11th Asia-Pacific Regional Conference of the International Society for Terrain-Vehicle Systems

Online
26-28 September 2022

ISBN: 979-8-3313-1248-0

Printed from e-media with permission by:

Curran Associates, Inc.
57 Morehouse Lane
Red Hook, NY 12571

Some format issues inherent in the e-media version may also appear in this print version.

Printed with permission by Curran Associates, Inc. (2025)

For permission requests, please contact the International Society for Terrain Vehicle Systems at the address below.

International Society for Terrain-Vehicle Systems
c/o Cold Regions Research and Engineering Laboratory
72 Lyme Road Hanover, New Hampshire 03755-1290
USA

http://www.istvs.org

Additional copies of this publication are available from:

Curran Associates, Inc.
57 Morehouse Lane
Red Hook, NY 12571 USA
Phone: 845-758-0400
Fax: 845-758-2634
Email: curran@proceedings.com
Web: www.proceedings.com

ISTVS

Proceedings of the 11th Asia-Pacific Regional Conference of the ISTVS

September 26-28, 2022 :: online
hosted by Harbin Institute of Technology, China
https://2022.istvs.org

Proceedings of the 11th Asia-Pacific Regional Conference of the ISTVS

Contents

Conference overview

The aim of the conference is to further advance knowledge in terrain-vehicle and road- vehicle system dynamics and design and, thus, to discuss and suggest new frontiers and research directions, facilitate engineering practice, and benefit the society at large in the area of the energy conservation, environmental sustainability, safety, cyber and physical security, application of artificial intelligence (AI) techniques in terramechanics and development of vetronics systems for civil and on ground-moving military vehicles.

- Terramechanics, locomotion, and soil/terrain modeling
- Recent accomplishments in terramechanics research and education
- Locomotion systems
- Soil and terrain modeling and characterization
- Advances in mobility, energy efficiency, ground vehicle dynamics, and high-fidelity simulation
- Agricultural, forestry, construction equipment, and machinery
- Mobile robotics for ground applications, planetary exploration, and other environments
- Design/control, sensing/actuation in cyber-physical systems for vehicle dynamics and mobility
- Innovative system designs for terrain and road-vehicle applications
- Application of artificial intelligence to vehicles

Conference Chair

Liang Ding
Harbin Institute of Technology | liangding@hit.edu.cn

Scientific Committee

Hosting Institution

Harbin Institute of Technology, affiliated with the Ministry of Industry and Information Technology of the People's Republic China, focuses on science and engineering and maintains the coordinated development of multiple disciplines including science, engineering, management, literature, economics, law, and art.

Since its founding in 1920, HIT has been well-known as the cradle of engineers today. HIT has three campuses in Harbin, Weihai, and Shenzhen.

Keynote Speakers

Baofeng Yuan
Locomotion and Exploration of Zhurong Mars Rover
Chief designer of the Zhurong Mars Rover mobility system
China Academy of Space Technology
Researcher, Space robot technology, China Academy of Space Technology | Beijing Institute of Space Systems Engineering

Baofeng Yuan has been engaged in the research of space robot technology since 2006. His research interests include space robot system design, precision transmission and mechanism design, drive control and dynamic analysis, perception and measurement technology. His research projects include Zhurong Mars rover mobile system, space station manipulator and Chang'e 5 sampling manipulator. He has published 5 SCI journal papers and 11 EI journal papers and applied 13 patents.

Salah Sukkarieh
Agriculture Robotics—Mechatronic and AI design for all terrain operations
Professor, Robotics and Intelligent Systems, University of Sydney | Sydney Institute of Agriculture | The University of Sydney Nano Institute | the Charles Perkins Centre | CEO of Agerris | C.V.

Dr. Salah Sukkarieh was the Director Research and Innovation at the Australian Centre for Field Robotics from 2007-2018. He was awarded the NSW Science and Engineering Award for Excellence in Engineering and Information and Communications Technologies in 2014, the 2017 CSIRO Eureka Prize for Leadership in Innovation and Science, and the 2019 NSW Australian of the Year nominee. He is a Fellow of Australian Academy of Technological Sciences and Engineering (ATSE) and has over 500 academic and industry publications in robotics and intelligent systems.

Guangming Chen
Development of a Biologically Inspired Walking Robot for Mars Exploration
Instructor, School of Mechanical and Electrical Engineering, Nanjing University of Aeronautics and Astronautics | Institute of Bio-Inspired Structure and Surface Engineering | C.V. | ResearchGate

Dr. Guangming Chen obtained his Ph.D. from Delft University of Technology, Netherlands. Since November 2017, he works at the Nanjin University of Aeronautics and Astronautics. His research interest is mobile robotics for unstructured environments. He has published 13 SCI journal papers and 10 conference papers and applied for 8 patents.

Ye Zhuang | on behalf of FISITA
Semi-Active Reinforcement Learning Suspension Control for Off-Road Vehicles
Professor | State Key Lab of Automotive Simulation and Control, College of Automotive Engineering | Jilin University | website

Professor Ye Zhuang obtained his Ph.D. from Jilin University, China. Since September 2005, he works at Jilin University. His research interest is vehicle multibody dynamics and control. He has published 20 SCI journal papers and 10 conference papers and been granted 12 patents.

Sponsor—Gold

Research Center of Aerospace Mechanism and Control relies on the mechanical engineering of Harbin Institute of Technology and carries out relevant basic theory and key technology research in view of the serious shortage of basic theory and basic technology in the development of Aerospace Science and technology. It carries out research on international cooperation projects by actively introducing and gathering international academic masters and backbones around seven specific academic directions, including planetary exploration robots, and space folding institutions.

Contact: yanghuaiguang@hit.edu.cn

Sponsor—Gold

Sunrise Instruments is a leading manufacturer of 6 axis force/torque sensor for robotics, automation, biomechanics, automotive crash dummies, and vehicle durability tests. It is specialized in the development of six axis force/torque sensors, multi-axis force/torque sensor, 6 axis load cell, 6 axis force transducer, and robot force-controlled grinding, robot grinding, robot polishing, and robot sanding.

Contact: +86-13061612685, sri@srisensor.com, www.srisensor.com

Sponsor—Silver

HRG International Institute (Hefei) of Research and Innovation (HRG (Hefei) for short) founded in August 2016, which is a new R&D institution of "city campus-enterprise cooperation." Its main business focuses on the research and development of high and new technologies and the transformation of scientific and technological achievements in the field of robot, artificial intelligence, intelligent equipment and the Internet of Things. HRG (Hefei) , which has set up six front-end scientific research industry research institutes, is committed to building an intelligent robot research system of "behavioral intelligence-brain-like intelligence-cloud brain intelligence. HRG (Hefei) has built an ecological chain of the upper, middle and lower reaches of the robot industry, covering the core key components of the robot, ontology, system integration and software services.

Contact: 861-0551-68996766, www.hiiri.cn, marketinghf@hitrobotgroup.com

Sponsor—Bronze

Weihai Tiante Intelligent Technology Co., Ltd., is a professional company of skateboard chassis + scene customized unmanned operation module. The core R&D team of the company comes from the Institute of Intelligent and Special Vehicles of Harbin University of Technology (Weihai) and has been engaged in the research of special vehicles and driverless vehicles for many years. The company is a well-known supplier of intelligent mobile platform. Its service scope covers logistics, fire protection, security, national defense, and other industries. It has formed a general chassis + scene customized R&D mode to serve the development of robot technology in various industries.

Contact: +86-18249758524 | wjfeee123@163.com

Conference program

All times are in Beijing time UTC+8. Check your local time here: everytimezone.com
View full paper abstracts: https://2022.istvs.org/papers
View abstract-only abstracts: https://2022.istvs.org/abstract-only
View advance video presentations: https://vimeo.com/showcase/istvs2022 *(password sent to registered attendees)*

Monday, September 26, 2022

19:00 - 19:40 Opening Ceremony

Chair: Liang Ding
Welcome by Conference Chair: Liang Ding
Welcome by Zhiwu Han, Executive Director of Chinese Society for Agricultural Machinery
President Address: Corina Sandu
Report on Journal of Terramechanics: Corina Sandu for Vladimir Vantsevich

19:40 - 19:45 5-minute break

19:45 - 20:35 Plenary — Keynote

Chair: Lutz Richter

Locomotion and Exploration of Zhurong Mars Rover
Baofeng Yuan
Chief designer of the Zhurong Mars Rover mobility system | China Academy of Space Technology

20:35 - 20:45 10-minute break

20:45 - 22:30

Session 1-1-A Best Paper Candidates	Session 1-2-A Mobility and Traction Characterization I	Session 1-3-A Control, Planning and Estimation I
Time: 20:45 - 21:25 **Chair: Péter Kiss** **Co-Chair: Dror Rubinstein**	**Time: 20:45 - 21:35** **Co-Chair: Meng Zou** **Chair: Chen Li**	**Time: 20:45 - 21:35** **Chair: Parish Nalavade** **Co-Chair: He Kong**
Acquisition of Flipper Motion of Tracked Robot in Step-Climbing Using Reinforcement Learning **Ryosuke Eto, Junya Yamakawa | 9913**	A Wheel and Vehicle Mobility Index Based on Traction and Velocity for Optimization of Mobility Performance **Vladimir Vantsevich, David Gorsich, Jesse Paldan, Masood Ghasemi, Lee Moradi | 0963**	Semi-active Reinforcement Learning Suspension Control for the Off-road Vehicles **Ye Zhuang, Haojie Sun, Yingchun Qi, Weiguang Fan, Hui Ye | 1128**
Research on Drag Reduction Performance of Sliding Plate of Rice Direct Seeding Machine Based on Non-smooth Structure of Loach Surface **Hongchang Wang, Zhen Jiang, Kaiquan Ding, Guozhong Zhang, Abouelnadar Salem, Yuan Gao | 5979**	A Comprehensive Lumped Parameter Approach for the Dynamic Simulation of Agricultural Tractors in Real Operating Conditions **Marco Polastri, Damiano Chiarabelli, Silvia Gessi, Massimo Martelli, Emiliano Mucchi, Pietro Marani | 2034**	Foothold Selection Considering Constraint and Slippage Evaluation for Legged Robots **Yufei Liu, Lei Jiang, Chong Tian, Boyang Xing, Zhirui Wang, Bo Su, Tong Yan, Liang Ding, Haibo Gao | 1534**
Factors Affecting Bevameter Soil Characterization **Schalk Els, Herman Hamersma, Ray Kruger | 6174**	Investigation of the Shear Stress Dynamics on Silty Loam Soil and Measurement of Traction-Wheel Slip Relationship of a Tractor Tire **César Andrés Arévalo-Montaña, Stefan Böttinger | 2149**	Model Predictive Control of a Robot Driven Vehicle for Testing of Advanced Driver Assist Systems **Mike Huang, Qiu Haixuan,Yang Chunyu, Xia Lian, Lin Zhaomin, Wang Yanqing, Xu Zongqing, Mingfu Tang | 2539**
	Research on High Traction Bionic Wheel Based on Traction Characteristics of Jerboa Foot **Hao Pang, Liangliang Zhao, Huailiang Wang, Rui Zhang | 2681**	Energy Consumption Analysis of Door Opening with a Mobile Manipulator in the Nuclear Power Plants **Changyou Ma, Binbin Yan, Xu Xiong, Haibo Gao, Liang Ding | 2632**
	The Running Gear Construction Impact on Obstacles Overcoming by Light High-Mobility UGV **Daniela Szpaczyńska, Marian Łopatka, Piotr Krogul | 4260**	An Improved Simultaneous Localization and Mapping Method Based on LeGO-LOAM and Motion Compensation **Mengyang Li, Xinsheng Wang, Xiyue Wang, Shuang Liu | 2643**

Session 1-1-B Best Student Paper Candidates	Session 1-2-B Mobility and Traction Characterization II	Session 1-3-B Control, Planning and Estimation II
Time: 21:30 - 22:30 **Chair: Péter Kiss** **Co-Chair: Shashank Agarwal**	**Time: 21:40 - 22:30** **Co-Chair: Chen Li** **Chair: Meng Zou**	**Time: 21:40 - 22:30** **Chair: Parish Nalavade** **Co-Chair: He Kong**
The Effect of Integrating a Bio-inspired Convex Structure with a Low-Surface Energy Polymer on Soil Adhesion and Friction **Abouelnadar Salem, Guozhong Zhang, Hongchang Wang | 8349**	Steadily Learn to Drive with Virtual Memory **Yuhang Zhang, Yao Mu, Shengbo Li, Yangang Ren, Liye Tang, Yujie Yang, Chen Chen | 4774**	Obstacle Avoidance of Mobile Robots Using Modified Artificial Potential Field Algorithm Based on Vortex and PID Adjustment **Zhuoran Sheng, Song Wang | 3004 (Abstract only)**
A Time Domain Passivity Controller for Teleoperation of Four-Wheeled Differential Mobile Robot on Slippery Surface **Hui Xi, Yanjing Li, Pengyu Sun, Weihua Li, Jianfeng Wang | 7092**	Research on Vehicle Running Performance on Paved Roads Covered with Falling Volcanic Ash **Junya Yamakawa, Ryosuke Eto, Yasuhiro Ichikado, Mitsuhiro Yoshimoto, Tatsuji Nishizawa, Tomohiro Kubo, Hiroyuki Yamada | 6718**	Perceptive Locomotion of Legged Robot Coupling Model Predictive Control and Terrain Mapping **Boyang Xing, Lei Jiang, Bo Su, Yufei Liu, Zhirui Wang, Tianqi Qiu,Jianxin Zhao | 6316**
Study of Passive Steering Mechanism for Mars Surface Exploration Rovers **Asahi Oe, Shin-Ichiro Nishida, Shintaro Nakatani | 7233**	Vehicle Dynamic Factor Characterized by Actual Velocity and Combined Influence of the Transmission and Driveline System **Vladimir Vantsevich, David Gorsich, Jesse Paldan, Jordan Whitson, Brian Butrico, Oleg Sapunkov | 7199**	Study of Passive Steering Mechanism for Mars Surface Exploration Rovers Vehicle Dynamic Factor Characterized by Actual Velocity and Combined Influence of the Transmission and Driveline System Research on Energy-Saving Underactuated Bionic Biped Robot **Youhao Diao, Zhiqiang Zhuang, Xuebo Wang | 7349 (Abstract only)**
Parameters Calibration of Red Clay Soil in Hilly Area of Southwest China for Discrete Element Simulation Based on Repose Angle Test **Le Yang, Qinghui Lai, Liangliang Zhao, Peihang Li, Zhihong Zhang, Zhaoyang Chen | 8131**	Determination of the Necessary Indicators of Vehicles and the Movement Surface for Calculating the Mobility Criterion **Alina Markovnina, Umar Vakhidov, Vladimir Belyakov, Vladimir Makarov | 8104**	
Design and Verification of a Creeping Mars Rover **Wangjun Zhang, Yang Jia, Zhuo Tao | 1491**	Force Chain Analysis on Wheel-Soil Interaction Using Photoelastic Method **Yuto Yoshida, Sota Yuasa, Kenji Nagaoka | 9273**	
A Review of Modeling and Validation Techniques for Tire-Deformable Soil Interactions **Varsha S Swamy, Rashna Pandit, Alba Yerro-Colom, Corina Sandu, Denise Rizzo, Katherine Sebeck | 7018**		

Tuesday, September 27, 2022

19:00 - 19:10 Welcome

Chair: Liang Ding

19:10 - 20:00 Plenary — Keynote

Chair: Corina Sandu

Agriculture Robotics—Mechatronic and AI Design for All Terrain Operations

Salah Sukkarieh

Professor, Robotics and Intelligent Systems, University of Sydney | Sydney Institute of Agriculture | The University of Sydney Nano Institute | the Charles Perkins Centre | CEO of Agerris

20:00 - 20:05 5-minute break

20:05 - 20:55 Plenary — Keynote

Chair: Lutz Richter

Development of a Biologically Inspired Walking Robot for Mars Exploration

Guangming Chen

Instructor, School of Mechanical and Electrical Engineering, Nanjing University of Aeronautics and Astronautics
Institute of Bio-Inspired Structure and Surface Engineering

20:00 - 20:05 5-minute break

21:05 - 22:30

Session 2-1-A Robotics and Space Applications I	Session 2-2-A Modeling and Simulation I	Session 2-3-A Modeling and Simulation III
Time: 21:05 - 21:45 **Chair: Huaiguang Yang** **Co-Chair: Jiayu Li**	**Time: 21:05 - 21:45** **Co-Chair: Schalk Els** **Chair: Zhenzhong Jia**	**Time: 21:05 - 21:45** **Chair: Junya Yamakawa** **Co-Chair: Kyeong Uk Kim**
Analysis on Climbing Mode of Zhurong Rover **Zhen Chen, Meng Zou, Dong Pan, Baofeng Yuan, Lining Chen | 0196 (Abstract only)**	Modeling of Lunar Rover Vehicle Wheel-Soil Interaction Using FEM-DEM Method **Kaidi Zhang, Yunqing Zhang, Junwei Shi, Weili Kong | 1655**	Numerical Analysis of Multi-Pass Effect Based on the Extended Terramechanics Theory **Shingo Nakano, Shingo Ozaki | 5636 (Abstract only)**
Review on the Reconfigurable Wheel-Tracked System **Jiaxuan Wang, Cheng Liu, Wei Wei, Qingdong Yan | 0861**	Tyre Parameteriaztion Tests: Dynamic Vs. Static **Carl Becker, Schalk Els | 2190**	Interaction Modeling and Dynamic Control Strategy for C-Shaped Leg with Sandy Terrain in Terradynamics **Chuanxiao Yang, Zhiyue Xin, Xiong Hu, Shibin Sun, Liang Ding, Dewei Tang | 5800**
Bearing Capacity Analysis of Bionic Walking Wheel for Manned Lunar Rover **Rui Zhang, Liang Liang Zhao, Yupei Du, Bin Zhao, Hao Pang, Lige Wen, Weijun Wang, Hua Zhang, Zhenyu Hu, Meng Zou | 2928 (Abstract only)**	Modeling Tire-Soil Compression Resistance on Artificial Soil Using a Scaling Law of Pressure—Soil Sinkage Relationship **Pius Jjagwe, Mehari Tekeste, Nisreen Alkalifa | 2398 (Abstract only)**	Tire-Soil Tangential Force Reinforcement Learning Modelling **Yingchun Qi, Jiaqi Zhao, Ye Zhuang | 7399**
Why We Need Alternative Ground Robots to Traverse Sandy and Rocky Extraterrestrial Terrain, and How We Can Progress Towards Them **Chen Li, Kevin Lewis | 3290 (Abstract only)**		Modelling of the Shear Displacement for Tracked Vehicles in Transient Maneuvers **Yang Jiao, Jozsef Kovecses | 7739 (Abstract only)**

Session 2-1-B Robotics and Space Applications II	Session 2-2-B Modeling and Simulation II	Session 2-3-B Modeling and Simulation IV
Time: 21:50 - 22:30 **Chair: Huaiguang Yang** **Co-Chair: Jiayu Li**	**Time: 21:50 - 22:30** **Co-Chair: Schalk Els** **Chair: Zhenzhong Jia**	**Time: 21:50 - 22:30** **Chair: Junya Yamakawa** **Co-Chair: Kyeong Uk Kim**
Field Validation of Egress Process for Planetary Rover **Dang Zhaolong, Chen Baichao | 4054**	Terramechanics Model Augmentation Using Machine Learning **Eric Karpman, Jozsef Kovecses, Marek Teichmann | 2960 (Abstract only)**	A Method for Fast Obtaining of Soil Shear Strength Index Based on DEM Free-Fall Cone Penetration Simulation **Jincheng Diao, Jingwei Gao, Xiaobo Song, Baojun Di | 7878**
Development and Traveling Performance Analysis of Driving Test Rover in Sandy Terrain **Masataku Sutoh, Yuji Katsumata, Sachiko Wakabayashi | 4170 (Abstract only)**	DEM Analysis of the Dynamics of Granular Media Considering with Interparticle Forces under Low Gravity Condition **Takuru Nishino, Kenta Takase, Shingo Ozaki, Takao Maeda, Mitsuhisa Baba, Masatsugu Otsuki | 4136 (Abstract only)**	A High-Fidelity Dynamics Simulation Method for Legged Robot Based on Foot–Terrain Interaction Model **Jianghua Ge, Jianghu Wu, Xiaofei Zhu | 8362 (Abstract only)**
Development and Demonstration of Robotic Investigation Tool for Terrain Mechanical Property **Yuzuki Morita, Genya Ishigami | 4323 (Abstract only)**	Multi-Fidelity Machine Learning Modeling for Wheeled Locomotion on Soft Soil **Vladyslav Fediukov, Felix Dietrich, Fabian Buse | 4812**	Examining the Simulation-to-Reality Gap of a Wheel Loader Interacting with Deformable Terrain **Koji Aoshima, Daniel Lindmark, Martin Servin | 9092 (Abstract only)**

Terrain Classification Using Mars Raw Images Based on Deep Learning Algorithms with Application to Wheeled Planetary Rovers **Junlong Guo, Xingyang Zhang, Yunpeng Dong, Zhao Xue, Bo Huang | 4744**		3D-DEM Simulation and Post-Process Method of Wheel-Terrain Interaction for Planetary Rovers **Qingning Lan, Zhengyin Wang, Huaiguang Yang, Liang Ding, Haibo Gao | 9352**

Wednesday, September 28, 2022

19:00 - 19:10 Welcome

19:10 - 20:00 Plenary — Keynote

Chair: Massimo Martelli

Semi-active Reinforcement Learning Suspension Control for Off-road Vehicles

Ye Zhuang | on behalf of FISITA

Professor | State Key Lab of Automotive Simulation and Control, College of Automotive Engineering | Jilin University

20:00 - 20:10 10-minute break

20:10 - 21:25

Session 3-1-A Soil and Ground Testing and Characterization I	Session 3-2-A Design I	Session 3-3-A Soil and Ground Testing and Characterization II
Time: 20:10 - 20:50 **Chair: Taizo Kobayashi** **Co-Chair: Junlong Guo**	**Time: 20:10 - 20:50** **Co-Chair: Linnea Hansson** **Chair: Mehari Tekeste**	**Time: 20:10 - 20:50** **Chair: Vilas M. Salokhe** **Co-Chair: Genya Ishigami**
Composite Beam Tests with Closed Cell Polyurethane and Aluminum Foam **George Mason, Ethan Salman, Shiraz Mujahid | 0303**	Design and Simulation Analysis of Intelligent Suspension for Manned Lunar Rover **Tao Li, Chongfeng Zhang, Weijun Wang, Junwei Shi | 0356**	Ride Comfort Comparison between Suspension Modes: Input towards Designing Difference Threshold Experiments during Driving **Cor-Jacques Kat, Kylian Praet, Miguel Dhaens, Schalk Els | 5408**
Benchmarking of Compression Testing Devices in Snow **Mohit Nitin Shenvi, Corina Sandu, Costin Untaroiu | 3351**	Comparative Analysis of Hydrodynamical Efficiency of Full-submerged Archimedes Screws of Rotary-Screw Propulsion Units of Snow and Swamp-Going Amphibious Vehicles with Single and Tandem Propulsor Design **Svetlana Karaseva, Aleksey Papunin, Vladimir Makarov, Dmitry Malahov | 1704 (Abstract only)**	Nonparametric Terrain Estimation Based on the Interaction Simulation between Planetary Penetrator and Soil **Xintao Yang, Han Huang, Zhixin Xiang, Qinghao Yan, Haozhe Wang, Shucai Xu | 6796**
Soil Compaction Monitoring Technique Using Deep Learning **Shota Teramoto, Taizo Kobayashi | 4243**	Design of Self-driving Bulldozer System **Yang Junhua, Zhang Biao, Tang Haokai, Shen Binghua, Ou Linlin, Yu Xinyi, Feng Yuanjing, Feng Yu, Zhou Libo | 4409**	Construction of a Soil Clods Recognition Bench-Scale Experiment for Discrete Element Method Modeling of Tilling Phenomena **Shuto Ishii, Isami Suto, Hiroaki Tabe, Keisuke Nagato, Moju Zhao, Yoshifumi Ueshige, Takashi Iritani, Masayuki Nakao | 8654**
Experimental Study of Track-Soil Interactions of the Steering Performance of Tracked Robots over Soft Deformable Terrains **Qiaowen Wang, Zhenzhong Jia | 4782 (Abstract only)**	Introducing Polibot: A High Mobility Tracked Robot with Innovative Passive Suspensions **Giulio Reina, Rocco Galati, Andrea Grazioso, Angelo Ugenti, Giacomo Mantriota | 4827**	Investigation of the Relationship between the Cone Index and the Physical and Mechanical Parameters of the Soil of Typical Surfaces of the Movement of Agricultural Tractors and Machines **Sergey Zhukov, Vladimir Makarov, Vladimir Belyakov | 8658**

Session 3-1-B Soil and Ground Testing and Characterization I	Session 3-2-B Design I	Session 3-3-B Soil and Ground Testing and Characterization II
Time: 20:55 - 21:25 **Chair: Taizo Kobayashi** **Co-Chair: Junlong Guo**	**Time: 20:55 - 21:25** **Co-Chair: Linnea Hansson** **Chair: Mehari Tekeste**	**Time: 20:55 - 21:25** **Chair: Vilas M. Salokhe** **Co-Chair: Genya Ishigami**

	Bionic Quadruped Robot for Mars Surface Exploration **Long Qiao, Guangming Chen, Lutz Richter, Aihong Ji | 5060**	Research on Bionic Anti-skid Tires Suitable for Icy and Snow-Covered Roads **Rui Zhang, Yu Han, Hao Pang | 5943 (Abstract only)**	Prominent Problems and Thoughts of "Paddy Soil-Terrain Machine System" Based on Disturbed Saturated Paddy Soil Conditions in South China **Guozhong Zhang, Hongchang Wang, Jun Du, Kaiquan Ding, Wanru Liu, Nanrui Tang, Yong Zhou | 1561**
	Penetration Dynamics of Asteroid Exploration Landing Based on DEM **Yongbin Wang, Shiqing Wu, Shutong Chen, Xinghua Liu, Huan Liu, He Jia, Xuyan Hou, Shaomin Liang, Shunying Ji, Zijian Zhang | 9219 (Abstract only)**	Design and Testing of Parallel Embedded Six-Axis Force Sensors for Paddy Walking Wheels **Zaiman Wang, Jianfei He, Erli Zhang, Miao Su, Yue Huang, Wenwu Yang, Minghua Zhang, Weiqin Jia, Yuanli Huang, Yifan Ma, Dongyang Yu, Peizhao Zhong, Zhihao Zeng, Ziyou Guo | 8685 (Abstract only)**	Experimental Analysis and Numerical Modeling of Bulldozing Force with Varied Soil Moisture Content **Naohiro Sato, Genya Ishigami | 9260 (Abstract only)**
	Configuration Design and Nonlinear Mechanical Properties Analysis of Tensegrity Robot **Ruiwei Liu, Xuntao Lin, Hongwei Guo, Yating Fang | 9943 (Permission to publish not granted)**	Design and Traction Performance Test of Bionic Paddy Wheel Based on Cattle Hoof **Lan Li, Jing Li, Baofeng Xie, Fei Lin, Long Xue | 9768**	
21:25 - 21:30	5-minute break		
21:30 - 22:05	Closing Ceremony **Chair: Liang Ding** **Appreciations by Conference Chair: Liang Ding** **Award Presentation: Best Paper & Best Student Paper: Péter Kiss** **Announcement of new ISTVS Fellows: Corina Sandu** **Introduction of 2022 ISTVS Americas Symposium, Montreal, October 5-7: Lutz Richter** **Introduction of ISTVS 2023—16th European-African Regional Conference: Jaroslaw Pytka**		

ISTVS

Conference papers

0303 / Composite Beam Tests with Closed Cell Polyurethane and Aluminum Foam

George L. Mason*, John E. Salmon, Jennie Daigler, and Shiraz Mujahid
Mississippi State University, Starkville MS, USA
**Corresponding author: mason@cavs.msstate.edu*

https://doi.org/10.56884/VCIX2361

Abstract

The frames of many commercial vehicles are composed of rectangular tubular steel and are a significant portion of the gross vehicle weight. The introduction of a composite frame would potentially reduce weight and increase structural properties. We compare a baseline frame section to new composite sections injected with two different types of foam: a polyurethane and aluminum foam. The objective was to increase the overall strength and stiffness, while leaving the overall weight of the frame unchanged or reduced by reducing the plate thickness of the tubular steel. Sections of rectangular tubular steel are tested individually with and without the inclusion of aluminum and closed cell polyurethane foam. The potential advantages include light weighting vehicles, increasing strength and durability, improving ride through a more ridged/stiffer un-sprung mass, and the ability to make quick structural repairs of a compromised vehicle frame. The study revealed that injecting of polyurethane foam produced increase in strength while the aluminum foam did not increase the strength.

Keywords: composites, structures, frames, foam

1. Introduction

Studies were conducted by the U.S. Corps of Engineers in the late 1990's (Mason, G., et al.,) examining the use of closed cell polyurethane foam in the field as a method to override obstacles by heavy tracked and wheeled vehicles. The strength of high-density polyurethane foam was latter considered as a method to repairing/strengthening the rectangular tubular steel frame of military HMMWVs in the field (Mason G.L., et al.,). The early HMMWV frames were compromised by excessive weight and significant rugged usage. The opportunity to strength the frame with a composite will decrease the weight of the unsprung mass and provide an improvement in ride (Jellnek et al., 2021) along with fuel economy. A modification by injection of a polyurethane or aluminum foam in a rectangular tubular steel frame was considered to increase the strength and stiffness. Several automotive manufacturers have considered the use of foam to increase the performance of their vehicles (Swart, 2004, Moodie et al., 2007, and Kim et al., 2002). In each of these studies the foam was injected into sections of the automotive panels to add rigidity, reduce noise/thermal signature, and add strength.

In this study, frame sections typical of structural support members of light trucks are evaluated, comparing baseline frame components to composite members created by the injection of various foam types and densities. The injection or placement of high-density structural foam into the steel frame hollow inner cores would hypothetically reduce buckling and increase stiffness. Procedures for the use of structural foam in vehicles are outlined by the Society of Automotive Engineers (SAE, 2004). However, this technology is directed primarily at acoustic foam inserted into vehicles at the factory for damping of outside noises. Several patents exist for the use of foam in vehicles such as Spaulding 2002 where foam is used to support concealment by mitigating heat and sound signatures. The purpose of this study is to determine if the addition of structural foam into a common tubular steel vehicle frame adds any structural support and/or potential for weight reduction.

We also consider a lower density foam and different thicknesses of plates. This study was limited in scope and did not consider possible corrosion that may occur due to trapped moisture. Plazankis et al., 2001 did an exhausted study on steel panels with inhibitors, this study could be extended to the foam application. A follow-on study is considered to evaluate the corrosion of the steel after application of the foam as compared to the baseline.

2. Test Design

The experiment consisted of several combinations of frame sections in different lengths, loading profiles, steel grades, foam types, and foam densities as seen in Figure 1. All the compression and bending tests were loaded and supported on the shorter width face so the longer face or sides were always vertical. Short frame sections were loaded in compression in a uni-axle manner. Mid-length and longer sections were loaded in three-point and four-point bending.

The frame sections were comprised of 80 KSI steel with a thickness of 0.138 *in.* [3.51*mm*]. The steel frame sections were made from two "C" shaped sections welded together as seen in Figure 2.

These tests were designed to show any increase in bending stiffness or buckling from the baseline frame to the composite frame sections.

Figure 1. Composite frame sections with aluminum (left) and polyurethane (right).

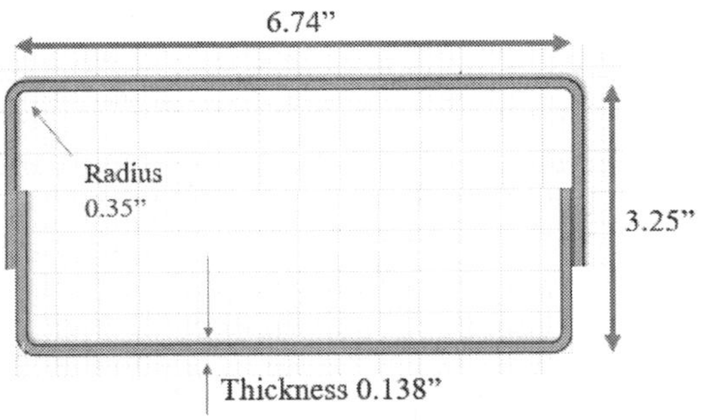

Figure 2. The cross-section of the hollow steel tubing used in experimental testing.

2.1 Uni-axle compression

The first type of tests were compression tests. 5 *in.* [127 *mm*] sections were used for compression tests. The different sections were loaded at a constant grass-head speed of 0.05 *in./sec.* [1.27*mm/sec.*] until a clear failure of the section occurred.

The density of the aluminum foam was 12 *lb./ft^3* [0.192 *g/cm^3*] and the density of the polyurethane foam was 13.3 *lb./ft^3* [0.213 *g/cm^3*]. The polyurethane foam showed successive shattering (see points P and Q in Figure 5) as well as delamination due to strain incompatibility. The aluminum foam was produced in prefabricated strips that were inserted into the frame section, while the polyurethane was sprayed and formed within the sections. The aluminum foam inserted in 'horizontal strips' showed negligible change from the hollow baseline. The aluminum foam inserted in 'vertical' orientation showed a small increase in both the first and second peak loads as shown in Figure 5 and Table 1.

Figure 3. An example of a baseline frame section with no foam (left) and a baseline frame section with no foam after compression (right).

Figure 4. Example of an aluminum (left) and a polyurethane (right) section after compression testing.

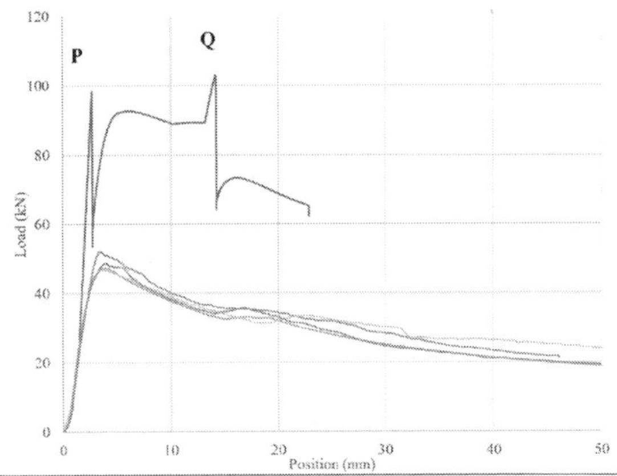

Figure 5. Compression tests, Load vs Displacement. Baseline (gray), polyurethane foam filled section (red), Al foam filled section vertical #9 (green), Al foam filled section vertical #8 (blue), Al foam filled section horizontal #7 (yellow)

Table 1. Compression test. 1st and 2nd load peaks for the baseline, polyurethane, and aluminum foams

Sample	Loading Slope [kN/mm]	1st Peak Load [kN]	2nd Peak Load [kN]	Foam Density [g/cm³]	Weight [kg]
Baseline Hollow	21.91	47.25	33.12	0	3
Al Horizontal	21.47	46.67	33.42	0.192	4
Al Vertical (a)	20.39	48.62	35.66	0.192	5
Al Vertical (b)	21.08	52.07	36.54	0.192	6
Polyurethane	61.7	96.77	102.54	0.213	7

2.2 Three-point bending

The second type of tests were three-point bending. Frame sections 3 *ft* [0.914 *m*] were loaded by a 1.5 in diameter steel rod in the center on the top and supported by two 1.5 in. steel rods 20 in apart centered as seen in Figure 6. During these tests the plate supporting the beam is shown to flex and may have created issues with the data. Supports were added later in the tests, but this interfered with the load cell measurements. While publishing these results, we expect to rerun these tests with a different support mechanism and reevaluate the results of the three-point bending.

Figure 6. Three-point bending

2.3 Four-point loading

The four-point bending was conducted on 59.5 *in.* [1.51 *m*] long sections and loaded to failure. Two 5 *in.* [0.127 m] loading surfaces were placed 17 *in* [0.43 *m*] apart from their centers, on center. The bottom supports were flat 6 *in* [0.152 *m*] wide supports 51 *in.* [1.295 *m*] apart from their centers, from the center of the section, as seen in Figure 7.

Figure 7. Four-point bending test on hollow section

Table 2 Four-point bending test results

59.5" Frame Rail Sections in 4-point bending				
	Foam Material	**Foam Density (g/cm^3)**	**Weight (g)**	**Max Load at Failure (kg)**
Section #1	N/A (hollow)	N/A	24.7	16,440
Section #2	Polyurethane	0.22	29.7	21,635
Section #3	Polyurethane	0.32	31.0	24,053
Section #4	Polyurethane	0.64	36.8	30,327

3. Experiment Results

The results showed that aluminum foam did not add any significant structural benefits. Part of this was most likely from the manner the foam was prefabricated and placed within the frame sections at both vertical and horizontal orientations. The polyurethane foam did increase structural properties in stiffness, buckling, and bending.

Figure 8. A 3-point bend polyurethane foam filled tube cut to show inside the center of loading area

4. Conclusions

The inclusion of aluminum foam overall did not add any significant structural benefits. Part of this was most likely from the manner the foam was prefabricated and placed within the frame sections at both vertical and horizontal orientations. The polyurethane foam did increase structural properties in stiffness, buckling, and bending. As to be expected, the higher the foam density, the stronger the composite section. The idea of reducing the steel frame section thickness in hopes of replacing its structural properties with an appropriate density of polyurethane foam needs to be further researched. From this experiment, weight was always added from the hollow baseline. In existing frames on vehicles, perhaps the injection of a dense, closed cell, polyurethane foam into a weak area of a frame would be practical.

5. References

Jellnek, Bohumir, Salmon, E., Mason George, Gibson, Michael, Hannis, Tyler, Pachel, Nathan, Walker, Jarrell, and Towne, Brennt, 2021. "Simulation Study of Light-Weighting Effects on Ride Quality and Mobility," 2021 NDIA Ground Vehicle Systems Engineering and Technology Symposium, Power & Mobility Technical Session, August 10-12, 2021, Novi, Michigan.

Esposito, S., Arias, N., and Soltisz, B. J. 1999. "Effect of Polyurethane Structural Foam on Vehicle Stiffness," Society of Automotive Engineers, Document Number 1999-01-1785. Noise & Vibration Conference & Exposition, May 1999, Traverse City, MI, USA, Session: Body/Chassis

Mason, G., Green, J., Hall, K., and Durst, B., "Ridged Polyurethane Foam Applications for Minefield Neutralization," Journal of Terramechanics, 2000, Vol. 37, pp. 87-98.

Mason, G. L., Towne, B., Eng, D., and Grey, C., "'Polyurethane Foam Injection for Strengthening the Frame of a HMMWV," Proceedings of the 18th International Conference of the International Society of Terrain-Vehicle Systems, September 2010, Sapporo, Hokkaido, Japan.

Spaulding, H. 2002. "Polyurethane Foam Concealment Panel," US Patent Number 6,441,799, August 27, 2002.

Placzankis, B. E., Miller, C. E., Buckley, P. F., and Beatty, J. H. 2001. "Accelerated Corrosion Analysis of E-coated Steel Box Panels to Determine Reapplication Intervals for Carwell Inhibitor," ARL-RP-34, Tri-Service Conference on Corrosion, Myrtle Beach, SC, 15-19 November 1999. U.S. Army Research Laboratory, Adelphi, MD.

Society of Automotive Engineers. 2004. "Surface Vehicle Recommended Practice, Qualifying Aftermarket Two-Component Structural Foams," SAE J2621, Warrendale, PA.

Moodie, J. R., Fowler, A.N., and Beck, W. 2007. "Composite Cross Member System," January 2, 2007, US Patent Number 7,156,421, January 2, 2007.

Swart, C. 2004. "Dow Automotive Materials Provide Performance Benefits for the 2004 Cadillac SRX," June 14, 2004, Howe Marketing Communications LLC, Kansas City, Mo.,

Kim, H-S, Chen, W., and Wierzbicki, T. 2002. "Weight and Crash Optimization of Foam-Filled Three-Dimensional "S" Frame," Computational Mechanics 28(5), 417-424.

0356 / Design and Simulation Analysis of Intelligent Suspension for Manned Lunar Rover

Tao Li[a]*, Chongfeng Zhang[b], Weijun Wang[a], Junwei Shi[a]

[a] Institute of Aerospace System Engineering Shanghai, China

[b] Shanghai Academy of Spaceflight Technology, China

**Corresponding author: 805686854@qq.com*

https://doi.org/10.56884/NSTJ2451

Abstract

As the assembly connecting the vehicle body and wheels, the suspension takes responsibility for transmitting force, ensuring fine wheel ground adhesion, turning portability, buffering impact load and attenuating vibration. Combining the intelligent control technology of ground vehicle with the operating environment and moving characteristics of manned lunar rover, this study expounds the feasibility and necessity of applying intelligent suspension to manned lunar rover.

Based on the requirements of the mobility performance, including handleability, stability, trafficability and ride comfort, a rotating electromagnetic intelligent suspension for manned lunar rover is designed. The complete suspension model is reasonably simplified and equivalent, based on the design of the intelligent suspension mechanism. We use various strategies to control the intelligent suspension for manned lunar rover and compare the control effect.

The effectiveness of the intelligent suspension for manned lunar rover is verified by using Adams and Matlab cooperative simulation. The results show that the evaluating indicators, including body acceleration, suspension dynamic displacement and tire dynamic deflection, are obviously better than the traditional passive suspension, the intelligent suspension can significantly improve the mobility performance of the manned lunar rover.

Keywords: manned lunar rover, intelligent suspension, control strategy, cooperative simulation, vehicle dynamics

1. Introduction

As the center of lunar exploration activities, the manned lunar rover takes the responsibilities for rapid transfer of the astronauts, emergency rescue and efficient material transportation. It plays an indispensable role in carrying out exploration missions, ensuring the safety of astronauts and improving work efficiency.

To ensure the good performance of the manned lunar rover, the crucial part is the reasonable design of the suspension system. Passive suspensions are designed to offer a reasonable compromise between comfort and passive safety, considering the character of the vehicle in question.

In Apollo program, which realized manned lunar exploration, the manned lunar rovers were equipped with passive suspension. According to the astronauts' reaction, when driving the manned lunar rover, the body comfort was poor and phenomena such as taking off and tilting often occur.

The intelligent suspension control technology of ground vehicles is becoming more and more mature. The active suspension has excellent damping control effect, but the structure is complicated, and the power consumption is large, the safety cannot be guaranteed on the complex and rugged lunar surface. The semi-active suspension has a similar structure to the passive suspension. The stiffness and damping coefficient of the shock absorber can be adjusted with a small amount of external energy. Using this technology, the posture of the vehicle body can be actively adjusted to improve the cushioning and damping performance. After the adjustment of the control system, a superior damping effect can also be obtained. It is highly feasible to apply it to the suspension system of manned lunar rover.

2. Model establishment and Analysis of Intelligent Suspension for Manned Lunar Rover

2.1 Characteristic analysis of suspension for Manned Lunar Rover

Similar to ground vehicles, the suspension for manned lunar rover is also responsible for connecting the vehicle body and wheels. Although the basic functions are similar, the suspension for manned lunar rover has unique characteristics.

In terms of structural design, the suspension for manned lunar rover, the mature experience of ground vehicle suspension can be applied. Due to the complex road conditions on the moon, the independent suspension of off-road vehicles with better ride comfort and trafficability is more suitable for manned lunar rovers.

In particular, due to the structural size requirements, the manned lunar rover and its suspension need to be foldable.

Based on the particularity of the lunar environment and the complex road conditions on the lunar surface, the design the suspension are more different than that of the ground vehicles. For example, due to the influence of low gravity on the lunar surface (the gravity constant is about

1/6 of the earth's surface), the manned lunar rover shows low positive pressure, low dynamic load and low handling force in the vertical direction and in the longitudinal directions of the road, which has an adverse impact on traction, trafficability and stability.

At the same time, the terrain of the lunar is rugged, with craters and rock blocks everywhere, the bumps and depressions are randomly distributed, which is prone to rollover and lift the wheel off the ground. Under the coupling action of low gravity and rugged lunar surface, the primary task of the suspension for manned lunar rover is to improve safety performance and optimize ride comfort on this basis. The driving setting of Adams can be used to simulate the road excitation, and the grounding performance can be expressed by recording the dynamics deflection of tires.

In Adams, the same sinusoidal excitation was set to qualitatively analyze the effect of low gravity. With same excitation of lunar surface 100sinωt, set normal gravity and 1/6 gravity respectively, the influence of low gravity on the grounding capacity is illustrated in Fig.1. If tire dynamic deflection is greater than zero, the wheel will be off the lunar surface.

According to the analysis of Fig.1, in low gravity, the probability of the dynamic deflection of the tire being greater than zero will suddenly increase, the ground separation phenomenon will occur, and the grounding performance and safety will greatly deteriorate.

Therefore, attention should be paid to the grounding performance in the matching design of stiffness and damping of suspension for manned lunar rover.

Combined with the functional requirements and characteristics, a collapsible double-wishbone suspension is predesigned, including structure, torsional stiffness and damping. The analysis and simulation of this paper are based on the suspension.

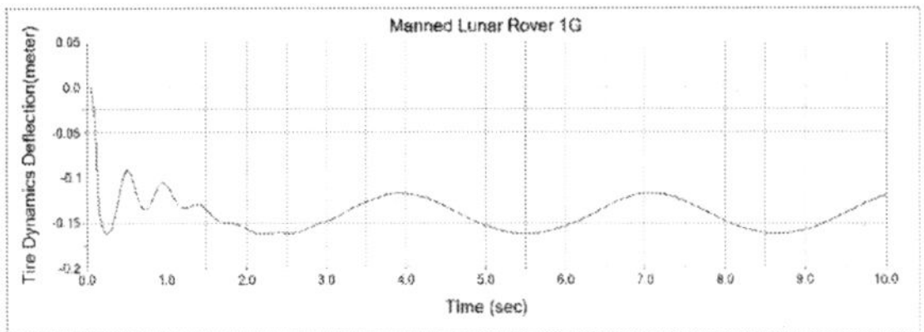

a) Tire dynamic deflection under 1 Gravity

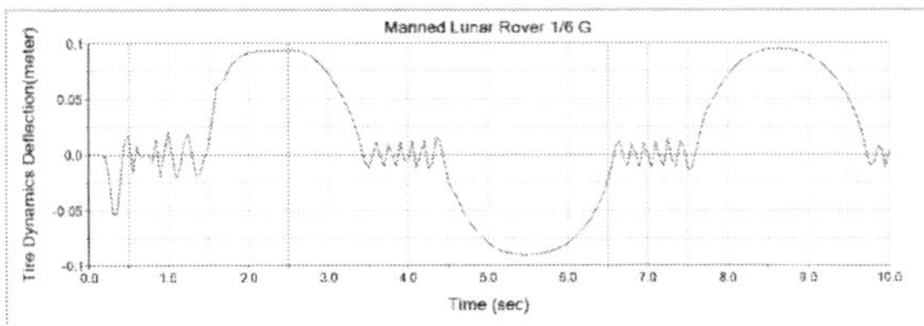

b) Tire dynamic deflection under 1/6 Gravity
Figure 1. Grounding performance under different gravity

2.2 Performance indicators of damping

Compared with the high-speed and plat conditions of most ground vehicles, the manned lunar rover mostly runs at lower speed on the rugged lunar surface.

When establishing the performance indicators of damping, the vertical acceleration of the vehicle body reflects the stability and driving comfort, the amplitude and root mean square value of the vertical body acceleration are commonly used to evaluate.

Suspension dynamic displacement reflects the change of body posture and trajectory; Tire dynamic deflection reflects the ground-contact performance, related to stability and safety; If suspension dynamic displacement is too large, the suspension will hit the limit block and there is a risk of breakdown. At the same time, it will change the wheel alignment parameters, making the ride comfort and handling stability of the vehicle worse. If the suspension dynamic displacement is too small, it will transmit more impact to the body, making the ride comfort worse.

As mentioned above, grounding safety performance is primary for manned lunar rover. It is reflected by the tire dynamic displacement under the action of the suspension

The main function of the suspension is to ensure stable tire load and acceptable ride comfort, while controlling the suspension travel within a reasonable range.

2.3 Vibration model of intelligent suspension

In order to analyze the vibration response of intelligent suspension, it is necessary to establish the corresponding vibration model. Based on the vehicle dynamics theory, for the consideration of safety and response frequency of the manned lunar rover, the form of actuator in parallel with elastic element and damping element is adopted. It can ensure that the passive suspension always exists and ensure a faster response speed.

In order to improve ride comfort, maneuverability and stability, and test the effectiveness of the control strategy, the 1/4 vehicle model of 2 degrees of freedom is established, as illustrated in Fig.2.

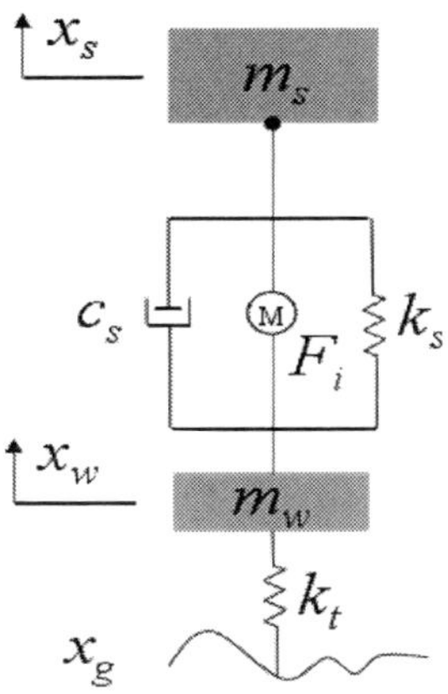

Figure 2. Vibration model of quarter vehicle model

In this figure, the m_s and m_w represent the sprung and unsprung masses; x_s and x_w represent the positions of the sprung and unsprung masses; c_s and k_s represents the damper coefficient and stiffness coefficient of the suspension; k_t represents the stiffness coefficient of the tire; x_g is an external signal representing the bumps and potholes of the road, generally called excitation of the rode; F_i is the control force provided by the actuator. Based on the mission requirements of manned lunar exploration and the pre-designed intelligent suspension model, these values can be determined.

According to Newton's second law of motion, the dynamic model with two degrees of freedom of body and wheel of intelligent suspension can be expressed, and its dynamic equation is as follows:

$$\begin{cases} m_s\ddot{x}_s = k_s(x_w - x_s) + c_s(\dot{x}_w - \dot{x}_s) + F_i \\ m_w\ddot{x}_w = k_t(x_g - x_w) - k_s(x_w - x_s) - \\ \qquad c_s(\dot{x}_w - \dot{x}_s) - F_i \end{cases} \quad (1)$$

Convert the dynamic equation into the equation of state, and take the state vector as follows:

$$X = [x_s, \dot{x}_s, x_w, \dot{x}_w]^T \quad (2)$$

The state equation of the system is obtained

$$\dot{X} = AX + BU \quad (3)$$

Take the control force provided by the actuator and the excitation of lunar surface as the inputs; Take the vertical body acceleration, suspension dynamic displacement and tire dynamic deflection as the output:

$$U = [F_i\ x_g]^T \quad (4)$$

$$Y = [\ddot{x}_s, (x_s - x_w), (x_w - x_g)]^T \quad (5)$$

The output equation of the system is obtained

$$Y = C(X) + DU \quad (6)$$

A is the system matrix; *B* is the control input matrix; *C* is the output matrix; *D* is the feedforward matrix.

$$A = \begin{bmatrix} 0 & 1 & 0 & 0 \\ -\frac{k_s}{m_s} & -\frac{c_s}{m_s} & \frac{k_s}{m_s} & \frac{c_s}{m_s} \\ 0 & 0 & 0 & 1 \\ \frac{k_s}{m_w} & \frac{c_s}{m_w} & -\frac{k_s + k_t}{m_w} & -\frac{c_s}{m_w} \end{bmatrix}$$

$$B = \begin{bmatrix} 0 & 0 \\ \frac{1}{m_s} & 0 \\ 0 & 0 \\ -\frac{1}{m_w} & \frac{k_t}{m_w} \end{bmatrix}$$

$$C = \begin{bmatrix} -\frac{k_s}{m_s} & -\frac{c_s}{m_s} & \frac{k_s}{m_s} & \frac{c_s}{m_s} \\ 1 & 0 & -1 & 0 \\ 0 & 0 & 1 & 0 \end{bmatrix}$$

$$D = \begin{bmatrix} \frac{1}{m_s} & 0 \\ 0 & 0 \\ 0 & -1 \end{bmatrix}$$

2.4 Multibody dynamics model of intelligent suspension

The intelligent suspension model for manned lunar rover can be simulated and analyzed by using the multi-body dynamics software Adams.

The intelligent suspension studied in this paper is based on the double wishbone independent suspension model for ground vehicles, as illustrated in Fig. 3, which is mainly composed of kingpin, upper cross arm, lower cross arm, pull arm, torsion bar spring, steering rod, steering knuckle, wheel, test platform and other objects.

The suspension and complete vehicle model are established in Adams, and the road excitation is applied to four tires through test platform. In order to meet the foldable requirements of manned lunar rover, the normal strut shock absorber cannot be adopted.

A rotating electromagnetic damper is designed, which is fixed at the end of the upper cross arm to generate damping force to suppress the suspension movement.

The concrete structure of the double wishbone intelligent suspension and the complete vehicle is illustrated in Fig. 4.

The variation range of wheel alignment parameters should be small, and the change trend should be reasonable. By adjusting the position of hard points, the wheel alignment parameters can be optimized, including caster angle, kingpin inclination angle, wheels toe angle and camber angle.

Caster angle is the angle between kingpin axis and ground vertical line in the longitudinal plane of the vehicle. The function of caster angle is to reset the wheels and improve the stability of straight-line driving.

Kingpin inclination angle is the angle between the kingpin axis and the vertical line in the transverse plane of the vehicle. The function of kingpin inclination angle is to make the wheels of the vehicle automatically align after steering. The larger the kingpin inclination angle, the stronger the alignment effect.

Toe angle is the angle between the horizontal diameter of the wheel and the longitudinal vertical plane of the vehicle. The function of toe angle is to balance the negative effect of the camber angle on the tire, and to reduce tire wear and rolling friction.

Figure 3. Double wishbone suspension for ground vehicles

Camber angle is the angle between the actual center line of the wheel and the vertical line. Its function is to make the steering portable and improve the safety of the wheel when working.

In ADAMS, adjust the wheel alignment parameters by changing the hard point parameters, they can be set within a reasonable range to achieve better steering portability and safety.

The variation range of wheel alignment parameters is shown in Fig. 5.

Based on the multibody dynamics model of the intelligent suspension, taking the road excitation and the active force provided by the actuator as the inputs, taking the vertical vibration acceleration of the vehicle body, the dynamic load or deflection of the wheel, the dynamic deflection of suspension as the outputs, the running condition of the manned lunar rover on the lunar surface can be simulated.

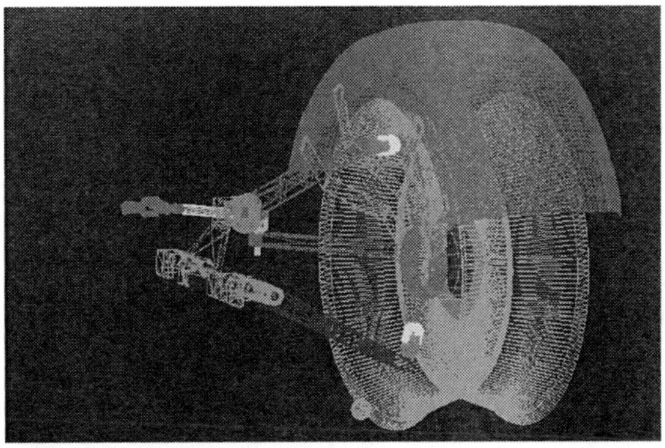

a) double wishbone intelligent suspension

b) complete vehicle model
Figure 4. Multibody dynamics model based on Adams

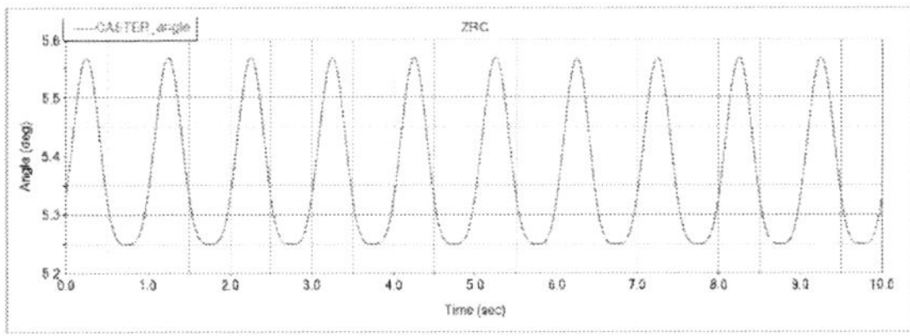

a) caster angle

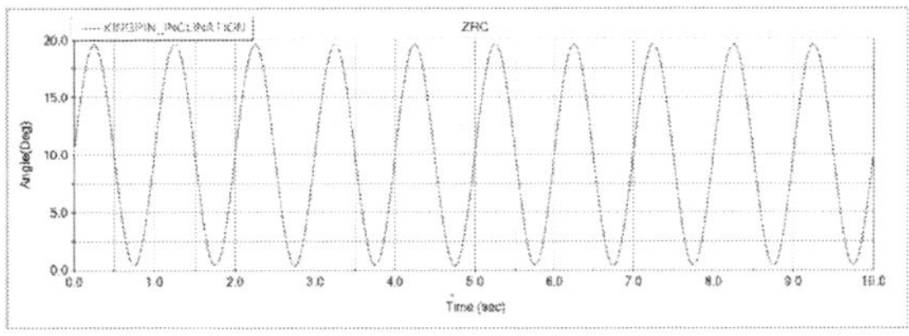

b) kingpin inclination angle

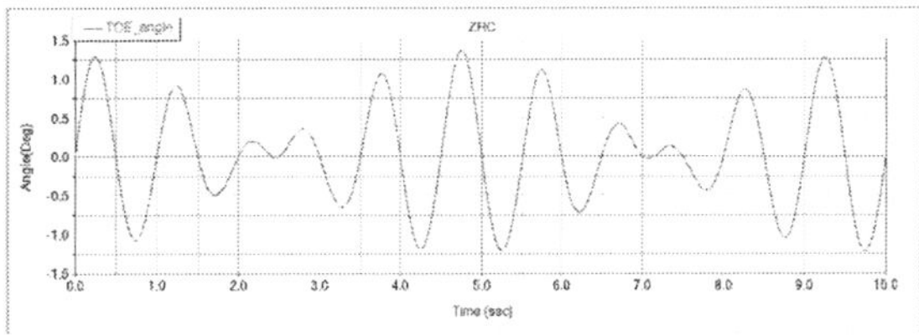

c) toe angle

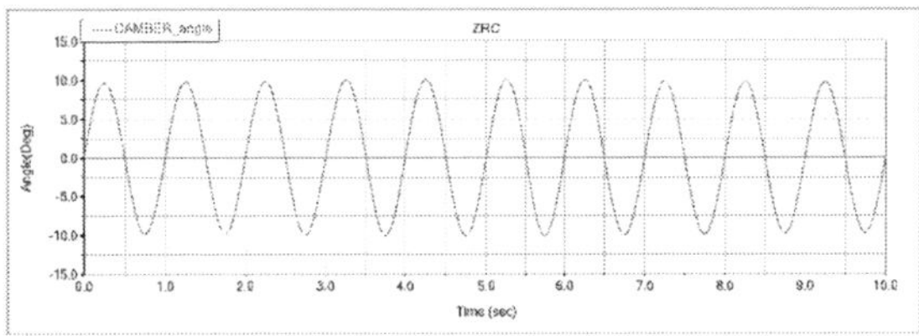

d) camber angle
Figure 5. Optimized wheel alignment parameters

3. Cooperative Simulation Based on Adams and Matlab/Simulink

Multibody dynamics model of intelligent suspension for manned lunar rover is established in Adams. Various control strategies are applied in Matlab. In order to realize cooperative simulation, Adams and Matlab/Simulink are connected through Adams/Control module.

In Adams, the input variables are set as excitation of lunar surface and the control force provided by the actuator, the output variables are set as body acceleration, suspension dynamic displacement and tire dynamic deflection.These variables are not only the response of suspension vibration, but also the feedback signal of the intelligent suspension control system, which determines the working state of the actuator.

Based on the multibody dynamic model of manned lunar rover, different control strategies are adopted to reflect the damping effect of intelligent suspension, as can be seen in Fig. 6.

The excitation of lunar surface is transmitted to the vehicle body through the wheel and suspension subsystem. The intelligent suspension can adjust the electromagnetic damping through the actuator, and the actuator can provide active control force to suppress the vibration of the suspension cross arm, The actuator provides active control force to suppress the vibration of the suspension cross arm, thus maintaining the stability of the vehicle body.

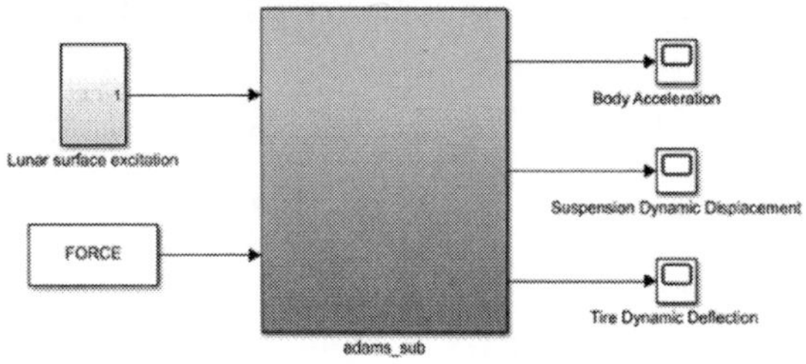

Figure 6. Block diagram of cooperative simulation

3.1 Excitation of lunar surface

As the external input, excitation of the lunar surface affects the stability and driving comfort of the manned lunar rover.

Typical road conditions on the lunar surface can be divided into two categories, impact pavement and random vibration pavement.

By using the signal simulation of triangular wave, rectangular wave or sine wave, the impact pavement in space domain can be transformed into that in time domain; The random vibration pavement on the lunar surface can be expressed by the change $q(I)$ of the road roughness (the height q of the actual road surface relative to the base plane along the length I of the road,

$$G_q(n) = G_q(n_0).(n/n_0)^{-W} \quad (7)$$

n is the spatial frequency, representing the number of waves per meter of length; n_0 is the reference spatial frequency, generally taken as 0.1 m^{-1}; $G_q(n_0)$ is the roughness coefficient of the lunar surface; W is the frequency index, generally taken as 2.

When the manned lunar rover drives on the lunar surface with spatial frequency n at the speed of $v(m/s)$, the spatial-frequency density function of the lunar surface is converted to obtain the time-frequency density function of the road spectrum.

$$G_q(f) = \frac{1}{v}G_q(n) = G_q(n_0){n_0}^2\frac{v}{f^2} \quad (8)$$

Based on the data of lunar surface provided by NASA, the power spectral density of lunar surface is illustrated in Fig. 6, including two typical working conditions: flat lunar surface and rugged lunar surface.

Taking n_0 as $0.1m^{-1}$, under these two working conditions, $G_q(n_0)$ is equal to 8.3×10^{-3} m^2/m^{-1} and 4.45×10^{-2} m^2/m^{-1} on flat lunar surface and on rugged lunar surface respectively, the driving speed of manned lunar rover is set as 18 km/h and 3 km/h respectively.

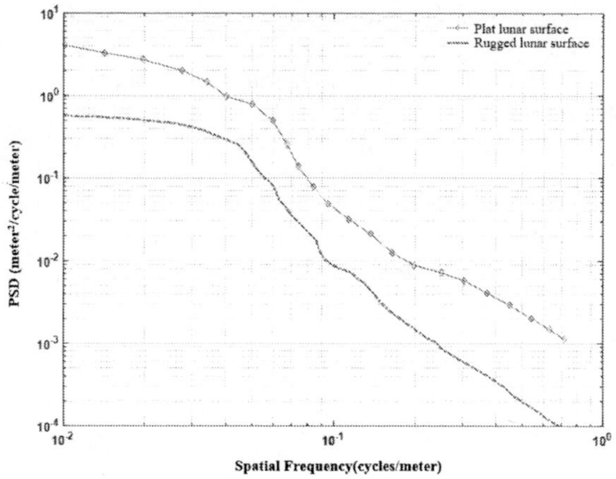

Figure 7. Power spectral density of lunar surface

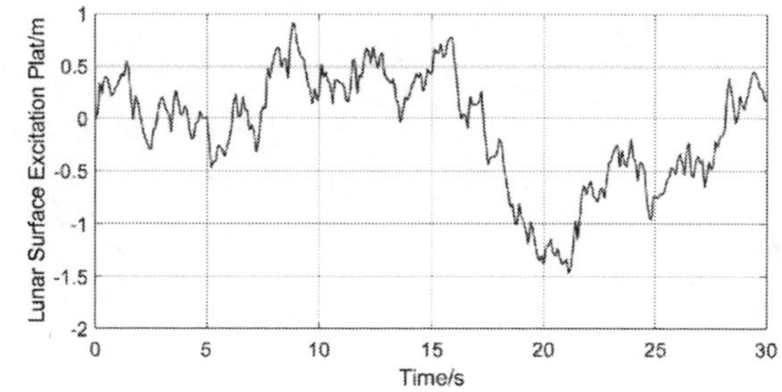

Figure 8. Excitation of plat lunar surface with speed of 18 km/h

3.2 Design of LQR control system

Linear quadratic regulator (LQR) is an optimum control method. Based on the idea of optimization, the best suspension damping effect can be achieved by setting different weighting factors.

The purpose of applying LQR control strategy is to improve the driving comfort of the manned lunar rover on the premise of ensuring the driving safety.

While ensuring the dynamic displacement of the intelligent suspension bearable, the vertical acceleration of the body and the tire dynamic deflection should be restricted as much as possible.

The objective function is created:

$$J = \int_0^\infty \left[q_1\ddot{x}_s^2 + q_2(x_s - x_w)^2 + q_3(x_w - x_g)^2\right]dt \quad (9)$$

q_1, q_2, and q_3 are the weighting factors of vertical body acceleration, suspension dynamic displacement and tire dynamic deflection respectively.

According to LQR control principle, the function can be converted to

$$J = \int_0^\infty (X^TQX + U^TRU + 2X^TNU)dt \quad (10)$$

Based on the previously described equation of state in section 2.2, weight matrix can be obtained.

$$Q=\begin{bmatrix} (\frac{k_s}{m_s})^2 q_1+q_2 & \frac{k_s c_s}{m_s^2} q_1 & -\left(\frac{k_s}{m_s}\right)^2 q_1-q_2 & -\frac{k_s c_s}{m_s^2} q_1 \\ \frac{k_s c_s}{{m_s}^2} q_1 & (\frac{c_s}{m_s})^2 q_1 & -\frac{k_s c_s}{m_s^2} q_1 & -\left(\frac{c_s}{m_s}\right)^2 q_1 \\ -\left(\frac{k_s}{m_s}\right)^2 q_1-q_2 & -\frac{k_s c_s}{m_s^2} q_1 & \left(\frac{k_s}{m_s}\right)^2 q_1+q_2+q_3 & \frac{k_s c_s}{m_s} q_1 \\ -\frac{k_s c_s}{m_s^2} q_1 & -\left(\frac{c_s}{m_s}\right)^2 q_1 & \frac{k_s c_s}{m_s^2} q_1 & \left(\frac{c_s}{m_s}\right)^2 q_1 \end{bmatrix}$$

$$R=\frac{q_1}{m_s^2}$$

$$N=\begin{bmatrix} -\frac{k_s}{m_s^2} q_1 & 0 \\ -\frac{c_s}{m_s^2} q_1 & 0 \\ \frac{k_s}{m_s^2} q_1 & -q_3 \\ \frac{c_s}{m_s^2} q_1 & 0 \end{bmatrix}$$

When the weight factors of vehicle parameters and performance indicators are determined, the LQR control feedback gain matrix can be obtained by Riccati formula.

The form is as follows:

$$\mathrm{PA}+\mathrm{A}^{\mathrm{T}}\mathrm{P}-(\mathrm{PB}+\mathrm{N})\mathrm{R}^{-1}\left(\mathrm{B}^{\mathrm{T}}\mathrm{P}+\mathrm{N}^{\mathrm{T}}\right)+\mathrm{Q}=0 \quad (11)$$

$$K=R^{-1}(B^T P+N^T) \quad (12)$$

After setting the weight factors q_1, q_2, and q_3, the control force provided by the actuator can be obtained.

$$U=-K.X \quad (14)$$

For the manned lunar rover, the primary goal is ensuring the safety rather than safety. When selecting the weight factors, it needs to be biased towards tire dynamic deflection to achieve better grounding performance.

3.3 Comparison of control effect

PID and LQR control strategies are used to control the intelligent suspension of manned lunar rover, and the comparison of vibration reduction effect is shown in Fig. 9:

In order to compare the effect of different control strategies more clearly, the time domain diagram is converted into the root mean square value of each evaluation index, as shown in Table1. (a_B represents vertical body acceleration; d_S represents suspension dynamic displacement; d_T represents tire dynamic deflection.)

As illustrated in Table 1 and Fig.7, under the same excitation of lunar surface, compared with the passive suspension, the vertical body acceleration of the intelligent suspension controlled by PID and by LQR is reduced by 18.6% and 14.5% respectively, more importantly, the peak value of body acceleration is significantly reduced. It means that the stability and comfort of the manned lunar rover have been improved.

Compared with passive suspension, the suspension dynamic displacement of the intelligent suspension controlled by PID and by LQR is reduced by 7.73% and 18.8%, the optimal effect of LQR control is better than that of PID control. It represents better body posture and less danger of suspension overstroke.

Compared with passive suspension, the tire dynamic deflection of the intelligent suspension controlled by PID and by LQR is reduced by 3.58% and 19.1%, representing better grounding performance and safety performance.

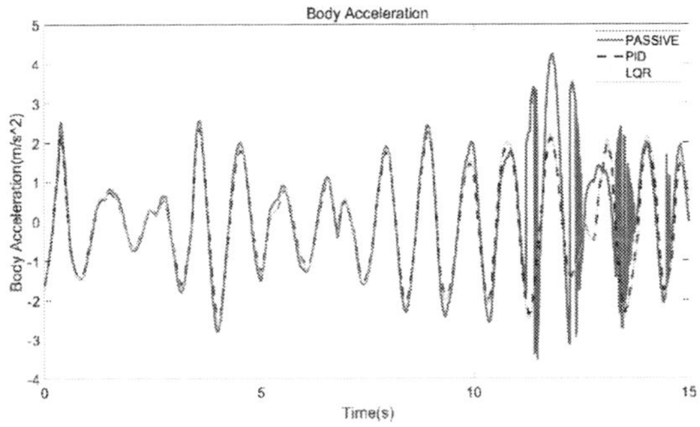

a) Control effect of vertical body acceleration

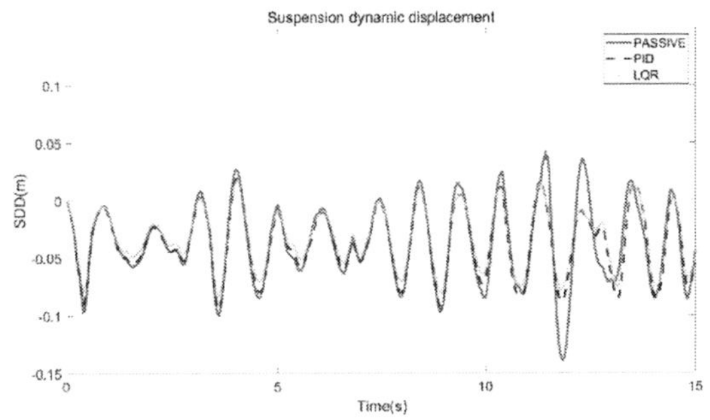

b) Control effect of suspension dynamic displacement

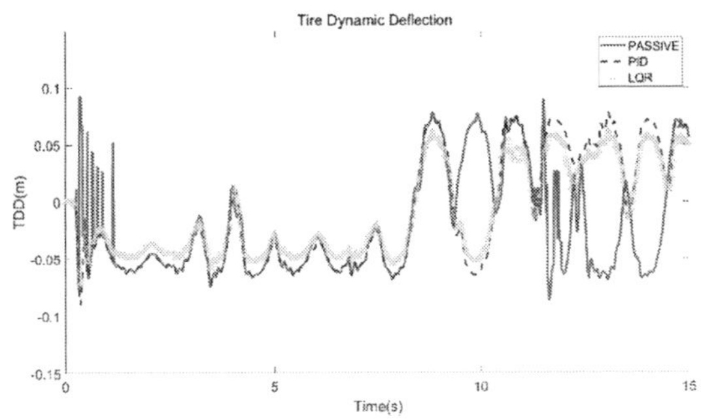

c) Control effect of tire dynamic deflection

Figure 9. Comparison of control effects

Table 1. Root mean square value of evaluation index

RMS	a_B[m/s^2]	d_S[m]	d_T[m]
PASSIVE	1.449	0.0504	0.0502
PID	1.179	0.0465	0.0484
LQR	1.238	0.0409	0.0406

4. Conclusions

Based on the designed structure intelligent suspension for manned lunar rover, the vibration model and multi-body dynamics model are established.

The research conducts cooperative simulation, PID and LQR control strategies are applied to improve vibration damping performance. By comparing and analyzing the control effect, the conclusions are as follows:

Compared with passive suspension, the intelligent suspension controlled by PID and by LQR have better mobility performance, and the evaluating indicators have certain improvement. In terms of the vertical body acceleration, the most significant improvement is the sharp reduction of peak value, representing better stability and driving comfort in specific extreme conditions. However, there is still room for improvement in the body acceleration during normal driving.

Based on the particularity and complexity of the driving environment of the manned lunar rover, safety should be taken as a priority. For the consideration of improving grounding performance, LQR control strategy can improve the weight factor of tire dynamic deflection. Therefore, the safety and grounding performance is greatly improved.

Although the reduction of suspension dynamic deflection is not emphasized when designing the controller, it still has a good optimization effect when applying LQR control strategy. In the subsequent research, other optimization algorithms can be used to select more appropriate weight factors to achieve the optimal overall performance.

In conclusion, the application of intelligent suspension technology in manned lunar rover is effective and necessary. However, further adjustment is still needed to achieve better mobility performance.

5. References

E. G. Cowart, 1973. Lunar Roving Vehicle: Spacecraft. J. Proceedings of the Institution of Mechanical Engineers. 187, 463-491

Giancarlo Genta, Lorenzo Morello, 2009. The Automotive Chassis, first ed., Springer US, New York.

Gysen B, Van D, Paulides J, et al. 2011.Efficiency of regenerative direct-drive electromagnetic active suspension. J. IEEE Transactions on Vehicular Technology, 60(4): 1384-1393.

Hsu H W., Horanyi M., 2012. Ballistic motion of dust particles in the lunar roving vehicle dust trails. J. American Journal of Physics. 80,452-456.

J. Blanco, L. Munoz. 2013.Methodology for the Design of a Double- wishbone Suspension System by Means of Target Cascading: Model-geometry Interactions and Case Study. SAE 2013 World Congress and Exhibition.

Jie Chen, Yangjun Wu, Limin Zhang, 2020.Suspension parameter design of underframe equipment considering series stiffness ofshock absorber. J. Advances in Mechanical Engineering, 12(5), PP1687814020922647

Liang Z C., Wang Y F., Chen G., 2015. A mechanical model for deformable and mesh pattern wheel of lunar roving vehicle. J. Advances in Space Research. 56, 2515-2526.

Lund T., 2018. Lunar Roving Vehicle and Exploration of the Moon: Ranger to Apollo, Luna to Lunniy Korabl. J. Early Exploration of the Moon.11, 303-324.

Montazeri-Gh M, Soleymani M, Hashemi S. 2013.Impact of traffic conditions on the active suspension energy regeneration in hybrid electric vehicles. J. IEEE Transactions on Industrial Electronics, 60(10):4546-4553.

Rajesh Rajamani, 2012. Vehicle Dynamics and Control, second ed., Springer US, New York.

Thompson A G, Davis B R. 2007.Technical note on the lotus suspension patents. J. Vehicle System Dynamics, 20(6):381-383.

XI Jiasheng, NIE Hong, CHEN Jingbao, 2013. Optimization analysis of steering breaking point position of torsion bar folding suspension for manned lunar rover. J. Science Technology and Engineering, 13(28): 8080-8085+8395.

YU Fan, LIN Yi. 2016.Automobile System Dynamics, second ed., China Machine Press, Beijing.

0861 / Review on the Reconfigurable Wheel-Tracked System

Jiaxuan Wang[a,b], Cheng Liu [a,b]*, Wei Wei[a,b] Qingdong Yan[a]
[a]*Beijing Institute of Technology, Beijing, China*
[b]*Beijing Institute of Technology Chongqing Innovation Center, Chongqing, China*
**Corresponding author: liucheng@bit.edu.cn*

https://doi.org/10.56884/CXRT8456

Abstract

Reconfigurable wheel-tracked system, which enables a vehicle to have the high speed of wheels and the increased mobility of tracks, is one of the key technologies on improving the off-road performance of vehicles in extreme environment. A series of such systems has already been developed and equipped on robots and medium or heavy ground vehicles. The paper reviews the research work, prototypes and applications of reconfigurable wheel-tracked systems. To start with, different systems are classified according to their mechanical structures. Development and current research interests are then reviewed. Triangle track configuration is the most popular layout of the wheel-tracked system, so this paper highlights its structure characteristics, research contents and applications. Following the review, key technologies, including design of power transmission system and elastic track, are analyzed. Finally, prospective of the reconfigurable wheel-tracked locomotion system is discussed in terms of transformation principle innovation, depth of research and width of application.

Keywords: wheel-tracked system, mobility, power transmission layout, elastic track

1. Introduction

It has been difficult for single-wheeled or tracked locomotion mechanisms to meet the requirements of special vehicles taking into account high speed, economy, and mobility. A high-mobility chassis compatible with wheels and tracks has attracted a large amount of interest. Vehicles with a combined chassis[0] are driven by tires and tracks at the same time, but the locomotion system cannot be switched in order to adapt to road conditions. Although an alternate wheel-tracked locomotion system[0] can be transitioned between wheel and track mode depending on road conditions, generally, two locomotion systems need to be implemented on the same chassis, leading to a relatively large mass and long mode transition time. A reconfigurable wheel-tracked locomotion system integrates two locomotion devices, namely wheels and tracks, into a system. This mechanism can maximize the utilization of existing vehicle chassis. The wheel/track mode can be switched at any time during travel, greatly improving a vehicle's mobility performance in off-road environments such as beaches, ramps, wetlands, and ravines, and great interest has been reported in the research of wheel-tracked locomotion systems.

2. Primary Structural Forms of Reconfigurable Wheel-tracked Systems

There have been many studies on reconfigurable locomotion system. The results of these studies can be divided into two categories, as shown in *0*, in terms of the track shape and shaft relationship. According to the track shape in the track mode, the locomotion systems can be divided into three types: triangular tracks, polygonal tracks, and arched tracks. According to the positional relationship between the track sprocket axis and the wheel main shaft in the crawler mode, these systems can be divided into the coaxial shaft type and the parallel shaft type. A locomotion system in which the axis of the sprocket and the main shaft of the wheel coincide is a coaxial-shaft locomotion system. Likewise, a locomotion system for which the two axes are parallel but not coincident is a parallel-shaft locomotion system.

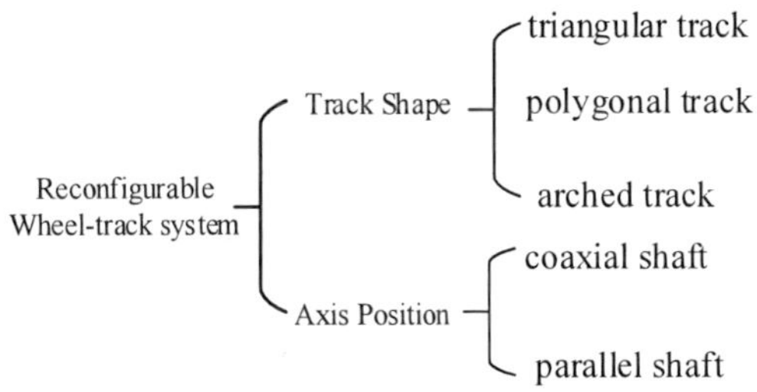

Figure 1. Main structure of reconfigurable wheel-tracked system

2.1 Triangular Track

2.1.1 Coaxial Shaft Type

A coaxial-shaft reconfigurable wheel-tracked system was implemented for the first time by Israeli researchers. In 2003, the Israeli company Galileo Mobility Instruments patented a reconfigurable wheel locomotion mechanism. As shown in *0*, the locomotion system used hydraulic cylinders and leaf springs as the wheel-to-track deformation mechanism to implement the mode conversion. The system achieved three reconfigurations: wheeled, long contact patch track mode, and short contact patch track mode. Vehicles using this locomotion system could change the height of the chassis from the ground during transformation[0]. In 2007, the Israeli military developed a small robot called VIPeR based on this technology.

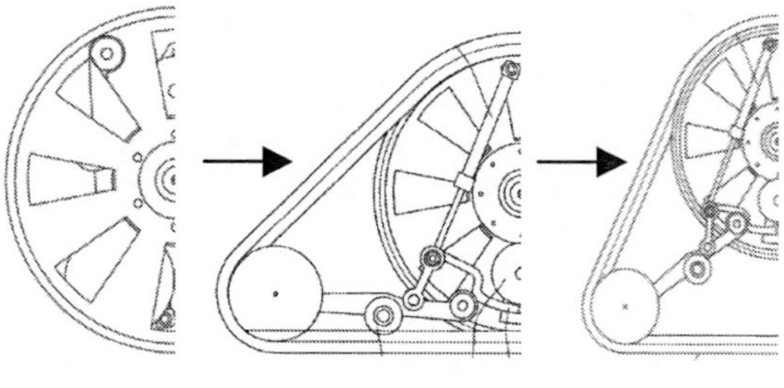

Figure 2. Reconfigurable locomotion system (Galileo Mobility Instruments)

Figure 3. VIPeR

0 shows a mini type robot of Beijing Institute of Technology (BIT). Researchers started early in studying reconfigurable wheel-tracked systems. Gao et al. designed a tail-rod type small search and reconnaissance robot[0] as well as a wheelchair based on this with high mobility[0,0]. The design method of a four-shaft wheel-to-track deformable device and a screw drive was proposed. The position of each hinge point of the deformable structure device was determined based on a graphical method[0] , and then the force state of the transformation device was analyzed in depth[0] . Because the tail rod greatly affected the robot's mobility, the process of climbing steps in both the forward and reverse directions in the wheel/track configuration with/without tail rod assistance was studied in detail. The static and dynamic mechanics models of each climbing stage were established. Obstacle-surmounting rules with respect to step height were established. For terrains such as flat ground, steps, ravines, and slopes, the geometrical mobility, steering, and emergency stop stability of the robot in wheel and track modes were analyzed with the centroid method[0]. Then, with manipulators as the typical load of a robot, the effects of their attitude and dynamic characteristics on the obstacle-surmounting performance of the robot in the track mode were studied [0-] . The research established a relatively complete theoretical method for the design and development of a small wheel-tracked robot.

Figure 4. Wheel-tracked robot (Beijing Institute of Technology)

Luo et al. of the National University of Defense Technology (NUDT) established the kinematics and dynamics models of a wheel-to-track conversion mechanism, derived the motion trajectory of a support wheel and the driving torque during mode switch, and obtained and optimized the driving torque as a function of the driving angle through simulation[0,]. The optimized driving torque was significantly smaller and has better continuity, effectively reducing the load of the driving motor (*0*). Zhang et al. of Yanshan University accurately predicted the speed characteristics, acceleration characteristics, and driving torques required for the support wheels during transformation. The research provided an important reference for the design and optimization of the same type of wheel track conversion mechanism[0-].

Liu et al. at Tianjin University of Technology and Kang et al. of Army Military Transportation University applied a wheel-tracked system (*0*) on a distributed driving military ATV, and they conducted a field test of the first domestic wheel-tracked prototype vehicle[0]. As shown in *0*, the metal segmented track were elastic using built-in leaf springs. A self-locking mechanism was designed within each segment to prevent the track from being loosened in the wheeled high-speed mode. This design led to a complex and heavy track structure. One single system weighs high up to 56 kg. The prototype was considered with weight of 1.2 t and wheel radius of 280 mm. The maximum speed in the wheel mode was 25 km/h. The maximum step height was 180 mm. The maximum climbing angle was 28°. The maximum width across the trench was 400 mm. The wheel-to-track mode transition could be completed within 10 s[0]. The system could also be used on off-road vehicles such as wheel-tracked rescue robots[0], agricultural machinery[0], and trucks[0].

Researchers at Harbin Institute of Technology and Chinese Academy of Sciences also conducted in-depth research on similar locomotion systems[0,].

To summarize, the coaxial-shaft triangular track reconfigurable systems represented by the VIPeR robot started to be published early on. Their principles were relatively simple. In the wheel and track modes, the same mechanism was used to drive the tracks, and there was no need to consider power switching. Due to the relatively large change in the track length when the configuration was changed, it was necessary to design suitable elastic tracks. These mechanisms have been primarily applied to mobile platforms below 1.5 t, and have been widely used on small mobile platforms such as two-wheeled robots and intelligent wheelchairs. For light vehicles such as military vehicles, rescue vehicles, and trucks, these locomotion systems have primarily been studied using computational simulation and tests. Applications of these mechanisms to heavy vehicles have not been studied yet. In structural design, most designs use hydraulic cylinders or linear motors to drive deformable mechanisms for wheel-to-track conversion, and thus have similar power transmission layouts. The research in China primarily focuses on the kinematics analysis, dynamic analysis, obstacle-surmounting ability analysis, dynamic parameter matching, suspension system optimization, and optimal design of the mode transition mechanisms of mobile platforms.

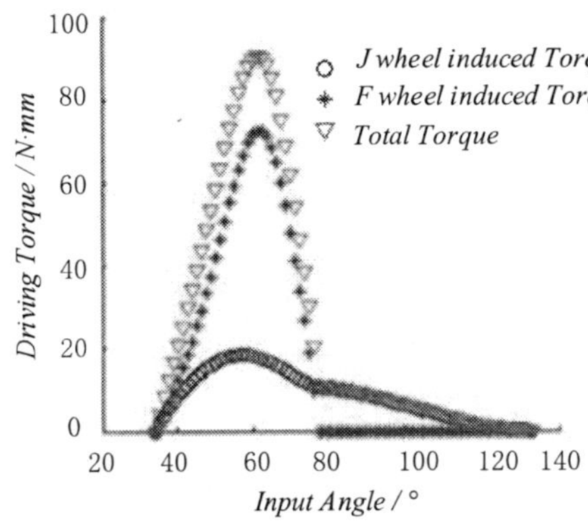

a) Original driving torques

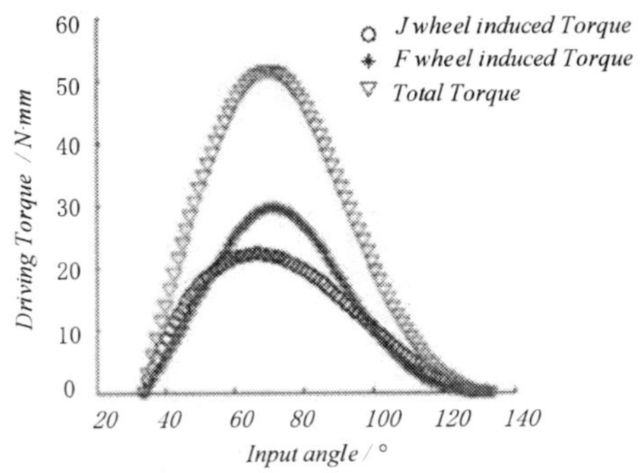

b) Optimized driving torque

Figure 5. Optimal design of the shape conversion mechanism

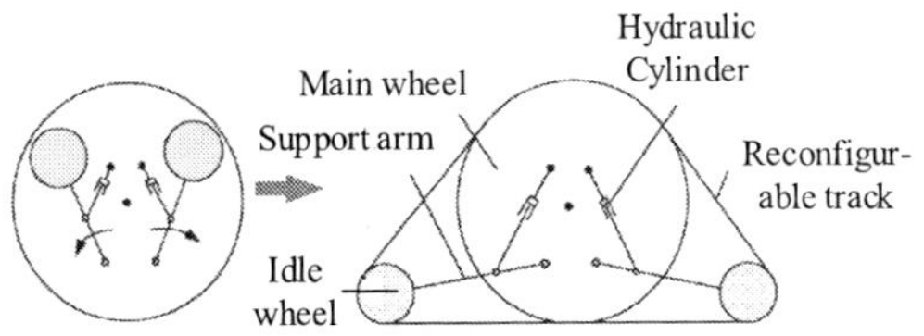

Figure 6. Reconfigurable wheel-tracked system (Tianjin University of Technology)

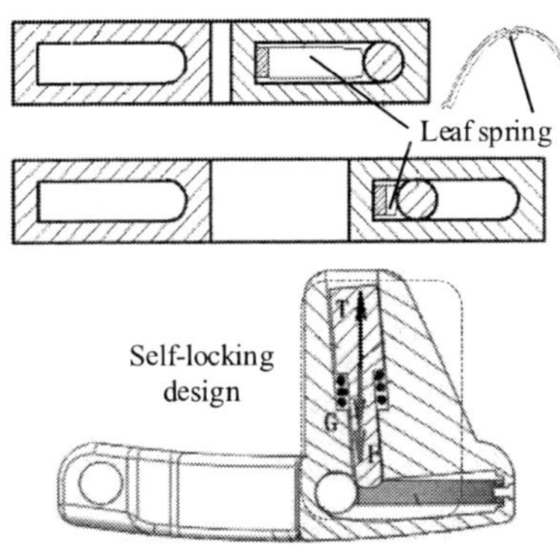

Figure 7. Metal reconfigurable track

2.1.2 Parallel Shaft Type

The parallel-shaft reconfigurable wheel-tracked system originated from the Ground X-Vehicle Technology (GXV-T) project proposed by the US Defense Advanced Research Projects Agency around 2014. Hannah et al. from Carnegie Mellon University (CMU) had an in-depth research on locomotion system of this kind. *0* shows a US military vehicle equipped with the system. During transition, the sprockets are radially extended and retracted with the action of the electric actuator to switch between wheel and track modes. A clutch was installed between the wheel and the frame. In the wheel mode, the clutch locked the wheel and the main shaft of the vehicle, and the two rotated together. In the track mode, the clutched started to engage, locking the main structure of the assembly in place, and the deformed wheels moved in translation with the vehicle as a whole. In the track mode, the power from the shaft was transmitted to the sprockets through the planetary gears inside the wheels so that the speed of the vehicle remained unchanged during the wheel-to-track transition[0]. Tests showed that when the vehicle speed was in the range of 5~35 km/h, the relative error between the vehicle speed and the target speed could be controlled within 6% during the wheel-to-track mode transition, and the entire transition process lasted 5~60 s[0].

Chen et al. of the China Northern Vehicle Research Institute (NOVERI) studied the application of this type of reconfigurable locomotion mechanism on heavy-duty vehicles. The drawing of the structure is shown in Figure 8. The locomotion mechanism drew on the idea of a distributed drive. An integrated drive system with a motor series reducer was set in the track and wheel drive systems. This method could decouple the sprocket's translation along the radial axis and the rotation around the axis[0]. With performance matching, the performance parameters of the legged unit, such as the grounding specific pressure and the driving and braking ability, were close to the standard for traditional military vehicles[0].

Figure 8. U.S. military vehicles equipped with the RWT system

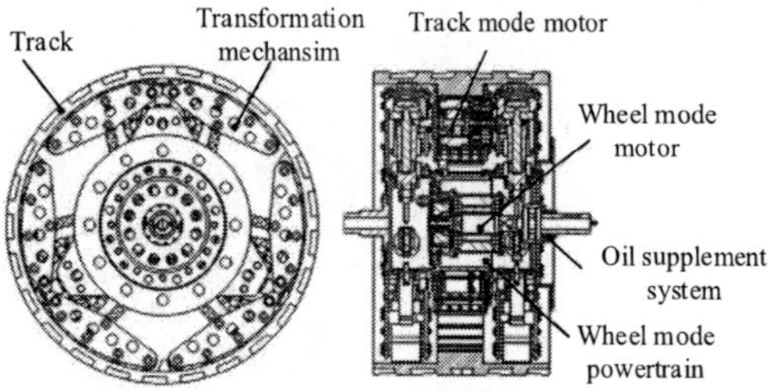

Figure 9. Reconfigurable wheel-tracked system (China Northern Vehicle Research Institute)

Liu et al. of the China General Nuclear Power Group applied deformable wheels and control systems to a disassembly robot[0]. During the transition process of the deformable wheels, the sprockets did not translate in the radial direction, effectively decoupling the moving pairs and the rotating pairs on the sprockets and reducing the design difficulty. The deformable wheel adopted two sets of drive mechanisms that were responsible for the drive control in wheel and track modes, with a simple structure, coherent mode transition, and direct and effective control[0].

Figure 10. Deformable wheel (China General Nuclear Power Group)

The researchers of the Chinese Academy of Sciences, Shaanxi University of Science and Technology, and Jilin University have also conducted research on similar locomotion mechanisms using in-wheel motors[0], shifting forks[0], and sliding spline shafts[0], respectively, as power switching mechanisms in the process of wheel-to-track transition.

The parallel-shaft triangular track reconfigurable locomotion mechanism represented by the RWT system was released later on. The principle of the mode transition mechanism was complex. The track length is almost unchanged during the wheel track conversion process. It was difficult to achieve power switching during conversion. Thus, the research started with a high threshold. Research on the GXV-derived reconfigurable wheel-tracked locomotion system in China is still in its infancy, and the research mainly focuses on conceptual design, theoretical verification of geometric mobility, and strength verification of structural parts.

2.2 Polygon Track

Figures 11 and 12 show the polygonal wheel-tracked mechanism designed by Richard et al. of Carnegie Mellon University. The locomotion mechanism shown in Figure 10 implements the conversion between the wheel and track modes by rotating a total of four inner and outer support plates around the main axis of the wheel[0]. In the wheel mode, the locomotion mechanism drives the integral rotation through the wheel spindle. In the track mode, the mechanism translates with the vehicle through the clutch and vehicle body engagement. The two ends of a sliding spline shaft in the locomotion mechanism are connected by a sprocket and a wheel main shaft through bevel gears, which are used to transmit power in the track mode. The locomotion mechanism shown in Figure 12 is in an arched shape in track mode[0], and its deformable structure device is similar to that of the coaxial triangular system described above.

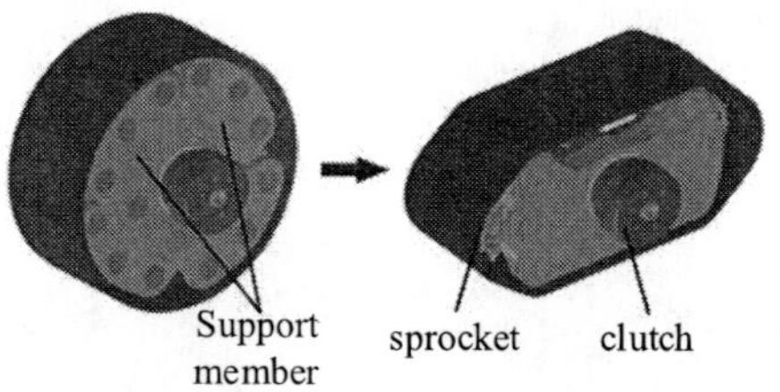

Figure 11. Diamond-shape(polygonal) track[0]

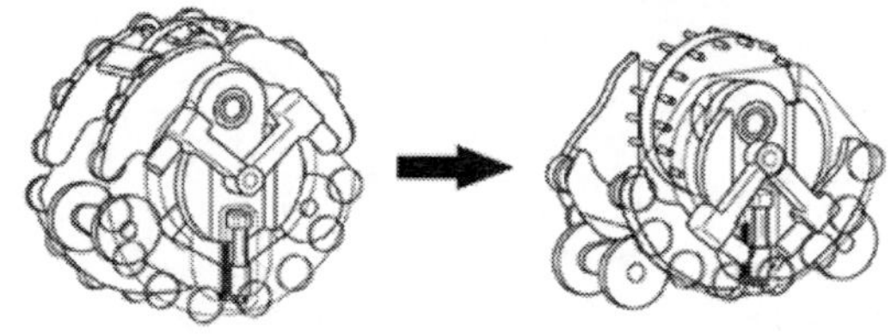

Figure 12. Arch-shape(arched) track[0]

The Mattracks Inc. developed an arched coaxial-shaft system that included a collapsible bladder, a track, pneumatic cylinders, and idler wheels (0). In the track mode, the bladder deflates and retracts, and the idler gears push outward, driven by the cylinders to support the track and support the vehicle body. The teeth on the track mesh with the sprockets. In the wheel mode, the bladder is inflated to form the casing of the tire, and separated from the sprockets, the track becomes the tire threads. The two work together in the wheel mode. The prototype vehicle (*0*) equipped with deformable wheels weighed about 1 ton. The results of the power, maneuverability, and mobility tests showed that the wheel-tracked system could improve the traction power, climbing ability, and mobility of the vehicle on a soft surface. However, due to the heavier weight of the mechanism, the maximum speed was lower than that for the same type of wheeled vehicle[0].

Zhou et al. of National Defense University of Science and Technology proposed a multi-mode motion robot based on wheel rim deformation. The robot effectively integrated wheel, track, and leg locomotion modes by folding the wheel rim. It had a compact structure and high terrain adaptability[0] . Qu et al. of South China University of Technology used contractible wheel rims to design a wheel-tracked mechanism which ensured that the vehicle floor clearance remained unchanged before and after deformation, and they manually operated the crank to drive bevel gears, worm gears, and sliding splines to achieve configuration conversion[0].

The wheel/track manned lunar rover developed by Jiang et al. of the China North Vehicle Research Institute adopted a track module for its rear wheels. The module could transition between full track, half-track, and wheel modes. Thus, the ground-engaging length, approach angle, and ground-engaging position of the track of the locomotion mechanism were changed to improve the obstacle-surmounting performance of the lunar rover[0]. Bai et al. applied a similar design to a small unmanned mobile platform with wheel-track-leg coupling locomotion system (*0*) and adopted different obstacle-surmounting strategies for different obstacle heights to enable the platform to autonomously create an obstacle-surmounting maneuver plan[0]. This configuration focused on the difference in the contact patch of the tire and the track. Although it did not involve wheel-to-track shape conversion, the same effect was achieved functionally.

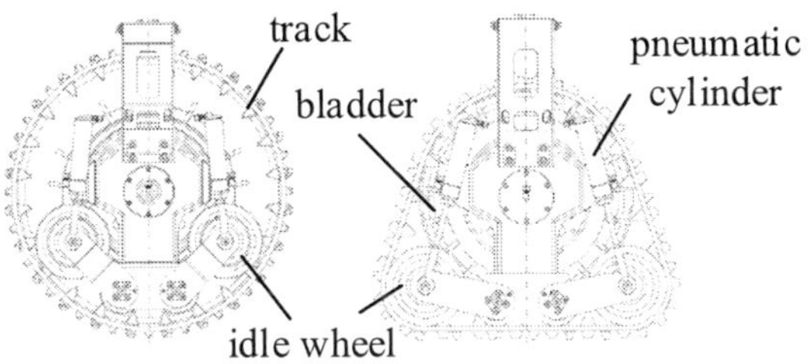

Figure 13. Mattracks Inc. wheel-tracked system

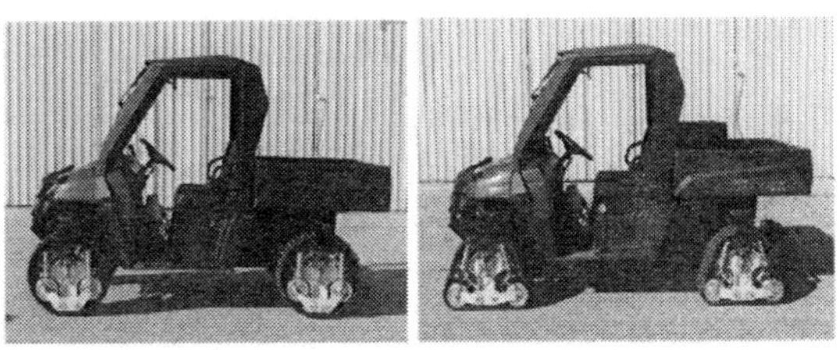

Figure 14. Mattracks Inc. tested vehicle

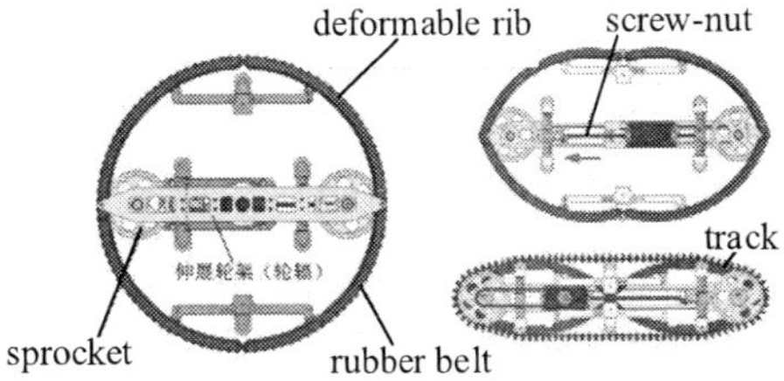

Figure 15. Deformable wheel[0] (National Defense Science and Technology University)

Figure 16. Wheel-track-leg coupling robot

3. Technical Difficulties of Reconfigurable Wheel-tracked Systems

3.1 Power Transmission Layout Design

The two prevalent triangular track reconfigurable locomotion system represent two power transmission layouts. Represented by reference 0, the coaxial locomotion system connects itself to the vehicle chassis and translates with the chassis in both wheel and track modes. In both modes, the sprocket drives the track. Therefore, in the process of mode transition, it is only necessary to change the shape of the track with a deformable device, and there is no need to deal with power switching. Represented by reference 0, the parallel-shaft locomotion system is driven by the wheel main shafts in the wheel mode and the sprockets in the track mode. At present, the highly integrated solutions use a clutch as the power switching mechanism, as shown in 0, 0, and 0. Each locomotion system includes two clutches. Clutch A works to lock the assembly to the chassis, and B locks the whole mechanism to the wheel main shaft. In wheel mode, clutch A is disengaged, and B is engaged, and the entire system rotates as a whole driven by the wheel main shaft, operating in the same manner as ordinary wheels. In track mode, clutch A is engaged, and B is disengaged, which mounts the frame of the locomotion mechanism to the vehicle chassis. A set of mechanisms including planetary gears, bevel gears and worm gears transmits the power from the wheel's main shaft to the sprockets. The driving sprockets are wound around the track, operating in a way similar to the driving wheels on a tracked armored vehicle. *Table 1* compares several typical powertrain layouts.

Table 1. Comparison of powertrain layouts

Institute	Type of the system	Driving Mechanism	Power Switch Method
BIT	Triangular Track, Coaxial shaft	Inner/outer positioning ring[0]	no power switch
CMU	Triangular Track, parallel shaft	wheel mode: wheel main shafts track mode: sprockets[0]	clutch A clutch B
NOVERI	Triangular Track, Parallel Shaft	wheel/track motor connected with a reducer[0]	motor control system, track brakes
NDUT	Polygonal Track, Parallel Shaft	driven by motors in wheel/track mode respectively[0]	motor control system

3.2 Design of Elastic Track

In designing a deformable track, the decision to deploy metal tracks or non-metal tracks requires comprehensive consideration of the driving conditions, ground type, reliability, total weight, maneuverability, and mobility requirements. Metal tracks have high mobility, and good stability in crossing vertical walls and climbing longitudinal slopes and transverse slopes. They are also strong, wear-resistant, and protective. However, the elasticity of a metal track prevents it from meeting some reconfiguration requirements, and reconfiguration from a triangular track to a circular track can only be achieved by adding elastic elements0. At present, only the prototype presented in reference 0 uses metal tracks. Friction between the driving wheel and the track is used as the main connection mechanism0, aided by plate-hole meshing as the secondary connection, resulting in a low power transmission efficiency. Fish-scale segmented tracks[0] adopt wear-resistant rubber as the track grousers and connect the aluminum track shoes with springs and hinges, which are somewhat elastic and improve reliability.

Non-metal tracks possess high elasticity and can easily achieve automatic reconfiguration during mode transition. However, the operating environments for mobile platforms are harsh, and thus the tracks need to be retracted frequently, creating strict requirements for wear resistance, elasticity, corrosion resistance, and tensile fatigue strength for the tracks. The key aspect of non-metal track design is elasticity control. That is, the track must be highly elastic so that it is able to continue to fasten the locomotion mechanism after frequent expansion and contraction; additionally, the track should avoid been overstretched to failure. The primary implementation method is to directly make tracks out of high elastic materials, add elastic elements to enhance elasticity [0,0,0] , and use a limit control measure for protection[0].

The change of the track length during mode conversion affects the tension of a track, which in turn affects the meshing loss of the sprockets and the traction performance of the locomotion system [0]. When operating in the track mode, the additional friction force F_f generated between the loose edge and the tight edge of the track can improve the power transmission efficiency of the system. Increasing the wrap angle of a sprocket, using meshing transmission, increasing the friction coefficient of the track, and increasing the meshing depth are all beneficial for the increase of additional friction [0].

4. Summary and Outlook

A reconfigurable wheel-tracked system combines the cost-effectiveness and mobility of a wheel with the high mobility of a track. It can adopt different locomotion methods according to the terrain characteristics and improve the adaptability of a vehicle to an off-road environment. This technology is of great significance for advancing the development of locomotion technology for high-mobility ground mobile platforms. At present, the research and application of this system are still in the development stage. The future major development of this technology is described below.

4.1 Principle Innovation

At present, there are only two configurations of mainstream reconfigurable locomotion mechanisms. They were first developed outside China and then widely studied in China. Most research institutions do not use the guidance of scientific methodologies such as the kinematics of machinery and topology in designing reconfigurable mechanisms, and thus systematic design methods are lacking. Zhou[47] of National University of Defense Technology combined AD theory (Axiomatic Design Theory) and TRIZ theory (Theory of Inventive Problem Solving) to guide the design of rim-deformable mobile robots. This approach was instructive. In the future, it will be necessary to further strengthen independent innovation at the principle level and own completely independent intellectual property rights for all design aspects such as the design of the wheel-to-track deformation mechanism, elastic track, and power switch method/mechanism.

4.2 In-Depth Study of the Performance Characteristics of Locomotion Systems

Until now, the studies on the performance of reconfigurable locomotion mechanisms have primarily focused on the system level, such as mobility, maneuverability, and power system parameter matching. These studies lack the optimal design of the performance of the mechanism itself. In the future, researchers should continue to conduct in-depth studies to tackle the technical difficulties such as the driving force/torque and control strategy of the mode transition mechanism, overall dynamic balance of the system, suppression of the belt-loosening phenomenon in the high-speed wheel mode, and design of the power switch mechanism. These studies allow researchers to explore the intrinsic performance characteristics of a reconfigurable mechanism.

4.3 Expansion in Engineering Application

In China, the VIPeR-based locomotion mechanism has only been widely used on small mobile platforms. It has not been widely equipped on light off-road vehicles, let alone medium and heavy vehicles. Chinese made RWT systems are still in their infancy and their application on small mobile platforms is scarce. Future activities should expand the scope of application of the reconfigurable systems, independently explore their integration with medium and heavy mobile platforms and promote their application to special vehicles.

5. References

XIE Xia, KANG Shao-hua, SHI Cai-hong. A Study on Mobile Chassis Traveling in Complex Terrains [J]. Construction Machinery and Equipment, 2015, 46(03):45-52+8. (in Chinese)

LIU Yang, XIE Xia, ZHOU Xuan, et al. Summary of Research on Wheel-Track Composite Chassis [J]. Modern Manufacturing Technology and Equipment, 2019 (04):87-90. (in Chinese)

Spector Y, Raz D, Novoplanski A, et al. Adaptable traction system of vehicle: Israel, WO03/029070 [P]. 2003-04-10.

GAO Xue-shan, LI Ke-jie. Mini type wheel/caterpillar structure-variable mobile-search reconnaissance robot [P]. China: CN101570216, 2009-11-04. (in Chinese)

LI Jing-yi, BAI Yang, JIN Fei, et al. Mechanics analysis of a wheelchair robot with wheel-track coupling mechanism [J]. Journal of Beijing Institute of Technology, 2013, 22(03):301-307. (in Chinese)

GAO Xue-shan, BAI Yang, WANG Dong-xiao, et al. Wheel/track coupled type self-balancing barrier-free wheelchair [P]. China: CN102614055B, 2014-06-25. (in Chinese)

GUO Wen-zeng, JIANG Shi-gong, DAI Fu-quan, et al. Design of a Small Wheel-Track Mobile Robot and Obstacle-Surmounting Analysis[J]. Transaction of Beijing Institute of Technology, 2015, 35(02):144-148+165. (in Chinese)

Guo Wen-zeng, Jiang Shi-gong, Zong Cheng-guo, et al. Development of a transformable wheel-track mobile robot and obstacle-crossing mode selection [C]. IEEE International Conference on Mechatronics & Automation. IEEE, 2014.

Cui Deng-qi, Gao Xue-shan, Guo Wen-zeng, et al. Design and Stability Analysis of a Wheel-Track Robot [C]. 2016 3rd International Conference on Information Science and Control Engineering (ICISCE). IEEE, 2016.

Guo Wen-zeng, Yu Mu, Gao Xue-shan. Step-climbing ability research of a small scout wheel-track robot platform [C]. 2015 IEEE International Conference on Robotics and Biomimetics (ROBIO). IEEE, 2015.

Cui Deng-qi, Gao Xue-shan, Guo Wen-zeng, Li Jian. Multimode obstacle-crossing analysis of a wheel/track mobile robot [C]. Proceedings of 2017 IEEE International Conference on Unmanned Systems (ICUS). 2017:211-217.

CUI Deng-qi. Study on the Key Technology of Special Wheel-tracked Mobile Robot with Movement Transformation Function [D]. Beijing Institute of Technology, 2018. (in Chinese)

GUO Wen-zeng. Study on a Small Special Ground Mobile Robot Coupling Wheel and Track[D]. Beijing Institute of Technology, 2016. (in Chinese)

CUI Yu-ning, LUO Zi-rong, SHANG Jian-zhong, et al. Machine design of a reconfigurable wheel-track hybrid mobile robot with multi-locomotion [J]. Journal of Harbin Institute of Technology, 2018, 50(07):80-86. (in Chinese)

LUO Zi-rong , SHANG Jian-zhong , WEI Guo-wu, et al. A reconfigurable hybrid wheel-track mobile robot based on Watt II six-bar linkage [J]. Mechanism & Machine Theory, 2018, 128:16-32.

ZHANG Shuo, YAO Jian-tao, XU Yun-dou, et al. Design and Analysis of Moving Mechanism of Shape Reconfigurable Mobile Robot [J]. Transactions of the Chinese Society for Agricultural Machinery, 2019, 50(8):9. (in Chinese)

ZHAO Yong-sheng, ZHANG Shuo, HAN Bo, et al. High-rigidity reconfigurable wheel/track deformation walking mechanism [P]. China: CN209534639U, 2019-10-25. (in Chinese)

ZHAO Yong-sheng, ZHANG Shuo, ZHU Hai-xiao, et al. Wheel and pedrail shape-reconfigurable mobile rescue robot [P]. China: CN110466628A, 2019-11-19.

WU Xing, SHI Cai-hong, KANG Shao-hua, et al. Matching Design of the Powertrain of a New Type Wheel-track Vehicle [J]. Journal of Agricultural Mechanization Research, 2014, 36(09):65-71. (in Chinese)

LIU Zong-hao.Design and Analysis of a Kind Wheel - Tracked Composite Variant Wheel[D]. Tianjin University of Technology, 2014. (in Chinese)

LI Hao, SHI Cai-hong, KANG Shao-hua, et al. Analysis and Optimization of Ride Comfort for a Wheel-tracked Rescue Robot[J]. China Mechanical Engineering, 2015, 26(11):1444-1449. (in Chinese)

LIU Zong-hao, SHI Cai-hong, KANG Shao-hua, et al. Design and Obstacle Analysis of a Novel Agricultural Machinery Walking Mechanism[J]. Journal of Agricultural Mechanization Research, 2014, 36(07):220-224. (in Chinese)

WU Xing, SHI Cai-hong, KANG Shao-hua, et al. Study on the Parameter Matching and Simulation of the Powertrain of a FWD Wheel-track Carrier[J]. Journal of Mechanical Transmission, 2014, 38(05):1-6. (in Chinese)

LI Hia-hong. Design and Research of Anti-Terrorism Robot with Wheel-Tracked Variable Structure [D]. Harbin Institute of Technology, 2011. (in Chinese)

Hu J , Peng A , Ou Y , et al. On study of a wheel-track transformation robot[C]. 2015 IEEE International Conference on Robotics and Biomimetics (ROBIO). IEEE, 2015.

Apostolopoulos Dimitrios, Muecke Karl, Pol Nishant, Mutschler Edward, Glisson Matthew, Pantaleo Richard, Novick Stephen. Reconfigurable wheel-track for all-terrain mobility: America, WO2020223295 [P]. 2020-11-05.

Lyness Hannah, Apostolopoulos Dimitrios. Validation of two reconfigurable wheel-track testbeds for military vehicles [J]. Journal of Terramechanics, 2021.

CHEN Yu, LI Chun-ming, SONG Hu-xin, et al. Crawler wheel walking device capable of realizing wheel-crawler conversion[P]. China: CN110182269B, 2022-02-22. (in Chinese)

CHNE Yu, SONG Hui-xin. Design of Track Wheel with Shape Conversion Function [J]. Vehicle and Power Technology, 2020(01):15-19. (in Chinese)

CHEN Shao-nan, DENG Zhi-yan, LIU Shuai. Research on the system of an intelligent disassembly robot under nuclear radiation[J]. Robot Technique and Application,2021(06):23-28. (in Chinese)

WU Feng-qi, CHEN Shao-nan, DENG Zhi-yan. Movable Chassis and Mobile Movable [P]. China: CN113371085B, 2022-07-15. (in Chinese)

XUN Lin-sen, LIU Lei, LIU Jin-fu, et al. Reconfigurable wheel [P]. China: CN210455005U, 2020-05-05. (in Chinese)

ZHANG Peng-chao, LU Hao-nan, ZHANG Zheng-wu, et al. Reconfigurable wheel-track composite wheel applied to trolley robot [P]. China: CN113492619A,2021-10-12.(in Chinese)

WANG Jia-hua, SHI Yu, XIE Fei, et al. Wheel-track combined type reconfigurable wheel [P]. China: CN112848792A,2021-05-28. (in Chinese)

Pantaleo Richard, Apostolopoulos Dimitrios, Mutschler Edward, Glisson Matthew. Reconfigurable wheel-track for all-terrain mobility: America, WO2020223298[P]. 2020-11-05.

Pantaleo Richard, Apostolopoulos Dimitrios, Mutschler Edward, Glisson Matthew. Reconfigurable wheel-track for all-terrain mobility: America, WO2020223305[P]. 2020-11-05.

Laux, D., Alef, J., 2016. Tire to track transforming system, ISTVS 8th Americas Regional Conference, Detroit, MI.

ZHOU Fa-liang. Configuration Design and Passibility Study of a Multimodal Mobile Robot Based on a Transformable Wheel Rim[D]. National University of Defense Technology, 2019. (in Chinese)

QU Jie, ZHONG Wie-bin. Design and Obstacle-Surmounting Performance Analysis of Wheel-Track Transformable Wheel[J]. Journal of South China University of Technology (Natural Science Edition),2013,41(05):119-124. (in Chinese)

ZHONG Wei-bin, QU Jie. Wheel-track switching mechanism[P]. China: CN202358210U,2012-08-01. (in Chinese)

JIANG Lei, QI Ying-chun, DENG Qin-dan, et al. Simulation and Analysis of the Moving Performance of the Wheel-track Combined Manned Lunar Rover [J]. Vehicle and Power Technology,2020(03):1-6+48. (in Chinese)

Renggui B, Runxin N, Huawei L, et al. Motion Planning and Performance Analysis of a Novel Six-Tracked Robot[C]. 2021 IEEE International Conference on Robotics, Control and Automation Engineering. IEEE, 2021.

KANG Shao-hua, HUANG Lin, SHI Cai-hong, et al. Design and analysis on reconfigurable rubber-track variant wheel [J].Chinese Journal of Construction Machinery, 2014, 12(04):320-324. (in Chinese)

XIE Xia, ZHANG Xue-ling, DUAN Xiu-bing, et al. Research on obstacle-surmounting performance of wheel-track vehicle chassis [J]. Transaction of Beijing Institute of Technology, 2018, 38(S1):107~111.(in Chinese)

WONG JY. Theory of ground vehicles, 4th ed. New York: John Wiley & Sons Inc, 2008.

WANG Ming-de, ZHAO Yu-qin, ZHU Jia-guang. Driving Principle of Tank. Beijing: National Defense Industry Press,1983,7.

Design and Kinetic Analysis of Anti-terrorism Robot with Wheel-tracked Variable Structure[J]. Machinery and Electronics,2011(10):66-69. (in Chinese)

LIU Zong-hao, SHI Cai-hong, KANG Shao-hua. Design and Analysis of the Pose-Varied Wheel-Tracked Robot Travel Mechanism[J]. Machinery Design and Manufacture,2014(04):175-177+181. (in Chinese)

GAO Xue-shan, ZONG Cheng-guo, WANG Dong-xiao. Fish scale-type segmenting elastic caterpillar[P]. China: CN102530118B,2013-09-25. (in Chinese)

LIU Bin. Study on Dynamics of Driving Device of High-Speed Tracked Vehicle[D]. Beijing Institute of Technology, 2015.(in Chinese)

SU Jie, LI Chun-ming, FAN Zhi-you. Analysis on Sprocket and Track Mesh Power Loss of Wheel-tracked Vehicle [J]. Vehicle and Power Technology,2017(01):15-20.(in Chinese)

PENG An-si, Zhou Yi-min, Hu Jian-bing, et al. Mechanical design for wheel/track transform mobile platform-search and rescue robot[C]. IEEE International Conference on Robotics & Biomimetics. IEEE, 2014.

0963 / A Wheel and Vehicle Mobility Index Based on Traction and Velocity for Optimization of Mobility Performance

Vladimir V. Vantsevich[a], David J. Gorsich[b], Jesse R. Paldan[c]*, Masood Ghasemi[c], Lee Moradi[a]
[a]*Worcester Polytechnic Institute, Worcester, Massachusetts, United States*
[b]*U.S. Army DEVCOM Ground Vehicle Systems Center, Warren, Michigan, United States*
[c]*University of Alabama at Birmingham, Birmingham, Alabama, United States*
**Corresponding author: jpaldan@uab.edu*

https://doi.org/10.56884/ZOFE8973

Abstract

In this paper, two new indices for wheel and vehicle mobility performance suitable for mobility performance optimization are proposed based on the wheel circumferential forces and the vehicle actual linear velocity. The indices are intended for on-the go assessment of mobility for autonomous real-time optimization of wheel power distribution. The wheel mobility performance (WMP) characterizes the mobility performance of a wheel in terms of its traction and velocity with regard to a potential but unachievable maximum. The index is extended to a vehicle with a given number of drive wheels, where the traction characteristics and theoretical velocities may be different at each wheel, as the vehicle mobility performance (VMP). The VMP index is demonstrated through a simulated comparison of multiple driveline configurations of a 4x4 vehicle on different terrains to the optimal value. The optimal mobility is computed through an objective function with the use of Lagrange Multipliers which determine the optimal distribution of vehicle forces and slippages which achieve optimal mobility.

Keywords: terrain mobility, traction performance, mobility index

1. Introduction

A commonly accepted definition of vehicle mobility is the vehicle's ability to move from one place to another place and perform a mission or task. Different empirical and physics-based methods have been used to predict ground vehicle mobility. The Vehicle Cone Index (VCI) has been a widely used metric for mobility featured in the NATO Reference Mobility Model. VCI is the minimum strength of the critical layer of soil that permits a given vehicle to successfully make a specific number of passes, usually one pass or 50 passes (Wong, 2008). The VCI is empirically related to a Mobility Index (MI) which is a function of a specific set of vehicle design factors (Wong et al., 2020).

For next generation mobility models, a goal is to move from empirical models to physics-based models capable of assessing mobility quantitatively (Wong et al., 2020). The drawbar pull coefficient at a slippage of 20% for optimal tractive efficiency and the drawbar pull coefficient at 80% as the upper limit of traction was proposed by (Wong et al., 2020) as a new mobility metric for next generation mobility models. The speed-made-good, or maximum possible speed of a vehicle in straight path under steady-state conditions, is used in generating mobility maps to show trafficability and max speed in a mapped region; a goal of research is to utilize physics-based models without requiring complex dynamic simulations with high computation cost (Karpman et al., 2021).

Recently, alternative driveline configurations such as in-wheel motors have received investigation for improving off-road mobility through control of individual motors (Savitski et al., 2017). Additionally, there is an interest in utilizing autonomous ground vehicles, which may be used to perform missions too dangerous for human drivers and developing means to perform terrain estimation for autonomous vehicles in off-road environments (Parker et al., 2021, Dallas et al., 2020). Assessment of autonomous vehicle mobility has unique needs which differ from manned vehicles. For their control systems to function in off-road environments with changing terrain conditions and environmental challenges, the vehicle needs means of assessing mobility on the go in real time. The above-given definition of mobility may be considered as two components:

1) Assessing the mobility state of a vehicle relative to its immobilization state (mobility margin), and
2) Estimating the mobility performance on terrain while the vehicle still maintains its margin of mobility.

A set of indices to assess the mobility margin of a wheel or vehicle relative to its immobilization state was developed in (Gray et al., 2013). This paper is concerned with item (2) by defining a means of first quantifying mobility performance of an individual wheel and extending that assessment to the vehicle level.

2. Mobility of a Single Wheel

In this section, the mobility of a single wheel is derived which combines the wheel's traction and velocity. The traction of a wheel is closely related to the tire slippage, defined as a ratio reflecting the velocity reduction V_δ from the wheel's theoretical velocity V_t (without slip) to the actual linear velocity of the wheel center V_x.

$$s_\delta = \frac{V_t - V_x}{V_t} = \frac{V_\delta}{V_t} \quad (1)$$

The tire's rolling radius at nonzero wheel torque is conventionally defined as the ratio of the linear velocity of the center of the wheel to the wheel's rotational velocity ω_w.

$$r_w = \frac{V_x}{\omega_w} \quad (2)$$

The theoretical velocity in Eq. (1) can be defined as

$$V_t = \omega_w r_w^0 \quad (3)$$

where r_w^0 is the rolling radius in the driven mode, the mode of wheel rolling where no torque is applied and the slippage is zero.

The applied wheel torque results in a circumferential force F_x which overcomes the resistance to motion to move and accelerate the wheel as shown in Fig. 1. Equation (4) is the wheel force balance equation which relates circumferential force, wheel torque, and resistances to motion (Andreev et al., 2010).

$$T_w = F_x r_w^0 + I_x \varepsilon_w = (R_x + W_w sin\theta_n + F_{frame.x} + F_a) r_w^0 + I_x \varepsilon_w \quad (4)$$

In Eq. (4), R_x is the rolling resistance, W_w is the wheel load, θ_n is the slope angle, $F_{frame.x}$ is the force exerted by the vehicle frame on the wheel, F_a is the longitudinal inertia force in acceleration, I_x is the wheel moment of inertia, and ε_w is the angular acceleration.

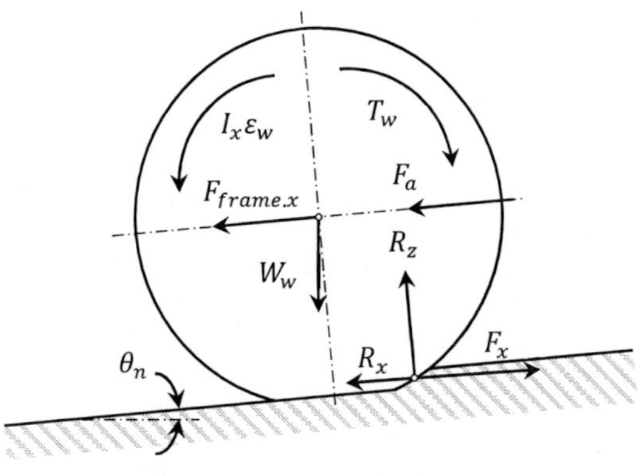

Fig. 1. Force balance of driving wheel; direction of forward motion toward the right.

When nonzero torque is applied to the wheel, the wheel starts to gain slippage. Slippage is related to the circumferential force in a nonlinear relation. The exponential function for $F_x = f(s_\delta)$ in Eq. (5) derived in (Andreev and Vantsevich, 2017) using the shear stress-strain relationship proposed by (Janosi and Hanamoto, 1961) is used in this paper.

$$F_x = \mu_{px} R_z \left(1 - \frac{s_{\delta c}}{2s_\delta}\left(1 - exp\left(-\frac{2s_\delta}{s_{\delta c}}\right)\right)\right) \quad (5)$$

$s_{\delta c}$ is the characteristic slippage, which is the value of s_δ past which the rate of slippage growth with further increase in F_x begins to drastically rise. μ_{px} in Eq. (5) is the peak friction coefficient, a maximum grip coefficient defining the maximum of the current friction coefficient, μ_x. The current friction coefficient μ_x is F_x normalized by the normal reaction R_z of the wheel load W_w.

$$\mu_x = \frac{F_x}{R_z} \quad (6)$$

From the form of Eq. (5), F_x approaches, but never reaches, a maximum value that is approximately

$$F_{x,max} = \mu_{px} R_z \quad (7)$$

Thus, higher values of the product $\mu_{px} R_z$ allow a greater maximum traction. The implication of Eqs. (6-7) is that a fraction of the wheel's normal reaction R_z, limited by the terrain condition, can be realized as traction. This fact leads to one half of the index for mobility assessment; the second is the velocity, which from Eq. (1) is related to the slippage.

While slippage results in a speed reduction and drop in the wheel's rolling radius, from Eq. (5) it can be seen that some slippage is necessary for the wheel to produce a tractive force. Therefore, the technical problem in mobility optimization becomes how to determine how much slippage is necessary for the optimal condition and how it is defined. Mobility control is often performed by controlling the slippage as the optimal slippage will correspond to a wheel circumferential force.

Figure 2 shows computed mobility characteristics for a single wheel module with a 335/65R22.5 tire and 14kN wheel load. The product of force and velocity (Fig. 2A and B) have only one combination which gives the maximum (Fig. 2C). Furthermore, the maximum $F_x V_x$ occurs at a different value than the maximum tractive energy efficiency η_w^t. The tractive energy efficiency is the ratio of wheel input to output power and is calculated with Eq. (10) as the product of rolling resistance efficiency η_{fw} and slip energy efficiency $\eta_{\delta w}$ (Andreev et al., 2010).

$$\eta_{fw} = \frac{F_w}{F_x} = \frac{F_x - R_x}{F_x} = 1 - \frac{R_x}{F_x} \quad (8)$$

$$\eta_{\delta w} = \frac{V_x}{V_t} = \frac{V_t - V_\delta}{V_t} = 1 - \frac{V_\delta}{V_t} = 1 - s_\delta \quad (9)$$

$$\eta_w^t = \frac{P_w^{out}}{P_w^{in}} = \frac{F_w V_x}{F_x V_t} = \eta_{fw}\eta_{\delta w} \quad (10)$$

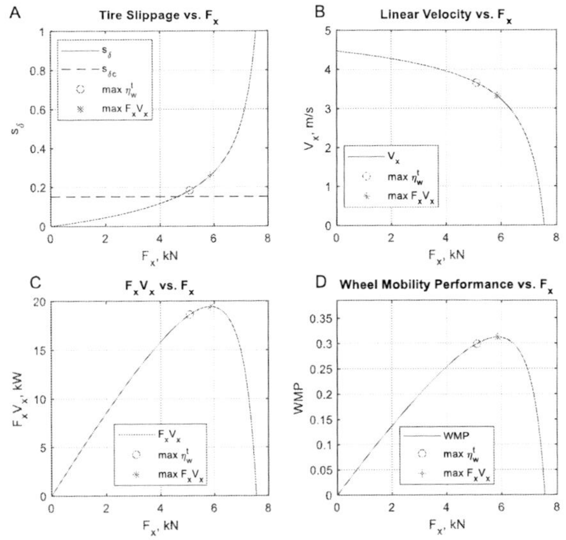

Fig. 2. Product (C) of tire slippage (A) and actual velocity (B); Wheel mobility performance Index (D) for single wheel module with tire model 335/65R22.5, 14kN load.

F_xV_x is a mobility index to mobility performance index to characterize the maximum power spent to support vehicle's movement, and therefore to complete a task and fulfill its mission.

$$F_xV_x \rightarrow max\ \{F_xV_x\} = F_{x*}V_{x*} \qquad (11)$$

The actual velocity V_x can be expressed in terms of the slippage and theoretical velocity by rewriting Eq. (1).

$$V_x = V_t(1 - s_\delta) \qquad (12)$$

Normalizing the actual velocity from Eq. (12) by the theoretical velocity yields

$$V_x/V_t = 1 - s_\delta \qquad (13)$$

The Wheel Mobility Index, WMP, is then introduced as F_xV_x from Eq. (11) normalized by the product R_zV_t.

$$WMP = \frac{F_xV_x}{R_zV_t} = \frac{F_x(1-s_\delta)}{R_z} \qquad (14)$$

In Eq. (14), the upper bound for WMP and thus the wheel mobility is the product R_zV_t, a hypothetical case where the complete potential of the wheel load can be converted into traction without any loss of velocity. This would require $\mu_{px} = 1$ and $s_\delta = 0$. In actuality, since the application of wheel torque will always result in some slippage and realistic values of μ_{px} are < 1, the WMP will take values < 1. To make the index suitable for autonomous self-observation and control of wheel mobility performance, the WMP index was built from components that can be measured or observed in real time without knowledge of the terrain characteristics. The rolling radius in the driven mode r_w^0 needed to computed the theoretical velocity from Eq. (3) can be experimentally determined as a function (or look up table) of the wheel normal reaction and inflation pressure.

To avoid excessive slippage which leads to immobilization, the optimal slippage should be limited to a maximum close to the characteristic slippage.

$$0 < s_{\delta*} < s_{\delta c} + \Delta \qquad (15)$$

Δ is an assigned value defining how far the slippage is allowed to exceed the characteristic slippage.

3. Mobility Performance of the Vehicle

The WMP presented in Eq. (14) provides an index for assessing mobility of a single wheel, which can be used as part of a mobility control system based on slippage control. Therefore, to be useful for this purpose, the index needs to be extended to a multi-wheel vehicle such that optimal slippages of all tires can be established. This is because components of the WMP may not be the same for all wheels. Reasons for differences include different rotational velocities and rolling radii in the driven mode, different inflation pressures or sizes of tires, traction characteristics from non-uniform terrain conditions, multi-pass effect, and weight distribution. The vehicle mobility index, VMP, is introduced as

$$VMP = \frac{\sum_{i=1}^{n} F'_{xi}V_x + F''_{xi}V_x}{\sum_{i=1}^{n} R'_{zi}V'_{ti} + R''_{zi}V''_{ti}}, i = 1,2 \qquad (16)$$

Here, superscript $'$ and $''$ indicate left and right wheels and i is the index of an axle, taking values $i = 1,2$ for the 4x4 vehicle considered in this paper. V_x is assumed to be the same for each wheel while each can have a different theoretical velocity V_t. Substituting the theoretical velocities in the VMP with their components (see Eq. (12)) gives

$$VMP = \frac{\sum_{i=1}^{n} F'_{xi} + F''_{xi}}{\sum_{i=1}^{n} R'_{zi}/(1-s'_{\delta i}) + R''_{zi}/(1-s''_{\delta i})}, i = 1,2 \qquad (17)$$

From Eq. (17), an objective function for maximization of mobility performance is established.

$$\frac{\sum_{i=1}^{n} F'_{xi} + F''_{xi}}{\sum_{i=1}^{n} R'_{zi}/(1-s'_{\delta i}) + R''_{zi}/(1-s''_{\delta i})} \rightarrow max \qquad (18)$$

The summation of the circumferential forces of the drive wheels, $\sum_{i=1}^{n} F'_{xi} + F''_{xi}$, balances the total resistance to motion of the vehicle in its planar motion. In straight line motion of the vehicle,

$$\sum_{i=1}^{n} F'_{xi} + F''_{xi} = R_{m\Sigma} \qquad (19)$$

where, $R_{m\Sigma}$ is the total resistance to vehicle straight motion, which includes the rolling resistance of the wheels, vehicle gravity components on hills, draw bar pull, acceleration/deceleration force, air drag forces, etc. Assuming these forces are known (computed or assigned), $R_{m\Sigma}$ becomes a known, non-controllable constant.

$$\frac{R_{m\Sigma}}{\sum_{i=1}^{n} R'_{zi}/(1-s'_{\delta i}) + R''_{zi}/(1-s''_{\delta i})} \rightarrow max \qquad (20)$$

The objective function in Eq. (18) can then be substituted with the minimization of

$$\sum_{i=1}^{n} R'_{zi}/(1 - s'_{\delta i}) + R''_{zi}/(1 - s''_{\delta i}) \rightarrow min \qquad (21)$$

The reason to use Eq. (21) is that, for the vehicle to meet its assigned motion requirements, the sum of F_x forces which result in $R_{m\Sigma}$ are fixed. However, the slippages can be changed through the driveline's allocation of F_x forces among the wheels without changing $R_{m\Sigma}$ because of different gripping and load conditions at each wheel.

Following Eq. (15), user-definable constraints for the optimal slippages should be considered to restrict approaching the region of mobility loss.

$$0 < s_{\delta i*}^{\prime(\prime\prime)} < s_{\delta ci}^{\prime(\prime\prime)} + \Delta_i^{\prime(\prime\prime)}, i = 1,2 \quad (22)$$

4. Tire Slippages for Optimal Mobility

The optimal slippages for mobility are determined by applying the method of Lagrange multipliers to the objective function given in Eq. (21). The Lagrange multiplier method is formulated starting from Eq. (23)

$$L(\bar{s}_\delta, \lambda) = F(\bar{s}_\delta, \bar{c}) + \lambda G(\bar{s}_\delta, \bar{b}, \bar{c}, \overline{k_c}) \quad (23)$$

where F is the objective function, G is a constraint function, λ is a scalar multiplier, and the vectors $\bar{s}_\delta, \bar{b}, \bar{c}$, and $\overline{k_c}$ are introduced as

$$\bar{s}_\delta = [s_{\delta 1}', s_{\delta 1}'', \dots, s_{\delta n}', s_{\delta n}'']^T \quad (24)$$

$$\overline{k_c} = \left[\frac{2}{s_{\delta c1}'}, \frac{2}{s_{\delta c1}''}, \dots, \frac{2}{s_{\delta cn}'}, \frac{2}{s_{\delta cn}''}\right]^T \quad (25)$$

$$\bar{b} = \left[\mu_{px1}', \mu_{px1}'', \dots, \mu_{pxn}', \mu_{pxn}''\right]^T \quad (26)$$

$$\bar{c} = [R_{z1}', R_{z1}'', \dots, R_{zn}', R_{zn}'']^T \quad (27)$$

Function $F(\bar{s}_\delta, \bar{c})$ is the objective function defined in Eq. (21).

$$F(\bar{s}_\delta, \bar{c}) = \sum_{i=1}^{n} c_i^{\prime(\prime\prime)} \left(1 - s_{\delta i}^{\prime(\prime\prime)}\right)^{-1}, i = 1,2 \quad (28)$$

The constraint function $G(\bar{s}_\delta, \bar{b}, \bar{c}, \bar{k})$ specifies that the sum of all wheels circumferential forces, expressed in terms of slippage using Eq. (5), must add up to the total resistance to the vehicle's motion $R_{m\Sigma}$.

$$G(\bar{s}_\delta, \bar{b}, \bar{c}, \bar{k}) = \sum_{i=1}^{n} b_i^{\prime(\prime\prime)} c_i^{\prime(\prime\prime)} \left(1 - \frac{1}{k_i^{\prime(\prime\prime)} s_{\delta i}^{\prime(\prime\prime)}} \left(1 - exp\left(-k_i^{\prime(\prime\prime)} s_{\delta i}^{\prime(\prime\prime)}\right)\right)\right) - R_{m\Sigma}, i = 1,2 \quad (29)$$

where $R_{m\Sigma}$ is the sum of air drag D_{ad}, inertia force F_a, drawbar pull force F_d, gravity component on a hill $W_a sin\theta_n$, and rolling resistance $R_{xi}^{\prime(\prime\prime)}$.

$$R_{m\Sigma} = \sum_{i=1}^{n} R_{xi}^{\prime(\prime\prime)} + W_a sin\theta_n + F_d + F_a + D_{ad}, i = 1,2 \quad (30)$$

The optimal tire slippages, $s_{\delta i*}'$ and $s_{\delta i*}''$, $i = 1, \dots, n$, come from the system that corresponds to the necessary conditions of an extremum of the objective function. Those conditions are determined from the following equations.

$$\frac{\partial L(\bar{s}_\delta, \lambda)}{\partial s_{\delta i}'} = 0, i = 1,2 \quad (31)$$

$$\frac{\partial L(\bar{s}_\delta, \lambda)}{\partial s_{\delta i}''} = 0, i = 1,2 \quad (32)$$

$$G(\bar{s}_\delta, \bar{b}, \bar{c}, k_c) \equiv 0 \quad (33)$$

Using Eqs. (29-30), Eqs. (31-33) become

$$c_i'(1 - s_{\delta i}')^{-2} + \lambda b_i' c_i' \left(\frac{1 - exp(-k_{ci}' s_{\delta i}')}{k_{ci}'(s_{\delta i}')^2} - \frac{exp(-k_{ci}' s_{\delta i}')}{s_{\delta i}'}\right) = 0, i = 1,2 \quad (34)$$

$$c_i''(1 - s_{\delta i}'')^{-2} + \lambda b_i'' c_i'' \left(\frac{1 - exp(-k_{ci}'' s_{\delta i}'')}{k_{ci}''(s_{\delta i}'')^2} - \frac{exp(-k_{ci}'' s_{\delta i}'')}{s_{\delta i}''}\right) = 0, i = 1,2 \quad (35)$$

$$\sum_{i=1}^{n} b_i' c_i' \left\{1 - \frac{1}{k_{ci}' s_{\delta i}'}[1 - exp(-k_{ci}' s_{\delta i}')]\right\} + b_i'' c_i'' \left\{1 - \frac{1}{k_{ci}'' s_{\delta i}''}[1 - exp(-k_{ci}'' s_{\delta i}'')]\right\} = R_{m\Sigma}, i = 1,2 \quad (36)$$

Eqs. (34-36) form a system of equations with unknown slippages $s_{\delta i}^{\prime(\prime\prime)}$ and multiplier λ, the solution of which gives the optimal slippages $s_{\delta i*}'$ and $s_{\delta i*}''$ for maximum mobility.

5. Mobility Simulation Results

Computer simulations of the mobility optimization were performed for a model of a 4x4 vehicle with a gross mass of 53.8kN. The vehicle parameters for the model are based on the FED Alpha Fuel Efficient Ground Vehicle Demonstrator; full data on the vehicle's characteristics is available at (KRC, 2021).

The optimal slippages were determined by solving the system of equations given by Eqs. (34-36). The mobility performance index VMP was then calculated for the optimal slippages using Eq. (17). Additionally, the mobility of two mechanical driveline configurations were computed and simulated for comparison. In total, the following three cases are considered:

1. Optimal slippages computed from solving Eqs. (34-36).
2. a differential driveline with an open symmetric differential in the transfer case which splits torque equally (labeled OD in figures).
3. a positive engagement driveline where the transfer case differential is locked providing a fixed ratio of angular velocities (labeled PE).

The distribution of circumferential forces and slippages for the mechanical drivelines was computed using a system of vehicle generalized parameters which builds a system of equations to solve for individual wheel slippages and a vehicle generalized slippage based on the number and configuration of power dividing units in the driveline. A detailed explanation of this method is available in (Vantsevich and Paldan, 2021).

The vehicle is simulated in a steady, straight motion at slow speed (zero F_a and neglecting air drag in Eq. (30)). The theoretical velocity V_a of the vehicle (velocity at zero tire slip) is assigned as a constant 5 MPH. The traction characteristics are simulated using Eq. (5) where parameters $s_{\delta c}$ and μ_{px} were determined from a curve fitting of experimental data from three terrain conditions: a meadow, a spring field with naturally packed soil after the field was tilled in the previous fall (plowed, harrowed and cultivated), and a snow dirt road. The meadow with grass cover of 0.05m to 0.10m has a loam soil sublayer with a moisture content varies from 8% to 12%. The field soil is sandy loam with a moisture content of about 10%. The dirt road is covered with fresh snow. The rolling resistance forces are computed using the Bekker-Wong model. The modeling includes the wheel multi-pass effect.

Figure 3 illustrates the change in the VMP index as the front and rear slippages take any combination of values which satisfy Eq. (29). Locations of the optimal VMP, as well as VMP of the positive engagement and differential driveline, are marked. As seen in Fig. 3, the optimal slippage values produce the maximum value of the VMP index. For the maximum mobility, the slippages of the front and rear wheels are not the same, differing from the requirements for maximum energy efficiency (Andreev et al., 2010). The VMP values of the positive engagement and differential drivelines are less than, but close to, the maximal VMP while having different combinations of slippages and circumferential forces.

A set of simulations was performed for all vehicle driveline configurations on flat smooth terrains loaded with a drawbar pull, F_d, which varies from zero up to the point of immobilization of the vehicle. In Fig. 4, the slippages of the front and rear tires along with their circumferential forces are shown for the meadow terrain.

The positive engagement driveline stays close to the optimal slippages throughout the simulation. The front and rear slippages are not the same since the rolling radii, r_{w1}^0 and r_{w2}^0, are different due to the normal load distribution among the axles influenced by the vehicle's weight distribution and drawbar pull. Immobilization occurs for the positive engagement drivelines at drawbar pull of around $F_d = 26.5$ kN. The front tires of the differential driveline reach immobilization quicker around $F_d = 18$ kN, which becomes its limit for mobility on this terrain; increasing F_d further for this driveline is not possible once the front wheels reach 100% slip due to the symmetric characteristic of the interaxle differential, which does not allow any increase of traction at the rear wheels.

The Mobility Performance Index and actual velocity of the vehicle on the meadow, soil field and snow road terrains are shown in Fig. 5 for the mechanical driveline configurations and the optimal slippages.

The mobility performance of the vehicle with the differential driveline on the meadow drops to zero at a drawbar pull force of $F_d = 18$ kN. The positive engagement and optimal drivelines more closely follow the optimal slippages and maintain better mobility performance, maintaining mobility up to a drawbar pull of $F_d = 26.5$ kN. A similar result is seen on the other two terrains with the differential driveline losing mobility at a lower drawbar pull (9.9 vs. 12.36 kN on the snow road and 5.9 vs. 6.8 kN on the soil field). On all terrains, past the maximum value of VMP, the actual velocity begins to drastically decrease and the VMP decreases accordingly until the vehicle is immobilized. Expanded modeling situations, including heterogeneous and sloped terrain, as well as a verification with experimental data, will be presented in other papers scheduled for publication.

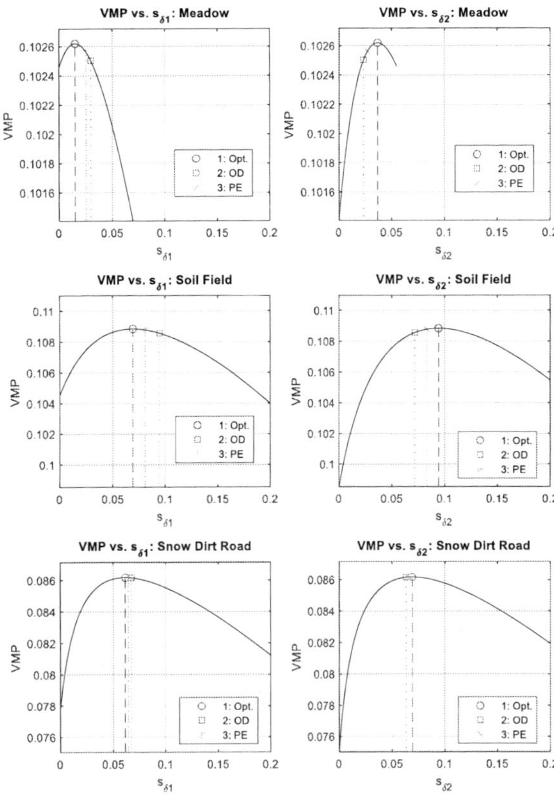

Fig. 3. Mobility performance index of FED Alpha at different combinations of front and rear wheel slippages.

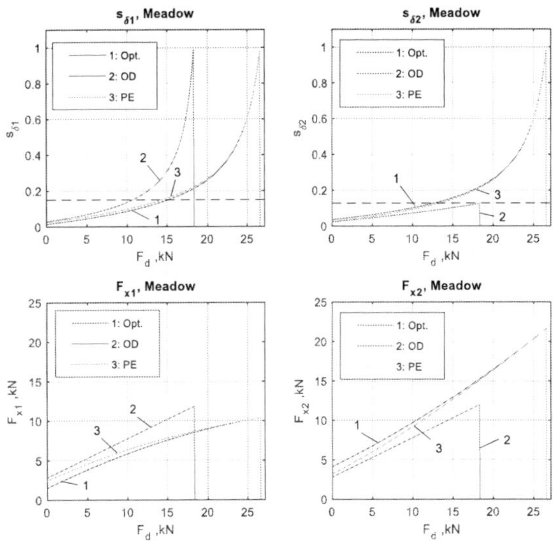

Fig. 4. Slippages and circumferential forces on meadow terrain.

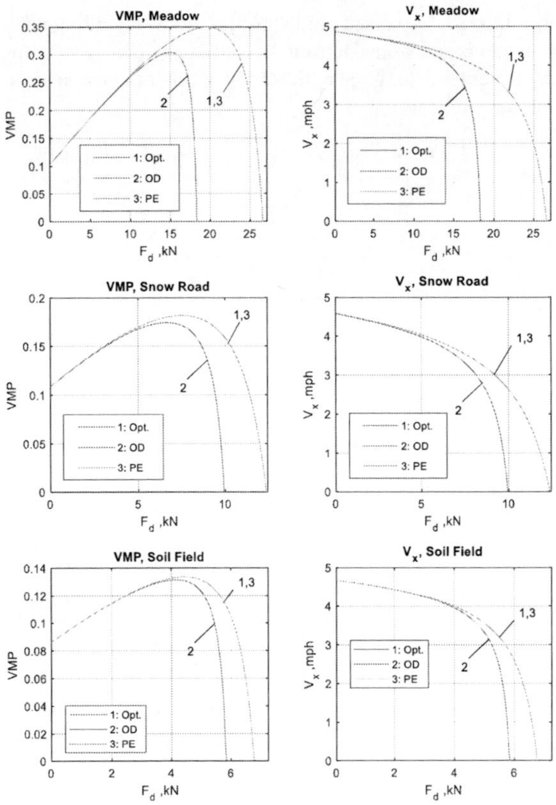

Fig. 5. Mobility performance index and actual velocity on meadow, snow road, and soil field.

6. Conclusions

In this paper, new indices for assessment of wheel and vehicle mobility was presented. The indices are based on parameters obtained from the interaction of the tire and terrain. They include components that can be measured and controlled in real time, making them suitable for on-the-go mobility assessment required by autonomous vehicle in challenging terrain conditions.

The wheel mobility performance (WMP) index evaluates wheel mobility performance in terms of traction and velocity relative to the theoretical maximum performance. The vehicle mobility performance (VMP) index relates the mobility performance of all wheels to the theoretical maximum performance of the vehicle.

An optimization problem was formulated to solve for the combination of wheel slippages and circumferential forces which provide the maximum mobility performance. Computer simulations of a 4x4 vehicle demonstrate the use of VMP in comparing the mobility performance of different configurations of driveline on different terrain conditions and at different levels of drawbar pull.

7. Nomenclature

D_{ad}	Aerodynamic drag force	[kN]
F_a	Linear inertia force	[kN]
F_d	Drawbar Pull	[kN]
$F_{frame.x}$	Longitudinal frame force	[kN]
F_x	Wheel circumferential force	[kN]
i	Index of an axle	
I_x	Wheel moment of inertia	[kgm^2]
m_a	Vehicle mass	[kg]
n	Number of axles	
$R_{m\Sigma}$	Total resistance to motion	[kN]
r_w	Rolling radius	[m]
r_w^0	Rolling radius (driven mode)	[m]
R_x	Rolling resistance	[kN]
R_z	Normal reaction	[kN]
s_δ	Tire slippage	
$s_{\delta c}$	Characteristic slippage	
T_w	Wheel torque	[Nm]
W_a	Vehicle weight	[kN]
W_w	Wheel load	[kN]
VMP	Vehicle mobility performance	
WMP	Wheel mobility performance	
V_t	Theoretical linear velocity	[mph]
V_x	Actual linear velocity	[mph]
V_δ	Slip velocity	[mph]
θ_n	Ground slope angle	[rad]
ε_w	Wheel angular acceleration	[rad/sec^2]
η_{fw}	Wheel Rolling resistance efficiency	
η_w^t	Wheel tractive energy efficiency	
$\eta_{\delta w}$	Wheel slip energy efficiency	
μ_x	Current friction coefficient	
μ_{px}	Peak friction coefficient	
ω_w	Wheel angular velocity	[rad/s]
$'$	Left wheels	
$''$	Right wheels	
$'('')$	Includes both left and right wheels	

8. Acknowledgements

This research is supported by Automotive Research Center (ARC) under federal award No. W56HZV-19-2-0001 and ARC subaward No. SUBK00013995.

9. References

Andreev, A.F., Kabanau, V., Vantsevich, V.V., 2010. Driveline systems of ground vehicles: theory and design, CRC Press.

Andreev, A.F., Vantsevich, V.V., 2017. Tyre and Soil Contribution to Tyre Traction Characteristic. The Dynamics of Vehicles on Roads and Tracks. CRC Press, 149-155.

Dallas, J., Jain, K., Dong, Z., Sapronov, L., Cole, M.P., Jayakumar, P., Ersal, T., 2020. Online terrain estimation for autonomous vehicles on deformable terrains. J. Terramechanics 91, 11-22.

Gray, J.P., Vantsevich, V.V., Overholt, J.L., 2013. Indices and Computational Strategy for Unmanned Ground Wheeled Vehicle Mobility Estimation and Enhancement. ASME 2013 International Design Engineering Technical Conferences and Computers and Information in Engineering Conference. V06AT07A068.

Janosi, Z.J., Hanamoto, B., 1961. The analytical determination of drawbar pull as a function of slip for tracked vehicles in deformable soils. ISTVS 1st International Conference. Turin, Italy.

Karpman, E., Huang, W., Kövecses, J., Teichmann, M., 2022. Speed-made-good: Mobility map generation for wheeled vehicles on soft terrain, J. Terramechanics 101. 11-22.

Michigan Technological University Keweenaw Research Center (KRC), 2021. NATO Applied Vehicle Technology (AVT) Panel Cooperative Demonstration of Technology CDT-308, Next-Generation NATO Reference Mobility Model (NG-NRMM) Demonstration. https://www.mtu.edu/cdt/ (accessed 2021).

Parker, M., Quinn, B., Bates, J., Hodgdon, T., Bodie, M., Shoop, S., Stott, A., 2021. Exploring cold regions autonomous operations. J. Terramechanics 96. 159-165.

Savitski, D., Schleinin, D., Ivanov, V., Augsburg, K., Jimenez, E., He, R., Sandu, C., Barber, P., 2017. Improvement of traction performance and off-road mobility for a vehicle with four individual electric motors: Driving over icy road. J. Terramechanics 69, 33-43.

Wong, J.Y. 2008. Theory of ground vehicles, John Wiley & Sons.

Wong, J.Y., Jayakumar, P., Toma, E., Preston-Thomas, J. 2020. A review of mobility metrics for next generation vehicle mobility models. J. Terramechanics 87, 11-20.

Vantsevich, V.V., Paldan, J.R., 2021. A Generalized Approach to Virtual Driveline Systems for E-Vehicle Operation Improvements. The 27th IAVSD Symposium on Dynamics of Vehicles on Roads and Tracks. Springer.

1128 / Semi-Active Reinforcement Learning Suspension Control for Off-Road Vehicles

Ye Zhuang**[a], Haojie Sun**[a]**, Yingchun Qi**[a]**, Weiguang Fan**[b]**, and Hui Ye**[a]*

[a] *Jilin University, Changchun, Jilin, China*

[b] *China North Vehicle Research Institute, Beijing, China*

**Corresponding author: zhuangye@jlu.edu.cn*

https://doi.org/10.56884/BCOR8152

Abstract

Vertical vibration of the terrain vehicle involves its comfort, maneuverability, and fatigue life of the key components. The type of road surface encountered by the vehicle is very complex and difficult to measure. Besides, the damper system of such vehicle has strong non-linearity, which makes the design of the passive suspension and active controller difficult. Compared with the active suspension, the semi-active suspension has the advantages of low energy consumption and high safety. The reinforcement learning technology is introduced to the semi-active control of a magneto-rheological damper, which is expected to improve the comfort of the terrain-vehicle with intelligent, low-power control method.

This paper applies Gaussian Process (GP) technology to learn and model the nonlinear part of the terrain vehicle system dynamics, and then carries out the design of the control strategy, establishes a linear control law with the feedback of the suspension deflection, the sprung mass velocity, and the unsprung mass velocity. The introduced reinforcement learning algorithm learns the appropriate feedback gain during the interaction process with the system, and then uses the learned feedback gain to carry out system simulation and experimental analysis. The reinforcement learning algorithm is compared with other classical semi-active control algorithms under the input of random, bump and sinusoidal pavement with a scaled suspension test facility. The simulation and experimental results show that the reinforcement learning algorithm introduced in the paper has excellent control effect in the full frequency band and can provide good comfort for terrain vehicles with low power consumption.

Keywords: terrain vehicle, gaussian process modelling, reinforcement learning, semi-active suspension

1. Introduction

This paper introduces the reinforcement learning control method to improve vehicle performance on-line by controlling a semi-active suspension. A teaching signal is designed to evaluate system performance and to update the internal state of the system with the aim of improving system closed-loop performance.

Compared with passive suspension, active and semi-active suspension can greatly improve vehicle performance, but due to the high cost of active suspension and high energy consumption, it has not been widely used in vehicle market (Kaldas 2015). The magnetorheological shock absorber's external characteristics can be changed by varying the input current. However, vehicle dynamics will become very complex when considering the actual frequency range (typically below 30Hz for ride comfort (Szymański et al. 2015). For this reason, model-based controller design strategies can be time-consuming or inefficient in practice. These considerations inspired the design of real-time learning control system. Reinforcement learning has been applied in many areas, such as games (Vinyals et al. 2017), robotics (Kober et al. 2013), and power systems (Glavic et al. 2017). However, most reinforcement learning algorithms are designed based on discrete actions, which is a restrictive problem for current dynamic control. There remains a need for a method that can be applied over a continuous action space, although some methods have been proposed to overcome this limitation, such as generalized approximation techniques (Thrun and Schwartz 1993), and homing in techniques (Frost et al. 1994).

In a typical reinforcement learning system, the learning subsystem sends an action or a set of actions to the environment, which then returns a scalar through the performance evaluation function indicating the quality of the action. The performance evaluation function implicitly encodes the learning system's objective. The environment is stochastic, so the action must be applied multiple times to decide which action yields to the biggest reward from the environment. The algorithm mentioned in this paper can be implemented by connecting multiple learning subsystems and can be applied across a continuous range of actions. Besides, it can operate in stochastic environments, be used on-line and be adaptable to changing environmental conditions.

2. Preparation

2.1 Damper external (F-v) characteristic test

The Gaussian process model is a data-driven algorithm, and large amounts of accurate data have an important impact on model accuracy. This paper will first describe the training data acquisition for hyperparameter optimization, then take the vehicle shock absorber as the modeling object to obtain the damper characteristic data through experiments for model training.

The damper selected for the test is a magneto-rheological shock absorber of a vehicle, which allows the hysteresis characteristics to be changed by changing the current applied to the damper. It has a maximum input control current of 1.0A and uses the MTS shock absorber test bench to test its external characteristics, which uses force and displacement sensors to obtain data such as the position, speed, and damping force in real time.

According to the specific characteristics of the damper, under the four highest speed sinusoidal displacement input conditions, the control current input is designed to be 0-1.0A, the interval is 0.1A, and the amplitude of the sinusoidal input is 20mm. (See *Table 1* for detailed experimental conditions)

Figure 1 shows the damper force - displacement (F-S) curve and the shock absorber force - speed curve (F-V) under the condition that the maximum speed of the sinusoidal input is 0.524m/s.

Table 1. Test condition of damper characteristics

Excitation type	Displacement (mm)	Velocity (m/s)	Current
sine	40	0.052 0.131 0.262 0.524	0-1.0A (interval is 0.1A)

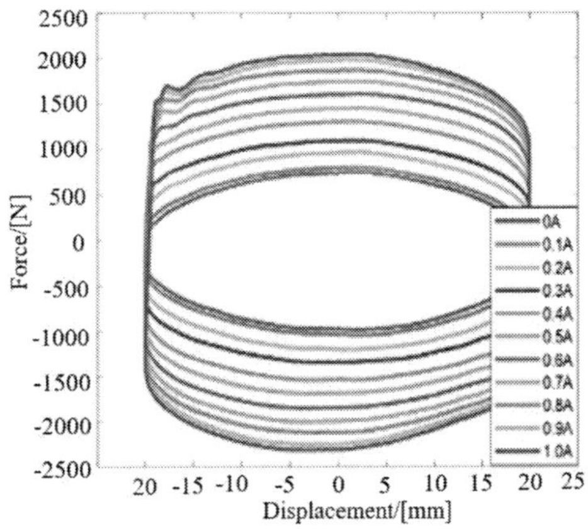

a) F-S curve

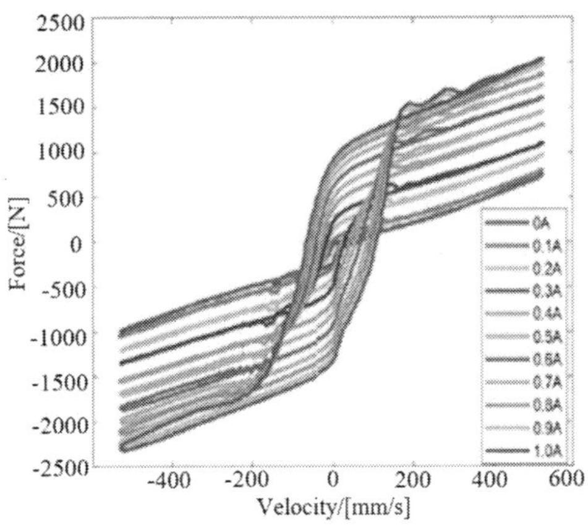

b) F-V curve

Figure 1. Damper external characteristic curve

2.2 Establishment of GP model for damper

The observed data of the damping force of the damper is related to the control current, position and speed. A certain measurement deviation exists in the data, so the following assumption is defined (Rasmussen et al. 2004):

$$\begin{cases} F_d = F_{true}(x_k) + \varepsilon, & F_{true}: \mathbb{R}^{n_x} \rightarrow \mathbb{R} \\ \varepsilon \sim \mathcal{N}(0, \sigma_n^2) \end{cases} \tag{1}$$

It is assumed that the damping force is a function of controlling current i_k, position s_k and velocity v_k , $\boldsymbol{x_k} = [i_k \ \ s_k \ \ v_k]$ is thus defined. The measurement deviation follows a normal distribution, and the variance is σ_n^2. F_{true} is a shock absorber model function, which is a Gaussian prior model:

$$F_{true} \sim \mathcal{GP}(0, \ \ k(x, x')) \tag{2}$$

Here, let $m(x) = \mathbf{0}$. Given this prior assumption and training data set $\mathcal{D}$, set the input of the predicted point to be $\boldsymbol{X_*}$, the corresponding output is $\boldsymbol{F_{true*}}$, and the training point set is:

$$X = [x_1, x_2, \dots, x_m]^T \in \mathbb{R}^{m\times 3}.$$
$$F_d = [F_{d1}, F_{d2}, \dots, F_{dm}]^T \in \mathbb{R}^{m\times 1}$$

Then the posterior distribution of $\boldsymbol{F_{true}}$ is:

$$F_{true*}|X_*, X, F_d \sim \mathcal{N}(\mu_*, \Sigma_*) \tag{3}$$

Where

$$\mu_* = \mathbb{E}[F_{true*}|X, Y, X_*]$$
$$= K(X_*, X)[K(X, X) + \sigma_n^2 I]^{-1} F_d$$
$$\Sigma_* = K(X_*, X_*) - K(X_*, X)[K(X, X) + \sigma_n^2 I]^{-1} K(X, X_*). \tag{4}$$

Smooth and continuous input-output characteristics are expected, so the ardSE kernel function is used here:

$$k(\boldsymbol{x}, \boldsymbol{x}') = \sigma_f^2 exp\left(-\tfrac{1}{2}(\boldsymbol{x} - \boldsymbol{x}')^T \boldsymbol{M}^{-1}(\boldsymbol{x} - \boldsymbol{x}')\right) \tag{5}$$

where

$$M = \begin{bmatrix} l_1 & 0 & 0 \\ 0 & l_2 & 0 \\ 0 & 0 & l_3 \end{bmatrix}$$

Unlike the squared exponential (SE) kernel function, the M matrix in the automatic correlation decision SE (ardSE) kernel function uses different hyperparameters $l_i (i = 1,2,3)$ according to the relative importance of different input features, which is a property called automatic correlation decision (ARD) that optimizes the structure of the Gaussian process model, i.e., the selection of input features (Csató and Opper 2002). The free parameters of the model, i.e., hyperparameters $\boldsymbol{\theta} = [\sigma_f^2, \sigma_n^2, l_1, l_2, l_3]$.

The training process of the Gaussian process model is also the process of hyperparameter optimization, and the maximum likelihood estimation method is applied. The Marginal likelihood of the observed quantity appears as follows:

$$F_d \,|X, \ \theta \sim \mathcal{N}(0, \ K(X, X) + \sigma_n^2 I)$$

Its likelihood function is:

$$log\ p(F_d\,|X, \theta) =$$
$$-\frac{1}{2} F_d{}^T (K(X, X) + \sigma_n^2 I)^{-1} F_d$$
$$-\tfrac{1}{2} log|K + \sigma_n^2 I| - \tfrac{n}{2} log(2\pi) \tag{6}$$

By maximizing the log-likelihood function above, the optimal hyperparameters can be obtained,

$$\theta^* = \underset{\theta}{argmax} \log p(F_d \mid X, \theta) \quad (7)$$

And the optimized hyperparameter results are shown in *Table 2.*

Figure 2 shows the comparison between the test results and the prediction results under a certain condition (the solid line represents the predicted value, and the point represents the test data). The results show that the prediction curve is located within the 95% confidence interval (shaded area), which matched well with the experimental data, which demonstrates the model accuracy and its capability for control application.

Table 2. Optimized hyperparameters

σ_f	σ_n	l_1	l_2	l_3
1021	98.76	1.084	3.249	37.051

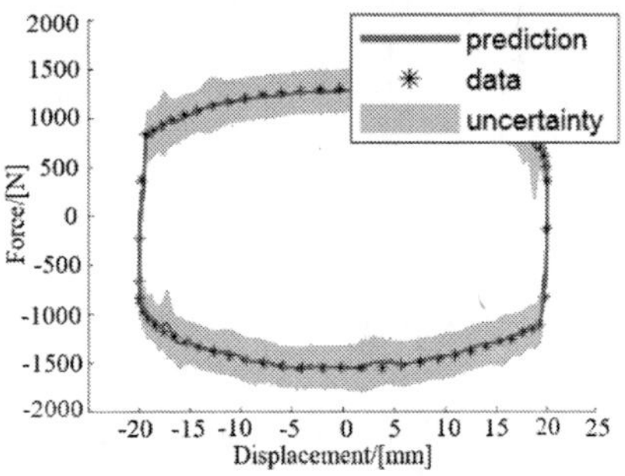

a) F-S curve

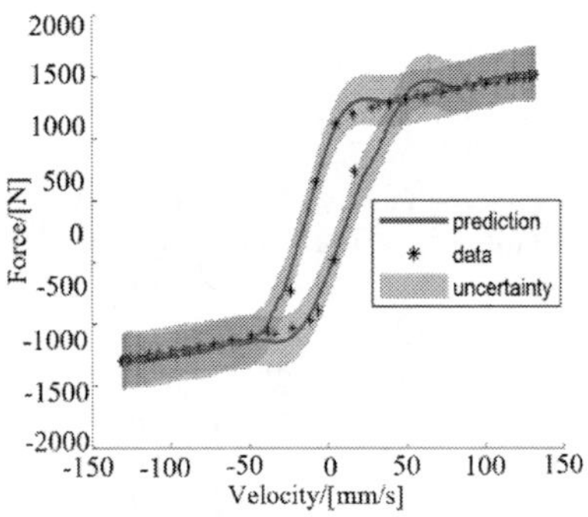

b) F-V curve

Figure 2. Comparison between experimental data with prediction data

2.3 Magnetorheological damper inverse model

It can be seen from the experimental data of the external characteristics of the damper that the hysteretic oscillator has obvious hysteresis characteristics, and the compression and recovery strokes are not much different, so here is based on the Hysteretic model (Kwok et al. 2006), combined with the experimental data, the parameters in the positive model are fitted. Then the Hysteretic positive model is inversely solved to obtain the inverse model. The following is a brief description of the establishment process of the Hysteretic model.

The expression for the Hysteretic model is shown in the equation:

$$F_d = c\dot{x} + kx + \alpha z + f_0$$

$$z = \tanh[\beta\dot{x} + \delta sign(x)] \quad (8)$$

Where the parameter c is the damping coefficient, k is the stiffness coefficient, z is the hysteresis variable, the δ and $\beta$are the form factors, the parameters are the compensatory forces, and their main role is to adjust the curve to move up and down along the Y axis.

The process of establishing a positive model is:

1. Test the external characteristics of the magnetorheological shock absorber and obtain the fitting data.

2. Write Matlab programs based on positive model expressions and external characteristic experimental data and use *lsqnonlin* fitting functions to identify the parameters. The goal of the fitting is to minimize the mean squared error between the prediction data and the experimental data:

$$J = min\sqrt{\frac{1}{N}\sum_{n=1}^{N}\left[F_{exp}(n) - F(x,\dot{x},n)\right]^2} \quad (9)$$

$F_{exp}(n)$ represents the damping force measured by the experiment at the nth point, and F represents the damping force predicted by the model based on the current speed and displacement of the damper.

3. Analyze the fitted parameter $\Theta = [c, k, \alpha, f_0, \delta, \beta]$, and the relationship between each parameter and the current according to the principle of least squares. The analysis shows that the current change has little effect on parameter δ and β, so they are taken as fixed values:

$$c(I) = c_1 I^2 + c_2 I + c_3$$

$$k(I) = k_1 I^2 + k_2 I + k_3$$

$$\alpha(I) = \alpha_1 I + \alpha_2$$

$$f_0(I) = f_{01} I^2 + f_{02} I + f_{03}$$

4. Bring the parameters obtained in the previous step into the Hysteretic model, then the relationship between the damping force, current, displacement and velocity can be obtained. (The final parameter fitting results are shown in *Table 3.*)

Table 3. Hysteretic model parameter identification result

Parameter	Value	Parameter	Value
c_1	-1.113	α_2	1.3
c_2	3.591	f_{01}	-6.134
c_3	1.021	f_{02}	21.691
k_1	1.341	f_{03}	-1.472
k_2	-1.979	δ	0.8108
k_3	0.746	β	0.09
α_1	1.121		

The highest order of current term in the damping force expression is 2, so the inverse model can be solved according to the knowledge of the quadratic equation. Its expression is as follows:

$$I_{ide}(\dot{x}, x, F) = \begin{cases} \frac{-N+\sqrt{N^2-4MP}}{2M} & \dot{x} \geq 0 \\ \frac{-N-\sqrt{N^2-4MP}}{2M} & \dot{x} < 0 \end{cases} \quad (10)$$

Where,

$$M(\dot{x}, x) = c_1\dot{x} + k_1 x + f_{01}$$
$$N(\dot{x}, x) = c_2\dot{x} + k_2 x + \alpha_1 Z + f_{02}$$
$$P(\dot{x}, x, F) = c_3\dot{x} + k_3 x + \alpha_2 Z + f_{03} - F$$

3. Suspension system modelling

When studying the suspension control problem, the system model could be 7 DOF (degrees of freedom), 4 DOF, and 2 DOF (Jeong et al. 2022). The higher the degrees of freedom are, the more accurately the model can reflect the response, but model will become more complex, which will increase the difficulty of algorithm design. This paper mainly studies the comfort performance of suspension, while the riding comfort is mainly related to the vertical vibration of the vehicle, so the 2 DOF vehicle model with only two vertical vibrations is selected, which is shown in Fig. 3.

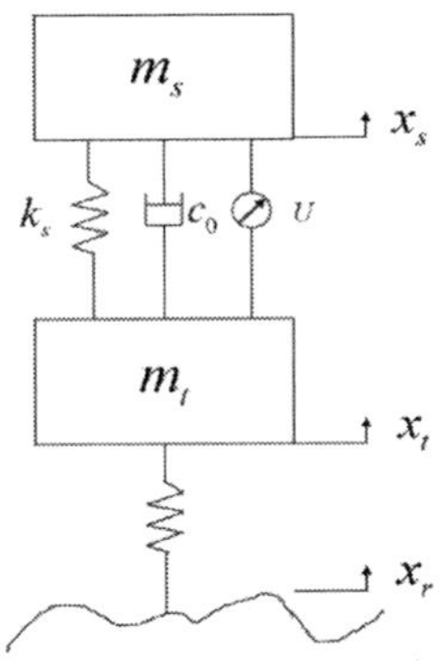

Figure 3. Quarter car suspension model

Fig 3 shows the basic structure of a quarter car model, and x_s , x_t , x_r respectively represents the vertical position of the sprung mass, the vertical position of the unsprung mass and the displacement of the road. $m_s = 40\text{kg}$ is the spring mass (mainly the body), $m_t = 10\text{kg}$ is the non-spring mass (wheel, brake disc and suspension connection rod system, etc.), and $k_s = 200\text{N/m}$, $k_t = 2500\text{N/m}$ represents the suspension stiffness and the tire stiffness respectively. The maximum and minimum damping are $c_{max} = 4000\text{Ns/m}$ and $c_{min} = 300\text{Ns/m}$, respectively, and the passive damping is $c_0 = 1500\text{Ns/m}$. Suspension systems can be described by the following differential equations:

$$\begin{cases} m_s\ddot{z_s} = -k_s(z_s - z_t) - c_0(\dot{z_s} - \dot{z_u}) + U \\ m_t\ddot{z_u} = k_s(z_s - z_t) + c_0(\dot{z_s} - \dot{z_u}) - k_t(z_s - z_r) - U \end{cases} \quad (11)$$

U is the damping force input, and the semi-active suspension system state variables are suspension deflection, sprung mass velocity, tire deformation and road speed as follows.

$$X = [z_s - z_t, \dot{z_s}, z_t - z_r, \dot{z_r}]$$

The output variables are sprung mass acceleration, suspension deflection, the dynamic tire loads, and the unsprung mass velocity:

$$Y = [\ddot{z_s}, z_s - z_t, k_t(z_t - z_r), \dot{z_t}]$$

Let $W = \dot{z_r}$ be the road speed input, so there is a semi-active suspension state space equation:

$$\begin{cases} \dot{X} = AX + B_u U + B_w W \\ Y = CX + DU \end{cases} \quad (12)$$

where the state matrix is:

$$A = \begin{bmatrix} 0 & 1 & 0 & -1 \\ -\frac{k_s}{m_s} & -\frac{c_0}{m_s} & 0 & \frac{c_0}{m_s} \\ 0 & 0 & 0 & 1 \\ \frac{k_s}{m_s} & \frac{c_0}{m_t} & -\frac{k_t}{m_t} & -\frac{c_0}{m_t} \end{bmatrix}$$

The damping force input matrix and the road input matrix respectively are:

$$B_u = \begin{bmatrix} 0 \\ \frac{1}{m_s} \\ 0 \\ -\frac{1}{m_t} \end{bmatrix}, \quad B_w = \begin{bmatrix} 0 \\ 0 \\ -1 \\ 0 \end{bmatrix}$$

The output matrix is:

$$C = \begin{bmatrix} -\frac{k_s}{m_s} & -\frac{c_0}{m_s} & 0 & \frac{c_0}{m_s} \\ 1 & 0 & 0 & 0 \\ 0 & 0 & k_t & 1 \\ 0 & 0 & 0 & 1 \end{bmatrix}$$

The transfer matrix is as follow:

$$D = \begin{bmatrix} \frac{1}{m_s} \\ 0 \\ 0 \\ 0 \end{bmatrix}$$

4. Control and simulation

4.1 Algorithm design

Reinforcement learning is a branch of machine learning that maximizes a scalar reward value during interaction with the environment. The reward only indicates that an action is good or bad in interacting with the environment, and does not indicate that the action is optimal, which is very different from supervised learning, in which the algorithm is learned through a series of input and output samples. Environments in reinforcement learning include known environments and unknown environments. For known environments, the algorithm will be programmed over a period time based on the dynamics of the environment to obtain the optimal control sequence; For unknown environments, reinforcement learning algorithms require a lot of interaction with the system. Since the suspension system faces random road input and is unpredictable, so the suspension system becomes a random environment. Reinforcement learning algorithm introduced in this chapter is to solve the problem of semi-active suspension optimal control in a random environment.

In a typical reinforcement learning system, learning subsystem acts on the environment with a set of actions, which then returns a scalar value through the performance evaluation function to reflect the performance impact of the action on the system. The design of the performance evaluation function needs to consider the control objectives of the system and can return a scalar value β (between $0 \sim 1$) that reflects the quality of the action, and the larger the value, the greater the contribution. Because of the random nature of the environment, the action must be applied multiple times to determine which action gets the highest average return from the environment. In the learning process, a balance must be maintained between "exploration"—that is, exploring more information about the system and "exploitation"—that is, exploiting the best action that has been already explored (Johnson et al. 2000). The algorithm interacts with the environment by acting on multiple actions on the environment (in this case, referring to the suspension system) and obtaining performance evaluation scalar values from the environment, and has good control effect under different input conditions of the system.

This section will introduce the stochastic optimal reinforcement learning control algorithm based on the control problem of a single continuous action variable, and the multi-action version can be constructed easily by connecting multiple learning subunits.

Action variable x is defined as a continuous random variable within a certain range $[x_{min}, x_{max}]$, then at the n th iteration, the probability distribution should satisfy the following condition:

$$\int_{x_{min}}^{x_{max}} f(x,n)dx = 1, f(x,n) \geq 0 \ \forall n, \forall x.$$

The pseudocode of the single-action continuous space stochastic optimal reinforcement learning algorithm is shown in *Table 4.*

The initial probability density distribution takes the following form:

$$f(x,0) = \begin{cases} \frac{1}{x_{max}-x_{min}} for\ x \in [x_{min}, x_{max}] \\ 0\ otherwise. \end{cases} \quad (13)$$

In this case, the random selection of the single action x at the beginning is obvious, i.e., $x \sim U[x_{min}, \ x_{max}]$. More generally, $x(n)$ needs to be sampled through a non-uniformly distributed probability density function, which can be achieved by introducing a uniformly distributed variable $z(n)$. Given a uniformly distributed random variable $z(n) \sim U[0,1]$ during iteration, the sampled action values need to satisfy the following equation:

$$\int_{x_{min}}^{x(n)} f(x,n)dx = z(n). \quad (14)$$

Apply the action value $x(n) = \xi$ to the system, and the system will return a cost value or performance index $J(n)$ to the performance evaluation function to evaluate the quality of the selected action. Stores current and past performance metric scalars in the collection R to determine the reward value $\beta(n)$:

Table 4. Single-action RL algorithm pseudocode

Initialize the probability density function to be uniformly distributed.

Loop:

1. Select an action based on the probability density distribution function;
2. Perform this action on the system;
3. Receive reward after the execution of the previous action;
4. Update performance assessment function β;
5. Update the probability density function;

Until stopping condition.

$$\beta(n) = max\left\{0, \frac{J_{med}-J(n)}{J_{med}-J_{min}}\right\} \quad (15)$$

J_{med} and J_{min} are the median and minimum values in the set R, respectively. Reward values $\beta(n) \in [0,1]$ relate the performance of recent actions to the results of past actions. To avoid the problem of infinite storage and enable the algorithm to adapt to the changing working conditions, only the last m values are saved.

The probability density function $f(x,n)$ will be updated according to the following rules:

$$f(x,n+1) = \begin{cases} \alpha[f(x,n) + \beta(n)H(x,r)]\ if\ x \in X \\ 0\ otherwise. \end{cases} \quad (16)$$

α is implicitly determined by the following expression:

$$\int_{x_{min}}^{x_{max}} f(x,n+1)dx = 1$$

Here, $H(x,r)$ is a Gaussian neighbor function centered on $r = x(n)$:

$$H(x,r) = \lambda\, exp(-\frac{(x-r)^2}{2\sigma^2})$$

λ and σ are parameters that affect Gaussian neighborhood functions, respectively, and they can be defined by the range of action values:

$$\lambda = \frac{g_h}{(x_{max} - x_{min})}$$

$$\sigma = g_w(x_{max} - x_{min})$$

The learning rate and learning accuracy are controlled by the parameter g_h and the parameter g_w, respectively. If the change in the cost function is assumed to be smooth, then the generation values with similar values should be able to generate similar reward values based on the performance evaluation function. The Gaussian neighborhood function $H(x,r)$ has the effect of spreading the reward value of the selected action to a similar action.

In addition, the learning process continues after the optimal value is found, and the system action value does not converge to a certain point but converges to a certain range.

For the programming implementation, the value of the probability density distribution function $f(x,n)$ is stored in the discrete form of equal interval ∇f, and the action probability density within the interval is calculated by interpolation. This implementation allows the probability density curve of the most important regions to be better defined.

Each learning subsystem can be thought of as a separate unit associated with a single continuous action variable.

4.2 Simulation results

In this section, the algorithm is built in the Matlab/Simulink environment to control the suspension system, and the action space dimension of the system is 3, that is, there are three continuous actions acting on the system. The control goal of the learning system is to minimize the root mean square value of sprung mass acceleration while keeping the suspension stroke within a safe range, so the design cost function J has the following form(Poussot-Vassal et al. 2012):

$$J = \sqrt{\frac{1}{N}\sum_{i=1}^{N}\ddot{x}_s(i)^2} + \alpha_1\sqrt{\frac{1}{N}\sum_{i=1}^{N}|x_s(i) - x_u(i)|^2} + \alpha_2\sqrt{\frac{1}{N}\sum_{i=1}^{N}|x_u(i) - x_r(i)|^2} \qquad (17)$$

$\ddot{x}_s$,$x_s(i) - x_u(i)$与$x_u(i) - x_r(i)$ represents sprung mass acceleration, suspension travel and tire deformation, respectively ; α_1 and α_2 represent the weights of the RMS values of suspension travel and tire deformation, respectively.

The following suspension system states will be used for feedback control: suspension relative displacement $x_s - x_u$, tire speed $\dot{x}_u$, and sprung mass velocity $\dot{x}_s$. The controller will calculate the desired control force based on above state values and feedback gains, which is expressed in the following equation:

$$F_s = k_1(x_s - x_u) + k_2\dot{x}_u + k_3\dot{x}_s \qquad (18)$$

The learning system is used to determine the controller parameters k1, k2, and k3, and the specified parameter range is determined by the previous literature on active and semi-active suspension control. The action variables of each subsystem correspond to one controller parameter, and the damping of the shock absorber can be determined by desired damping force and suspension speed. The specific implementation details of the control algorithm are shown in pseudocode *Table 5*, and the parameters of the algorithm are shown in *Table 6*. Setting the learning interval to 16s is necessary, not only for the sake of learning rate, but also for the dynamic characteristics of the controlled system, which can eliminate the effect of input fluctuations.

The learning system is used to determine the controller parameters k1, k2, and k3, and the specified parameter range is determined by the previous literature on active and semi-active suspension control. The action variables of each subsystem correspond to one controller parameter, and the damping of the shock absorber can be determined by desired damping force and suspension speed. The specific implementation details of the control algorithm are shown in pseudocode *Table 5*, and the parameters of the algorithm are shown in *Table 6*. Setting the learning interval to 16s is necessary, not only for the sake of learning rate, but also for the dynamic characteristics of the controlled system, which can eliminate the effect of input fluctuations.

Table 5. Stochastic optimal RL algorithm pseudocode

1. Set the number of iterations $n = 0$;
2. Define the action space:

   $A(n) = [k_1(n), k_2(n), k_3(n)]$,
   where $k_i \in [k_{i,min},\ k_{i,max}]$;
3. Define the probability density function:
   $f_i(k_i, n), i = 1,2,3;$
4. Initialize the probability density function $f_i(k_i, n)$ as a uniform distribution within the action range.
5. Repeat
   - In each iteration, use pseudo-random number generator to select a z_i value, $z_i \in U[0,1]$
   - Select $k_i \in A(n)$ to make it satisfy the following equation:
   - $\int_{k_{i,min}}^{k_i(n)} f(x,n)dx = z_i(n)$
   - Run the system for Δ seconds
   - Evaluate the cost J via *Eyq. 16*
   - Append to and evaluate the minimum, J_{min} and median, J_{med} ,values of R.
   - Evaluate $\beta(n)$ via *Eq. 14*
   - Update the probability density functions $f_i(k_i, n)$ via *Eq. 15*
   - Increment iteration number n

Table 6. Control system parameter list

$\Delta = 16$s	$\nabla f = 0.01$
$m = 500$ (size of set R)	$k_{2,max}$=4000
$g_w = 0.02$	$k_{2,min}$= -6000
g_h=0.3	$k_{3,max}$=20000
$k_{1,max}$=40000	$k_{3,min}$= 0
$k_{1,min}$= -20000	——

Figure 4 shows how the action values change during iteration. The learning system continuously selects the action based on the feedback gain probability density, then acts on the system, and then updates the probability density according to the system return value.

It can also be seen that the three learning subsystems converge after about 2100, 1300, and 1800 iterations, and k1, k2, and k3 converge at 13197.8, -529.9, and 5019.8, respectively. Set the number of iterations to 3000, and the shape of the feedback gain probability density function gradually becomes sharp during iteration.

Figure 5 shows the change of the probability density function of the parameters during the iteration. As can be seen from the figure, at the end of the iteration, the probability density function converges. The probability density function after its convergence is shown in *Figure 6*, and the action value corresponding to the spike position will be used as the controller parameter. *Figure 7* shows that during the iteration process, the cost value is decreasing, and when the final parameters are stable, the cost value is located at a very low position.

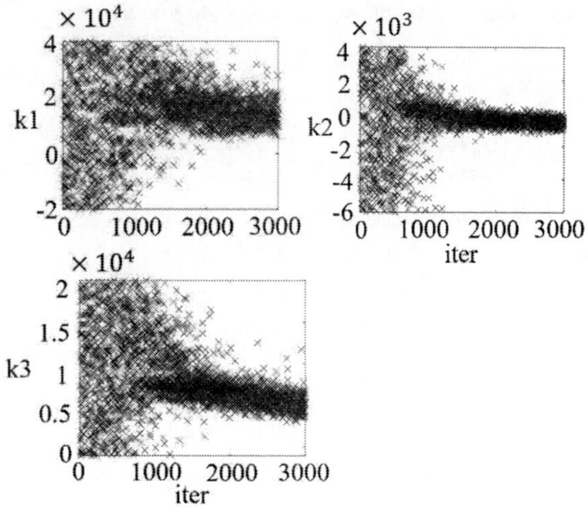

Figure 4. Control iteration process

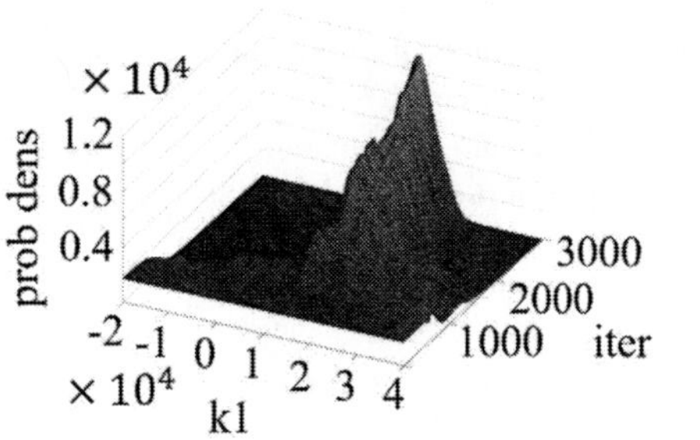

Figure 5. Controller parameter probability density

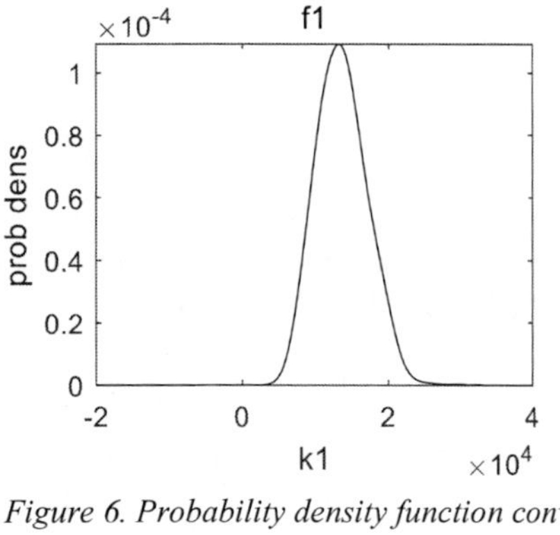

Figure 6. Probability density function converge

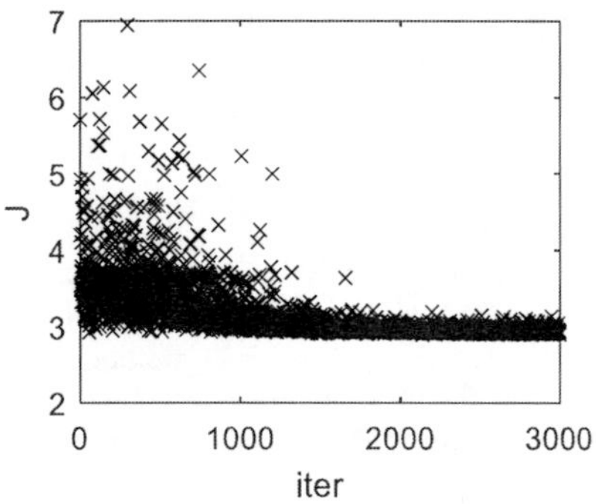

Figure 7. Cost change during iteration

To compare the control effects between different algorithms, passive control(Elahi et al. 2016), skyhook control (Emura et al. 1994), and this reinforcement learning algorithm (RL) simulations are carried out. The excitation is a random pavement, simulation step is 0.01s, and the maximum excitation frequency is 20Hz.

Figure 8 shows the graph of the amplitude-frequency characteristic function estimated by taking the acceleration of the sprung mass as the signal output and the road excitation as the signal input under different control algorithms. As can be seen from the figure, the two-order resonance frequency of the suspension is roughly located at 1 to 3Hz and 8 to 11 Hz, respectively. In the low frequency band, skyhook corresponding to the curve is the lowest, has the most significant effect on improving suspension comfort, followed by RL algorithm; In the high-frequency band, Cmin algorithm corresponds to the lowest curve, which can most effectively reduce the acceleration of the sprung mass, followed by RL algorithm. In the mid-band, skyhook and Cmin still work best, but in this band the curve is very close to the curve of the reinforcement learning algorithm. In summary, the skyhook and Cmin algorithms can achieve good comfort improvement effects in the low and high frequency bands respectively, but the reinforcement learning algorithm can achieve the optimal control effect from the view of the entire frequency band.

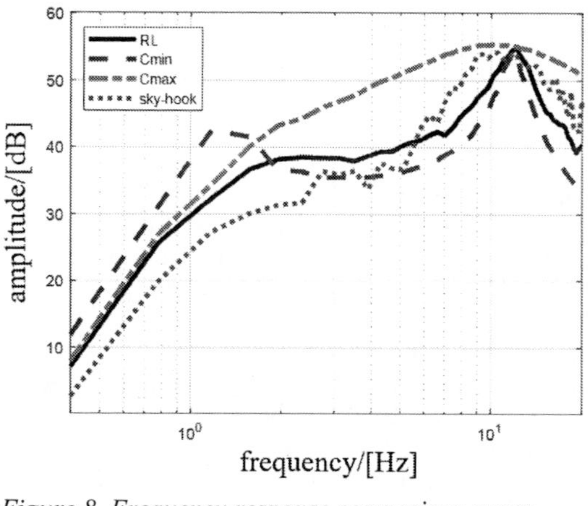

Figure 8. Frequency response comparison curve

5. Experiment

The semi-active suspension experimental system is mainly composed of power system, data acquisition and control system, bench system and sensor system.

As can be seen from *figure 9*, both the sprung mass and the unsprung mass are arranged with acceleration sensors for measuring acceleration, and the acceleration signal can be integrated to obtain the speed signal. An encoder is installed between the sprung mass and the unsprung mass to measure the suspension dynamic stroke, and the displacement signal is differential to obtain the relative speed of the suspension.

The suspension experimental data acquisition and control system is built through Matlab/Simulink Real-time software, which can realize model writing, C code compilation, data acquisition and real-time control.

The control current of damper is provided by the self-developed current drive board, and the supply voltage of the drive board is 18-24V. The driver board can output a control current equal to its numerical size according to the input voltage, and the maximum output current is 2A. The current damping control can be achieved by sending a voltage signal to the drive board.

To illustrates the control effects of the proposed algorithm under random pavement input conditions. The acceleration responses of the sprung mass under RL, Skyhook control and Cmax, Cmin are compared and shown in *Figure 10*.

Under the condition of random road, the comfort evaluation can be performed by calculating the RMS value of the sprung mass acceleration response under each algorithm. After calculation, under the passive suspension, skyhook, Cmin and RL algorithms, the RMS values of acceleration are: 0.959, 0.844, 0.886, 0.814 (m/s^2). Compared with passive suspension, the acceleration RMS values of the three algorithms have decreased by 12.02%, 7.59% and 15.1%, respectively, and the reinforcement learning algorithm has the most obvious effect on comfort improvement, followed by skyhook. The Cmin algorithm controls a single current range with only one control current, keeping the suspension damping at a low level. The sky-hook algorithm has two kinds of control currents, switching between the minimum current and the maximum current according to the suspension state, and the control effect is better than Cmin algorithm. The reinforcement learning algorithm calculates the continuous control input in the minimum and maximum current ranges according to the optimal state feedback gain learned during interaction with the system, whose control range is wider than that of Cmin and skyhook algorithm, and the control effect is better.

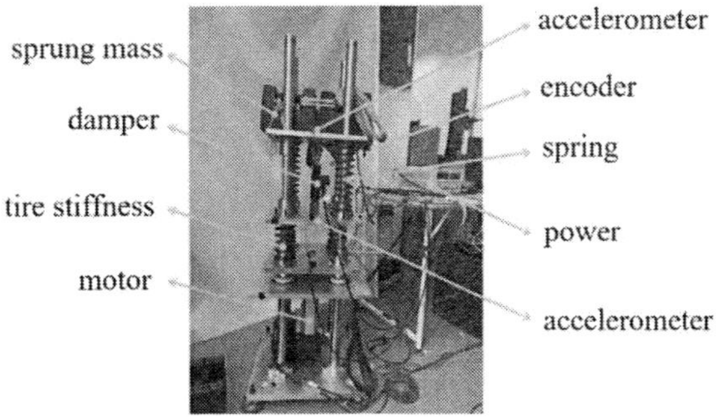

Figure 9. Experimental bench structure

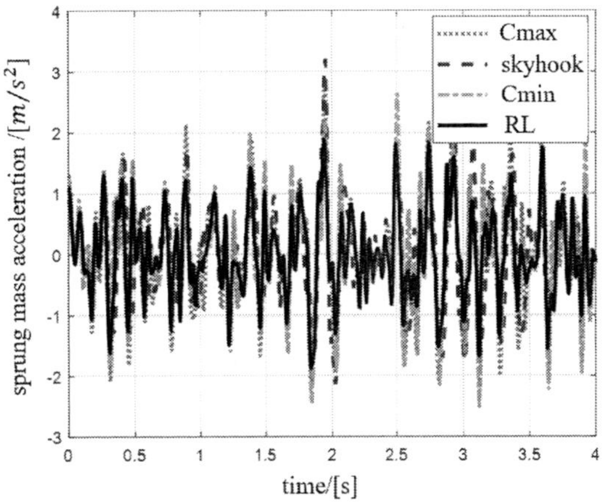

Figure 10. Sprung mass acceleration response

6. Conclusion

Based on the hysteresis characteristics and current characteristics of magneto-rheological damper, this paper explores the semi-active suspension reinforcement learning control algorithm of terrain vehicles to improve and balance the suspension performance in each frequency band.

The continuous action space reinforcement learning algorithm introduced in this paper has a good control effect in improving comfort performance. The algorithm has relatively few hyperparameters: the most important parameters are only the height and width parameters in the Gaussian neighborhood function, and the actual learning process is not very sensitive to these two parameters, and only affects the training speed to a small extent. The algorithm structure is simple but can be applied to complex systems with more action dimensions. This simple and general algorithm structure makes it easy to apply to control systems with random characteristics, and easy to implement in hardware. In addition, due to the randomness of the action selection process, the algorithm has the potential to avoid local optimal solutions.

7. References

Csató, L. and M. Opper (2002). "Sparse On-Line Gaussian Processes." Neural Computation 14(3): 641-668.

Elahi, H., A. Israr, M. Z. Khan and S. Ahmad (2016). "Robust vehicle suspension system by converting active & passive control of a vehicle to semi-active control system analytically." Journal of Automation and Control Engineering 4(4).

Emura, J., S. Kakizaki, F. Yamaoka and M. Nakamura (1994). "Development of the semi-active suspension system based on the sky-hook damper theory." SAE transactions: 1110-1119.

Frost, G., T. Gordon and Q. Wu (1994). The Application of Learning Automata to Advanced Vehicle Suspension, Control of Vibration, 5-8 Sept, Bath.

Glavic, M., R. Fonteneau and D. Ernst (2017). "Reinforcement learning for electric power system decision and control: Past considerations and perspectives." IFAC Papers OnLine 50(1): 6918-6927.

Jeong, Y., Y. Sohn, S. Chang and S. Yim (2022). "Design of Static Output Feedback Controllers for an Active Suspension System." IEEE Access 10: 26948-26964.

Johnson, J., J. Li and Z. Chen (2000). "Reinforcement Learning: An Introduction: R.S. Sutton, A.G. Barto, MIT

Press, Cambridge, MA 1998, 322 pp. ISBN 0-262-19398-1." Neurocomputing 35: 205-206.

Kaldas, M. M. S. (2015). Improvement of semi-active suspensions through fuzzy-logic and top mount optimization, Shaker-Verlag.

Kober, J., J. A. Bagnell, and J. Peters (2013). "Reinforcement learning in robotics: A survey." The International Journal of Robotics Research 32(11): 1238-1274.

Kwok, N. M., Q. P. Ha, T. H. Nguyen, J. Li, and B. Samali (2006). "A novel hysteretic model for magnetorheological fluid dampers and parameter identification using particle swarm optimization." Sensors and Actuators A: Physical 132(2): 441-451.

Poussot-Vassal, C., C. Spelta, O. Sename, S. M. Savaresi, and L. Dugard (2012). "Survey and performance evaluation on some automotive semi-active suspension control methods: A comparative study on a single-corner model." Annual Reviews in Control 36(1): 148-160.

Rasmussen, C. E., O. Bousquet, U. V. Luxburg, and G. Rtsch (2004). Gaussian Processes in Machine Learning, Springer Berlin Heidelberg.

Szymański, G., M. Josko, F. Tomaszewski, and R. Filipiak (2015). "Application of time–frequency analysis to the evaluation of the condition of car suspension." Mechanical Systems and Signal Processing 58: 298-307.

Thrun, S., and A. Schwartz (1993). Issues in using function approximation for reinforcement learning. Proceedings of the 1993 Connectionist Models Summer School Hillsdale, NJ. Lawrence Erlbaum.

Vinyals, O., T. Ewalds, S. Bartunov, P. Georgiev, A. S. Vezhnevets, M. Yeo, A. Makhzani, H. Küttler, J. Agapiou, and J. Schrittwieser (2017). "Starcraft ii: A new challenge for reinforcement learning." arXiv preprint: 1708.04782.

1491 / Design and Verification of a Creeping Mars Rover

Wangjun Zhang, Yang Jia, and Zhuo Tao*
Beijing Institute of Spacecraft System Engineering, Beijing, China
**Corresponding author: taozhuo90@foxmail.com*

https://doi.org/10.56884/BJJS7767

Abstract

Aiming at the possible wheel sinkage, difficulty in climbing and underpinning of Mars rover with traditional passive suspension due to the complex terrain of Mars surface, the characteristics of the inchworm motion are imitated, and a creeping Mars rover is designed, include the design of sinkage and extrication, slope-climbing, anti-underpinning, and wheel-lifting. The key parameters of the mobile system of creeping rover are identified and the key parameter values of different rover lift heights and the displacement value of a creeping cycle are calculated. and the results show that the creeping rover can solve the above problems encountered by the passive suspension rover. Ground tests and flight tests have been carried out on the designed creeping rover, and the ground tests have verified the ability of wheel sinkage and extrication, obstacle-overcoming, and slope-climbing. As an example of the application of the creeping rover, the "Zhurong" Mars rover has undergone flight tests, and its successful operation on the surface of Mars shows that the design measures of the creeping are effective.

Keywords: Mars rover, creeping motion, active suspension, ground tests, flight tests

1. Introduction

As Mars exploration missions continue to progress, Mars rovers, as the most used tool for Mars surface in-situ exploration, have become the focus of Mars mission by relevant countries.

The three generations of Mars rovers have been successfully launched and put into operation by the United States are all passive suspension rovers, which have experienced problems such as wheel sinkage, difficulty in climbing and underpinning during the operation on the Mars surface. Wheel sinkage and extrication has always been a difficult point of the design of Mars rover (Zhou et al., 2014; Reina et al., 2006; PAN et al., 2022). Reina, Pan, and others have conducted in-depth research on the mechanism of sinkage. Weathering causes the Martian soil to be loose inside and hard outside, which is visually deceptive. During the driving process of the Mars rover, the sand and soil collapsed under the action of the wheels, which could easily lead to the danger of deep sinkage without warning. According to statistics, the "Spirit" Mars rover had a total of 6 times of wheel sinking dangers and was trapped for a total of about 26 months. Eventually, the Mars rover was trapped in the Troy Dunes and permanently lost its ability to move.

The complex terrain on the surface of Mars is not only easy to cause the sinkage of the rover, but also causes difficulty in climbing and underpinning. The Martian surface is rocky and impact cratered, making the Martian surface rough. Under the action of sandstorms, impact craters, dunes, etc. will form slopes with small but very soft slopes. The soil cohesion and internal friction angle of such slopes are very small, and the wheel slip rate is very high, which makes the Mars rover unable to climb the slope normally. "Opportunity" was trapped in the ring pit for 3 months, and after choosing dozens of routes, it finally climbed out of the pit. According to calculations, the diameter of the sand pit is about 20m, the depth is about 3m, and the slope is about 15° (WANG et al., 2020). The surface of Mars has many rocks and high protrusions, and the distribution density of rocks is about twice that of the lunar surface. Large rocks have a higher probability of underpinning. The bottom of the "Spirit" rover was once held up by stones. Because the passive suspension could not actively adjust the height of the rover, it could only drive out. This kind of strong driving can easily damage the rover structure and destroy the thermal control measures on the rover surface.

In addition, both "Opportunity" and "Spirit" had wheel drive failures on orbit. The faulty wheels could not turn and could only be dragged by other wheels, which caused a bulldozing effect on the sandy ground and greatly increased driving resistance, making it difficult to drive, greatly reducing the efficiency and range of exploration, and in severe cases, it will lead to inability to drive.

The various dangers encountered by the US Mars rovers on the surface of Mars are not only related to the complex Martian surface environment, but also directly related to the passive suspension which cannot adjust the rover height in addition to the complex terrain on the Mars surface. Therefore, in recent years, the active suspension has become an important direction of research because the rover height can be lifted or descended.

The Jet Propulsion Laboratory of the United States proposed a microminiature four-wheel active suspension, which gives the Mars rover a variable structure with a good self-correction function, and the disadvantage is that it is applicable to the microminiature Mars rover (BRIAN et al., 1997, TUNSTEL et al.,1999). The Jet Propulsion Laboratory also proposed an SRR (Sample Return Rover) four-wheel active suspension, which can adjust the rover height and improve the Mars rover's ability to overcome obstacles and climb slopes but cannot solve the problem of wheel sinkage (IAGNEMMA et al., 2003). The European Exo-Mars rover adopts a six-wheeled active suspension form with three identical parallel frames, which has better terrain adaptability, but the wheel sinkage descent capability is poor (ELLERY et al., 2005). In China, Harbin Institute of Technology proposed a tandem articulated multi-joint suspension for six-wheeled lunar rovers, which is a mixed active-passive system that can improve the extrication ability by creeping and can also actively cross larger ditches and steps by lifting the wheels, and the shortcomings of this suspension are the complex mechanisms and the difficulty of control (TAO, 2009). Inspired by the natural inchworm motion, the author team of this paper proposes an inchworm-like active suspension system based on the active suspension Mars rover to solve the above problems with minimal resources. So far, no relevant literature has been retrieved on the application of inchworm motion to planet rover.

In this paper, we will design a kind of creeping Mars rover, as a kind of Mars rover with active suspension, which adopts active suspension modified by rocker type passive suspension, which can creep to improve the driving force to realize wheel sinkage and extrication and can lift the height of the rover to realize large obstacles overcoming and lower the height of the rover to realize large angle slope climbing. Ground tests are to be conducted to validate the creeping Mars vehicle.

2. The working principle of the creeping Mars rover

The inchworm is the larva of the geometer moth, which is an arthropod. The inchworm has a slender body and is mainly composed of three parts: head, tail, and body. It bends and stretches like an arch bridge when in motion and stretches like a branch when resting (SHI et al., 2019; LIU et al., 2019). The front wheel of the creeping rover can be regarded as the head of the inchworm, the rear wheel and middle wheel can be regarded as the tail of the inchworm, and the front main rocker arm and the back main rocker arm can be regarded as the inchworm's body, and the active suspension structure of the Mars rover is close to the inchworm in shape, as shown in *Figure 1*.

The most important function of the creeping rover is to imitate the inchworm motion through the lifting and descending of the rover. One cycle of creeping motion is divided into four stages. In the first half cycle of creeping motion, the angle adjustment mechanism moves, the rover body lifts, while the front wheel does not move, and the middle wheel and rear wheel roll forward, as shown in *Figure 2a*; the second half cycle of creeping motion, the angle adjustment mechanism moves, the body descends, and at the same time the front wheel rolls forward, the middle wheel and the rear wheel do not move, as shown in *Figure 2c*. The above-mentioned motion is reciprocated to realize the creeping motion. A Mars rover that can imitate the inchworm motion is called a creeping rover.

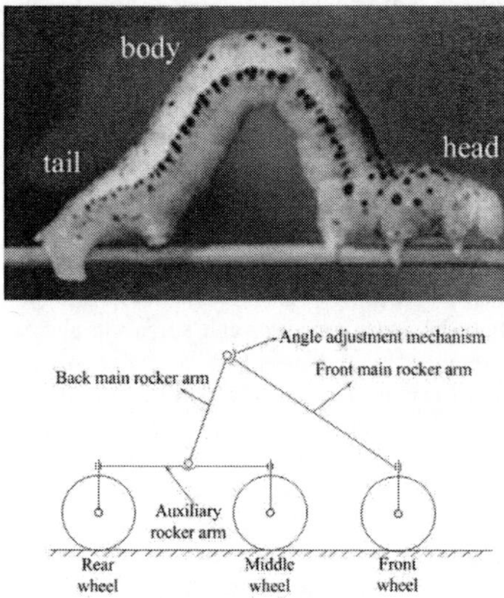

Figure 1. Schematic diagram of the active suspension of the bionic inchworm of Mars rover

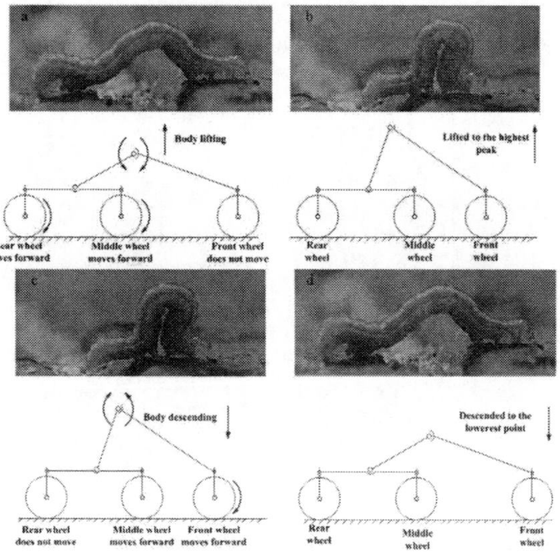

Figure 2. Schematic diagram of the rover imitating the inchworm motion

The principle of the creeping rover movement is that in the process of lifting or descending the rover body, the wheels need to rotate in accordance with the movement of the rover body, so that the wheel spacing changes periodically to realize the creeping motion. When the wheel of the Mars rover sinks in the soft soil, the creeping motion can improve the traction of the wheel and improve the ability to extrication (SUN et al., 2022). *Figure 3* and *Figure 4* show the stress state of the rover during normal motion and creeping motion. It can be seen from the figures that in the creeping motion, the change of the force is realized through the transformation of the driving wheel and the braking wheel, and the resistance of wheel is converted into the driving force, thereby improving the passing ability of the whole rover.

The drawbar pull of the normal motion can be expressed as *Eq. 1*, and the drawbar pull in the first half cycle and the second half cycle of the creeping motion can be expressed as *Eq. 2* and *Eq. 3*.

$$F_{\mathrm{PD,p}} = \frac{6rb}{\theta_{\mathrm{m}}(\theta_1-\theta_{\mathrm{m}})}\{t_{\mathrm{m}}[\theta_1\cos(\theta_{\mathrm{m}})-\theta_{\mathrm{m}}\cos(\theta_1)-\theta_1+\theta_{\mathrm{m}}]- \sigma_{\mathrm{m}}[\theta_1\sin(\theta_{\mathrm{m}})-\theta_{\mathrm{m}}\sin(\theta_1)]\} \tag{1}$$

$$F_{\mathrm{PD,w1}} = \frac{rb}{\theta_{\mathrm{m}}(\theta_1-\theta_{\mathrm{m}})}\{6t_{\mathrm{m}}[\theta_1\cos(\theta_{\mathrm{m}})-\theta_{\mathrm{m}}\cos(\theta_1)-\theta_1+\theta_{\mathrm{m}}]- 2\sigma_{\mathrm{m}}[\theta_1\sin(\theta_{\mathrm{m}})-\theta_{\mathrm{m}}\sin(\theta_1)]\} \tag{2}$$

$$F_{\mathrm{PD,w2}} = \frac{rb}{\theta_{\mathrm{m}}(\theta_1-\theta_{\mathrm{m}})}\{6t_{\mathrm{m}}[\theta_1\cos(\theta_{\mathrm{m}})-\theta_{\mathrm{m}}\cos(\theta_1)-\theta_1+\theta_{\mathrm{m}}]+ 2\sigma_{\mathrm{m}}[\theta_1\sin(\theta_{\mathrm{m}})-\theta_{\mathrm{m}}\sin(\theta_1)]\} \tag{3}$$

Where r is wheel radius(in this paper r =0.15m), b is wheel width(in this paper r =0.15m), σ_{m} is normal stress of the wheel, t_{m} is shear stress of the wheel, θ_1 is entry angle, θ_{m} is the angle at which maximum normal and shear stress located.

Only the static sinkage is considered, the entry angle θ_1 is:

$$\theta_1 = \arccos\left(\frac{r-z}{r}\right) \tag{4}$$

where, z is wheel sinkage.

The angle at which maximum normal and shear stress located θ_{m} is:

$$\theta_m = (c_1 + c_2 s)\theta_1 \tag{5}$$

Where c_1、c_2 is related to soil property, usually taken as $c_1 \approx 0.4$, $c_2 \in [0,0.3]$。

The slip rate s is defined as :

$$s = \begin{cases} \dfrac{\omega r - v}{\omega r}, & \omega r \geq v; \\ \dfrac{v-\omega r}{v}, & \omega r < v. \end{cases} \tag{6}$$

Where ω is angular velocity of the wheel, v is forward speed of the wheel.

According to *Eq. 1-Eq. 6*, the relationship curve between wheel sinkage and drawbar pull can be drawn, as shown in *Figure 5*. It can be seen from the curve that in the creeping motion when the wheel sinkage is 0.05m, the drawbar pull is increased by 80%; and in the normal motion wheel sinkage exceeds 0.12m, the rover will not be able to extrication, while the rover can be extrication when the whole wheel sinks 0.3m in the creeping motion.

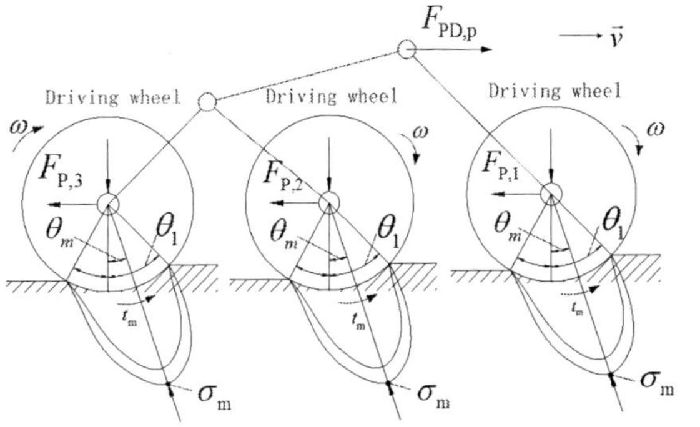

Figure 3. Stress state of the Mars rover during the normal motion

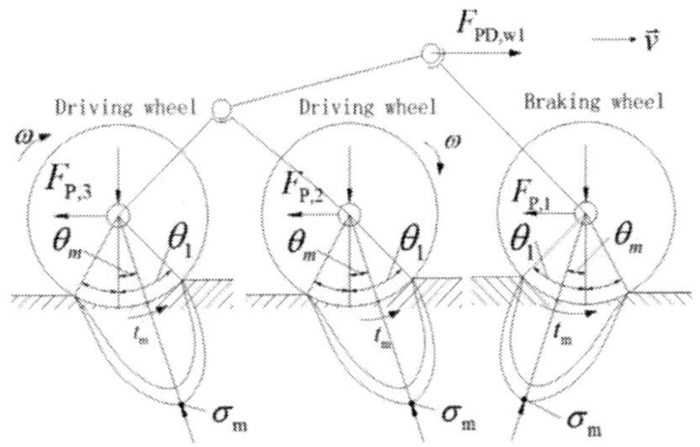

a) The first cycle

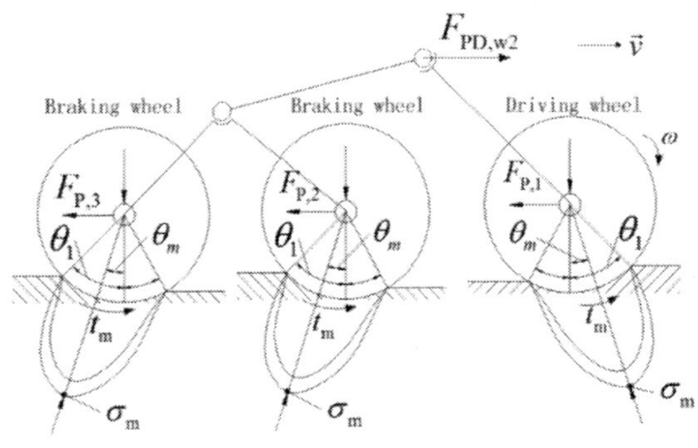

b) The second cycle

Figure 4. Stress state of the Mars rover during the creeping motion

The creeping Mars rover proposed in this paper is a Mars rover with active suspension (*Figure 6a*). The core difference between it and the Mars rover with passive suspension (*Figure 6b*) is whether the front main rocker arm and the back main rocker arm can move. If the two are fixedly connected, it becomes Mars rover with a passive suspension. It can be seen from *Figure 6* that each side of the suspension system of the creeping rover has one more set of angle adjustment mechanisms than the passive suspension rover.

The comparison between the creeping Mars rover and Mars rover with passive suspension is shown in Table 1, the creeping Mars rover can solve the four problems encountered by the Mars rover on the surface of Mars.

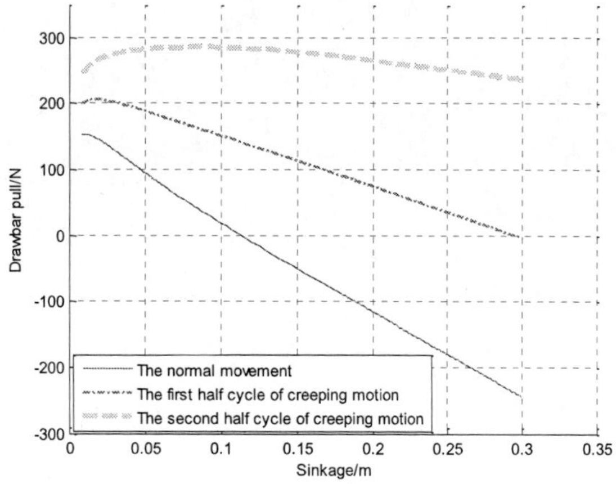

Figure 5. Curve of relationship between wheel sinkage and drawbar pull in different motion modes

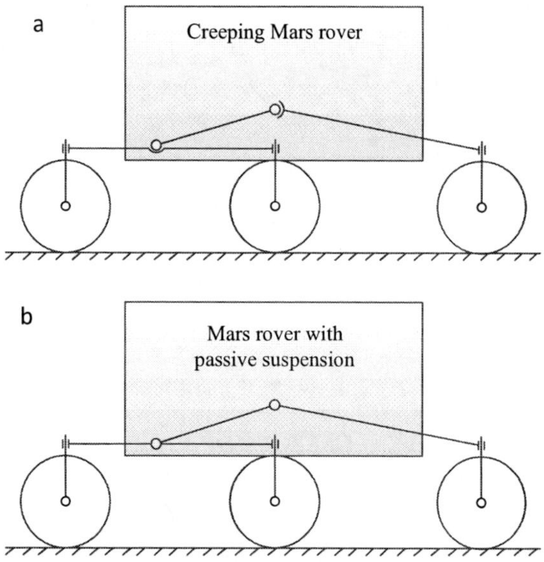

Figure 6. Comparison of creeping Mars rover and Mars rover with passive suspension

Table 1. Comparison of performance of the creeping Mars rover and Mars rover with passive suspension

	Creeping Mars rover	Mars rover with passive suspension
ability of wheel sinkage and extrication	ability of wheel sinkage and extrication can be enhanced by means of front and rear wheel creeps, wheel load adjustment and wheel lift.	The wheel is sinking, and the wheel can only be driven to get out of the predicament, and the ability to get out of the predicament is limited
ability of obstacle crossing	The height of the rover body can be lifted, the bearing capacity of the front and rear wheels can be distributed, and the obstacle-crossing performance can be improved.	Depending on the friction coefficient between the wheel and the rocks on the Martian surface, the driving torque and the height of the rover body, once the sinkage causes the rover body to underpin or the wheels get stuck, it is difficult to improve the obstacle-overcoming ability.
ability and stability of climbing	It can reduce the center of mass of the rover body, reduce the driving torque and load of the rear wheels for climbing, balance the load of each wheel, reduce the deviation distance of the center of mass, and improve the driving stability	Depending on the inherent structure size of the suspension and the size of the wheels, the driving torque of the rear wheel increases greatly when climbing, and the center of mass changes, which affects the stability
adaptability of wheel failure	Through load adjustment, the wheel can be lifted to reduce the load of the faulty wheel and avoid the bulldozing effect, and the driving torque of the remaining wheels is sufficient for long-term driving.	After the wheel fails, the wheel is only dragged to driving, which will cause a bulldozing effect on the sand, making it difficult to perform long-term driving.

3. Design of creeping Mars rover

3.1 Design of key parameters of active suspension

Based on the working principle of the creeping rover, in order to solve the demands of the rover to better adapt to the complex terrain on the surface of Mars, an creeping rover with active suspension is designed, as shown in *Figure 7,* and its active suspension system is shown in *Figure 6*. The suspension system mainly includes the front main rocker arm, the back main rocker arm, and the auxiliary rocker arm. When the angle adjustment mechanism is locked, the front main rocker arm and the back main rocker arm remain fixed, and the suspension becomes a rocker-type passive suspension; when the rover needs to be lowered or lifted, the angle adjustment mechanism moves, the angle between the front main rocker arm and the back main rocker arm is adjusted, and the suspension becomes an active suspension (YUAN et al., 2021; TAO et al., 2016).

To facilitate the analysis and calculation, the active suspension is simplified. The simplified active suspension configuration is shown in *Figure 8*. The key parameters of the simplified active suspension of the Mars rover are: the distance between the axis of the angle adjustment mechanism and the axis of the front wheel is equivalent to the first half of the main rocker arm, taken as l_1; the distance between the brake and the angle adjustment mechanism is equivalent to the second half of the main rocker arm, taken as l_2; Assuming that the distance between the locking mechanism and the middle and rear wheel axes is equal, and is equivalent to half length of the auxiliary rocker arm, taken as l_3; the distance between the front wheel and the middle wheel is l_4; the distance between the middle wheel and the rear wheel is l_5; the distance between the outermost edges of the front and rear wheels is *L,* the angle between the main rocker arm is α, the angle between the rear half of the main rocker arm and the horizontal line is γ, and the angle between the auxiliary rocker arm is β. The key parameters of the rover are calculated at different heights are shown in *Table 2*.

3.2 Design of wheel sinkage and extrication

Aiming to solve the problem of the wheel sinkage of the Mars rover, the creeping Mars rover was designed to realize the extrication.

The rover can creep forward like an inchworm. The specific process is as follows: the rover body descends from the initial state (the height of the rover is 300mm), the front wheel rotates forward, the middle wheel and the rear wheel do not move, and the wheel distance between the front wheel and the middle and rear wheels becomes larger, the front wheel moves forward under the push of the front main rocker arm, as shown in *Figure 9a*; then the rover body is lifted, the front wheel does not move, the middle wheel and the rear wheel rotate forward synchronously, the wheel distance becomes smaller, and the middle and rear wheel moves forward under the push of the back main rocker arm and the auxiliary rocker arm, so one time forward creeping is completed, as shown in *Fig. 9b*. After several repetitions, the rover will leave the sinkage area. It can be seen from *Table 2* that the rover will move forward by 209.3mm in a forward creeping cycle from 300mm → 200mm →500mm → 300mm.

According to the characteristics of the active suspension, the order and direction of wheel rotation can be controlled during the descent and lift of the rover, and the backward creeping of the rover can also be realized.

The rover can extrication from the sinkage area by creeping forward or backward, and the forward or backward movement is mainly determined according to the sinkage of the rover and the terrain characteristics of the front and rear of the rover.

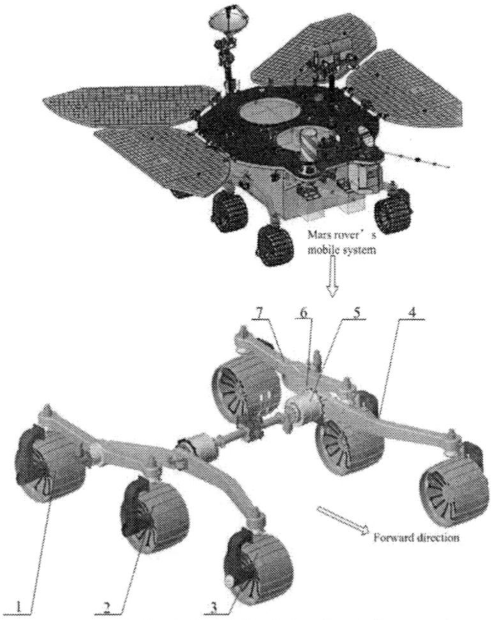

1-Rear wheel; 2-Middle wheel; 3-Front wheel; 4-Front main rocker arm; 5-Angle adjustment mechannism; 6-Back main rocker arm; 7-Auxlary rocker arm

Figure 7. Design of the mobile system of the Mars rover using active suspension

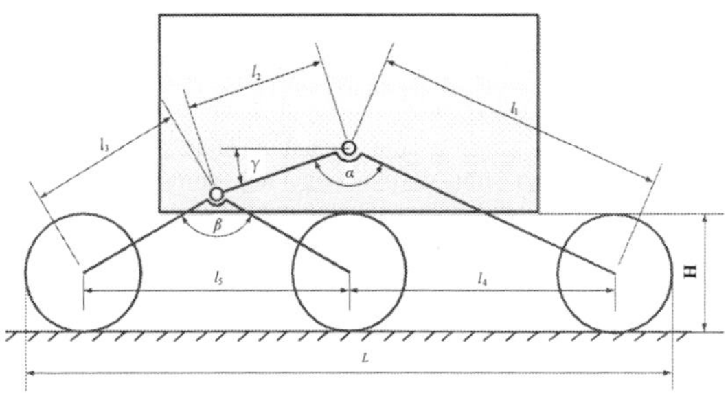

Figure 8. Key parameters of simplified mobile system of Mars rover

Table 2. Key parameters of mobile system under different rover body lift heights

PARAM.	0mm	200mm	300mm	500mm
l_1[mm]	814.3	814.3	814.3	814.3
l_2[mm]	390.7	390.7	390.7	390.7
l_3[mm]	436.1	436.1	436.1	436.1
l_4[mm]	725.5	800.8	775	591.5
l_5[mm]	775	775	775	775
L[mm]	1800.5	1875.8	1850	1666.5
α[°]	-22.364	176.715	154.769	106.673
β[°]	125.4	125.4	125.4	125.4
γ[°]	-39.783	-7.353	7.352	39.781

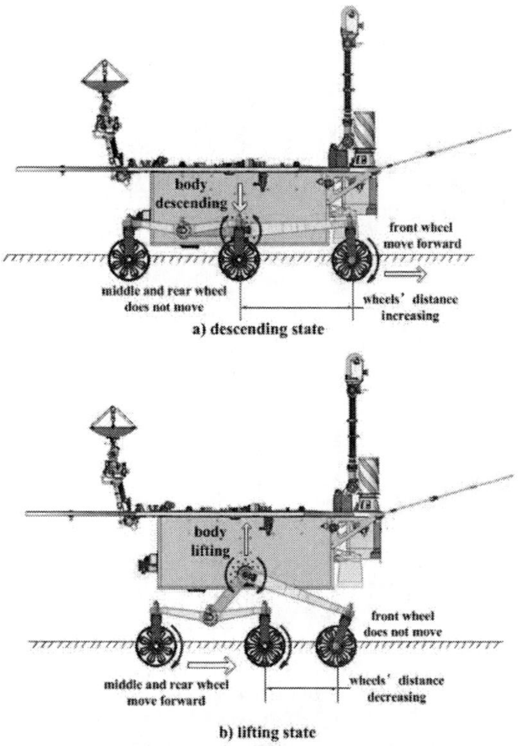

Figure 9. Schematic diagram of the forward creeping motion process of Mars rover

3.3 Design of climbing and anti-underpinning

The design of the Mars rover for climbing is mainly by reducing the height of the rover body, which can make the force of the front, middle and rear wheels more balanced, thereby reducing the maximum ground pressure value of the wheels and improving the climbing ability. The maximum climbing angle of the Mars rover is 20°, as shown in *Figure 10a.*

The design of the Mars rover for underpinning is mainly to increase the height of the rover body to increase the ground clearance of the rover body, to pass through large obstacles. The maximum obstacle height that the Mars rover can cross is 500mm, as shown in *Figure 10b.*

3.4 Design of lifting wheel

The solution to the wheel failure of the Mars rover is to lift the wheel. Different wheels have different wheel lifting strategies. The most important thing is to adjust the center of mass. When lifting the front wheel, lift the rover body and move the center of mass backward; When lifting middle and rear wheel, the center of mass needs to be moved forward. When the wheel is lifted by 300mm, the state of the rover is shown in *Figure 11* and the key parameters are shown in *Table 3.*

4. Verification of Mars rover

To verify the correctness of the design of the creeping rover, ground tests verification and flight tests verification were carried out.

4.1 Verification of ground tests

The state of the rover in ground tests is shown in *Figure 12,* the displacement curve of the rover movement is shown in *Figure 13* , and the height of the rover's descend and lift during the movement is shown in *Figure 14.* The results of the two movements are shown in *Table 4.* The test shows that the rover has the ability of creeping, but the movement speed is slow.

4.2 Verification of flight tests

As an example of the application of the Creeping Mars rover, the "Zhurong" Mars rover has carried out a flight test to verify the effectiveness of the design measures of the Mars rover. The work is normal, all telemetry parameters are within the normal range, and no abnormal phenomenon occurs. Through the telemetry information such as the current and angle of the mechanism, it can be judged that the current state of the mechanism is good, as shown in *Figure* 15.

5. Conclusion

1) The creeping rover can effectively solve the problems such as wheel sinkage, climbing, large obstacles encountered during the movement of the Mars surface.

2) The rover can improve the traction force of the whole vehicle through the creeping motion. When the sinkage exceeds 0.05m, the traction force is increased by more than 80%.

3) By lifting the rover body, the rover can pass through large obstacles, reduce the probability of underpinning, and improve the passing performance by more than 60%.

4) By lowering the rover body, the Mars rover can improve its climbing ability, and can pass a maximum slope of 20°.

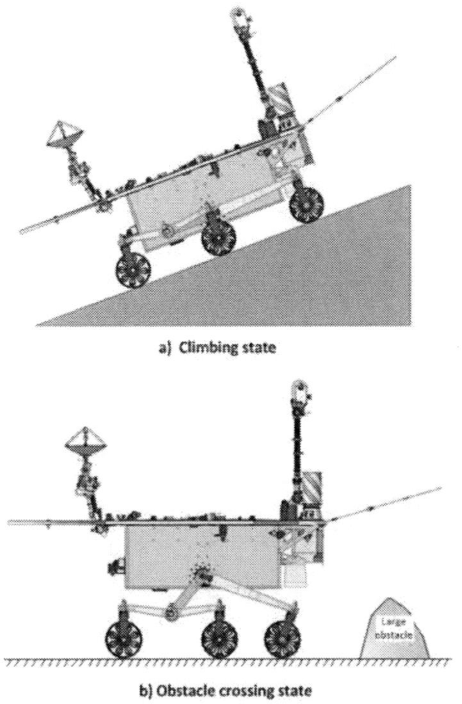

Figure 10. Schematic diagram of the working state of Mars rover in complex terrain

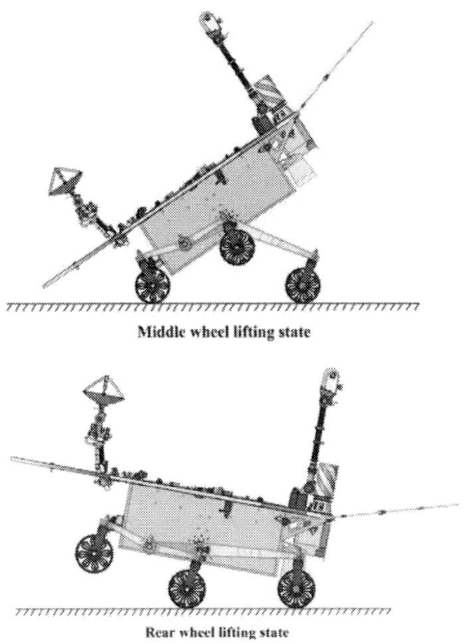

Figure 11. Schematic diagram of the attitude of the Mars rover when lifting different wheels

Table 3. Key parameters of mobile system under different wheel lifting attitudes

Param.	Front wheel	Middle wheel	Rear wheel
α[°]	129.604	132.614	179.775
β[°]	125.4	125.4	125.4
γ[°]	39.781	7.3524	7.3524

Figure 12. Mars rover in ground tests

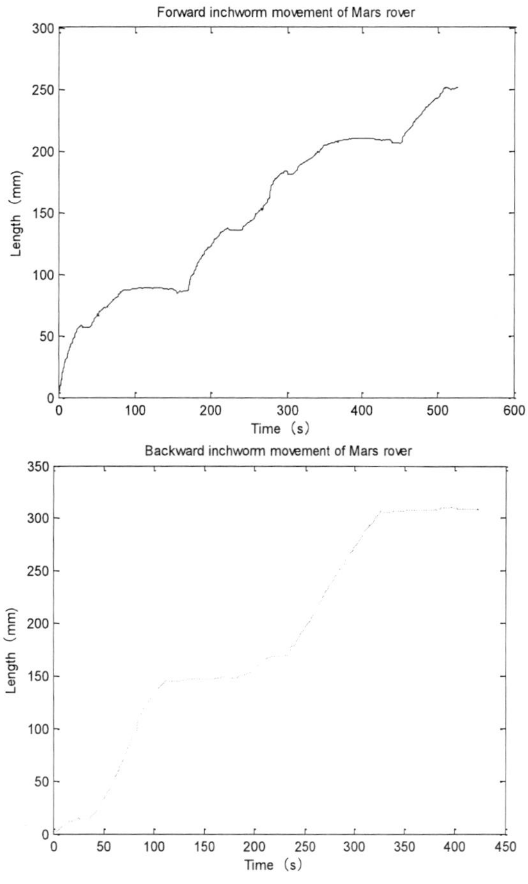

Figure 13. Displacement curve of the creeping motion of the Mars rover in ground tests

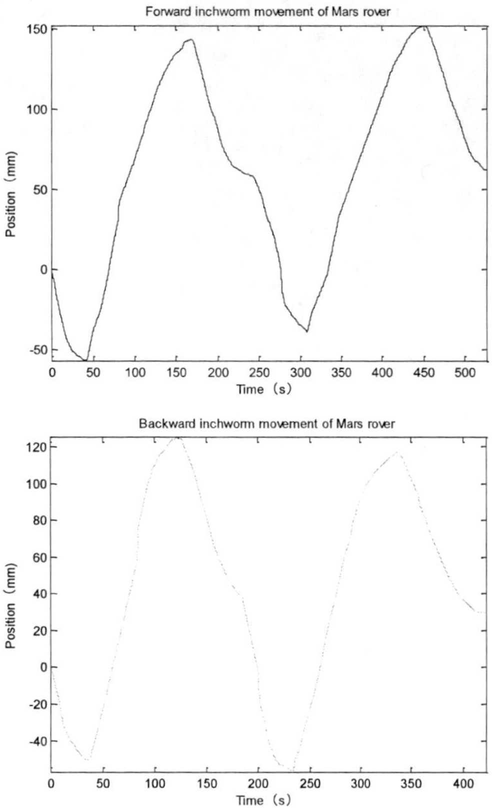

Figure 14. Height curve of the creeping motion of the Mars rover in ground tests

Table 4. Comparison of results of the creeping motion of Mars rover in the ground tests

	Forward [mm]	Backward [mm]
First cycle descend	56	52
First cycle lift	199	176
Second cycle descend	187	184
Second cycle lift	196	172
Total displacement	254	305

Figure 15. Rover working on Mars surface

6. References

ELLERY A, PATEL N, RICHTER L, et al., 2005. ExoMars rover chassis analysis and design. Proceedings of the 8th International Symposium on Artificial Intelligence, Robotics and Automation in Space (iSAIRAS).

F Zhou, RE Arvidson, K Bennett, et al., 2014. Simulation of Mars Rover Traverses. Journal of Field Robotics, 31(1), 141-160.

Giulio Reina, Lauro Ojeda, Annalisa Milella, et al., 2006. Wheel slippage and sinkage detection for planetary rovers. IEEE/ASME Transactions on Mechatronics, 11(2), 185-195.

IAGNEMMA K, RZEPNIESKI A, DUBOWSKY S, 2003. Control of robotic vehicles with active articulated suspensions in rough terrain. Autonomous Robots, 14(1): 5-16.

LIU Yanwei, WANG Limeng, LIU Sanwa, et al., 2019. Design and analysis of an inchworm-inspired wall-climbing robot[J]. Mechanical Transmission, 43(8), 87-91.

PAN Dong, CHEN Zhen, YUAN Baofeng, et al., 2022. Sinkage mechanism and extrication strategy of Mars rover. ROBOT, 44 (1), 2-8.

SHI Z Y, PAN J, TIAN J, et al., 2019. An Inchworm-Inspired crawling robot. Journal of Bionic Engineering, 16, 582-592.

SUN Zezhou, CHEN Baichao, JIA Yang, et al., 2022. The Tianwen-1 roving exploration technology for the Martian surface.Sci sin tech, 52(2), 214-225.

TAO Zhuo, CHEN Baichao, JIA Yang, 2016. Optimization of geometric parameters for Mars rover active sus-pension. Spacecraft Engineering, 25(6), 48-54.

TAO Jianguo, 2009. Research on six-wheeled rover mobile system with series multi-articulated suspension and its key technology. Harbin: Harbin Institute of Technology.

TUNSTEL, E, 1998. Evolution of autonomous self-righting behaviors for articulated nanorovers. Proceedings of the 5th International Symposium on Artificial Intelligence Robotics and Automation in Space.

YUAN Baofeng, WANG Chengen, ZOU Meng, et al., 2021. Design active suspension system and creeping control strategy for Mars rover of China. Journey of Jilin University (Engineering and Technology Edition), 51(1), 154-162.

WANG Z Y, YANG H G, YUAN B F, et al., 2020. Wheels performance of Mars exploration rover: experimental study from the perspective of terramechanics and structural mechanics. Journal of Terramechanics, 30(5), 23-42.

WILCOX BRIAN, NASIF ANNETTE, WELCH RICK, 1997. A nanorover for Mars. Space Technology, 3(17): 163-172.

1534 / Foothold Selection Considering Constraint and Slippage Evaluation for Legged Robots

***Yufei Liu*[a], *Lei Jiang*[a,b]*, *Chong Tian*[a], *Boyang Xing*[a], *Zhirui Wang*[a], *Bo Su*[a]*, *Tong Yan*[a], *Liang Ding*[c], and *Haibo Gao*[c]**

[a]*Unmanned center, China North Vehicle Research Institute, Beijing, China*

[b]*College of Computer Science and Technology, Zhejiang University, Hangzhou, China*

[c]*Harbin Institute of Technology, Harbin, Heilongjiang, China*

**Corresponding author: 11921167@zju.edu.cn*

https://doi.org/10.56884/ZEES3819

Abstract

Legged robots over a rugged terrain can be accomplished by taking the environment and control constraint models into account when planning foothold and robot movements. In this paper, the foot-terrain slippage evaluation model for 3-DOF leg based on dynamics model is proposed. The authors evaluate the geometrical characteristics of each cell on the local elevation map, checks slippage state constraints and kinematic constraints. Then, we employ reinforcement learning and imitation learning to develop the relationship between the local elevation map, robot state, and robot behavior. The method can help to select the foothold considered geometrical characteristics on the elevation map and the robot state can be determined. The experiments were carried out on legged robots walking over rough terrain in both simulation and real robotic platforms. And the experimental results have demonstrated the effectiveness of the proposed method.

Keywords: slippage evaluation model, geometrical characteristics, reinforcement learning, legged robots

1. Introduction

The quadruped robot could be utilized to complete tasks in a hazardous environment. Compared with tracked/wheeled robots, quadruped can freely choose their contact points with the environment, making them more agile on rough terrain (Raibert, 2008). Due to recent advances in control theory and hardware, the development of quadruped robots has made significant progress (Katz, 2019). Many organizations, including Boston Dynamics, MIT, and ETH, have released quadruped robots for research or commercial applications (Hutter, 2016).

Precise velocity and slippage estimation of legged robots are important for machine learning, and the control strategy can update based on collected data samples and estimates autonomously (Wawrzyński, 2013, 2014). The orientation of the legged robot body is estimated using the joint variables and high-order sliding-mode observer. Rotella et al. (2014) presented an approach for velocity and body pose estimation by joint encoders and the IMU data calculation based on state estimator.

The control algorithm is crucial for agile locomotion. Using a modular controller to operate the quadruped robot is a popular method. This technique decomposes the control problem into smaller manageable subproblems, which are subsequently solved independently (Raibert, 1986). Using the modular controller, the quadruped robot could traverse challenging terrain and smoothly switch between various dynamic gaits (Kalakrishnan, 2010). Alternatively, there is a trajectory optimization method. Typically, this strategy consists of two steps: planning and tracking. Using optimization techniques, the controller determines the optimal path throughout the planning step (Bellicoso, 2018). Then, the controller directs the robot to follow the path in the tracking step. The trajectory optimization method has been applied to control the quadruped robot without requiring prespecified contact switches and gait patterns (Neunert, 2017).

The aim of the study is to propose the foothold selection method based on reinforcement learning system. The main contributions of the work are as follows: 1) the models of slippage evaluation and geometrical characteristics are presented; 2) reinforcement and imitation learning are employed in quadruped robot control system; and 3) extensive simulation and real experimental studies of the proposed method have been carried out.

This paper is organized as follows: Section 2 introduces the foot-terrain slippage evaluation. The geometrical characteristics evaluation model is formulated in Section 3. Section 4 introduces the robot control system based on reinforcement learning. Finally, comprehensive simulation and real experiments are presented in Section 5. The conclusions are given in Section 6.

2. Foot-terrain Slippage Evaluation

Foot terrain interaction for slippage is mainly divided into normal and tangential directions. The contact model is analyzed for the foot terrain dynamic contact, and the impact model of the foot terrain mechanics is obtained. The process of foot terrain interaction for tangential directions is as shown in Fig. 1.

The normal force along the vertical contact direction can be calculated as:

$$F_N = \int_L F_p \, ds - G_{RA} \quad (1)$$

where G_{RA} carries the gravity of the foot active area.

The velocity vector v of tangential movement has a velocity component in the normal direction of the tangential contact line. The passive pressure generated by each vertical edge element is F_N. The tangential force F_T along the lateral contact direction can be obtained as:

$$\sum F_T = \iint_\kappa \frac{\sum F_N}{\kappa} \tan\varphi \cdot \mathrm{th}(j, \frac{3K}{2}) \, d\kappa \quad (2)$$

where κ is the contact surface between the flat foot and soft terrain, j is the tangential displacement of the foot, K is the shear deformation of the soft terrain.

According to the normal and tangential force equations, the authors can analyze the influence of normal load and slip velocity load on the experimental results of foot-terrain slippage mechanics, and solve the problem of difficulty in establishing the theoretical model.

In order to evaluate for the invalid slippage of the foot terrain contact process, the slippage estimation method for the mammal type robots based on the dynamic model. The body velocity estimation value and the foot tip velocity estimation value are fused based on the EKF, and the foot slippage of the robot can be estimated.

The two control variables of a single leg with three links are defined as shown in Figure 2. The angle between the hip joint and the normal direction is q_H, and the angle between the knee joint and the normal direction is q_K.

The foot terrain contact point C_s is defined as supporting position, the mass of coxa is M_A, the length of coxa is L_A, and the moment of inertia based on the center of mass of the coxa is J_A. The mass of thigh is M_H, the length of thigh is L_H, and the moment of inertia based on the center of mass of the thigh is J_H. The mass of calf is M_K, the length of calf is L_K, and the moment of inertia based on the center of mass of the calf is J_K.

The position of the foot tip for the supporting phase represents the slippage displacement. According to the single-leg slippage model for mammal-type legged robot walking, the position vector of the three-link can be obtained as:

$$\mathbf{p}_K = \begin{bmatrix} {}^L P_{sy} + L_K \sin q_K \\ {}^L P_{sz} + L_K \cos q_K \end{bmatrix}$$
$$\mathbf{p}_H = \begin{bmatrix} {}^L P_{sy} + L_K \sin q_K + L_H \sin q_H \\ {}^L P_{sz} + L_K \cos q_K + L_H \cos q_H \end{bmatrix} \quad (3)$$

The kinetic energy of the robot's single-leg movement for mammal-type legged robot walking is:

$$T = \frac{1}{2} M_K \dot{\mathbf{p}}_K^{\mathrm{T}} \dot{\mathbf{p}}_k + \frac{1}{2} M_H \dot{\mathbf{p}}_H^{\mathrm{T}} \dot{\mathbf{p}}_H + \frac{1}{2}(J_K \dot{q}_K^2 + J_H \dot{q}_H^2) \quad (4)$$

The potential energy of the legged robot for the mammal type legged robot walking can be expressed as:

$$V = M_K g({}^L P_{sz} + L_K \cos q_K) + M_H g({}^L P_{sz} + L_K \cos q_K + L_H \cos q_H) \quad (5)$$

The general form of the equation for single leg movement according to Lagrange equation:

$$\mathbf{D_d(q)\ddot{q} + B_d(q,\dot{q})\dot{q} + G_d(q) - C_d u - E_d F} = 0 \quad (6)$$

where the $\mathbf{D_d(q)}$ is Inertia matrix, and the $\mathbf{B_d(q)}$ is Coriolis matrix, and the $\mathbf{G_d(q)}$ is Gravity matrix and $\mathbf{u}$ is joint torque.

According to the single leg Lagrange equation, the forward slippage estimation model is obtained as:

$$ {}^L\ddot{P}_{sy} = ((M_K L_K + L_K M_H)(\dot{q}_K^2 \sin q_K - \ddot{q}_K \cos q_K) + M_H L_H (\dot{q}_H^2 \sin q_H - \ddot{q}_H \cos q_H) + F_T) / (M_K + M_H) \quad (7)$$

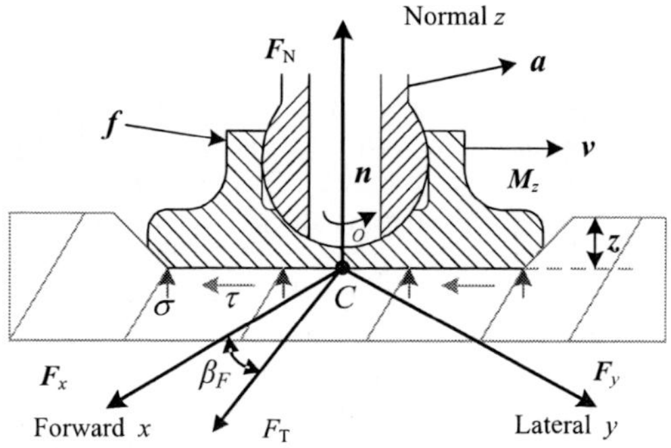

Figure 1. The foot-terrain interaction for forward and lateral directions

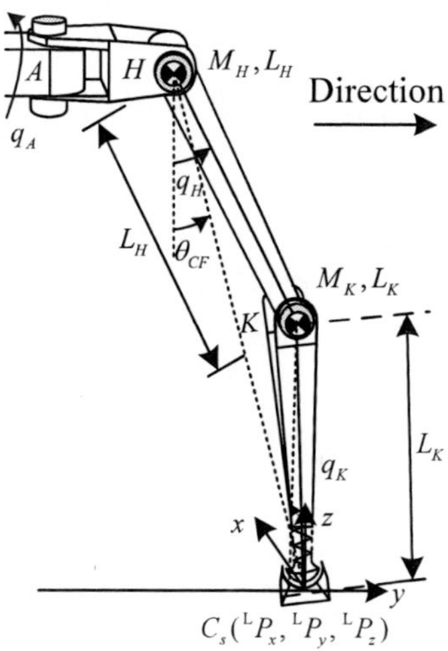

Figure 2. The foot-terrain interaction for forward and lateral directions

3. Geometrical Characteristics Evaluation

In order to select foothold based on the topographic elevation map, the authors considered the foot slippage and safe landing area. The geometric feature of terrain based on visual information is obtained. The convex and concave degree of the terrain and inclination of the terrain can be described, and the foothold selection can be determined. so as to determine the foothold position. In order to define the coordinate position in the grid, the 8 coordinate points around the target point is signed. The height value of topographic map based on the geometric characteristics of the terrain is as shown in Fig. 3.

The grid map includes convex and concave terrains, and the grid target positions for the different terrains are defined, and the point cloud data around the target position is calculated. The evaluation function of the target position $K_1(i,j)$ can be calculated as:

$$K_1(i,j)=\sum_{m=-1}^{1}\sum_{n=-1}^{1}(z_{(i,j)}-z_{(i+m,j+n)}) \tag{8}$$

where $z_{(i,j)}, z_{(i+m,j+n)}$ are the height information for the convex and concave grid map.

When $K_1(i,j)$ is positive, the terrain is defined as convex; and when $K_1(i,j)$ is negative, the terrain is defined as concave. However, the evaluation function can not distinguish between flat and sloped terrains. Thus, the supplementary evaluation function of the target position $K_2(i,j)$ can be calculated as:

$$K_2(i,j)=\sum_{m=-1}^{1}\sum_{n=-1}^{1}\left|z_{(i,j)}-z_{(i+m,j+n)}\right| \tag{9}$$

Assumed that the feet are flat or spherical shapes, the triangle area is fitted based on the point cloud height coordinates. The triangle area of the foothold selection can be calculated as:

$$S_{\Delta ABC}=\frac{1}{2}(\left|\overrightarrow{AB}\times\overrightarrow{AC}\right|) \tag{10}$$

The inscribed circle area of the triangle can be calculated as:

$$S_{circle}=\pi\left(\frac{\left|\overrightarrow{AB}\times\overrightarrow{AC}\right|}{\left|\overline{AB}\right|+\left|\overline{AC}\right|+\left|\overline{BC}\right|}\right)^2 \tag{11}$$

The schematic diagram of foot hold selection based on height geometric characteristics calculation is as shown in Fig. 4.

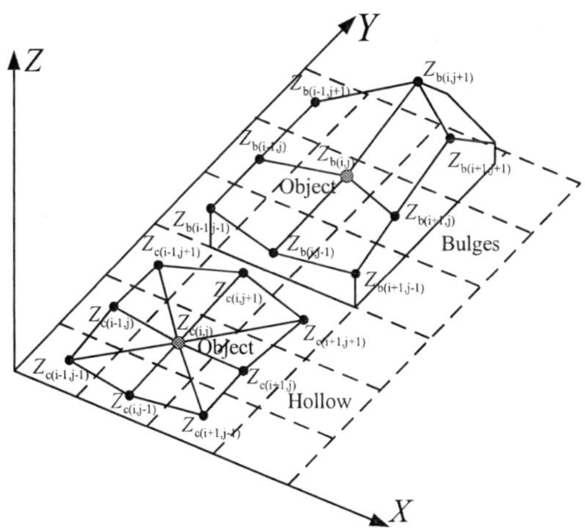

Figure 3. Geometric features of bulge and hollow in topographic elevation map

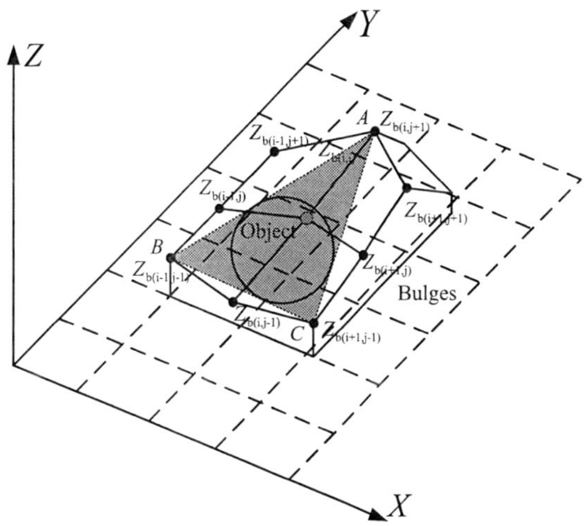

Figure 4. Schematic diagram of foothold selection

4. Robot Control based on Reinforcement Learning

We utilize reinforcement learning to learn the robot control policy. The overall framework is shown in Fig. 5. To improve the robustness and optimality of the policy, the policy is trained using the teacher-student training approach. The trained policy takes upper-level command and sensors' data as input and outputs joint level commands, which will be converted to torque using PD controllers.

We utilize PPO as the training algorithm to train the teacher Policy π_T. Because of the reality gap between the real world and simulation, the controller trained in the simulation often can not be transferred to the real world. There are two common approaches to reducing the gap: domain randomization and external disturbance. However, because the policy π does not have access to the domain and disturbance parameters, the agent can not adapt its behavior accordingly. The optimality will be sacrificed. To improve the robustness and decrease the influence on the optimality, we incorporate the domain parameters α_t^{dom} and disturbance states α_t^{dist} in the teach policy training. The setup follows.

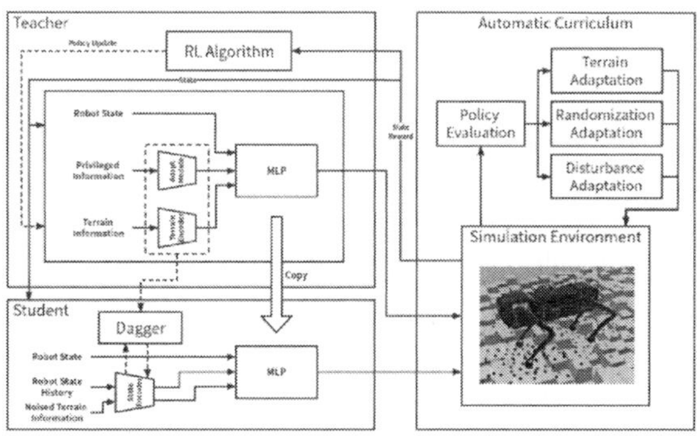

Figure 5. Framework of reinforcement learning

Observation Space: The agent has access to all the information $s_t = [o_t, s_t^{pr}, s_t^{tr}]$, where o_t is the robot observation including robot state and the upper-level command, s_t^{pr} is privileged information, and s_t^{tr} is terrain information. The privileged information s_t^{pr} contains domain parameters α_t^{dom} and disturbance states α_t^{dist}. The privileged information s_t^{pr} and terrain information s_t^{tr} are encoded using the adaptation encoder and terrain encoder:

$$l_t^{pr} = g\ (s_t^{pr}) \quad (12)$$

$$l_t^{tr} = h\ (s_t^{tr}) \quad (13)$$

where l_t^{pr} and l_t^{tr} are encoded privileged information and encoded terrain information, respectively.

Action Space: The action $a = \pi_T(s_t), a \in R^{12}$ is the desired joint position $\theta_{d,i}, i \in [1,12]$ for the PD controller. The desired joint velocity $\dot{\theta}_{d,i}, i \in [1,12]$ is set to zero.

Reward: The reward function $R = \sum c \cdot r$, where c is the reward coefficient, r is the reward for different term, includes three main components: linear velocity tracking r_{lv}, angular velocity tracking r_{av}, and stability of the body r_{st}, and six minor components, including foot clearance r_{fc}, body collision r_{bc}, smoothness r_s, energy consumption r_e, foot slippage r_{fs}, and joint limitation r_j. Because the robot in the real world cannot precisely sense the terrain information s_t^{tr} and privileged information s_t^{pr}, the teacher policy π_T can not be applied directly. One possible solution is leveraging the robot's past observable state $s_o = [o_h, s_t^{tr\_n}]$ to infer this information, where o_h is the robot state history, $s_t^{tr\_n}$ is the noised terrain information. In the student policy π_S, We feed the robot's past observable state s_o into a state encoder m to infer the encoded privileged information l_t^{pr} and the encoded terrain information l_t^{tr}:

$$\bar{l}_t^{pr}, \bar{l}_t^{tr} = m\ (s_o) \quad (14)$$

where $\bar{l}_t^{pr}$ and $\bar{l}_t^{tr}$ are inferred encoded privileged information and inferred encoded terrain information, respectively. Then $\bar{l}_t^{pr}$ and $\bar{l}_t^{tr}$ are fed to the MLP copied from π_T to obtain action a. The state encoder is trained using the Dagger algorithm. The loss function is the mean squared distance between the inferred and encoded accurate information:

$$L = (\bar{l}_t^{pr} - l_t^{pr})^2 + (\bar{l}_t^{tr} - l_t^{tr})^2 \quad (15)$$

One of the most complex challenges for reinforcement learning is exploration. If the agent cannot obtain a meaningful reward, the algorithm will fail. The locomotion problem without any ambiguity is already challenging to explore. If we start the training with hard terrain, extremely uncertain domain parameters, and severe external disturbance, the agent will most likely learn to suicide. To overcome this challenge, we built an automatic curriculum. This curriculum will first assess the policy performance $P\ (\tau \sim \pi)$ with a policy evaluation module, where $\tau \sim \pi$ is the trajectory genereted by the policy π, then alter the terrain parameters β_{tr}, domain randomization parameters β_{dm}, and disturbance parameters β_{dt} using the terrain adaptation module, randomization adaptation module, and disturbance adaptation module accordingly. This automatic curriculum guarantees that the environment is neither too hard nor too easy for the agent.

5. Experiments Verification

The experimental systems of quadruped simulation and hexapod prototype robots include proprioceptive and external measurement equipments. The accelerations and angular velocities can be integrated over time to derive the position, velocity of the trunk body. The acceleration in the IMU and kinematic frames have been periodically measured in the experiments; and it is adopted to evaluate the slippage and geometrical characteristics to further determine the walking performance of legged robots. The simulation environment of quadruped robot is as shown in Fig. 6.

The quadruped robot can walk on the step terrains autonomously based on the slippage estimation and terrain map geometrical characteristics of terrain map. And the foot force of the stance and swing phase for two legs are as shown in Fig. 7 and Fig. 8.

The tracking speed of quadruped robot based on the evaluation models is plotted as shown in Fig. 9. And the robot can track the desired tracking velocity of trunk body as shown in the Fig. 9. For the hexapod prototype robot, the terrain height map is obtained from the elevation map information. A square obstacle areas 16*21cm and 30*28cm are placed in the passing area of the robot as shown in Fig. 10. The foothold selection method is based on geometrical characteristics evaluation as shown in Fig. 11. Due to the red points do not satisfy the foothold supporting area, and the inclination of the terrain will cause the slippage of the landing. However, the green points satisfy the terrain geometrical evaluation function, thus the points will be the safe foothold selection.

The authors applied the continuous tracking gait for walking on the unstructured terrain based on the foothold selection method. It is proved that the hexapod robot can select the appropriate foothold points according to the terrain height map, and the effectiveness of continuous gait tracking is also proved, which ensures the hexapod robot to select the foothold landing point movement in the unstructured terrains automatically as shown in Fig. 12.

6. Conclusions

In this study, we have investigated the foothold selection method of legged robot based on reinforcement learning system. And the foot-terrain slippage evaluation model and geometrical characteristics evaluation model have been analyzed. The validity of the proposed method is evaluated by using a quadruped simulation prototype and real hexapod robot. Several conclusions are drawn as follows: (1) the models of slippage evaluation and geometrical characteristics evaluation are presented; (2) reinforcement learning and imitation learning are employed in quadruped robot control system for walking on random unstructured terrains; (3) comprehensive simulation and real experiments for different terrains are carried out, and the results validate the theoretical analysis. The software deployment and high tracking performance for the real control system will be considered in the future.

Figure 6. Step terrains for foothold selection

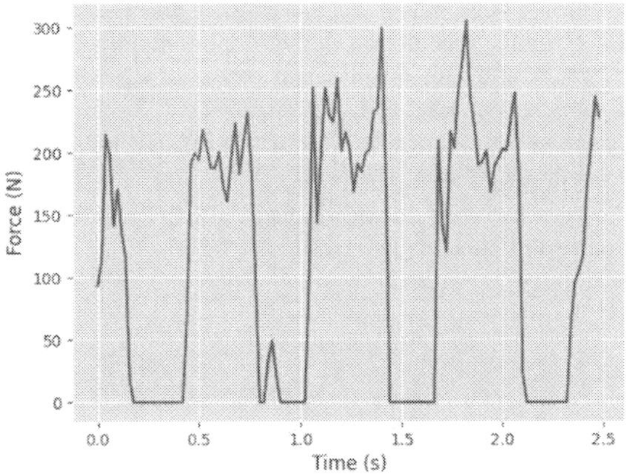

Figure 7. Foot force of leg1 for the simulation

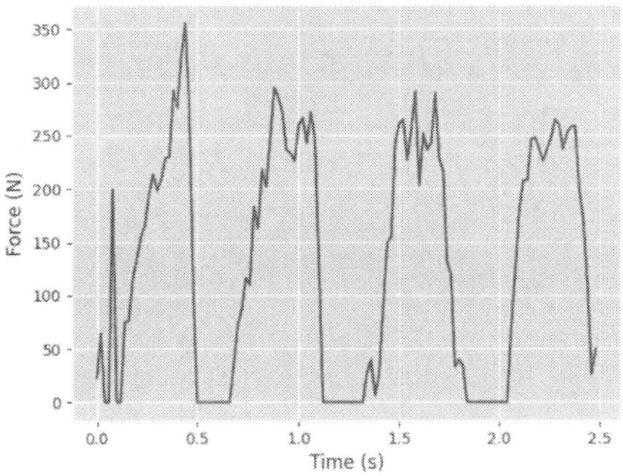

Figure 8. Foot force of leg3 for the simulation

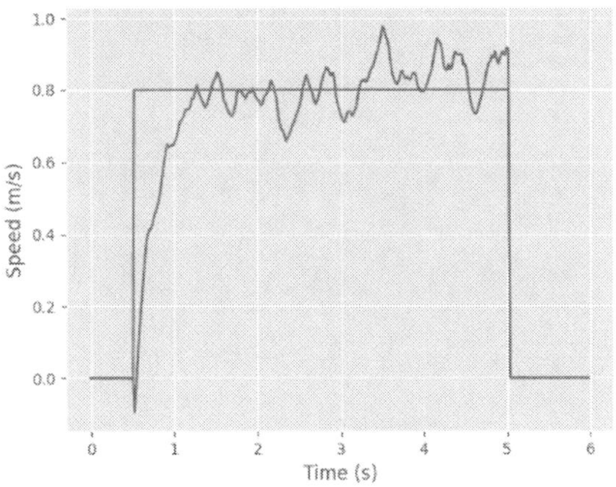

Figure 9. Tracking speed of body velocity

Figure 10. Obstacle terrains for foothold selection

Figure 11. Verification of foothold geometrical characteristics

Fig 12. Process of foothold selection

7. Acknowledgements

This work is supported by the National Key Research and Development Program of China (Grant No. 2019YFB1309500). State Administration of Science, Technology and Industry for National Defence, PRC under Grant JCKY2019208B024.

8. References

Raibert, M., Blankespoor, K., Nelson, G. and Playter, R., 2008. Bigdog, the rough-terrain quadruped robot. IFAC Proceedings Volumes. 41(2), 10822-10825.

Katz B., Di Carlo J., Kim S., 2019. Mini cheetah: A platform for pushing the limits of dynamic quadruped control. 2019 International Conference on Robotics and Automation (ICRA). 6295-6301.

Hutter, M., Gehring, C., Jud, D., Lauber, A., Bellicoso, C.D., Tsounis, V., Hwangbo, J., Bodie, K., Fankhauser, P., Bloesch, M. and Diethelm, R., 2016. Anymal-a highly mobile and dynamic quadrupedal robot. In 2016 IEEE/RSJ international conference on intelligent robots and systems (IROS). 38-44.

Wawrzyński, P., Tanwani, A. K., 2013. Autonomous reinforcement learning with experience replay. Neural Networks. 41(5), 156-167.

Wawrzyński, P., 2014. Reinforcement Learning with Experience Replay for Model-Free Humanoid Walking Optimization. International Journal of Humanoid Robotics. 11(3), 137-2187.

Rotella, N., Bloesch, M., Righetti, L., 2014. State estimation for a humanoid robot. IEEE/RSJ International Conference on Intelligent Robots and Systems (IROS). 952-958.

Raibert, M.H., 1986. Legged robots that balance. MIT Press.

Kalakrishnan, M., Buchli, J., Pastor, P., Mistry, M. and Schaal, S., 2010. Fast, robust quadruped locomotion over challenging terrain. In 2010 IEEE International Conference on Robotics and Automation. 2665-2670.

Bellicoso, C.D., Jenelten, F., Gehring, C. and Hutter, M., 2018. Dynamic locomotion through online nonlinear motion optimization for quadrupedal robots. IEEE Robotics and Automation Letters. 3(3), 2261-2268.

Neunert, M., Farshidian, F., Winkler, A.W. and Buchli, J., 2017. Trajectory optimization through contacts and automatic gait discovery for quadrupeds. IEEE Robotics and Automation Letters. 2(3), 1502-1509.

1561 / *Prominent Problems and Thoughts of "Paddy Soil-Terrain Machine System" Based on Disturbed Saturated Paddy Soil Conditions in South China*

Guozhong Zhang, Hongchang Wang, Jun Du, Kaiquan Ding, Wanru Liu, Nanrui Tang, and Yong Zhou
Huazhong Agricultural University, Wuhan, Hubei, China
**Corresponding author: zhyong@mail.hzau.edu.cn*

https://doi.org/10.56884/CHSD4959

Abstract

There is a large number of paddy fields in South China. During rice sowing and transplanting, agronomy requires the topsoil to be soft, viscous, muddy, mixed with water and disturbed-saturated mud, after soaking, tilling and rotary harrowing. At this time, the soil has a high fluidity, high adhesion and low shear strength. Thus, it would be productive to research the contact relationship between the soil contacting parts and the soil surface, so as to improve the operational efficiency of the tillage and planting machinery of paddy field in this soil condition. For this purpose, the disturbed-saturated soil formation process is described, the coupling mechanism between soil-engaging components and such soil conditions is clarified and the current state and prospects of research on soil-tool interactions under paddy field conditions is analyzed. This paper is expected to serve as a useful reference for future work in studying the surface characteristic parameters of soil contacting parts; the coupling correlation mechanism of "contact-rheology-resistance" among the structural shape, motional parameters and the disturbed paddy field soil, creating favorable conditions at the soil-tool contact interface for drag force reduction in the paddy field soil, the design and the optimization of paddy field equipment in South China, and the research on the technology of reducing viscosity and drag force of paddy field machinery.

Keywords: paddy field, disturbed-saturated paddy soil, soil contacting parts, contact-rheology-resistance, viscosity reduction and drag reduction.

1. Introduction

In order to ensure that rice seeds or seedlings have a suitable growth environment with gentle softness, high adhesion and sufficient fertilizer. Paddy soil is often subject to rotary tillage with water or plowshare with water, followed by harrowing and appropriate sedimentation (Chen et al., 2015, Wang et al., 2007). When the rotary tiller, plowshare and harrow roll are operating, high-speed rotation or movement strongly impacts, squeezes and breaks the soil, so as to soften the topsoil and promote the uniform mixing of soil, moisture, straw and fertilizer (Zhang et al., 2015). The changes of structure and mechanical properties of typical paddy soil after the above treatments are shown in Figure 1.

a) undisturbed state before tillage

b) burying the previous rice residue in the presence of water

c) disturbed saturation state after harrowing and sedimentation

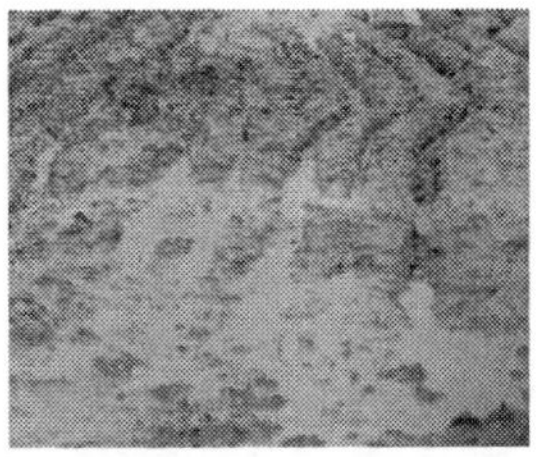

d) slurry overflow state during machine operating

Figure 1. Typical treatment of paddy field soil and changing process

After the above treatments, the arable soil presents a "disturbed-saturated state," which is characterized by a small amount of free water on the surface and a slurry mixed with plant residues inside. Its moisture content tends to be saturated, its mechanical strength is low, and its adhesion and fluidity are significantly enhanced, thus greatly affecting the passing and working performance of paddy field operating machinery. Take rice planting machinery as an example. In recent years, with the continuous and in-depth research at home and abroad, the working performance of key working parts of rice planting machinery has been greatly improved, such as rice transplanter (Zhu et al., 2016), pot seedling transplanter (Zhao et al., 2016, Na et al., 2016, Zuo et al., 2014), seedling transplanting part, pot seedling transplanting part or rice seed sowing part of precision seeder (Zhang et al., 2017, Wang et al., 2016).

By observing the operation of the existing rice planting machinery in the disturbed-saturated paddy soil environment, it can still be found that the working performance of the seedling transplanting part, pot seedling transplanting part or rice seed sowing part is stable and reliable respectively. The quantitative seedlings, pot seedlings or seeds can be taken out from the seedling tray or seed box and planted and placed in the designated position in the soil. However, the auxiliary working parts of the planting machinery, such as the furrow sliding plate, the seeding sliding plate and other flat soil contacting parts, squeeze and push the muddy paddy soil, which will cause the mud to overflow, flow and block, and then push down and bury the seedlings and bowl seedlings in the operation line, causing them to fall down and even die. The seed surface after sowing is covered with too thick mud, which makes it unable to germinate. The machines and tools sink and move difficultly due to the mud overflow and adhesion, as a result, the working efficiency and operating performance of rice planting machinery are reduced (as shown in Figure 2).

In view of the bottleneck of planting mechanization in the whole process mechanization of rice production, in addition to the theoretical and technical research on the key working parts of rice planting machinery, such as the seedling transplanting part of rice transplanter, the bowl seedling transplanting part of pot seedling transplanter, and the rice seed sowing part of precision seeder, more attention should be paid to the overall working performance of rice planting machinery under the disturbed-saturated paddy soil environment. In particular, attention should be paid to the impact on the performance of the working tool caused by the interaction between the auxiliary working parts of the planting machinery and the soil, such as the furrow sliding plate, the seeding sliding plate.

a) machine sinking

b) surging mud

c) seedling lodging

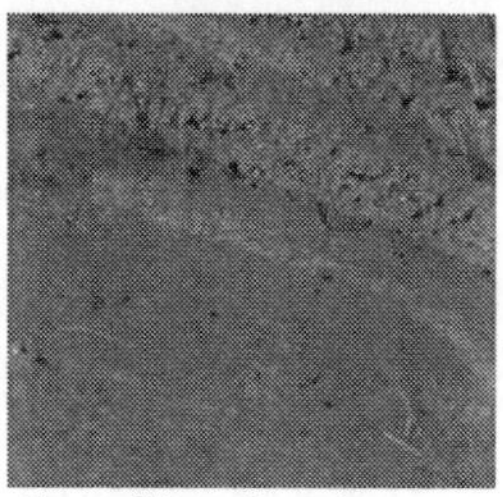

d) seedlings burying

Figure 2. Rheology of disturbed-saturated paddy field soil and its influence

2. Principal question

The influence of furrow and seeding sliding plates on the performance of work tools mainly comes from the interaction between soil contacting parts and soil. In other words, under the thixotropic, lubrication, adhesion and other interaction relations among the soil contacting parts, soil and soil moisture, the muddy paddy soil is disturbed again after tillage and sedimentation, and then flows. These interaction relations are coupled with each other, which can be shown schematically in Figure 3.

These coupling relationships are mostly closely related to soil particle size distribution, moisture content, water distribution, adhesion characteristics, surface characteristics of soil contacting parts, contact process with soil and other factors. They are the main source of mud overflow, machine subsidence, increased migration resistance, decreased trafficability of planting machinery, and increased power consumption.

Therefore, the following scientific problems arise:

(1) What is the contact relationship between the flat plate soil contacting parts such as ship bottom plate and seeding slide plate and the disturbed paddy soil? What are the relationships among these contact relationships and soil flow, deformation, and the changes of machine subsidence, movement resistance, and adhesion resistance, and what factors affect them?

(2) Which of these surface contact relationships is conducive to minimizing soil flow and deformation, limiting machine subsidence, reducing resistance and viscosity of the soil contacting working parts of rice planting machinery, decreasing movement resistance and adhesion resistance, and controlling the flow range of disturbed paddy soil?

(3) What is the mechanism of these favorable drag reduction and viscosity reduction effects? How are these favorable surface contact relationships established and maintained? What are the relationships between these favorable surface contact relations and the external structure shape, surface characteristic parameters and motion parameters of the soil contacting parts?

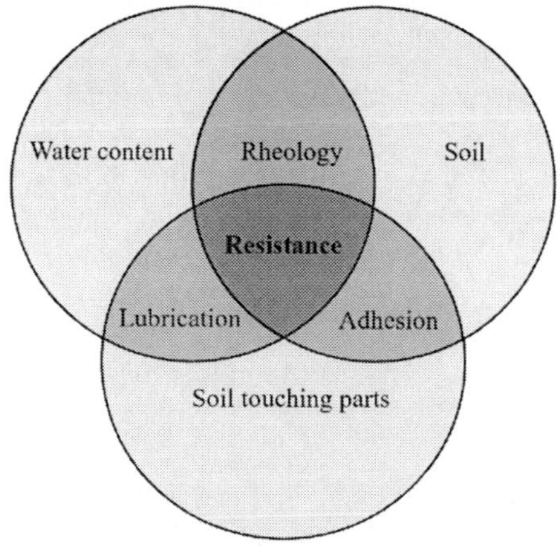

Figure 3. Schematic diagram of contact-rheology-resistance coupling relationship between soil contacting parts and soil

3. Research state and future directions

3.1 Research progress on moving resistance of flat plate soil contacting parts

The bottom plate of a ship tractor is a typical flat plate soil contacting part. When the ship tractor operates in paddy field, the data shows that the sliding resistance of the hull accounts for about 20%-40% of the total resistance (Shi et al., 1979). In order to reduce this resistance, Dinghua Qian, Qinye Lao, Zhen Zhuge, and others studied the contact relationship between the flat plate soil contacting parts and the soil (Qian et al., 1965, Lao et al., 1990, Zhuge et al., 1989). They believed that the contact among the moldboard, hull bottom and metal can be divided into five layers when plowshares and boat tractors are operating: metal surface layer, water and metal surface bonding interface layer, water film or liquid layer, water and soil surface interface layer, soil surface layer. Among them, the water film or liquid layer is a high-energy layer, and the separation of the contact interface generally occurs in this layer, and the magnitude of the separation force depends on the thickness and continuity of this layer. Therefore, the presence of thick mud-water film between the hull of the boat-type tractor and soil is a conducive condition for reducing the sliding resistance at the contact interface. Based on this theory, they also stated that the sliding resistance of the ship tractor is primarily caused by the deformation and subsidence of the soil, and the comprehensive resistance formed by compaction, friction, adsorption, push and shear. They also pointed out that the sliding resistance increases with the increase of the grounding specific pressure of ship tractor hull, the increase of subsidence and the intensification of soil deformation. The average grounding specific pressure of ship tractor hull is between 2.5 to 7.0 kPa, which is more appropriate. Zhenyong Li (Li et al., 1995, Li et al., 1997) and others believe that if there is mud water film lubrication between the bottom of the ship and the soil, the bottom plate of the ship will not have direct friction with the soil, so as to reduce the sliding resistance of the ship. This concept was verified through the field test of adding baffles on both sides of the ship to force mud water to flow out from the stern and reduce leakage from both sides of the ship. The experimental research of Chuzhou Tang (Tang et al., 1998) et al. shows that the running resistance of the large carriage on the paddy field tillage machine is approximately in a linear relationship with the vertical load, and the movement resistance increased as subsidence increased. Wan Zhiyuan (Wan et al., 1998) conducted an experimental study on the traction characteristics of the seeding slide plate of the rice direct seeding machine mounted to a tractor. The results revealed that the material and shape of the ship plate have a great impact on the traction performance of the chassis. It can significantly reduces the friction between the soil and the ship plate by selecting low frictional materials to make the ship plate, improving the shape of the ship plate or adding auxiliary devices to form a complete mud water film lubrication. The research of Rashid Qaisrani and Li Jian qiao (Rashid et al., 2010) shows that the usage of bionic bulldozing plate and moldboard plough wall in dryland soil can reduce the adhesion resistance. B. Azadegan, J. Massah (Azadegan et al., 2012) carried out experimental research on the influence of clayey soil on steel adhesion at different temperatures in the range of 5°C-30°C. The results show that the adhesion decreases with the increase of temperature. This relatively new result may indicate that temperature causes the change of surface contact state between soil and soil contact parts, especially the change of surface water distribution, thus causing the change of soil adhesion to steel.

3.2 Research progress on testing technology of "paddy soil-terrain machine system"

It is an important method and technology to evaluate various ground machine systems by using modern testing technology to study the relationship between "paddy soil-terrain machine system." For example, GuoZhen Wu, Zuan Feng, etc. (Wu et al., 1984) used a self-made ultra-low pressure diaphragm type specific pressure sensor to measure the size and distribution of the specific pressure at the bottom of the boat-type tractor when it was moving freely, rotating and plowing, which provided a basis for determining the general layout of the tractor boat and the design of the bottom shape. In order to test the adhesion between bionic surface and soil, Lu Sun (Sun et al., 2005) of Jilin university designed a soil adhesion testing system based on virtual instrument technology and universal material testing machine and studied its sampling frequency and control parameters. Li Jianqiao, Yang Zhiqiang, et al. (Li et al., 2005) developed and validated the Micro Soil Adhesion Test System (RMSACS), which broke through the minimum soil adhesion measurement of the previous soil adhesion testing systems and provided the basis for measuring the micro adhesion on the surface of small soil animals such as earthworms in bionic research. Chihang Zhao (Zhao et al., 2001) developed a paddy field soil rheological parameter measuring instrument under dynamic load based on the virtual measurement principle. The field test shows that the measuring instrument can be used for actual detection. Lixin Zhai (Zhai et al., 2011) sealed 11 pressure sensors with self-made isolation boxes and arranged them on the surface of the plough to establish a sensor array for specific pressure at the soil-plow contact interface. The constructed system was tested in the rheological cohesive soil. The results indicated that it is feasible to use the sensor array to test the surface of the plough. In order to analyze the dynamic change law of wheel sand settlement and traction force accurately under different working conditions, Zhongjia Chen (Chen et al., 2009) used a torque sensor installed in the transmission system and a tension sensor installed on the vehicle body to measure the torque of the wheel and the traction force generated by the wheel in real time. Juanli Zhang (Zhang et al., 2004) used CTM-4F automobile tractor comprehensive tester to test and study the traction adhesion performance of dry land driving impeller with DongFangHong-150 tractor as the load. Meanwhile, Mingu Zhou (Zhou et al., 2005) used a lever system to test and study the traction adhesion performance of tire and impeller. The measurement error was corrected by compensating the lever length. Guozhong Zhang, Xinglin Yao, Zhishan Wang, etc. used the torque test system to study the working torque and power consumption of the double spiral rotary tillage straw burying blade roller (Zhang et al., 2012, Yao et al., 2012, Wang et al., 2012).

3.3 Research progress on prediction and simulation of soil tillage mechanics

In recent years, the application of finite element (FEM) virtual simulation technology or model to predict soil tillage mechanics has been attracting much attention (Xu et al., 2005, Roul et al., 2009, Formato A. et al., 2005). For example, using the finite element method, Godwin R J (Godwin et al., 2007) and Kannakara S (Kannakara et al., 2007) established the interaction force model between soil and soil contacting working parts, predicted and simulated the tillage resistance of plowshares and other working parts; Kheiralla A F (Kheiralla et al., 2004) analyzed the energy demand model of tillage equipment under sandy soil conditions; Abdul Mounem Mouazen (Abdul et al., 1999, Abdul et al., 2002) analyzed the rheology of plastic soil; Susila E. (Susila et al., 2003) studied the permeability of consolidated soil; Ming Zhou (Zhou et al., 2009), Junfang Xia (Xia et al., 2013) and Aili Ma (Xia et al., 2009) simulated the cutting process of the cross blade for straw burying and the rotary tillage blade for ditching to soil in the double spiral rotary tillage straw burying blade roller respectively. Yuanshuang Lu and Guozhong Zhang (Lu et al., 2014, Lu et al., 2013) studied the shape of the double screw straw burying roller and its working process of cutting soil by combining the method of proportional model. Mootaz Abo-Elnor (Mootaz et al., 2003), Nidal H. (Nidal et al., 2003), S. Karmakara (Karmakara et al., 2007), A. mardani (A. mardani et al., 2010) respectively analyzed the interaction between soil and working parts and the force of disk plow on soil. Yong Zhou and Jian Yang (Zhou et al., 2006) of Guangxi University combined the dynamic simulation software ADAMS with the finite element analysis software ANSYS and optimization technology to study the influence of the inclination of the front cross baffle of the ship bottom plate, the shape of the bottom plate section and the depth of the bottom plate section on the damming and walking resistance when the ship bottom plate is walking and turning, and optimized the design of the ship bottom plate which can reduce the damming and walking resistance. Ikuya ono (Ikuya et al., 2013) studied the resistance of a narrow blade cutter in soil by using the method of three-dimensional discrete element. Qi long, Ma Xu, et al. (Qi et al., 2015) used ALE (Arabic Lagrange-Euler) multi-material coupling algorithm to establish a soil water two material coupling finite element model and used fluid solid coupling algorithm to analyze the interaction process between weeding wheel and soil water model, and obtained the influence law of soil water model coupling stress and soil disturbance rate.

3.4 Research Progress on drag reduction mechanism and technology of soil contacting parts

The researchers found that many soil animals have excellent soil adaptability, which provides a new idea for the research of bionic soil touching working parts with the functions of drag reduction, desorption, wear resistance and consumption reduction (Cheng et al., 2002). For example, Luquan Ren, academician of Jilin University (Ren et al., 2001) and others prepared the bionic plough surface with pit shape according to the surface of stinking dung beetles on which the pits can reduce attachment and carried out the farmland test. The results showed that under the same experimental conditions, the resistance of 20-steel bionic plough with the same material was reduced by 12.7% compared with that of ordinary 20-steel smooth plough; Compared with common 35-steel smooth plow, the resistance of 35-steel bionic plow with the same material is reduced by 18%. Jun Jin et al. (Jin et al., 2014) studied the problems of serious soil adhesion to the plough wall during field operation of paddy field plough, resulting in increased ploughing resistance, increased fuel consumption, and reduced farming quality and production efficiency. Field comparative tests showed that the bionic plough body could reduce ploughing resistance by 15.9%-18.0% compared with the ordinary smooth paddy field plough wall. Referring to the surface morphology of the earthworm head when it shrinks, Yunhai Ma et al. (Ma et al., 2014) added ultra-high molecular weight polyethylene coating on the basis of the core moldboard ditcher and manufactured a bionic ditcher with corrugated non smooth surface, which can reduce drag and viscosity by 9%. Jin Tong, Qingzhu Zhang et al. (Tong et al., 2014) based on the geometric structure of the ventral side of the stinking dung beetle and combined with the rib bionic pressing roller designed with ultra-high molecular weight polyethylene material, achieved good drag reduction and anti-sticking effect, with the maximum drag reduction rate of 41.08% and the maximum drag reduction rate of 39.40%.

4. Analysis and consideration

At present, the research on soil-terrain machine system of paddy field in South China mainly starts with the contact relationship among the ship bottom, plough wall and the soil, and analyzes the influence of the moisture content, fluidity, vertical load and mud water lubrication film on the subsidence and moving resistance of the flat soil contacting parts such as the ship hull during paddy field operating. In order to reduce the moving resistance of the plate type soil contacting parts in the disturbed paddy soil, it is necessary to establish the surface contacting state system between the soil and the soil contacting parts, and comprehensively consider the effects of soil moisture content, soil moisture distribution, soil properties, load on the soil contact parts, surface shape, center of gravity position, grounding pressure distribution and other factors. Finite element analysis and dynamic simulation can effectively analyze the influence law, help to clarify the test scope and design the test scheme more reasonably. It is an important auxiliary method for the study of the relationship between soil and soil-contacting parts. With the continuous clarity of soil characteristics and the emergence of more reasonable soil structure and mechanical models, its application will be increasingly in-depth. The adhesion of paddy field soil has a great impact on the passing performance of paddy field planting machinery. In order to achieve the functional requirements of effective desorption and drag reduction, the research on the desorption and drag reduction technology of paddy field terrain machinery based on the multi coupling bionic theory and technology of surface structure, morphology and materials has great potential development and broad prospects (Ren et al., 2008, Li et al., 2008, Ren et al., 2008).

In view of the mutual coupling mechanism of the movement resistance formed by the contact and rheology of the typical flat paddy field contacting parts such as the hull and the seeding slide plate in the disturbed-saturated paddy field soil, the following scientific problems need to be studied:

(1) To obtain the rheological model of disturbed-saturated paddy soil, and preliminarily clarify the influence of moisture content, sedimentation time, load and loading rate on soil mobility.

(2) To clarify the influence of the rheological characteristics of the disturbed-saturated paddy field soil on the motion characteristics of the plate type soil contacting parts, and obtain the preferable results of the structure parameters, motion parameters and grounding specific pressure distribution of the plate type soil contacting parts with the motion resistance as the evaluation index.

(3) To clarify the law of adhesion resistance of flat soil contacting parts with different external characteristics in disturbed paddy soil environment, to obtain the conditions for the establishment of mud water lubricating film on non smooth surface, and to obtain the typical surface structure to reduce adhesion resistance.

(4) To establish the motion and dynamic model of the bottom and seeding slide plate of the ship type rice direct seeder in the typical disturbed-saturated paddy soil, and raise the ways and methods to optimize the shape and dynamic characteristics of the bottom and seeding slide plate of the ship type rice direct seeder.

5. Acknowledgements

This study was supported by the National Natural Science Foundation of China (No. 51775220).

6. References

A. mardani*, K. shahidi, A. rahmani, B. mashoofi, H. karimmaslak. Studies on a long soil bin for soil-tool interaction[J]. Cercetări Agronomice în Moldova Vol. XLIII, No. 2 (142) / 2010:5-10.

Abdul Mounem Mouazen, Herman ramon. A numerical statical hybrid scheme for evalution of draught requirements of a subsoiler cutting a sandy loam soil, as affected by moisture content, bulk density and depth [J]. Soil and Tillage Research, 2002, 63(1): 155-165.

Abdul Mounem Mouazen, Mikles Nemenyi. Tillage Tool Design by the Finite Element Method:Part I , Finite Element Modelling of Soil Plastic Behaviour [J]. J.Agr Engin. Res.,1999.72(1):37-51.

B.azadegan J. massah. Effect of temperature on adhesion of clay soil to steel [J].Cercetari Agronomice in Moldova, Vol. XLV, No. 2 (150) / 2012: 21-27.

Chen, Q., Shi, Y., Ding, Q., Ding, W., Tian, Y., 2015. Comparison of straw mixed burying effect between positive and negative rotary tillage. Journal of Agricultural Engineering, 31 (9): 13-18

Chen, Z., 2009. Experimental study on traction performance of shrub stubble leveler and its influence on sandy land. Beijing Forestry University

Cheng, H., Sun, J., Li, J., Ren, L., 2002. The structure of the body wall surface of dung beetle and its relationship with the function of reducing adhesion and desorption. Journal of Entomology. 45(2):175-181

Formato A, Faugno S, Paolillo G. Numerical simulation of soil-Plough mould board interaction [J]. Biosystems Engineering, 2005, 92(3):309-316.

Godwin R J, O'Dogherty M J, Saunders C, et al. A force Prediction model for mould board Ploughs incorporating the effects of soil characteristic properties, plough geometric factors and plough speed [J]. Biosystems Engineering, 2007 (97):117-129.

Guozhong Z., Yong Z, Junfang X., Qichuan X., Research of the torque testing system in the spiral rotary tillage and stubble-mulch roller based on virtual instrument [C]. 2012 ASABE Annual International Meeting, Dallas, Texas, July 2012.

Ikuya Ono, Hiroshi Nakashima, Hiroshi Shimizu, Juro Miyasaka, Katsuaki Ohdoi. Investigation of elemental shape for 3D DEM modeling of interaction between soil and a narrow cutting tool [J]. Journal of Terramechanics. 50 (2013) 265–276.

Jin, J., Li, J., Zhang, G., Li, Y., Pu, M., Huang, H.,2015. Design of bionic non smooth paddy field plough wall and field application experiment. Research on agricultural mechanization. 12:160-165

Kannakara S, Kushwalla RL, Lague C. Numerical modeling of soil stress and pressure distribution on a fiat tillage tool using computational fluid dynamics [J]. Bio-systems Engineering, 2007, 97(1):407-414.

Kheiralla A F, Yahya A, Zohadie M, et al. Modelling of Power and energy requirements for Tillage implements operating in Serdang sandy clay loam, Malaysia [J]. Soil & Tillage Research, 2004 (78):21-34.

Lao, Q., Li, Y., Wang, J., Li Z., 1990. Theoretical analysis of sliding resistance of towing hull lubricated by mud water film. Journal of Zhejiang Agricultural University. 16(2): 125-128

Li, J., Huang, H., Wang, Y., Tian, L., Ren, L., 2015. Research progress of machine system on soft ground. Journal of Agricultural Machinery. 46(5):306-320

Li, J., Yang, Z., Liu, G., Xu, Y., Sun, J. Experimental study on micro soil adhesion testing system [EB/OL] http://www.paper.edu.cn

Li, Z, Li Y., Li J., et al.1997. Study on the mechanism and measures of waterproof wave of high-speed rice transplanter. Journal of Agricultural Machinery. 28(4):33-39

Li, Z., Li, J., Li Y., Hu, H., 1995. Application and experimental study of "water cushion" formation principle in ship towing. Journal of Agricultural Machinery. 26 (4): 41-44

Lu, Y., Zhang, G., Zhou Y., Ji W., Li, Z., Li, Y., 2014. Reverse reconstruction of the spiral line of the blade edge of the cross blade of the blade roller of the double spiral rotary tillage model. Journal of Huazhong Agricultural University (SCIENCE EDITION). 33(3):1-7

Lu, Y.,2013. Experimental study on dynamic characteristics of straw burying blade roller in double spiral rotary tillage based on similarity method. Huazhong Agricultural University

Ma, A., Liao, Q., Tian, B., 2009. Numerical simulation of soil cutting with spiral cutter based on ANSYS/LS-DYNA. Journal of Huazhong Agricultural University. 28(2):248-252

Ma, Y., Ma, S., Jia, H., Liu, Y., Peng, J., Gao, Z., 2014. Test and analysis of viscosity and drag reduction performance of bionic corrugated ditcher. Journal of Agricultural Engineering. 30 (5):36-41

Mootaz Abo-Elnor, R. Hamilton, J.T. Boyleb., 2003. 3D Dynamic analysis of soil-tool interaction using the finite element method. Journal of Terrain mechanics. 40(1):51-62.

Na, M., Song, Z., Zhou, M., Zhu, H., Wang, Q., Zhao, Y., 2016. Design and experiment of ratchet gear type rice pot seedling transplanting longitudinal seedling feeding mechanism. Journal of Agricultural Machinery. 46(11): 43-48

Nidal H. Abu-Hamdeha, Randall C. Reederb. A nonlinear 3D finite element analysis of the soil forces acting on a disk plow. Soil & Tillage Research 2003 (74) :115-124.

Qi, L., Liang, Z., Ma, X., Tan, Y., Jiang, L., 2015. Fluid structure coupling simulation analysis and verification of the interaction between rake pressure weeding wheel and paddy soil. Journal of Agricultural Engineering. 31(5):29-37

Qian, D., 1965. Study on the adhesion characteristics of white cast iron to heavy clay. Journal of Agricultural Machinery. 8(2):145-150

Rashid Qaisrani, LI Jian-qiao, M. Azam Khan, Iram Rashid. Soil Adhesion Preventing Mechanism of Bionic Bulldozing Plates and Mouldboard Ploughs[J]. Advances in Natural Science. Vol. 3, No. 2, 2010: 100-107. (The 3rd International Conference of Bionic Engineering)

Ren L Q, Tong J, Li J Q, et al. Soil Adhesion and Biomimetics of Soil-engaging Components: A Review [J]. J Agriculture Engineering, 2001, 79(3): 239-263.

Ren, L., 2011. Soil adhesion mechanics. Beijing, China Machine Press

Ren, L.2008. Progress in bionic research on drag reduction of ground mechanical desorption. Chinese Science. 38(9): 1353-1364

Roul A K, Raheman H, Pansare M S, etal. Predieting the draught requirement of tillage implements in sandy clay loam soil using an artificial neural network [J].Biosystems Engineering, 2009(104):476-485.

S. Karmakara, R.L.Kushwaha, C.Lague Numerical modelling of soil stress and pressure distribution on a flat tillage tool using computational fluid dynamics[J]. Biosystems Engineering. 2007, 97(1): 407-414.

Shi, L., Deng, J., 1979. Development of tractor boats and analysis of their paddy field performance. Journal of Agricultural Machinery.10(2): 64-74

Sun, L., Study on sampling frequency and control parameters of soil adhesion characteristic test system. Jilin University

Susila E, Hryciw R D. Large displacement FEM modeling of the cone penetration test (CPT) in normally consolidated sand [J]. Int.J.Numer.And Meth. Geomech, 2003, 27(7): 585-602.

Tang, C., Liu, H., Wu, Q. 1998. Experimental study on the running resistance of the large carriage of paddy field cultivator. Journal of Agricultural Machinery. 29(2):27-31

Tong, J., Zhang, Q., Chang, Y., et al., 2014. Experiment on reducing viscosity and resistance of rib bionic roller [j] Journal of agricultural machinery. 45(5):135-140

Wan, Z., 2007. Traction characteristic test and structure optimization design of Rice Direct Seeder Based on virtual technology. Nanjing Agricultural University

Wang, C., 2016. Design and test of a simple precision hole direct seeding machine for rice. South China Agricultural University.

Wang, J., Yin, D., Han, Y., Ji, W., Sun Y., Zhao, T., 2007. Design and test of rice straw whole plant returning machine. Journal of Agricultural Machinery. 38 (10):54-56

Wang, Z., Xia, J., Xu, Q., Hu, R., 2010. Power consumption test of screw cutter roller of ship type rotary cultivator. Journal of Agricultural Machinery. 41(12), 44-47

Wu, G., Feng, Z., Li, W., 1984. Preliminary study on grounding specific voltage test of tractor boat. Journal of Zhejiang Agricultural University. 10(3):351-355

Xia, J., He, X., Yu, S., Zhou, Y. Finite element simulation of soil cutting with spiral cutter roller based on ANSYS/LS-DYNA. Journal of Agricultural Engineering. 29(10): 34-41

Xu, Z., Wang, J., 2005. Research progress in finite element analysis of soil cutting. Journal of Agricultural Machinery. 36(L): 134-137

Yao, X., Zhang, G., Xia, J., Zhou, Y., Shen, M., 2010. Design and test of torque detection system for straw burying blade roller of double spiral rotary tillage. Research on Agricultural Mechanization. 12:23-26

Zhai, L., 2011. Study on the influence of plough structure and working parameters on tillage resistance of convective soil. Nanjing Agricultural University

Zhai, L., Ji, C., Yu, Y., Ding, Q., 2011. Sensor array detection of plow surface load distribution. Journal of Agricultural Machinery. 42(8):50-53

Zhang, G., Zhou, Y., Fan, Q., Ji, W., Huang, H., Li, Y., Li, Z., Zhang, Y., Zhai, K., Zhang, Q., Zhang, S., 2015. The utility model relates to a double chamber pneumatic seed metering device, June 10, 2015, Chinese Invention Patent, zl201310132979.3

Zhang, J., 2004. Experimental study on traction adhesion performance of dry land driven impeller. Northwest University of Agriculture and Forestry Science and Technology

Zhang, S., Xia, Junfang, Zhou Yong, Wu Delin, caochengmao, Xia Ping Performance analysis and field experiment of pneumatic roller direct seeding precision metering device for rice [j] Journal of agricultural engineering, 2017, 33 (3):14-23

Zhang, X., Xia, J., Zhang J., He X., Liang S., Zhang, S., Wu, H. Wan, S., 2016. Operation performance test of straw returning combined knife roller for both water and drought. Journal of Agricultural Engineering. 32 (9): 9-15

Zhao, C.,2001. Development of a dynamic load measuring instrument for rheological parameters of paddy soil. Nanjing Agricultural University

Zhao, Y., Zhu, H., Xin, L., Zhou, M., Feng, J. Zhang, M., 2016. Mechanism analysis and experiment of fitting gear five bar rice pot seedling transplanting mechanism. Journal of Agricultural Engineering. 32(01): 12-21

Zhou, M., 2005. Experimental study on traction adhesion performance of tire and impeller. Northwest University of Agriculture and Forestry Science and Technology

Zhou, M., Zhang, G., Xu, Q., Xia, J., 2009. Finite element simulation of soil right angle cutting. Journal of Huazhong Agricultural University. 28(4):491-494

Zhou, Y., 2006. Optimization design and vibration research of walking mechanism of 2bd-6 rice direct seeder. Guangxi University

Zhu, D., Xiong, W., Jiang, R., Wu, L., Wang, C., Zhu, H., 2016. Design and test of 2ZGK-6 adjustable width narrow row high speed rice transplanter. Journal of Agricultural Engineering, 32 (21): 37-45

ZhuGe, Z., Jiang, C., Development, performance research and structure analysis of ship tractor. Journal of Agricultural Machinery. 2:46-52

Zuo, Y., Cao, P., Zhao, Y., Yu, X., Lu, C., Zhang, Z., Zhang, H., 2014. Design and optimization of rice bowl seedling transplanting mechanism with B-spline noncircular gear planetary gear system. Journal of Agricultural Engineering. 30(10): 10-17

1655 / Modeling of Lunar Rover Vehicle Wheel-Soil Interaction Using Fem-Dem Method

Kaidi Zhang[a], Yunqing Zhang[a]*, Junwei Shi[b], and Weili Kong[a]

[a]*Huazhong University of Science and Technology, Wuhan, Hubei, China*
[b]*Shanghai Institute of Astronautical Systems Engineering, Shanghai, China*
**Corresponding author: zhangyq@hust.edu.cn*

https://doi.org/10.56884/OTLT5367

Abstract

Apollo Lunar Rover Vehicle (LRV) used wire mesh wheels to adapt to the loose and soft soil conditions on the lunar surface, aimed to meet the mobility performance of LRV on the moon. Therefore, it is of great significance to study the wheel-soil interaction characteristics of the wire mesh wheel to improve the maneuverability and traction performance of LRVs. This paper proposed a coupled analysis method of wire mesh wheel-soil based on the DEM-FEM method. Firstly, a lunar terrain DEM model was established, which conforming to the physical characteristics of the lunar soil. Then, a FEM model of the flexible wire mesh wheel with large deformation characteristics was developed, which was verified by stiffness tests. In addition, the mobility performance of the wheel model developed were studied under different slip rates and sideslip angles, respectively. The simulation results were compared and validated with NASA experiment data. The results can provide technical support for the coupling simulation of wire mesh wheels and the traction characteristics of LRV.

Keywords: lunar rover vehicle, wire mesh wheel, FBM-DEM, lunar soil

1. Introduction

Manned lunar rover is a hot research topic in the field of lunar exploration in recent years. As the main auxiliary tool for short-term exploration and long-term stationing on the lunar surface, the manned lunar rover can efficiently carry astronauts to reach far away from the lunar module and accomplish various exploration tasks more effectively. The lunar surface is covered with a loose, fine-grained, powdered, weathered layer of lunar soil, and the gravity of the Moon is 1/6 of that of the Earth, so manned lunar vehicles need to have high traction characteristics in the lunar surface operating environment.

The wheel is one of the most important components of the lunar rover, which directly affect the traction and comfort performance of the vehicle. At present, the developed manned lunar rovers mainly used elastic wheels. The Russian Institute of Vehicle Engineering has developed various elastic wheels with internal elastic spokes, slatted spokes, spring rings, etc. (Berkelman et al., 1995; Matrossov et al., 2004a, 2004b); the Apollo LRV used elastic mesh wheels woven with galvanized high-carbon steel wire and successfully conducted exploration missions on the lunar surface (Asnani et al., 2009); NASA designed a drum-shaped Mars wheel with spiral spokes to form elastic wheels (Lindemann et al., 2005); ESA has developed a variety of structures of exploration wheels (Lee et al., 2006). NASA's latest development for NASA's newest flexible wheel for Mars exploration, named 'Superelastic' (Stefano et al., 2021), used a nickel-titanium alloy mesh with shape memory, which allows it to travel well on rocky roads. Therefore, the wire mesh wheel is the most successful and proven wheel structure for LRV, which is suitable for the extreme environment on the Moon and Mars. It is the preferred solution for future countries to develop manned lunar wheels. So it is important to study the mobility performance of the wire mesh wheel on the lunar surface.

In order to accurately simulate the mechanism of wheel-soil interaction, researchers have proposed many methods. Bekker (1956) developed a classical analytical model including wheel and soil parameters based on experiments. Hambleton and Drescher (2008) established a 2D and 3D wheel-soil interaction simulation system using ABAQUS to analyze the sinkage and driving performance of the wheels when walking on elastic and plastic soils; Li and Schindler (2013) established a FEM model of Bridgestone diagonal tires and analyzed the effect of inflation pressure on tire deformation and soil compaction process. The FEM method treats the soil as a continuous medium and ignores the important individual properties such as the shape, size, and friction force of soil grains. Therefore, more researchers are currently using the discrete element (DEM) method to study the microscopic properties of soil and lunar soil and the complex interactions with wheels. Fujii (2002) and Nakashima (2007) analyzed the interaction between the wheels of lunar deep space exploration vehicles and the lunar terrain using the DEM method, and investigated the effects of wheel structure parameters such as the width of wheel teeth, number of wheel teeth, and height of wheel teeth on the traction performance and sinkage of wheels.

Smith and Peng (2013) used 3D DEM to study the running behavior of wheels on granular surfaces. Although DEM can simulate the granular characteristics of sandy soil well, the elastic wheels with large deformation characteristics is a difficult problem for DEM method.

In addition, some researchers have combined discrete elements and finite elements to analyze the wheel-soil interaction. Nakashima and Oida (2004, 2005) combined discrete elements and finite element methods to treat the surface soil in contact with the wheel using the discrete element method and the deep soil far from the wheel action region using the finite element method; Mark et al. (2015) analyzed the interaction between wheel tread and wheel teeth of rubber tires and soil particles based on the DEM-FEM method. Yang et al. (2020a, 2020b) studied the dynamic interaction simulation between commercial vehicle tires and cobblestone ground, and the simulation results were verified by bench tests, which proved the effectiveness of the coupling analysis method.

The FEM method is specialized in the simulation of the large deformed wheels when interacting with the ground, while the discrete element method has a more accurate representation of discrete materials such as lunar soil. Therefore, the DEM-FEM analysis method is a great way to simulate the mobility characteristics of elastic wheels on lunar surface accurately.

In the numerical simulation of the wire mesh wheel, the LRV wire mesh wheel is different from other wheels and has a more complex structure, mainly including galvanized piano wire, titanium alloy treads, internal elastic restrictor, load bearing reeds, and hub (Giancarlo, 2006). Most of the current studies on wire mesh wheels are based on finite element analysis; Fan et al. (2014) established a finite element model of a wire mesh wheel and analyzed the stress characteristics of the wheel under various loading conditions; Xiao and Zhang (2016) studied the traction characteristics of a wire mesh wheel on elastic ground based on single FEM method. Due to the complexity of the wheel structure and the peculiarities of the physical properties of lunar soil, we have not yet seen any case of studying the simulation of the traction characteristics of the wire mesh wheel based on the DEM-FEM method. Therefore, this paper combines the three-dimensional FEM method and the DEM method to study the mobility characteristics of the wire mesh wheel of LRV under lunar surface conditions.

In this paper, firstly, a DEM lunar surface model was established, and the DEM parameters were calibrated based on the triaxial test and the pressure plate test. Secondly, a flexible large-deformation finite element model of the LRV wheel was established based on the Timoshenko beam elements and shell elements, which is verified by stiffness tests; then a virtual soil bin experiment system (VSBE) was established to study the traction characteristics of the wire mesh wheel under different slip conditions. The simulation results are in high agreement with the NASA soil bin experiments. The results show that the wheel-soil interaction analysis method proposed in this paper has high accuracy and is of guidance for the wheel design and traction characteristics study of the manned lunar rover.

2. Modeling of soil–soil interaction

An accurate DEM model of lunar soil is the basis of wheel-soil coupling simulation. In this section, a discrete element model of weathering layer soil was established, which has the most significant interaction with the LRV wheels. And the key contact parameters were calibrated for the physical properties of the lunar soil.

2.1 DEs equation of motion

In discrete element theory, each discrete element particle has six degrees of freedom, and the equation of motion for each particle is expressed as (Yang et al., 2020a):

$$m_i(\mathrm{d}^2\boldsymbol{u}_i/\mathrm{d}t^2) = \boldsymbol{F}_i^c + \boldsymbol{F}_i \quad (1)$$

$$I_i(\mathrm{d}^2\boldsymbol{\theta}_i/\mathrm{d}t^2) = \boldsymbol{r}_i \times \boldsymbol{F}_i^c + \boldsymbol{M}_i + \boldsymbol{M}_{ir} \quad (2)$$

where m_i is the particle mass; $\boldsymbol{u}_i$ is the three-directional translational displacement of the particle; $\boldsymbol{F}_i^c$ is the contact force between the particles, including normal elastic force, tangential elastic force, normal damping force and tangential damping force; $\boldsymbol{F}_i$ is the external and field forces on the particle, including gravity, electromagnetic force, etc.; I_i is the rotational inertia of the particle; $\boldsymbol{\theta}_i$ is the rotation angle of the particle in the three rotation directions; $\boldsymbol{M}_i$ is the external moment caused by the external force $\boldsymbol{F}_i$; $\boldsymbol{r}_i$ is the position vector from the particle center of mass point to the contact point; $\boldsymbol{M}_{ir}$ is the resistance moment of the particle.

2.2 Interaction between DEs

As demonstrated (Giancarlo, 2006), the soil particles in the weathered layer of the lunar surface have strong adhesive properties. To better characterize the cohesion, van der Waals forces, and hardness between lunar soil particles, the Hertz-Mindlin with JKR contact model is used in this paper to express the contact properties between lunar soil particles. This model introduces the concept of interparticle surfactant energy in the calculation, takes into account the influence of van der Waals forces in the contact region, allows the construction of strong adhesive systems (Santos et al., 2015), and can well characterize the adhesion properties between particles, which are mostly used to represent the bonding between small particles.

The tangential elastic force, tangential damping force and normal damping force of the JKR model are calculated based on the Hertz-Mindlin contact model theory, while the normal elastic contact force is based on the Johnson-Kendall-Roberts theory (Johnson et al., 1971), and the normal contact force is calculated using the normal overlap, interaction parameters, and the particle surface energy as:

$$\boldsymbol{F}_{JKR}^c = -4\sqrt{\pi E^*}a^{3/2} + \frac{4E^*}{3R^*}a^3 \quad (3)$$

$$\delta_{n1,2} = -\sqrt{\frac{4\pi\Delta\gamma\alpha}{E^*}} + \frac{\alpha^2}{R^*} \quad (4)$$

Where: $\boldsymbol{F}_{JKR}^c$ is the normal elastic contact force; $\delta_{n1,2}$ is the normal overlap; a is the radius of the contact point; E^* and R^* are respectively the equivalent Young's modulus and equivalent radius of the contact spheres defined in terms of shear modulus, E_1 and E_2, Poisson's ratio v_1 and

v_2, radius R_1 and R_2. $\Delta\gamma$ is the contact particle surface energy.

$$\frac{1}{E^*} = \frac{1-v_1^2}{E_1} + \frac{1-v_2^2}{E_2} \quad (5)$$

$$\frac{1}{R^*} = \frac{1}{R_1} + \frac{1}{R_2} \quad (6)$$

$$\Delta\gamma = \gamma_1 + \gamma_2 - \gamma_{1,2} \quad (7)$$

Where, γ_1 and γ_2 are the adhesion surface energy per unit area of the two interacting particle surfaces, $\gamma_{1,2}$ is the interface surface energy, when the particle materials in contact are the same, $\gamma_1 = \gamma_2$ and $\gamma_{1,2}$= 0, so there is $\Delta\gamma = 2\gamma$.

Without JKR gravitational force $\Delta\gamma = 0$, the normal contact force becomes Hertz normal contact force.

$$\boldsymbol{F}_{Hertz}^{c} = \frac{4E^*}{3}\sqrt{R^*}\delta_{n1,2}^{3/2} \quad (8)$$

The tangential contact force, normal damping force and tangential damping force between the particles are expressed as follows (Baran et al., 2009).

$$\boldsymbol{F}_{t1,2}^{c} = K_t \delta_{t1,2} \quad (9)$$

$$\boldsymbol{F}_{n1,2}^{d} = 2\sqrt{\frac{5}{6}}\beta\sqrt{K_n m^*} v_n \quad (10)$$

$$\boldsymbol{F}_{t1,2}^{d} = 2\sqrt{\frac{5}{6}}\beta\sqrt{K_t m^*} v_t \quad (11)$$

where $\delta_{t1,2}$ is the tangential displacement of the particles; v_n and v_t are the normal and tangential components of the relative velocities of the two particles.

$$K_n = \frac{4}{3}E^*\sqrt{R^*\delta_{1,2}} \quad (12)$$

$$K_t = 8G^*\sqrt{R^*\delta_{1,2}} \quad (13)$$

$$\beta = \ln e / \sqrt{\ln^2 e + \pi^2} \quad (14)$$

$$\frac{1}{G^*} = \frac{2-v_1}{G_1} + \frac{2-v_2}{G_2} \quad (15)$$

$$\frac{1}{m^*} = \frac{1}{m_1} + \frac{1}{m_2} \quad (16)$$

All forces on a given particle include contact force, damping force and gravitational force, the combined force of which is expressed as

$$\boldsymbol{F}_{all1,2} = \boldsymbol{F}_{n1,2}^{c} + \boldsymbol{F}_{t1,2}^{c} + \boldsymbol{F}_{n1,2}^{d} + \boldsymbol{F}_{t1,2}^{d} + G \quad (17)$$

The product of the tangential force and the vertical distance from the contact point to the center of mass represents the moment M caused by the tangential contact force.

$$\boldsymbol{M}_{1,2} = R^*\left(\boldsymbol{F}_{t1,2}^{c} + \boldsymbol{F}_{t1,2}^{d}\right) \quad (18)$$

The additional moment $\boldsymbol{M}_{a1,2}$ caused by rolling friction is expressed as

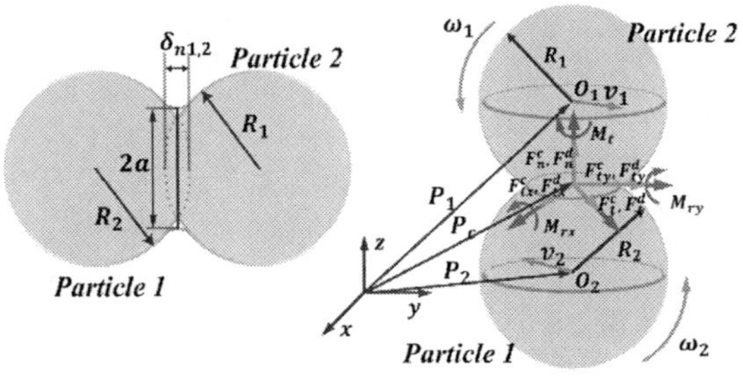

Figure 1. Schematic diagram of the interaction forces between particles

$$\boldsymbol{M}_{a1,2} = -\mu_r \boldsymbol{F}_{n1,2}^{c} R^* \quad (19)$$

Where, μ_r is the rolling resistance coefficient between particles in contact. The synthetic moment $M_{r1,2}$ is

$$\boldsymbol{M}_{r1,2} = \boldsymbol{M}_{1,2} + \boldsymbol{M}_{a1,2} \quad (20)$$

2.3 Geometric features of lunar surface soil

The geometric features of lunar soil grains are the main properties which determine the physical and mechanical properties of the lunar soil, mainly including the grain size and shape. The size distribution of the lunar soil grains is wide, and the general diameter is less than 1 mm, the vast majority of the diameter is ranged from 30 μm to 1 mm, and the median particle size is between 40 μm and 130 mm, which is smaller than the resolution ability of the astronaut's eye.

The particle morphology of lunar soil grains is diverse, ranging from spherical shapes to extreme angularity. However, the elongated, sub-angular, and angular grain shapes are more common. The occlusion of particles of other shapes is somewhat stronger than that of spherical particles. Therefore, the shear resistance of non-spherical particles is more significant under the same conditions of material friction coefficient. Based on the weathered layer lunar soil samples brought back by Luna 16 and 20 (Slyuto, 2014), this paper classifies lunar soil particles into three categories: 1. regular shaped glass and metal spherical grains (in figure 2a); 2. sharp angular fragments of different rocks, minerals, and glasses (in figure 2b-2e); 3. dendritic shaped agglomerates and slag-like particles (in figure 2f). In order to better characterize the different morphological features of lunar soil particles and to take into account the computational efficiency reduction caused by non-spherical particles, this study used the spherical (30%), cubic (30%) and elongated (40%) particles to express the morphological features of real lunar soil, as shown in figure 3.

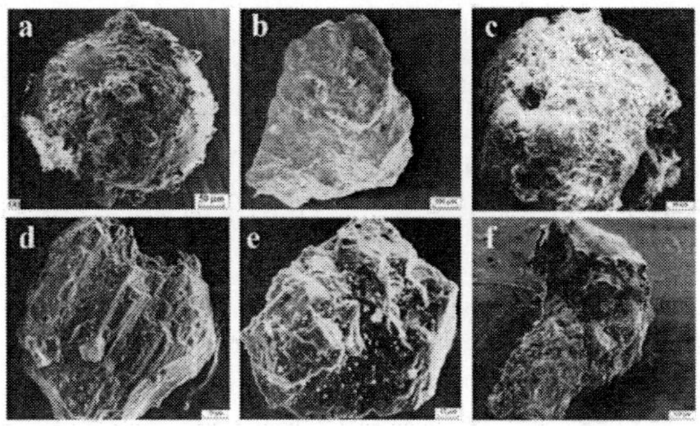

Figure 2. Shape of the lunar soil particles

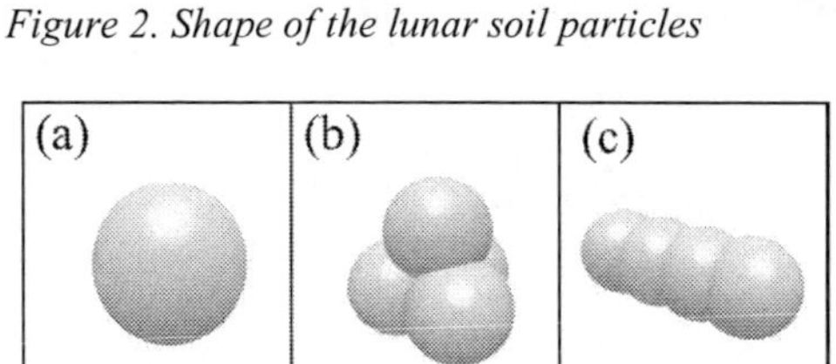

Figure 3. Lunar soil discrete element simulation particles

The median size of lunar soil particles ranges from 40 to 130 μm, with an average value of 70 μm(Carrier, 1973). Based on data from 21 samples of weathered layers returned by the Apollo 17 expedition, the cumulative weight content of particles larger than a specified size (N(>d) /%, expressed as a percentage) is described by *Eq. (1)*:

$$N_{>d}\% = -23.205\ln(d) + 143.586 \quad (21)$$

To improve the computational efficiency and to ensure that the lunar soil particles can enter the wheel through the tire gap of the wire mesh wheel, the particle diameter in the DEM is scaled down to 3.0 mm. In order to reduce the error between the simulation results and the experimental results, five particle sizes are adjusted during the simulation, which are 0.5, 0.75, 1.0, 1.25, and 1.5 times the nominal particle size, respectively.

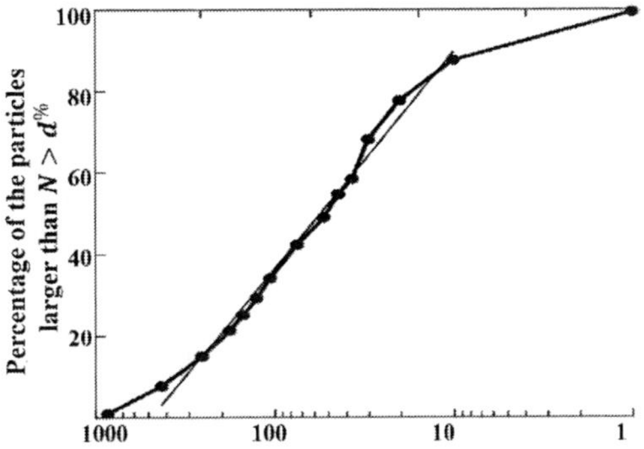

Figure 4. The cumulative weight content of particles larger than the specified size (Slyuto, 2014)

3. Modeling of wheel–soil interaction

The wire mesh wheels of Apollo LRV have flexible and large deformation characteristics. A large wheel load will cause a significant deformation. How to build an accurate model of woven wire mesh wheel is a difficult modeling problem. In this paper, the wire model is built by using the beam element based on the Timoshenko beam theory, and the shell element is used to build other wheel components.

3.1 FEs equation of motion

In the finite element theory, the modified Lagrange algorithm is applied (Yang et al., 2020b). The motion equation of FEs is expressed as:

$$\boldsymbol{M}\ddot{\boldsymbol{U}} = \boldsymbol{F}^{Ext} - \boldsymbol{F}^{Int} - \boldsymbol{F}^{C} \quad (22)$$

where $\boldsymbol{M}$ is the mass matrix; $\boldsymbol{U}$ is the finite element node displacement vector in the global coordinate system; $\boldsymbol{F}^{Ext}$ is the external force of the system; $\boldsymbol{F}^{Int}$ is the external force of the system; and $\boldsymbol{F}^{C}$ is the contact force generated with other objects.

The woven wire of the wheel is modeled using beam elements, which is based on the Timoshenko beam theory in which the cross section remains plane in deformation. Its displacements and section angles are interpolated independently, instead of being derived from the derivatives of the displacements. The Timoshenko beam considers not only the bending deformation of the beam, but also the shear deformation and rotational inertia of the beam, as shown in Fig. 5.

The force balance equation of the beam element is expressed as:

$$\boldsymbol{Q} = \frac{dM}{dx} = \boldsymbol{M}', \frac{dQ}{dx} - \boldsymbol{q} = \boldsymbol{M}'' - \boldsymbol{q} = 0 \quad (23)$$

The geometric relationships need to satisfy:

$$\boldsymbol{\varphi} = \frac{dw}{dx} - \boldsymbol{\gamma} \quad (24)$$

The physical equation is:

$$\boldsymbol{M} = -\frac{d\varphi}{dx}EI, \boldsymbol{Q} = \kappa GA\boldsymbol{\gamma} \quad (25)$$

where E is the elastic modulus; I is the second moment of inertia; G is the shear modulus; A is the cross-sectional area; κ is the Timoshenko shear coefficient, which is related to the cross-sectional shape; and $\boldsymbol{\gamma}$ is the shear angle.

The control equation for the Timoshenko beam element is as follows:

$$EI\frac{d^4\boldsymbol{w}}{dx^4} = \boldsymbol{q}(x) - \frac{EI}{\kappa GA}\cdot\frac{d^2\boldsymbol{q}}{dx^2} \quad (26)$$

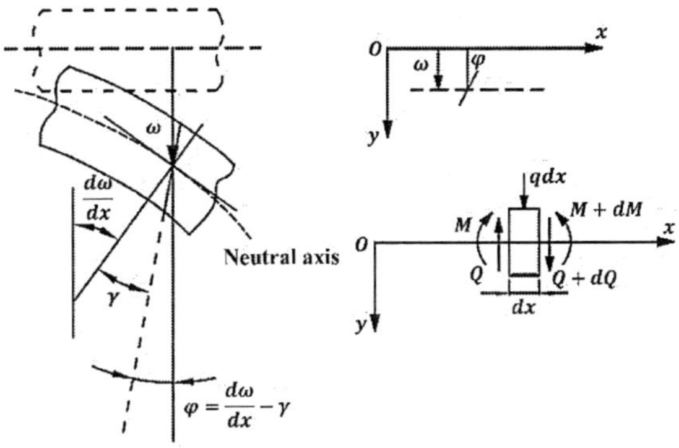

Figure 5. Timoshenko beam model

3.2 DEM-FEM contact model

In actual interaction, the tread of the wheel is in direct contact with the soil and provides traction, while there is almost no significant force between the wire mesh and the soil. In this study, the interaction between the lunar soil and the metal wire is neglected, and the lunar soil particles are considered to be in direct contact with the shell elements of the tread only. The contact forces between the two are calculated based on the Hertz-Mindlin contact model. As shown in Fig. 6, the contact forces between DEs and FEs mainly include the normal force F_{nij} and the tangential force F_{tij}.

As analyzed in Section 2.2, the normal force $\boldsymbol{F}_{n1,2}$ between DEs and FEs consists of normal elastic force $\boldsymbol{F}^{c}_{n1,2}$ and normal damping force $\boldsymbol{F}^{d}_{n1,2}$; the tangential force $\boldsymbol{F}_{t1,2}$ between DEs and FEs consists of tangential elastic force $\boldsymbol{F}^{c}_{t1,2}$ and tangential damping force $\boldsymbol{F}^{d}_{t1,2}$.

$$\boldsymbol{F}_{n1,2} = \boldsymbol{F}^{c}_{n1,2} + \boldsymbol{F}^{d}_{n1,2} \quad (27)$$

$$\boldsymbol{F}_{t1,2} = \boldsymbol{F}^{c}_{t1,2} + \boldsymbol{F}^{d}_{t1,2} \quad (28)$$

According to Hertz-Mindlin theory, each component of the contact force is expressed as

$$F_{n1,2}^{c} = \frac{4E^*}{3}\sqrt{R^*}\delta_{n1,2}^{3/2} \tag{29}$$

$$F_{t1,2}^{c} = 8G^*\sqrt{R^*\delta_{1,2}}\delta_{t1,2} \tag{30}$$

$$F_{n1,2}^{d} = 2\sqrt{\frac{5}{6}}\beta\sqrt{K_n m^*}v_{n1,2} \tag{31}$$

$$F_{t1,2}^{d} = 2\sqrt{\frac{5}{6}}\beta\sqrt{K_t m^*}v_{t1,2} \tag{32}$$

Where, $m^* = \frac{m_1 m_2}{m_1+m_2}$, m_1 and m_2 are the mass of DEs and FEs. E^* and R^* are the equivalent Young's modulus and equivalent radius of DEs and FEs. $\sigma_{1,2}$ is the penetration distance. $\sigma_{n1,2}$, $\sigma_{t1,2}$ are the normal and tangential components of $\sigma_{1,2}$. $v_{n1,2}$ and $v_{t1,2}$ are the normal and tangential relative velocities on the contact point.

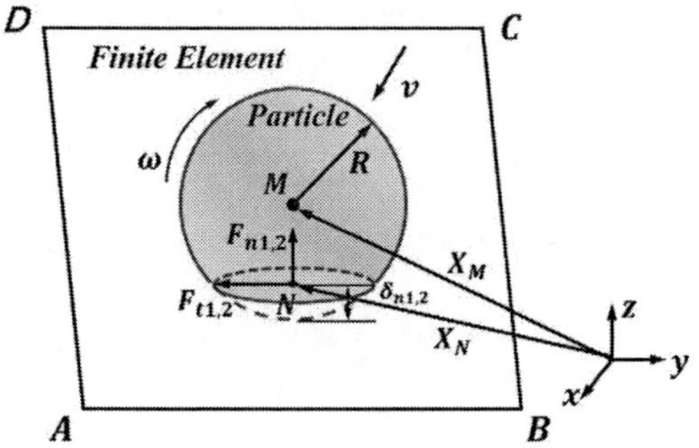

Figure 6. Contact model between DEs and FEs

3.3 Finite element model of wheel

Apollo LRV wheel is a metal wire mesh wheel, the tread is made of galvanized high-carbon steel wire woven, connected to the aluminum alloy hub through the titanium alloy frame. The wheel has a diameter of 0.82 m and a width of 0.23 m. The outer surface of the wheel is fitted with titanium treads, evenly distributed around the circumference of the tire, covering 50% of the wheel surface. The material parameters are shown in Table 2.

In this paper, a finite element model of the wire mesh wheel considering large deformation characteristics was developed (see Fig. 8). The woven wire needs to consider the deformation of tensile, torsion, and bending. Therefore, the wire was modeled using the Timoshenko beam elements, and the nodes were merged between the crossed beam elements. The inner frame and the hubs were built with shell elements. Considering that the titanium alloy tread was attached to the wire mesh by rivets, nodes were merged between the wire beam element and the tread shell element, and fixed constraints are applied among the wires, the inner frame, and the rim. Define a contact friction model between the inner frame and the woven wire mesh.

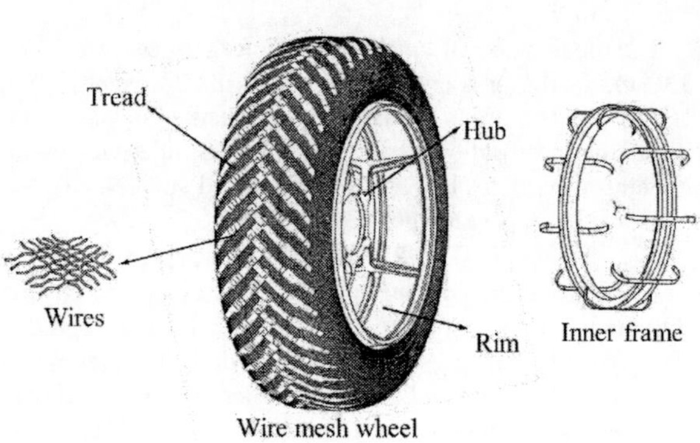

Figure 7. The structure of LRV wire mesh wheel

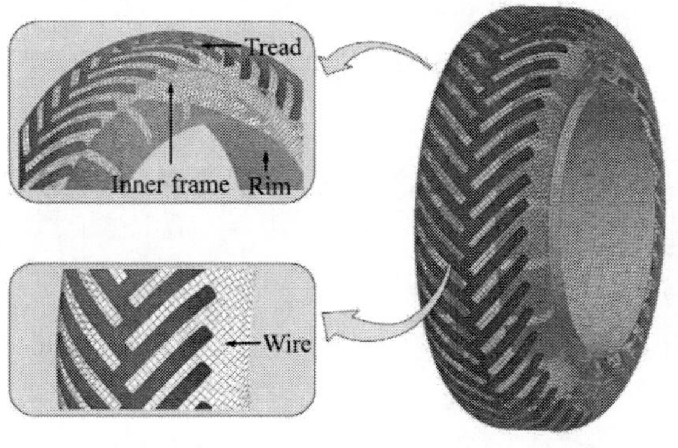

Figure 8. Finite element model of wire mesh wheel

Table 1. Structural parameters of Apollo LRV wire mesh wheels

Parameters	Values
Total mass [kg]	5.48
Outside diameter [mm]	818
Wheel width [mm]	229
Diameter of metal wire [mm]	0. 838
Number of metal wires	800
Spacing of metal wires [mm]	4.7
Tread thickness [mm]	0.279
Tread coverage area	50%

Table 2. Material parameters of wheels

Components	Density [kg/m^3]	Young modulus [Gpa]	Poisson ratio
Metal wires	7850	210	0.3
Tread	4510	108	0.34
Inner frame	4510	108	0.34
Rim	2750	72	0.33

4. Numerical Validation

4.1 Calibration of the DEM parameters

Scott (1987) performed a miniature triaxial test on the real lunar soil brought back by Apollo 12 and obtained the stress-strain curve of the real lunar soil, as shown in Fig. 9 and 10. In this paper, the 3D DEM model of the lunar soil was calibrated using triaxial tests and pressure plate tests under quasi-static loading conditions.

This work is only for the loose soil of the lunar surface. The soil sample size of the constructed virtual triaxial test is 15mm×15mm×40mm, the initial porosity is set to 0.465, and the adjusted confining pressures are 53kPa, 78kPa, and 103kPa, respectively. The working conditions are consistent with the loose lunar soil triaxial test done by Scott (Scott R.F., 1987). The axial stress difference curves of the lunar soil DEM simulant at different envelope pressures are shown in Fig. 11. The Mohr's circle and shear strength envelope of the simulated lunar soil are calculated based on the Mohr-Coulomb damage criterion, as shown in Fig. 12.

The lunar soil exhibits special properties of high internal friction angle and low cohesion relative to the Earth soil, as illustrated in table 3. The cohesion of DEM simulant is 0.33 kPa and the internal friction angle is 40.18°, which is in the allowable range of real lunar soil.

The lunar soil is a typical soft ground, and the pressure-bearing model proposed by Bekker is used to describe the pressure-bearing characteristics of the lunar soil. The relationship between the vertical load and the simulated lunar soil deformation under the action of the pressure plate is expressed as:

$$p = \left(\frac{k_c}{b} + k_\varphi\right) z^n \qquad (33)$$

Where, p is the pressure acting on a unit area; z is the sinkage of the simulated lunar soil; b is the width or radius of the plate; n is the deformation index of the simulated lunar soil; k_c is the cohesion modulus; k_φ is the friction modulus. These three parameters, including deformation index n, cohesion modulus k_c and friction modulus k_φ, are obtained by simulated lunar soil pressure plate test.

The pressure plate test is shown in Fig. 13, the plate used for the test is round, the radius is 50, 70, and 90mm. The pressure plate loading speed is 3mm/s, to prevent the loading speed is too fast to have an impact on the test results.

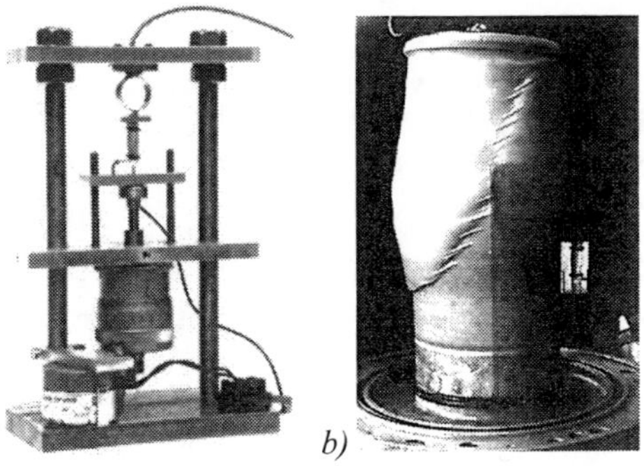

a) b)

Figure 9. Micro triaxial test. a) the test equipment; b) the tested lunar soil samples

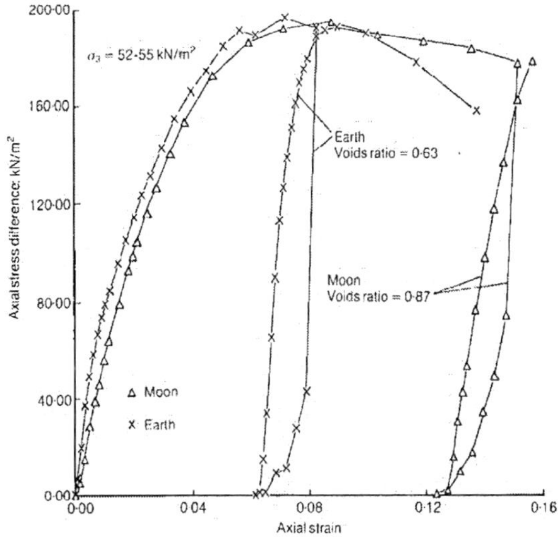

Figure 10. Trial test result of loose lunar soil conducted by Scott

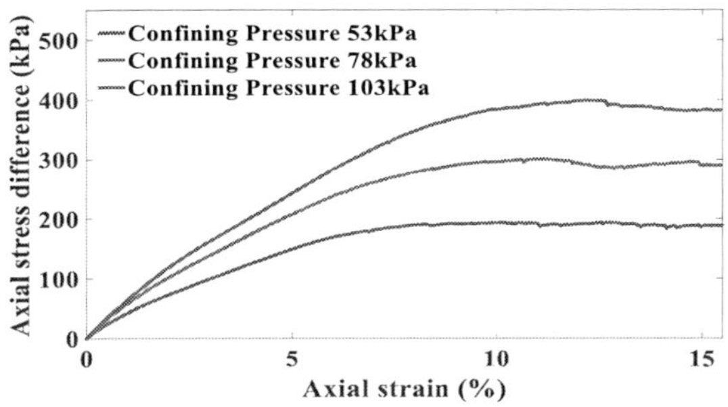

Figure 11. Stress-strain curve under different confining pressures

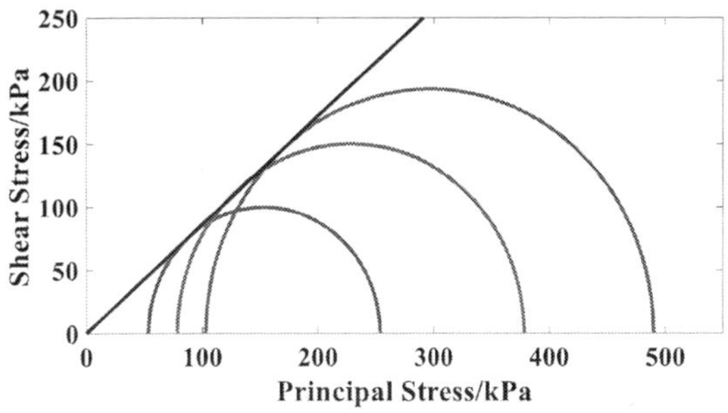

Figure 12. Mohr's circle and shear strength envelope

Table 3. Shear parameters of lunar soil at Apollo landing site

Landing site	Cohesion [kPa]	Internal friction angle [°]
Apollo 11	0.75-2.1	37-45
Apollo 12	1.2-4.8	25-50
Apollo 14	0.03-0.1	35-45
Apollo 15	1.0	47.5-51.5
Apollo 16	0.6	46.5
Apollo 17	1.1-1.8	30-50
DEM simulant	0.33	40.18

During the simulation test, the vertical pressure of the particle to which the platen is subjected and the compression distance under the plate are monitored. The final variation curve of pressure versus compression distance is obtained. Three simulation tests were conducted for each size of plane. As shown in Table 4, the pressure-sinkage characteristics of the calibrated DEM simulant meet the requirements of the lunar regolith guidance values.

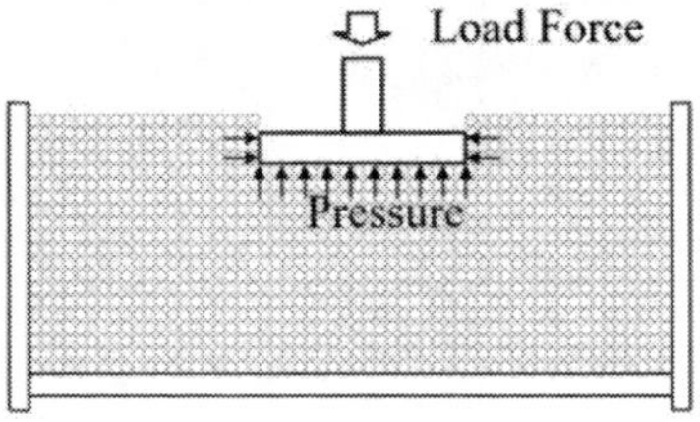

Figure 13. Schematic diagram of pressure plate test

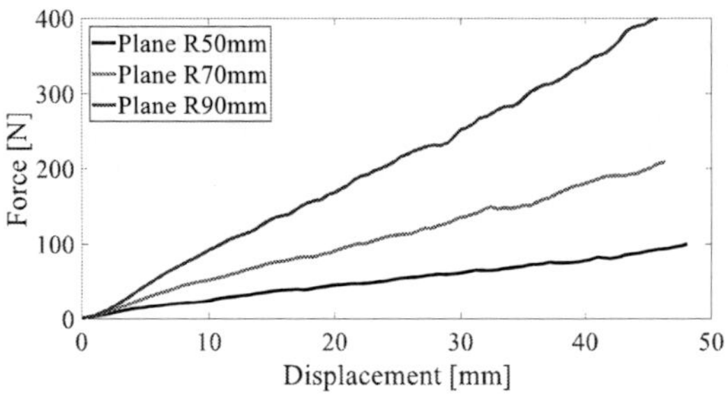

Figure 14. Pressure curve of plate test.

Table 4. Reference values of pressure-sinkage parameters

Environment	n	k_c $[kPa \cdot m^{1-n}]$	k_φ $[kPa \cdot m^{-n}]$
Dry sand	1.1	0.99	1528.43
Lunar regolith	1.0	0.52	820
DEM lunar soil	0.96	0.36	644.45

4.2 Calibration of the wheel stiffness

Literature (Vivake et al., 2009) states that the radial stiffness of the Apollo LRV wire mesh wheel is about 15 kN/m. Based on the FEM model of the wire mesh wheel, the material properties of each component were assigned as in Table 2, and vertical loading tests were performed to verify the vertical stiffness of the wheel.

Based on the vertical loading test, the radial stiffness of the proposed model is about 12.2 kN/m, which is approximate to the Apollo LRV wheel. The model can be considered sufficient to express the physical properties of the Apollo lunar wheel.

5. Numerical simulation of wheel-soil interations

5.1 Soil bin experiment

During the design phase of the Apollo LRV, the U.S. Army Engineer Waterways Experiment Station (WES) conducted a single-wheel soil bin experiment for the LRV wire mesh wheel (Melzer K.J., 1971), as shown in Fig. 17. The lunar soil simulant used for the test was close to the lunar soil samples brought back by Apollo 11 and Apollo 12, and its range of cohesive and frictional properties was considered to satisfy the real lunar soil properties.

In this paper, a virtual soil bin experiment (VSBE) was constructed, as shown in Fig. 18. The soil used in the bin was the calibrated DEM lunar soil. The VSBE system can monitor the wheel condition parameters such as drawbar pull D_P, sinkage z, translational speed v, rotational velocity ω and driving torque M of the wheel. In addition, it is able to adjust wheel load, wheel slip rate, sideslip angle, and camber, etc.

The slip rate s of elastic wheel is expressed as:

where r_e is the wheel effective radius, which is defined as the average value between the radius of the undeflected wheel and the radius of the deflected wheel on hard surface under the action of load W.

In this work, the wheel-soil interaction in a 1/6 gravity environment on the lunar surface is simulated based on the established virtual soil bin experiment. The pull coefficient P_C, torque coefficient T_C, power number P_N, efficiency η and sinkage z are used to evaluate the traction performance of the drive wheel on the lunar surface.

- Pull coefficient P_C

$$P_C = D_P/W \tag{35}$$

- Torque coefficient T_C

$$T_C = M/Wr_e \tag{36}$$

- Power Number P_N

$$P_N = \frac{M\omega}{Wr_e} \tag{37}$$

- Efficiency η

The efficiency η reflects the ratio of the input power converted by the wheel into useful power. the larger the T_E, the greater the share of the output power, the less power is used to overcome the drag losses.

$$\eta = \frac{D_P v}{M\omega} \tag{38}$$

Where D_P is drawbar pull; F_x is the soil thrust and F_r is the soil resistance; W is the load on the wheel; r_e is the wheel effective radius; M is the wheel torque; v is the translational speed; ω is the angular velocity of the wheel.

5.2 Mobility performance simulation

The relationship between the drawbar pull and the slip rate of the wheel is investigated. 25.92 kg, which is about quarter of the Apollo LRV body mass. The gravity was set as g/6. The sideslip angle was set to 0°, and the slip rate was range of 5%-80%. The test procedure is shown in Fig. 19.

As shown in Fig. 20, the pull coefficient increases rapidly with the increase of slip rate, reaches the ideal value at the slip rate of 15%-30%. The distribution of the torque coefficient is similar to that of the pull coefficient. The power number increases rapidly with the increase of the pull coefficient. The sinkage is positively correlated with the slip rate. In summary, the VSBE simulation test results have high accuracy compared with the WES experimental data.

As shown in Fig. 21, the efficiency of the wheels reaches its maximum at pull coefficients between 0.2 and 0.3, corresponding to a range of slip rates from 15% to 25%. The efficiency of the wheels gradually decreases when the pull coefficient is greater than 0.3, which means that the power consumption of the wheels increases greatly. Therefore, the performance at range of 15%-25% slip rate is the most efficient and also has a high mobility, which is considered to be the ideal state of the wheel.

5.3 Effect of sideslip on mobility performance

Sideslip phenomena occurs when the translational velocity direction deviates from the rolling direction of the wheels. The soil exerts a lateral force on the wheel. In this experiment, the wheel load was set to be 25.92 kg, the gravity is g/6, the translational velocity of the wheel is 1 m/s, and the slip rate is 20%. The wheel sideslip angles are set to 3°, 6°, 9°, 12°, and 15°. The test procedure is shown in Fig. 22.

As shown in Fig. 23, as the sideslip angle of the wheel increases, the lateral force of the wheel becomes larger. As we all know, the lateral force of rubber tires on hard ground increases with the increase of sideslip angle and then decreases. However, the elastic wheel on soft ground is different, and the lateral force will always increase as the sideslip angle increases. At the same time, as the lateral force increases, the drawbar pull decreases, because the lateral force takes away part of the ground adhesion force. This is shown by the fact that the traction force of the wheels will be weakened when the lunar vehicle turns forward.

6. Conclusion

In this paper, a wheel-soil coupled modeling and analysis method for wire mesh wheels based on the FBM-DEM method is proposed. Firstly, a lunar soil discrete element model is established, and the contact parameters are calibrated using the triaxial test and pressure plate test; then, a FEM model of a large deformation wire mesh wheel is established in this paper based on the Timoshenko beam element and shell element. On this basis, a virtual soil bin experiment was established. The mobility performance of the large deformation wire mesh wheel was studied under different slip rates and sideslip angles, respectively. The simulation results are in good agreement with the NASA soil bin experiment results. This study will be a guide for the design and simulation analysis of the elastic wheel of the manned lunar vehicle. In the future, we will further study the mobility performance and climbing characteristics of the wheel under different loads and apply them to the simulation of the whole vehicle dynamics.

Figure 15. Wheel vertical loading test

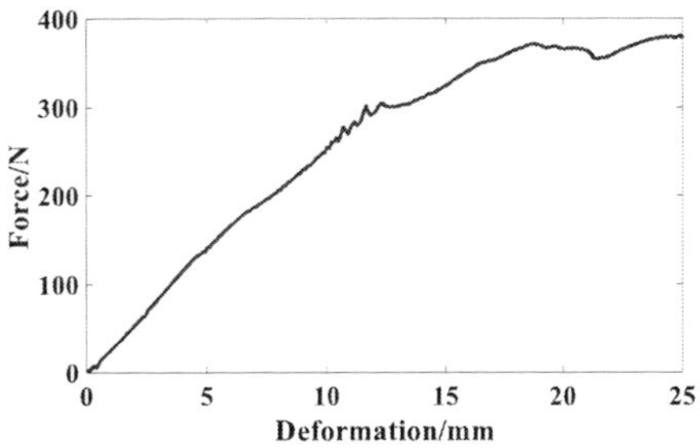

Figure 16. Curve of wheel radial stiffness

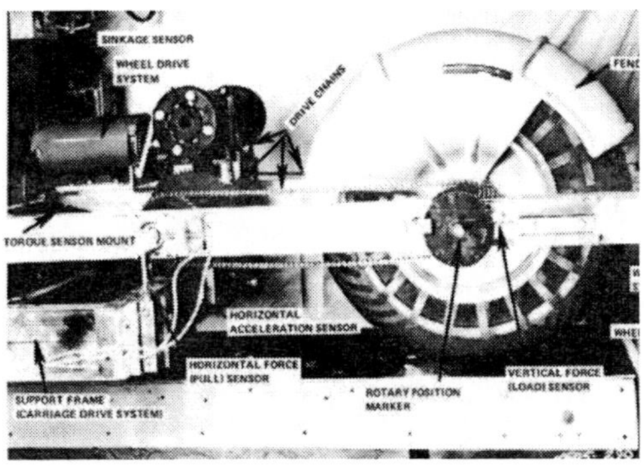

Figure 17. WES soil bin experiment

$$s = \frac{\omega r_e - v}{\omega r_e} \tag{34}$$

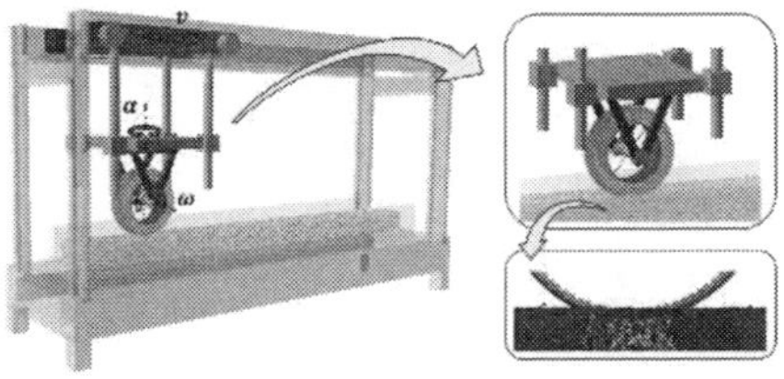

Figure 18. Virtual soil bin experiment (VSBE)

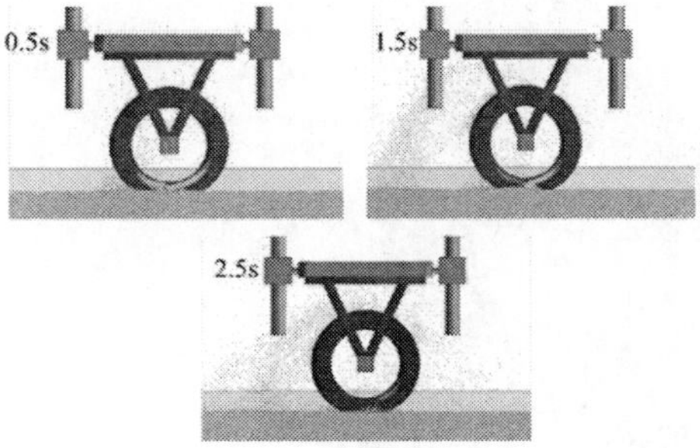

Figure 19. Screenshots of wheel traveling process given a 20% slip condition

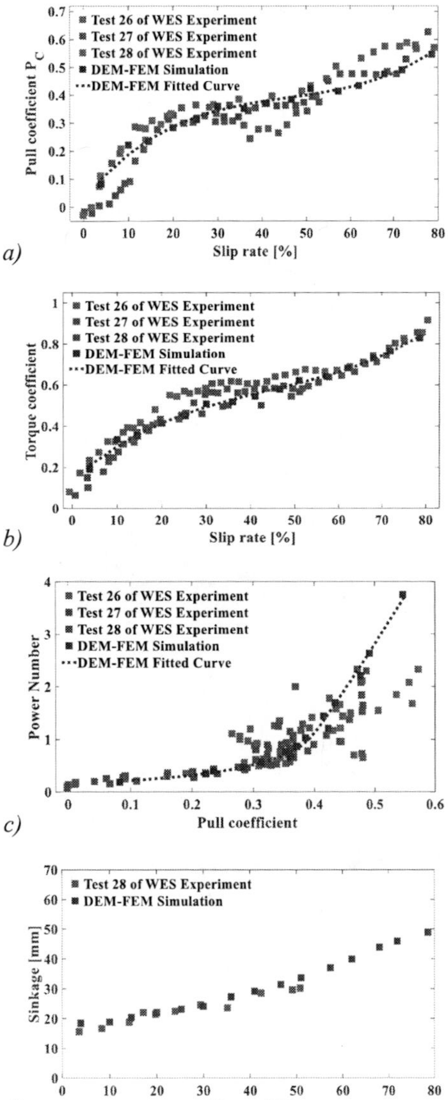

Figure 20. Results of the mobility performance experiment: a) Pull coefficient; b) Torque coefficient; c) Power number; dd) Sinkage

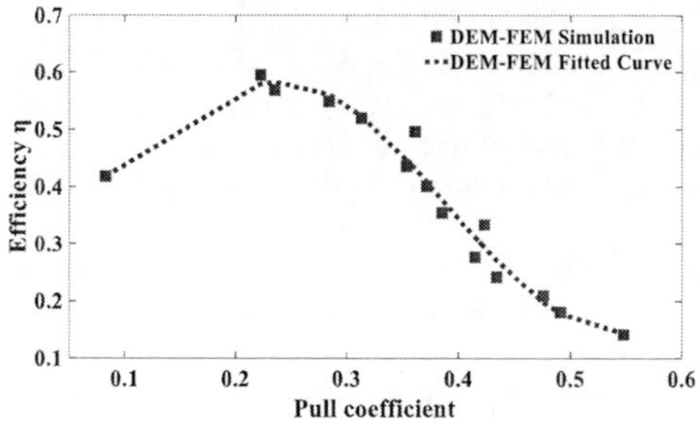

Figure 21. Results of efficiency vs pull coefficient

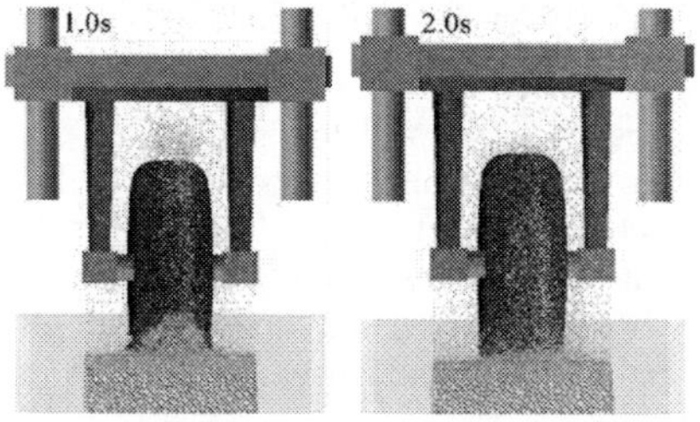

Figure 22. Screenshots of wheel traveling process given a 3° sideslip condition

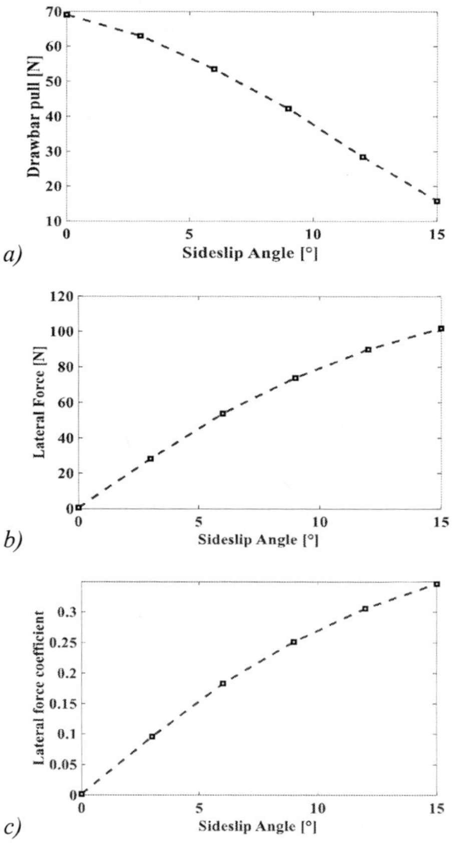

Figure 23. Results of the sideslip characteristics experiment: a) Drawbar pull; b) Lateral force; c) Lateral force coefficient.

7. Acknowledgements

This research was supported by the National Natural Science Foundation of China (11772136). This financial support is gratefully acknowledged.

8. References

Berkelman, P., Chen, M., Easudes, J., et al., 1995. Design of a Day/night Lunar Rover. Pittsburgh: Carnegie Mellon University.

Matrossov, S., Bogatchev, A., Kutcherenko, V.,et al, 2004. Summary the Russian Finnish Cooperation in Robotics. *Proceeding of International Workshop on Adaptive Robots & GSLT Petersburg Russia*, 521-526.

Malenkov, M., Gromov, V., Matrossov, S., et al, 2004. Results of ESA/ESTEC and VN Transmash & RCL Joint Activities Aimed at Planetary Rovers Development. *In Proceedings of the 8th ESA Workshop on Advanced Space Technologies for Robotics and Automation Noordwijk*, 1-10.

Asnani, V., Delap, D., Creager, C., 2009. The Development of Wheels for the Lunar Roving Vehicle. *Journal of Terramechanics*, 46: 89-103.

Lindemann, R.A., Voorhees, C.J., 2005. Mars Exploration Rover Mobility Assembly Design Test and Performance. *In Proceeding of IEEE Conference on Systems Man and Cybernetics Hawaii*, 1-6.

Lee, C., Dalcolmo, J., Klinkner, S., 2006. Design and Manufacture of Full Size Breadboard Exomars Rover Chassis. *9th ESA Workshop on Advanced Space Technologies for Robotics and Automation*, 236-241.

Bekker, M.G., 1956. Theory of land locomotion. The Univ. of Michigan Press, Ann Arbor, Michigan.

Hambleton, J.P., Drescher, A., 2008. Modeling of wheel induced rutting in soils: Interaction. *Journal of Terramechanics*, 45:201-211.

Li, H., Schindler, C., 2013. Investigation of tire-soil interaction with analytical and finite element method. *Mech. Based Des. Struct.*, 41, 293-315.

Fujii, H., Oida, A., Kanamori, H., et al., 2002. Analysis of interaction between lunar terrain and treaded wheel by distinct element method. *14th International Conference of the International Society for Terrain-Vehicle Systems. Vicksburg*, MS USA, 1-10.

Nakashima, H., Fuji, H., Oida, A., 2007. Parametric analysis of lugged wheel performance for a lunar microrover by means of DEM. *Journal of Terramechanics*, 44:153-162.

Nakashima, H., Oida, A., 2004. Algorithm and implementation of soil-tire contact analysis code based ondynamic FE-DE method. *Journal of Terramechanics*, 41:127-137.

Nakashima, H., Oida, A., Kawase, Y., 2005. Simulation of soil-tire system interaction by a combined Finite Element-Discrete Element Method, *15th International Conference of the International Society for Terrain Vehicle Systems 2005*, 709-718.

Mark, M., Frank, V., Bernhard, P., 2015. DEM–FEM coupling simulations of the interactions between a tire tread and granular terrain. Comput. Methods Appl. Mech. Engrg., 289:227-248.

Yang, P., Zang, M., Zeng, H., Guo, X., 2020a. The interactions between an off-road tire and granular terrain: GPU-based DEM-FEM simulation and experimental validation. *International Journal of Mechanical Sciences*, 179:105634.

Yang, P., Zang, M., Zeng, H., 2020b. An efficient 3D DEM-FEM contact detection algorithm for tire-sand interaction. *Powder Technology*, 360:1102-1116.

Giancarlo, G., 2006. Design of planetary exploration vehicles. *8th Biennial ASME Conference on Engineering Systems Design and Analysis*, Torino, Italy.

Fan, X.B., Deng, Z.Q., Gao, H.B., Ding, L., 2014. Design & Analysis of Flexible Wire Mesh Tire for Manned Lunar Roving Vehicle [J]. *Journal of Astronautics*, 35(02): 235-244.

Xiao, W.S., Zhang, Y., 2016. Design of manned lunar rover wheels and improvement in soil mechanics formulas for elastic wheels in consideration of deformation[J]. *Journal of Terramechanics*, 65:61-71.

Santos, K., Campos, A., Oliveira, O., et al, 2015. DEM simulations of dynamic angle of repose of acerola residue: a parametric study using a response surface technique. *Blucher Chem. Eng. Proceed*, 1:11326–11333.

Johnson, K.L., Kendall, K., Roberts, A.D., 1971. Surface Energy and the Contact of Elastic Solids. *Proc. R. Soc. Lond. A*, 324(1558):301-313.

Baran, O., Degennaro, A., Rame, E., Wilkinson, A., 2009. DEM Simulation of a Schulze Ring Shear Tester. *International Conference on Micromechanics of Granular Media*, 409-412.

Carrier, W.D., 1973. Lunar soil grain size distribution. *Moon*, 6(3-4):250-263.

Slyuto, E.N., 2014. Physical and Mechanical Properties of the Lunar Soil (A Review). *Solar System Research*, 48(5):330-353.

Scott, R.F., 1987. Failure. *Geotechnique*, 37(4), 423-466.

Vivake, A., Damon, D., Colin, C., 2009. The development of wheels for the Lunar Roving Vehicle. *Journal of Terramechanics*, 46(3): 89-103.

Melzer, K. J., 1971. Performance of the Boeing LRV Wheels in a Lunar Soil Simulant. Report 2. Effects of Speed, Wheel Load, and Soil.

2034 / A Comprehensive Lumped Parameter Approach for the Dynamic Simulation of Agricultural Tractors in Real Operating Conditions

Marco Polastri[a], Damiano Chiarabelli[a,b], Silvia Gessi[b], Massimo Martelli*[b], Emiliano Mucchi[a], and Pietro Marani[b]

[a]Department of Engineering, University of Ferrara, Ferrara, Italy
[b]Institute of Sciences and Technologies for Sustainable Energy and Mobility, CNR-STEMS, National Research Council, Ferrara, Italy
**Corresponding author: massimo.martelli@stems.cnr.it*

https://doi.org/10.56884/ZLTY2074

Abstract

Vehicle dynamics is of primary importance for the determination of the vertical load on wheels and consequently on their traction capability. This is even more true if the vehicle is travelling on an uncompacted soil and influenced by a variable load applied to the hitch, as it is for a ploughing tractor. In this framework, the authors present a comprehensive lumped parameter approach for performance assessment of agricultural tractors in real operating conditions. The proposed methodology integrates in a modular context different numerical models related to the main subsystems of a modern tractor, i.e., diesel engine, hydro-mechanical transmission, full multibody frame and tire mechanics. In particular, the engine and transmission modules reproduce powertrain characteristics and control strategy, the multibody module characterizes the dynamic behaviour of the vehicle detailing the interaction between the tractor rigid bodies, and the tire model predicts tractive capability and resistance to motion on soft soil. It also provides the possibility to properly reproduce real load cycles and their influence on the vehicle setup. The presented lumped parameter model is intended as a powerful simulation tool, capable of considering a large number of phenomena affecting tractor performance, both in terms of fuel consumption and longitudinal response due to load distribution. The predictive capabilities of the proposed modelling approach are presented by simulating a realistic ploughing operation, focusing on tire-soil interaction. Considering the cascade phenomena from the wheel-ground interaction to the engine, passing through the dynamic of vehicle bodies and their mass transfer, numerical results are presented in terms of tractive capability and its effect on fuel consumption.

Keywords: tire-soil interaction, lumped parameter simulation, agricultural tractor simulation, multibody, powertrain.

1. Introduction

Modern agricultural tractors are the result of an ever-increasing demand for power density, efficiency, reliability, and safety. The growing design complexity combined with the challenging demand for reduction of fuel consumption and pollutant emissions requires the adoption of advanced numerical tools for the design and simulation of modern tractors. Within this framework, many numerical approaches have been presented to analyse the tractor behaviour by focusing on the performance of specific vehicle subsystems. Many papers present lumped and distributed parameter models to explore possible solutions for fuel consumption reduction by introducing novel powertrain design alternatives. For instance, Macor and Rossetti (2013) investigated the performance of a vehicle with a dual stage hydromechanical transmission by means of a lumped parameter model. Zahidi *et al.* (2020) presented a simplified numerical model to evaluate the energy consumption of a tractor equipped with a parallel hybrid engine considering three different farming operations. Many researchers addressed efficiency increase of agricultural tractors by focusing on hydraulic systems. Zardin *et al.* (2018) proposed a modelling and analysis tool for the simulation of a hydrostatic steering system for agricultural tractors to assess the performance of the system and determine the influence of its main design parameters, allowing a proper prediction of the energy dissipation. Gaiola *et al.* (2020) investigated new and flexible solutions to increase the efficiency of the hydraulic system through a lumped parameter method. A wide variety of studies dealt with handling and comfort characteristics by modelling suspension systems and tire-soil interactions. Panetta *et al.* (2015) integrated a full car model and the hydraulic model of the suspension system of an agricultural tractor to understand the influence of several design and control parameters on the dynamic behaviour of the vehicle. He and Jing (2019) proposed an optimized air brake system to improve the braking stability of tractor and semitrailer vehicles based on a lumped parameter approach. Zheng *et al.* (2019) presented a dynamic model for the analysis of the vibrational characteristics of the wheeled tractor system with implement and front axle hydropneumatic suspension. Leanza *et al.* (2021) investigated vertical ground forces exchanged between a rigid wheel and a soft terrain, obtaining valuable information on off-road vehicles behaviour in terms of vibrations, dynamic wheel sinkage, terrain hardness. Despite the availability of a wide range of studies and theoretical dissertations detailing the characteristics of specific tractor components, these detailed approaches are always proposed as single technical insights and never collected in an integrated context.

Some authors proposed broader approaches to combine engine and transmission modules with a vehicle dynamics model, but all the introduced methodologies are based on simplified hypotheses. Hong *et al.* (2009) developed a numerical tool for parallel hybrid electric vehicles to foresee the drivability issues for the hybrid system. It includes a module dedicated to the engine and electric motor, another for the transmission and a third simplified module for vehicle dynamics. In a similar fashion, Macor *et al.* (2017)

presented a model dedicated to the simulation of hybrid transmission vehicles based on a simplified lumped parameter approach. However, the cited works share the common characteristic of being based on very simplified models of the vehicle, which do not provide the chance to consider variable load distribution on the axles and, therefore, the significant effects of variable vertical forces acting on the tires.

In this paper, the authors present a comprehensive lumped parameter approach for the simulation of agricultural tractors in real operating conditions. Developed in the Simcenter Amesim environment, it integrates in a modular scheme the detailed submodels representing the main tractor subsystems involved in the field operations of the vehicle. In particular, the model takes into account engine performance, hybrid transmission, the multibody characteristics of the vehicle and the longitudinal dynamics of tires on soft soil. Therefore, it provides the fundamental possibility to accurately predict the performance of the whole system by considering a large number of design parameters and to delve into the dynamic effects due to the interaction of the subsystems. The potential of the model is presented by applying the proposed methodology to a test-case vehicle and simulating a ploughing operation and the results are presented in terms of fuel consumption, dynamic behaviour, and tractive capabilities. The target of this work is to present the features and the capability of the model through the result of the simulations; an experimental validation will regard a further research development for model parameter tuning, in order to validate this approach as a valuable tool for tractor performance prediction.

The following Section is dedicated to the presentation of the proposed lumped parameter approach of agricultural tractors, by portraying all the modules and the model setup. In the third Section the numerical tool is adopted to simulate a ploughing operation by applying a standardized load cycle, and the main outcomes are discussed to outline the potentiality of the proposed methodology. Finally, the last Section is dedicated to the concluding remarks.

2. Model description

The flowchart representing the overall concept of the introduced lumped parameter approach is depicted in Fig. 1. Each block is dedicated to a functional subsystem, and all submodels constituting the whole model are linked together according to a power port approach. *Internal combustion engine* reproduces the dynamic characteristics of the power unit accounting for a simplified behaviour of the control unit. By simply receiving target speed information, it delivers torque output to the transmission block, providing also fuel consumption data. *Transmission* block reproduces driveline dynamics, offering the fundamental chance to account for hybrid architectures. Based on target speed and gearing information, it delivers torque to front and rear axles. *3-D multibody* block reproduces the dynamic behaviour of the tractor moving in a 3-D domain by discretizing the whole vehicle mass into several rigid bodies linked together by means of constraint submodels. It allows to account for spatial distribution of inertial properties, hence providing a better prediction of the forces acting on chassis and tires than the common 1-D simplified strategy. Since 3-D multibody module also includes the proper reproduction of rear and front hitch, *load cycle* can be applied to the tractor by faithfully reproducing the variable load both in terms of magnitude and direction. Finally, *tire-soil interaction* module calculates the tractive capabilities and torque feedback based on vertical load, soil properties and tire features.

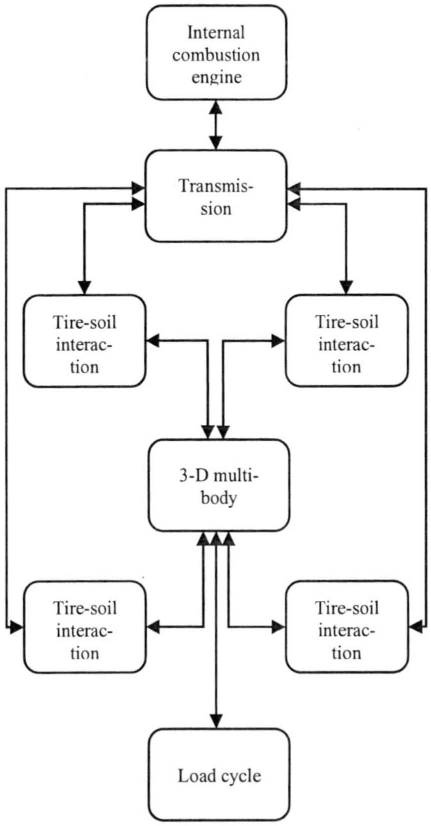

Figure 1. Workflow of the model

It is worth highlighting that this modelling strategy provides the possibility to build a comprehensive digital twin of any agricultural tractor as well as comparing the performances of different design solutions for the same vehicle.

The detailed description of the modules is presented in the following Subsections. The main data of the test-case tractor are shown in Table 1 and are available to the authors thanks to the TASC project (2020) aimed at studying and developing efficiency improvements in agricultural tractors.

Table 1. New Holland T7 main data

Vehicle	Vehicle mass [kg]	10410
	Front tire	540/65 R30
	Rear tire	650/65 R42
Engine	Maximum power [kW]	167
	Rated speed [rpm]	1800

2.1 Internal combustion engine

The engine is modelled according to Zarotti (2010) by defining its operational envelope, shown in Fig. 2 (top). This can be imagined as a region of the torque-speed diagram, where T_e and ω_e are the engine torque and speed, respectively. The envelope area is delimited as follows: on the left by the minimum speed of the engine that allows regularity and on the other side by the maximum speed in order the have the structural integrity. Then the top limit is the maximum torque of the engine (T_M), corresponding to an injection rate (z) of the fuel equal to 1. The curve associated with an injection rate equal to zero, i.e. the minimum torque (T_m), is the brake torque of the engine when no fuel is injected and it is negative. Normally, the only information which is available is the maximum torque curve and the two limits of speed come with it. So that, the minimum torque curve is assumed to be the 25% of the maximum torque at maximum speed, but negative. Since the limit curves are known, the actual engine torque T_e for an intermediate load condition is defined by a series of intermediate curves, defined by the linear combination in Eq. 1:

$$T_e = z \cdot T_M - (z - 1) \cdot T_m \qquad (1)$$

The engine control logic (the 'governor') changes the injection rate to deliver the torque necessary to maintain the requested engine speed (ω_s), by means of a dedicated proportional-integral controller, according to Eq. 2:

$$z = \int_0^t G_2 \left[G_1 (\omega_s - \omega_e) - z + 1 \right] dt; \; 0 < z < 1 \qquad (2)$$

where G_1 and G_2 are the static and dynamic gain, respectively. The schema of this typical engine governor equation is shown in Fig. 2 (bottom). The real time signals of T_e and ω_e are used as inputs to the fuel consumption map, provided by the engine manufacturer, to calculate the real time consumption.

2.2 Transmission

The test-case tractor is a New Holland T7 equipped with a Continuous Variable Transmission (CVT) with four synchronized forward gears (F1, F2, F3, F4) and two reverse gears (R1, R2). In Fig. 3 a schema of the architecture shows the two volumetric units, HP and HM (hydraulic 'pump' and 'motor') and the gearings. In relation to the latter, S1 is the sun of the first stage planetary gear and R is its ring, S2 is the sun of the second stage and P is the planet carrier. The C shaft can be considered as the output of the variable ratio part of the transmission. In a CVT transmission the power coming from the engine is split in two paths, a purely mechanical one, with fixed ratios and a hydraulic one, that allows to adjust the transmission ratio over a continuous range. These two paths are then combined via a double stage planetary gear in order to achieve continuously variable speed. The mechanical path sets the gear ratio and for any of the gears the hydraulic variable input allows to continuously change the transmission ratio. To better understand how the system works, the achievable ratios for any gear are shown in Table 2; they are expressed as ω_{out}/ω_{in}, where ω_{in} is the the engine speed and ω_{out} is the speed of the pinion, so that the final reduction of the bevel gear and the planetary gear inside the wheel are not considered. Due to confidentiality reasons, the listed ratios are normalized to the maximum value and the engine reference speed is omitted.

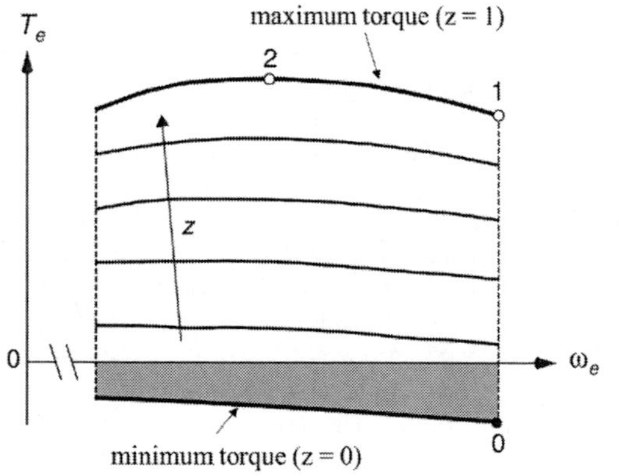

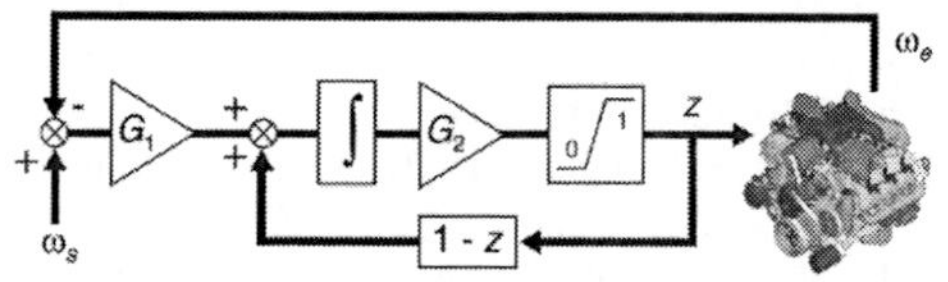

Figure 2. Diesel engine operational envelope (top); governor block diagram (bottom)

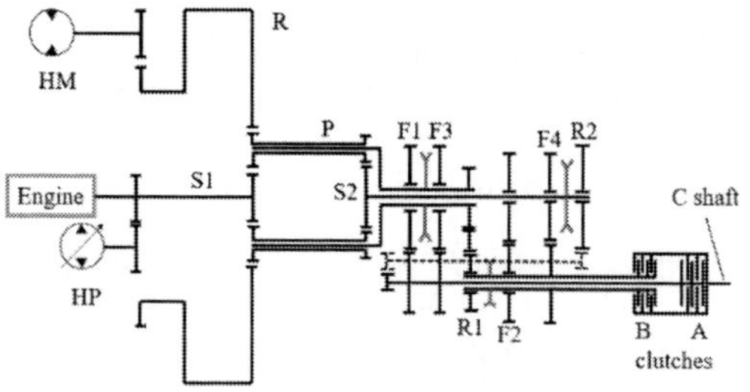

Figure 3. Schematic diagram of the input coupled transmission architecture

As it can be seen from Table 2, considering an acceleration manoeuvre, all the gears are combined to obtain the end of a gear with the same ratio of the beginning of the next gear. Moreover, since the first gear allows a ratio equal to zero, it is possible to achieve the so-called 'power-zero' condition, which is a smooth start from standstill with the engine operating at its maximum power. The transmission consists of an Input coupled architecture. The first hydraulic unit (HP), presenting variable displacement, is directly connected to the engine through a fixed ratio; on the other hand, the second unit (HM) features a fixed displacement. The transmission ratio of the hydrostatic subsystem, defined as $\omega_{mot}/\omega_{pump}$, can then be changed by adjusting a single control variable, the pump displacement setting (α_P). This, in turn, changes the output speed of the overall transmission. The architecture has two outputs: the planet carrier is common between the two planetary stages and the second internal ring. The first output is the odd gears shaft (F1, F3, R1) and the second is the even gears shaft (F2, F4, R2). As it can be seen in Fig. 3, each of these output shafts is connected via a clutch (A or B) to the single output shaft (C) that is connected to the differential and finally to the wheels. The six gears are selected by means of three synchronizers, which are red coloured in Fig. 3. So, the set of control variables needed to achieve a desired vehicle speed is composed as follows:

- Engine speed
- Synchronizers position
- Clutches engagement (A or B)
- Pump displacement setting

Given the engine speed and the desired vehicle speed, the control system implemented in the model generates the command signals for clutches, synchronizers and pump via lookup maps, with additional logic to prevent issues such as 'gear hunting'.

Both quasi-constant and variable vehicle speed manoeuvres can be effectively simulated, for typical field (work) and road (driving) conditions.

Table 2. CVT ratios (normalized) and tractor speed

Gear	Ratio range [-]	Tractor speed (engine @ reference ω_e) [km/h]
R2	-0.259 ÷ -0.175	-11.6 ÷ -7.7
R1	-0.175 ÷ 0	-7.7 ÷ 0
F1	0 ÷ 0.175	0 ÷ 7.7
F2	0.175 ÷ 0.259	7.7 ÷ 11.6
F3	0.259 ÷ 0.538	11.6 ÷ 23.7
F4	0.538 ÷ 1.000	23.7 ÷ 44.7

A standardized open field duty cycle, characterized by constant engine speed and variable load, is applied to the drawbar: the average vehicle speed is prescribed to be within a narrow range (2.5 km/h), according to DLG Standards. An important part of the model is the representation of energy losses in its components. In this architecture, in order to simplify the modelling, all the losses of the 'purely mechanical' part, *i.e.* gears, shafts and their lubrication are represented by a set of selected viscous friction losses located on specific shafts of the transmission. Friction parameters are tuned to match the losses measured experimentally on an equivalent transmission. The largest contribution to the losses comes from the hydraulic components, which endow the transmission with an infinite variation of the ratio at the cost of having some operating points in which the efficiency of the power transmitted through the "hydraulic path is about 70%. The volumetric and mechanical efficiency maps of the hydraulic units are implemented based on the manufacturer technical documentation. This results in a full load driveline efficiency which is about 80 ÷ 84 % at rated engine speed, where the best performance is reached in the speed range of 6 ÷ 12 km/h. These are the desired efficiency targets for CVTs of larger tractors above 100 kW, according to Rhenius (2020).

2.3 3-D multibody

3-D multibody model discretizes the massive body of the vehicle into several rigid parts interconnected through constraint elements. Figure 4 depicts the whole module, making use of solid and dashed arrows to indicate rigid bodies and constraint submodels, respectively.

Driveline represents the main component to which rear hitch, cab, tank, rear tires, and front axle are linked. Steering system is reproduced in detail: front tires are connected to the knuckles, which are connected to the front axle through pivot joints and linked together through the tie rod by means of spherical joints. Steering manoeuvre is carried out thanks to 3-D actuator submodels, which are connected to mechanical or hydraulic jacks. Then, front suspension, characterized by a Panhard architecture is faithfully reproduced with pivot, spherical and actuator joints connecting the spar and the front axle to the driveline block. Likewise, rear hitch is built taking into account the geometry of structural components, mechanical connections and actuation devices, providing the chance to exactly replicate the operating setup of the vehicle during farming operations with attached equipment.

All the connections between body parts are modelled as ideal constraints. However, since this module is based on the 3-D multibody library of Simcenter Amesim software, it is possible to account for more realistic linkage between rigid bodies by adopting yielding constraints, whose stiffness and damping properties can be defined as finite values for the main directions. In this fashion, cabin suspension is modelled making use of custom constraints reproducing the dynamic properties of triaxial bushings and dampers, which are computed as functions of both relative position and velocity for the main directions.

Each rigid body considers both the geometry of the component defined by the coordinates of the communication ports with the constraints, and the inertial properties expressed in the local reference frame defined by the position of the centre of gravity. 3-D body submodels compute the forces exchanged between solid components in compliance with constraint characteristics. As a result, the whole module calculates both the internal forces acting between the components and the external forces acting on the vehicle and transmitted to the wheels.

2.4 Tire-soil interaction

Tire-soil interaction module is responsible for the calculation of the variables related to the tire dynamics. The Amesim block for a single wheel is depicted in Fig. 5: it is connected to the multibody model through the dedicated constraint submodels shown in Fig. 4. Inertial properties of the unsprung mass (spindle + wheel) are set in the *wheel mass* element, and rotary inertia of the wheel is transferred to the tire kinematic model through the dedicated *wheel inertia* submodel, which handles the rotational degree of freedom of the wheel relative to the spindle. Along its axis, it outputs the wheel rotary speed relative to a Galilean frame on 1-D rotary ports and receives the brake and engine torques through those ports. It combines those with the torque along its axis developed by the tire to compute the rotary acceleration and update the rotary speed.

The *tire kinematic model* computes all kinematic elements of the centre of tire contact at the tire-ground interface. According to Fig. 6, these elements are computed with reference to the kinematics of the real wheel centre A_2. Furthermore, it allows the transfer of complete motion from tire-ground contact point B to the wheel centre. Vertical tire stiffness and damping are defined through a *spring and damper* submodel: different kinds of spring and damper can be implemented, *i.e.* linear, non-linear, visco-elastic, from table. This submodel plays the role of interface between 3-D multibody vehicle model and global tire model.

The *tire belt model* allows the computation of characteristic inputs of tire models according to Pacejka (1996), and Pacejka and Besselink (1997). Characteristic frames and geometric elements of tire kinematic and tire belt model are summarized in Fig. 6. *Tire belt model* also allows the transfer of elements relative to the road model and used in the tire kinematics model. On the other hand, it allows the transfer of absolute velocity of the wheel centre expressed in the Galilean frame and the absolute position of centre of wheel A_2 to the road model. Since tire forces correspond to only the longitudinal and lateral forces, longitudinal slip K is calculated according to Eq. 3:

$$K = \frac{R_{dyn}\left(\frac{\vec{\Omega}_{wheel}}{R_0} \cdot \vec{J_2}\right) - V(\vec{B}_{dyn})\big|_{R_0} \cdot \vec{X_w}}{V(\vec{B}_{dyn})\big|_{R_0} \cdot \vec{X_w}} \quad (3)$$

R_{dyn} is the dynamic rolling radius, $V(\vec{B}_{dyn})\big|_{R_0}$ is the absolute velocity of the tire contact centre computed with dynamic rolling radius from the absolute velocity of the wheel centre, $\vec{\Omega}_{wheel}/R_0$ is the absolute rotary velocity of the wheel.

Tire model generates the contact force at the tire-soil interface. Based on a modified Brixius model for the longitudinal behaviour (Zoz, 1979), this submodel predicts tractive capability and resistance to motion on soft soil focusing on modelling the resulting rolling resistance and tractive capacity of the tire.

It also implements a simplified approach for the lateral behaviour proposed by Dugoff *et al.* (1970) to set up the vehicle on uneven ground (ruts and bumps) and analyse all the effects of the resulting uneven load repartition on the longitudinal response. However, the lateral response does not take into account the soft soil nature of the ground for lateral phenomena effects like bulldozing, therefore the model prioritises the longitudinal behaviour of the tire dynamics point of view.

Since the interaction between tire and soft soil is extremely complex, the tire model implements a simplified method based on few easily obtainable tyre parameters: tire dimension and a measure of soil hardness. The ground features are described simply by defining the cone index parameter in kPa. Then, longitudinal behaviour is calculated according to Brixius (1979).

Road grip model is an adherence generator, an interface between ground and tire implemented in order to obtain the local road grip. Despite its usage is not mandatory, it enables to set a variable road grip in case the parameter is not constant and equal to 1 for all the wheels.

Finally, *road model* assesses the contact between the tire and the road. It queries the road with the wheel centre position and velocity and outputs the height, normal and apparent vertical road speed (rate of altitude change).

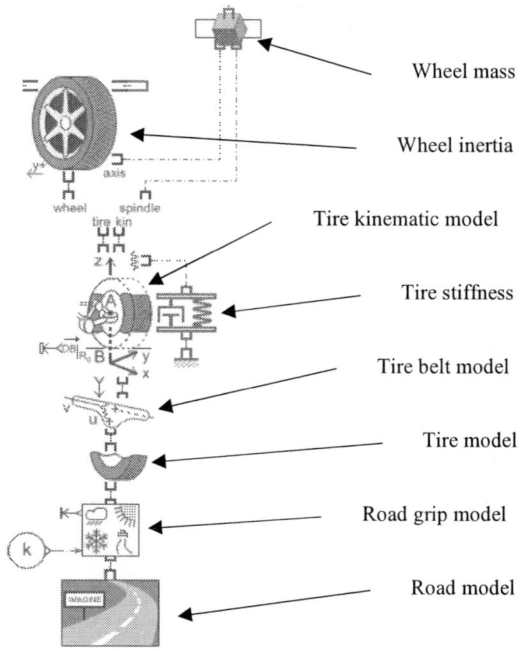

Figure 4. Tire-soil interaction module

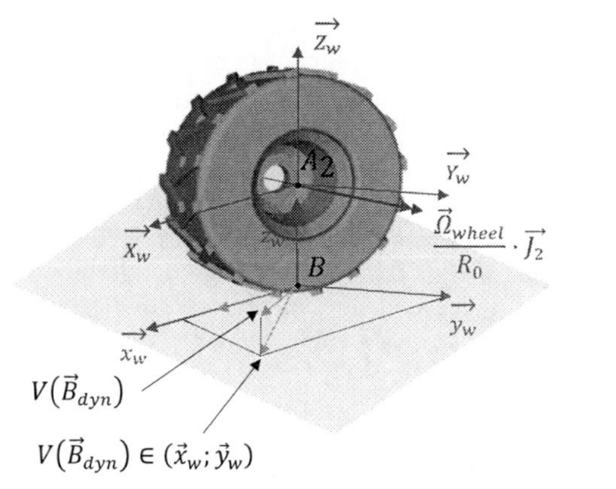

Figure 5. Reference frames of the tire kinematic model

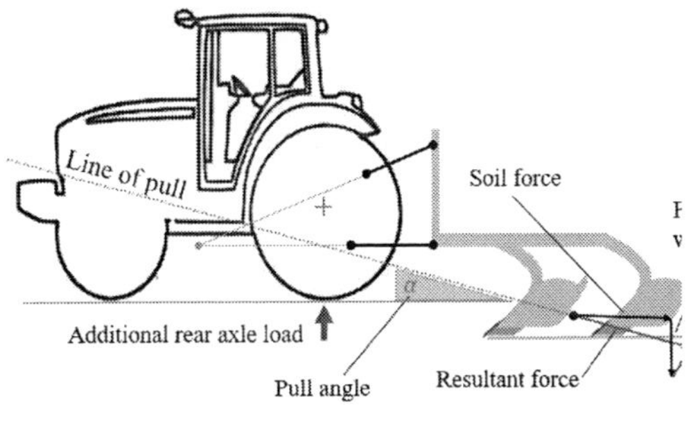

Figure 6. Definition of the pull force due to a ploughing work

$$\begin{cases} F_a = F \cdot \cos(\alpha) \\ F_b = F \cdot \sin(\alpha) \end{cases} \quad (4)$$

3. Simulations

3.1 Test conditions

Model capabilities are tested by applying standardized load cycles for tractors, the 'PowerMix' defined by DLG Standards. These cycles are time-based curves that reproduce suitable drawbar pull forces for specific work tasks, *e.g.* 'plough' and 'cultivator', some of them including also a PTO (Power Take Off) and a hydraulic power contribution. The 'shape' of the load is fixed and has a standard duration; the magnitude is determined by the size of the tractor because the cycle is scaled with the power of the engine in order to saturate the engine power when cycle peaks occur. These tests are normally performed on a chassis dynamometer or by using a load car on a flat track to standardize the test conditions and avoid all the variations that could occur on the field. This work is primarily focused on the '100% Plough' test. The main simplifications introduced by the DLG methodology are related to the direction of the load force and the possible variation of the soil type. These aspects affect the traction capability, and this is exactly what the comprehensive model developed within this work can evaluate. As previously described, the multibody of the tractor includes also the three-point hitch that determines, together with the shape of the plough, the direction of the force transmitted to the tractor body. As it can be seen from Fig. 7, derived from the work of Rhenius (2020), the resultant load force felt by the tractor is inclined and tends to develop an additional load on the rear wheels while unloading the front axle. Since the target of this work is to show the capability of the model and the detailed geometry of a real plough was not available, the angle of the resultant force with respect to the horizontal, *i.e.* the pull angle α, was estimated and applied to the centre of gravity of the rear ballast. So that the force coming from the DLG cycle (F) is split in a vertical component (F_b) and a horizontal component (F_a), as follows:

It is important to observe that the three-point hitch controls are not considered. Normally there is a logic aimed to keep the pull force almost constant thanks to a force sensor based on the pressure measured on a hydraulic valve. If the field soil density increases during operation, the system lifts the implement reducing the tillage depth and the pull on the tractor. According to Rhenius (2020), even if such a system is present there are still some variations on the pull force. Therefore, the three-hitch point controls have not been considered and the DLG cycle variable force has been kept as a reference. Using this load configuration, a suitable way to present the capabilities of the model is the simulation of different settings of the tractor in terms of ballasting and see how the tire-soil interaction is influenced and its impact on fuel consumption. Three settings are considered, increasing the front ballast mass in 500 kg steps, as shown in Table 3.

The simulated ploughing work cycle has a duration of 320 seconds; the pull force has a peak value of 85 kN, to have an idea of its profile it is sufficient to look at any of the quantities calculated during the cycle, *e.g.* Fig. 9.

3.2 Results

For each test condition, the vehicle performance has been characterized for efficiency, in terms of average fuel consumption and average driveline efficiency, and traction capability, using the travelled distance as a simple yet effective measure (Table 3).

Table 3. Fuel consumption and performance

Front ballast mass [kg]	Average fuel consumption [l/h]	Average driveline efficiency [%]	Travelled distance [m]
1064	30.30	79.47	528.81
1500	30.42	79.59	530.25
2000	30.56	79.72	531.75

The driveline efficiency (η_{drv}) is calculated as in Eq. 5, where P_{out} is the output power at the C shaft (Fig. 3) and P_{in} is the input power at the engine shaft:

$$\eta_{drv} = \frac{P_{out}}{P_{in}} \quad (5)$$

Moreover, the traction capability corresponds to the ability to convert the wheel torque into an effective traction force at the wheel-ground contact, which is directly related to the wheel longitudinal slip (S_l) and calculated by the tire model as reported in Eq. 3. This variable can be simplified as follows:

$$S_l = \frac{\omega r - v}{v} \quad (6)$$

where ω is the wheel angular velocity, r is the rolling radius of the tire and v is the vehicle speed. The formulation of the longitudinal slip given by Eq. 6 is provided by He *et al.* (2020).

The resulting longitudinal slips are shown in Fig. 8. Both the rear axle and the front axle benefit from the front ballast increasing mass due to the decreasing power dissipation for tire slipping, which leads to an increasing driveline efficiency. Conversely, the fuel consumption increases, as expected, because of the increase in the total mass of the vehicle.

As it can be seen in Table 3 a more efficient traction results in a longer travelled distance but also a higher load going back through the transmission and up to the engine, because of that the transmission works in slightly different operating conditions.

It is also worth noting how the multibody reacts to different settings, showing the actual interaction of external loads with the rigid bodies of the tractor and as a consequence, with the tires. Fig. 9 shows the displacement of the axles in the three cases and, as expected, the axle centre of gravity moves lower on the front, and vice versa on the rear, as the ballast mass increases. It can be noted that another way to compare these settings was at equal work done, *i.e.* considering the same distance of tilled soil with the same pull force. In this case it would have been necessary to model a resistance force as a function of speed; therefore, the more practical standardized DLG cycle has been preferred. Finally, it needs to be remarked that the target of this work is to show the potential of the model and this is achieved thanks to the coherent results of the simulations, nevertheless it can be noted that the variations of all the variables are very small, actually they are below 1%. As a first step, the quality of the model is evaluated by comparison of the simulations results and in this sense, the model responds in a coherent way to the changing of physical parameters and the absolute values of fuel consumption are realistic. Nonetheless, an experimental validation of the simulation results is needed in order to make this model a valuable tool to predict vehicle performance in real operating conditions.

The software provides an animation tool that makes it easier to visualize the multibody assembly and see the advancement of the simulation, Fig. 10 shows the graphic result of this modelling.

4. Conclusions

The present paper introduces a comprehensive lumped parameter approach for the simulation of agricultural tractors under real farming operations. Developed in the Simcenter Amesim environment, it integrates in a modular architecture the detailed models of the main vehicle subsystems involved during the vehicle functioning. In particular, the internal combustion engine is modelled as a functional torque generator. Based on the engine characteristics map, it implements a calibrated controller to replicate the control logic, in order to compute both the delivered power and fuel consumption information. Transmission block replicates the functioning of the hybrid driveline, detailing both the mechanical and the oil hydraulic paths. It includes a control unit which generates the command signals for clutches, synchronizers and pump via lookup maps based on engine speed and target velocity.

A key feature of the model is the introduction of the 3-D multibody module, a model that discretizes the mass of the vehicle in different solid bodies connected through proper constraint submodels. It provides the possibility to calculate the variable forces applied to the vehicle and then the proper distribution of the vertical forces acting on the wheels. Finally, tire-soil interaction module computes tire kinematics and dynamics, providing information regarding the tractive capability of the vehicle. Implementing a simplified approach for soft soil ground based on few easily available parameters, it enables to assess the tractive performance under representative operating conditions.

By integrating several modules describing in detail the characteristics of many subsystems, the proposed modelling approach represents a very powerful numerical tool since it covers a large number of phenomena and design parameters in the evaluation of vehicle performance. The potential of the proposed methodology is outlined by applying the Amesim model to a sample tractor in order to simulate a ploughing operation on soft and irregular soil, achieving valuable results in terms of tractive capabilities and overall performances.

A standardized DLG ploughing cycle is applied to the model and three values for the front ballast mass are considered. Results show a slight dependence of the tractor performances on front weight. In particular, both axles benefit from the front ballast increasing mass, which leads to a sensible decrease of the power dissipation due to tire slipping. As a consequence, an increasing driveline efficiency is observed. On the other hand, a slight increase of the fuel consumption is observed, and it is due to the increase in the total mass of the vehicle.

Tractive performance and fuel consumption are strictly related to the motion of the vehicle and to the resulting vertical loads. The model provides the chance to estimate the dynamic response of the vehicle, timely detailing the load distribution during the operation on uneven soil and variable load.

Generally, considering the level of detail of the modelled system combined with the straightforward implementation provided by the modular architecture, the proposed modelling approach reveals to be a powerful and effective simulation tool for agricultural tractors.

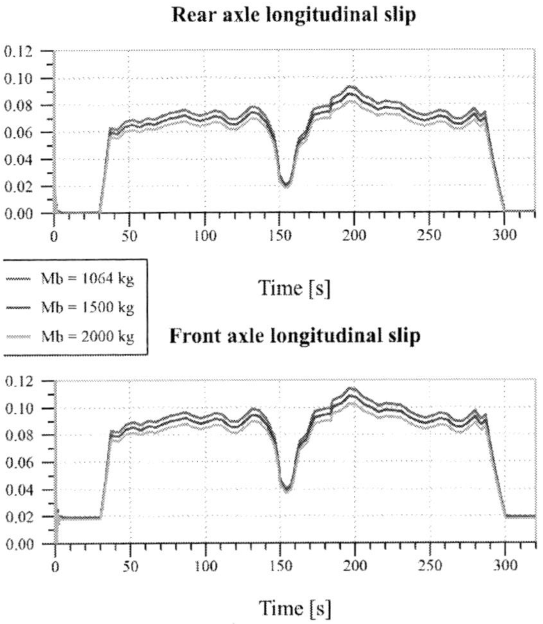

Figure 7. Effect of the front ballast mass (M_b) on the tire longitudinal slip

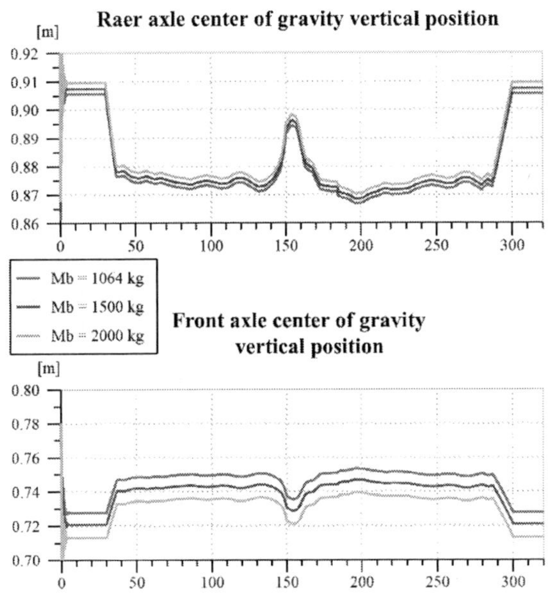

Figure 8. Effect of the front ballast mass (m) on the axles position

Figure 9. Partial view of the 3D multibody model in the Amesim animation tool

5. Nomenclature

A_2	*Wheel centre*
B	*Centre of tire contact*
F	*Pull force magnitude*
F_a	*Pull force horizontal component*
F_b	*Pull force vertical component*
G_1	*Engine governor static gain*
G_2	*Engine governor dynamic gain*
J_2	*Wheel rotary inertia*
K	*Wheel longitudinal slip*
M_b	*Front ballast mass*
P	*Mechanical power calculated in a point of the transmission*
r	*Tire rolling radius*
R_0	*Tire free radius*
R_{dyn}	*Tire dynamic rolling radius*
S_l	*Wheel longitudinal slip (ISTVS standards)*
T_e	*Engine torque*
T_M	*Maximum engine torque*
T_m	*Minimum engine torque*
v	*Vehicle velocity*
$V(\overrightarrow{B_{dyn}})\vert_{R0}$	*Absolute velocity of the tire contact center computed with dynamic rolling radius from the absolute velocity of the wheel center*
x_W, y_W, z_W	*Tire frame*
X_W, Y_W, Z_W	*Wheel frame*
z	*Engine injection rate*
α	*Pull angle*
α_P	*Pump displacement setting*
η_{drv}	*Driveline efficiency*
ω	*Wheel angular velocity*
ω_e	*Engine speed*
ω_s	*Desired engine speed*
Ω_{wheel}/R_0	*Wheel absolute rotary velocity*

6. References

Brixius, W.W., 1987, Traction Prediction Equations for Bias Ply Tires, American Society of Agricultural Engineers.

DLG Standards, Deutsche Landwirtschafts-Gesellschaft – German Agricultural Society, https://www.dlg.org/en/agriculture/tests/dlg-powermix

Dugoff, H., Fancher, P. S., Segel, L., 1970. An Analysis of Tire Traction Properties and Their Influence on Vehicle Dynamic Performance, SAE Technical Paper, DOI: 10.4271/700377.

Gaiola, A., Zardin, B., Casoli, P., Borghi, M., Mazzali, F., Pintore, F., Fiorati, S., 2020. The Hydraulic Power Generation and Transmission on Agricultural Tractors: Feasible architectures to reduce dissipation and fuel consumption - Part I, E3S Web of Conferences, 197.

He, R., Jing, Z., 2019. Study on braking stability of commercial vehicles: An optimized air brake system. Advances in Mechanical Engineering. DOI:10.1177/1687814019848593.

He, R., Sandu, C., Mousavi, H., Shenvi, M.N., Braun K., Kruger, R., Els, P.S., 2020. Updated Standards of the International Society for Terrain-Vehicle Systems, Journal of Terramechanics, ISTVS, Elsevier.

Hong, J., Kim, S., Min, B., 2009. Drivability Development Based on CoSimulation of AMESim Vehicle Model and Simulink HCU Model for Parallel Hybrid Electric Vehicle. SAE Technical Paper. DOI:10.4271/2009-01-0725.

Leanza, A., Mantriota, G., Reina, G., 2021. On the vertical dynamics modelling of rigid wheels on soft soil. Paper presented at the Proceedings of the 20th International and 9th Americas Conference of the International Society for Terrain-Vehicle Systems, ISTVS 2021.

Macor, A., Rossetti, A., 2013. Fuel consumption reduction in urban buses by using power split transmissions. In: Energy Conversion and Management 71, pp. 159–171. DOI: 10.1016/j.enconman.2013.03.019.

Macor, A., Benato, A., Rossetti, A., Bettio, Z., 2017. Study and Simulation of a Hydraulic Hybrid Powertrain, Energy Procedia, 126, DOI:10.1016/j.egypro.2017.08.279.

Pacejka, H.B., 1996. The tyre as a vehicle component, XXVI FISITA Congress, Prague, 1-19.

Pacejka, H.B., Besselink, I.J.M., 1997. Magic formula tyre model with transient properties, Proceeding of the 2nd Int. Colloquium on Tyre Models for Vehicle Dynamic Analysis, Berlin, 234-249.

Panetta, G., Mancarella, F., Borghi, M., Zardin, B., Pintore, F., 2015. Dynamic Modelling of an Off-Road Vehicle for the Design of a Semi-Active, Hydropneumatic Spring-Damper System., Proceedings of the ASME 2015 International Mechanical Engineering Congress and Exposition. Volume 4B: Dynamics, Vibration, and Control. Houston, Texas, USA.

Renius, K.T., 2020. Fundamentals of Tractor Design, Springer Nature Switzerland.

Simcenter Amesim 2020.2, Siemens Industry Software NV, 2022.

TASC – Trattrici Agricole Smart & Clean, 2020. Italian regional project, POR FESR 2014-2020, https://www.tascproject.eu/en/

Zahidi Y., El Moufid M., Benhadou S., Medromi H., 2020. An Assessment of Low-Cost Tractor Motorization with Main Farming Implements. World Electric Vehicle Journal. DOI:10.3390/wevj11040074.

Zardin, B., Borghi, M., Gherardini, F., Zanasi, N., 2018. Modelling and Simulation of a Hydrostatic Steering System for Agricultural Tractors. Energies, 11, 230. DOI: 10.3390/en11010230.

Zarotti, G.L., 2010. Trasmissioni idrostatiche - nozioni e lineamenti introduttivi (Seconda edizione). Quaderni tematici n. 5, IMAMOTER-CNR.

Zheng, E., Cui, S., Yang, Y., Xue, J., Zhu, Y., Lin, X., 2019. Simulation of the Vibration Characteristics for Agricultural Wheeled Tractor with Implement and Front Axle Hydropneumatic Suspension. Shock and Vibration. 1-19. DOI:10.1155/2019/9135412.

Zoz, F.M., Brixius, W.W., 1979. Traction prediction for agricultural tires on concrete. Summer Meeting of ASAE and CSAE.

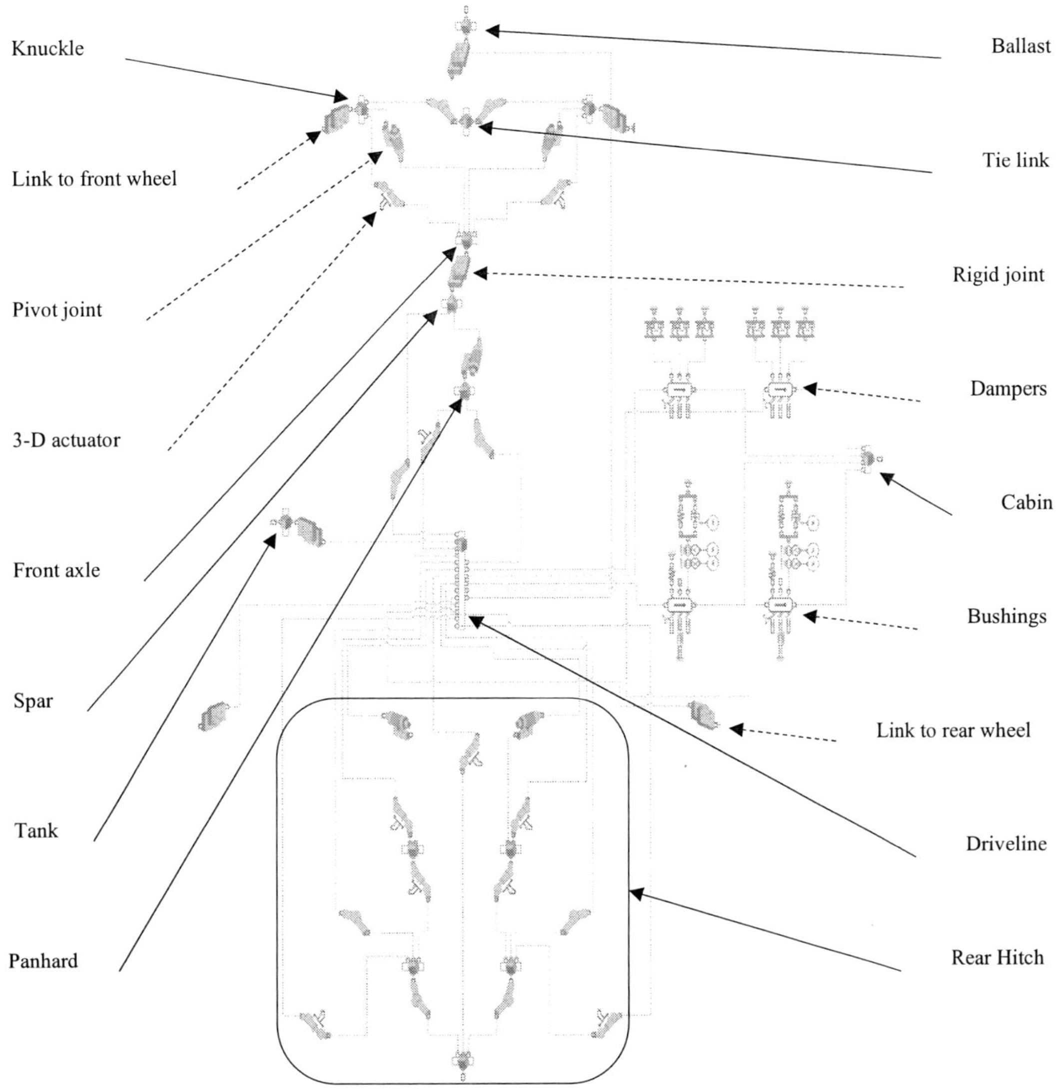

Figure 10. 3-D multibody model

2149 / Investigation of the Shear Stress Dynamics on Silty Loam Soil and Measurement of Traction-Wheel Slip Relationship of a Tractor Tire

César Andrés Arévalo-Montañaa* and Stefan Böttinger
Hohenheim University, Department of Agricultural Engineering (440), Stuttgart, Germany
**Corresponding author: cesar.arevalo@uni-hohenheim.de*

https://doi.org/10.56884/WVQN3391

Abstract

The Hohenheim Tire Model (HTM) was developed and validated at the University of Hohenheim. It simulates the specific behavior of large-volume agricultural tires and is used in conjunction with multi-body models of tractors and other agricultural machines. Special focus is placed on driving dynamics and comfort. Currently the HTM is being expanded for implementation on soft soils. Therefore the soil's shear response using Bevameter tests and the relationship traction-wheel slip using a single wheel tester are investigated. In field tests, the effect of shear rate on soil mechanical parameters was examined. The shear device of the Bevameter was used to conduct experiments in a soil bin with silty loam at eight shear rates and five vertical loads. The Mohr-Coulomb failure criterion was calculated and the measured values were fitted to the Wong-Preston and the Janosi-Hanamoto approaches. The shear stress-shear displacement curves show two different behaviors with respect to shear rate. At high shear rates, the shear stress reaches a peak and then drops to a residual value, which is typical for cohesive soils. At low shear rates, the trend of the curves is exponential, which is typical for granular soils. The cohesion and angle of internal friction values are comparable to those from literature. The angle of internal friction has no correlation with shear rate while the cohesion shows a low correlation. Traction-wheel slip curves for the tractor tire 480/70 R24 with two tire loads (17 and 22 kN) and at 2 km·h-1 were measured under identical conditions as the shear experiments. For the higher tire load, there is a tendency toward a larger tractive force. The curves of both wheel loads have equal gross traction coefficients. For the prediction of the traction force, the parameters obtained in the shear tests and the method of Wong and Preston-Thomas are used.

Keywords: tire-terrain interaction, shear behavior, bevameter, tractor tire, traction-wheel slip curve

1. Introduction

The performance of agricultural machines is predicted with the use of simulation and thus their operation can be optimized with less expenditure of time and resources than with real trials. This can be used for both, operation on the field and during the development process. When comparing the simulation of agricultural machinery to that of passenger cars or trucks, several differences must be considered: tire behavior, the absence of suspension systems, the need for sufficient traction in the field, the effect on soil compaction, and the versatility for both on-road and off-road operation. The primary focus of this work is the tire-terrain interaction.

At the Institute of Agricultural Engineering of the University of Hohenheim, numerous research works have been conducted with the goal of simulating the specific behavior of agricultural tires. One outcome of these investigations is the Hohenheim Tire Model, which simulates the dynamic three-dimensional tire behavior on firm ground (asphalt). The model has been used in the research of vehicle dynamics and ride comfort. Witzel extended it from a three-dimensional single-point contact model to a three-dimensional spoke model (Witzel, 2018a, 2018b, 2015). This last version enables the modeling of obstacle passages and represents an advance toward the simulation of soft soil interaction. The Hohenheim Tire Model reproduces the tire through multiple nonlinear Voigt-Kelvin elements (spokes) and uses a stick-slip contact to depict the transfer of forces between the ground and the contact patch. The parameterization is performed in the single wheel tester and in the flat belt test rig of the Institute of Agricultural Engineering at the University of Hohenheim.

The Hohenheim Tire Model is currently being expanded for implementation on soft soil and for this a soil model is necessary that provides the deformation of the soil profile and the stress distribution in the tire contact patch. Since the Hohenheim Tire Model works close to real time, finite element and discrete element based methods for modeling the soil response are discarded due to the high computation requirements associated with these methods and therefore semi-empirical models are considered for the calculation of soil forces (He et al., 2019; Sandu et al., 2020).

The first step in the further development of the Hohenheim Tire Model and the development of a soil model is the description of the soil response to shear forces. This paper reports on the determination of soil mechanical parameters with a Bevameter and of tire-soil characteristics (traction force- wheel slip curve) with a single wheel tester on a soil bin, which will serve as a test surface to parameterize and validate a future soil model. The effect of shear rate on the soil parameters is investigated and the traction force is predicted using the shear parameters.

2. Materials and Methods

2.1 Bevameter

The tractor-mounted Bevameter (Figure 1) is composed of a hydraulic linear shear device and a hydraulic pressure-sinkage device and is used in the determination of shear stress-shear displacement and pressure-sinkage relationships. The shear device was originally built by Barrelmeyer (Barrelmeyer, 1996) but modified for the use in this project and the sinkage device is a new assemble attached to it.

Figure 1. Bevameter consisting of linear shear device and pressure-sinkage device

2.1.1 Linear shear device

The main element of the shear device is the shearing carriage, to which the shear plate (rectangular shear plate of 0.05 m^2, with grousers) is attached (Figure 2). The shearing plate penetrates into the ground and on it are placed the force sensors. The shearing carriage is supported with bearings on guide rails and can be started from two positions. The first position follows a circular path, which is followed by a straight rail section parallel to the ground. Thus, the shear plate is loaded incrementally as it penetrates the soil to shear it. This type of test is called without preload. In the second start position, the shearing carriage is already in the straight rail section and parallel to the ground. The shear plate penetrates the ground when the unit is lowered and so a preload takes place. Then the hydraulic cylinder moves the shearing carriage only horizontally and the shearing of the soil starts at zero speed. The results in this work were obtained in shear tests with preload.

The drive of the hydraulic cylinder, which moves the shear carriage back and forth, is carried out directly from the hydraulic connection of the tractor. With this drive and with a 3-way flow control valve, shearing speeds up from approx. 0.015 m·s^{-1} and with a high fine adjustment can be realized.

For recording the forces on the shear plate, the shear device has two octagonal elements with strain gauges connected in Wheatstone bridge. These force sensors can separately record the vertical and horizontal, *i.e.,* shear forces. The voltages from the octagonal elements are recorded with an amplifier (MGC from HBM Company). For the measurement of the shear, displacement a laser sensor with measuring range of 750 mm is used (Fa. Micro-Epsilon optoNCDT 1710-750). An angle sensor (Elobau model 424A) was installed to register the vertical movement of the shearing carriage or its support frame during the test. Sensing this movement is necessary to determine the validity of the test, since excessive vertical motion of the shearing carriage can distort the shear force. The analog signals of all sensors are acquired in a National Instruments data acquisition device (NI USB 6216), digitized, and then transferred to a laptop running DASYLab software.

(1) Shearing carriage
(2) Force sensors
(3) Guide rails
(4) Hydraulic cylinder
(5) Weight-charged frame
(6) Three-point frame

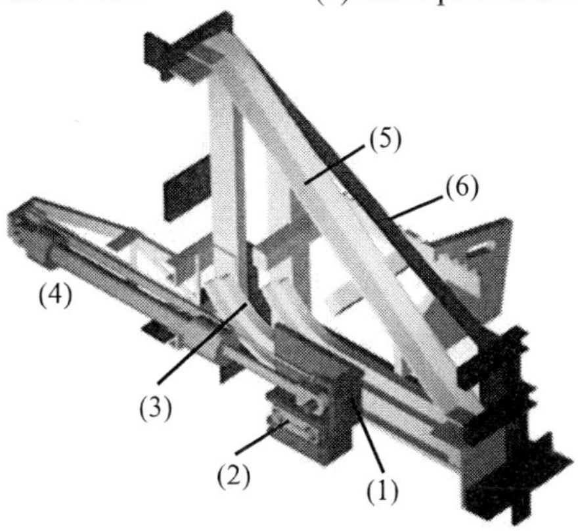

Figure 2. Cutaway drawing of the linear shear device (Barrelmeyer, 1996)

2.1.2 Pressure-Sinkage Device

The pressure-sinkage device consists of a double-acting hydraulic cylinder (50 cm travel) driven by the tractor hydraulics, a load cell and a displacement transducer. One of two circular pressure plates with areas 0.05 m^2 and 0.025 m^2 can be installed. A load cell (Type K-12, Lorenz Messtechnik) records the vertical force between the pressure plate and the piston rod of the hydraulic cylinder. The sinkage is measured with a displacement transducer LWG450 (Company Novotechnik) parallel to the cylinder. An inductive sensor senses the contact between the frame of the device and the soil surface. This measurement is necessary to stop the test as soon as the hydraulic cylinder lifts the device together with the tractor. Namely, from this moment the measurement is no longer valid. The signals from the sensors of the pressure-sinkage device are also received in the DaqBook NI USB 6216 and processed with the DasyLab software.

2.2 Single wheel tester

With single wheel tester, the behavior of agricultural tractor tires on any soil type can be researched. The main components are the towing tractor (MB-Trac 1500) and the measuring frame in which the measuring wheel is installed. The following operating conditions of the measuring wheel can be set: Drive power up to 52 kW, slip angle up to 16°, camber angle up to 20° and wheel load up to 40 kN. The maximum tire diameter is 2 m. The variables that can be measured with the single wheel tester are: the tire forces and torques by a 6-component dynamometer, the wheel slip by means of a fifth wheel for the driving speed and a rotary incremental encoder for the angular velocity of the measuring wheel. The distance between the wheel hub and the soil is recorded using a laser sensor.

Figure 3. Single wheel tester (Schlotter, 2006)

2.3 Experimental area

The experiments of this work were carried out at the soil bin (110 m long and 4 m wide) of the Heidfeldhof Experimental Station of the University of Hohenheim. The soil type is silty loam. This experimental area was prepared as follows: Single pass of cultivator to a depth of 40 cm. Prior to the Bevameter and to the single wheel tester trials, a rotary harrow to a depth of approx. 25 cm was used. This was followed by a pass with a farm roller to firm and smooth. Soil moisture and penetration resistance were measured before each test with the single wheel or Bevameter.

2.4 Experiment on shear stress dynamics

In the shear test, the shear stress is measured with respect to the soil deformation (shear displacement). Curves of shear stress versus shear displacement under five vertical loads and at eight shear rates were recorded using the Bevameter's shear device. The combination of the following vertical loads and shear rates results in a number of 40 tests:

Vertical loads in kN: 0.77 | 2.3 | 3.8 | 5.74 | 7.27
Shear velocities in cm·s^{-1}: 2.8 | 5.6 | 11.1 | 16.7 | 22.2 | 27.8 | 33.3 | 38.9

The normal stress generated by the vertical loads corresponds to a range from low to medium of typical loads during farm operations (field or transportation). The shear rates tested were derived from the sliding velocity between the tire and the ground at 2 km·h^{-1} and wheel slip from 5 to 70 %.

For each vertical load, the maximum shear stress is determined and the Mohr-Coulomb shear strength diagram is generated from these pairs of data. In the diagram, the intersection with the y-axis corresponds to the cohesion and the slope of the trend line corresponds to the tangent of the angle of shearing resistance φ.

2.5 Driving tests with the single wheel tester

Under soil conditions in the soil bin comparable to the shear tests, traction force-wheel slip curves for the tire 480/70 R 24 were recorded with the single wheel tester. Wheel loads of 22 kN and 17 kN were investigated at a driving speed of 0.556 m·s^{-1} and 120 kPa tire pressure. The recording of traction force-wheel slip curves should include the determination of the zero-slip point. For this purpose, the method proposed by Grečenko was used (Barrelmeyer, 1996). With free-rolling and towed wheel, rolling resistance is determined both on asphalt and in the soil bin, and for each wheel load. It is assumed that the internal rolling resistance in the soil bin is equal to the internal rolling resistance on asphalt. It is also assumed that the zero-slip point is located by traction force equal to the negative value of the external rolling resistance.

2.6 Modeling approaches for soil response to shear stress

Fitting of the soil response was performed using Wong and Janosi-Hanamoto approaches (Wong, 2010):

2.6.1 Wong approach

This model equation is applied for cohesive soils that have residual stress after reaching a maximum shear stress (Wong, 2010):

$$\frac{\tau}{\tau_{max}} = K_r \cdot \left[1 + \left[\frac{1}{K_r\left(1 - \frac{1}{e}\right)}\right] e^{\left(1 - \frac{j}{K_w}\right)}\right] \cdot \left[1 - e^{-\frac{j}{K_w}}\right] \quad (1)$$

With:

τ:	Shear stress
τ_{max}:	Maximum shear stress
K_r:	Ratio of the residual shear stress τ_r to the maximum shear stress τ_{max}
K_w:	Shear displacement at τ_{max}

A weighted least squares method is used to calculate the value for K_r.

2.6.2 Janosi-Hanamoto approach

This approach is used for soils, which have an increasing shear stress up to a constant value:

$$\tau/\tau_{max} = 1 - e^{(-j/K)} \quad (2)$$

With:

τ:	Shear stress
τ_{max}:	Maximum shear stress
K:	Shear deformation coefficient
j:	Shear displacement

From the *Eq. (*1) for Wong's shear stress and with the parameters determined for it, the traction force-wheel slip curve is calculated. The following model equation is used to determine the traction force in relation to the soil properties and to the state of motion of the tire (Wong and Preston-Thomas, 1983):

$$F_T = b\tau_{max}K_r \int_0^l \left(1 - e^{\left(\frac{-ix}{K_w}\right)} + \left[\frac{1}{\left(K_r\left(1-\frac{1}{e}\right)\right)} - 1\right] \cdot \left[e^{\left(1-\frac{ix}{K_w}\right)} - e^{\left(1-\frac{2ix}{K_w}\right)}\right]\right) dx \qquad (3)$$

With:
b: Width of the contact patch
l: Length of the contact patch
i: Wheel slip
x: Distance from the front of the contact area

3. Results and discussion

3.1 Shear tests

From the results of the shear tests, the dynamic behavior of the shear strength of the soil was analyzed. As the normal stress increases, the shear strength increases, and a similar behavior shows the K_w value. For example, Figure 4, depicts the curves for shear rates 0.36 m·s^{-1} and 0.01-m·s^{-1}. Appendix A shows the curves for all 40 measurements. The shear stress-shear displacement curves show two different behaviors with respect to shear rate: above 0.16 m·s^{-1}, the shear strength increases to a maximum value and then decreases to a residual stress. For shear rates below 0.16 m·s^{-1}, the shear stress increases up to the maximum value and remains unchanged with increasing shear displacement.

The cohesion and angle of shearing resistance were calculated from the Mohr-Coulomb shear strength diagrams obtained from the shear tests (Appendix B). Table 1 lists the parameters for the 8 shear rates. The angle of shearing resistance varies between 25° and 31.6°. Thus, it agrees with the parameters of Wong (Wong, 2010) for loams, but it does not show any relationship with the shear rate. Except for the shear rates 0.22 and 0.29 m·s^{-1}, the cohesion shows a slight increase with increasing shear rate. On the other hand, the obtained cohesion values differ with Wong's results but are similar to Barrelmeyer's (Barrelmeyer, 1996) This may be due to the type of shearing device used. The results published by Wong were obtained namely with a ring shear device and the results of the present work with a linear shear device.

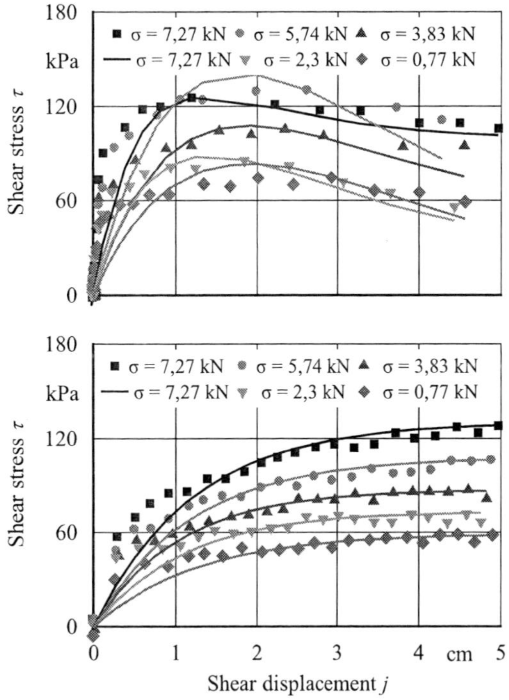

Figure 4. Shear stress-shear displacement relationship: measuring data and fit curves according to Wong (Wong, 2010) for shear rate of 0.36 m·s^{-1} (top) and fit curve according to Janosi-Hanamoto (Wong, 2010) for shear rate of 0.01 m·s^{-1} (bottom)

Table 1: Soil mechanical parameters in relation to shear rate.

Shear rate, m·s^{-1}	**Angle of shearing resistance tan φ, °**	**Cohesion c, kPa**
0.4	30.5	28.6
0.36	30	25.9
0.29	33.9	13.6
0.22	31.3	17.4
0.16	27.4	25.5
0.09	26.8	25.5
0.03	28.6	17.6
0.01	30.9	11.8

3.2 Soil reaction modeling

At shear rates above 0.16 m·s^{-1}, the shear stress shows the typical behavior for loams, but the observed behavior at shear rates below 0.16 m·s^{-1} suggests a sandy soil. In the last case, the Janosi-Hanamoto model equation is used. Figure 4 shows the Wong model fit for 0.36 m·s^{-1} shear rate and the Janosi-Hanamoto model fit for 0.01 m·s^{-1}. Appendix A contains the fits for all 40 measurements. The values of shear strength (τ_{max}), K_r, K, K_w and angle of shearing resistance, as well as the goodness of fit, are consistent with the parameters published by Wong (Wong, 2010). Similar deviations of the fits of the shear stress-shear displacement relationship to Wong and Janosi-Hanamoto approaches are also observed in Barrelmeyer's work (Barrelmeyer, 1996). Here again the shear device type used can have an influence.

3.3 Traction force-wheel slip curve

Figure 5 shows the traction force-wheel slip curve measured with the single wheel tester on the soil bin. Under identical test conditions, the higher wheel load of 22 kN produces the higher traction force. The gross traction coefficient (Figure 6) behaves similarly for the two wheel loads.

Using the soil mechanical parameters determined in the shear tests and the approach suggested by Wong and Preston (Wong and Preston-Thomas, 1983) the traction force-wheel slip and the gross traction-wheel slip curve are predicted.

Since the soil mechanical parameters do not show a clear dependence with the shear rate, the average value of the cohesion and the angle of shearing resistance was taken for the calculation. K_r and K_w correspond to the maximum calculated values from the shear tests with normal stresses of 130 kPa and 100 kPa. Assumed contact patch are 0.17 m^2 and 0.19 m^2 for 17 and 22 kN wheel loads, respectively.

The precision of the prediction is insufficient. For both wheel loads, there is a systematic deviation of the rise of the curves. The predicted traction force reaches its maximum very soon with wheel slips under the 10 %. The predicted traction force is only acceptably for the wheel load of 17 kN. For the wheel load of 22 kN, the maximum traction force is underestimated.

The difficulties in predicting the traction force may be because the approach of Wong and Preston is not optimal neither for the tested conditions (tire and soil) nor for the measuring equipment used. In this work, the soil parameters were measured with a linear shear instrument. In this regard, the difference with the results from a ring shear instrument such as that reported by Wong's publications is not known. Specific cohesion values determined in this work are significantly greater than those reported by Wong. Experimental errors or errors on the application of the methods cannot be excluded and therefore further tests are planned.

4. Conclusions

Shear tests with a Bevameter were performed on a soil bin to characterize its soil and to investigate the influence of shear rate on soil mechanical parameters. The measured values were fitted to the Wong and Janosi-Hanamoto approaches to model the shear behavior of the soil. The shear strength increases clearly with the increase of the normal stress. Variation of the shear rate shows shear stress-shear displacement behaviors of different soil types. At high shear rates, the soil behaves like a cohesive soil. At low shear rates like a sandy soil. In this work, the shear rate has no strong influence neither on the angle of shear strength nor on the cohesion.

Driving tests with a single wheel tester with two wheel loads on the soil bin were performed to determine traction force-wheel slip relationships. To determine the zero slip condition driving tests with the single wheel tester were conducted. The prediction of the traction force-wheel slip curve based on the determined soil parameters was not satisfactory. Systematic discrepancies between the modeled traction force and traction force-wheel slip curve, measured in the field, suggest further experiments and revision of the method and its application.

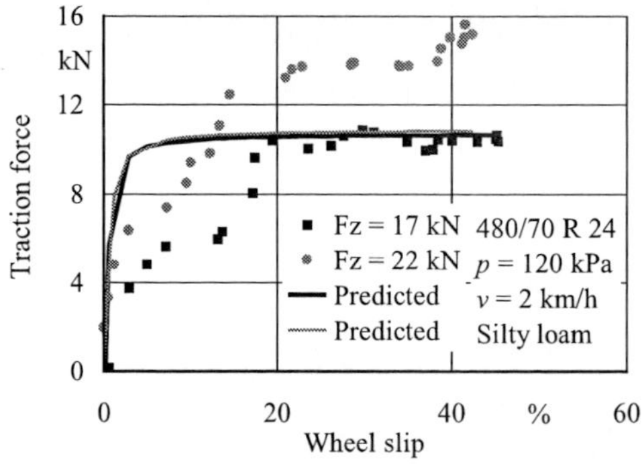

Figure 5. Measured traction force-wheel slip curves and prediction according to Wong and Preston for the tire 480/70 R 24 at 2 km·h-1 and 120 kPa tire pressure

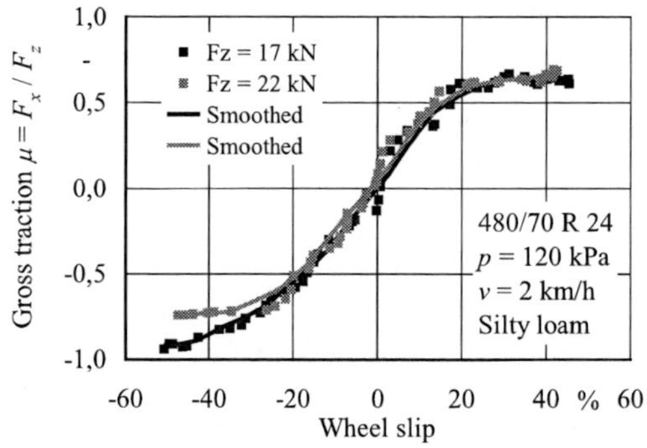

Figure 6. Gross traction coefficient-wheel slip curves for the tire 480/70 R 24 at 2 km·h^{-1} and 120 kPa tire pressure

5. References

Barrelmeyer, T., 1996. Untersuchung der Kräfte an gelenkten und angetriebenen Ackerschlepperrädern bei Gelände- und Straßenfahrt (Investigation of forces on steered and driven farm tractor wheels during off-road and road travel). PhD Thesis, University of Stuttgart. VDI, Düsseldorf, Stuttgart, Germany.

He, R., Sandu, C., Khan, A.K., Guthrie, A.G., Schalk Els, P., Hamersma, H.A., 2019. Review of terramechanics models and their applicability to real-time applications. Journal of Terramechanics 81, 3–22. https://doi.org/10.1016/j.jterra.2018.04.003.

Sandu, C., Taheri, S., Taheri, S., Els, S., Jimenez, E., 2020. Hybrid soft soil tire model (HSSTM). Part III: Model parameterization and validation. Journal of Terramechanics 88, 1–15. https://doi.org/10.1016/j.jterra.2019.11.002.

Schlotter, V., 2006. Einfluss dynamischer Radlastschwankungen und Schräglaufwinkeländerungen auf die horizontale Kraftübertragung von Ackerschlepperreifen (Influence of dynamic wheel load variations and slip angle changes on the horizontal power transmission of farm tractor tires). PhD Thesis, University of Stuttgart. Shaker, Aachen, Stuttgart, Germany.

Witzel, P., 2015. Ein validiertes Reifenmodell zur Simulation des fahrdynamischen und fahrkomfortrelevanten Verhaltens von Ackerschleppern bei Hindernisüberfahrt (A validated tire model for simulating the driving dynamic and ride comfort behavior of farm tractors during obstacle passages). PhD Thesis, University of Stuttgart. Shaker, Aachen, Stuttgart, Germany.

Witzel, P., 2018a. The Hohenheim Tyre Model: A validated approach for the simulation of high-volume tyres—Part I: Model structure and parameterisation. Journal of Terramechanics 75, 3–14. https://doi.org/10.1016/j.jterra.2017.07.002.

Witzel, P., 2018b. The Hohenheim Tyre Model: A validated approach for the simulation of high-volume tyres—Part II: Validation. Journal of Terramechanics 75, 15–24. https://doi.org/10.1016/j.jterra.2017.09.004.

Wong, J.Y., 2010. Characterization of the Shearing Behaviour of Terrains, in: Wong, J.Y. (Ed.), Terramechanics and off-road vehicle engineering: Terrain behaviour, off-road vehicle performance and design, 2nd ed. Else-vier, Amsterdam, pp. 115–128.

Wong, J.Y., Preston-Thomas, J., 1983. On the characterization of the shear stress-displacement relationship of terrain. Journal of Terramechanics 19, 225–234. https://doi.org/10.1016/0022-4898(83)90028-9.

Appendix A

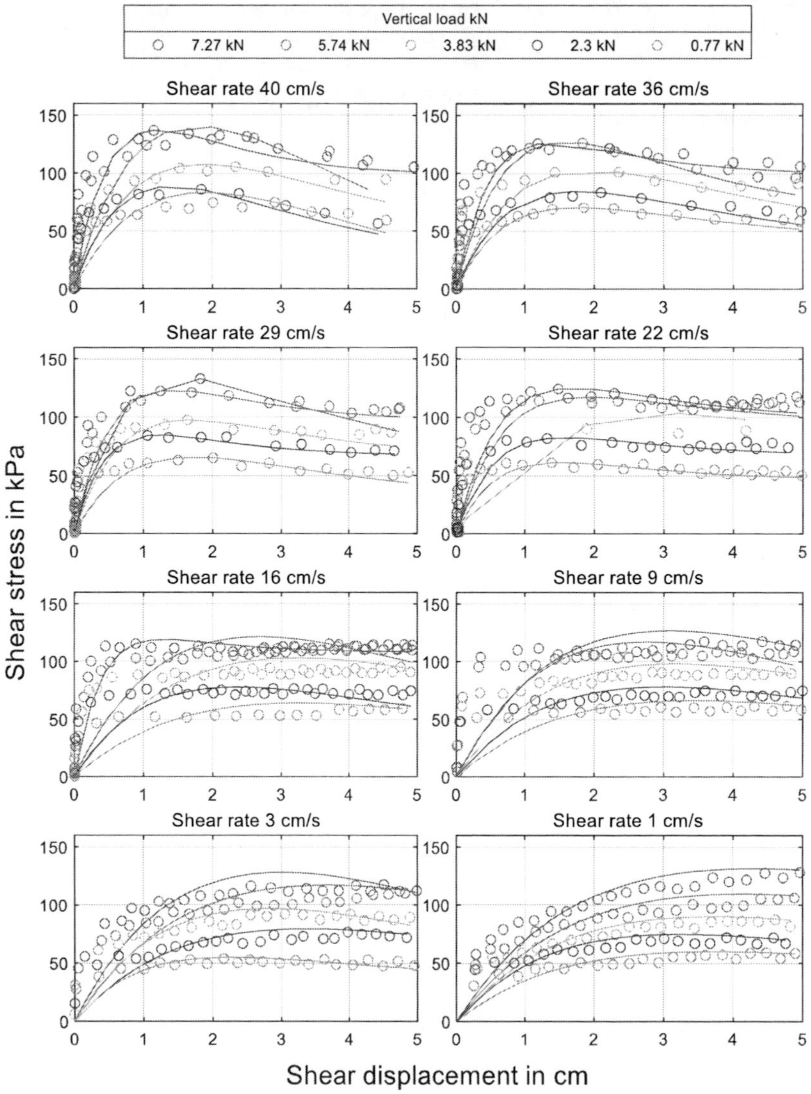

Figure A1. Shear stress-shear displacement relationships: Measuring points and fitting curves according to Wong

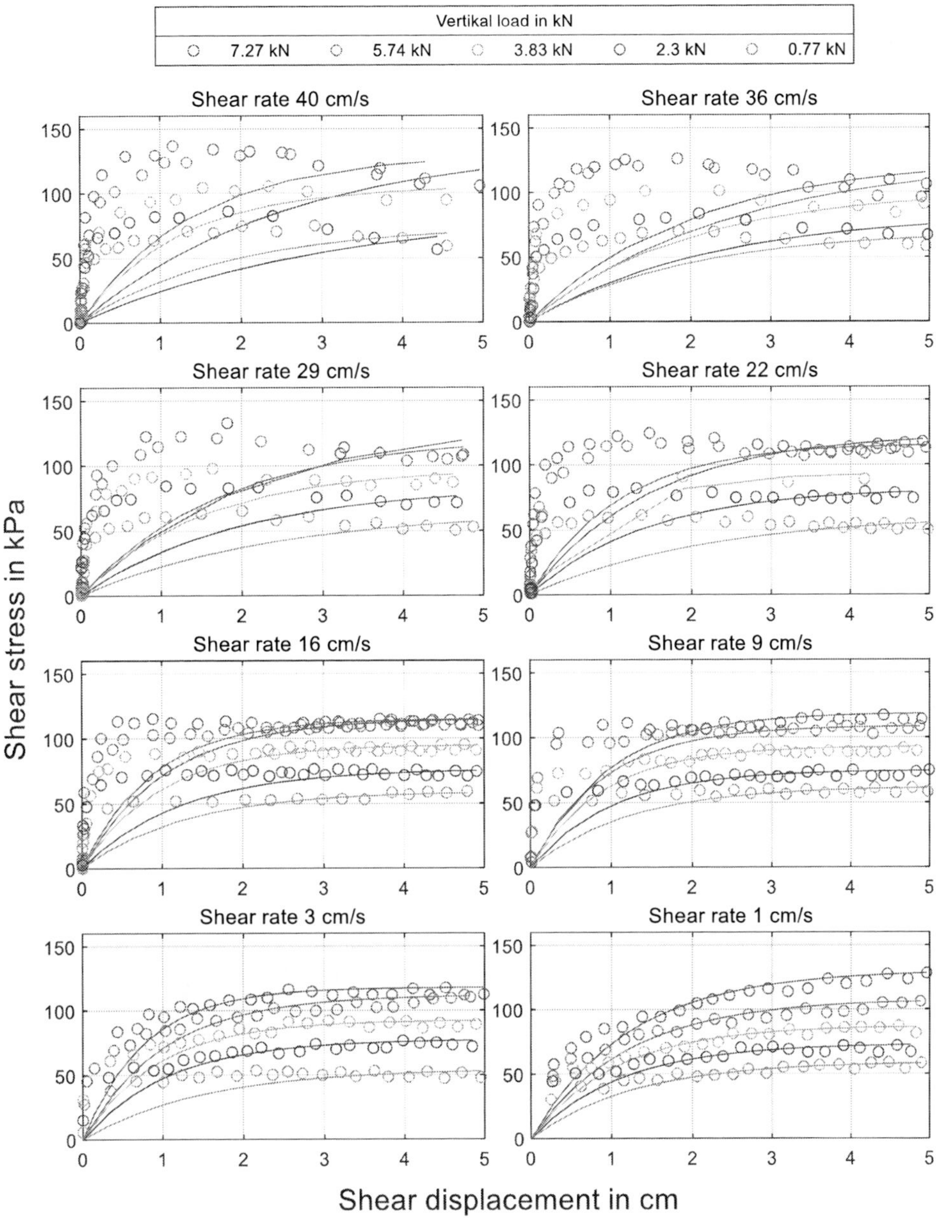

Figure A2. Shear stress-shear displacement relationships: Measuring points and fitting curves according to Janosi-Hanamoto

Appendix B

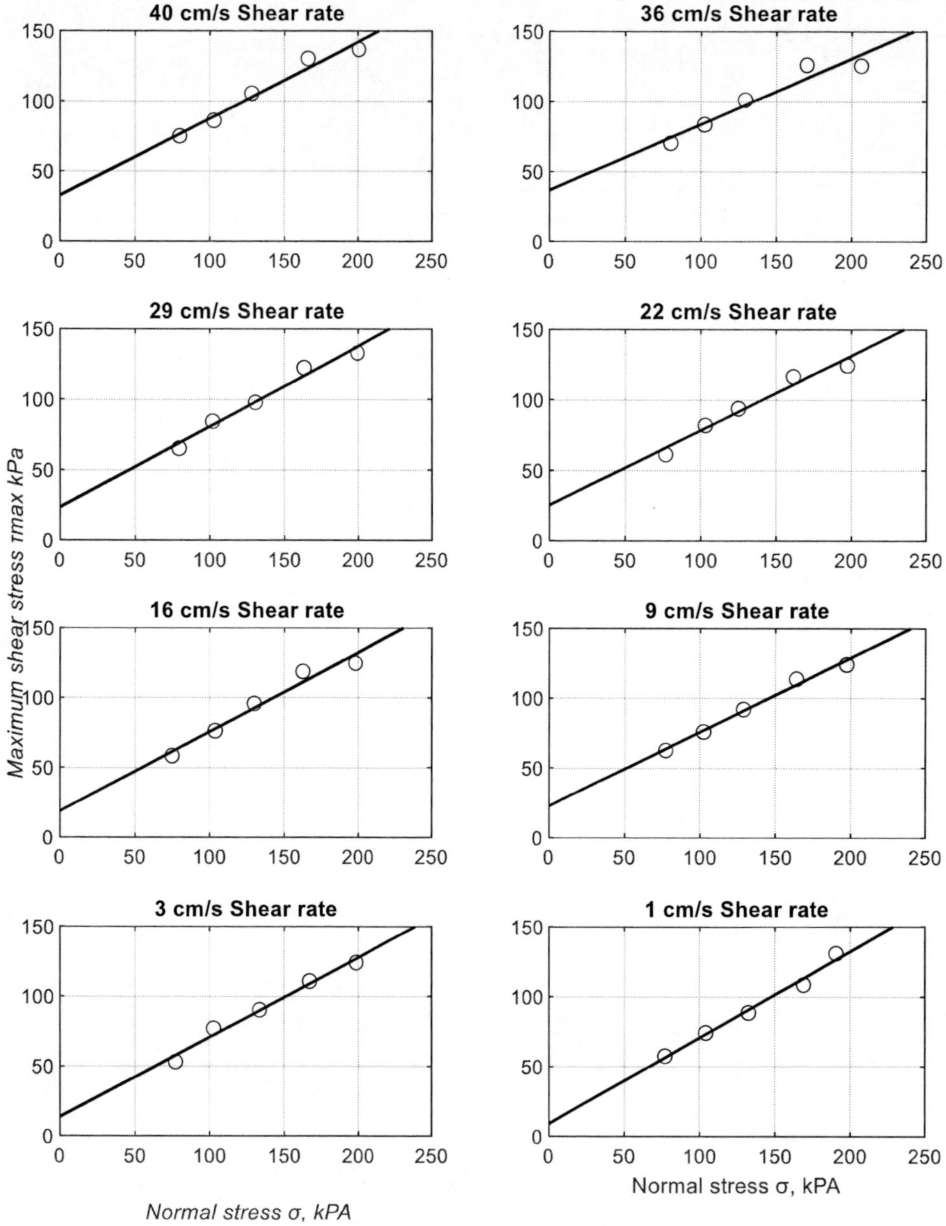

Figure B.1: Mohr-Coulomb shear strength diagrams for the determination of cohesion and angle of shearing resistance.

2190 / Tyre Parameterization Tests: Dynamic vs. Static

Carl Becker* and Schalk Els
University of Pretoria, Pretoria, South Africa
**Corresponding author: carl.becker@up.ac.za*

https://doi.org/10.56884/VTFC7095

Abstract

This is abstract text. Agricultural vehicles are used more on asphalt roads and needs to be safe in an evasive maneuver or during an emergency brake scenario. The designers of these vehicles require tyre parameterization data to obtain accurate and representative simulation results of these vehicles. The standard laboratory tyre testing procedures are not always possible on very large tyres. The limitations in the form of the actual size of the tyre and the required loads on the tyre cannot always be accommodated by the current equipment available to conduct dynamic tests.

This study investigates how the tyre parameterization results compare between a dynamic/rolling tyre test and a static/non-rolling tyre. This is in an attempt to parameterize tyre models using largely or only static test data. Tests are conducted on a smaller agricultural tyre where dynamic tyres test results are compared to static tests.

Keywords: tyre parameterization, tyre models, static test, dynamic test, agricultural tyre

1. Introduction

Over the last two decades considerable advances have been made in the levels of detail design and simulation of large off-road vehicles. This includes construction, mining and agricultural vehicles amongst others. Evolving vehicle design technology incorporates simulation up to the point where homologation of new vehicles is possible based on simulation results, instead of real-world testing (TüV SüD, 2021). Simulation results are however only as reliable as the input data. In vehicle ride or handling simulations the primary excitation into the vehicle is via the tyres, thus the importance of having reliable tyre data for accurate tyre modelling and vehicle simulation. With the tyres being the only suspension on many agricultural and construction vehicles, the importance of accurate tyre models is even more critical. Agricultural vehicles are used more on asphalt roads and needs to be safe in an evasive maneuver or during an emergency brake scenario. As a result, designers need to simulate these maneuvers to improve the safety of these vehicles. Stayner et al., (1984). Lines (1991) indicated that poor information about the agricultural tyre's behavior over any terrain has a very large effect on the simulation errors. The advances made in tyre technology to date have placed a large question mark on the reliability on the limited historical collection of tyre data available especially in the agricultural and construction industry. Other vehicle factors that are affected by the tyre characteristics range from off-road traction and soil compaction to on-road handling, ride comfort, braking and motion resistance.

Obtaining tyre data is not a trivial or inexpensive exercise, as a result tyre manufactures are reluctant to freely supply tyre data. Tyre manufactures only supply high-end clients with limited tyre data or they simply do not have the required data. Test equipment suitable for large agricultural and off-road vehicles are very limited due to the size of the tyres, the required operating load and logistics around testing these large tyres at the designed operating conditions.

2. Tyre models

Every tyre model used in simulations requires different sets of parameterization data. Different tyre parameterization test methods exist and vary from laboratory tests to field tests on different surfaces. Tyre parameters include longitudinal, lateral and vertical stiffness which are dependent on the inflation pressure and carcass construction. The tyre-terrain interaction parameters are dependent on the friction coefficient between the tyre and terrain, which is dependent on the contact area which directly effects the motion resistance and maximum traction available. Each test surface has different surface roughness and will result in different measured tyre characteristics.

Tyre models fall broadly into two types of models namely:

- Empirical or semi-empirical models, for example a Pacejka/ Magic Formula model, Hybrid Soft Soil Tire Model, etc.

- Physics based models, for example FTire, RMOD-K, Hohenheim, etc.

The Pacejka tyre model, also known as the Magic Formula, MF, was formulated by Dr. H.B Pacejka in the 1980s (Bakker et al., 1987). The model is widely known and used for handling simulations on smooth terrains but has limitations when it comes to ride comfort simulations (Becker, 2009). The model is semi-empirical and completely dependent on experimental data for parameterization. The following data sets are the minimum required to parameterize a useful Pacejka model:

- Longitudinal force vs. % longitudinal slip at three vertical loads
- Lateral force vs. slip angle at three vertical loads
- Vertical stiffness and damping at three vertical loads

Curve fits are made to the test data and the curve fit coefficients are used in the simulation model. Changes in tyre operating conditions, for example a different inflation pressure, requires that a new MF model needs to be parameterized.

The FTire model is a physics-based flexible structure tyre model developed by cosin scientific software (2017). The model can be used for handling and ride comfort simulations on rough terrain. The FTire model can be parameterized at two inflation pressures with the use of static tests to obtain vertical stiffness (on flat surface and cleats), lateral stiffness and longitudinal stiffness together with footprints and tyre geometry. Modal analysis will contribute to model parameterization but it is not an essential requirement. Damping over cleats and dynamic tests in the form of lateral force vs. slip angle and longitudinal force vs. % longitudinal slip are used for validation. The required tests can be conducted on a flat track or drum tester. FTire is a very flexible/adjustable tyre model and enables the user to adjust the amount of tread on the belt and inflation pressure amongst other parameters.

3. Tyre parameterization tests

Tyre characteristics can be measured using multiple methods. Each method has different constraints and have their own set of positives and negatives features. This is the case for passenger cars, sports cars, motorcycles, commercial trucks, heavy duty trucks, agricultural and construction vehicles. Many test laboratories are spread worldwide and all have the capability to parameterize relatively small tyres. Testing larger tyres have its own challenges as one needs large amounts of driving power and infrastructure to test these tyres at the designed loads. Typical indoor laboratory test rigs include drum test rigs and flat track rigs. Some test rigs can only be used for traction characterization, where other test rigs can conduct lateral force vs. slip angle tests, motion resistance measurements or even simulate track racing conditions. These laboratory test rigs tend to have different test surfaces (corundum surfaces/sandpaper or steel belts) and on some of these test rigs the test surface can be changed.

The Vehicle Dynamics Group (VDG) in the Department of Mechanical and Aeronautical Engineering at the University of Pretoria in South Africa, has developed a Dynamic Tyre Test Trailer (DTTT) and a Static Tyre Test Rig (STTR) as described by Becker and Els (2018). The DTTT tests rolling tyres with diameters larger than 800mm and up to 2000mm, where two of the same tyres are tested in opposed lateral slip directions up to a maximum vertical load of 50kN per wheel. The DTTT can measure tyre forces and moments for lateral force vs. slip angle measurements, longitudinal brake force vs. longitudinal slip measurements as well as combined lateral and longitudinal force measurements during braking at pre-determined lateral slip angles. The STTR can measure the static stiffness of a tyre in the longitudinal or lateral direction for a predetermined vertical load on test surfaces that are interchangeable. The STTR is equipped with 1MN actuators and can accommodate 4000mm diameter tyres.

The University of Hohenheim, has multiple large tyre test rigs in the form of a flat belt test stand, single wheel test trailer and an instrumented test tractor as presented by Witzel (2018). These test rigs/platforms can be used to measure longitudinal forces, lateral forces and motion resistance on larger agricultural tyres. Other single wheel test trailers have been built for agricultural tyres, used and upgraded from early 1960's to present. These trailers can be used on any terrain. Billington (1973) introduced the NIAE MkII single wheel tester in 1973. Ambruster and Kutzbach (1989), developed a single wheel tester which parameterized a driven wheel at pre-set slip angles. Shmuleviuch et al. (1996) presented a field single wheel tester. These trailers can be used on any terrain, however the size of the test tyres and the vertical load applied to them, with the use of these test rigs are limited. It is however possible to obtain static parameterization data on larger tyres with the use of the STTR at the University of Pretoria. This raised the question of how static parameterization data compares to dynamic parameterization data. Tests were conducted on a Trelleborg TM700 280/70R16 agricultural tyre with 100% tread. Although this is a relatively small tyre, the investigated principles can be used on large diameter tyres. The findings for this analysis are discussed in this paper.

4. Parameterization Test Results

In this section the different parameterization test results are presented and compared.

4.1 Static/non-rolling tyre

A large number of factors influence the characteristics of a tyre. This includes, but are not limited to, operating temperature, inflation pressure, rubber compound, terrain/test surface, vertical load, tread condition etc. Some of these factors can be controlled by the operator, designed and tyre manufacture, however none of them has control over all of these factors. As an example, Figure 1 shows the difference in measured longitudinal tyre characteristics, with the use of the STTR, by only changing the roughness of the test surface during static testing. The tests conducted on concrete will be used during the analysis in this study as the roughness of the concrete corresponded to the surface on the outdoor test track.

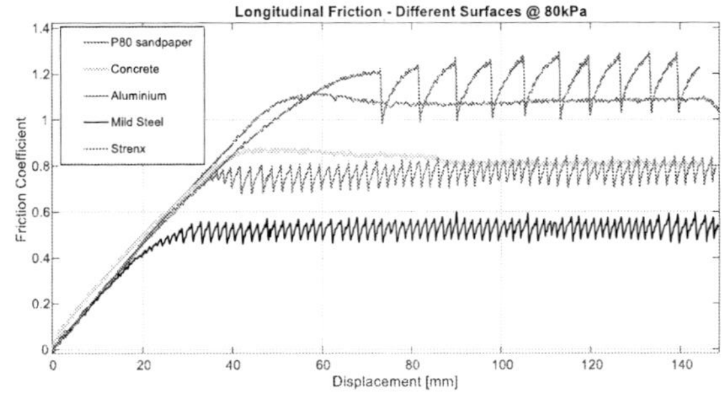

Figure 1. Longitudinal tyre characteristics for different test surfaces

4.2 Dynamic/rolling tyre

Dynamic/rolling tyre characterization tests were conducted with the use of the DTTT in order to obtain the longitudinal and lateral characteristics of a tyre on concrete. Test were conducted at inflation pressures of 80kPa and 200kPa and a speed of 11km/h, with a vertical load of 5.68kN. This was representative operating conditions of the vehicle on which the tyre of interest is typically fitted to.

Unfiltered data for a typical rolling tyre, at 11km/h, longitudinal friction coefficient vs. longitudinal slip under braking test is shown in Figure 2. The graph shows the longitudinal forces normalized with the vertical force to give the friction coefficient at 80kPa and 200kPa inflation pressure. It can be seen that for a rolling tyre, the longitudinal stiffness is very consistent and independent of the inflation pressure.

The lateral friction coefficient vs. slip angle of a rolling tyre is as repeatable as the longitudinal test. An example of the unfiltered lateral test data is shown in Figure 3, where it can be seen that the lateral stiffness of the rolling tyre is very consistent. This is caused by the stiffening of the carcass belt as the tyre rotates. The effect of the lugs in the tread pattern and radial runout are clearly seen in the vertical force oscillations in the data as seen in Figure 4.

During the friction envelop tests the tyre is rolling at a set slip angle with the brakes applied up to wheel lockup. Figure shows the state of the lateral deformation of the tyre before and during lockup, respectively. This is a clear representation of why a vehicle, that is operating at the limit, can only generate a maximum steering force or only apply a maximum braking force. The maximum combined forces will always be lower.

The friction envelop describes the combination of the applied longitudinal and lateral forces during different slip angles and longitudinal wheel slip as shown in Figure 5. It can be seen that at 80kPa inflation pressure a larger longitudinal friction coefficient can be generated due to a larger footprint at 80kPa, also noted in Figure during pure longitudinal slip tests. The increased side wall stiffness, due to the increased inflation pressure, correlates to the increase maximum lateral friction coefficient generated at 200kPa.

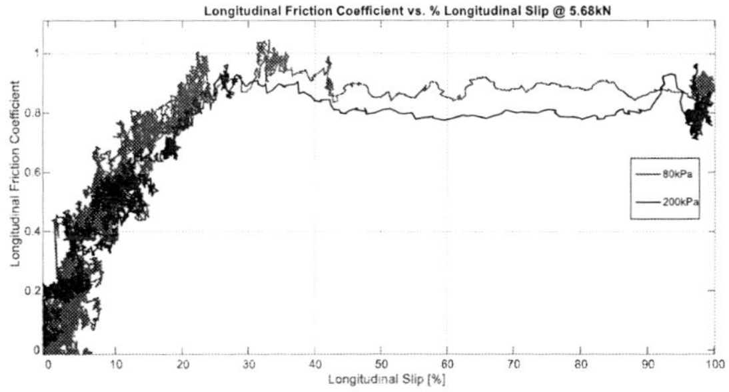

Figure 2. Unfiltered longitudinal friction coefficient vs. longitudinal slip measurements under braking on Trelleborg TM700 280/70R16 at a speed of 11km/h at 80 kPa and 200kPa inflation pressures

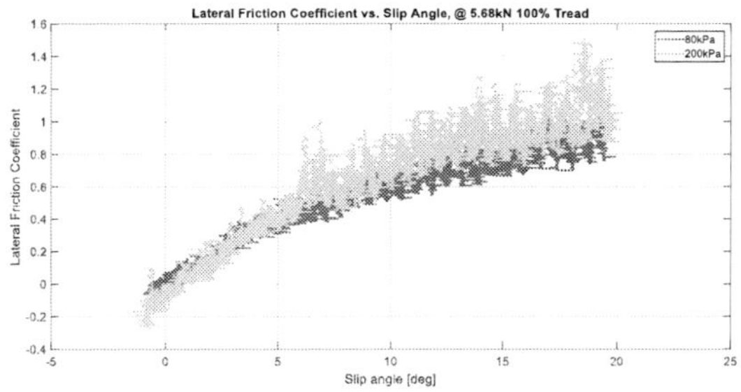

Figure 3. Unfiltered lateral friction coefficient vs slip angle measurements on Trelleborg TM700 280/70R16 at 11km/h, at 80kPa and 200kPa inflation pressure

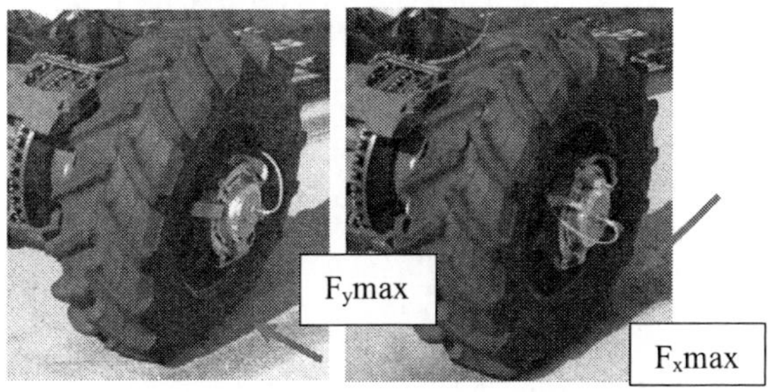

Figure 4. Friction envelope tests on Trelleborg TM700 280/70R16 at 11km/h, vertical load at 5.68kN before and during lockup at a constant lateral slip angle, respectively

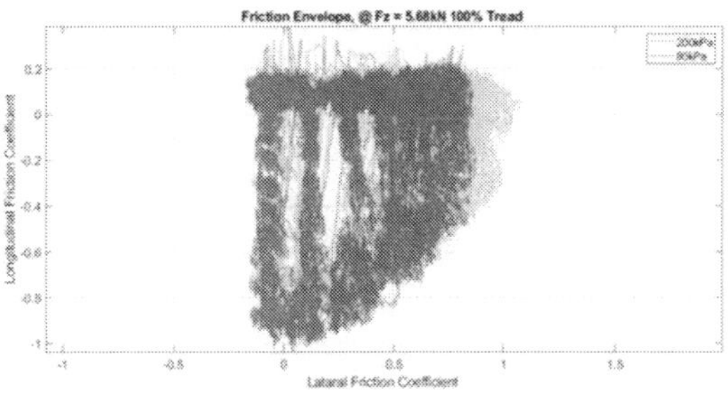

Figure 5. Friction envelope unfiltered measurement on Trelleborg TM700 280/70R16 at 11km/h, at 80kPa and 200kPa inflation pressure, respectively

4.3 Static/Non-rolling vs. Dynamic/Rolling

Outdoor dynamic tyre tests are time consuming, expensive and less repeatable over different times of year testing than tests conducted in a laboratory. Also, tyre size/load limitations are a major factor when conducting dynamic/rolling tests on large tyres. If static test rig results could be used to, at least estimate, tyre parameters that represent dynamic/rolling characteristics, it can become more affordable to obtain tyre characteristics in order to parameterize more tyre models. A method to compare the longitudinal friction coefficient as a function of % longitudinal slip of a rolling tyre to static/non-rolling longitudinal friction coefficient as a function of longitudinal displacement, is to translate longitudinal slip of the contact patch to longitudinal displacement, $D_{longStatic}$.

For the dynamic longitudinal slip tests this can be achieved by calculating the longitudinal displacement of the contact patch, D_{wheel}, relative to the displacement of the vehicle ($D_{vehicle}$) and dividing it by the number of wheel rotations, N_{rotate}, during the longitudinal slip period from 0% to 100% slip as shown in Eq. 1:

$$D_{longStatic} = \frac{(D_{vehicle} - D_{Wheel})}{N_{rotate}} \quad (1)$$

The result of this transformation from % longitudinal slip to relative longitudinal displacement is shown in Figure 6.

Figure 7 shows the comparison between the static tests conducted on the STTR on concrete and the translated DTTT tests.

Relatively good correlation is seen between the two test methods. The initial longitudinal stiffness correlates very well within the first 30mm of longitudinal stiffness, with the maximum sliding friction coefficient within 10 to 20% for the tests conducted on the outdoor test track with the DTTT. The difference in boundary conditions between the static and dynamic test methods will also contribute to the difference in the results.

The lateral friction coefficient vs. slip angle tests can be translated to the lateral friction coefficient vs. 90-degree lateral displacement with the use of trigonometry to calculate the lateral displacement component, $Slip_{Lat\_mm}$, at the specific lateral slip angle, β, and the contact patch length, L, as shown in Eq. 2:

$$Slip_{Lat\_mm} = L\tan\beta \quad (2)$$

The result of the comparison between the static tests on the STTR on concrete with the lateral translation of the friction envelope tests and lateral force vs. slip angle tests, as conducted with the DTTT at 200kPa inflation pressure is shown in Figure 8.

Very good correlation is seen at the high inflation pressure tests. This shows consistency for all three tests methods being static or dynamic.

5. Conclusions

The results from the dynamic/rolling tyre outdoor tests at a 100% tread condition is successfully compared to the static parameterization tests, with limited success at low inflation pressures due to the tyre belt stiffening for a rolling tyre. This indicates that it is possible to obtain accurate tyre characteristics from static tests which are representative of a rolling tyre at lower velocities and high inflation pressures. This is an important observation as most agricultural vehicles operate at low speeds and construction vehicles operate with higher inflation pressures. This will allow engineers to obtain valuable tyre parameterization data at a lower cost with the use of static tyre testing. This data can also be used to parameterize physics-based tyre models for rolling tyre simulations, only when using carefully executed static tests on representative test surfaces. The difference in boundary conditions can be eliminated during tyre model validation as the physics-based tyre model can simulate the boundary conditions and movements of the STTR which is used to measure the tyre characteristics.

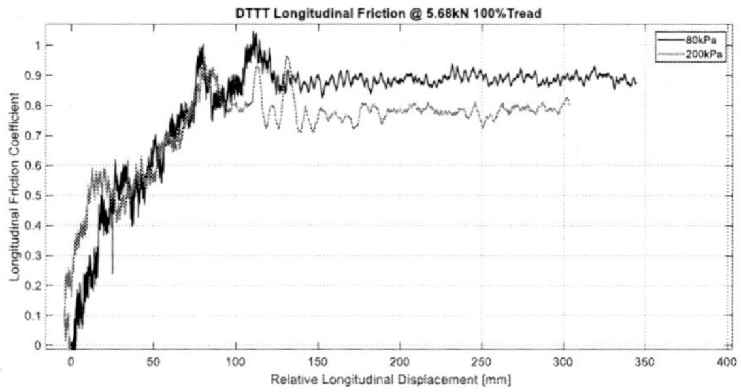

Figure 6. Translated Rolling Longitudinal Friction Coefficient vs. Relative Longitudinal Displacement at 80kPa and 200kPa inflation pressure, respectively

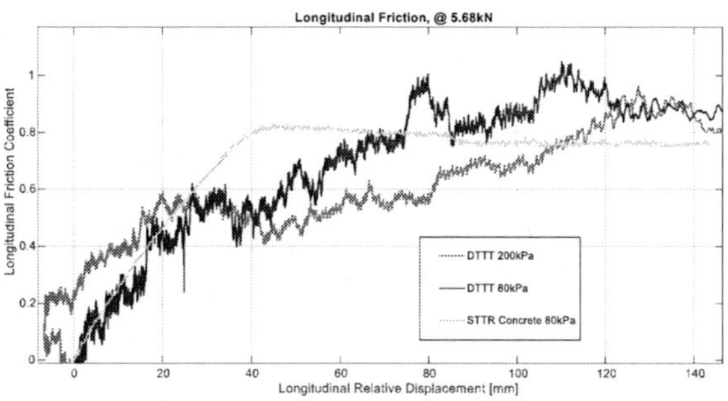

Figure 7. Longitudinal Friction Coefficient vs. Longitudinal Displacement from different test methods

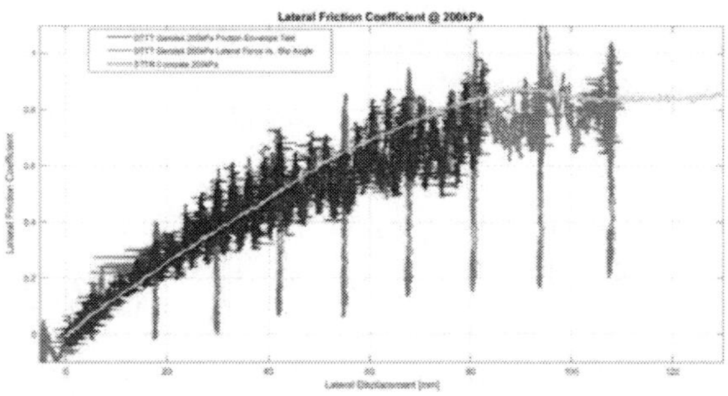

Figure 8. Dynamic lateral friction coefficient tests compared to static testes on concrete at 200kPa inflation pressure for 100% tread

6. Nomenclature

$D_{longStatic}$	*Longitudinal displacement of static wheel*	*[mm]*
$D_{Vehicle}$	*Longitudinal displacement of vehicle*	*[mm]*
D_{Wheel}	*longitudinal displacement of the contact patch on rolling wheel*	*[mm]*
L	*Contact patch length*	*[mm]*
N_{rotate}	*Number of wheel rotations*	
$Slip_{Lat\_mm}$	*lateral displacement component*	*[mm]*
β	*Lateral slip angle*	*[°]*

7. References

Ambruster, K, Kutzbach, H.D., 1989, Development of a single wheel tester for measurement on driven angled wheel. In: Proceedings of 5th European conference of the ISTVS, Wageningen, The Netherlands; p. 8–14.

Bakker, E, Nyborg, L and Pacejka, H.B., 1987, Tyre modelling for use in vehicle dynamics studies, Society of Automotive Engineers, Warrendale, PA.

Becker, C.M, 2009, Profiling of rough terrain, Master of Engineering Thesis, University of Pretoria, South Africa, http://hdl.handle.net/2263/29833.

Becker, C.M. and Els, P.S., 2018, Static and Dynamic Parameterization Test Rigs for Large Tyres, Proceedings of the 10th Asia-Pacific Conference of the ISTVS, Kyoto, Japan, July 11-13.

Billington, P.W., 1973 The NIAE Mk II single wheel tester. J Agric Eng Res; 18:67–70.

Cosin scientific software, 2017, FTire, Modelization and Parameter Specification. 2017-1-r14630. [pdf]. Software documentation and user guide.

Lines, J.A., 1991, The Suspension Characteristics of Agricultural Tractor Tyres. PhD Thesis, Cranfield Institute of Technology Silsoe College.

Shmuleviuch, I., Ronai, D., Wolf, D., 1996. A new field single wheel tester. Journal of Terramechanics, Vol. 33(3), p133–41.

Stayner, R.M., Collins, J.A. and Lines, J.A., 1984, Tractor Ride Vibration Simulation as an Aid to Design. Journal of Agriculture Engineering Res. 29. 345-355.

TüV SüD, 2021, Virtual Simulation to Ensure the Safety of Vehicles, website: https://www.tuvsud.com/en/e-ssentials-newsletter/automotive-essentials/e-ssentials-3-2017/virtual-simulation-to-ensure-the-safety-of-vehicles, accessed 5 October 2021.

Witzel, P., 2018. The Hohenheim tyre model: a validated approach for the simulation of high-volume tyres—Part I: Model structure and parameterisation. Journal of Terramechanics. 75, 3–14.

2539 / Model Predictive Control of a Robot Driven Vehicle for Testing of Advanced Driver Assist Systems

Mike Huang[a]*, Haixuan Qiu[b], Chunyu Yang[b], Lian Xia[b], Zhaomin Lin[b], Yanqing Wang[b], Zongqing Xu[b], and Mingfu Tang[c]
[a]Sunrise Instruments USA, Canton, Michigan, USA
[b]SAIC Motor R&D Innovation Headquarters, Shanghai, China
[c]Sunrise Instruments, Nanning, China
**Corresponding author: mhuang@srisensor.com*

https://doi.org/10.56884/WZVY2027

Abstract

Abstract: Advanced Driver Assist Systems (ADAS) are becoming more prevalent and more sophisticated in passenger vehicles, with features such as automatic lane keeping, pedestrian detection, and emergency breaking. In line with the increased production deployment of ADAS, testing of these systems are becoming more rigorous with more scenarios needing to be considered every year, see, for example, the ADAS testing conducted by Euro NCAP. To fit the need of placing test vehicles and environment factors in very specific and repeatable scenarios, physical driving robots are commonly used. While most current tests set up the test vehicle in near steady state conditions, e.g., constant speed and straight, in the future, more complex, possibly dynamic scenarios will need to be tested. This paper presents a Model Predictive Control (MPC) strategy for controlling a vehicle along dynamic paths and is easily deployable across different vehicles. Experiment results demonstrating the controller capability for both an electric and a conventional vehicle is presented.

Keywords: model predictive control, optimal control, robot driver, advanced driver assist systems

1. Introduction

In the following, a Model Predictive Control (MPC) for a driving robot will be developed for the purpose of testing ADAS systems. The robot consists of motors and linkages for pressing the test vehicle's pedal and brake and for steering and will drive the vehicle at specific speeds along specific paths, e.g., as defined by Euro NCAP test scenarios (Euro NCAP, 2020). Tight tracking of the test plan, as defined by target positions and speeds or target positions in time, is pertinent to successful and repeatable tests, particularly since the tests often involve coordinating multiple objects in addition to test vehicle, e.g., leading and following vehicles, cross traffic, and pedestrians.

MPC is a control methodology which uses a plant model to predict states and outputs into the future as a function of a sequence of future controls and current state. This sequence of future controls is then optimized with respect to a cost function, e.g., predicted tracking error, and constraints. Then the first element of the sequence is applied to the plant. The next sampling instant, the optimization problem is updated with the new state measurements / estimates and new targets, the future control sequence is re-optimized, and the first element of the control sequence is again applied to plant.

In the context of control of vehicle dynamics, most work tend to focus on lateral control and tire modelling. Rarely is lateral and longitudinal control considered together, possibly because tight speed tracking may be considered well handled by inner loop controllers. For example, Sun et al. (2019) and Wang et al. (2019) use bicycle and tire models for prediction but only optimize for a steering command while assuming speed is constant over the horizon. Tang et al. (2020) uses a bicycle and tire model for prediction and optimizes for a yaw rate target which is sent to a lower level PID controller for steering. In dynamic scenarios, e.g., accelerating while turning, these decoupled strategies will be suboptimal because yaw rate is fundamentally a nonlinear function of both the steering angle and vehicle speed.

For multi-variable control, Alcalá et al. (2019) uses a multi-layer control structure, first using a Linear Parameter Varying (LPV)-MPC for kinematic planning of speed and yaw rate targets, then using an LPV-LQR controller to obtain steering angle and acceleration from the speed and yaw rate targets. Ugo and Borelli (2020) built a MPC simultaneously controlling steering and acceleration with the ability learn optimal racing trajectories over a sequence of laps and was demonstrated on a RC car platform. Liao-McPherson et al. (2020) considers how to solve the constrained multi-variable (steering and drive force demand) MPC optimization problem in real time using a Fischer-Burmeister function to transform complementarity conditions into nonlinear equations.

None of these prior works consider the powertrain dynamics, i.e., pedal to vehicle acceleration, which we observed in vehicle testing to significantly degrade tracking performance and even stability, particularly if the test vehicle is a conventional gasoline vehicle. It may be that for autonomous vehicles, tight speed tracking is not as important as steering around obstacles. Additionally, autonomous vehicles are commonly built on top of hybrid or electric vehicle platform, thus, not as exposed to delays in power generation. However, in other developments, e.g., for emissions and fuel economy testing, tight tracking of speed profiles is explicitly required. Pedal robots are sometimes used for these tests. A strategy using a recorded human drive for feed-forward combined with a PID could be used to control the vehicle speed. A more sophisticated strategy (Park et al., 2022) uses Reinforcement Learning (RL) to find a policy for adjusting time varying PI gains. However, to accommodate vehicle variations, e.g., drive modes or entirely different vehicles, a P gain table as a function of vehicle speed must be made first for each vehicle variation.

We have seen in our experiments that the lag / delay between the pedal and vehicle acceleration can be large, e.g., up to 400msec. on conventional vehicle tests. This greatly limits the performance of controllers if the model has no ability to express that lag. Adding on top of this, braking dynamics and acceleration dynamics are vastly different. A model free approach, e.g., PID, would also be sensitive to the lag, greatly limiting its performance. Furthermore, unlike MPC, PID cannot take advantage of knowing future speed targets, and, as a result, requires high gains which are not feasible in the presence of substantial lag. One may suggest a simple model of the powertrain, e.g., a first order filter. However, our experiments also indicate that this is not sufficient to obtain good, closed loop performance.

Our final MPC design, described herein, takes a data driven approach to modelling the vehicle. First, a neural network model of the powertrain is made. Specifically, we use a Long Short-Term Memory (LSTM) network (The MathWorks, Inc., 2022), which can represent general nonlinearities and has internal states capable of representing lags. We then co-optimize for the steering wheel angle and pedal and brake position as a function of the GPS-IMU measurements and target path.

The remainder of this section is organized as follows. Section 2 describes the vehicle model and model validation. Section 3 describes the MPC optimization problem to be solved at every sample time. Section 4 shows closed loop experiment results. Finally, Section 5 contains concluding remarks.

2. Vehicle Model

The vehicle model has the following physical states: longitudinal acceleration, a, longitudinal velocity, v, eastward position, x_E, northward position, y_E, yaw rate (clockwise), ω_z, and yaw, ψ. The controls are the normalized pedal, $0 \leq u_p \leq 1$, brake, $0 \leq u_b \leq 1$, and steering, $-1 \leq u_s \leq 1$. The evolution of the acceleration is modeled with a neural network, f_{nn},

$$\begin{bmatrix} a_{k+1} \\ c_{k+1} \\ h_{k+1} \end{bmatrix} = f_{nn}(a_k, v_k, u_{p,k}, u_{b,k}, c_k, h_k), \tag{1}$$

where k denotes a discrete time step (we utilize a 10Hz sampling rate) and c and h denote cell and hidden state vectors for the LSTM network respectively. The function, f_{nn}, consists of an input saturation layer, LSTM layer, and a fully connected output layer. The first layer l_1 adds a bias, β_{in}, and saturation to the brake and pedal inputs. This is used to capture a dead band arising from a non-rigid contact between the robot's push rod and the pedals,

$$l_{1,k} = \begin{bmatrix} a_k \\ v_k \\ max(u_{p,k} + \beta_{in,1}, 0) \\ max(u_{b,k} + \beta_{in,2}, 0) \end{bmatrix}. \tag{2}$$

The second layer is the LSTM layer with weights W_{LSTM} and R_{LSTM} and bias β_{LSTM} where

$$W_{LSTM} = \begin{bmatrix} W_i \\ W_f \\ W_g \\ W_o \end{bmatrix}, R_{LSTM} = \begin{bmatrix} R_i \\ R_f \\ R_g \\ R_o \end{bmatrix}, \beta_{LSTM} = \begin{bmatrix} \beta_i \\ \beta_f \\ \beta_g \\ \beta_o \end{bmatrix}. \tag{3}$$

Let $\sigma_c(x) = \tanh(x)$ and $\sigma_g(x) = (1 + e^{-x})^{-1}$ denote cell and gate activation functions and let $.\times$ denote an element-wise multiplication. The cell and hidden state update are computed from the following sequence of equations.

$$\begin{aligned} i_k &= \sigma_g(W_i l_{1,k} + R_i h_k + \beta_i), \\ f_k &= \sigma_g(W_f l_{1,k} + R_f h_k + \beta_f), \\ g_k &= \sigma_c(W_g l_{1,k} + R_g h_k + \beta_g), \\ o_k &= \sigma_g(W_o l_{1,k} + R_o h_k + \beta_o), \\ c_{k+1} &= f_k.\times c_k + i_k.\times g_k, \\ h_{k+1} &= o_k.\times \sigma_c(c_{k+1}). \end{aligned} \tag{4}$$

Finally, the acceleration at time $k+1$ is computed from a fully connected layer with weights, W_{fc}, and bias β_{fc},

$$a_{k+1} = W_{fc} h_{k+1} + \beta_{fc}. \tag{5}$$

Fitting of the weights $\beta_{in}, W_{LSTM}, R_{LSTM}, \beta_{LSTM}, W_{fc}$ and β_{fc} is executed with MathWork's Deep Learning Toolbox (The MathWorks, Inc., 2022).

The remaining state equations are simple and natural, with a discretization step size, Δt (100msec.),

$$\begin{aligned} v_{k+1} &= v_k + \Delta t a_k, \\ x_{E,k+1} &= x_{E,k} + \Delta t\, v_k cos\,\psi_k, \\ y_{E,k+1} &= y_{E,k} + \Delta t\, v_k sin\,\psi_k, \\ \omega_{z,k+1} &= \alpha_1 \omega_z + \alpha_2 v_k u_{s,k}, \\ \psi_{k+1} &= \psi_k + \Delta t \omega_{z,k}, \end{aligned} \tag{6}$$

where α_1 and α_2 are coefficients, fit with linear least squares, giving the yaw rate response the ability to represent first order dynamics from the steering input.

The model was trained on less than 10 min. of driving data from a combination of human driving and an old version of a MPC controller. The driving scenarios include straight line drives, circular drives and weaving through a series of cones. The maximum speed was 60kph due to limited size of the testing site. The number of cell and hidden states is eight each, i.e., $c, h \in \mathbb{R}^8$.

In addition to the neural network acceleration model, model results from a first order acceleration model and a static model will be shown for comparison. The first order acceleration model, a_{fo}, has the form

$$a_{fo,k+1} = c_0 + c_1 a_{fo,k} + c_2\, max\left(u_{p,k} + \beta_{in,1}, 0\right) + c_3\, max\left(u_{b,k} + \beta_{in,2}, 0\right) + c_4 v_k + c_5 v_k^2, \quad (7)$$

where c_0 through c_5 are fitted coefficients. The static model is the same as the first order model except with $c_1 = 0$. Figure 1 shows a comparison of the models against data from a straight-line drive scenario with a conventional vehicle. As expected, the static model greatly leads the acceleration response, e.g., see the time difference between the acceleration peaks between 38 sec. and 45 sec. While the first order model can roughly match the phase of the oscillations between 38 sec. and 45 sec., the neural net model is able to match the peak-to-peak amplitude better. The neural net model can also express a different brake response, between 55 sec. and 60 sec., from the pedal response whereas the first order model must use the same time constant for both pedal and brake.

Figure 2 shows the modeled yaw rate versus data where the vehicle is weaving through a series of cones. As can be seen, the simple yaw rate model is effective for matching the test data and, as will be shown in later, is sufficient for achieving low lateral path tracking error when used by the MPC.

3. Model Predictive Controller Design

In the following, the optimization problem associated with the MPC will be given and the solver strategy will be discussed.

First, because the pedal and brake should not be pressed at the same time, they can be combined into a single actuator for the purposes of prediction and optimization. Let the combined pedal and brake be denoted u_{pb} and let u_p and u_b in acceleration model, eq. (1), be replaced with $u_p = \max(u_{pb}, 0)$ and $u_b = \max\left(-u_{pb}, 0\right)$. Let u be a concatenation of controls, $u = \left[u_{pb}\ u_s\right]^T$. For prediction, $\beta_{in} = 0$ is utilized, i.e., removing the dead band in prediction, knowing that integral action will take care of static offsets and avoiding loss of local sensitivity to the controls in the MPC computations. Let x_k be a concatenation of discrete time states,

$$x_k = \left[a_k\ v_k\ x_{E,k}\ y_{E,k}\ \omega_{Z,k}\ \psi_k\ c_k^T\ h_k^T\ u_{k-1}^T\right]^T, \quad (8)$$

where u_{k-1} is the control applied at the previous time step and will be used to facilitate evaluating the rate of change of the control in the cost function to be minimized. The equations (1) – (6) and (8) can be compactly written into a system of nonlinear difference equations,

$$x_{k+1} = f(x_k, u_k). \quad (9)$$

We will define the number of states and number of controls as N_s and N_c respectively.

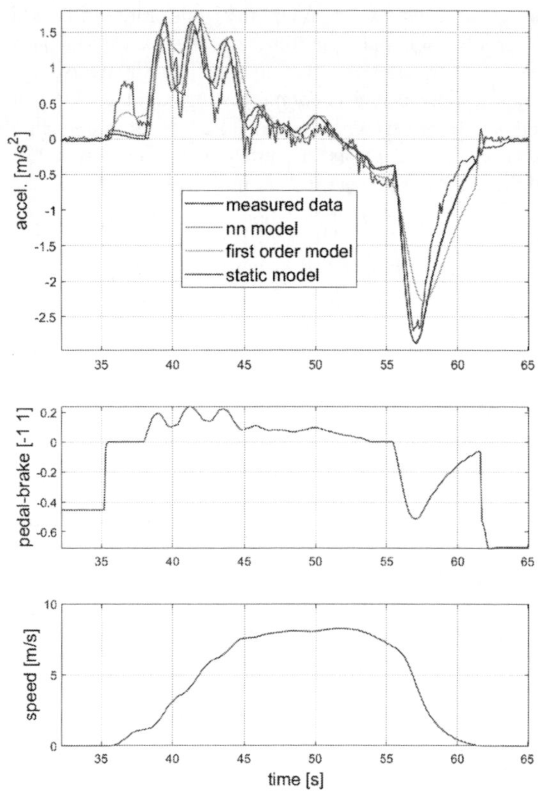

Fig. 1. Acceleration models versus measured data

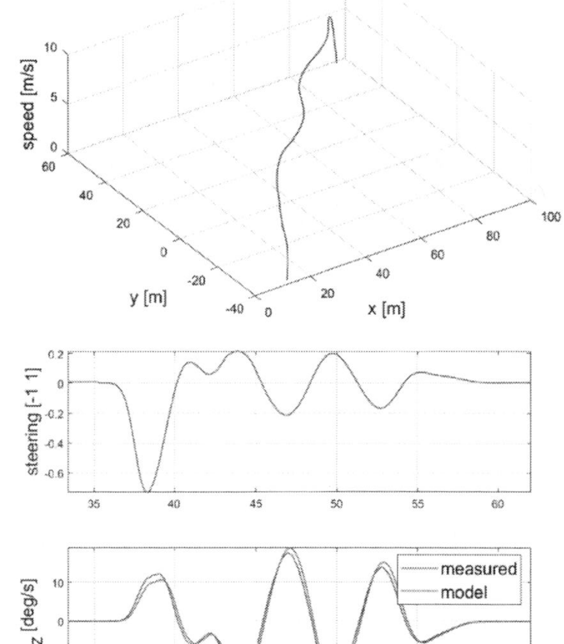

Fig. 2. Yaw rate model versus measured data

The objective of the MPC is to track a sequence of positions as a function of time (or sample), i.e., the test plan is defined through functions $x_E^{target}(t)$ and $y_E^{target}(t)$ which may be stored in a lookup table and interpolated. The optimization problem to be solved at each sample time is

$$\min_{u_0,\ldots,u_{N-1}} \sum_{k=0}^{N-1} \left(x_{E,k} - x_{E,k}^{target}\right)^2 + \left(y_{E,k} - y_{E,k}^{target}\right)^2 + (u_k - u_{k-1})^T R (u_k - u_{k-1})$$

$$\textit{subject to the constraints:}$$
$$x_0 = x(t),$$
$$x_{k+1} = f(x_k, u_k),$$
$$-1 \le u_k \le 1,$$
$$v_k \ge 0, \qquad (10)$$

where N is the horizon length and $x(t)$ is the measured state at sample time t with cell and hidden states obtained from evaluating equations (2) and (4) once per sample. The targets, $x_{E,k}^{target}$ and $y_{E,k}^{target}$ are obtained from sampling $x_E^{target}(t)$ and $y_E^{target}(t)$. The cost function in (10) penalizes the control effort through the control rate of change, $u_k - u_{k-1}$, giving a smooth control action and rider comfort and removes the need to compute a control target.

An alternative to the cost function in (10) would be to instead have terms to track speed and yaw targets (or x-y velocity components instead of yaw) and penalize lateral path offset and control effort. In our simulations and experiments, we have observed that this alternate version has difficulties at low speed, e.g., less than 10-15kph. Also, compared to (10), this alternate version has more tuning parameters, and requires additional integrators / adaptors / estimators for speed tracking, which also need to be tuned. In contrast, the MPC, (10), only has tuning on the control effort and speed tracking is inherited because position error is the integral of speed error, assuming the speed targets and position targets are consistent, i.e., $v^{target}(t) = \left[\dot{x}_E^{target}(t)\ \dot{y}_E^{target}(t)\right]^T$.

To assist the solving of the optimization problem (10) in real time, the inequality constraints are converted to an exterior penalty function

$$\gamma(u_k, v_k) = \gamma_1 \left|\left|max(u_k - 1, 0)\right|\right|_2^2 + \gamma_1 \left|\left|max(1 - u_k, 0)\right|\right|_2^2 + \gamma_2 \left|\left|v_k\right|\right|_2^2, \qquad (11)$$

where $\left|\left|x\right|\right|_2^2 = x^T x$ and γ_1 and γ_2 are tuning variables which can be set large relative to R, i.e., $\gamma_1, \gamma_2 \gg \left|\left|R\right|\right|$. In practice, values for γ_1 and γ_2 are easily selected in simulations and does not need to be changed afterward.

For convenience, let the incremental cost, l, be defined as

$$l(x_k, u_k) = \left(x_{E,k} - x_{E,k}^{target}\right)^2 + \left(y_{E,k} - y_{E,k}^{target}\right)^2 + (u_k - u_{k-1})^T R (u_k - u_{k-1}) + \gamma(u_k, v_k). \qquad (12)$$

The exterior penalized optimized problem to be solved in real time is

$$\min_{u_0,\ldots,u_{N-1}} \sum_{k=0}^{N-1} l(x_k, u_k)$$

$$\textit{subject to the constraints:}$$
$$x_0 = x(t),$$
$$x_{k+1} = f(x_k, u_k). \qquad (13)$$

From the optimization problem (13), necessary conditions for optimality are obtained through applying Karush-Kuhn-Tucker conditions and are as follows,

$$x_{k+1} = f(x_k, u_k)$$
$$\forall k \in \{0, \ldots, N-1\}, x_0 = x(t), \qquad (14)$$

$$p_k = l_x(x_k, u_k) + f_x^T(x_k, u_k) p_{k+1}$$
$$\forall k \in \{1, \ldots, N-1\},\ p_N = 0, \qquad (15)$$

$$l_u(x_k, u_k) + f_u^T(x_k, u_k) p_{k+1} = 0$$
$$\forall k \in \{0, \ldots, N-1\}, \qquad (16)$$

where l_x, l_u, f_x and f_u are $\partial l/\partial x$, $\partial l/\partial u$, $\partial f/\partial x$ and $\partial f/\partial u$ respectively. The co-states, $p_k \in \mathbb{R}^{N_s}$, are the Lagrange multipliers associated with state equation equality constraints. Equation (15) is referred to as the co-state equation and equation (16) is referred to as the stationarity condition.

Let X be the sequence of future states, $X = [x_0^T, \ldots, x_{N-1}^T]^T \in \mathbb{R}^{(N-1)N_s}$ and P be the sequence of co-states, $P = [p_1^T, \ldots, p_N^T]^T \in \mathbb{R}^{(N-1)N_s}$. The states, X, can be computed as a function of U and initial condition $x(t)$ through propagation of the state equations, i.e., $X = X(U, x(t))$. The co-states can be computed as a function of U and X through back-propagation of the co-state equations, i.e., $P = P(X, U)$. We define the residual, $F \in \mathbb{R}^{(N-1)N_c}$, as the left-hand side of the stationarity conditions,

$$F(X, U, P) = \begin{bmatrix} l_u(x_0, u_0) + f_u^T(x_0, u_0) p_1 \\ \cdots \\ l_u(x_{N-1}, u_{N-1}) + f_u^T(x_{N-1}, u_{N-1}) p_N \end{bmatrix}. \qquad (17)$$

Note that $F(X, U, P) = F\big(X(U), U, P(X(U), U)\big) = F(U)$. Newton's method can now be used to find a sequence of U's such that $F(U) \to 0$, i.e., satisfying the necessary conditions. Let $U^i(t)$ denote the i-th iteration of U at sample time t. The Newton update is

$$U^{i+1}(t) = U^i(t) + \delta U, \qquad (18)$$

where δU is the Newton step. Specifically, the Forward Difference Conjugate Gradient (FDCG) algorithm (Algorithm 2.5.1 in Kelley, 1999) is used calculate δU starting from an initial point $U^i(t)$ and residual $F(U^i(t))$. The advantage of the FDCG algorithm is that it does not require computing the Jacobian, $J(U^i(t))$, of the residual around the point $U^i(t)$. Rather the Jacobian times vector product in the standard Conjugate Gradient (CG) algorithm (Algorithm 1.5.1 in Kelley, 1999) can be approximated by a forward difference calculation,

$$J(U^i(t))\rho \approx \frac{\left(F(U^i(t) + \epsilon\rho) - F(U^i(t))\right)}{\epsilon}. \qquad (19)$$

Finally, a Real Time Iteration (RTI) strategy (Diehl et al. 2005) will be used, meaning that the optimization problem (13) will not be solved to completion, but rather, a fixed, finite number, N_{New}, of Newton steps (18) will be taken each sampling instant. The initial guess for the sequence of controls at the next sampling instant is the final iterate from the current sampling instant, i.e., $U^0(t + \Delta t) = U^{N_{New}}(t)$.

4. Experiment Results

The MPC used in experiments utilize a prediction horizon of 4.5 sec. (N = 45 steps), the sampling rate is 10Hz and 10 Newton steps are taken every sample. Figure 3 shows experiments with MPC controlling a conventional vehicle along a straight-line path. The target positions versus time were obtained from a human drive. Note that the red pedal-brake line is not used as a target for the control and is only there to show how the human drove the path. Both a MPC with the neural net powertrain model and a MPC with a first order model are shown, where the MPC with a first order model struggles to track the speed target while the MPC with neural net can track the target closely. Different tunings of the MPC with the first order model were not able to achieve satisfying speed tracking performance. A different configuration with a velocity and lateral offset type cost function was somewhat more successful but still did not perform as well as the MPC with neural net model.

The MPC with neural net in Fig. 3 gains some speed at the beginning of the experiment when the target is zero. This is due to down-sampling of the target drive and target interpolation and is not an issue of the MPC itself. At the transition between acceleration and braking, the MPC looks like it is a little slow to brake. This is due to the deadband as the pedal comes off and as the brake engages which the MPC is currently configured to smoothly pass through.

Figures 4 and 5 show the MPC (with neural net powertrain model) controlling the conventional vehicle along a cone weaving path and along a circular path respectively. In these cases, the speed tracking is also satisfactory and the path error, as defined as the distance to the closest point on the target path, is less than 30 cm after the initial condition response. Note that the MPC softly stops to zero speed due to the $v \geq 0$ constraint. This causes some overshoot of the final target position (and is why the path error becomes large at the end of the path).

Finally, Fig. 6 shows the MPC driving an electric vehicle along a cone weaving path (with the model refit to electric vehicle data). As expected, the speed tracking and path error is good. The main thing to note is that after model refitting, no other parameters/tunings of the MPC were changed from the conventional vehicle settings. This shows the general applicability of the approach.

5. Conclusions

This paper describes a MPC for a driving robot to be used for testing ADAS systems. Specifically, this MPC utilizes a neural network, with LSTM layer, to model the acceleration dynamics of the vehicle. Being able to easily adapt to a wide variety of vehicles and powertrains is necessary for the driving robot, a capability not clearly seen in other robot driver or autonomous driving developments. Despite the model being data driven, little data is needed to fit the model, e.g., 10 min of driving data. Furthermore, no tuning of the controller was required when swapping between the conventional and electric vehicles in our experiments.

In future work, we will be testing the controller in additional scenarios and higher speeds, specifically ensuring that Euro NCAP test requirements are met. We will also be considering the coordination of the test vehicle with other robot-controlled vehicles and objects.

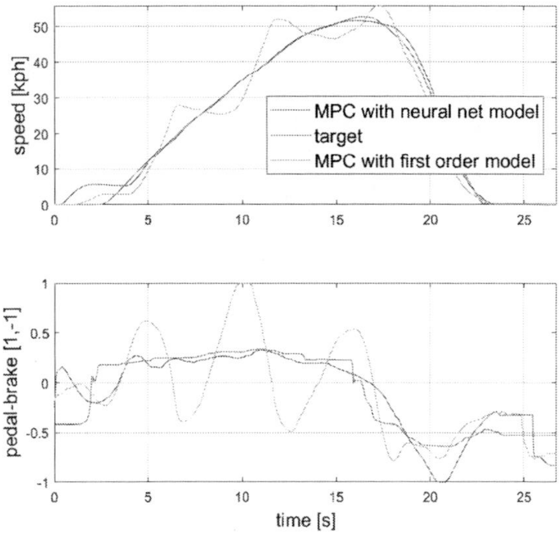

Fig. 3. Speed tracking performance on a conventional vehicle comparison between MPC with neural network model and MPC with a first order powertrain model on a straight-line drive

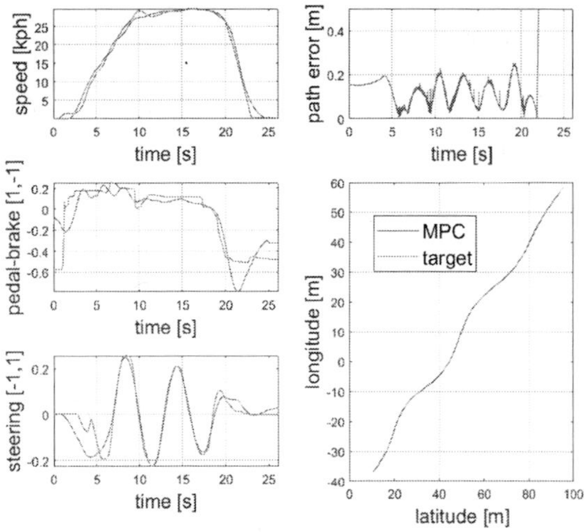

Fig. 4. Speed and path tracking performance on a conventional vehicle with the MPC when weaving through cones

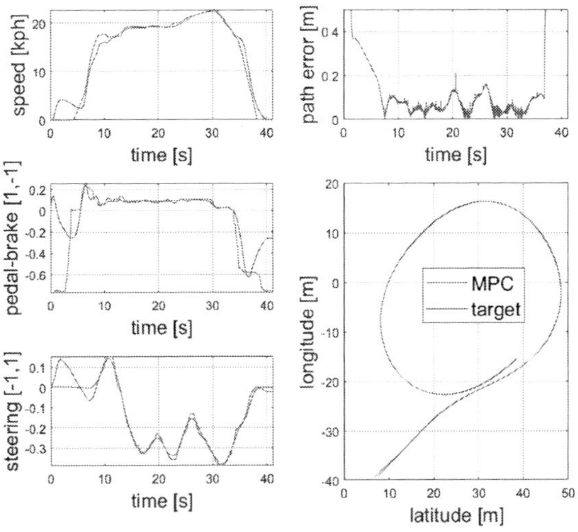

Fig. 5. Speed and path tracking performance on a conventional vehicle with the MPC when driving in a circle

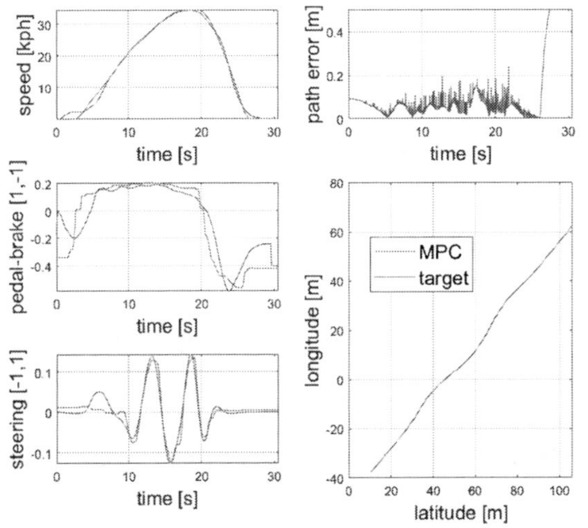

Fig. 6. Speed and path tracking performance on a electric vehicle with the MPC when weaving through cones

6. References

Alcalá, E., Puig, V., Quevedo, J., 2019. LPV-MPC Control for Autonomous Vehicles. *IFAC PapersOnline*. Vol. 32, No. 28, pp. 106-113.

Diehl, M., Findeisen, R., Allgöwer, F., Bock, H. G., Schlöder, J. P., 2005. Nominal Stability of the Real-Time Iteration Scheme for Nonlinear Model Predictive Control. *IEEE Proceedings-Control Theory and Applications*, Vol. 152, No. 3, pp. 296-308.

Euro NCAP, 2020. 2020 Assisted Driving Tests, Retrieved from https://www.euroncap.com/en/vehicle-safety/safety-campaigns/2020-assisted-driving-tests/, July 13, 2022.

Kelley, C. T., 1999. Iterative Methods for Optimization. *Society for Industrial and Applied Mathematics*, Philadelphia.

Liao-McPherson, D., Huang, M., Zaseck, K., 2020. Smoothed and regularized Fischer-Burmeister solver for embedded real-time constrained optimal control problems in autonomous systems. *United States Patent*, No. US 10,739,768 B2.

The MathWorks, Inc. 2022. Deep Learning Toolbox. Retrieved from https://www.mathworks.com/products/deep-learning.html, July 14, 2022.

Park, J., Kim, H., Hwang, K., Lim, S., 2022. Deep reinforcement learning based dynamic proportional-integral (PI) gain auto-tuning method for a robot driver system. *IEEE Access*. Vol 10., pp. 31043-31057.

Sun, C., Zhang, X., Zhou, Q., Tian, Y., 2019. A model predictive control with switched tracking error for autonomous vehicle path tracking. IEEE Access. Vol. 7, pp. 53103 – 53114.

Tang, L., Yan, Fuwu, Y., Zou, B., Wang, K., Chen L., 2020. An improved kinematic model predictive control for high-speed path tracking of autonomous vehicles. *IEEE Access*. Vol 8., pp. 51400-51413.

Ugo, R., Borelli, F., 2020. Learning how to autonomously race a car: a predictive control approach. Direct control acceleration and steering angle. *IEEE Transactions on Control Systems Technology*, Vol. 28, No. 6, pp. 2713-2719.

Wang, H., Liu, B., Ping, X., An Q., 2019. Path tracking control for autonomous vehicles based on an improved MPC. *IEEE Access*. Vol. 7, pp. 161064-161073.

2632 / Energy Consumption Analysis of Door Opening with a Mobile Manipulator in the Nuclear Power Plants

Changyou Ma[a]*, Binbin Yan[a], Xu Xiong[a], Haibo Gao[b], and Liang Ding[b]

[a]Jiamusi University, Jiamusi, Heilongjiang, China
[b]Harbin Institute of Technology, Harbin, Heilongjiang, China
**Corresponding author: mcyazh@126.com*

https://doi.org/10.56884/FPQM8003

Abstract

Abstract: The problem of rescue is necessary for the mobile manipulator in nuclear power plants (NPPs), however, the research on energy consumption is of great importance to the mobile manipulator in the process of rescue. In this paper, we focus on the energy consumption of nuclear power plant rescue robot about the task of door-opening and present a method of path planning of the door opening. The energy consumption model is derived to evaluate the performances of modular manipulators. Extensive door-opening capture experiments have been carried out. The compares of the manipulator summation energy consumption of the different holding positions of knob radius rk, and the energy consumption of each joint in the same path planning are studied. Experimental results show that the minimum summation of manipulator energy consumption is the end-effector grasps the position of doorknob, which is the value of rk (40mm), and the third joint of the average energy consumption is the maximal in the same path planning.

Keywords: energy consumption, door opening, mobile manipulator, nuclear power plants

This work was supported by the National Key Basic Research Development Plan Project of China (973) (2013CB035502), Heilongjiang Province Higher Education Project of Basic Scientific Research (2021-KYYWF-0558).

1. Introduction

The Fukushima Daiichi NPP accident in Japan on March 11, 2011 was triggered by an earthquake of magnitude 9.0 and the resultant tsunami marked the beginning of the worst nuclear accident of the last two decades (Ma, 2014).After the Fukushima Daiichi NPP accident, the Defense Advanced Research Projects Agency (DARPA) initiated a new challenge in 2012, the DARPA Robotics Challenge (DRC) (DeDonato, 2015), including eight rescue tasks to test the capabilities of teams, as well as of individuals. The tasks were ranked by DARPA in terms of difficulty as Valve (easiest); Terrain and Hose (easier), Door, Debris, Wall, and Ladder (harder) and Vehicle (hardest) (Yanco, 2015).

At present, there are two kinds of research methods about the problem of door opening manipulator. One is to study how the robot can accurately recognize the door handle in unknown environment. The other is to study the movement control of the robot base, especially the obstacle avoidance in the movement process. However, in some special occasions, the energy consumption of the robot door opening needs to be considered. For example, in the nuclear power plant accident treatment, the use of energy needs to be reasonably arranged to complete the rescue work faster.

Nagatani (Nagatani, 1996) proposed a method for opening a door and passing through a doorway using a mobile manipulator called a YAMABICO-10 robot. Sachin Chitta managed to overcome the high-dimensionality planning problem during the opening process by identifying a graph-based representation that is small enough for efficient planning yet rich enough to contain feasible motions that open doors (S. Chitta, 2010). Peterson et al. presented the design and implementation of a door-opening controller using a hybrid dynamic system model, when using this simple controller, the radius and center of rotation of the door are estimated online. The results of their experiments demonstrated that off-the-shelf algorithms for force/torque control are very effective for solving the task of grasping the handle and opening the door (Peterson L, 2000). Chung et al. (Chung W, 2009) proposed a control strategy for the door-opening procedure executed by a service robot, PSR1, which utilized a three-fingered robot hand for grasping the door handle. Hongjun Xing user teleoperation system, consisting of two local robots and an omnidirectional WMM, are conducted to verify the proposed approach's feasibility and effectiveness (Hongjun Xing, 2022).

Tomas proposed a general framework for opening doors and drawers in unknown environments (Ruhr T, 2012). Their framework can be divided into four parts: detecting and localizing the handle, operating the handle while learning the kinematics model, integrating the learned model into robot's semantic map and re-use the model for future encounters. The framework has been well tested on PR2 platform and the approach was successful in 51.9% of all 104 trails.

Yiannis and Christian designed a adaptive controller to deal with the kinematics and dynamics uncertainties of doors (Karayiannidis, 2012). This controller explicitly included the uncertainty estimates and can be proved to be stable even with initially large errors

A general framework for door opening task with a Kinect as a vision sensor was proposed by Zhou Li and Guodong Chen (Zhou L, 2013). A grasp pattern for various kinds of door handles and a door-opening pattern were proposed. Then they focused mainly on getting and processing 3D point cloud with which the different types of door handles can be recognized and the location of the door axis can be computed. The framework was tested on Gazebo simulator with a SCHUNK robot arm and a Kinect imported as a vision sensor.

The rest of paper is organized as follows. Section 2 provides the description of the mobile modular robot and introduces the forward kinematic and the inverse kinematics model of the 6-DOF modular manipulators. In the Section 3 we address the method and path planning of door opening. The energy consumption model is derived to evaluate the performances of modular manipulators in Section 4. The results of experiments are discussed in Section 5, followed by some conclusions in Section 6.

2. Description of the Mobile Modular Robot

The kinematics model configuration of the 6-DOF Schunk modular manipulator is shown in Fig. 1.

The forward kinematic is developed by using D-H formalism and represents by ${}^{0}_{6}T$.

$$ {}^{0}_{6}T = \begin{bmatrix} R_{3\times3} & & & T_{1\times3} \\ 0 & 0 & 0 & 1 \end{bmatrix} = \begin{bmatrix} r_{11} & r_{12} & r_{13} & T_x \\ r_{21} & r_{22} & r_{23} & T_y \\ r_{31} & r_{32} & r_{33} & T_z \\ 0 & 0 & 0 & 1 \end{bmatrix} \qquad (1) $$

Where $R_{3\times3}$ and $T_{1\times3}$ are the rotation and translation matrix. Based on Eq. 1, the joint angles by inverse kinematics calculations can be derived as follows:

$$ \begin{bmatrix} \theta_1 \\ \theta_2 \\ \theta_3 \\ \theta_4 \\ \theta_5 \\ \theta_6 \end{bmatrix} = \begin{bmatrix} \operatorname{atan2}(T_y, T_x) - \operatorname{atan2}(0, \pm\sqrt{T_x^2 + T_y^2})_1 \\ \theta_{23} - \theta_3 \\ \operatorname{atan2}(K, \pm\sqrt{d_4^2 - K^2}) \\ \operatorname{atan2}(-r_{13}s_1 + r_{23}c_1, -r_{13}c_1c_{23} - r_{23}s_1c_{23} + r_{33}s_{23}) \\ \operatorname{atan2}[-r_{13}(c_1c_{23}c_4 + s_1s_4) - r_{23}(s_1c_{23}c_4 - c_1s_4) \\ r_{33}s_{23}c_4, -r_{13}c_1s_{23} - r_{23}s_1s_{23} - r_{33}c_{23}] \\ \operatorname{atan2}(s_6, c_6) \end{bmatrix} \qquad (2) $$

$$ where: K = \frac{T_x^2 + T_y^2 + T_z^2 - a_2^2 - d_4^2}{2a_2}, $$

$$ \theta_{23} = \operatorname{atan2}[-a_2c_3T_z - (c_1T_x + s_1T_y)(d_4 - a_2s_3), (a_2s_3 - d_4)T_z - (a_3 + a_2c_3)(c_1T_x + s_1T_y)], $$

$$ c_i = \cos(\theta_i), and\ s_i = \sin(\theta_i) $$

The joint angles own eight available analytical solutions based on the intersection of the last three joints of the arm. An algorithm which can minimize the Euler distance in joint space from the initial state should be chosen from the eight solutions (Li J, 2015). Jacobian matrix of 6-DOF manipulator as Eq. 3 showed:

$$ J = \begin{bmatrix} J_{11} & J_{12} & J_{13} & J_{14} & J_{15} & J_{16} \\ J_{21} & J_{22} & J_{23} & J_{24} & J_{25} & J_{26} \\ J_{31} & J_{32} & J_{33} & J_{34} & J_{35} & J_{36} \\ 0 & 0 & 0 & J_{44} & J_{45} & J_{46} \\ 0 & 0 & 0 & J_{54} & J_{55} & 0 \\ 0 & 0 & 0 & 0 & 0 & 1 \end{bmatrix} \qquad (3) $$

Based on Eq. 3, angular velocity $\dot{\theta}_i$ and acceleration $\ddot{\theta}_i$ of the joint can be derived as follows:

$$ \left(\dot{\theta}_1\ \dot{\theta}_2\ \dot{\theta}_3\ \dot{\theta}_4\ \dot{\theta}_5\ \dot{\theta}_6\right)^T = J^{-1}\sum\dot{T} \qquad (4) $$

$$ \left(\ddot{\theta}_1\ \ddot{\theta}_2\ \ddot{\theta}_3\ \ddot{\theta}_4\ \ddot{\theta}_5\ \ddot{\theta}_6\right)^T = J^{-1}\ddot{P} + J^{-1}\dot{T} \qquad (5) $$

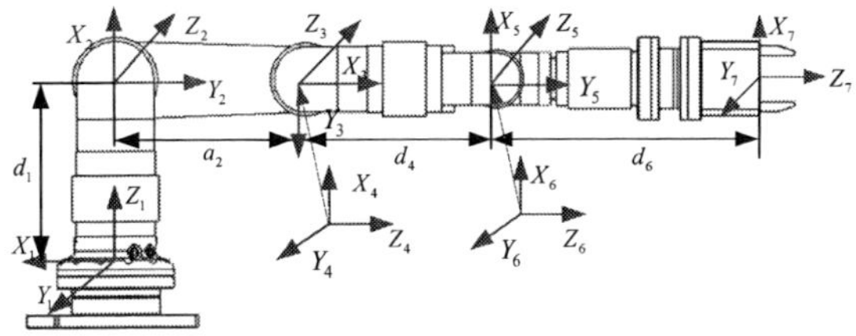

Figure 1. The kinematics configuration of the 6-DOF manipulator

3. Door-Opening

In this paper, we focus on the door opening and have accomplished the searching/reaching task of the door (Gao H, 2019). In the proposed door-opening method, after the mobile manipulator robot has grasped the doorknob, and rotated, position control is used to rotate the doorknob and open the door using only the mobile manipulator. During this process, the home position of the mobile base, the doorknob radius, the door radius, and the height of the doorknob are measured using the API T3 laser tracker system. These measured parameters are then used for planning the path of the mobile manipulator that will allow rotating the doorknob and the opening the door to the desired angle (Xing H, 2019).

3.1 Door-Opening Method

In this section, we propose a door-opening method. The following assumptions were made: 1) The door opening and door handle rotating directions are known (the door is opened toward the left-side and, the door handle is rotated toward the right-side); 2) the door axis of rotation is perpendicular to the floor; 3) the door moves in the horizontal *xy* plane; 4) the mobile platform travels on the ground, which can always be adjusted for a structured laboratory environment; and 5) the axis of rotation of the first manipulator module is perpendicular to the ground (Ma C, 2018).

3.2 Path Planning

The model of the door-opening process using the mobile manipulator is depicted in Fig. 2. It shows the origin of the reference frame $\{O_d\}$ is set at the intersection point of the door hinge, the origin of the reference frame $\{O_k\}$ is set in the center point of the doorknob axis, the horizontal plane that crosses the origin of the reference frame of the mobile manipulator $\{O_b\}$ and the reference frame $\{O_e\}$ is placed at the position the manipulator end-effector has grasped the doorknob.

In order to plan the path of the mobile manipulator, we firstly need to know the accurate value of the door knob radius (r_k) and the door radius (r_d) , the initial base position of the mobile manipulator (x_b, y_b, z_b), and the position of end-effector holds the door knob firmly (x_e, y_e, z_e). The doorknob motion is conformed to follow the doorknob trajectory in the *xz* plane with a center of rotation at (x_k, y_k) and a radius r_k in the Fig. 2. With the assumption that the end-effector grasps the doorknob firmly, the end-effector's position forms the trajectory of doorknob during twisting doorknob process. The rotating trajectory radius of doorknob is derived as follows:

$$[x_e(t)-x_k]^2+[z_e(t)-z_k]^2=r_k^{\,2} \qquad (6)$$

The door motion is conformed to follow the door trajectory in the *xy* plane with a center of rotation at (x_d,y_d)and a radius rd in the Fig. 2. With the assumption that the end-effector grasps the doorknob firmly, the end-effector's position forms the trajectory of door knob during twisting door knob process. The trajectory radius of pulling door is derived as follows:

$$[x_e(t)-x_d]^2+[y_e(t)-y_d]^2=r_d^{\,2} \qquad (7)$$

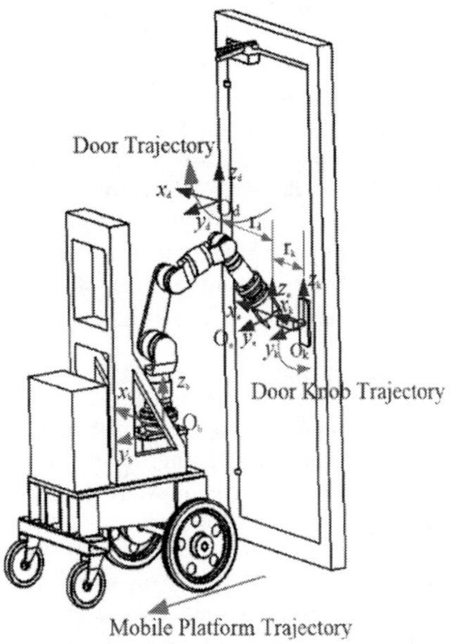

Figure 2. The model of the door-opening process using mobile manipulator

3.2.1 Rotating Doorknob

The mobile manipulator realizes rotating the doorknob by tracking the circle arc represented by the doorknob trajectory as shown in Fig. 3.

The arc of doorknob trajectory can be acquired by interpolation using points Q_1, Q_2, and Q_3. In the Fig. 3, r_k denotes the length of the rotating radius which the end-effector grasps doorknob, θ_k denotes the rotation angle of the doorknob when reaching points *Q1, Q2*, and *Q3*, and the coordinate position of *Q1 (x1, z1*) in the *xz* plane. Above all, the points can be calculated by the following formula:

$$Q_{1,2,3}=\begin{bmatrix} Q_x \\ Q_y \\ Q_z \end{bmatrix}=\begin{bmatrix} r_k(1-\cos\theta_k) \\ y_1 \\ z_1-r_k\sin\theta_k \end{bmatrix} \qquad (8)$$

In this paper, we choose the orientations of the end-effector to be consistent with the rotating doorknob during the twisting process. When posture of the end-effector stays constant with door handles as shown in Fig. 3, the rotation matrix varies denotes ${}^0_eR_k^{(t)}$ during the process. ${}^0_eR_k^{(0)}$ denotes the initial rotation matrix of the end-effector holds the doorknob can be get using forward kinematics. Then at time *t*, this matrix is multiplied by a rotation matrix parameterized by the rotating angle of the doorknob $\theta_k^{(t)}$, as shown in (9).

$${}^0_eR_k^{(t)}=\begin{bmatrix} \cos\theta_k^{(t)} & 0 & \sin\theta_k^{(t)} \\ 0 & 1 & 0 \\ -\sin\theta_k^{(t)} & 0 & \cos\theta_k^{(t)} \end{bmatrix}\cdot {}^0_eR_k^{(0)} \qquad (9)$$

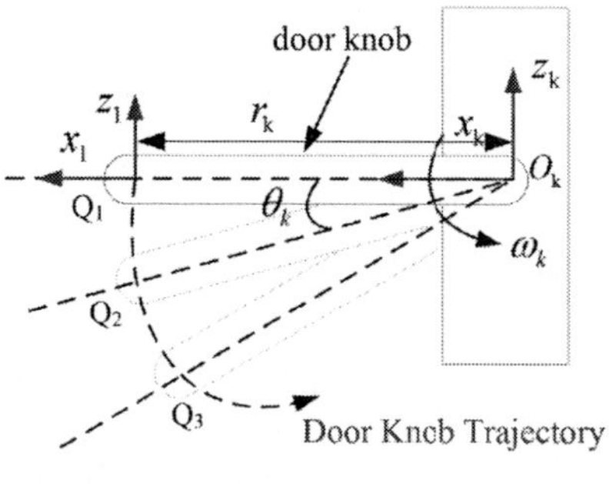

Figure 3. Planar model of rotating doorknob

3.2.2 Pulling Door

The mobile manipulator achieves pulling the door by tracking the circle arc represented by the door trajectory as shown in Fig. 4. The arc of door trajectory can be acquired by interpolation using points *Q4, Q5*, and *Q6*. In the Fig. 4, r_d denotes the length of the rotating radius which the end-effector grasps door, θ_d denotes the rotation angle of the door when reaching points *Q4, Q5*, and *Q6*, and the coordinate position of *Q4* (x_4, y_4) in the *xy* plane.

The position of end-effector on the points *Q4, Q5*, and *Q6.* are computed as follows:

$$Q_{4,5,6}=\begin{bmatrix}Q_x\\Q_y\\Q_z\end{bmatrix}=\begin{bmatrix}r_d(1-\cos\theta_d^{(t)})\\y_4-r_d\sin\theta_d^{(t)}\\z_4\end{bmatrix} \quad (10)$$

In this section, the orientations of the end-effector and the door are the same during the pulling door process. When posture of the end-effector stays constant with doorknob as shown in Fig. 4, the rotation matrix varies denotes ${}_e^0R_d^{(t)}$ during the process. ${}_e^0R_d^{(0)}$ denotes the initial rotation matrix of the end-effector after twisting doorknob can be get using forward kinematics. Then at time t, this matrix is multiplied by a rotation matrix parameterized by the rotating angle of the doorknob $\theta_d^{(t)}$, as shown in (11).

$${}_e^0R_d^{(t)}=\begin{bmatrix}\cos\theta_d^{(t)} & -\sin\theta_d^{(t)} & 0\\ \sin\theta_d^{(t)} & \cos\theta_d^{(t)} & 0\\ 0 & 0 & 1\end{bmatrix}\cdot {}_e^0R_d^{(0)} \quad (11)$$

Above analyses, the end-effector trajectory of the door-opening process is shown in Fig. 5. This Cur consisting of the trajectory of close to holding the handle, the rotating door knob and pulling door. Fig. 6 shows the manipulator joints angle of the door-opening process which is based on Eq. 2, in the Section 2.

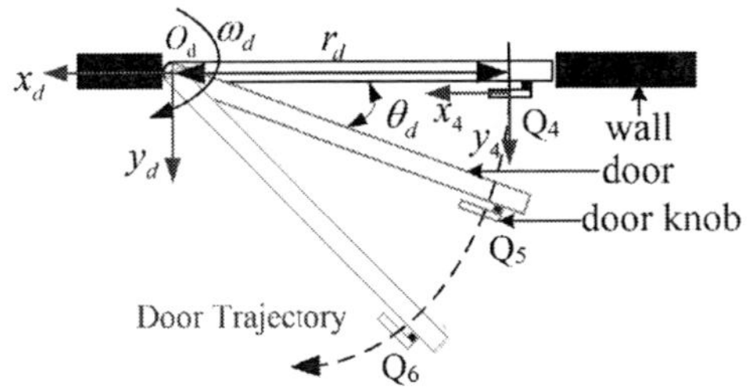

Figure 4. Planar model of pulling door

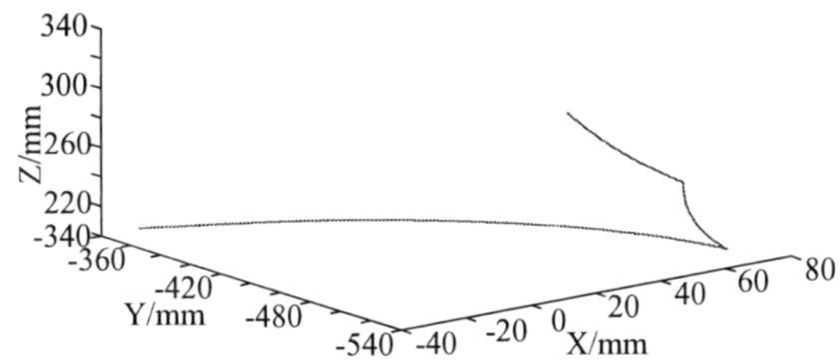

Figure 5. The end-effector trajectory of the door-opening process

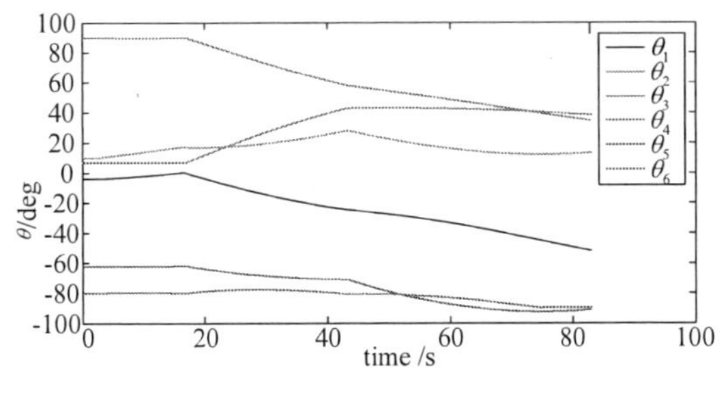

Figure 6. The manipulator joints angle of the door-opening process

4. Formulation of Power Consumption

The energetic cost is an important metric for energy consumption for the manipulator. We experimentally evaluate how different grasp the different position of the doorknob and wave gait patterns affect the power consumption of the manipulator. The input energy E_m of the motor is as follows:

$$E_m=\int_0^T U_{in}\cdot I_m dt\cdots \quad (12)$$

Where U_{in} is the voltage of power supply. I_m is the instantaneous current of DC motor. The total energy consumed by the robot used by the actuators include the generating mechanical power (P_{mech}), heat power (P_{heat}) and power losses (Δ_{loss}). This can be described as:

$$P_{in}=P_{mech}+P_{heat}+\Delta_{loss} \quad (13)$$

The torque τ_m of each actuator is related to the torque constant K_m, i is the gear ratio of joint, η is the efficiency of transmission mechanism and instantaneous current I_m can be given by

$$\tau_m=K_m\cdot I_m\cdot i\cdot\eta \quad (14)$$

The mechanical power generated by each actuator is related to the rotation speed and torque (Zhuang H, 2013). Therefore, the total instantaneous mechanical power during movement of the manipulator can be stated by

$$T_s=60$$
$$P_{mech}=\sum_{j=1}^{6}\tau_{mj}\cdot\left(\frac{n_j}{T_s}\right) \quad (15)$$

where P_{mech} is the total mechanical power of actuators, T_s is unit conversion constant, τ_{mj} is the torque of each actuator in $N{\cdot}m$, $n_j$is the rotation speed of motors in rpm, j is the number of manipulator joints.

The heat dissipated by the motors can be given by

$$P_{heat}=\sum_{j=1}^{6}\frac{R_i}{K_{mj}^{2}}\cdot\tau_{mj}^2 \quad (16)$$

The total power consumed of the manipulator can be computed by

$$P_{mot}=\sum_{j=1}^{6}\tau_{mi}\cdot\left(\frac{n_j}{T_s}\right)+\sum_{j=1}^{6}\frac{R_i}{K_{mj}^{2}}\cdot\tau_{mj}^2 \quad (17)$$

The performance parameters of motors and actuating devices of joints are showed in Table 1.

Table 1. Performance parameters of drive devices and actuating devices of joints

Joint of ARM j	U_{in}/(V)	K_{mj}/(mN·m)	i_j	R_j/(Ω)	η_j/(%)
Joint1、2	24	31.4	596	1.5	0.85
Joint3、4	24	38	625	0.6	0.85
Joint5、6	24	16	552	0.1	0.85

5. Experimental Results

5.1 Experimental set up

The mobile robot system (see Figure 7) consists of a 6-DOF modular manipulator produced by the Schunk Company, a four-wheeled platform with two drive wheels, and a two free wheel sand two-finger gripper mounted on the arm wrist module. The 6-DOF modular manipulator comprises three types of modules, and each joint module consists of a brushless DC motor, a harmonic drive, a braking system, and an encoder. The mobile robot system is equipped with various types of sensors, including two six-axis force sensors and a camera. The two six-axis force sensors are mounted on the manipulator wrist module and the base module, respectively, to measure the mechanics date of the door-opening procedure. A joystick with associated force feedback control from the base-mounted 6-axis force/torque sensor is used for teleoperation. A camera is mounted on top of the frame to support 3D displays for continuous teleportation and object recognition (Ma C, 2018).

At this stage, two kinds of experiments are carried out: (1) comparative power consumption of the different path and (2) joint power consumption of the same path. During the comparative power consumption of the different path experiments, the end-effector holds the doorknob on the different positions according to the planning trajectory of the section3 in order to find the minimum power consumption of the manipulator. The positions are 10, 25, 35, 50, and 80mm, which are the length of the doorknob rotating radius r_k. In the joint power consumption of the same path experiment, the end-effector grasps the doorknob on the position of the minimum summation of power consumption in the experiment of comparative power consumption of the different path. The end-effector holds doorknob moves along the planning trajectory of the section 3.

The door-opening experiment is shown by sequential pictures in the Fig. 8. The mobile robot system is used to measure the instantaneous current Im and the rotation speed of the each joint ni of manipulator in all experiments. The tested data is imported in real-time during the experimental progress and stop the mobile robot system in the end. The experiments with the same condition are performed for more than three times to ensure the repeatability and effectiveness of the experimental results.

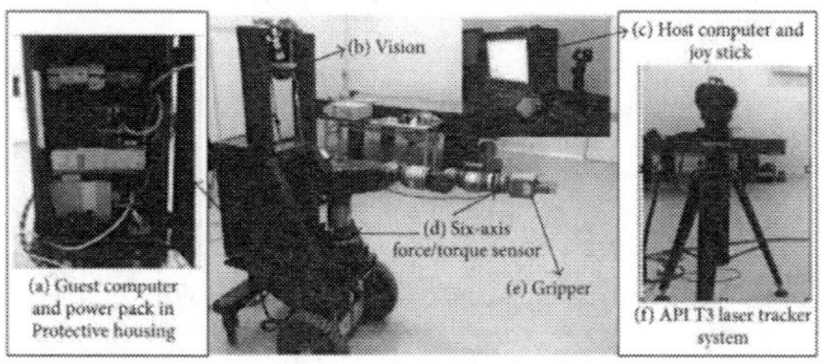

Figure 7. Software and hardware of the mobile manipulator

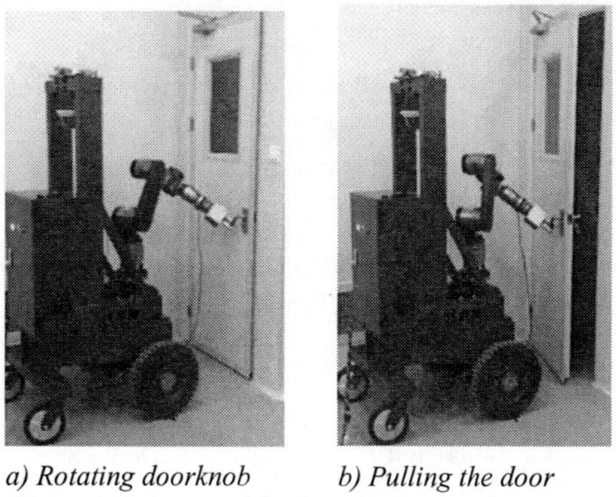

a) Rotating doorknob *b) Pulling the door*
Figure 8. Pictures of the door opening experiment

5.2 Comparative Power Consumption of Different path

Fig. 9 shows the comparison of the sum power consumption of the 6 joints with the end-effector holding the five different positions of the doorknob, which are the length of the doorknob rotating radius r_k, 70, 55, 40, 25, and 10mm, move along the door-opening trajectory. G1, G2, G3, G4, and G5 denote the group number corresponding to the value of rotating radius r_k, 70, 55, 40, 25, and 10mm. In the Fig. 8, the end time of door-opening is 82s, the rotating doorknob time is 20-44s, and the pulling door time is 44-82s. The max power consumption is G5 during the rotating doorknob because the end-effector grasping the point closed to the rotating axis of doorknob. The max power consumption is G2 during the pulling door.

Fig. 10 shows the average of the sum of the joints power consumption with the five different doorknobs rotating radius r_k, 70, 55, 40, 25, and 10mm, during the door-opening process. The x-axis denotes the number of G1, G2, G3, G4, and G5. In the Fig. 10, we could obtain the max value is G5(103.33W), the min value is G3(81.02W), G1 is 85.68W, G2 is 89.29W, and G4 is 90.37W. Based the analysis, the end-effector grasps on the value of r_k (40mm), is the minimum summation of manipulator power consumption during the door-opening process.

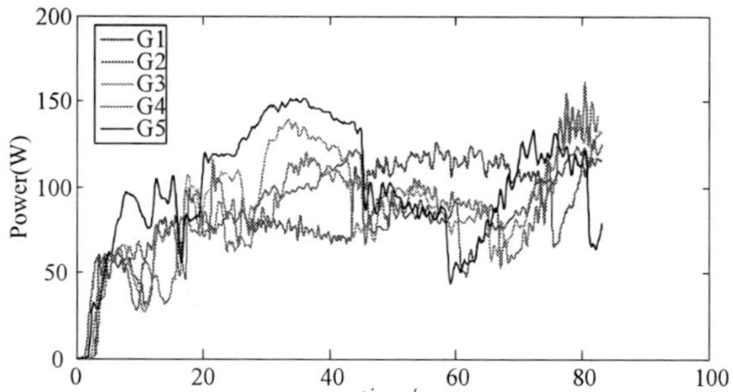

Figure 9. Comparison of the sum power consumption of the joints with the different rk

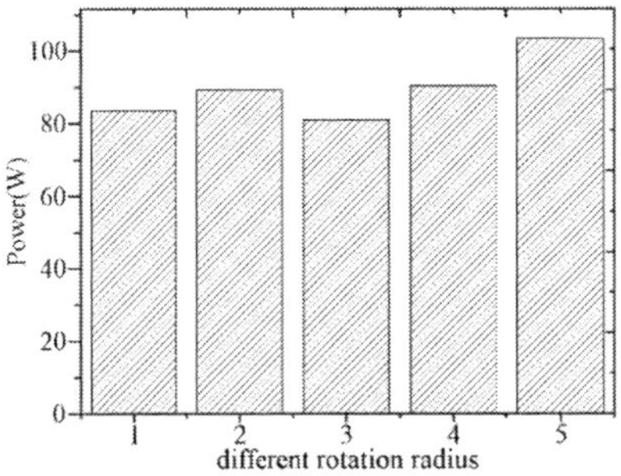

Figure 10. The average of the sum of the joints power consumption

5.3 Joint power consumption of the same path

Fig. 11 shows the average of each joints power consumption with the doorknob rotating radius r_k, 40mm, along the planning door-opening trajectory. The x-axis denotes the number of the modular joint. In the Fig. 11, we could find the max value is joint3 (18.20W), the min value is joint4 (11.26W), joint1 is 17.54W, joint2 is 12.31W, joint5 is 16.02W and joint6 is 15.04W.

The maximum power consumption of each joint is shown in table2. The maximum power consumption is joint2 (52.58W) during the door-opening process.

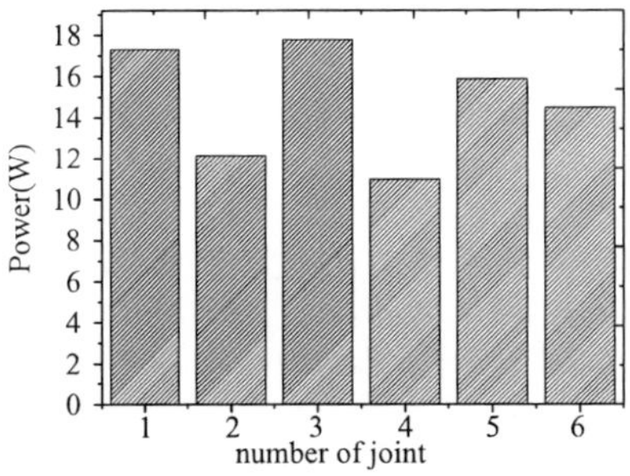

Figure 11. The average power consumption of each joint

Table 2. The maximum power consumption of each joint

Joint of ARM n	Joint1	Joint2	Joint3	Joint4	Joint5	Joint6
Maximum power (W)	40.96	52.58	26.07	29.69	47.19	35.49

6. Conclusion

In this paper, we have investigated the power consumption analysis model and experiment of a mobile manipulator door-opening and presented a method of path planning of the door opening. Several conclusions are drawn:

(1) The trajectory of door-opening, which contains the circle arc of rotating the doorknob and pulling the door, has been planning. In the door-opening trajectory, the posture and position matrix of the end-effector have been obtained.

(2) A novel model of power consumption is derived. The tested parameters of models are instantaneous current I_m and the rotation speed of motors n_j by using the door-opening experiments.

(3) Experiments of several groups door-opening have been carried out, including the comparative power consumption of the different path and joint power consumption of the same path. The experimental results show that the minimum summation of manipulator power consumption is 81.02W, which is the end-effector grasps on the value of r_k (40mm), and the value of the average power consumption of the third joint is 18.2W, which is the maximal in the same path planning.

7. References

Ma, J., et al., 2014. Design, Simulation and Manufacturing of a Tracked Robot or Nuclear Accidents. IEEE International Conference on Robotics and Biomimetics, Bali, Indonesia, December 5-10.

DeDonato, M., et al., 2015. Human-in-the-loop Control of a Humanoid Robot for Disaster Response: A Report from the DARPA Robotics Challenge Trials. Journal of Field Robotics, Vol. 32 No. 2, 275-292.

Yanco, H.A., et al., 2015. Analysis of Human-robot Interaction at the DARPA Robotics Challenge Trials. Journal of Field Robotics, Vol. 32 No. 3, 420-444.

Nagatani, Keiji, and S. Yuta., 1996. "Designing strategy and implementation of mobile manipulator control system for opening door," Robotics and Automation, 1996. Proceedings, IEEE International Conference on, Vol. 3, pp. 2828-2834.

S. Chitta, B. Cohen, and M. Likhachev, 2010. "Planning for autonomous door opening with a mobile manipulator," in Robotics and Automation (ICRA), 2010 IEEE International Conference, pp. 1799-1806.

Peterson, L. Austin, D., Kragic, D., 2000. High-level control of a mobile manipulator for door opening. in Proc. IEEE/RSJ Int. Conf. Intelligent Robots and Systems, Takamatsu, Japan, Oct. 31-Nov. 5, pp. 2333-2338.

Chung, W., et al., 2009. Door-Opening control of a service robot using the multifingered robot hand. IEEE Transactions on Industrial Electronics, Vol. 56, No. 10, 3975-3984.

H. Xing, L. Ding, H. Gao, W. Li, and M. Tavakoli, "Dual-User Haptic Teleoperation of Complementary Motions of a Redundant Wheeled Mobile Manipulator Considering Task Priority," in IEEE Transactions on Systems, Man, and Cybernetics Systems, doi: 10.1109/TSMC.2022.3144009.

Ruhr T., et al., 2012. "A generalized framework for opening doors and drawers in kitchen environments," in Robotics and Automation (ICRA), 2012 IEEE International Conference, 3852-3858.

Karayiannidis Y., et al., 2012. "Open sesame! Adaptive force/velocity control for opening unknown doors," in Intelligent Robots and Systems (IROS), 2012. IEEE/RSJ International Conference, 4040-4047.

Zhou L., et al., 2013. "Kinect-based robotic manipulation for door opening," in Robotics and Biomimetics (ROBIO), 2013 IEEE International Conference, 656-660.

Li J., et al., 2015. "Twisting door handles and pulling open doors with a mobile manipulator," IEEE International Conference on Robotics and Biomimetics, 686-691.

Gao H., et al., 2019. Dynamic Modeling and Experimental Valida-tion of Door-Opening Process by a Mobile Manipulator. IEEE Access (99):1-1.

Xing H., et al., 2019. Unknown geometrical constraints estimation and trajectory planning for robotic door-opening task with visual teleoperation assists. Assembly Automation, Vol. 39, No. 3, 479-488.

Ma C., et al., 2018. Optimal Energy Consumption for Mobile Manipulators Executing Door-Opening Task. Mathematical Problems in Engineering, 1-11.

Zhuang H., et al., 2013. "Method for analyzing articulated torques of heavy-duty six-legged robot," Chinese Journal of Mechanical Engineering, Vol. 26 No. 4, 801-812.

2643 / An Improved Simultaneous Localization and Mapping Method Base on LeGO-LOAM and Motion Compensation

Mengyang Li, Xinsheng Wang*, Xiyue Wang, and Shuang Liu
Author affiliations
**Corresponding author: xswang@hit.edu.cn*

https://doi.org/10.56884/HHVH4446

Abstract

With the rapid development of mobile robot, the premise of all decisions and planning is to perceive the surrounding environment, especially in complex environments such as wild and mountainous areas. The mainstream Simultaneous Localization and Mapping (SLAM) algorithm Lightweight and Ground-Optimized Lidar Odometry and Mapping (LeGO-LOAM) can be well adapted to this complex environment, but it does not take into account the motion compensation of the point cloud, which leads to a decrease in perception accuracy. LeGO-LOAM adopts the way of feature point and feature point matching for posture estimation. Due to the vertical launch angle of every scanning laser radar being fixed, the radar motion will lead to distortion between the matching feature points, which can cause incorrect pose estimation. Therefore, based on LeGO-LOAM, this paper proposes a motion compensation algorithm (LeGO-LOAM-MC). Compared with the current mainstream of laser slam algorithm LeGO-LOAM, the result shows LeGO-LOAM-MC has a smoother mapping effect, smaller path drift, the maximum error is reduced by 29.1%, the mean error is reduced by 31.0%, the median error is reduced by 31.3%, the standard deviation is reduced by 37.0%, the root mean square error is reduced by 32.1%, and the sum of squares due to error is reduced by 53.9%. The experimental results show the superior performance of the proposed algorithm.

Keywords: simultaneous localization and mapping, LeGO-LOAM, motion compensation

1. Introduction

The problems of how a mobile robot to determines its position in an unknown environment and how to describe the surrounding environment are the primary problems to be solved in the studies of autonomous motion robots. The current mainstream SLAM algorithms are mainly divided into several types according to the primary sensor types. The first is the vision-based slam. Due to the relatively cheap visual sensor, large amount of information, rich texture, and other characteristics, it developed rapidly in the early stages. Davison et al. [1] proposed a MonoSLAM algorithm based on a monocular vision camera, which used EKF to build a map for feature points. It solved the problem of monocular feature initialization. Due to the established map being too sparse, it is easy to degenerate when the features are insufficient. Sim et al. [2] presented and analyzed an implementation of a Rao-Blackwellized particle filter (RBPF) that used stereo vision to localize a camera and 3D landmarks as the camera moves through an unknown environment. It improved the robustness and accuracy of vision-based SLAM, but the time complexity of the algorithm was increased. Klein et al. [3] designed a monocular PTAM SLAM algorithm based on key frames, which separated odometry and mapping. The trajectory and map are optimized, but the lack of global optimization makes it easy to fail in large scenes. Mur-Artal et al. [4] proposed a real-time robust monocular ORBSLAM algorithm that estimated the pose and built the map in real time, with high positioning accuracy.

The second is the lidar-based slam. Early laser SLAM was mainly two-dimensional. Grisetti G et al. [5] presented the Gmapping algorithm, which improves the SLAM algorithm based on the particle filter algorithm. It improves particle distribution and selective resampling and can be used for two-dimensional plane positioning and mapping. The slam algorithm is decomposed into front-end and back-end components. The front-end matches the grid map, the back-end is optimized, and the bilinear interpolation method is used to improve the accuracy of the algorithm. Cartographer, presented by Olson E. B [6], used the frame-to-map matching method to reduce the accumulated error. The front-end performs loop-close detection and optimizes the local map of the front-end. In the 3D lidar SLAM field. J Zhang et al. [7] proposed LOAM to extract points with particularly large and small curvatures as corner points and surface points. For the corner points of the current frame, the line composed of adjacent corner points in the previous frame was searched. For the surface points, the plane formed by the adjacent plane points of the previous frame was searched, and the distance from the corner point to the line and the distance from the surface point to the plane are used as constraints to construct a least squares problem and optimize it with the Levenberg-Marquardt method. In addition to frame-to-frame matching, LOAM also matches the current frame to the map, improved global accuracy. T Shan et al. [8] proposed LeGO-LOAM. It added a point cloud segmentation algorithm on the basis of LOAM. It reduced the number of feature points and improved the operation speed and efficiency.

In recent years, the fusion SLAM algorithms of vision, Lidar and Inertial Measurement Unit (IMU) have emerged. IMU can be used to improve performance in vision and lidar SLAM algorithm, Hanzhang Xue et al. [9] proposed a method that the IMU, wheel encoder, and lidar odometry are utilized together to estimate the ego-motion of an unmanned ground vehicle, improved odometry accuracy. Tong Qin et al. [10]. presented VINS-Mono: a robust and versatile monocular visual-inertial state estimator with highly accurate visual-inertial odometry by fusing preintegrated IMU measurements.

2. LeGO-LOAM

The overall framework of the system is shown in Fig. 1. The input is a lidar point cloud. First, the point cloud is mapped to the plane as an image for segmentation. The ground points are marked, the bad points and occlusion points are removed, and then the feature extraction module performs a rough calculation on the segmented point cloud. The feature matching module transforms the current feature to the previous frame or map coordinate system to find matching features and sends the matching features to the pose estimation module. The transformation fusion of the front-end and back-end pose estimation results as the final output.

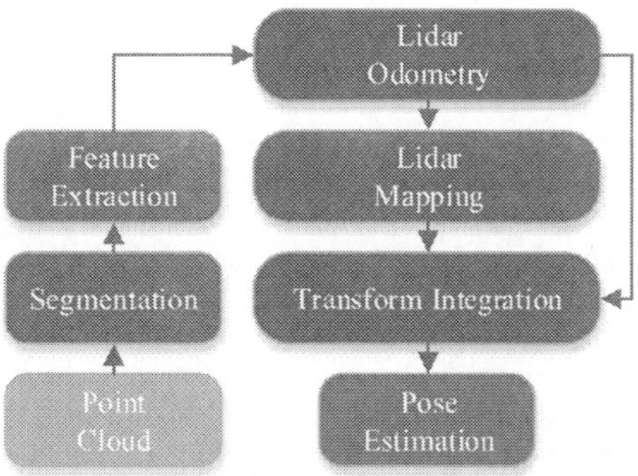

Figure 1. System overview of LeGO-LOAM

2.1 Segmentation

Due to the large number of lidar point clouds, unstable feature points are easily generated at the edges of adjacent objects. Usually, the lidar installation position is always close to the ground. Therefore, the ground point cloud accounts for the majority of the overall point cloud. While the ground is relatively flat and aligned, posture estimation is not binding enough. Therefore, according to the lidar vertical and horizontal resolution, N_{scan} is the number of lidar scan lines, H_r is the lidar horizontal resolution, the lidar point cloud is projected into the depth image with size N_{scan} $*(360/H_r)$, and the pixel intensity is the Euclidean distance of each point cloud to the lidar center, uses an image-based segmentation algorithm to segment and cluster the point cloud, assigns a specific label to each cluster, and marks the ground points at the same time.

2.2 Feature extraction

When performing feature extraction on the segmented point cloud clusters, ignore the point cloud clusters with less than 30 points, because they are usually unstable and unreliable small object features, which may not be continuously scanned, such as leaves and branches on the road. For each point in the point cloud cluster, as shown in Eq. 1, calculate its roughness c. S is the set of continuous points on the same horizontal scan line in the point cloud cluster, and LeGO-LOAM sets the number of them to 10. Define $P_t = \{p_1, p_2, \ldots, p_n\}$ be the point cloud data obtained by lidar at time t, where p_i is a point in P_t and its Euclidean distance to the lidar is indicated by r_i.

$$c = \frac{1}{|S| \cdot \|r_i\|} \left\| \sum_{j \in S, j \neq i} (r_j - r_i) \right\| \quad (1)$$

Where, r_i and r_j are the respective Euclidean distances from the points p_i and p_j in set S to the center of the lidar.

In order to prevent feature points from clustering, LeGO-LOAM divides each horizontal scan line into 6 uniform sub-images, and extracts feature points from each sub-image to ensure that reliable features can be extracted in each direction.

According to the size of c, the roughness threshold c_{th} is set. For the points with c greater than c_{th}, they are called edge feature points, and for the points with c less than c_{th}, they are called plane feature points. At the same time, the set F_e is defined as the edge point set with the largest c, and it cannot be a ground point. F_s is defined as the set of plane points with the smallest c, selected from the ground points. f_e is defined as the set of edge points of the second largest c, f_s is the set of the next largest plane points. F_e belongs to f_e, F_s belongs to f_s. Each in the sub-images, the number of F_e, f_e, F_s, f_s is set to 2, 40, 4, 80.

2.3 LiDAR odometry

The lidar odometry is used to estimate the motion of the lidar between two adjacent frames in time and provides the motion prior to the mapping module. Motion estimation relies on the matching of edge points to edge lines and plane points to planes. For every edge point F_e in the current frame, find the edge points on the upper and lower adjacent lines of the previous frame to form the edge line. For the plane point F_s of the current frame, find the two nearest plane points on the same line of the previous frame and find a nearest plane point on the adjacent line. Those three plane points form a plane. In order to improve the matching accuracy, LeGO-LOAM require the matching point and the target point to have the same point cloud label. After finding the corresponding match, in order to improve the calculation speed, it adopts a two-step Levenberg-Marquardt algorithm, taking the distance from the edge point to the edge line and the distance from the plane point to the plane to be the smallest as the optimization goal, and rotates angle θ_{roll}, θ_{pitch}, θ_{yaw}, and translation t_x, t_y, t_z in three axes are optimization variables. First use plane matching to estimate t_z, θ_{pitch}, θ_{yaw}, and then use edge matching to estimate t_x, t_y, θ_{roll}. The two-step Levenberg-Marquardt algorithm is in the similar estimation results, which saves about 35% of the computation time compared to the one-step Levenberg-Marquardt algorithm.

2.4 LiDAR mapping

Match the feature point cloud of the current frame with the map composed of historical key frames. Compared with frame-frame matching, the frame-map has more abundant and reliable features, which reduces the accumulation error of the frame-frame matching odometry to a certain extent and further optimizes the posture variables. The optimized posture variables and historical key frame posture variables constitute constraints; build a factor graph; and use GTSAM [11] for optimization. At the same time, the current frame and the historical key frame are compared. If a loopback is detected, the point cloud of the current frame and key frames of the loopback are performed ICP [12] to correct the posture of the loopback path. Compared with LOAM, the constructed map of LeGO-LOAM is sparse because it limits the scale of feature points. The pose estimated by the lidar odometry and the mapping module is fused as the final pose estimate.

3. Motion Compensate

As shown in the Fig.2, let the lidar move from t_1 to t_2, observe corner points A_1 and A_2 at t_1, and observe points B_1 and B_2 at t_2, A_1 and B_1 are points on the same corner, A_2 and B_2 are the same corner, A_1 and B_1, A_2 and B_2 are usually regarded as matching points by matching strategy. In fact, due to movement and the lidar observation angle θ is fixed, A_1 and B_1, A_2 and B_2 will not be aligned, and in the next stage pose estimation, in order to make the matching points closer, the wrong rotation will be estimated, and the accuracy of the pose estimation will be reduced.

However, if the true position of the feature point of the current frame in the matching point of the previous frame can be found, the matching point can be compensated, thereby reducing the estimation error. It is assumed in this paper that this error is mainly caused by the magnitude and direction of motion between two adjacent frames of motion. Therefore, the magnitude and direction of the motion between two frames are estimated first, and different compensation strategies are formulated according to the relative position of the feature points and the motion direction.

As shown in Fig.3, after motion compensate, the pose estimate result is more realistic. Based on this, this paper proposes the following tightly coupled motion compensation algorithm.

3.1 Estimate Motion Vector

Let P_i^L represent the ith frame under the lidar coordinate system, note the transformation from the lidar coordinate system to the world coordinate system as T_L^W, and define the transformation from the current frame to the previous frame of the lidar coordinate system as T_{L-1}^L, T_{L-1}^L is the function of three-axis rotation R_x, R_y, R_z and three-axis translation t_x, t_y, t_z. Define the point in the lidar coordinate system converted to the point in the world coordinate system as P_i^W, $P_i^W = T_L^W(P_i^L)$, Define the point in the current lidar coordinate system converted to the point under the last lidar coordinate system as P_i^{L-1} ,$P_i^{L-1} = T_L^{L-1}(P_i^L)$, define P_{ti}^{L-1} as the matching point of the current frame of the lidar coordinate system in the previous frame, and the distance between the two points is defined as d_i.

For all corner feature points P_i^L of the current frame, use T_{L-1}^L to convert them to P_i^{L-1} in the previous frame coordinate system, use kd-tree [13] to find the nearest feature point P_{ti}^{L-1} and calculate the distance d_i from P_i^L to P_{ti}^{L-1}. Equation. 2 defined the current frame feature matching loss as D.

$$D = \sum_{1}^{n} d_i \tag{2}$$

n is the number of corner feature in current frame. With the goal of minimizing the matching point loss D, use T_{L-1}^L as the optimization object to build a least squares problem. Equation. 3 defined the jacobian of the loss function with respect to the rotation and translation as $\boldsymbol{J}$.

$$\boldsymbol{J} = \frac{\partial D}{\partial T_L^{L-1}} = \left[\frac{\partial D}{\partial R_x}, \frac{\partial D}{\partial R_y}, \frac{\partial D}{\partial R_z}, \frac{\partial D}{\partial t_x}, \frac{\partial D}{\partial t_y}, \frac{\partial D}{\partial t_z} \right]^T \tag{3}$$

Equation. 4 defined the optimization variables as $\boldsymbol{x}$.

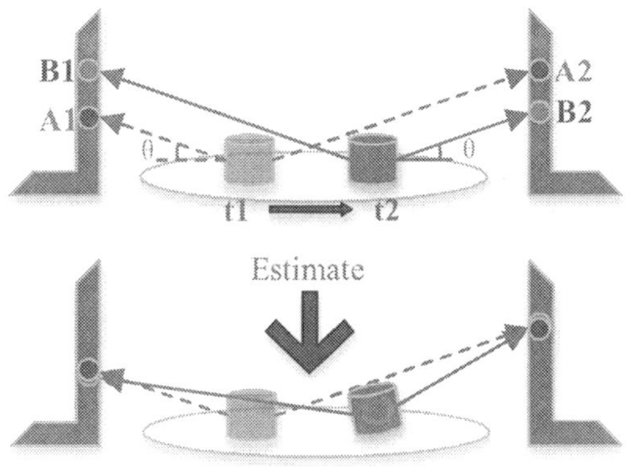

Figure 2. Distortion of point cloud leads to wrong rotation

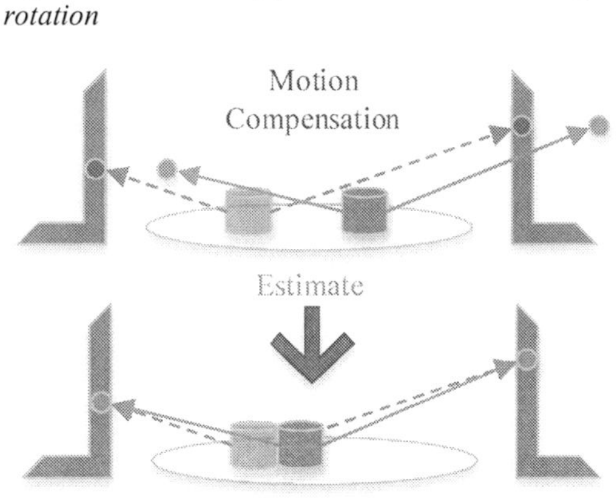

Figure 3. The compensated point cloud can be estimated correctly

$$x = [R_x, R_y, R_z, t_x, t_y, t_z]^T \quad (4)$$

δx is the tiny variable of x.

As shown in Eq. 4 and Eq. 5, for the t-th iteration, solve using Gauss-Newton's method.

$$JJ^T \delta x_t = -JD \quad (5)$$

$$x_{t+1} = x_t + \delta x_t \quad (6)$$

Iterate Eq. 4, Eq. 5 until δx converges. Shown in Eq. 7, take the translation part of x as the motion vector V_m from the current frame to the previous frame.

$$V_m = [t_x, t_y, t_z]^T \quad (7)$$

3.2 Compensate strategy

Usually, the feature points that are more parallel to the V_m direction have greater distortion, while the more perpendicular to the V_m direction feature points are more stable, so a compensation factor is defined as μ, shown in Eq. 8.

$$\mu = \frac{V_m \cdot P_i^{L-1}}{|V_m| * |P_i^{L-1}|} \quad (8)$$

where, P_i^{L-1} represent the vector form of P_i^{L-1}.

Equation. 9 defined the compensate vector as V_c.

$$V_c = P_{ti}^{L-1} + V_m - P_i^L \quad (9)$$

Compensate strategy shown in Eq. 10 :

$$P_i^L = \begin{cases} P_i^L + \mu * V_c & \text{if } \mu > 0 \text{ and } |P_{ti}^{L-1}(z)| > |P_i^L(z)| \\ P_i^L - \mu * V_c & \text{if } \mu < 0 \text{ and } |P_{ti}^{L-1}(z)| < |P_i^L(z)| \\ P_i^L & \text{default} \end{cases} \quad (10)$$

Where, $P_{ti}^{L-1}(z)$, $P_i^L(z)$ represent the Z-axis value of P_{ti}^{L-1}, P_i^L.

Then, the pose estimation is performed again according to the compensated corner feature points, and the process is repeated until the algorithm converges.

For frame-map motion compensation, sometimes the movement distance from current frame to map is too large, and the motion compensation will cause wrong compensation when the rotation angle is too large, so set a threshold r_{th}, do the motion compensation only if the rotation angle θ is less than r_{th}.

The frame-map motion compensate strategy is similar to frame-frame motion compensate with the exception that the coordinate system is changed to the world.

The process of tightly coupled motion compensation is shown in Fig.4. First, a rough motion direction is obtained according to the pose estimation, and then motion compensation is performed according to the direction. The compensated point cloud is subjected to pose estimation again, and the iteration is repeated until the algorithm converges, finally getting a highly accurate pose estimation result.

Finally, the presented LeGO-LOAM-MC algorithm structure is shown in Fig.5. A tightly coupled motion compensation module is added to the lidar odometry module and lidar mapping module to replace the original corner feature point to line matching optimization, and the pose estimation results are continuously optimized.

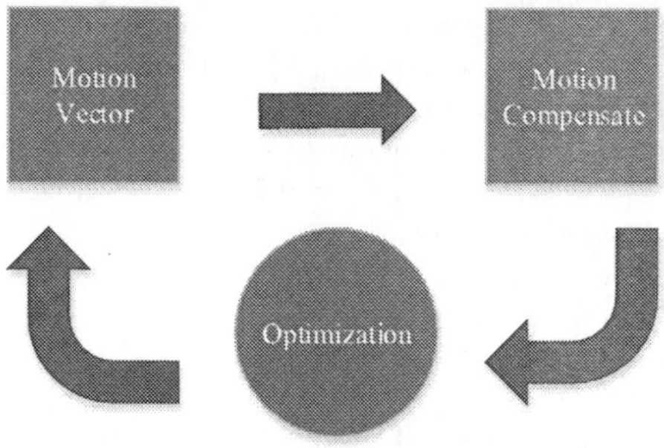

Figure 4. Tightly coupled motion compensation

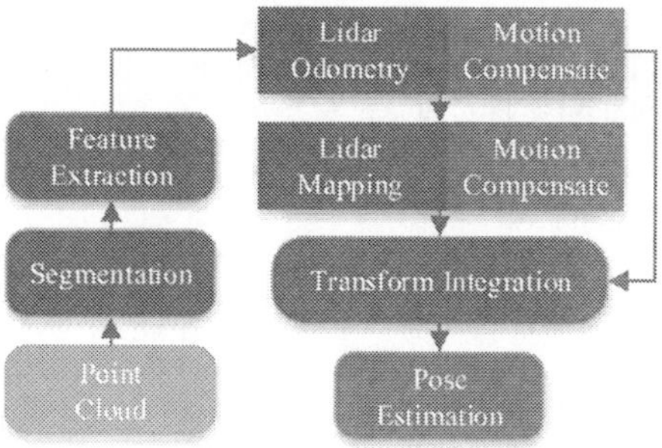

Figure 5. System overview of LeGO-LOAM-MC

4. Experiment

4.1 Experimental Platforms and Datasets

The test hardware platform is Intel i5-6200U CPU, 8GB RAM, and NVIDIA GeForce 940MX Video Card. The test software platform is Ubuntu 18.04 with ROS melodic, gtsam 4.0.0 and ceres. The radar data in KITTI is collected by Velodyne HDL-64E which is loaded on a car, the parameters of the Velodyne HDL-64E are: 10 Hz acquisition frequency, 64 beams, 0.09 degree angular resolution, 2 cm distance accuracy, collecting 1.3 million points/second, field of view: 360° horizontal, 26.8° vertical, 120 m detect range. The source code can be found at https://github.com/1475015695/lego-loam-mc.

4.2 Experiment procedure

To verify the effectiveness of the proposed method, this paper tests on the public dataset KITTI, and compares the obtained results with LeGO-LOAM. All test environments and algorithm parameters are consistent, except for the algorithm itself, the KITTI data set is converted into a ROSBag file and played back during the test, which can ensure the consistency of the input data of the algorithm. In ROS, the compensation process is visualized, as shown in

Fig. 6, the blue squares represent the feature points in the previous frame, the green squares represent the feature points of the current frame, the red squares represent the compensated feature points, and the white arrows represent the direction of radar movement. As expected, the feature points along the motion direction have greater compensation, and the feature points perpendicular to the motion direction are relatively stable. The output of the algorithm is the trajectory of the radar movement and a map of the surrounding environment.

4.3 Track Compare

Use open-source tools evo [14], evo is a Python package for the evaluation of odometry and SLAM. This package provides executables and a small library for handling, evaluating, and comparing the trajectory output of odometry and SLAM algorithms. It supports 'TUM' trajectory files, 'KITTI' pose files, 'EuRoC MAV' (.csv ground truth and TUM trajectory file), ROS and ROS2 bag files.

As shown in Fig. 7. It can be seen that as the moving distance increases, the Z-axis drift of the LeGO-LOAM continues to accumulate. When returning to the original path, the Z-axis drifts greatly, while the LeGO-LOAM-MC is very close to the original path. The Z-axis drift is smaller in the long term and without loopback trip, so it can be seen that the LeGO-LOAM-MC is more robust and reliable.

4.4 Comparison of length of trajectory

As shown in Table 1, the trajectory length ground truth of the KITTI 00 sequence is 3724.187*m*, and the LeGO-LOAM result is 3726.925*m*, the LeGO-LOAM-MC result is 3725.760*m*. Accuracy improved by 42.5%. It can be seen that with motion compensation, the trajectory length estimation is more accurate.

4.5 Comparison of absolute errors of trajectory

Define the ith ground truth pose as $Pose_{gt,i}$, define the ith estimate pose as $Pose_{est,i}$. Equation. 11 defined the ith absolute error as E_i.

$$E_i = \left\| Pose_{gt,i}^{-1} Pose_{est,i} - I \right\| \quad (11)$$

Here I is an identity matrix.

LeGO-LOAM-MC in the KITTI 00 sequence test, compared with the trajectory generated by LeGO-LOAM, the maximum error (max) is reduced by 29.1%, the mean error (mean) is reduced by 31.0%, the median error (median) is reduced by 31.3%, the standard deviation (std) is reduced by 37.0%, the root mean square error (rmse) is reduced by 32.1%, and the sum of squares due to error (sse) is reduced by 53.9%. It can be seen that compared with LeGO-LOAM, LeGO-LOAM-MC has less error accumulation in long-term motion and is more reliable and better. The detailed results are shown in Table 2.

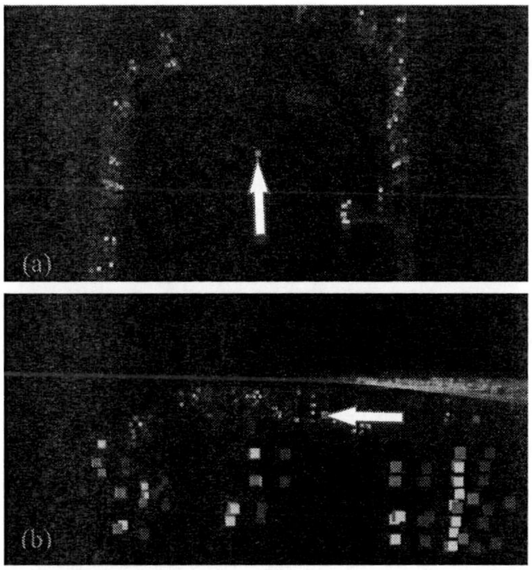

Figure 6. Compensation process.
(a) Top view; (b) Left view

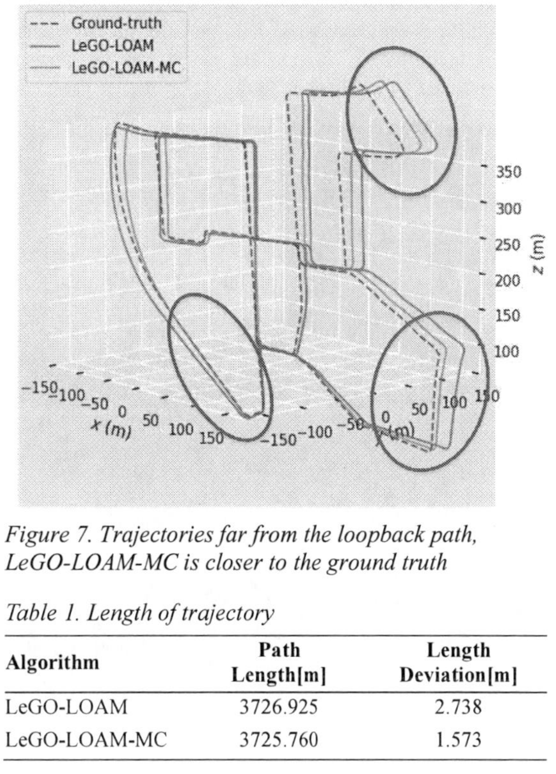

Figure 7. Trajectories far from the loopback path, LeGO-LOAM-MC is closer to the ground truth

Table 1. Length of trajectory

Algorithm	Path Length[m]	Length Deviation[m]
LeGO-LOAM	3726.925	2.738
LeGO-LOAM-MC	3725.760	1.573

Table 2. Absolute errors of trajectory

Absolute Error	LeGO-LOAM	LeGO-LOAM-MC
max[m]	8.451	5.986
mean[m]	3.367	2.323
median[m]	3.172	2.179
std	1.643	1.036
rmse	3.746	2.544
sse	63728.68	29387.55

4.6 Mapping Effect

Due to the limited computing power of the computing platform, usually in the KITTI 00 sequence test, when LeGO-LOAM returns to the starting point, the accumulated Z-axis drift is too large that it exceeds the closed-loop search radius and the loop cannot be closed, while LeGO-LOAM-MC can close the loop here every time. The specific results are shown in Fig. 8.

5. Conclusions

A motion compensate based algorithm LeGO-LOAM-MC is proposed. The results show that the LeGO-LOAM-MC has higher accuracy and lower drift than the LeGO-LOAM algorithm on the public dataset.

Using the 00 sequence of the KITTI dataset to test LeGO-LOAM-MC and LeGO-LOAM, the results show that compared with LeGO-LOAM, LeGO-LOAM-MC has a smoother mapping effect, smaller path drift, the maximum error is reduced by 29.1%, the mean error is reduced by 31.0%, the median error is reduced by 31.3%, the standard deviation is reduced by 37.0%, the root mean square error is reduced by 32.1%, and the sum of squares due to error is reduced by 53.9%.

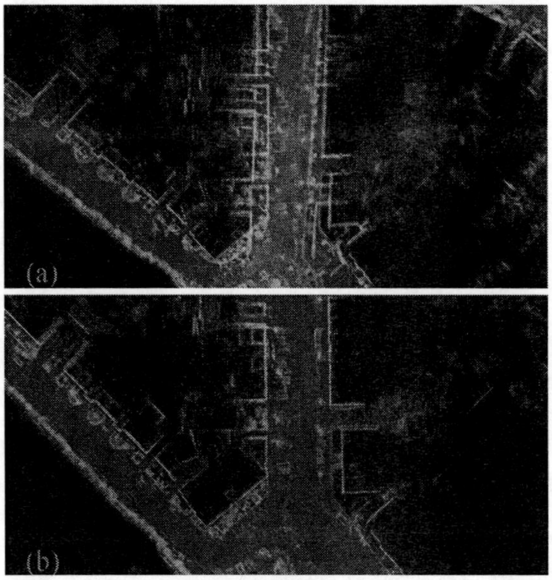

Figure 8. Map constructed when returning to the starting point. a) Map constructed by LeGO-LOAM; b) Map constructed by LeGO-LOAM-MC

6. References

[1] Davison A J, et al. 2007. MonoSLAM: Real-time single camera SLAM. *IEEE Trans. Pattern Anal. Mach. Intell*, 29(6): 1052-1067.

[2] Sim R, et al. 2006. Design and analysis of a framework for real-time vision-based SLAM using Rao-Blackwellised particle filters//*The 3rd Canadian Conference on Computer and Robot Vision (CRV'06)*. IEEE, 21-21.

[3] Klein G, Murray D. 2007. Parallel tracking and mapping for small AR workspaces//*2007 6th IEEE and ACM international symposium on mixed and augmented reality*. IEEE, 225-234.

[4] Mur-Artal R, et al. 2015. ORB-SLAM: a versatile and accurate monocular SLAM system. *IEEE Trans Robot*, 31(5): 1147-1163.

[5] Balasuriya B, et al. 2016. Outdoor robot navigation using Gmapping based SLAM algorithm//*2016 Moratuwa Engineering Research Conference (MERCon)*. IEEE, 403-408.

[6] Turchi P. 2011. Maps of the imagination: The writer as cartographer [M]. Trinity University Press.

[7] Zhang J, Singh S. 2014. LOAM: Lidar odometry and mapping in real-time//*Robotics: Science and Systems*. 2(9): 1-9.

[8] Shan T, Englot B. 2018. Lego-loam: Lightweight and ground-optimized lidar odometry and mapping on variable terrain//*2018 IEEE/RSJ International Conference on Intelligent Robots and Systems (IROS)*. IEEE, 4758-4765.

[9] Xue H, et al. 2019. IMU-aided high-frequency lidar odometry for autonomous driving. *Applied Sciences*, 9(7): 1506.

[10] Qin T, et al. 2018. Vins-mono: A robust and versatile monocular visual-inertial state estimator. *IEEE Trans Robot*, 34(4): 1004-1020.

[11] Kaess M, et al. 2012. iSAM2: Incremental smoothing and mapping using the Bayes tree. *Int J Rob Res*, 31(2): 216-235.

[12] Zhang J, et al. 2021. Fast and robust iterative closest point. *IEEE Trans. Pattern Anal. Mach. Intell.*

[13] Ram P, Sinha K. 2019. Revisiting kd-tree for nearest neighbor search//*Proceedings of the 25th acm sigkdd international conference on knowledge discovery & data mining*. 1378-1388.

[14] Evo: Python Package for the Evaluation of Odometry and SLAM. Available online: https://github.com/MichaelGrupp/evo

3351 / Benchmarking of Compression Testing Devices in Snow

Mohit Shenvi[a]*, Corina Sandu[b], and Costin Untaroiu[b]
[a]*Terramechanics, Multibody, and Vehicle Systems Laboratory, Virginia Tech, Blacksburg, VA, USA*
[b]*Center for Injury Biomechanics, Virginia Tech, Blacksburg, VA, USA*
**Corresponding author: mshenvi@vt.edu*

https://doi.org/10.56884/OTSP3025

Abstract

Traction testing of tires for "severe snow use" certification is performed according to ASTM F1805 in the Unites States and Canada. The ASTM standard provides guidelines for the preparation and compaction of the snow required for the testing and its evaluation using the CTI (Compliance Testing Incorporated) penetrometer. However, the CTI penetrometer provides a measure of the degree of compactness of the snow. From a computational perspective, this information in itself is not sufficient for modeling and/or validation of a snow model for virtual snow traction testing, which is important in the design stage of tires. This work attempts to compare the relative compression testing performance of three devices, namely a CTI penetrometer (rounded cone tip impactor), a standard Clegg Hammer (flat surface impactor), and an in-house developed device inspired by the Russian Snow Penetrometer (cone tip impactor). The approach used here includes a comparison of the available direct and indirect outputs of all the devices and a statistical analysis of these. The observed differences may be due to differences in the mass and drop height between the devices. The findings of this work provide a baseline for further research into the optimal mass and drop height for the evaluation of compacted snow properties. From a modeling perspective, the values generated using the Clegg impact hammer could be useful. However, it tends to underestimate the elastic modulus, when its results are compared to the laboratory technique, and to overestimate the ram resistance when compared to the in-house device. Further improvements to the in-house device may improve its ability to measure the snow properties most relevant to the modeling of snow.

Keywords: compacted snow, tire-snow traction, penetration testing, CTI penetrometer, Clegg Impact Hammer

1. Introduction

Snow mechanics has been a field of interest for many researchers; however, the field of compressed snow mechanics has seen less interest, and there exist fewer devices that could be used for compressed snow characterization (Shenvi et al., 2022). The density of snow can theoretically vary from fresh snow (70-150 kg/m^3) up to ice (917 kg/m^3) (Russell-Head and Budd, 1989). Automotive tire mobility on compacted snow has been a major focus area for about five decades for institutions like the U.S. Army's Cold Regions Research and Engineering Laboratory (CRREL), with a focus on snow-covered roads with snow densities ranging from 370 kg/m^3 to 600 kg/m^3 (Shoop et al., 2010).

From a mobility perspective, the role of the tire is critical as it affects the interfacial forces between the vehicle and the underlying terrain. The terrain being snow further adds to the complexity due to the snow's temperature-sensitive nonlinear viscoelasticity (Mellor, 1975). The testing of winter tires for research and development has primarily been done outdoors at testing facilities designed for the purpose. In North America, snow traction testing is most commonly performed according to the ASTM F1805-20 standard (ASTM, 2020). The standard specifies procedures for testing and evaluation of straight-line driving traction of passenger cars and light truck tires on snow and/or ice-covered surfaces.

ASTM F1805 recommends the usage of the CTI Snow Compaction Gauge to measure snow compaction and shear strength, as these are known to significantly affect tire traction. From a snow modeling point of view, the single index value generated by the CTI device is not very helpful. An accurate model of snow for tire traction would enable tire manufacturers to optimize tire designs before prototype manufacturing and testing. For this purpose, the measurement of compacted snow properties with outputs relevant to the computational modeling of snow is important.

This work studies the measurements of three compressive testing devices in compacted snow. Section 2 introduces the reader to the background of the devices used in this work. Section 0 details the procedures used for the collection and analysis of measured data by considering the assumptions related to the devices. Section 4 presents results from the analysis and inferences derived from the same. Lastly, Section 5 presents the conclusions of this work and the scope of future work.

2. Compressive Testing Devices

Interaction between the snow and the tire is affected by both the compressive and shear strength of the snow. The testing methodologies for each are different. Compressive properties are generally evaluated by vertical penetration or indentation using a plate or cone-type apparatus, but most of the devices were originally developed for testing of soil. The various devices used in this work are explained in the upcoming sections.

2.1 CTI Snow Compaction Gauge

The "CTI Snow Compaction Gauge" also known as the "CTI Penetrometer," was developed by a division of Smithers that was formerly called Compliance Testing, Incorporated (CTI). It was designed to provide a combined measurement of compacted snow's compressive and shear properties as a single index value (ASTM, 2020). The device employs a cone with a rounded tip with the hypothesis that driving the tip into the snow provides both compression and a lateral shift of the surface snow, thereby causing a penetration which is also indicative of the shear properties of the snow. A schematic is shown in Figure 1.

The outputs of the device are indicative of the degree of compactness with 50 corresponding to a penetration distance of approximately 50 mm (the softest snow) and 100 being no penetration (indicative of ice or some other non-deformable surface). ASTM F1805 recommends conducting the traction tests on compacted snow having CTI penetrometer values between 70 and 84. Though the device provides a measurement of the compaction of the snow, its index cannot be directly converted to known properties like penetration resistance (Shoop et al., 2019), or contact pressure. The applicability of the device in compacted snow is suitable (Shoop et al., 2019), but its outputs cannot be easily applied to characterize snow for purposes of computational modeling directly and research in this aspect could be useful for the research community (Shenvi et al., 2022).

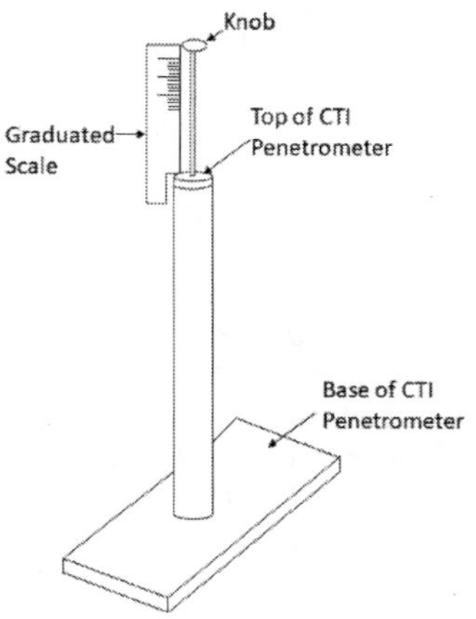

Figure 1. Schematic of CTI Penetrometer (Shenvi et al., 2022)

2.2 Clegg Impact Hammer

The Clegg Impact Hammer was developed by Baden Clegg (Clegg, 1976) as a device for evaluating the base of pavements. The device consists of a flat-faced hammer with an attached accelerometer to measure the peak deceleration experienced by the hammer. Various hammer masses can be chosen depending on the surface to be measured, ranging from 0.5 kg to 20 kg. The advantages and disadvantages of the device were presented by the authors (Shenvi et al., 2022) and a schematic is presented in Figure 2.

ASTM D5874 specifies a standard practice for use of the Clegg device (ASTM, 2016). The manufacturers and the ASTM standard consider the 4.5 kg variant as the "standard" or nominal version. The procedure involves dropping the hammer 5 consecutive times and using the impact value of the 4th drop (S D Instrumentation Ltd., 2022) to evaluate the strength of the material.

The earliest use of the Clegg device for snow was in 1989 (Lee et al., 1989) when it was used for strength evaluation of snow roads in Antarctica. However, the use of Clegg Impact Hammer for snow has primarily been performed by CRREL ever since. The findings of a CRREL group (Shoop et al., 2010) suggested that the standard (4.5 kg) hammer was too heavy to be used in compacted snow, contributing to a higher variation in the measured data. Further work by the same group (Shoop et al., 2012) suggested that the 2.25 kg size was better for measuring compacted snow properties. The 4.5 kg device was used for this work.

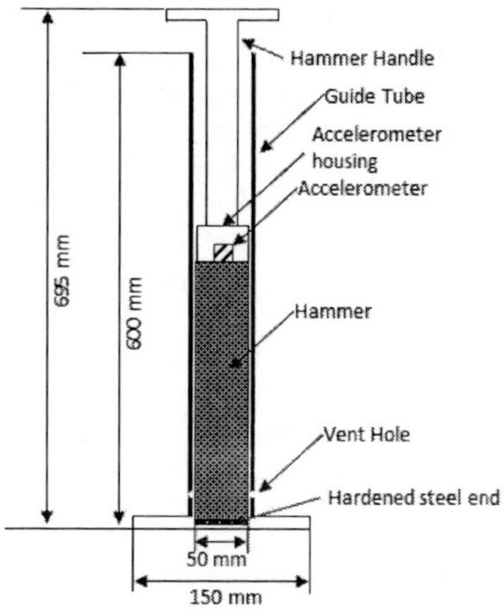

Figure 2. Schematic of Clegg Hammer (Shenvi et al., 2022)

2.3 In-House Developed Device

The Rammsonde penetrometer was found to have limitations in evaluation of hard compressed snow properties (Niedringhaus, 1965; Shoop et al., 2019). For better evaluation of compressed snow properties, a modified Rammsonde known as the Russian Snow Penetrometer (RSP) was created. It consists of a cone with an angle of 30°, a drop weight of 1.75 kg, and a base diameter of cone of 11.5 mm (Shoop et al., 2019).

As part of our current research, an in-house device was created in phase one as a further improvement of the RSP with a similar cone angle but with the drop weight reduced to 1.5 kg, the base diameter of the cone increased to 20 mm, and a maximum drop height of 400 mm. The developed device can be seen in Figure 3. The shaft of the device has gradations, allowing its drop height to be varied in 100 mm increments.

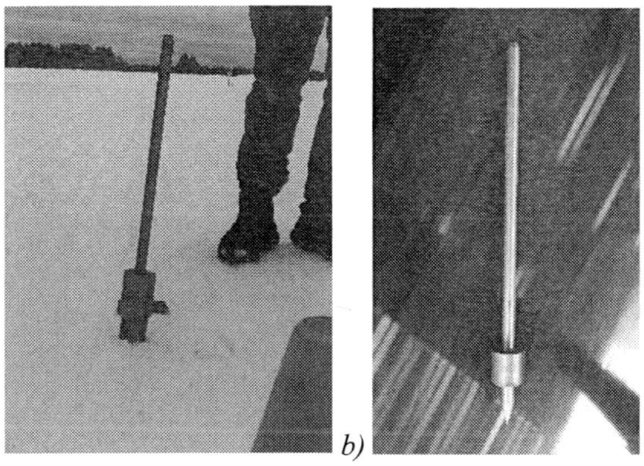
a) b)

Figure 3. In-house developed device (a) in use, (b) entire device

3. Procedures for Data Collection and Analysis

For this work, testing was conducted at the Smithers Winter Test Center, which is located in Michigan's Upper Peninsula. The surface chosen for the collection of data was part of a snow traction test surface and consisted of six locations over an area of about 40 m by 200 m. The locations were marked with traffic cones and measurements were performed within 2 feet of the cones. This specific part of the test surface is subjected to variable sunlight and wind effects due to its positioning. Multiple CTI readings (3) according to the ASTM procedure (ASTM, 2020) were made to evaluate the statistical significance of the measurements. A single Clegg test consists of 5 drops at the same location (ASTM, 2016), and hence only a single set of measurements at each location were performed each time data was collected. The measurements with the in-house device also consisted of 5 hammer drops at a single location, and so, like the Clegg, a single set of readings were performed each time data was collected. Thus, each dataset consisted of 3 CTI measurements and 5 drops each of the Clegg and the in-house device. Several datasets were collected over three days. In addition, density measurements were performed using a cylindrical cutter and an approximate snow density of 545.8 kg/m^3 was found. For analysis of the collected data, it is essential that the measured and/or evaluated outputs are comparable. The procedures for analysis of measured data are detailed in the following subsections.

3.1 CTI Penetrometer

The CTI penetrometer measurements were performed three times at each location for each dataset. The outputs are index values between 100 and 50 which are in a nonlinear relation to the sinkage. An index value of 100 denotes zero sinkage – as when the device is placed on a hard surface (ASTM, 2020). Lower values represent deeper penetration into the surface. Thus, for analysis of the measurements of the CTI, a statistical analysis of the mean and standard deviation of the measurements at the 6 locations was performed.

3.2 Clegg Impact Tester

The Clegg Impact Tester measures the resistance offered by the terrain in terms of deceleration when a hammer of a specific weight is dropped from a predetermined height. The ease of conversion to the California Bearing Ratio (He et al., 2020), the evaluation of the penetration, and the calculation of the Clegg Hammer Modulus (analogous to Young's Modulus) are important parameters that can be derived from the outputs (Shoop et al., 2019). The inherent assumptions in the derivations of the formulations (Clegg, 1994) are:

- The drop of the hammer is vertical, and the clearances between the guide tube and the hammer do not affect its movement.
  There is no energy loss on the impact of the hammer on the terrain.
- The retardation curve is constant with time.

The Clegg Impact Value (CIV) recorded by the device could be used to evaluate the Clegg Hammer modulus and penetration (Shoop et al., 2012). The formula for evaluation of Clegg Hammer Modulus as shown in eq. 1 is specific to the 4.5 kg Clegg. Penetration/sinkage of the device can also be evaluated using eq. 2. Further, it is also possible to evaluate an equivalent ram resistance force for the drops using eq. (3) (Shoop et al., 2010).

$$E = 0.088 * CIV^2 \qquad (1)$$

$$z = \frac{h}{10 * CIV} \qquad (2)$$

$$R = \sqrt[0.48]{\frac{(0.24 * CIV + 1)^2}{1.44}} \qquad (3)$$

where E is Young's modulus in MPa, z is penetration in mm, h is drop height in mm, and R is ram resistance force in kg$_f$. To analyze the consistency of the evaluations, a statistical analysis is attempted.

3.3 In-house Developed Device

The in-house device developed is a modification of the Rammsonde and RSP but works on the same principle. The inherent assumptions in the working of the device are:

- The friction between the guide tube and hammer is negligible.
  There is no energy loss on impact.
- The usage of a single type of cone, thereby considering the size effect to be negligible.

It was ensured that the drop of the hammer was vertical, and the device was not tilted at the end of the drop. Since the working principle is the same as the Rammsonde, the evaluation of the resistance to penetration offered by the snow and the pressure can be calculated using eq. 4 and eq. 5 respectively (Wong and Irwin, 1992). Thus, to evaluate the sinkage exponent and sinkage modulus (He et al., 2020), eq. 6 and 7 respectively were used. These values in turn can quantify the modified pressure-sinkage relation shown in eq. 8.

$$R = \frac{W*h*n_{blows}}{z} + W + Q \quad (4)$$

$$p = R/A \quad (5)$$

$$n = \frac{\Sigma p^2 \Sigma p^2 ln(p) ln(z) - \Sigma p^2 ln(p) \Sigma p^2 ln(z)}{\Sigma p^2 \Sigma p^2 (ln(z))^2 - (\Sigma p^2 ln(z))^2} \quad (6)$$

$$k = \frac{\Sigma p z^n}{\Sigma z^{2n}} \quad (7)$$

$$p = k z^n \quad (8)$$

where R, W, and Q represent the resistance of snow to penetration, the weight of drop hammer, and the weight of penetrometer in Newtons respectively, h is the height of drop in mm, z is sinkage in mm, n_{blows} is number of blows required to achieve sinkage z, A is base diameter of the cone in mm^2 and p is pressure in MPa.

As the measurement methodology employed involved measurement of the height of a fixed point on the device with respect to the terrain, a cumulative of 6 readings (initial one and one after every drop) were conducted which allows evaluation of 5 sinkage values for the hammer drops at each location in each dataset. However, to employ the use of eq. 4, the second and further sinkages need to be considered as cumulative sinkages instead and thus the number of blows in the evaluation needs to be changed accordingly.

4. Results and Discussion

The analysis of collected measurements was performed according to the methods described in Section 0. The results described in this section will highlight the findings of data collected on the 3rd day as the number of datasets were more and provide a better insight from statistical perspective.

4.1 CTI Penetrometer

The CTI index value was measured three times at each location, about 2 hours apart. The statistical analysis of the mean and standard deviations of the data on day 3 yielded the results shown in Table 2.

From Table 2, it is clear that the mean value of the CTI of each location was nearing the upper limit of the medium-pack snow defined by the ASTM F1805. However, on closer observation of the evolution of measurements through the datasets of the day, as shown in Figure 4 (error-bar plot showing the variation of one standard deviation from the mean), it can be deduced that the measurements of CTI over the 6 locations tended to increase as time elapsed from grooming (dataset number). This implies that snow hardening occurs over time. Further, the choice of locations led to points D and E being in direct sunlight for a shorter time than the other points, which could also lead to hardening of the surface snow at those points as suggested by a higher mean and overall range of measurements for these points in the later datasets. In addition, point F is situated at a location at the end of the track and in the path of the wind direction and faces wind loading which may affect the snow bonding. The average coefficient of variation (COV), which is a measure of the variation with respect to the mean is 0.0384.

Table 2. Mean and Standard deviation of CTI data collected on day 3

Location	Mean of CTI	Standard Deviation of CTI	Coefficient of Variation (COV)
A	82.0	2.82	0.034
B	81.2	2.11	0.026
C	80.8	3.99	0.049
D	80.9	3.20	0.040
E	81.6	4.10	0.050
F	80.9	2.50	0.031

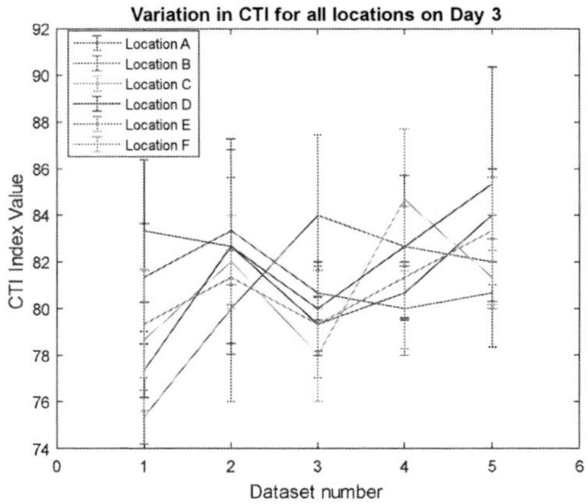

Figure 4. Plot of the mean and standard deviation of measured values of CTI Penetrometer on day 3

4.2 Clegg Impact Tester

The Clegg Impact Tester was used to measure deceleration values as described in section 3.2. The deceleration values were used to calculate the Clegg Hammer modulus, the sinkage per drop, and ram resistance force in Newtons from the mean of 3rd to 5th drop, as shown in Table 3, Table 4, and Table 5, respectively.

Table 3. Mean and Standard deviation of Clegg Hammer Modulus (MPa) evaluated using Clegg data of day 3

Location	Mean of CHM *(MPa)*	Std. Dev. of CHM *(MPa)*	Coefficient of Variation (COV)
A	22.3	3.58	0.161
B	22.0	5.69	0.259
C	19.5	4.82	0.247
D	18.1	5.80	0.320
E	20.9	2.37	0.113
F	18.5	8.25	0.446

Table 4. Mean and standard deviation of sinkage (z) evaluated using Clegg data of day 3

Location	Mean of sinkage of 3-5 drops *(mm)*	Std. Dev. of sinkage of 3-5 drops *(mm)*	Coefficient of Variation (COV)
A	2.9	0.22	0.076
B	2.9	0.32	0.110
C	3.1	0.39	0.126
D	3.3	0.67	0.203
E	3.0	0.18	0.06
F	3.3	0.66	0.2

Table 5. Mean and standard deviation of resistance force (R) evaluated using Clegg data of day 3

Location	Mean of R *(N)*	Std. Dev. of R *(N)*	Coefficient of Variation (COV)
A	3250	880	0.271
B	3270	1490	0.456
C	2660	1050	0.395
D	2400	1100	0.458
E	2920	520	0.178
F	2600	1880	0.723

The typical values of the Clegg hammer modulus were around 20 MPa. The other device that can be used to evaluate the Young's modulus in-situ: Snow MicroPenetrometer (Schneebeli and Johnson, 1998; Schneebeli et al., 1999) underestimates the value of Young's modulus (Gerling et al., 2017). Upon comparison, it was noted that the calculated modulus using the Clegg method is between that measured by the SMP and results from lab techniques like micro-computed tomography (μCT) or acoustic pulse technique, however, the calculated modulus should be treated as a dynamic modulus. The sinkage values evaluated in this work are similar to findings in the literature (Shoop et al., 2012) however on the addition of the individual sinkage values, it was found that the total penetration evaluated using this method underestimated the sinkage as the total sinkage measured experimentally after 5 drops was around 27-30 mm. Hence, the evaluation of pressure-sinkage relation using these values would not provide accurate data for comparison with the in-house device. An evaluation of the equivalent ram resistance (Table) was performed to compare with the findings from the in-house device in section 4.3. The average coefficients of variation for CHM, sinkage, and resistance forces are 0.26, 0.13, and 0.41 respectively, which could be considered higher compared to the CTI.

4.3 In-house Developed Device

The outputs from the in-house device consisted of height measurements by a laser distance sensor from a fixed point, yielding 5 sinkage values for each test approximately at 2-hour intervals. The analysis procedure described in section 3.3 was used. The results in Table 6 and Table 7 show the statistics of the evaluated resistance force and pressure-sinkage parameters for the data collected on day 3.

Comparison of the mean values of resistance force at each location shows that the resistance force offered at locations D, E, and F to the penetration of the device were lower. The ambient and weather factors discussed in section 4.1 also applied to these measurements. Closer observation of the resistance forces over the datasets of day 3, shown in Figure 5, found that subsequent drops of the hammer led to more sinkage at these locations (i.e., lesser resistance to penetration was offered by the terrain). Further, from the evaluated pressure-sinkage parameters of Table , there seems to be a higher consistency in the sinkage modulus values (albeit in two groups) in comparison to the sinkage exponent values. This observation, correlated with higher variation in the evaluated value of the sinkage exponent, matches an observation of Wong and Irwin (Wong and Irwin, 1992). The values in this work do not match the values of Wong and Irwin for two reasons. First, the density of the snow being tested in this work is higher and second, the base diameter of the cone used in this work is smaller and has a different cone angle, so a direct comparison of the evaluated pressure-sinkage parameters is not possible.

Table 6. Mean and Standard deviation of resistance force (R) evaluated using sinkage measurements of day 3

Location	Mean of *R (N)*	Std. Dev. of *R (N)*	Coefficient of Variation (COV)
A	730	200	0.274
B	780	210	0.269
C	770	280	0.364
D	610	190	0.311
E	600	190	0.317
F	660	200	0.303

Table 7. Mean and standard deviation of pressure-sinkage parameters evaluated from data of day 3

	Mean		Std. Dev.	
Location	k	n	k	n
A	0.182	1.030	0.038	0.699
B	0.208	1.043	0.034	0.730
C	0.192	0.846	0.076	0.700
D	0.133	1.289	0.033	0.500
E	0.132	1.475	0.025	0.333
F	0.154	1.191	0.042	0.437

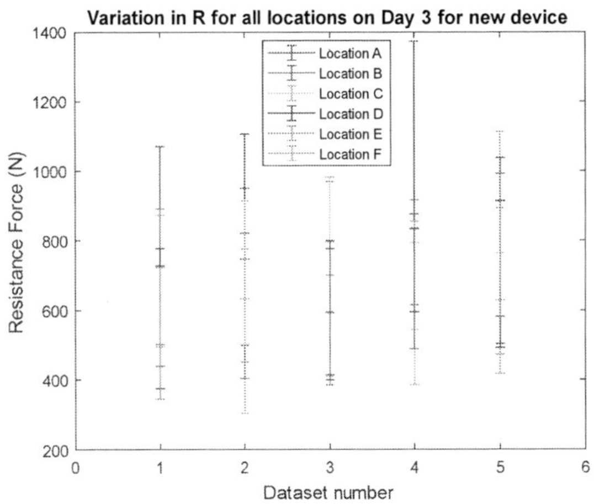

Figure 5. Plot of the mean and standard deviation of penetration resistance to new device on day 3

5. Conclusions and Future Work

In this work, a comparative study of three devices: the CTI penetrometer, the Clegg impact hammer, and an in-house device inspired by the RSP, was performed. Measurements were conducted on a compacted snow surface at the Smithers Winter Test Center that is used for winter tire traction measurement.

It was found that the CTI readings would not be useful for snow surface modeling perspective as it is difficult to accurately measure the sinkage in-situ after a specific drop of the device. Further, the low weight of the CTI (about 0.22 kg) is a limitation as even the 0.5 kg Clegg hammer was found to be lighter than required for evaluation of compacted snow properties (Shoop et al., 2012). The evaluation of physical properties using the Clegg hammer matched the data found in the literature, however, some discrepancies were observed in comparison to the measurements made in-field. The comparison of ram resistances using the in-house device and the Clegg leads to the finding that the Clegg hammer, by virtue of its greater weight and the plate-type surface, predicts a higher value of ram resistance. On comparison of the impacting element, it is found that while the mass of the Clegg hammer is three times that of the in-house device, the value of its ram resistance is approximately 4.5 times that of the in-house device.

The findings of this work could be used as a baseline for further analysis of the optimum mass and drop height for the evaluation of compacted snow properties. In literature (Wong and Irwin, 1992), it was found that a Rammsonde cone on a bevameter which would provide a more-controlled penetration rate that tended to override the size effects due to changes in the geometry of the cone and this could be a possible direction for future research into an improved device for measuring compacted snow properties. In addition, a more in-depth analysis of the theoretical formulations used in this work with datasets from other days would help to understand the specific over/underestimation of properties by a specific device.

6. Nomenclature

The symbols used in this work are:

A	*Base area of cone*	$[mm^2]$
E	*Young's Modulus/Clegg Hammer Modulus*	*[MPa]*
h	*Drop Height*	*[mm]*
n_{blows}	*Number of blows required to achieve sinkage z*	*[]*
P	*Pressure*	*[MPa]*
Q	*Weight of Penetrometer*	*[N]*
R	*Penetration resistance of the snow*	*[N]*
W	*Weight of drop hammer*	*[N]*
z	*Sinkage/Penetration*	*[mm]*

7. Acknowledgements

This study was supported by the Center for Tire Research (CenTiRe), an NSF-IUCRC (Industry University Cooperative Research Center) at Virginia Tech. The authors wish to thank the project mentors and the members of the Industrial Advisory Board of CenTiRe for their support and guidance and Smithers Winter Test Center for providing access to the test site.

8. References

ASTM (2016). D5874-16: Standard Test Method for Determination of the Impact Value (IV) of a Soil. West Conshohocken, PA, ASTM International: 1-9.

ASTM (2020). F1805-20: Standard Test Method for Single Wheel Driving Traction in a Straight Line on Snow. West Conshohocken, PA, ASTM International: 1-13.

Clegg, B. (1976). An impact testing device for in situ base course evaluation. Australian Road Research Board Conference Proc.

Clegg, B. (1994). Calculation of Penetration and Elastic Modulus from CIV. Technical Note. Jolimont, Western Australia, Australia, Dr. Baden Clegg Pty. Ltd.: 1-5.

Gerling, B., H. Löwe and A. van Herwijnen (2017). "Measuring the Elastic Modulus of Snow." Geophysical Research Letters 44(21): 11,088-011,096.

He, R., C. Sandu, H. Mousavi, M. N. Shenvi, K. Braun, R. Kruger and P. S. Els (2020). "Updated standards of the international society for terrain-vehicle systems." Journal of Terramechanics 91: 185-231.

Lee, S. M., W. M. Haas, R. L. Brown and A. F. Wuori (1989). Improving snow roads and airstrips in Antarctica. Hanover, New Hampshire, U. S. Army Corps of Engineers, Cold Regions Research and Engineering Laboratory: 1-24.

Mellor, M. (1975). A review of basic snow mechanics, U.S. Army Cold Regions Research and Engineering Laboratory: 1-42.

Niedringhaus, L. (1965). Study of the Rammsonde for Use in Hard Snow. Hanover, New Hampshire, U.S. Army Cold Regions Research and Engineering Laboratory: 23-23.

Russell-Head, D. S. and W. F. Budd (1989). Compacted-Snow Runways: Guidelines for their Design and Construction in Antarctica. Hanover, New Hampshire, U.S. Army Cold Regions Research and Engineering Laboratory.

S D Instrumentation Ltd. (2022). CIST/882: 4.5 Kg Clegg Impact Soil Tester (IV Display): Operating Manual. Top Farm, Tellisford, Bath, England: 1-30.

Schneebeli, M. and J. B. Johnson (1998). "A constant speed penetrometer for high resolution snow stratigraphy." Annals of Glaciology 26.

Schneebeli, M., C. Pielmeier and J. B. Johnson (1999). "Measuring snow microstructure and hardness using a high resolution penetrometer." Cold Regions Science and Technology 30: 101-114.

Shenvi, M. N., C. Sandu and C. Untaroiu (2022). "Review of compressed snow mechanics: Testing methods." Journal of Terramechanics 100: 25-37.

Shoop, S., M. Knuth and J. Crandell (2012). Using a Clegg Impact Hammer to measure snow strength. Cold Regions Engineering 2012: Sustainable Infrastructure Development in a Changing Cold Environment Quebec City, Canada, American Society of Civil Engineers.

Shoop, S., G. Phetteplace and W. L. Wieder (2010). Snow Roads at McMurdo Station, Antarctica. Hanover, New Hampshire, U.S. Army Cold Regions Research and Engineering Laboratory: 1-112.

Shoop, S. A., W. L. Wieder, B. C. Elder, S. A. Beal and E. J. Deeb (2019). Assessment of Field Methods for Measuring Mechanical Properties of Snow. Hanover, New Hampshire, U.S. Army Cold Regions Research and Engineering Laboratory.

Wong, J. Y. and G. J. Irwin (1992). "Measurement and Characterization of the Pressure-Sinkage data for Snow obtained using a Rammsonde." Journal of Terramechanics 29(2): 265-280.

4054 / Field Validation of Egress Process for Planetary Rover

Zhaolong Dang*, Jirong Zhang, and Baichao Chen
Beijing Institute of Spacecraft System Engineering, Beijing, China
**Corresponding author: dangzhaolong@126.com*

https://doi.org/10.56884/UOUD6258

Abstract

After the landing of planetary spacecraft on the surface, the planetary rovers will be powered up and tested in designed conditions. The technical systems of field validation for egress process of planetary rover are shown in this paper. Firstly, the typical rover egress procedures are described such as unlocked the fixed mechanisms of rover body through the lander, powered up the rover through the lander, unlocked the rover typical mechanism after the rover can be communicated directly with the earth station or orbiter, etc. Secondly, the field validation systems are given including the low gravity system, the integrated testing system, the rover models with assistant equipment, the simulated lander, etc. Thirdly, the field tests are described with test conditions and results. Finally, the egress process of Chinese lunar and Martian rovers were described. The technical systems of field validation work well and can be used for the planetary robots in the future.

Keywords: planetary rover, Mars rover; mobility; Mars terrain

1. Introduction

Before the planetary rover starts to explore the planetary surface, it has to leave the lander or landing platform which carried the rover. For example, the Chinese Yutu Lunar rover[1] and Zhurong Mars rover[2] were mounted on the lander and landing platform correspondingly. Yutu rover was mounted on the lander in Chang'e-3 or Chang'e-4 lunar probe. After the lander safely landed on the lunar surface, the Yutu rover was powered and self-inspected, and then the rover was transferred from the top surface of the lander to the lunar surface through one special transfer mechanism. Zhurong Mars rover was carried on the landing platform in Tianwen-1 probe. Zhurong Mars rover passed through the ramped transfer mechanism to Martian surface after rover mast & solar array deployment, rover body lifting by mobility system[3]. As for the success of rover mission, the departure process of the planetary rover should be carefully tested in proving ground.

This paper discusses the field validation of egress process for planetary rover, such as Zhurong Mars rover. First, the typical departure process of the planetary rover is described. Secondly, the ground systems of the departure test are given, including the state of the test model, the simulated planetary surface environments, the control system of the test model and the ground test instruments. Thirdly, the design of the verification conditions and the specific test conditions for Zhurong Mars rover are listed. Finally, the performances of Zhurong Mars rover on Martian surface was described and the validation of the ground verification system was illustrated.

2. Release Procedures for Planetary Rover

When the lander or landing platform with the planetary rover safely landed on the planetary surface, the planetary rover starts its release procedures after the lander or landing platform completes its initialization actions.

Take as an example, Tianwen-1 Mars probe, a combination orbiter, landing platform and Zhurong rover. The release procedures of Zhurong Mars rover are shown in Fig. 1.

1) Fig. 1a shows that the lander with rover is landed successfully on Martian surface. The deploy direction (forward or backward) of ramp transfer mechanism [4] are decided by the attitude value of the landing platform. The default deploy direction is forward direction.

2) Fig. 1b shows that the ramp transfer mechanism is deployed by its own motor.

3) Fig. 1c shows that the mechanisms of Zhurong Mars rover such as mast, solar array, high-gain antenna are deployed in series, the mobility system lift off the rover body height from the stowed configuration to the normal one which means the height between the rover body and Martian surface is 300mm. The Martian surface around the landing is sensed by the navigation cameras. The path is planned by the mission support center including the walking mode on the ramp and from the top of landing platform to the first rover station on Mars.

4) Fig. 1d shows that Zhurong Mars rover releases from the landing platform to the first rover station on Martian surface which means that the release procedures of Zhurong Mars rover are finished.

a) Landing on Martian surface

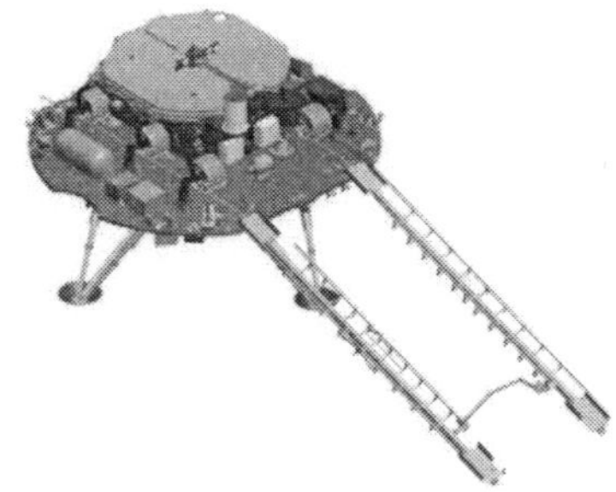

b) Deployment of ramp transfer mechanism

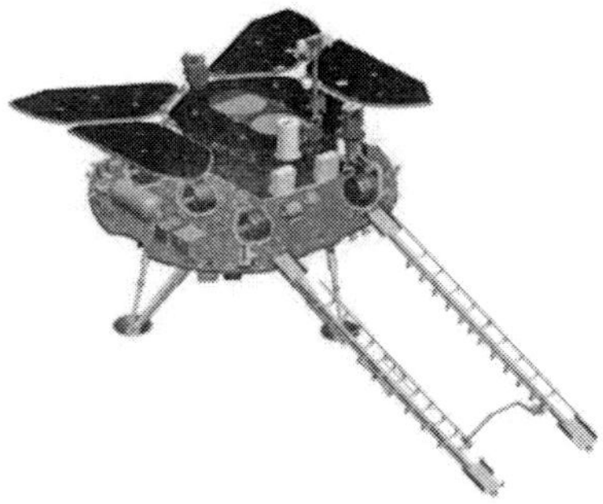

c) Rover mechanism deployment

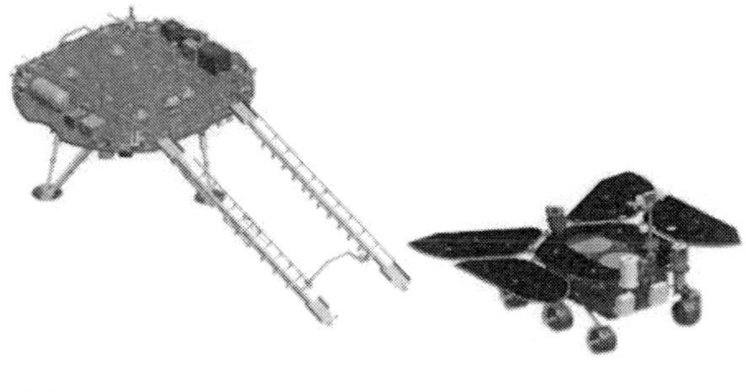

d) Rover release
Figure 1. Mars Rover Release Procedures

As compared with the Chang'e-3 or 4 Lunar probe, the release procedures of Yutu lunar rover was different with that of Zhurong Mars rover. But we can find some common stages as reference release procedures for planetary rovers:

1) Unlock the rover from the lander or landing platform. The connection between the planetary rover and its lander or landing platform is separated. In the launching state, the rover body and mobility subsystem should be locked, usually fixed on the lander or landing platform, and the corresponding pyrotechnic cutters are fired by command.

2) Self powered up and mechanism deployment of the planetary rover. In the launching phase, the planetary rover is often in power-off state. After the lander lands safely on planetary surface, the planetary rover will be powered up, unlocked the pyrotechnic cutters and deployed the mechanisms such as mast, solar array, and other mechanisms, established the telemetered and control link between the planet rover and the ground station or relay satellite. The status of planetary rover should be confirmed by the mission control center.

3) Got the surrounding environmental information for planetary rover egress. The planetary rover sees the surrounding environment information through the optical sensor installed on the mast. The mission support center builds the Digital elevation model (DEM) of surrounding environmental information and plans the route for planetary rover egress.

4) Deployment of the transfer mechanism. According to the transfer scheme of the planetary rover, the transfer mechanism is unlocked and deployed to the designated position. Of course, for the Tianwen-1 Mars probe, the transfer mechanism was deployed firstly for the interference with the rover wheels.

5) The rover wheels on planetary surface. The release mode of planetary rover is decided by the transfer mechanism. For example, the departure of Chang'e No. 3/No. 4 cruiser relies on the transfer mechanism of the lander to transfer the Yutu lunar rover from high position to the lunar surface; The Zhurong Mars rover moves on the mating surface of the ramp transfer mechanism and travels directly to the Martian surface.

After the rover wheels are on the planetary surface, the release procedures of planetary rover are finished.

3. The Ground Verification System for Planetary Rover Release Procedures

3.1 Verification requirements

The requirements of the ground verification system for planetary rover release procedures are as follows:

1) Planetary rover test model. As for the aim of the verification test, the test model should have the following performance as the real planetary rover such as mobility, GNC, data handling, etc.

2) Environmental simulation of landing zone[5~7]. The environmental simulation of the landing zone mainly includes the terrain environments such as slopes, obstacles and the planetary soil. The slope angle affects the initial landing angle of the lander or landing platform and is related to the position where the wheels of the planetary rover on the planetary surface. Generally, the area of landing zone is no more than 8m×8m. The slope is no more than 8 degrees (the baseline is about 10m). When the landing leg encounters an obstacle, the inclination angle of the lander will be changed. To simulate the mechanical properties of

the "soil" of the planetary soil, the simulated planetary soil should be laid in test instruments with certain thickness.

3) Simulation of landing leg deformation. During the landing process of the planetary probe, the landing leg of lander or landing platform will deform because of absorbing energy and the orientation of lander will be changed. In order to fully verify the release and separation capability of the planetary rover, the combined effects of terrain and landing deformation should be considered.

4) Simulation of planetary gravity[8]. For the planetary rover, the wheel load on planetary surface should be simulated on ground test by special Low Gravity System. The Low Gravity System should be designed according to the rover mobility subsystem. As for the rocker-bogie suspension, the rocker arm counterweight should be considered. Because the planetary rover is roving in the field, the Low Gravity System should have the tracking ability to the planetary rover in indoor test field. The cable between the planetary rover and its control instruments should be placed along the Low Gravity System before the tests begin.

5) Control instruments for planetary rover. Generally, the control instruments are similar to the one in AIT phase. Because of the Low Gravity System, the cable between the planetary rover and the control instruments should be adjusted adaptively according to the actual situation of the indoor test field.

6) External measurement of the planet rover. It is necessary to provide external measurement means such as the position, attitude and speed of the planetary rover to evaluate the motion state and performance of the planetary rover during the release procedures. The external measurements equipment should be placed around the indoor test field before the tests begin. The reference coordinate of external measurements equipment should be the same to the coordinate of the planetary rover.

During the design of verification system, every requirement should be satisfied by some equipment or test cases.

3.2 Components of verification system

The components of verification system are the test rover model, the integrated test system, the proving ground, the assistant system, the monitoring system. The test rover model is based on the electrical one whose mobility performance is similar to the real rover. The function of integrated test system is controlling the rover model as the same to that on planetary surface. The proving ground can simulate the planetary gravity (Low Gravity System), soil, topography. The assistant system includes the Simulated Release Equipment for planetary rover, balloon for mast deployment, mast temporary fixed structure. The monitoring system are used for measuring the positions, attitudes and velocities of test rover model.

The low gravity system is composed of the crane system, monitoring platform, constant pull equipment, payload for rocker-bogie suspension and field control equipment. The function of crane system is anchoring position for monitoring platform. The function of monitoring platform is following the rover test model in the indoor field. The function of constant pull equipment is keeping the pull force on rover model unchanged in any situation. The function of payload for rocker-bogie suspension is keeping the contact pressure of each wheel similar to the one on Martian surface. The field control equipment can power on/off and control the monitoring platform, constant pull equipment which is placed in special room.

The special test instrument for release procedures of planetary rover is the Simulated Release Equipment which is shown in Fig. 3. The Simulated Release Equipment is composed of the simulated lander module and the simulated planetary soil module. Among them, the simulated lander module can realize the simulation of the lunar terrain slope and the height (above the surface) and attitude (pitch and roll, yaw is not required) of the lander; The ramp transfer mechanism is assembled in the simulated lander module. The simulated planetary soil module should be laid some soil simulant, the height of soil should be more than 0.5m.

Before the tests, the components of verification system are debugged in advances. Every system is debugged individually. Then, the full system is integrated together after the test rover model are involved.

4. Results of Mars rover verification tests

4.1 Test cases

The test cases of the release procedures for planetary rover should be divided into normal case and limit cases. In the cases, the terrain of the landing area and the attitude of the lander need to be considered. The terrain of the landing area includes the slope and obstacles. According to the release direction of the planetary rover, the terrain of the landing area can be divided into upslope, downslope and lateral slope; Obstacles are divided into craters and rocks. The attitude of the lander mainly considers the attitude inclination caused by the deformation of the landing leg.

Taking the Zhurong Mars rover as an example, on the basis of the simulation analysis, the normal and limit cases are shown in Table 1.

In normal case, the deformation of landing leg is not considered. The pitch and roll angles of the landing platform are horizontal. In the limit cases, the pitch and roll angles of the simulated lander module and the slope angle of landing zone are changed. The "limit" means that angle in that situation is the biggest for the rover. As for the ramp transfer mechanism, the ramp pitch angle and its difference are two different cases. The pitch angle of ramp effects the pitch angle of the rover, but the pitch angle difference between left-right ramp effects the pitch and roll angle of the rover. In addition, the release and separation scheme of the Mars rover also includes the forward and backward deployment of the ramp transfer mechanism.

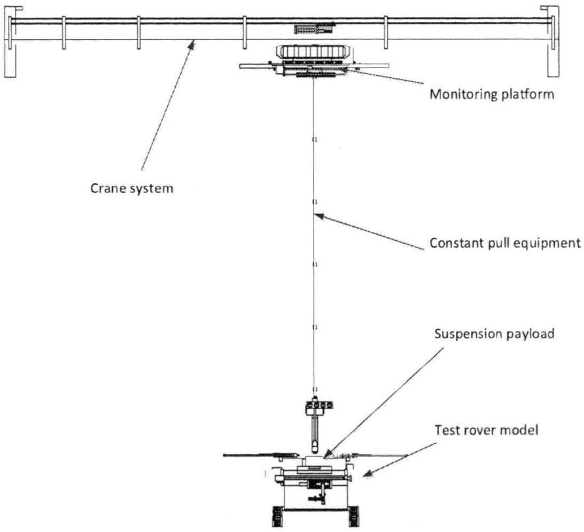

Figure 2. Low gravity simulation system for planetary rover in indoor field

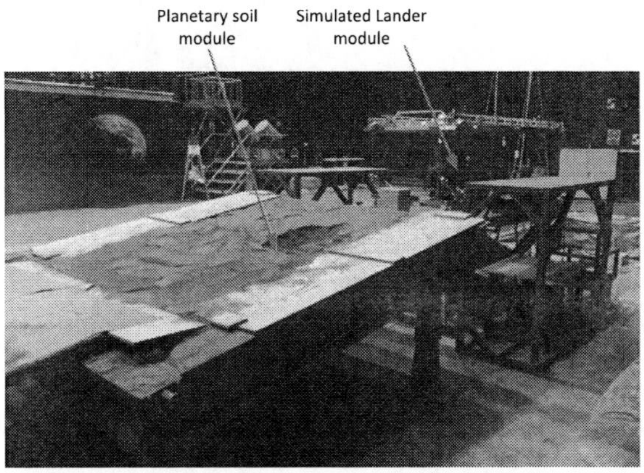

Figure 3. Release and separation test equipment for planetary rover

Table 1. Normal and limit test cases of release and separation for Mars rover

Cases	Lander pitch	Lander roll	Terrain angle
Normal case	0	0	0
Ramp pitch maximum case	8°	0	8° (down)
Pitch angle maximum difference	8°	6°	8° (down)
Ramp roll maximum case	0	8°	8° (lateral)
Roll angle maximum difference	0	2°	8° (lateral)

4.1.1 Normal case

The normal case is defined as follows: the simulated lander is in horizonal situation and the slope angle of ramp transfer mechanism is 20°. The aim of normal case is verification the release procedures. The normal case was tested several times using the rover structural model and electrical model. During the normal cases, the verification items including the mast deployment, solar array deployment, rover body lift-off, sense of NavCams, HazCams, Monitoring Cameras (loaded on landing platform). The picture of NavCams in tests are shown in Fig. 4.

The test results of model release procedures are shown in Fig. 5. After the rover model left the simulated lander module, the pictures took by front and rear Hazcams are shown in Fig. 6.

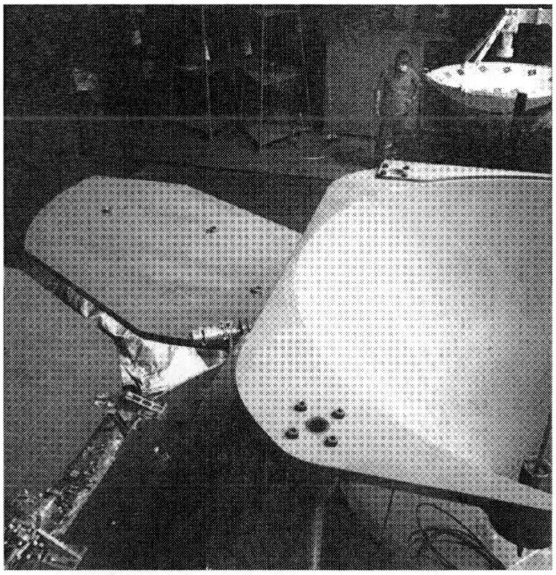

Figure 4. Picture by NavCams in tests

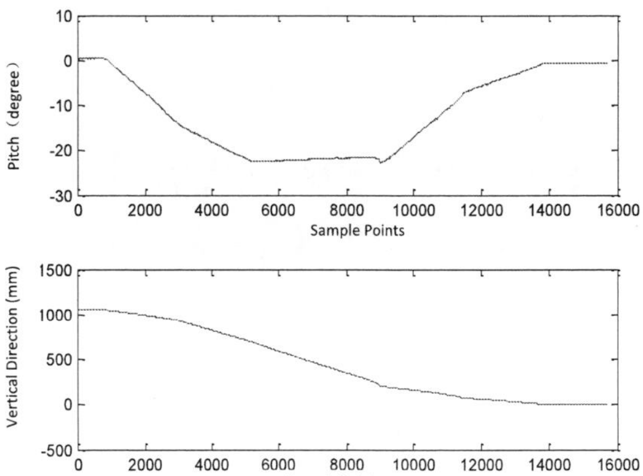

Figure 5. Measured curves of Mars rover release procedures

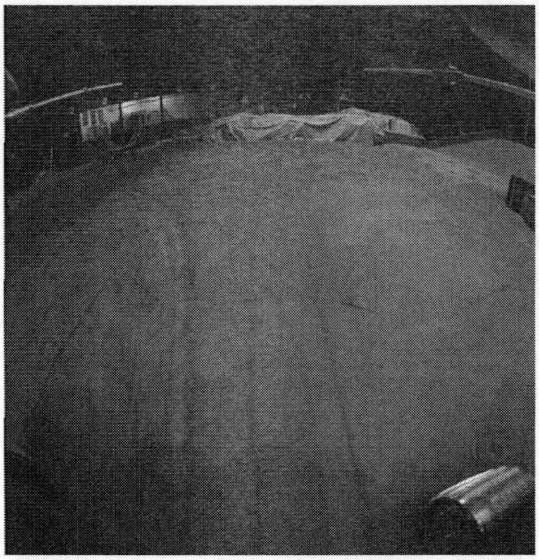

a) Front Hazcams

b) Rear Hazcams
Figure 6. Pictures of front and rear Hazcams of test rover model

4.1.2 Limit cases

In the limit cases, the results of the limit pitch angle maximum difference between left-right ramps are shown. The results of the other cases are similar, the test rover model travels successfully from the top of the simulated lander module to the surface. In the limit case of pitch angle maximum difference between left-right ramps, the pitch angle is 8.3°, the roll angle is 6.2°, and the ramp angle is 34.9°, and the pitch angle difference is 7.1°. The landing surface is 8° downhill. The change of attitude angle in the working condition of maximum pitch out of plane angle is shown in Fig. 7. During the departure of the rover, the maximum pitch angle of rover model is 31.8° and the maximum rolling angle is 6.0°. These values are less than the limit angle of the rover, so the rover can travel successfully.

Figure 7. Limit case of pitch angle maximum difference

4.1.3 Fault test cases

The aim of fault test cases is to test the ability of release in the condition that the separation plug between landing platform and rover fails to separate. In that condition, the rover has to use the drawpull to pull it. The fault test cases are divided into 4 types, as shown in Table 2. The difference between cases is the slope angle of landing platform. In the state of 18° upslope, the drawpull of rover model is tested by a special payload because the rover has to overcome the influence of Martian gravity which requires the maximum drawpull of the rover.

The mechanical detachment process of the rover is as follows: firstly the angle of the included angle adjustment mechanism in mobility subsystem changed from the stowed configuration to the angle of -30° that meets the requirement of the separation plug; secondly, rover move forward 200mm during which the separation plug will be "forced" separated; thirdly, the rover move backward 200mm in order to ensure that the front wheel is still in the "smooth" part of the guide rail of the ramp transfer mechanism; finally, the rover lift itself to the nominal position that the included angle adjustment mechanism is lifted to the included angle of 0°. In all 4 cases (horizontal, 18° downslope, 18° roll-slope and 18° upslope), the rover is lifted to the nominal position according to the strong detachment process. In the 4th case, the drawpull force of test rover model is no less than 181N, which is greater than the maximum detachment force 140N of the separation plug.

The results of fault test cases show that the Mars rover can finish its release procedures even in ill condition.

Table 2. Fault test cases of release and separation for Mars rover

Cases	Lander pitch	Lander roll
Horizon	0	0
Pitch 18° downslope	-18 °	0
Roll 18°	0	18 °
Pitch 18° upslope	18 °	0

5. Zhurong Mars rover on Martian surface

At 7:18 on May 15, 2021, Tianwen-1 Mars probe successfully landed in the preselected landing area in the south of the Martian Utopian plain. The tilting and pitching attitude of the landing platform are nearly horizontal, and the ramp mechanism is deployed in the forward direction. The Mars rover deployed independently its mechanisms such as the mast, solar wing and high-gain antenna, etc. According to the working procedures, after completing the global perception and self-testing of the mobile mechanism on the top of the landing platform, the Mars rover successfully reached on the Martian surface on May 23 and began to carry out the exploration. Images from the front hazard cameras and the rear hazard cameras during the departure of the Rover are shown in Fig. 8.

The condition provided by the ramp transfer mechanism is similar to the nominal case. During the development of the rover, two release procedures tests (nominal case) were conducted before and after the mechanical test by using the structure rover model; a full process release procedures test under nominal case was carried out in the indoor field using an electric rover model. These results of nominal case in different development phase show that the release procedures of Zhurong Mars rover have been verified fully.

6. Conclusions

This paper presents a ground verification system for egress process of the planetary rover taking the field test of Zhurong Mars rover as an example. The requirements and design results of ground verification system are given and the test results are shown. The Zhurong Mars rover work well on Martian surface based on the fully tested in development phase. Therefore, the following conclusions can be drawn:

1) The design of test cases should consider the terrain of the landing area (slopes and obstacles), the deformation of the landing leg, and the actual situation of the planetary vehicle and the transfer mechanism.

2) During the test, the control process of the planetary rover shall be the same as that on planetary surface. The similarity of the ground verification was ensured, and the correctness of the commands was also verified.

3) The test cases shall be divided into normal, limit, and fault ones. The normal case shall be fully verified, and the limit and fault cases shall also be verified to ensure that the planetary rover can complete the departure work under any conditions.

a) Front Hazcams before release

b) Rear Hazcams after release

Figure 8. Pictures from front and rear Hazcams on Martian surface

7. References

1) Jia Yang, Shen Zhenrong, Pang Yu, et al. Summary of ground test methods and technologies for lunar probe. *Spacecraft Environment Engineering*. 2014 (5):464-469.

2) Sun Zezhou, Chen Baichao, Jia Yang, et. al. The Tianwen-1 roving exploration technology for the Martian surface. *SCIETIA SINICA Technologica*. 2022. 52(2):214-225.

3) Pan Dong, Jia Yang, Yuan Baofeng, et al. Design and verification of the active suspension mobility system of the Zhurong Mars rover. *SCIETIA SINICA Technologica*. 2022, 52(2): 278-291.

4) Liu Wei, Ma Chao, Yan Qingqing, et al. Conceptual Design and Analysis of Ramp Type Transfer Mechanism for Martian Rover. *Spacecraft Engineering*. 2018. 27(3): 52-60.

5) Jia Yang, Li Ye, Ji Long, et al. Demands of Mars exploration missions on environmental simulation technologies. *Spacecraft Environment Engineering*. 2015. 32(5): 464-468.

6) Dang Zhaolong, Chen Baichao. Physical and mechanical characteristics analysis of Martian soil. *Journal of Deep Space Exploration*. 2016. 3(2):129-133.

7) Xue long, Dang Zhaolong, Chen Baichao, et al. Advances and prospects of Martian soil parameter identification based on terramechanics, *Journal of Astronautics*, Vol. 41 No. 2, 2020.

8) Deng Zongquan, Gao Haibo, Ding Liang. 2015. *Design of Mobility Systems for Lunar Rovers*. ISBN 978-7-04-042331-0. High Education Press.

4243 / Soil Compaction Monitoring Technique Using Deep Learning

Shota Teramoto[a]* and Taizo Kobayashi[b]
Author affiliations
[a] Graduate School of Science and Engineering, Ritsumeikan University, Shiga, Japan
[b] Faculty of Science and Engineering, Ritsumeikan University, Shiga, Japan
**Corresponding author: rd0084rv@ed.ritsumei.ac.jp*

https://doi.org/10.56884/ZBSA8929

Abstract

Abstract: It is commonly known that the dynamic behavior of a vibratory drum of a soil compaction machine changes with soil stiffness. Although real-time monitoring techniques of compaction quality by measuring the acceleration of the vibratory drum have already been put into practice use, their applicability depends on the soil type and condition. In this study, to extend the range of applicability and improve accuracy, we propose a deep learning-based technique that allows the regression estimation of soil stiffness from the acceleration responses of a vibration drum. To collect a large amount of noise-free training data, the acceleration responses of a vibratory drum were simulated by numerically solving the equations of the motion mass-spring-damper system. We also conducted a field experiment to verify the proposed technique. The experimental results show that the estimated values of soil stiffness correlate with the measured values, with the correlation coefficient of approximately 0.79. Thus, the proposed method has potential as a new real-time monitoring technique for soil compaction quality.

Keywords: soil compaction, vibratory compaction roller, subgrade reaction coefficient, intelligent compaction, deep learning

1. Introduction

For soil compaction quality control, soil density and moisture content measurements have been used for a long time. This traditional method offers direct quality evaluation but only a point-by-point evaluation (spot test). Furthermore, this approach causes construction delays during the testing and data analysis. To address these problems, real-time compaction monitoring techniques, called Intelligent Compaction (IC), have been developed. IC technology evaluates soil stiffness based on the dynamic response of the vibratory drum of a compaction machine, focusing on the fact that acceleration waveforms vary with the stiffness of the compacted soil. This concept was first implemented in Sweden (Thurner and Sandstrom, 1980). In the late 2000s, the U.S. Federal Highway Administration (FHWA) initiated the Intelligent Compaction Project and developed vibratory rollers that enable feedback control of the compaction force using real-time accelerometer measurements called IC measurement value (ICMV). ICMV is a generic term for index values that correlate acceleration data with soil stiffness. The ICMV used in the U.S. include Compaction Meter Value (CMV) (Thurner, 1980), Compaction Control Value (CCV) (Sakai, 2005), vibration modulus, E_{VIB} (Kröber, 2001), soil stiffness value, K_{b} (AMMANN, 2002), roller-integrated stiffness, K_{s} (Anderegg and Kauffmann, 2004) and Machine Drive Power (MDP) (White et al., 2005). These are in different forms of metrics with various correlations to the mechanical and physical properties of the compacted materials. IC technology has been applied; however, its applicability depends on the soil type and soil conditions. In this study, to extend the range of applicability and improve accuracy, we proposed a deep learning-based technique that allows the regression estimation of soil stiffness from the acceleration responses of a vibratory drum.

2. Soil stiffness evaluation by deep learning

Deep learning generally requires a large amount of labeled training data, the preparation process of which is labor-intensive. In this study, we propose a substitute for data collection using numerical simulations. As shown in Fig. 1, we first developed an equation of motion for a soil-machine system that incorporates the stiffness (spring constant) of compacted soil and prepared a large number of acceleration waveforms by numerically solving the equation with a wide variety of soil stiffness values. Then, we developed AI regression models to estimate the soil stiffness by learning the labeled training dataset.

2.1 Simulation of acceleration waveforms of vibratory drum of soil compactor

In this study, the dynamic behavior of the soil-machine interaction was modeled as a mass-spring-damper system (Fujiyama and Tateyama, 1996), as depicted in Fig. 2. The soil-drum system and drum-machine frame system were simplified into Voigt models and connected in series. The equations of motion for the frame and vibratory drum are expressed as:

$$m_1\ddot{x} + k_r(x - y) + c_r(\dot{x} - \dot{y}) = m_1 g, \quad (1)$$

$$m_2\ddot{y} + k_s y + c_s\dot{y} - k_r(x - y) - c_r(\dot{x} - \dot{y}) = m_2 g + F\,sin(2\pi f t) \quad (2)$$

where m_1: mass of frame, m_2: mass of vibratory drum, x: vertical displacement of frame, y: vertical displacement of vibratory drum, k_r: spring constant of anti-vibration rubber damper, k_s: spring constant of soil, c_r: viscous damping coefficient of rubber damper, c_s: viscous damping coefficient of soil, F: vibration force, and f: basic frequency of vibratory drum.

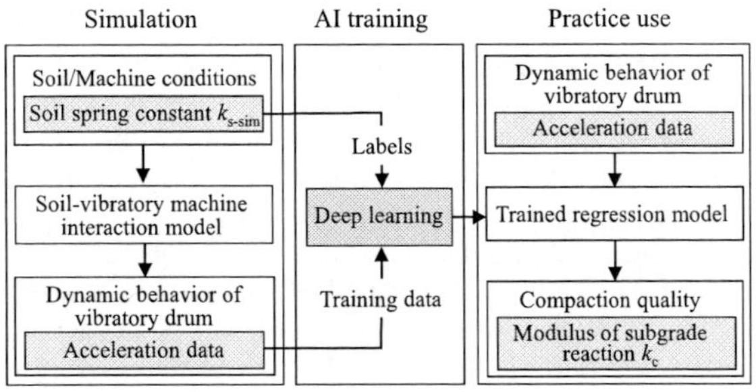

Figure 1. Flowchart of AI-based intelligent compaction (IC) technique

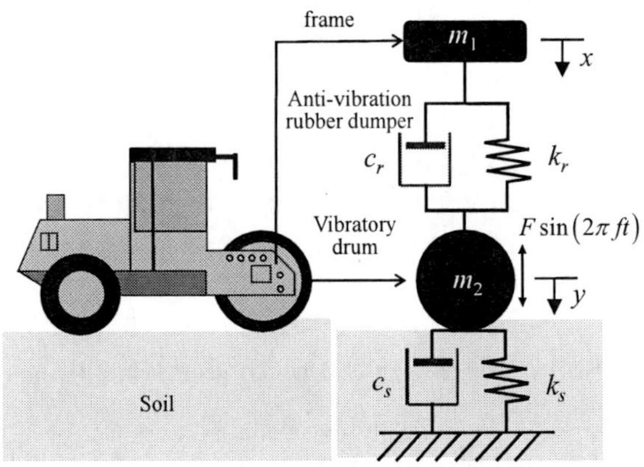

Figure 2. Soil-vibratory compactor interaction model

Table. 1. Parameters used in simulation

Parameter	Symbol	Quantity
Mass of frame	m_1	2,080 (kg)
Mass of vibratory drum	m_2	4,070 (kg)
Vibration force	F	172.0 (KN)
Frequency of vibratory drum	f	31.2-35.2 (Hz)
Spring constant of dumper	k_r	1,000 (N/m)
Viscous damping coefficient of soil	c_r	$2D_r\sqrt{m_1 k_r}$ (Ns/m)
Spring constant of soil	$k_{s\text{-sim}}$	0.1～250 (MN/m)
Viscous damping coefficient of soil	c_s	$2D_s\sqrt{m_2 k_s}$ (Ns/m)

As soil has no tensile strength, the soil reaction force, expressed as $k_s x + c_s\dot{y}$, will not be negative. To consider this, calculations were performed under the condition $k_s = c_s = 0$ when $k_s x + c_s\dot{y} < 0$. Acceleration waveforms corresponding to soil and machine conditions were obtained by numerically solving Equations (1) and (2).

Fig. 3 presents examples of the measured and simulated acceleration waveforms (upper figures) and their amplitude spectra (lower figures) obtained via Fast Fourier Transform (FFT). The measured data in Figs. 3(a), (b), and (c) represent the results obtained when the number of passes of the compactor was 2, 8, and 16, respectively. The corresponding moduli of the subgrade reaction measured in the field experiment were $k_{c\text{-mea}}$ = 77.2, 127.5, and 193.5 MN/m^3. The details of the field experiments will be described later. In the simulation, the spring constants of soil of $k_{s\text{-sim}}$ = 38.6, 63.8, and 96.8 MN/m and the parameters listed in Table 1 (assuming an 11-ton class vibratory compactor) were considered. The reason for using different parameters ($k_{s\text{-sim}}$ and $k_{c\text{-mea}}$) is described later.

As shown in Figs. 3(a) and (b), representing a condition when the soil is not yet compacted, the measured and

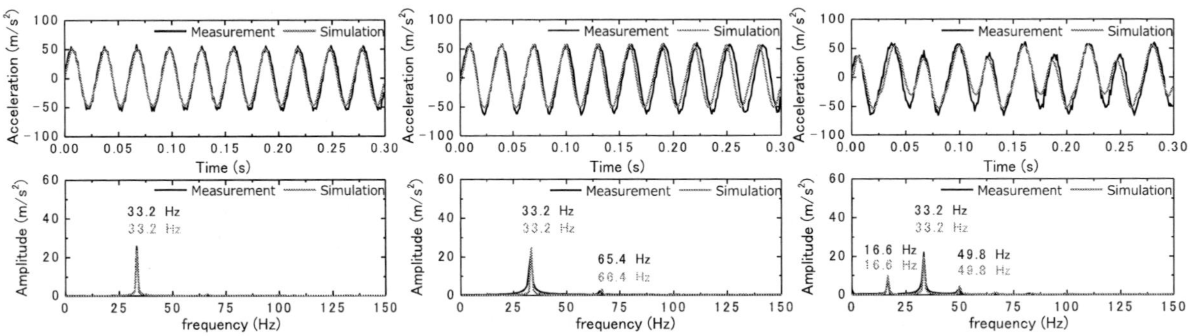

(a) $k_{s\text{-sim}}$=38.6 MN/m, $k_{c\text{-mea}}$=77.2 MN/m^3 (b) $k_{s\text{-sim}}$=63.8 MN/m, $k_{c\text{-mea}}$=127.5 MN/m^3 (c) $k_{s\text{-sim}}$=96.8 MN/m, $k_{c\text{-mea}}$=193.5 MN/m^3

Figure 3. Comparison between measured and simulated acceleration data simulated data are in good agreement, and both show sinusoidal behaviors with a regular amplitude of ±50 m/s^2. The dominant frequency also coincides with the frequency of the vibratory drum, indicating that low soil stiffness has no effect on the vibratory drum motion. In Fig. 3(c), representing the condition when the soil is well-compacted, the trends of the simulation and experimental results are consistent. It was observed that when the soil stiffness was high, the acceleration waveform was disturbed, and the peaks of different levels appeared alternately. The frequency spectra show the appearance of a 1/2-order subharmonic frequency component.

2.2 Deep learning for soil stiffness estimation

The conventional IC technology uses a theoretical/empirical model that correlates between acceleration data and soil stiffness. Deep learning is a subset of machine learning which uses multilayer neural networks. One of the biggest advantages of using deep learning approach is its ability to automatically extract useful features from input data. In this study, we consider that applying deep learning approach to IC technology will extend the range of applicability and improve the accuracy of existing methods.

This study employed a one-dimensional convolutional neural network (1D-CNN) to develop regression models for estimating soil stiffness. The neural network constituted an input layer, three sets of convolution and pooling layers with a ReLU activation function, and a fully connected layer with a squared error loss function. The trained CNN model allowed estimation of the numerical value of the soil spring constant $k_{s\text{-est}}$ from the acceleration data.

We generated 5,000 sets of labeled training data via simulation, with varying soil spring constant $k_{s\text{-}sim}$ of 0.1–250 MN/m in increments of 0.05 MN/m. The parameters used in the simulations are listed in Table 1. The vibration frequency of the compaction drum slightly varies according to the machine operation and the values of f were randomly set ranging from 31.2 to 35.2 Hz.

Additionally, in this study, two estimation models that were trained using different data forms were developed, and their generalization performances were compared.

(a) Time-series data: One-dimensional array of time-series data (numeric scalar) of acceleration responses. The sampling interval and frequency of each data point were 1.024 s and 1,000 Hz (1,024 points), respectively.

(b) Amplitude spectrum data: A one-dimensional array of amplitude spectra (numeric scalar) obtained through FFT analysis of the time-domain acceleration data (frequency of 1,000 Hz).

3. Verification of soil stiffness estimation model

3.1 Verification with simulation data

To confirm the feasibility of the proposed method, validation was performed using 50 sets of test data generated by additional simulations. Fig. 4 displays a comparison between the true (input values) and estimated values of the soil spring constant, $k_{s\text{-sim}}$ and $k_{s\text{-est}}$. The results show good agreement between $k_{s\text{-sim}}$ and $k_{s\text{-est}}$ in both cases of training data forms: time series data and amplitude spectrum data. Thus, the developed neural network learned the relationship between soil stiffness and acceleration behavior from the simulated training data, and the CNN model has the potential to estimate the soil spring constant from acceleration measurements of the compaction machine.

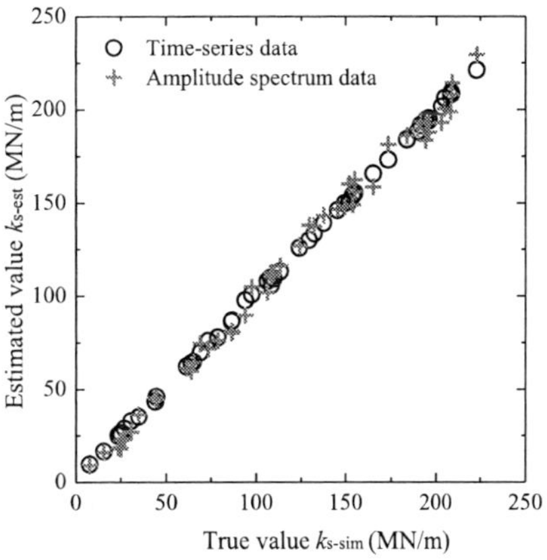

Figure 4. Comparison between $k_{s\text{-sim}}$ and $k_{s\text{-est}}$

3.2 Verification with experimental data

In this study, we conducted a field experiment at an embankment construction site to validate the proposed method. Soil compaction was demonstrated using an 11-ton class vibratory compactor (Fig. 5) in a rectangular area with length and width of 10 and 8 m, respectively (Fig. 6). The area was divided into four lanes, each of which was compacted using a different number of passes. The experiment was conducted in two phases: the first phase involved 2, 4, 6, and 8 passes on each lane, while the second phase involved 10, 12, 14, and 16 passes. The average travel speed of the compactor was approximately 4.0 m/s. A strain gauge accelerometer was attached to the bracket of the vibratory drum to collect acceleration data during compaction (sampling frequency of 1,000 Hz).

The soil material constituted 45.8% gravel, 23.4% sand, and 30.8% fine-grained content. The average moisture content during compaction was 12.1% (optimal moisture content of 15.7% and maximum dry density of 1.81 g/cm^3). The lift thickness was not controlled in this experiment and varied in the range of 30–50 cm.

Fig. 7 depicts the relationship between the number of passes and the modulus of the subgrade reaction $k_{c\text{-mea}}$. The modulus of the subgrade reaction shown here was measured using a portable in-situ soil testing tool called CASPOL, developed by the Kinki technical office of the Ministry of Land, Infrastructure, Transport and Tourism, Japan. This is a drop-weight impact tester that records the magnitude of maximum impact acceleration at the time when the rammer hits the ground, and the value is converted to $k_{c\text{-mea}}$ through an empirical equation. As shown in Fig. 7, $k_{c\text{-mea}}$ increases with the number of passes up to eight and remains constant for a higher number of passes, except for 16 passes. For the validation of the CNN model, from the records of acceleration data of the vibratory drum, we extracted sections while the compactor passed through the points where the drop weight impact test was performed.

Figure 5. Vibratory compactor used in this study

Figure 6. Test area

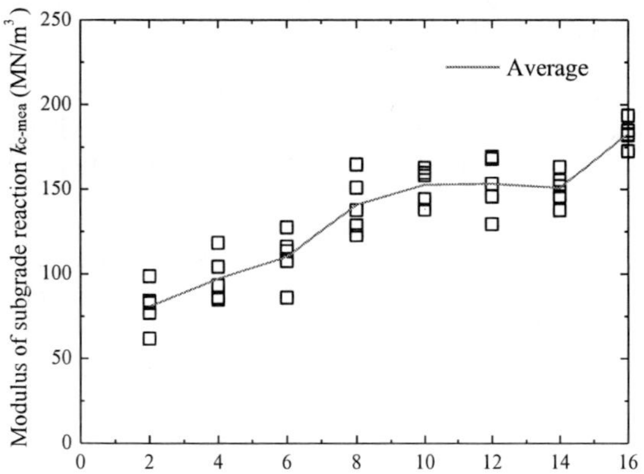

Figure 7. Soil stiffness results

Fig. 8 shows comparisons between the measurements of $k_{c\text{-}mea}$ and the estimated values of $k_{s\text{-}est}$. The CNN model trained using time-series acceleration data (Fig. 8(a)) exhibited a relatively high positive correlation with the correlation coefficient $r = 0.790$. The model trained using the amplitude spectrum data (Fig. 8(b)) also showed a positive correlation, while the correlation coefficient was rather low. Considering that there are substantial variations in the measurements of $k_{c\text{-}mea}$ obtained by the drop-weight impact tester, the correlation seems reasonably good, and we can conclude that the deep learning-based approach has the potential to estimate soil stiffness.

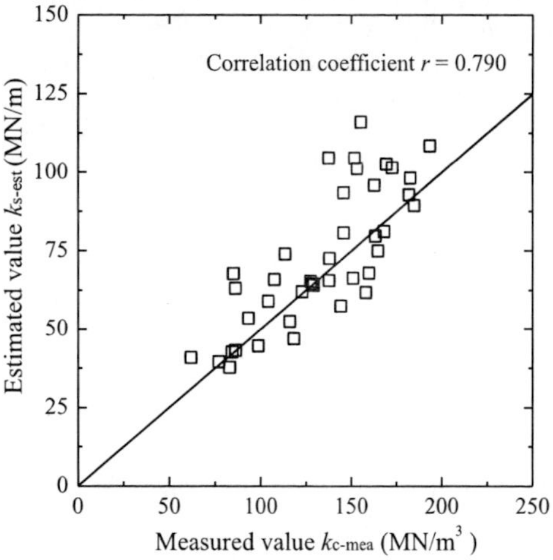

a) Trained using time-series data

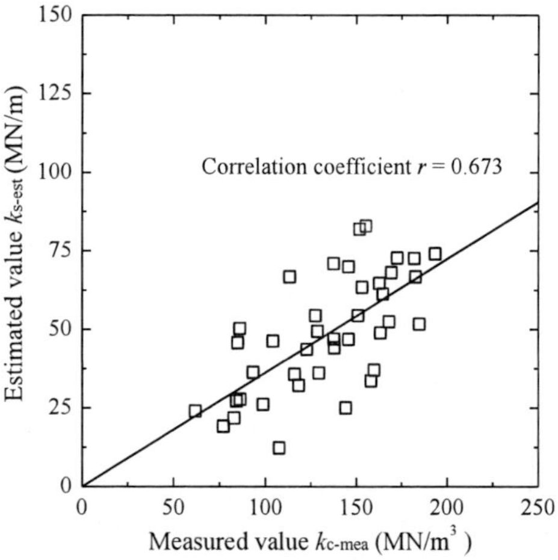

b) Trained via amplitude spectrum data

Figure 8. Correlation between measured and estimated values

It should be noted that the estimation model can estimate the soil spring constant and not the modulus of the subgrade reaction. The soil spring constant is defined as the ratio of the force acting on the ground to the settlement caused by it, whereas the modulus of the subgrade reaction is defined as the ratio of the pressure (stress) to the settlement. Therefore, to estimate the modulus of the subgrade reaction, the estimated soil spring constant $k_{s\text{-}est}$ must be divided by the contact area of the soil and the vibratory drum. It is difficult to measure and/or predict the contact area during compaction and no reasonable solution has been found to date. In Fig. 3, the comparisons were made using the simulation results for $k_{s\text{-}sim}$ = 38.6, 63.8, and 96.8 MN/m. The values of $k_{s\text{-}sim}$ used here were obtained by intentionally assuming the contact area of 0.5 m^2 (0.25m wide × 2.0m long).

4. Conclusions

In this study, we proposed a deep learning-based compaction monitoring technique that allows the estimation of soil stiffness from the acceleration responses of a vibratory drum. The results of the verification experiment showed a high positive correlation (r=0.79) between estimated soil stiffness and measured soil stiffness, indicating that deep learning-based approach is technically feasible.

Quality assurance and quality control (QA/QC) of soil compaction has been performed based on soil density and moisture content. Quality should be evaluated based on mechanical properties (design parameters), such as the stiffness and strength of soils. We believe the proposed method will contribute to the rationalization of QA/QC for soil compaction.

5. Acknowledgments

This research was supported by the Japan Society for the Promotion of Science (JSPS) [grant number 19K04605]. The authors would like to thank Shimizu Corporation, GeoSurf Corporation, and Soil and Rock Engineering Corporation for their cooperation and advice.

6. References

AMMANN, 2002. European and U.S. Patents on the ACE-System.

Anderegg, R., Kaufmann, K., 2004. Intelligent Compaction with Vibratory Rollers: Feedback control systems in automatic compaction and compaction control. Transportation Research Record, no. 1868, pp. 124-134.

Fujiyama, T., Tateyama, K., 1996. Study of The Evaluation of Compacting Ground Stiffness by the Acceleration Response of Vibratory Roller (in Japanese). Proceedings of the JSCE, No. 544, III-37, pp. 231–237.

Furuya, H., Fujiyama, T., 2011. Development of Soil Stiffness Evaluation Equipment "ALFA SYSTEM" Using Acceleration Response of Vibratory Roller. Proceedings of the 28th ISARC.

HAMM COMPACTION METER (HCM). https://mdx.vn/hamm-compaction-meter-hcm (accessed 18 July 2022).

Kröber, W., Floss, E., Wallrath, W., 2001. Dynamic soil stiffness as quality criterion for soil compaction. Geotechnics for Roads, Rail Tracks and Earth Structures, A.A. Balkema Publishers, pp. 189-199.

Kloubert, K., 2006. Intelligent Soil and Asphalt Compaction Technology. TRB Annual Meeting, Washington, D.C.

Ltd Sakai Heavy Industries, 2007. Vibrating roller type soil compaction quality controller CCV (Compaction Control Value). Operating Maint. Instr, pp. 1-5.

Richart, F. E., Hall, J. R., and Woods, R.D., 1970. Vibrations of soils and foundation. Englewood Cliffs: Prentice Hall.

Sandstrom, A., geodynamic AB., 1994. Numerical Simulation of a vibratory roller on Cohesionless Soil. Internal Report, Geodynamik.

Teramoto, S., Kobayashi, T., 2021. A Proposal of In-Situ Soil Compaction Quality Measurements Using Deep Learning. JSCE. Journal of AI and Data Science. No 2, vol 2. pp. 785-791.

The Transtec Group, Inc., 2018. Intelligent Compaction and Infrared Scanning Field Projects with Consulting Support Final Report. MoDOT Publication, No. cmr 18-003.

Thurner, H., Sandstrom, A., 1980. A New Device for Instant Compaction Control. Proceedings International Conference on Compaction, vlo2, pp. 611-614.

White, D.J., Jaselskis, E.J., 2005. Real-Time Compaction Monitoring in Cohesive Soils from Machine Response. Transportation Research Record, no. 1936, pp.173-180

White, D.J., 2009. Report of the Workshop on Intelligent Compaction for Soils and HMA. 2nd Annual Intelligent Compaction for Earthworks Workshop.

White, D.J., Vennapusa, P.KR., 2010. A Review of Roller-Integrated Compaction Monitoring Technologies for Earthworks. Earthworks Engineering Research Center (EERC), Final Report ER10-04.

4260 / The Running Gear Construction Impact on Obstacles Overcoming by Light High-Mobility UGV

Daniela Szpaczyńska*, Marian Łopatka, and Piotr Krogul
Military University of Technology, Warsaw, Poland
**Corresponding author: daniela.szpaczynska@wat.edu.pl*

https://doi.org/10.56884/QSJO7805

Abstract

Rubber tracked running gears are widely used in high-mobility Unmanned Ground Vehicles (UGV) to increase obstacle negotiation possibility in urban and rural terrain. The paper proposes a method of assessing the configuration of the light UGV`s tracked running gear in terms of the ability to overcome terrain obstacles. Five types of obstacles have been proposed: a road ditch, a lying log, a curb, a crop rows and a wavy meadow profile. Then designed four types of running gear with the same geometrical and mass parameters, differing in the way of applying the running wheels (rigid construction, consisted of sprocket, idler and four road wheels, two bogies solution, system with two bogies elastically mounted to frame and rocker-bogie construction). To compare the selected solutions, simulation in the multi-body environment program and the equations based on the Bekker theory was performed. The simulation results showed an improvement in the driving properties with the use of elastically suspended elements. Better copying of the shape of the obstacle by the running gear contributes to the required tractive effort at lower slip values. At the same time, the negative impact of elastically suspended elements on the platform's stability was indicated.

Keywords: terrain mobility, rubber track running gear, high-mobility unmanned ground vehicles, obstacles overcoming

1. Introduction

Lightweight UGVs (up to 300 kg) are increasingly used in EOD missions (Bartnicki et al., 2015; Grigore et al., 2021), military support(Khurshid and Bing-rong, 2004; Lopatka, 2020; Micire, 2008; Yamauchi, 2004), search and rescue operations (Choi et al., 2008; Davids, 2002). Effective tasks performance requires of them high terrain mobility, understood as the ability to overcome terrain with various ground types, as well as negotiate obstacles and rough terrain, with comparable to the human movement speed (Bruzzone and Quaglia, 2012). Despite the limited size of the platform and its running gear, lightweight UGVs should efficiently negotiate both urban and rural areas with encountered obstacles (Lopatka, 2020; Massey, 2016). For this purpose, most of UGVs are equipped with tracked locomotion system, which compared to wheeled, provide higher tractive forces and better stability (Guo, 2016; Rasool and Raheman, 2018; Wong and Huang, 2006).

Lightweight UGVs negotiate terrain differently than regular-sized vehicles. Especially terrain obstacles and surface roughness are a significant factor to restrict overcoming abilities of UGV. Their decreased weight and lower ground pressure limiting the available tractive force(Weidong et al., 2008), while the dimensions of the typical terrain obstacles become relatively larger than for regular vehicles. Articles (Grigore et al., 2021; Haueisen, 2003) highlight the scale impact as the major limitation of UGV mobility. It was estimated (Haueisen et al., n.d.)using the NRMM algorithm that when performing a probable tactical task tracked robot Packbot 510 (weight 31 kg) is not able to negotiate about 35% of the area (NO-Go zone), while for the 100 times heavier HMMVV wheeled vehicle No-Go zone is less than 10%. The report points out that for standard military vehicles the main no-go reason is insufficient obstacle clearance, while for lightweight UGV like Packbot the main difficulty is vegetation and obstacle override problems. At the same time, the mobility results in relation to the vehicle dimensions reach similar values for HMMVV and Packbot 510.

According to field experiments (Krotkov et al., 2007) terrain roughness, ditches and slopes are the cause of many emergency stops as well as loss of stability and rolled over of UGV. However, smaller terrain roughness becomes especially dangerous. Testing autonomous UGVs it was highlighted that none of the sensor systems was able to detect gentler slopes, therefore their unconscious overcoming attempt led to immobilization. Terrain slope is an inseparable component of tactical terrain—surfaces inclined by more than 10 degrees account for more than half of the studied tactical operations areas (Nuttall, 1979).

A special example of terrain obstacles with small but steep slopes are crop rows. According to the report (Lopatka et al., 2014) it may be one of the most demanding obstacles for light UGV in rural terrain. Fallen log trees are another characteristic object encountered in rural areas. Climbing them during autonomous missions also ended in many cases with emergency stops and the necessity to tow the UGV (Krotkov et al., 2007). Another publication (Yajima and Nagatani, 2017) highlights the specific problems of light tracked robots climbing cylindrical objects located on slopes. Even a small, unfixed object can cause sliding down of the UGV. The climbable obstacle height is reduced by 50 percent in a 10^0 slope.

Terrain profile may also be the result of its coverage. The report (Lopatka et al., 2014)indicated that while negotiating clumps of grass and vegetation, the terrain profile is formed in repeating undulations. The same study shows that UGVs below 200 kg are not able to effectively overcome the terrain covered with high grass due to the deformation resistance of the vegetation and insufficient tractive force (Lopatka et al., 2014). Poor effectiveness in negotiating Johnson's grass by the robot TALON V was described in the report (Richmond et al., 2009). The robot did not complete 75% of the runs due to stalled out. Research (Krotkov et al., 2007) on the performance of heavier (800 kg) autonomous UGV also points out that thick vegetation such as brambles and shrubs efficiently slowed the vehicle runs. Although, the presence of vegetation increases the strength of the soil (Wieder and Shoop, 2018) and effect positively on standard-sized vehicle mobility, for light UGVs grass can significantly reduce developed drawbar pull. In report (Richmond et al., 2009) Packbots drawbar performance tests were carried out on various surfaces - the drawbar pull value on sand was 10% lower than on asphalt surface, while the drawbar pull developed on grass, decreased by 30% compared to asphalt.

Insufficient drawbar pull can limit the size of negotiated obstacles. Packbot 510, despite satisfactory stability level, was able to climb the 380 concrete slope and only 30^0 sand slopes (Longoria et al., 2003). Another experiment (Richmond et al., 2009) points out that by lower the roughness of the concrete surface, it is possible for Packbot 510 to climb on 40^0 slopes, while for more rough concrete surface it can climb 37^0 slopes. The authors justify this by the reduced track-terrain contact surface in rougher materials. This indicates that small running gears are particularly sensitive to the track-terrain contact conditions and even a surface microstructure.

One of the most common methods of the drawbar pull estimation for heavy vehicles with steel tracks is Bekker theory (Bekker, 1969). Wong extend this theory with equations describing the strength of various types of soil (J. Wong, 2008). This approach was used by Okello to describe rubber tracked running gear performance (Okello et al., 1998). The possibility of use this method to describe small, tracked vehicles performance was verified in (Wong et al., 2015) with a track load of 155-190 N, with load 1300 N in (Wang et al., 2018), with load 5000 N in (Park et al., 2008), on respectively Quikrete Medium sand, clay and sandy land and loamy soil. Measurements were carried out on a flat, even surface with the speeds of range 0,5 - 0,12 m/s. The assumed types of ground, and especially the speed values, do not fully describe the typical UGV operating conditions.

According to the experiments, the vehicle's speed on a flat asphalt surface may have an influence on the developed drawbar pull. The maximum value of drawbar pull occurs at the minimum speed, and it decrease as speed increase (Richmond et al., 2009). Therefore, the NRMM algorithms predicted to overcome obstacles such as a curb or driveway by robots: Matilda, Talon and Packbot, with speeds value respectively 1,2 m/s, 2,4 and 2,6 m /s despite they higher value maximum speeds. Another articles (McBride et al., 2003; Wang et al., 2007) presents how dynamic effect can improve the ability to overcome ditches. Increasing Packbot 510 speed from 0,2 to 1,9 m/s is managed to overcome 50% wider ditch (McBride et al., 2003). Article authors also mention possible travel speed impact on climbing higher steps.

Another factor that significantly determines the mobility of the UGV is the shape and design of the running gear. The articles (Edlund et al., 2013; Grigore et al., 2021; Li et al., 2022) point out that changing the shape of tracked running gear can increase the ability to negotiate ditches, slopes and stairs. Evaluating various solutions of arctic machine locomotion system (Dudziński et al., 2012) the positive impact of elastically mounted running gear elements is acknowledged on the vertical acceleration of the vehicle frame, the distribution of pressures and the loads acting while climbing the slope. The simulation results (Kim et al., 2018) indicate that better running gear shape adaptation to the surface roughness improves obtained traction force.

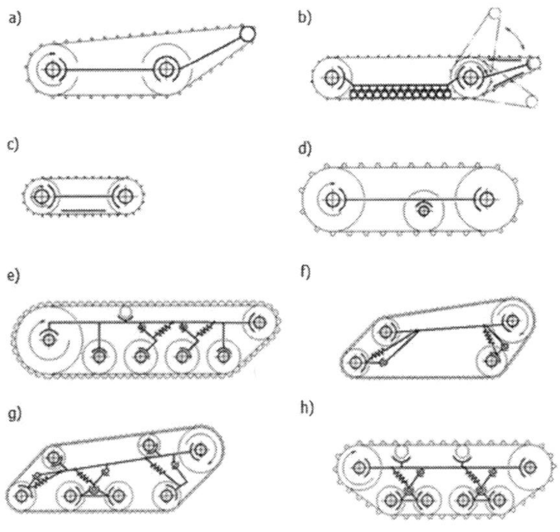

Figure 1. Running gear constructions schemes:
a) Dragon Runner 20 by Foster Miller; b) iRobot's PackBot 510; c) AEODRS increment 1 by Northrop Grumman Corporation; d) Foster Millers TALON V; e) thEODor by Telerob; f) RS1 by Howe and Howe; g) RS2H1 "Gryzly" by Howe and Howe; h) Aslan by Aselsan

The running gears of currently used UGVs differ in shape and structure—schemes of the most popular solutions are presented in Fig. 1. The need to reduce the weight is the main cause to affect the running gear construction. Backpackable (up to 20 kg) and portable (up to 75 kg) UGVs are characterized by a rigid suspension and a simple structure of track system. Obstacles overcome ability increase is realized by inserting additional, active elements (e.g., flippers), which allow changing the locomotion system shape while obstacle negotiating. In medium-sized

UGVs (up to 1000 kg) running gears usually contain flexible elements—bogies, rocker system and springs. Mobility improvement may be accomplished by increasing the number of flexible elements.

The design of the locomotion systems especially in lightweight UGVs is the main factor that determine their ability to overcome the terrain, and thus the utility of UGVs. Currently, very different running gears are adopted in widely used robots. However, there are no clear comparisons of how these designs affect mobility. Since there are no possibility to unify and directly compare individual UGVs tests results—the aim of the paper is to compare and evaluate the influence of the structure of the tracked running gear on mobility. This assumes ensuring unify simulation conditions, considering the operational speed of the robots and the characteristic obstacles in rural terrain.

2. Methodology and model

The aim of the research was to determine the impact of alternative lightweight UGV track systems solutions on their ability to overcome obstacles. The research was carried out by multibody simulation. Four undercarriage assemblies were selected for the simulations (Fig. 2): a) rigid construction, consisted of sprocket, idler and four road wheels, b) two bogies solution, c) system with two bogies elastically mounted to frame, d) rocker-bogie construction. The spring's stiffness and dumping value are selected to ensure even pressure of the wheels on the ground while driving on flat surface.

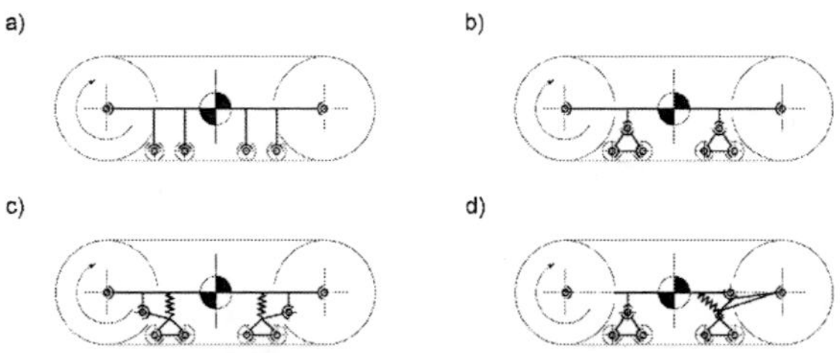

Figure 2. Selected undercarriage schemes

The tracked running gear model was built in the MSC ADAMS program. Models in this environment consist of simple-shaped rigid bodies, constraints and external forces acting on them. The program enables the performance of dynamics and kinematics analyzes by arranging and solving the equations of motion.

It was assumed that the analyzed undercarriage solutions are equipped with the same rubber track of the DTV Shredder high mobility vehicle. The rubber belt was modeled as a series of 56 links connected with each other by general force vectors. Their elasticity and damping characteristics (Fig. 3, Table 1) were determined during the identification tests. The identified resistance to track bending and sliding on wheels without load was 15% of the UGV weight. The model also takes into account the wheel rolling resistance. In result, the rolling resistance coefficient was about 0,25. Both values are within specified for similar tracked vehicles ranges (Lopatka et al., 2014; Morales et al., 2009; Rasool and Raheman, 2018).

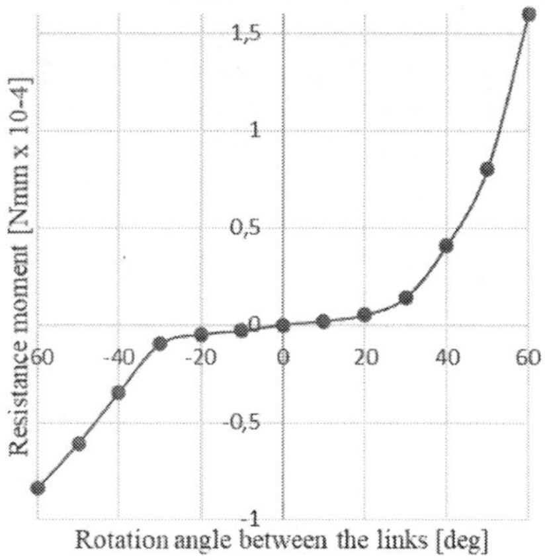

Figure 3. Determined rotational elasticity characteristic of track links

Main parameters of model are listed in tab.1. It was assumed that the center of gravity (CG) is in the running gear's symmetry axis at the height of the sprocket's centerline. There was the positive drive transmission on the track, also there was no track tensioning mechanism and a 5 mm sag was assumed. Ground response under loading conditions was determined by IMPACT function (Šulka et al., 2018):

$$F = \begin{cases} 0 \; if \; y > y_1 \\ k \cdot (y_1 - y)^e - c\dot{y} \cdot STEP(y, y_1 - d, 1, y_1, 0) \\ if \; y \leq y_1 \end{cases} \tag{1}$$

Where y and y_1 characterize the distances bodies (links and the ground) in contact. Other parameters are listed in tab.1. The STEP function is used in ADAMS to make a smooth transition between two value levels. The function characteristic is presented in the figure 4.

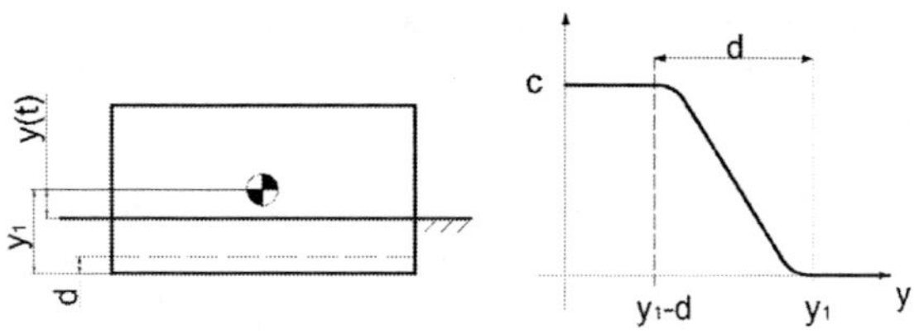

Figure 4. Function STEP plot and parameters scheme

The thrust required for tracked running gear driving is developed by the shearing action between the track and the terrain. Gross traction for each link was determined and summed up according to the equation (Wong, 2008):

$$F = b \int_0^l \tau(x) \cdot dx \tag{2}$$

Where b is the track width, l is the track contact length; x is the distance from the track-ground contact area beginning and τ(x) is the shear stresses determined by horizontal shear displacement according to the Wong function (Wong, 2008):

$$\tau(x) = c + \sigma \cdot \tan\emptyset \cdot K_r \cdot \left(1 + \left(\frac{1}{K_r\left(1-\frac{1}{e}\right)} - 1\right) \cdot e^{\left(1-\frac{j}{K_\omega}\right)}\right) \cdot \left(1 - e^{\left(\frac{-j}{K_\omega}\right)}\right) \quad (3)$$

Where: c is soil cohesion; σ is normal pressure; Ø is angle of internal shearing resistance of the terrain; K_r is the ratio of residual stress; K_ω shear displacement where the maximum shear stress occurs; j is horizontal shear displacement increasing along the contact length and defined by Eq. 4 (Wong, 2008):

$$j = i \cdot x \quad (4)$$

$$i = \begin{cases} \frac{V_t - V}{V_t} & if\ V_t > V \\ \frac{V - V_t}{V} & if\ V > V_t \end{cases} \quad (5)$$

Where x is the distance from the beginning of the track's contact with the ground; i is slip factor, which can be defined by Eq. 5 (Wong, 2008) depending on the difference in theoretical speed; V_t and V are actual forward vehicle speed.

Using of these equations allows including the shear strength enhancement of cohesive soil in the initial phase of ground deformation.

Table 1. Simulation characteristics

Parameter	Value
Running gear assembly mass [kg]	100
Rubber belt mass [kg]	6,4
Track width [mm]	160
Track length [mm]	2450
Length of the track on the ground [mm]	685
Sprocket, idler diameter [mm]	320
Drive wheels diameter [mm]	70
Sprocket peripheral speed [mm/s]	845
Vertical, longitudinal, and horizontal track belt stiffness [N/mm]	10000
Vertical, longitudinal, and horizontal track belt dumping [N·s²/mm]	1000
Yaw and roll rotational stiffness [N/rad]	10000
Yaw and roll rotational dumping [N·s²/rad]	1000
Pitch rotational dumping [N·s²/rad]	550
Ground stiffness k [N/mm]	100
Ground dumping c [N·s²/mm]	10
Contact force exponent e	2
Contact penetration depth d[mm]	10

In order to evaluate the ability to overcome terrain by the running gear solutions, simulations of negotiating 5 obstacles were carried out. Typically faced obstacles were selected (Fig. 5): a roadside drainage ditch, sinusoidal profile roughness reflecting vegetation clumps every 1 m, crop rows—typical for agricultural areas; a log with a diameter equal to idler radius and a curb 0,2 m high.

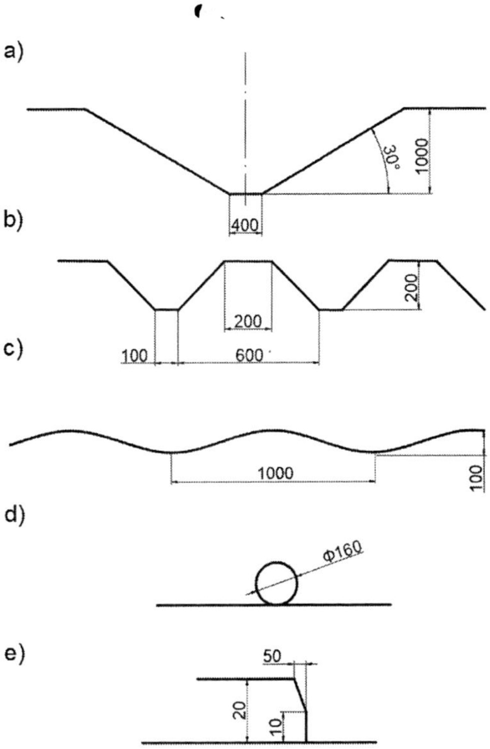

Figure 5. Shape and dimensions of selected terrain obstacles profiles

To compare the analyzed solutions, few parameters were used. Driving forces on sprocket necessary to overcome motion resistance allow determining energy consumption. Gross traction was compared to evaluate soil strength requirements and drawbar pull. In addition, to assess the stability of the UGV driving, the pitch angle and the Root Mean Square of vertical center of mass acceleration values were presented.

3. Results

While ditch negotiating, the highest track loads occur when climbing a slope. Driving forces characteristics are comparable for all analyzed solutions (Fig. 6a), however significant force differences occur at the beginning and ending of the climb. In these phases, solutions A and B are more advantageous. The slip time series does not show any significant differences (Fig. 6b). Also, no significant differences on the time to overcome the obstacle were indicated.

When negotiating the crop rows, it was assumed that the bearing capacity of the soil is twice as low as on other obstacles. As a result, sinking the running gear was up to 10 mm. Due to the dimensions of the rows and the limited length of the track, during the drive, there were intense longitudinal inclinations of the running gear (Fig. 7c) and cyclical driving force increases (Fig. 7a) and slip increases (Fig. 7b). The largest pitching occurred for solution C, while the smallest for solution D (Fig. 7c). Solution D, however, required developing the largest driving forces

(Fig. 7a). The greater slip occurred in solution C (Fig. 7b). Solution A made it possible to overcome obstacle profile without driving into furrows, with the lowest driving force and slip required.

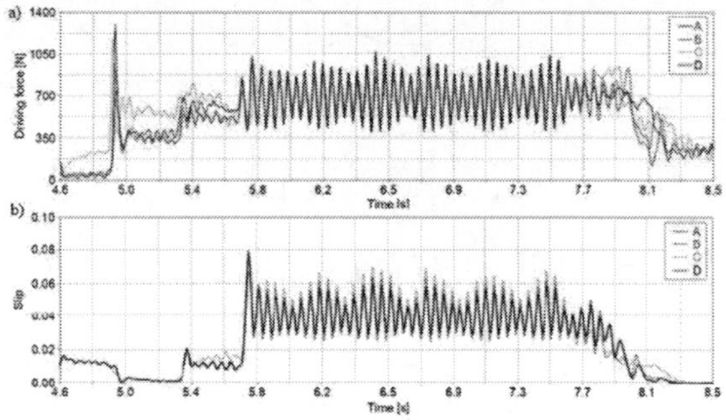

Figure 6. Ditch overcoming: a) Driving force time series; b) Slip time series

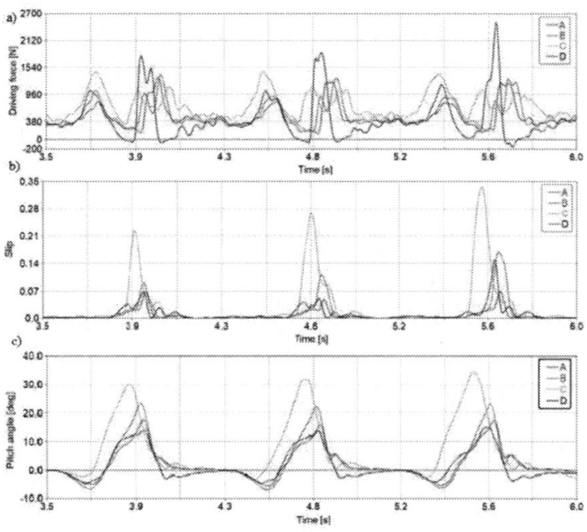

Figure 7. Crop rows overcoming characteristics: a) Driving force time series; b) Slip time series; c) Pitch angle time series

The sinusoidal surface profile reflecting the presence of clumps of vegetation, that are driven over by tracks. This obstacle requires the smallest driving forces in solution A (Figure 8a), however, major slip increases periodically occurs due to poor track-terrain contact ensured by rigid construction. There is also relatively large UGV pitching. In this case, solution C is the most advantageous.

While climbing a log, the contact between the track and the obstacle was characterized by the dynamic and static friction coefficient respectively: 0.7 and 0.8. The driving force time series (Fig. 9a) indicate that the maximum load occurs during attacking the obstacle. There are large slips, reaching 90% (Fig. 9b) and major pitch angle, reaching 28^0 (Fig. 9c). The most advantageous solution in terms of slip and pitching is solution C.

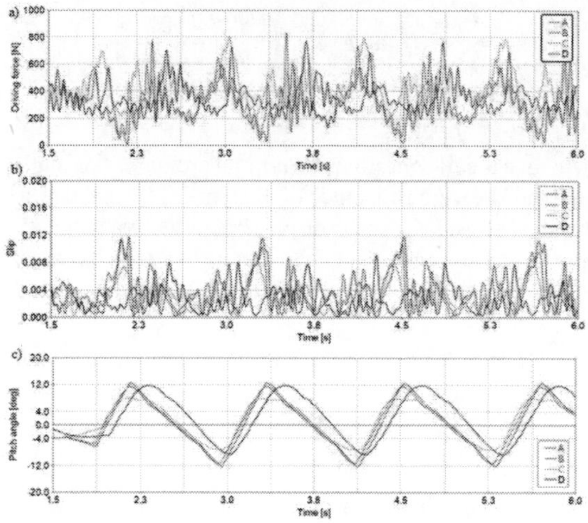

Figure 8. Grass clamps overcoming characteristics: a) Driving force time series; b) Slip time series; c) Pitch angle time series

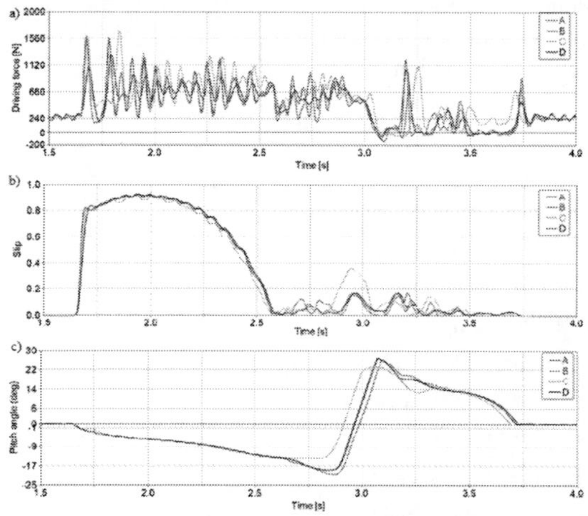

Figure 9. Log overcoming characteristics: a) Driving force time series; b) Slip time series; c) Pitch angle time series

Negotiating the curb, similar to climbing a log, the greatest required gross traction and slips occurred while attacking an obstacle (Fig.10a and Fig.10b). In this phase, no significant differences were observed between the analyzed solutions with only exception greater, prolonged slip in solution C. In the next phase, the smallest load variation was caused by solutions C and D. The advantage of using solution C is, however, 20% lower longitudinal inclinations. While overcoming this obstacle, solutions B and D have almost identical courses due to the small track clearance and the characteristic front bogie pivot location. This made it impossible to change the shape of the front part of the running gear.

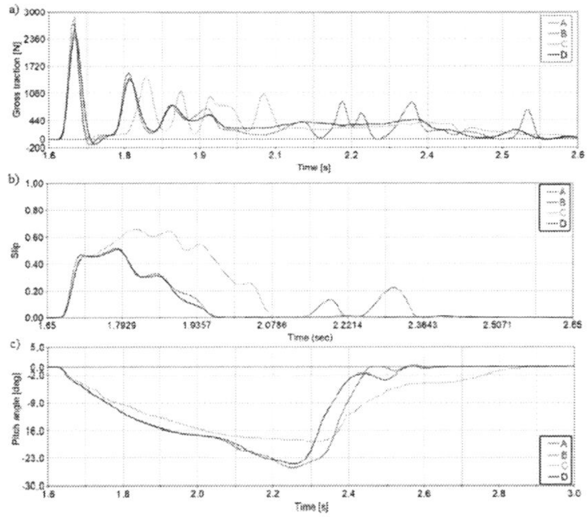

Figure 10. Curb overcoming characteristics:
a) Gross traction time series; b) Slip time series;
c) Pitch angle time series

4. Discussion

For the analyzed solutions, the easiest obstacle to overcome was the clumps of vegetation profile. It required the smallest gross traction and driving force (Figs. 11a-d). At that time, the smallest slip also occurred (Fig. 11e-f). Despite this, peak-to-peak value of pitch angle (Fig. 11g) achieved similar values as for crop rows. In the pitch angle criterion, in most cases, C-type running gear was the best solution (except for crop rows). Reduced pitching is an essential ability for remote controlled UGVs operating in teleoperation mode. Situational awareness of operator and steering efficiency can be truly limited by UGV pitching.

The maximum dynamic loads occurred when crossing the curb, but the highest average gross traction was required to ditch negotiation—it was a relatively long section of high slope. Due to the gentle profile changes, the lowest RMS value of acceleration occurred (Fig. 11h).

Analyzing slip time series, the hardest task was to climb a log, where the value of slip was approaching 1. There were also high of pitch angle, which may be a problem for obtaining adequate stability for teleoperation cameras work.

While crop rows overcoming, solutions containing flexible elements needed to develop higher gross traction and driving force. Driving through the ridges possibility indicated the advantage of the rigid constructions from solutions A and B. They required on 30% less mean driving force compared to the solution C (Fig. 11d) and their average slip was up to five times lower (Fig.11 f). In addition, solutions A and D reduced longitudinal inclination by approx. 35% compared to solution C. Solution A was also characterized by the lowest RMS acceleration value.

On the profile of the clumps of vegetation the most advantageous was solution C. It required 30% less gross traction and 25% lower slip, causing the smallest longitudinal inclination and the lowest vertical acceleration. However, the mean driving force was still 50% higher than in solution A.

Negotiating single obstacles, such as a log or curb it was shown that the type of running gear has low impact on required forces and slip of the track. Rigid solutions allowed overcoming obstacles with lower driving forces, but flexible solutions allow up to 25% less UGV pitching while obstacles negotiation.

There were no major differences between the solutions in negotiating time—each solution was driven at the same speed regardless of how much torque it required. Some solutions moved to the initial horizontal position faster, e.g., solution A climbed a curb—it led to loss of contact of a part of the track from the ground.

5. Conclusions

For the analyzed solutions it was indicated that the running gear of lightweight UGV may affect the ability to overcome obstacles. The simulation results shows that all-rigid solutions may require even 35% lower driving forces to overcome obstacles and can provide to negotiate wider ground recesses and furrows. In the case of limiting engine capabilities and the energy reserve, they may be an advantageous solution. However, estimating their energy consumption requires more detailed simulations. e.g., more specific sinkage evaluation, soil excavation and bulldozing effect consideration. In the future, it is planned to develop the simulation model on a more accurate depiction of the action at the extremities of the crop rows profile. The soil there may have less strength and lead to vehicle landslides and uncontrolled sinking into furrows.

On the other hand, the solution containing flexible elements are characterized by increase in the possibility of copping the terrain shape and developing the gross traction in terrain with slighter roughness. This means that this type of running gear will enable it to overcome obstacles with even 34% lower traction force needed, and also with 30% lower slip than all-rigid solution. Although this will be achieved at the expense of the higher required driving force. The use of flexible elements also allows to significantly reduce UGV pitching (up to 40%) and vibrations caused by rough terrain

For a more accurate assessment of overcoming objects such as a curb or a log, it should be tested the greater sag track, so the track tension cannot limit the kinematics of the flexible, moving running gear elements. The next step will also be testing dedicated to the obstacles undercarriage systems with a front axle raised, with smaller diameter wheels and in more parallelogram shape.

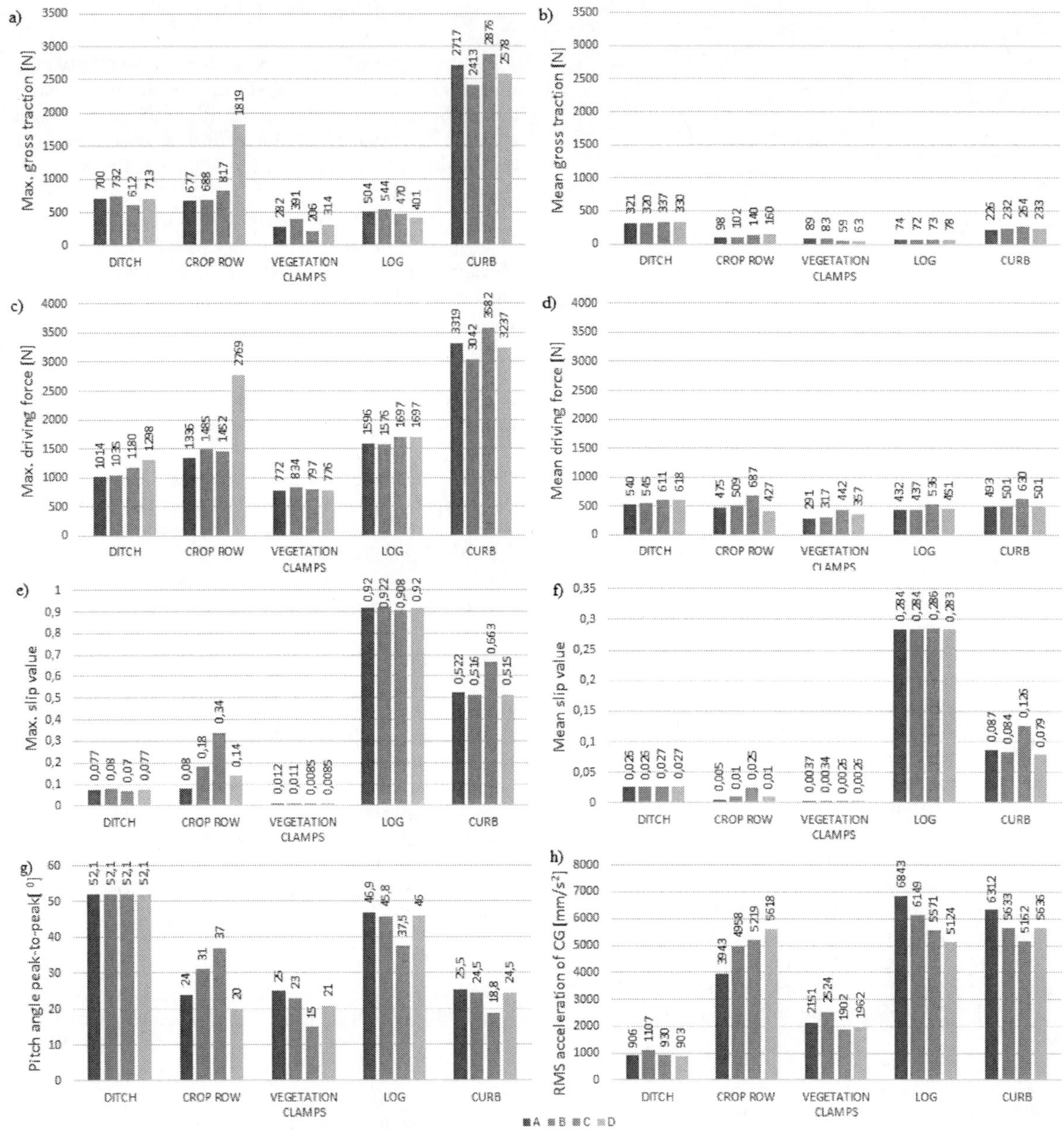

Figure 11. Results graph of the measured simulations parameters:
a) max gross traction
b) mean gross traction
c) max. driving force
d) mean driving force
e) max slip value
f) mean slip value
g) pitch angle peak-to peak value
h) RMS acceleration of CG

6. References

Bartnicki, A., Łopatka, J.M., Muszyński, T., Wrona, J., 2015. Concept of IED/EOD Operations (CONOPs) for engineer mission support robot team. Journal of KONES 22, 269–273.

Bekker, M.G., 1969. Introduction to terrain-vehicle systems. University of Michigan Press, Ann Arbor.

Bruzzone, L., Quaglia, G., 2012. Review article: locomotion systems for ground mobile robots in unstructured environments. Mechanical Sciences 3, 49 - 62.

Choi, Y., Jeong, K., Seo, Y., Lee, S., Cho, J., Jung, S., Kim, S., 2008. QuadTrack-II: A Remotely Operated Mobile Robot with Four Articulated Tracks. IFAC Proceedings 41, 3017–3020.

Davids, A., 2002. Urban Search and Rescue Robots: From Tragedy to Technology. IEEE Intelligent Systems 17, 81–83.

Dudziński, P., Kosiara, A., Konieczny, A., 2012. Virtual prototyping of a new generation undercarriage with rubber tracks for arctic applications (in Polish). Postępy Nauki i Techniki 64–74.

Edlund, J., Keramati, E., Servin, M., 2013. A long-tracked bogie design for forestry machines on soft and rough terrain. Journal of Terramechanics 50, 73–83.

Grigore, L., Ionica, H., Priescu, I., Joiţa, D., 2021. Development and Evaluation of the Traction Characteristics of a Crawler EOD Robot. Applied Sciences 11, 3757.

Guo, T., 2016. Power Consumption Models for Tracked and Wheeled Small Unmanned Ground Vehicles on Deformable Terrains.

Haueisen, B., 2003. Mobility analysis of small, lightweight robotic vehicles.

Haueisen, B., Hudas, G., Gorsich, D., Ahlvin, R., Jones, R., Priddy, J., Mason, G., Hulbert, G., n.d. Case Study of the Evaluation and Verification of a PackBot Model in NRMM.

Khurshid, J., Bing-rong, H., 2004. Military robots—a glimpse from today and tomorrow, in: ICARCV 2004 8th Control, Automation, Robotics and Vision Conference, 1, 771-777

Kim, Jihoon, Kim, Jongwon, Lee, D., 2018. Mobile robot with passively articulated driving tracks for high terrainability and maneuverability on unstructured rough terrain: Design, analysis, and performance evaluation. Journal of Mechanical Science and Technology 32, 5389–5400.

Krotkov, E., Fish, S., Jackel, L., McBride, B., Perschbacher, M., Pippine, J., 2007. The DARPA PerceptOR evaluation experiments. Autonomous Robots 22, 19–35.

Li, Y., Zang, L., Shi, T., Lv, T., Lin, F., 2022. Design and Dynamic Simulation Analysis of a Wheel–Track Composite Chassis Based on RecurDyn. World Electric Vehicle Journal 13.

Longoria, R.G., Krotkov, E.P., Mcbride, B., Longoria, R., Krotkov, E., 2003. Measurement and prediction of the off-road mobility of small, robotic ground vehicles.

Lopatka, M., 2020. UGV for Close Support Dismounted Operations—Current Possibility to Fulfil Military Demand. Challenges to national defense in contemporary geopolitical situation 2020, 16–23.

Lopatka, M. et al. 2014. State of the art of working capabilities and mobility of robots—"Medium platform (weight 800 kg)" project report (in Polish). Warszawa.

Massey, K., 2016. Squad Mission Equipment Transport (SMET)—Lessons Learned for Industry, in: NDIA Ground Robotics Capability Conference.

McBride, B., Longoria, R., Krotkov, E., 2003. Measurement and prediction of the off-road mobility of small, robotic ground vehicles.

Micire, M.J., 2008. Evolution and field performance of a rescue robot. Journal of Field Robotics 25, 17–30.

Morales, J., Martínez, J.L., Mandow, A., García-Cerezo, A.J., Pedraza, S., 2009. Power consumption modeling of skid-steer tracked mobile robots on rigid terrain. IEEE Transactions on Robotics 25, 1098–1108.

Nuttall, C.J., 1979. An Assessment of the Inland Water Crossing Performance of Selected Tactical Support Vehicles (WACROSS), Engineer Research and Development Center (U.S.).

Okello, J., Watany, M., Crolla, D., 1998. A Theoretical and Experimental Investigation of Rubber Track Performance Models. Journal of Agricultural Engineering Research 69, 15–24.

Park, W., Chang, Y., Lee, S., Hong, J., Park, J., Lee, K., 2008. Prediction of the tractive performance of a flexible tracked vehicle. Journal of Terramechanics 45, 13–23.

Rasool, S., Raheman, H., 2018. Improving the tractive performance of walking tractors using rubber tracks. Biosystems Engineering 167, 51–62.

Richmond, P.W., Mason, G.L., Coutermarsh, B.A., Pusey, J., Moore, V.D., 2009. ERDC TR-09-6, Mobility Performance Algorithms for Small Unmanned Ground Vehicles.

Šulka, P., Sapietová, A., Dekýš, V., Sapieta, M., 2018. Analysis and synthesis parameters influencing to the effects of impact. MATEC Web of Conferences 157,

Wang, W., Du, Z., Sun, L., 2007. Dynamic load effect on tracked robot obstacle performance. Proceedings of the 2007 4th IEEE International Conference on Mechatronics, 230, 12–17.

Wang, W., Zhiyuan, Y., Zhijiang, D., 2018. Experimental study of a tracked mobile robot's mobility performance. Journal of Terramechanics 77, 75–84.

Weidong, W., Lei, Z., Zhijiang, D., Lining, S., 2008. Track-terrain interaction analysis for tracked mobile robot. IEEE/ASME International Conference on Advanced Intelligent Mechatronics, 370, 126–131.

Wieder, W.L., Shoop, S.A., 2018. State of the knowledge of vegetation impact on soil strength and trafficability. Journal of Terramechanics 78, 1–14.

Wong, J., 2008. Theory of Ground Vehicles.

Wong, J.Y., Huang, W., 2006. "Wheels vs. tracks"—A fundamental evaluation from the traction perspective. Journal of Terramechanics 43, 27–42.

Wong, J.Y., Senatore, C., Jayakumar, P., Iagnemma, K., 2015. Predicting mobility performance of a small, lightweight track system using the computer-aided method NTVPM. Journal of Terramechanics 61, 23–32.

Yajima, R., Nagatani, K., 2017. Investigation of tip-over condition for tracked vehicles climbing over an obstacle on a slope.

Yamauchi, B.M., 2004. PackBot: a versatile platform for military robotics, in: Unmanned Ground Vehicle Technology VI. SPIE, 228–237.

4409 / Design of Self-Driving Bulldozer System

***Junhua Yang*[a], *Biao Zhang*[a], *Haokai Tang*[a], *Binghua Shen*[b], *Linlin Ou*[a], *Xinyi Yu*[a]*, *Yuanjing Feng*[a], *Yu Feng*[a], *and Libo Zhou*[a]**

[a]*Zhejiang University of Technology, Hangzhou, Zhejiang, China*
[b]*Hangzhou Dengyuan Technology, Hangzhou, Zhejiang, China*
**Corresponding author: yuxy@zjut.edu.cn*

https://doi.org/10.56884/CRUY6863

Abstract

Abstract: To improve the efficiency of manual operations in the coal bunker, a completely automatic system based on multi-sensor perceptual positioning and map-based clearance planning is proposed. Firstly, the perceptual positi*oning link combines semantic segmentation to improve the traditional feature extraction and restricts the loop keyframe by using brute force matching, which improves the localization ability of simultaneous localization and mapping (SLAM) in the coal bunker. Compared with the incremental map maintenance strategy, a neighboring strategy is proposed to update the map in the dynamic environment of the coal bunker. Secondly, in response to the requirements of the coal bunker cleaning operation, based on the SLAM positioning results, we formulate behavior planning combined with a hierarchical finite state machine (FSM) and a grid map-based motion planning. The proposed clearing plan combines the geometric characteristics of crawler rakes, reduces the error of trajectory tracking, and improves the smoothness of trajectory curvature. The system can realize accurate real-time positioning in the bunker environment, and successfully complete the requirements of cleaning. In addition, a high simulation environment based on the unreal engine is built to verify the effectiveness and robustness of the system.

Keywords: SLAM, behavior planning, motion planning

1. Introduction

Bulldozer is the key clear-cabin equipment for wharf operations. The traditional manual-driven bulldozer has some problems in its work environment, such as low work efficiency, unfavorable physical-mental health of the operators. To solve these problems, the research and development of a self-driving bulldozer are essential.

The bulldozer system involves two aspects: perceptual positioning and clearance planning. The simultaneous localization and mapping (SLAM) algorithm is widely used in unmanned vehicles. The visual SLAM algorithm has problems with positioning drift in the coal bunker due to degradation of image features and susceptibility to external interference. Zhang and Singh (2014) proposed a real-time LiDAR ranging and mapping (LOAM) method, which enables autonomous localization in unknown environments of vehicles. In common with LIO-SAM, most general lidar-based SLAM positioning algorithms assume static or locally dynamic structured environments. However, the coal bunker environment is unstructured and constantly changing, which impacts the effectiveness of traditional SLAM positioning.

To solve the positioning problems, an improved positioning algorithm is proposed to enhance the positioning ability of the bulldozer in the coal bunker. Unlike the traditional curvature-based feature extraction strategy, a feature extraction strategy combined with semantic segmentation based on linear fitting is adopted to enhance the adaptability of the positioning in the highly dynamic environment. As for the selection of loopback keyframes, an adaptive matching-based verification algorithm is proposed to ensure the accuracy of keyframes. In contrast to traditional incremental historical map methods, a nearest neighbor update method is proposed to maintain the map and provide a reliable basis for pose estimation.

Clearance planning, which includes behavior planning and motion planning, is an essential part of the design of the self-driving bulldozer system.

By fusing multi-sensor information, behavior planning makes decisions based on real-time mission requirements and sends them to the bulldozer for execution. As demonstrated (Mordechai and Francesco, 2018), the finite state machine (FSM) consists of states and transitions between pairs of states, where a state can be transitioned to another state by a corresponding transition. Based on the FSM and actual navigation process, Xiaoping Bai et al. (2020) establish relationships between the navigation events and navigation tasks with different priorities. Moreover, they realize the fast-grading response to the navigation events in a bus-based navigation control system. The structure of the decision system based on the FSM model is relatively simple and reliable when the environment is simple. However, it is not conducive to maintaining the whole model when too many states exist in the FSM.

Motion planning is responsible for specifying the sent commands into the movement of the vehicle. The point-to-point algorithm is commonly used for the motion planning part. The breadth first search (BFS) algorithm (Lee, 1961) is the basic algorithm for graph search and is particularly effective for solving the shortest or smallest problems with a small search depth. However, its aimless path search leads to inefficiency. Dijkstra (1951) proposed an algorithm to plan optimal solution paths using the cost function as a judgment criterion for prioritization. But the algorithm is still essentially a BFS algorithm with high space complexity and time complexity. Considering the heuristic cost function, the A^* algorithm (Hart, Nilsson, Raphael, 1968) calculates the generated value of expanding nodes to the target. When the heuristic distance is smaller than the true distance, the A^* algorithm can guarantee that the obtained path is the optimal solution. However, the A^* algorithm planning direction is limited and the planned paths do not conform to the vehicle kinematics. Saeid Sedighi et al. (2019) fused a Hybrid A^* search engine with visibility graph planning to find the shortest possible non-holonomic path in a hybrid environment. By maintaining a modified state update rule, The Hybrid A^* algorithm can capture the continuous state of the vehicle and is more consistent with vehicle kinematics than the A^* algorithm path.

The coal bunker is a special environment, and the uncertainties introduced by the environment are determinable. Thus, the FSM model is used in the behavior planning module. Considering the complex model brought by too many task states, we design a hierarchical FSM into the clearing system, consisting of three FSMs, namely system monitoring state machine, automatic stacking state machine, and task switching state machine. Through a hierarchical finite state mechanism, the complexity of the model and the coupling between task logics brought by a single FSM are reduced. In motion planning, a Hybrid A^* algorithm combined with bulldozer kinematic constraint is applied as a path search method, and the curvature of the search path is optimized using a cubic B-spline. Due to the structure of the modified bulldozer, we use two tracks to track the path separately to reduce the error instead of tracking the path with only a single vehicle center positioning information.

The remainder of this paper is structured as follows: The overall framework of the system is described in Section 2. We detail the realizations of the improved perceptual positioning in Section 3. Section 4 presents specific clearance strategies for bulldozers. In Section 5, we build the cabin and bulldozer models using the Unreal Engine to recreate the real environment. In Section 6, The effectiveness and robustness of the algorithm is verified using data from real environments and simulations.

2. Frame and Process of the System

The system framework of this article is shown in Fig. 1. The system operation process can be divided into perceptual positioning and clearance planning.

As shown in Fig. 1, the bulldozer and the sensors form the hardware structure of the system. In perceptual positioning, an improved SLAM positioning algorithm is proposed to calculate the pose estimation of the bulldozer in real-time and maintain a global map by acquiring sensor information. After acquiring the pose and map, the clearance planning continuously determines the current behavior pattern and plans the corresponding path, which is sent down to the bulldozer for execution. The sensory information is then continuously updated via sensors.

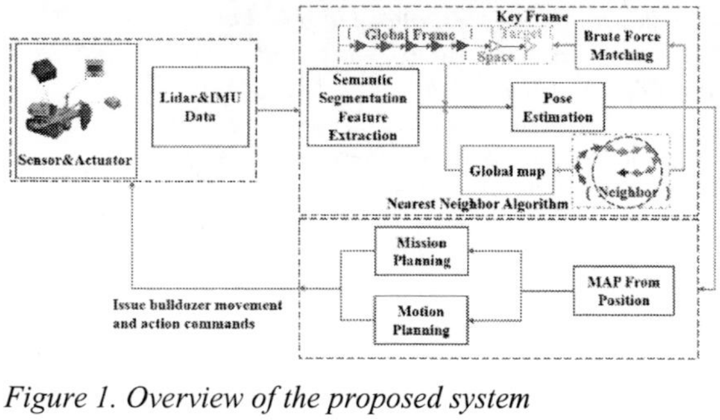

Figure 1. Overview of the proposed system

3. Improved Perceptual Positioning

3.1 Semantic Segmentation

After obtaining perceptual information about the sensor environment, the point cloud and imu information are matched with the calibration matrix and timestamps. Shan and Englot (2018) proposed a lightweight and ground-optimized lidar odometry and mapping method, LeGO-LOAM, for real-time 6DOF pose estimation with ground vehicles. In one of the preprocessing sections, they applied a spatial sampling approach to divide the point cloud in a reasonable way. We propose a semantic segmentation operation based on it. Due to this near-dense and far-sparse feature of the lidar information, uniform down sampling is not well adapted to the lidar scanning method.

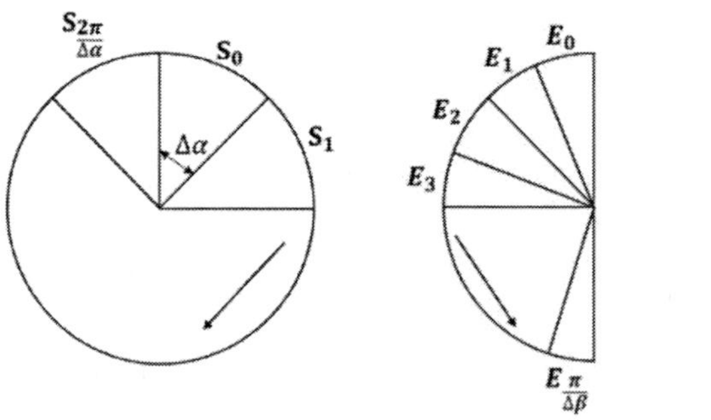

a) Horizontal sector division b) Vertical sector division
Figure 2. Basis of space division

A spatial coordinate division method is applied to down sample the point cloud, as in Fig. 2. Firstly, the horizontal coordinates of the points are divided in a clockwise direction with the angular resolution, as shown in Fig. 2a, where $\{S_0, S_1, \ldots, S_{2\pi/\Delta\alpha}\}$ denotes the divided set which can be divided into a total of K sectors ($K = 2\pi/\Delta\alpha$). The vertical section is next divided in an anticlockwise region, as shown in Fig. 2b, where $\{E_0, E_1, E_2, E_3 \ldots, E_{\pi/\Delta\beta}\}$ denotes the divided set which can be divided into a total of M sectors ($M = \pi/\Delta\beta$). Based on this spatial division method, a spatial point $\{x_i, y_i, z_i\}$ can be transformed into 2D coordinates $\{segment_i, bins_i\}$.

$$segment_i = \frac{\tan^{-1}\frac{y_i}{x_i}}{\Delta\alpha\pi} \quad (1)$$

$$bins_i = \frac{\tan^{-1} z_i / \sqrt{x^2 + y^2}}{\Delta\beta\pi} \quad (2)$$

The point cloud is divided into coal pile point cloud, wall point cloud, and ceiling point cloud by calculating the angle between adjacent point cloud in the same sector, the horizontal plane, and the distance in the XY plane.

The point cloud after initial processing is clustered by density-based spatial clustering of applications with noise (DBSCAN). After the DBSCAN cluster filtering, the point cloud is fed to the odometry module as an input of feature points.

The semantic segmentation results are shown in Fig. 3. Fig. 3a is the original point cloud, and Fig. 3b is the semantic segmentation result.

It can be concluded from Fig. 3a and Fig. 3b that the fusion of multiple sensors is adapted to the perception of the complex environment in the coal bunker, providing an accurate feature point cloud for the improved mapping algorithm.

Because of the dynamic feature of the coal pile, lidar odometry and mapping (LOAM) will reduce the accuracy of the positioning algorithm by using the points of the coal pile as the basis for positioning. As shown in Fig. 3c, the feature corner points extracted by the traditional LOAM are concentrated on the coal pile, and feature plane points are concentrated on the walls. However, for the positioning algorithm, the extracted feature corner points are dynamic, which affects the matching effect. Therefore, we use the wall point cloud as input of the plane points and the ceiling point cloud as input of the corner points, which avoids the influence of dynamic objects on the matching. As in Fig. 3d, the improved plane feature points are concentrated on the walls, while the corner points are concentrated on the ceiling, which can well constrain the 6DOF of the bulldozer.

3.2 Brute Force Matching

After inputting the improved feature points into the odometry module, the pose estimation of the bulldozer is obtained. The LOAM algorithm generally extracts keyframes at a fixed frequency or adds a threshold of rotation and translation to limit the keyframes. However, these verification algorithms do not adaptively match the different situations. Hess et al. (2016) proposed a new loopback detection method based on laser data, called Cartographer. However, the detection method in algorithms regarding keyframes is not adapted to 3D laser data and is restricted to local scan map, so an adaptive loopback keyframe verification algorithm is proposed to solve the keyframe verification problem for long-term localization in complex environments.

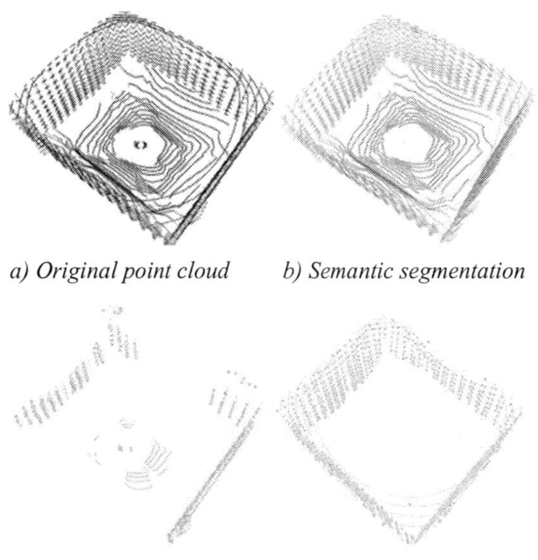

a) Original point cloud *b) Semantic segmentation*

c) Traditional feature *d) Improved feature*

Figure 3. Feature extraction results

As in the keyframe session in Fig. 1, the algorithm maintains a global grid map of appropriate resolution and size. Keyframes before the current frame are then projected in a 2D grid coordinate system to update the global grid map. As shown in Fig. 4, Fig. 4a is the global wall point cloud projected onto a 2D grid. Fig. 4b is the ideal current wall point cloud projected onto a 2D grid. Fig. 4c is the result of a grid with some positioning bias in a real environment. A cubic variable $\{T_x, T_y, T_{yaw}\}$ is used for brute force matching to calculate the score, which results in an optimal score with the map.

$$\begin{cases} F(T) = argmin\sum \| 1 - M(S_i(T)) \|^2 \\ T = (T_x, T_y, T_{yaw}) \end{cases} \quad (3)$$

Firstly, the rotation and translation of odometry estimation are extracted based on the grid coordinate system. Secondly, after dimensionality reduction, the wall point cloud is rotated and translated according to a specific resolution, and the subscript is set according to the amount of rotation and translation. Then, based on the interpolation principle, the matching score of each point cloud under the action of T is calculated. Finally, the optimal matching T and the minimum error result are calculated by constructing a least square problem such as Eq. 3, where $S_i(T)$ denotes the wall point cloud with (T_x, T_y, T_{yaw}) superimposed on it, and M represents the interpolation calculation score.

If the optimized translation and rotation are within the allowable range, then the currently optimized pose is accurate, and the current frame is set as a loopback keyframe.

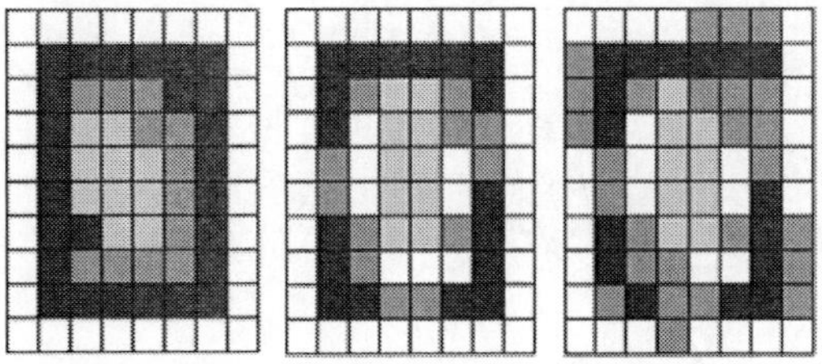

(a) Global grid b) Current grid c) Errors grid
Figure 4. Brute force matching map

3.3 Nearest Neighbor Strategy

Shan et al. (2020) proposed a tightly coupled lidar-inertial odometry method, LIO-SAM, which optimized multiple sensor factors through a factor graph to obtain the globally consistent pose of the robot. In the SLAM process, many keyframes are generated and multiple keyframes may be generated at the same location. The local scale of scan matching is used to replace the global scale which can effectively improve the real-time performance of LIO-SAM.

However, as the bulldozer works and the surroundings dynamically change, duplicate keyframes take up much memory, and part of the local map becomes out of date at some point in time and requires updating using new point cloud frames. Therefore, we propose a dynamic filtering-based map maintenance strategy.

As in the Nearest Neighbor session in *Fig. 1*, the nearest neighboring keyframes that can be searched around the current frame are $\{x_1, x_2, x_3, \dots, x_{pre}, x_{curr}\}$. Based on the timestamp comparison, where $\{x_1, x_2, x_3\}$ is the obsolete frame. However, it is unnecessary to thoroughly remove obsolete frames since some points remain static, such as walls, ceilings, and untransformed coal piles. Therefore, the non-obsolete frames are formed into a local map, and $\{x_1, x_2, x_3\}$ are iterated with the local map based on the point-to-line iterative closest point (PLICP) method to remove the point cloud with significant matching errors. As a result, the remaining plausible points and local map are reconstructed into a global map, which is used to match the current frame.

4. Clearing Planning

4.1 Behavior Planning for Bulldozer

The design requirement of the clearance system is to determine the different modes of the bulldozer in various situations. The requirements of the clearance system include the implementation of autonomous clearance tasks in automatic mode and the execution of the given path commands in navigation mode. Therefore, switching the task modes is crucial in the task decision. As a result of the complexity of the logic under the overall clearing task, a state machine is used.

Figure. 5a is a schematic diagram of the path of the bulldozer when it works automatically, the overall path when it works is similar to a fan, and we use the data in the coal bunker to obtain a sequence of target points to drive the bulldozer to finish traversing the fan-shaped area.

We design a task-switching state machine, a system-monitoring state machine, and an automatic-stacking state machine to enable the bulldozer to switch modes. Fig. 5b shows a task-switching state machine with an automatic state, a navigation state, and a waiting state. The automatic state indicates that the bulldozer performs a specific job task, the navigation state indicates that the bulldozer performs a specific navigation task, and the wait state indicates that the bulldozer waits for command input.

When the task-switching state machine is in the automatic state, the automatic state machine starts to work. As shown in Fig. 5c, the automatic stacking state machine consists of five states: generate-goal-queue state, forward-to-goal state, rake-material state, backward state, and target-area state. The generate-goal-queue state is used to generate goals for stacking through certain strategies according to the distribution of materials. The forward-to-goal state generates a path based on the path search algorithm and reaches the goal by path tracking. The rake-material state is used to control the bulldozer stack material. The backward state means that the bulldozer is backing up to deliver the material to the stacking area, and the target-area state means that the bulldozer is in the designated stacking area. There are corresponding transfer conditions between the states.

The system monitoring state machine is always in operation and responsible for receiving, processing, and feeding data to other modules. As shown in Fig. 5d, the system monitoring state machine comprises four states: normal state, risk state, adjust state and deal-failure state. The normal state refers to the system without special circumstances, the risk state refers to unexpected occurrences, the adjust state refers to the corresponding recovery measures when the unexpected situation can be handled, and the deal-failure state refers to the unexpected situation cannot be handled and feedback to other modules.

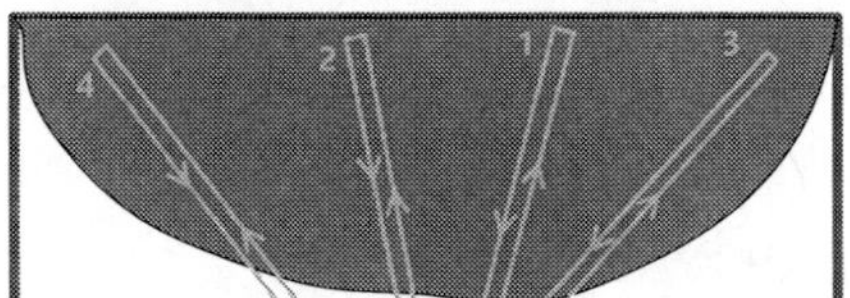

a) Automatic bulldozer working schematic

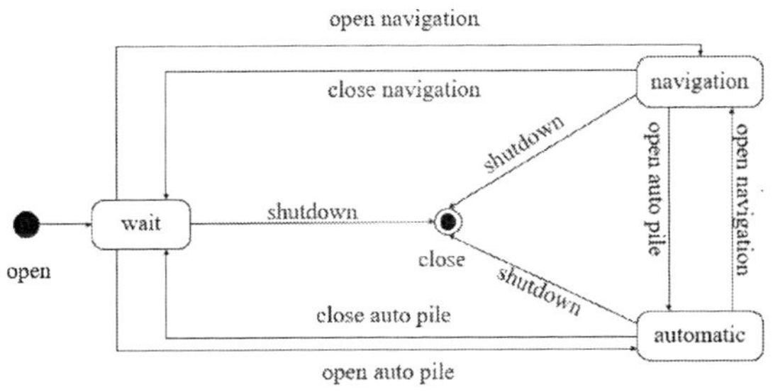

b) Task-switching state machine

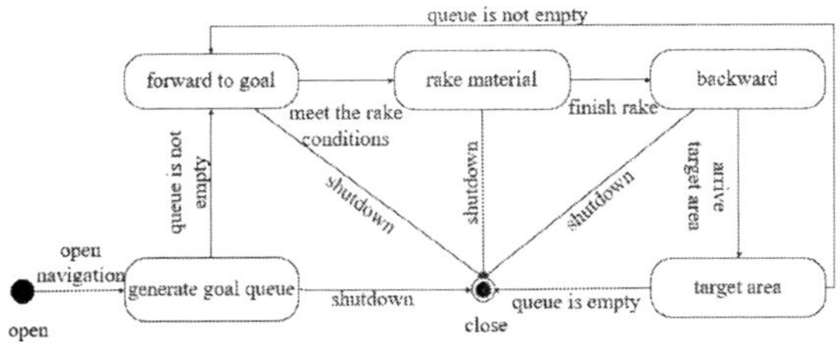

c) Automatic-stacking state machine

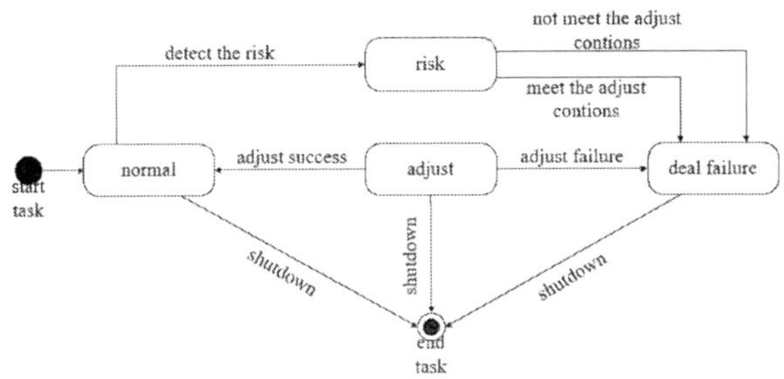

d) System-monitoring state machine
Figure 5. Clearing system

4.2 Motion Planning

We use a point-to-point motion planning method to execute the set of target points generated in behavior planning. The specific implementation of the system uses a Hybrid A^* algorithm with a cubic B-spline optimized curvature to calculate a path that matches the motion characteristics of the bulldozer with the cost function shown in Eq. 4.

$$f(x) = g(x) + h(x) \tag{4}$$

where the cost function $g(x)$ is an accumulation of expanded distances that match the movement profile of the bulldozer, the heuristic function $h(x)$ takes the weighted sum of the Euclidean distance and the target angle difference as the cost, and the sum of the costs of the two functions as the value of the total cost function f(x).

As shown in Fig. 6, in motion planning, considering that there are obstacles such as pillars or mechanical equipment in the coal bunker, the planned path should conform to the kinematics of the bulldozer as much as possible. Fig. 6a shows a 2D grid map of the coal bunker, with the x-axis of the coal bunker coordinate system in green and the y-axis in red.

Fig. 6b shows the effect of the grid path based on the Hybrid A^* algorithm in the coal bunker with the presence of obstacles, where the angle of the starting point is 145 degrees and the angle of the target point is 45 degrees, which shows that the path searched by Hybrid A^* is close to the vehicle kinematic characteristics. Fig. 6c shows the path after cubic B-spline smoothing, which makes the curvature of the planned path more continuous.

The steering method of the modified bulldozer is single-power flow steering type, which leads to a significant error in the vehicle center position in the real-world scenario. After sampling the forward target point, the bulldozer center position is not considered in the steering calculation at the specified distance after sampling the forward target point to reduce tracking errors, as shown in Fig. 7. Instead, the steering angle is calculated using the positions of the left and right wheels so that the problem of significant errors in reaching the target point is solved.

a) Coal bunker environment

b) Hybrid A^* result *c) B-spline result*
Figure 6. coal bunker environment and planning

5. Simulation Environment

Alexey Dosovitskiy et al. (2017) introduced CARLA, an open-source simulator that supported various specifications of environments for autonomous driving systems to control vehicles running tests in a virtual environment. Shah S et al. (2018) presented a new simulator, AirSim, to provide physically and visually realistic simulations for autonomous driving platforms.

However, to make the generated coal pile particles satisfy the physical properties, so we provide a quick solution based on the Unreal engine introduced by Sanders (2016), using AGX Dynamics for the multiphysics of bulldozer systems and small objects. As shown in Fig. 8, the controller and actuator are interconnected through the ROS-Integration plug to complete the construction of a high assurance simulation environment.

In this simulation environment, we use ray collision detection to write the lidar sensor plug -in, which can customize the level and vertical FOV, line number and point, point number, detection distance and other parameters, then lidar will get the data input SLAM module, the output of the six -freedom position then put into the path planning module, and get the corresponding control instruction finally.

The environment not only provides a variety of plug-ins but also solves the problem of complex, inefficient, and time-consuming functional testing in real-world environments, enabling simulation modeling with high fidelity, high interactivity, and scalability.

6. Result

To demonstrate the effectiveness of the algorithm, we test it by using data from highly simulated and real environments and compare it with the traditional LOAM algorithm. In the real environment, the radar model is Ouster M1, the imu model is Xesens Mti 1, and the IPC is NVIDIA Jetson AGX Xavier. The hardware and environmental configuration used in the simulation is as follows: the operating system is win10, the CPU is AMD Ryzen7 3700X, the graphics card is NVIDIA GTX1660s, and the memory is 16G. The imu and lidar are first calibrated, and then the sensor drive is activated, where the lidar data is published at 10 Hz and the inertial element data at 100 Hz.

For the semantic segmentation module in positioning, we validate our algorithm on data inside a real coal bunker environment. As shown in Fig. 9, Fig. 9a is the original point cloud, Fig. 9b is the point cloud segmented by LOAM, and Fig. 9c is the result of our semantic segmentation based on linear fitting, which ensures the reliability of feature point extraction for positioning.

Fig. 10 shows the different results of the traditional SLAM algorithm and the improved positioning algorithm in a simulated and real environment. The results of the ALOAM algorithm for positioning in the simulation environment are shown in Fig. 10a. The result of perceptual positioning combined with semantic segmentation is shown in Fig. 10b. ALOAM, with its concentration of feature points and unreliable feature points, is not accurate in terms of positioning. The improved algorithm can achieve accurate positioning in complex environments. And Fig. 10c shows that the unverified keyframes cause some drift in the long-term positioning. However, positioning with a combination of brute force matching can eliminate the matching error of keyframes and achieve an accurate match, as shown in Fig. 10d. As the work progresses, the global-based incremental map approach leads to a continuous decrease in map confidence, resulting in traditional SLAM positioning drift, as in Fig. 10e, and the program memory is constantly occupied, eventually leading to memory overflow problems. In contrast, using a nearest-neighbor-based update strategy ensures that the map is always guaranteed to be statically plausible, as in Fig. 10f, and the long-term positioning results remain accurate. Moreover, the update strategy ensures that computer memory never overflows.

After acquiring the exact pose and map, the clearing module decides on the behavior pattern based on the current state. It then uses the generated 2D raster map to plan the path information to complete the cleaning task.

To verify the effect of traversal planning in the cabin and the path planning algorithm, clearance planning experiments were carried out on the ship's cabin map to verify whether the task of traversing the area in a similar fan-shaped path could be accomplished using the state machine model and to verify the smoothness of the optimized path by comparing the path planned by Hybrid A^* with the path optimised based on the B-spline algorithm.

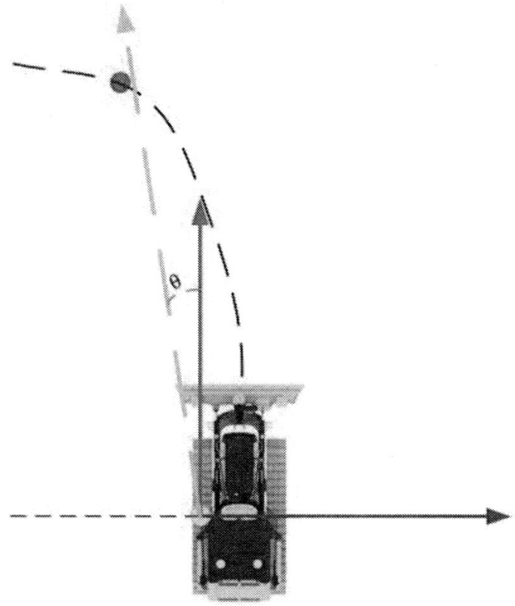

Figure 7. Path tracking diagram

Figure 8. Environment built on Unreal Engine

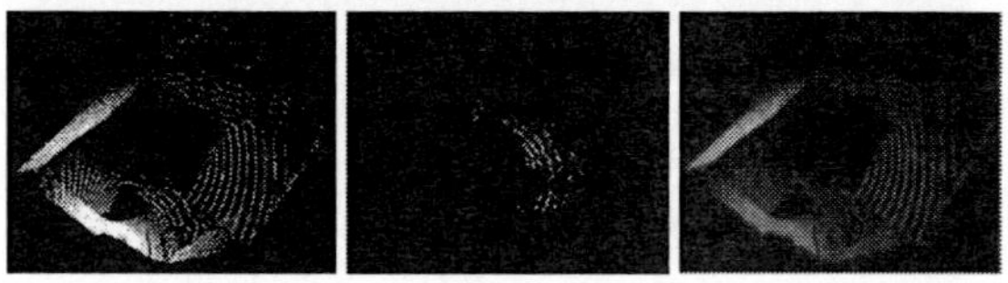

a) Original cloud b) Traditional res c) Improved res
Figure 9. Semantic segmentation results

In Fig. 11, the colored curves are the effect of traversing by fan-shaped paths, where the path planned by Hybrid A^* is shown by the green line, the result of optimization based on the B-spline algorithm is shown by the yellow line, and the black line indicates that the bulldozer rotates at an angle and backs up to a point within the stacking area to complete the raking task. From the results, it can be seen that the designed state machine can be used to traverse the region in a fan-like path, and the comparison of the paths of the non-overlapping yellow and green segments, especially the curvature of the green segment, shows that the optimization algorithm has smoothed out the Hybrid A^* to some extent.

7. Conclusions

In this paper, we design two critical modules of a self-driving bulldozer: positioning and planning. Under the experimental conditions reported in this paper, the main conclusions are as follows.

(1) In the positioning module, an improved multi-line SLAM algorithm is proposed for dynamic coal bunker positioning. The algorithm implements the filtering of dynamic coal bunker in the front end and the verification of brute force matching of keyframes in the back end. Finally, the nearest neighbor update principle is applied throughout the positioning phase to ensure the accuracy of the map. The algorithm satisfies the requirements of positioning in a complex environment for the long term in practical engineering.

(2) In the clearing planning, the behavior decision module can determine the current bulldozer behavior in different situations. The action planning plans the working path under specific behavior patterns and optimizes the forward path based on the bulldozer's kinematics, and finally sends it to the bulldozer for execution.

(3) Through the design of the above system, the new unmanned bulldozer achieves automatic clearance and accelerates the overall working speed in the actual dock work.

At the same time, there is space for further research in our work. The high inter-frame dependency in the localization module can bring the accuracy of positioning, but inevitably reduces the robustness of positioning. Future work can further reduce the coupling without sacrificing accuracy by using a high-precision map-based matching method. In future work, we will concretize the nodes in the planning module and further improve the overall system logic.

8. Acknowledgments

This research is supported by Zhejiang Provincial Natural Science Foundation of China under Grant No. LQ22F030021 and Zhejiang Provincial Natural Science Foundation of China under Grant No. GB21031030049.

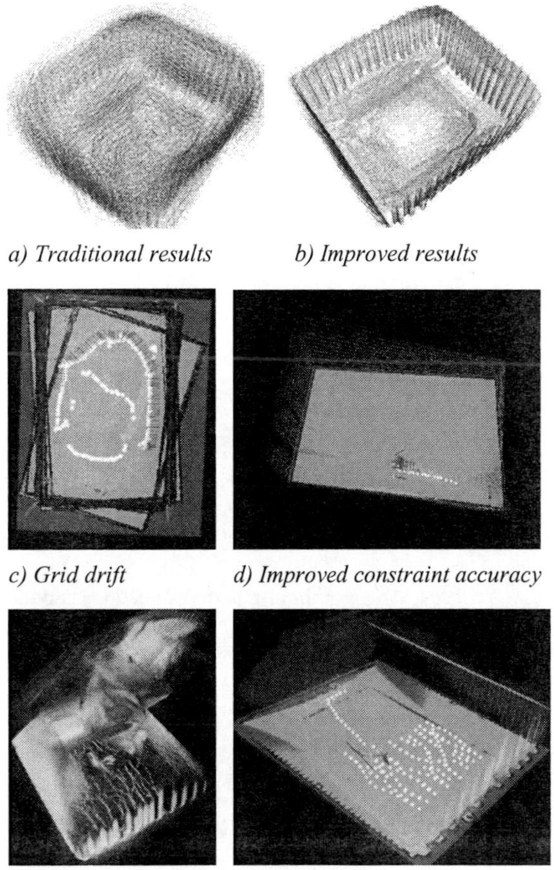

a) Traditional results *b) Improved results*

c) Grid drift *d) Improved constraint accuracy*

(e) Traditional drift *f) Improved accuracy*

Figure 10. Comparison of positioning effects

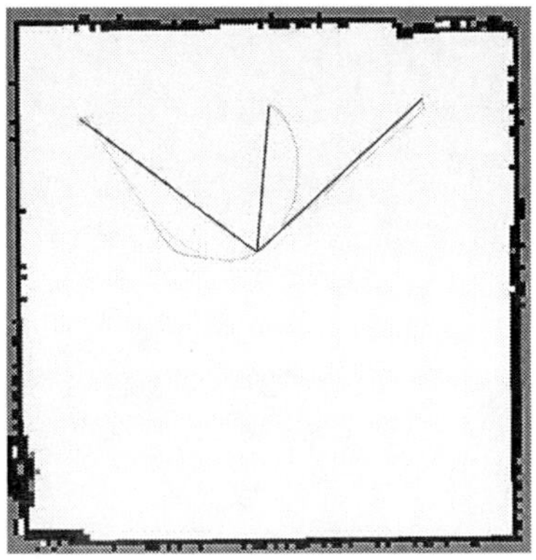

Figure 11. Fan-like path traversal map

9. References

Alexey, D., German, R., Felipe, C., et al., 2017. CARLA: An Open Urban Driving Simulator. *Proceedings of the 1st Annual Conference on Robot Learning*. 78, 1-16.

Bai, Xiaoping, Wang, Zhuo, et al., 2017. Multi-task Scheduling Research of Farm Machinery Navigation System Based on the FSM. *Journal of Agricultural Mechanization Research*. 39, 26-31.

Dijkstra, E.W., 1959. A note on two problems in connexion with graphs. *Numerische mathematic*. 1, 269-271.

Hart, P.E., Nilsson, N. J., Raphael, B., 1968. A formal basis for the heuristic determination of minimum cost paths. *IEEE transactions on Systems Science and Cybernetics*. 4, 100-107.

Hess, W., Kohler, D., Rapp, H., et al., 2016. Real-time loop closure in 2D LIDAR SLAM. *2016 IEEE International Conference on Robotics and Automation (ICRA)*. 1271-1278.

Lee, C.Y., 1961. An algorithm for path connections and its applications. *IRE transactions on electronic computers*. 3, 346-365.

Mordechai, Ben-Ari., Francesco, M., 2018. Elements of Robotics, *Springer*.

Shan, T, Englot, B., Meyers, D., et al., 2020. LIO-SAM: tightly-coupled Lidar Inertial Odometry via Smoothing and Mapping. *2020 IEEE/RSJ International Conference on Intelligent Robots and Systems*. 5135-5142.

Sedighi, S., Nguyen, D., Kuhnert, K., 2019. Guided Hybrid A-star Path Planning Algorithm for Valet Parking Applications. *2019 5th International Conference on Control, Automation and Robotics (ICCAR)*. 570-575.

Shah, S., Dey, D., Lovett, C., et al., 2018. Airsim: High-fidelity visual and physical simulation for autonomous vehicles. *In Field and service robotics. Springer, Cham*. 621-635.

Sanders, A., 2016. An Introduction to Unreal Engine 4. *CRC Press*.

Shan, T., Englot, B., 2018. Lego-loam: Lightweight and ground-optimized lidar odometry and mapping on variable terrain. *In 2018 IEEE/RSJ International Conference on Intelligent Robots and Systems (IROS). IEEE*. 4758-4765

Zhang, J., Singh, S., 2014. LOAM: Lidar odometry and mapping in real-time. *In Robotics: Science and Systems*. 2, 1-9.

4744 / Terrain Classification Using Mars Raw Images Based on Deep Learning Algorithms with Application to Wheeled Planetary Rovers

Junlong Guo, Xingyang Zhang*, Yunpeng Dong, Zhao Xue, and Bo Huang

State Key Laboratory of Robotics and System, Harbin Institute of Technology, Harbin 150001, China

**Corresponding author: xingyang.zhang@outlook.com*

https://doi.org/10.56884/WSDO4112

Abstract

Scene information plays a crucial role in motion control, attitude perception, and path planning for wheeled planetary rovers (WPRs). Terrain recognition is the fundamental component of scene recognition. Due to the rich information, visual sensors are usually used in terrain classification. However, teleoperation delay prevents WPRs from using visual information efficiently. End-to-end learning method of deep learning (DL) that does not need complex image preprocessing was proposed to deal with this issue. This paper first built a terrain dataset (consists of loose sand, bedrock, small rock, large rock, and outcrop) using real Mars images to directly support You Only Look Once (YOLOv5) to test its performance on terrain classification. Because the capability of end-to-end training scheme is positively correlated with dataset, the performance of YOLOv5 can be significantly improved by exploiting orders of magnitude more data. The best combination of hyperparameters and models was achieved by slightly tuning YOLOv5, and data augmentation was also applied to optimize its accuracy. Furthermore, its performance was compared with two other end-to-end network architectures. Deep learning algorithms can be used in the future planetary exploration missions, such as WPRs autonomy improvement, traversability analysis, and avoiding getting trapped.

Keywords: terrain classification, deep convolutional neural network, Mars raw images, wheeled planetary rover

1. Introduction

Due to the complex unstructured surface, autonomous mobility of wheeled planetary rovers (WPRs) plays a crucial role in Mars exploration. Safely accessing rough and steep terrain is essential to expand the search space to increase scientific value. Most of the recent review work on WPRs focuses on trajectory planning, localization, and obstacle avoidance (except terrain recognition).

The Opportunity got stuck in a sand dune in Meridiani Planum (Cowen, 2015) [1], and the engineers took five weeks to resume the rover from this entrapment. The twin mars exploration rover, Spirit, got trapped in late 2009, and the mission was declared concluded on May 24, 2011. While Curiosity's wheels were injured due to sharp terrain features (Rankin et al., 2020) [2].

Terrain classification can be used to avoid these aforementioned hazardous scenarios. Terrain classification focuses on safe navigation, andnavigation and deals with the problem of identifying terrain being traversed (or to be traversed) from a list of candidate terrains (Ojeda et al., 2006) [3].

There are two main bodies of research for terrain classification. Because of the intuitive, rich shapes, colors and textures, the most popular body of research uses visual cameras to classify ground surface by extracting color, texture, or objects like rocks (Bellutta et al., 2000; Manduchi et al., 2005) [4, 5]. The second body of research uses both proprioceptive and exteroceptive signals, such as acoustic signals (Brooks and Iagnemma, 2005; Valada et al., 2018) [6, 7], motor current (Ojeda et al., 2006) [3], accelerations (Giguere and Dudek, 2009) [8], radar (Walas and Nowicki., 2014) [9], and vibration (Sadhukhan et al., 2004) [10].

Furthermore, with the development of artificial intelligence, deep learning (DL) technique was also applied to conduct terrain recognition because of its powerful learning capabilities and scalability. Deep learning was first used in R-CNN (an important DL method) to detect target, while DL can realize terrain recognition on real-time maps (Chavez-Garcia et al., 2017) [11]. Combination of DL and Bayesian classifier can be used to classify the upcoming terrains (Cheng et al., 2014) [12]. The performance of DL in terrain classification can be improved through data augmentation methods (Li et al., 2020) [13].

However, there are some challenges that hinder the application of DL models. The first challenge is the lack of high-quality datasets for model training. Developing effective strategies to improve existing DL models for terrain classification is the second challenge. The authors used high resolution pictures of Mars surface to build high-quality datasets for model training. Hyperparameters optimization and data augmentation were applied to improve existing DL models. Considering the training cost, three typical DL model architectures, YOLOv5, CenterNet, and Faster-RCNN were selected to detect terrains.

This paper is organized as follows: Section II introduces three typical DL model architectures. The collection, annotation, and enhancement of Martian terrain dataset are presented in Section III. Section IV presents a series of experiments to investigate factors influencing the optimal performance of different models. Finally, Section V summarizes this study.

2. Selected deep learning frameworks

Current mainstream target detection algorithms are developed based on anchor box, such as You-Only-Look-Once (YOLO; Redmon et al., 2016) [14], Single Shot MultiBox Detector (SSD; Liu et al., 2016) [15] and Region-CNN (R-CNN; Girshick et al., 2014) [16].

YOLO divides an image into $n \times n$ sub-grids, and then places a fixed-size anchor box at the center point of each grid, and finally uses a regression model to predict bounding boxes and classes. The single-stage detector represented by the YOLO series only uses a network to determine category and location, and thus it is very fast.

Faster-RCNN (Ren et al., 2015) [17] relies on the region proposal network, which can identify plenty of bounding boxes (Bboxes) to localize objects. Meanwhile, Faster-RCNN uses region of interest pooling to refine Bboxes. Figure 1 shows the schematic diagram of Faster-RCNN architecture.

Faster-RCNN and YOLO are representatives of one-stage and two-stage target detection networks, respectively. Both candidate frame and complex calculation are needed in these two networks. CenterNet that is an anchor-free and NMS-free model can be used to solve these aforementioned issues (Zhou et al., 2019) [18]. It estimates key points to find center point and regresses to object to simplify training process and inference process. CenterNet extracts image features through CNN operations, and uses three head layers to determine category number, center point positions, and bias terms. Its structure is shown in Fig. 2.

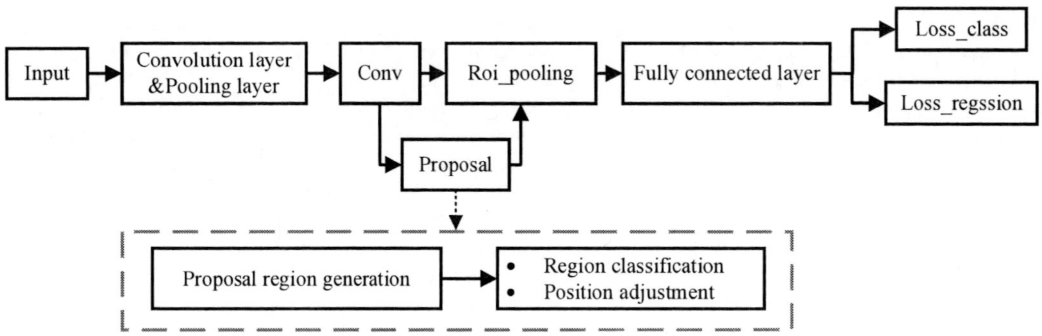

Figure 1. Schematic diagram of Faster-RCNN architecture

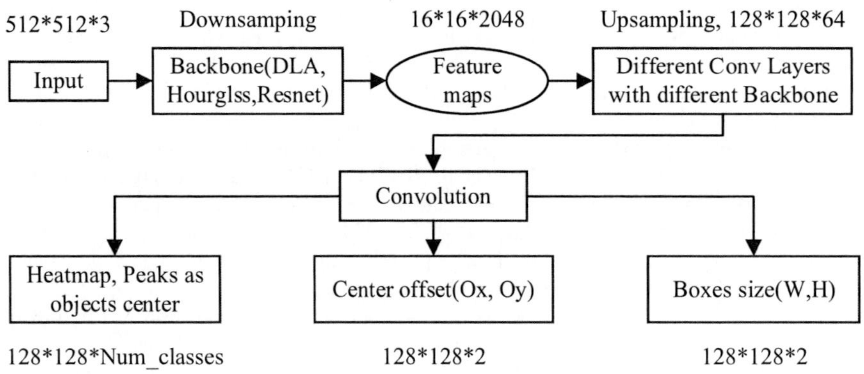

Figure 2. Schematic diagram of CenterNet architecture

3. Methodology

3.1 Training data collection

The performance of DL depends on high-quality image datasets. Although open access datasets, such as ImageNet and MS COCO, can be used to detect object, neither can be used to do terrain recognition. The training dataset in this work was built using NASA's Planetary Data System (PDS) Imaging Atlas. The Martian surface was divided into five categories (loose sand, large rock, small rock, outcrop, and bedrock) in this study, and Table 1 lists the definitions of five terrains.

One thousand two hundred and fifty images were selected from more than 30,000 images in the PDS, and each category contains 250 images. Target detection annotation tool was then used to annotate these images, as shown in Fig. 3. Finally, high-quality terrain dataset was obtained using data cleaning methods.

Table 1. Detailed definition of terrain classes

Class	Definition
Loose sand	Without any visible rocks, prone to high slip and wheel sinkage
Large rock	Large geometric obstacles
Outcrop	Hard and rough outcrop with loose rocks on top
Bedrock	Relatively flat exposed bedrock
Small rock	Do not interfere with wheeled rover and can easily be mistaken for large rock

a) Loose sand b) Bedrock c) Small rock

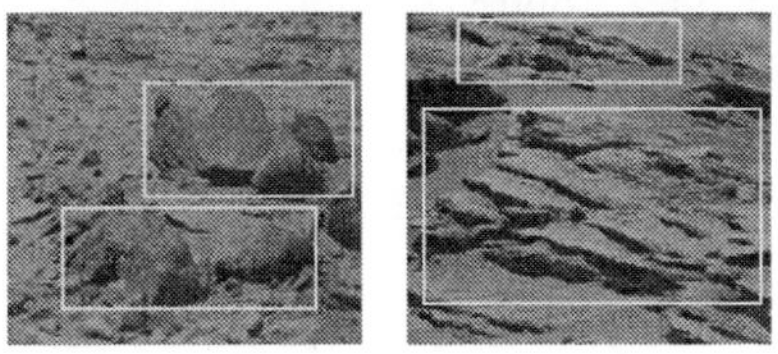

d) Large rock e) Outcrop
Figure 3. Examples of manually labeled terrain

3.2 Data augmentation

The size of the dataset will greatly influence the performance of the DL model. To further expand the size of the dataset, data enhancement techniques are performed on each image. These techniques include inversion, noise addition, rotation, grayscale conversion and brightness adjustment. The inversion technique uses 255 minus the corresponding RGB value to determine the RGB value of the new images.

Although the color of the target has changed, the color relationship between the object and the background still exists. The pictures obtained by the camera and the data processing process will be mixed with some noise. Adding salt and pepper noise can imitate the real data and increase the diversity of the data.

Rotation technology rotates images at various angles (90°, 180°, and 270°), and then the rotated images are resized and filled with black pixels. The rotated images need to be manually re-labeled. Grayscale conversion is the weighted summation of the RGB values. The phenomenon that camera is sensitive to brightness can be improved using brightness adjustment technology. Fig. 4 shows examples of data augmentation.

4. Experiment results

Two experiments were conducted to investigate the influence of various parameters and optimization methods on model performance. Fine-tuning and model hyperparameters (learning rate, number of iterations, mini-batch size, etc.) tuning through orthogonal method were carried out in experiment 1 to improve the performance of DL models. The performance of various DL models was then compared, and finally a relatively superior model was selected. Influence of various data augmentation techniques on the selected model performance was investigated in experiment 2.

The dataset was divided into training dataset, validation dataset, and testing dataset (70%, 15%, and 15%), while the augmented dataset was also divided into three parts (80%, 10%, and 10%). The selected three DL models were deployed on Linux Ubuntu 18.04 LTS using Pytorch framework, and the experiments were carried out on NVIDIA Tesla P40 GPUs. The iteration number of each experiment was selected to ensure DL models' performance, and each experiment was performed twice to ensure repeatability.

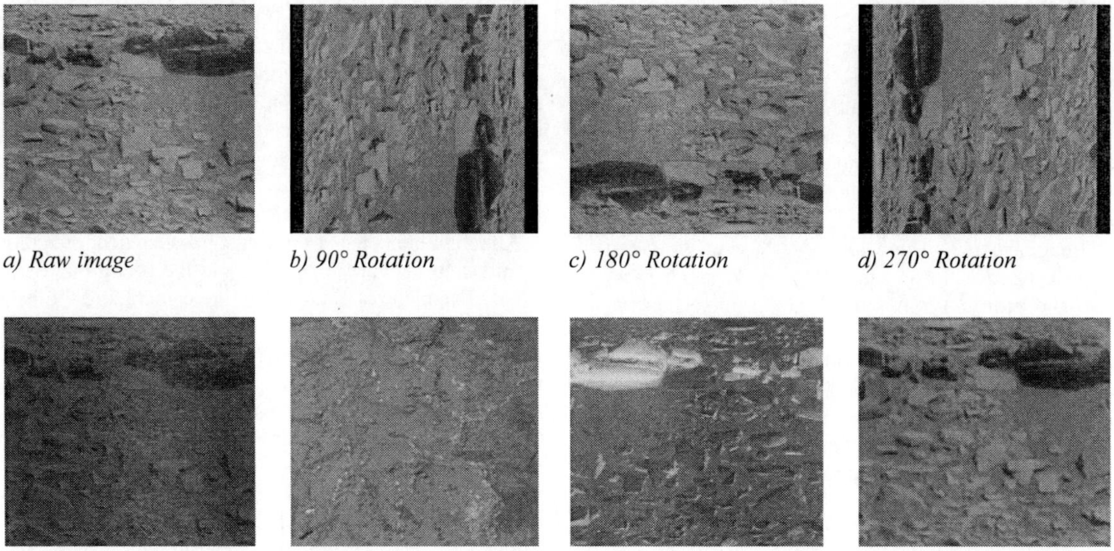

a) Raw image *b) 90° Rotation* *c) 180° Rotation* *d) 270° Rotation*

e) Brightness adjustment *f) Grayscale conversion* *g) Inversion* *h) Noise addition*

Figure 4. Examples of image augmentation

4.1 Comparison of model performance using fine-tuning

The fine-turned hyperparameters in this study contain learning rate, batch size, and image size. Metrics (mAP@.5 and mAP@[.5, .95]) and frames per second (FPS) were selected as performance indicators.

To compare their performance, each model was trained by running different iterations with a similar learning rate. The resolution of input images was about 600×600 pixels. During training process, the validation dataset was used to evaluate the intermediate model, and the model moved in a favorable direction.

Table 2 lists the performance of YOLOv5, CenterNet, and Faster-RCNN. In terms of averaged accuracy and FPS, YOLOv5 was selected to classify terrain. This is because YOLOv5 has a deeper structure and powerful design strategy. In addition, in terms of FPS, YOLOv5 and CenterNet perform much better than Faster-RCNN that is a two-stage model.

Table 2. Performance of YOLOv5, CenterNet, and Faster-RCNN

Metric / Model	mAP@.5	mAP@[.5, .95]	FPS
YOLOv5	0.776	0.485	76.9
CenterNet	0.771	0.479	45.3
Faster-RCNN	0.765	—	15.4

In terms of per unit training time, Faster-RCNN is about 3-times slower than CenterNet, and CenterNet is about 1.6-times slower than YOLOv5.

Figure 5 illustrates the curves of prediction accuracy versus epoch. The learning rate and resolution were adjusted through fine-tuning technology at the training stage. The initial value of learning rate plays a key role in the performance of DL model, while a comparatively high resolution may obtain a better result.

Figure 6 compares the labeled and predicted Bboxes using the optimal trained YOLOv5 model. Both terrain classes and Bboxes can be predicted correctly, and the confidence coefficient ranges from 0.5 to 0.9. The large rocks at far distance were confused with the small ones in near distance, and thus the Bbox of large rock in Fig. 6f) was predicted with a confidence of 0.5. As shown in Figs.s 6e) and 6f), both the large rock and the loose sand that was not labeled can also be predicted, which demonstrates that YOLOv5 has powerful generation capacity.

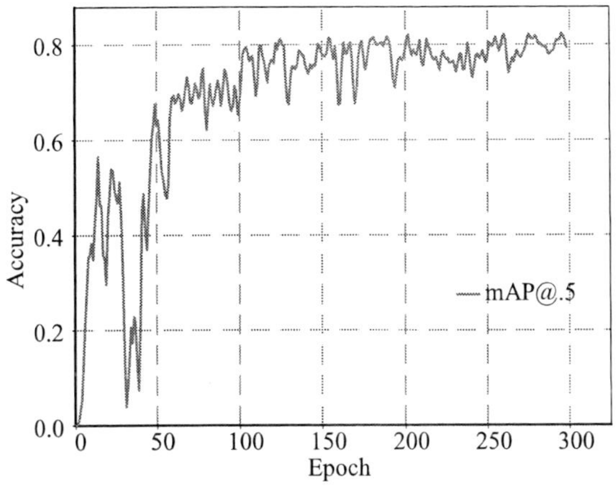

a) mAP@.5

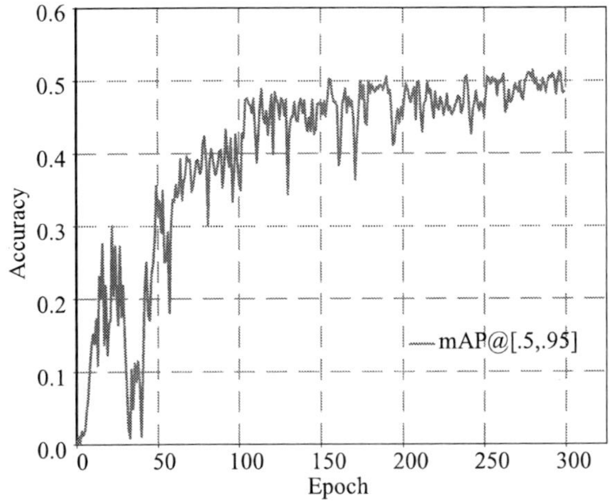

b) mAP@[.5, .95]
Figure 5. Accuracy of YOLOv5

4.2 Effect of data augmentation on model performance

Data augmentation extracts more information from the original dataset to build a more comprehensive dataset. Augmented dataset can be used to strengthen the generation capacity of a DL model in unknown dataset. Thus, experiment 2 was carried out to investigate the influence of data augmentation on terrain classification. As mentioned in Section 3.2, augmentation technologies were used to rebuild a larger dataset (containing 10,000 pictures). The experimental results are listed in Table 3, where "Aft. Aug." stands for "After Data Augmentation,", and "Bef. Aug." for "Before Data Augmentation.".

Table 3. Prediction accuracy using data augmentation

Metric \ Classes	Loose sand	Bedrock	Small rock	Large rock	Outcrop
mAP@.5 (Aft. Aug.)	0.945	0.928	0.852	0.872	0.900
mAP@.5 (Bef. Aug.)	0.853	0.858	0.697	0.731	0.741
mAP@[.5, .95] (Aft. Aug.)	0.743	0.682	0.484	0.678	0.653
mAP@[.5, .95] (Bef. Aug.)	0.634	0.553	0.343	0.385	0.510

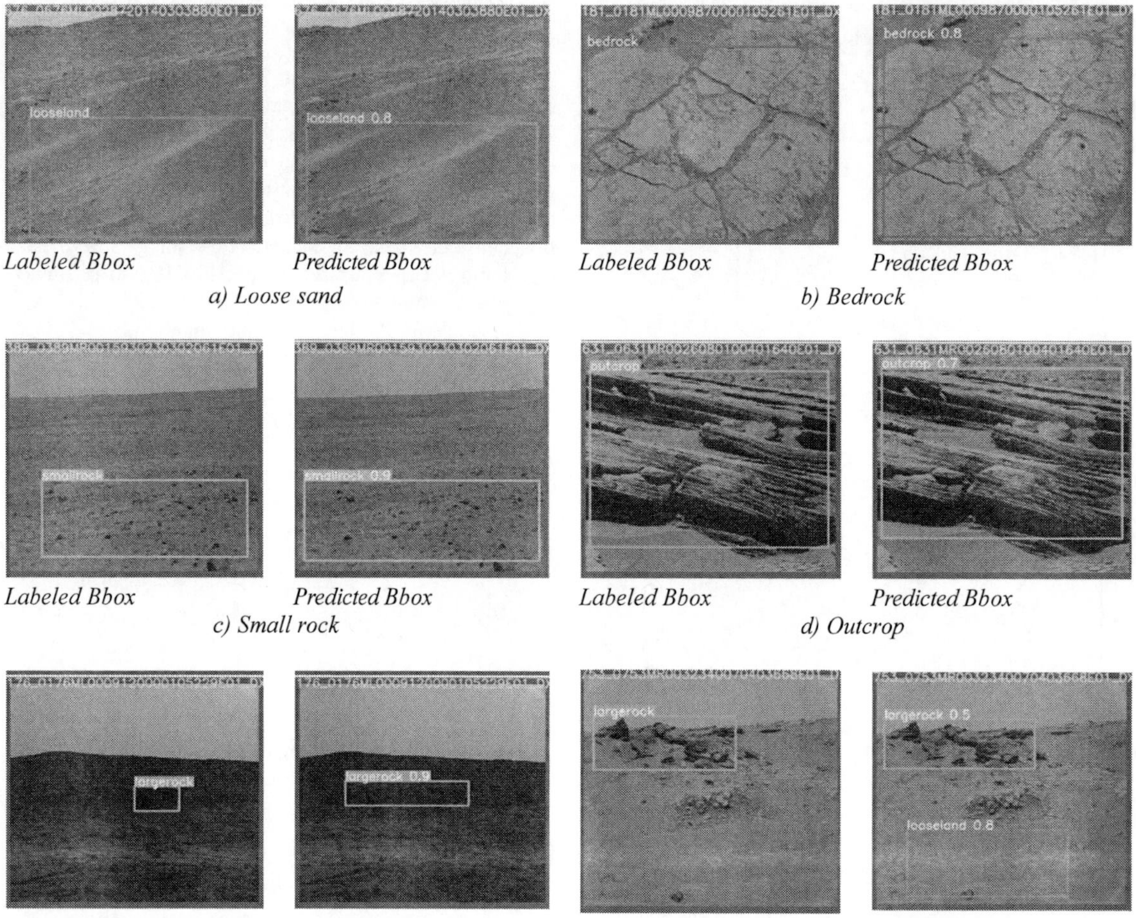

Labeled Bbox *Predicted Bbox* *Labeled Bbox* *Predicted Bbox*

a) Loose sand *b) Bedrock*

Labeled Bbox *Predicted Bbox* *Labeled Bbox* *Predicted Bbox*

c) Small rock *d) Outcrop*

Labeled Bbox *Predicted Bbox* *Labeled Bbox* *Predicted Bbox*

e) Large rock *f) Large rock*

Figure. 6. Comparison of labeled and predicted Bboxes using the optimal trained YOLOv5 model.

The overall performance of terrain classification can be strengthened by data augmentation. After data augmentation, YOLOv5 can detect the classified five terrains with high accuracy. The mAP@.5 ranges from 0.852 to 0.945, while mAP@[.5, .95] ranges from 0.484 to 0.743.

Loose sandy terrain can be predicted with the highest accuracy, which can form a hazardous obstacle for WPRs. The detection accuracy of large rock is lower than most of the other terrains. Compared with other terrains, the size range of large rock is larger according to the classification criterion listed in Table 1, which increases the classification difficulty. Small rocks in near distance are frequently misclassified as large rock, and thus its prediction accuracy is the lowest.

In conclusion, data augmentation can be used to improve the performance of DL models in terrain classification.

5. Conclusions and discussion

Real-time and accurate terrain identification can be used to prevent WPRs from getting trapped and to slow down wheel wear. A Martian terrain classification method with DL models was proposed in this paper. (1) A dataset of Mars scenes was first developed. (2) Three typical DL models (YOLOv5, CenterNet, and Faster-RCNN) were then selected to demonstrate their performance in Martian terrain classification. (3) Transfer learning and data augmentation were finally deployed to strengthen model's generation capacity. Comprehensive experiments show that YOLOv5 is the best choice to conduct terrain classifycation.

Though the five terrain classes can be identified correctly using YOLOv5, a more robust terrain recognition framework should be developed to identify terrain in real time with a higher accuracy by combining DL model with exteroceptive sensors and proprioceptive sensors. In addition, terrain recognition research based on unsupervised learning methods is also emerging, which will provide more possibilities for the development of intelligent terrain analysis and scene interpretation science.

6. Acknowledgements

This study was supported in part by the National Natural Science Foundation of China [Grant No. 51905119], Fundamental Research Funds for Central Universities [Grant No. HIT.NSRIF.2020090], and Self-Planned Task of State Key Laboratory of Robotics and System [Grant No. SKLRS202003C].

7. References

[1] Cowen R. Opportunity rolls out of purgatory. Science News. 2005; 167:413.

[2] Rankin A, Maimone M, Biesiadecki J, Patel N, Levine D, Toupet O. Driving curiosity: Mars rover mobility trends during the first seven years. 2020 IEEE Aerospace Conference: IEEE; 2020. p. 1-19.

[3] Ojeda L, Borenstein J, Witus G, Karlsen R. Terrain characterization and classification with a mobile robot. Journal of field robotics. 2006; 23:103-22.

[4] Bellutta P, Manduchi R, Matthies L, Owens K, Rankin A. Terrain perception for DEMO III. Proceedings of the IEEE Intelligent Vehicles Symposium 2000 (Cat. No. 00TH8511): IEEE; 2000. p. 326-31.

[5] Manduchi R, Castano A, Talukder A, Matthies L. Obstacle detection and terrain classification for autonomous off-road navigation. Autonomous robots. 2005; 18:81-102.

[6] Brooks CA, Iagnemma K. Vibration-based terrain classification for planetary exploration rovers. IEEE Transactions on Robotics. 2005; 21:1185-91.

[7] Valada A, Spinello L, Burgard W. Deep feature learning for acoustics-based terrain classification. Robotics research: Springer; 2018. p. 21-37.

[8] Giguere P, Dudek G. Clustering sensor data for autonomous terrain identification using time-dependency. Autonomous Robots. 2009; 26:171-86.

[9] Walas K, Nowicki M. Terrain classification using laser range finder. 2014 IEEE/RSJ International Conference on Intelligent Robots and Systems: IEEE; 2014. p. 5003-9.

[10] Sadhukhan D, Moore C, Collins E. Terrain estimation using internal sensors. Proc of the IASTED Int Conf on Robotics and Applications, 2004.

[11] Chavez-Garcia RO, Guzzi J, Gambardella LM, Giusti A. Image classification for ground traversability estimation in robotics. International Conference on Advanced Concepts for Intelligent Vision Systems: Springer; 2017. p. 325-36.

[12] Cheng H, Chen H, Liu Y. Topological indoor localization and navigation for autonomous mobile robot. IEEE Transactions on Automation Science and Engineering. 2014; 12:729-38.

[13] Li W, Hsu C-Y. Automated terrain feature identification from remote sensing imagery: a deep learning approach. International Journal of Geographical Information Science. 2020; 34:637-60.

[14] Redmon J, Divvala S, Girshick R, Farhadi A. You only look once: Unified, real-time object detection. Proceedings of the IEEE conference on computer vision and pattern recognition2016. p. 779-88.

[15] Liu W, Anguelov D, Erhan D, Szegedy C, Reed S, Fu C-Y, et al. Ssd: Single shot multibox detector. European conference on computer vision: Springer; 2016. p. 21-37.

[16] Girshick R, Donahue J, Darrell T, Malik J. Rich feature hierarchies for accurate object detection and semantic segmentation. Proceedings of the IEEE conference on computer vision and pattern recognition2014. p. 580-7.

[17] Ren S, He K, Girshick R, Sun J. Faster r-cnn: Towards real-time object detection with region proposal networks. Advances in neural information processing systems. 2015;28.

[18] Zhou X, Wang D, Krähenbühl P. Objects as points. arXiv preprint arXiv:190407850. 2019.

4774 / Steadily Learn to Drive with Virtual Memory

Yuhang Zhang[a], Yao Mu[b], Shengbo Li[a]*, Yangang Ren[a], Liye Tang[a], Yujie Yang[a], and Chen Chen[a]
[a]State Key Lab of Automotive Safety and Energy, School of Vehicle and Mobility, Tsinghua University, Beijing, China
[b]The University of Hong Kong, Hong Kong, China
**Corresponding author: lishbo@tsinghua.edu.cn*

https://doi.org/10.56884/ONWH8454

Abstract

Reinforcement learning has shown great potential in developing high-level autonomous driving systems. However, for high-dimensional tasks, current RL methods suffer from low data efficiency and oscillation in the training process. This paper proposes an algorithm called Learn to drive with Virtual Memory (LVM) to overcome these problems. LVM compresses the high-dimensional information into compact latent states and learns a latent dynamic model to summarize the agent's experience. Various imagined latent trajectories are generated as virtual memory by the latent dynamic model. The policy is learned by propagating gradient through the learned latent model with the imagined latent trajectories and thus leads to high data efficiency. Furthermore, a double critic structure is designed to reduce the oscillation during the training process. The effectiveness of LVM is demonstrated by an image-input autonomous driving task, in which LVM outperforms the existing method in terms of data efficiency, learning stability, and control performance.

Keywords: autonomous driving, reinforcement learning, representation learning

1. Introduction

With the development of artificial intelligence technologies, autonomous driving has become an important tendency in the automotive industry for the potential to improve road safety, reduce fuel consumption, and improve traffic efficiency (Yurtsever et al., 2019; Keshav, 2015). There are two kinds of schemes widely employed in the autonomous driving system: the hierarchical scheme (Urmson et al., 2008; Montemerlo et al., 2008; Duan et al., 2020; Mu et al., 2020) and the end-to-end scheme (Jaritz et al., 2018; Pomerleau, 1989; Bojarski et al., 2016). The hierarchical scheme divides the entire autonomous driving system into several modules, including environment perception, decision-making, and motion control (Tian et al., 2016). Each module is manually designed and tuned to ensure satisfying driving performance. Moreover, these modules are hierarchically structured, making errors easily propagated to downstream modules. The end-to-end scheme employs an integrated module that directly generates control commands from raw sensor inputs. It has gained more and more attention in recent years due to its straightforward system design and better overall performance compared with its hierarchical counterparts theoretically (Bojarski et al., 2016).

In the end-to-end scheme, Reinforcement Learning (RL) has great advantages over supervised learning because of insufficient natural driving data with such high-dimensional inputs (Chen et al., 2020; Chen et al., 2021). It learns from interactions with an environment or an analytical model, achieving incredible success in several fields varying from games to robotics (Volodymyr et al., 2015; Vinyals et al., 2019). Yu et al. (2016) trained an agent to master the turning operation and navigation in the game JavaScript Racer by Deep Q-Network (DQN). Jaritz et al. (2018) utilized Asynchronous Advantage Actor-Critic (A3C) to learn a vehicle controller in a physically and graphically realistic rally game in an end-to-end manner. Kendall et al. (2019) trained a Deep Deterministic Policy Gradient (DDPG) controller in the lane-keeping task, in which a single front-view image was used as input. However, these methods still suffer from low data efficiency and poor driving performance when applied in real autonomous driving scenarios.

Several studies in RL have been proposed to deal with high-dimensional sensory inputs. Ha and Schmidhuber (2018) proposed the World Model, which divided the algorithm into three parts. First, a Variational Auto-Encoder (VAE) (Kingma and Welling, 2014; Higgins et al., 2016) is trained to extract low-dimensional latent features of the image inputs by variational inference. Second, an RNN-based prediction model is learned to predict the state transition in the latent space. Third, the evolutionary strategy algorithm uses the learned representation model and prediction model to train a policy in the latent space. However, such a separated training scheme limits the prediction accuracy and the final policy performance because of the mismatch of the training data distribution for the three processes. In order to learn a better representation model and prediction model, Hafner et al. (2018) proposed to use

Recurrent Stochastic State Model (RSSM) to represent the environment dynamics. The representation and prediction model are trained together by variational inference exhibiting better performance than World Model. RSSM employs both stochastic variables and deterministic variables as latent states, which has the ability of uncertainty representation and achieves higher prediction accuracy at the same time. Lee et al. (2019) presented Stochastic Latent Actor-Critic (SLAC), which used a latent dynamic model with only stochastic latent states and achieved comparative performance with algorithms based on RSSM. After mapping the high-dimensional images into the low-dimensional latent space, the performance of RL algorithms in image-input tasks is improved effectively compared with those standard RL methods.

However, when applied in autonomous driving tasks, these methods still suffer from low data efficiency and oscillation in the training process. The interaction data collected by the exploration in the real physical world is expensive. Moreover, the unstable policy during the training process may lead to safety accidents and make the training process dangerous. Hafner et al. (2019) proposed a novel model-based algorithm with latent representation, Dreamer, to improve data efficiency. Meanwhile, the overestimation of the state value caused by the model error leads to unstable policy during the training process. Hasselt et al. (2015) argued that any estimation errors caused by system noise, function approximation, etc., will induce an upward bias in the value estimation.

In this paper, we propose an end-to-end autonomous driving algorithm, **L**earn to drive with **V**irtual **M**emory (LVM), which combines representation learning and model-based reinforcement learning to achieve high data-efficiency training for end-to-end driving policy. Besides, inspired by the Clipped Double-Q technique in model-free methods (Fujimoto et al., 2018; Haarnoja et al., 2018; Duan et al., 2021), a double critic structure is utilized to make the training process more stable than its counterparts by reducing the value overestimation caused by the model errors and noise. The contributions of this paper are as follows:

(1) A high data-efficiency end-to-end autonomous driving algorithm, LVM, is proposed. LVM learns a representation model in a self-supervised manner that maps the high-dimensional camera data into low-dimensional latent states. Besides, a transition model is synchronously trained to represent the dynamics of the latent states. LVM uses the learned dynamic model to generate imagined trajectories in the latent space, which are recorded as virtual memory. The policy is then optimized by the model-based RL method with the virtual memory and does not need to directly interact with the real environment, significantly improving the data efficiency.

(2) A double critic structure of state value estimation is designed for model-based RL. LVM learns two critics independently to reduce the overestimation of the state value caused by the model errors and make the training process more stable than its counterparts.

This paper is organized as follows. Section 2 describes the autonomous driving problem. In section 3, the proposed algorithm, LVM, is presented. Section 4 discusses the experiment and training results. Section 5 concludes this paper.

2. Preliminaries

Nowadays, self-driving vehicles are equipped with a significant number of expensive sensors and high energy-consuming chips, such as LiDAR and GPUs, to achieve high-level autonomous abilities. Here, we try to build an autonomous driving system with a single cheap camera. Historically, researchers have investigated end-to-end autonomous driving systems, which use camera images as input (Yu et al., 2016; Jaritz et al., 2018; Kendall et al., 2019). However, these methods suffer from low training efficiency due to high-dimensional state space. Different from aforementioned methods, the autonomous driving task is described as a Partial Observable Markov Decision Process (POMDP) (Shengbo, 2022) with high-dimensional observation $o \in \Omega$, low-dimensional latent state $s \in \mathcal{S}$, action $a \in \mathcal{A}$, transition model $P(s'|s,a): \mathcal{S} \times \mathcal{A} \to \mathcal{P}(s')$, reward function $r(s,a): \mathcal{S} \times \mathcal{A} \to \mathbb{R}$, observation function $O(o|s,a): \mathcal{S} \times \mathcal{A} \to \Omega$, and discounted factor $\gamma \in (0,1)$.

In this problem, the observation o is a three-channel RGB image generated by the camera attached to the front part of the vehicle. The state s represents the dynamic state of the vehicle and surrounding environments, such as lane position and obstacle position. The state is learned by a representation learning algorithm, in which its ground truth is not accessible. The transition model P determines the distribution of the next state given the current state and action. The action a is given by the policy $\pi(a|o_{\le t}, a_{<t})$ which maps the direct product space of $\mathcal{S}$ and $\mathcal{A}$ to action space $\mathcal{A}$. The subscript $\le t$ stands for all the information before or at time step t, while $< t$ does not include time step t. The policy definition is different from standard RL problems because of the lack of Markov property in the observation states. Reward function r is given by the environment, which measures how good an action is. Good actions, such as getting closer to the centerline of the road, lead to higher rewards, while bad actions, such as crashing out of the road, lead to lower rewards. At each step, the agent takes an action based on historical observations and actions, receives a reward r, and arrives at a new state s'. However, the agent cannot get access to the ground truth state and only receives a new observation o' generated by the observation function.

The goal of the agent is to find the optimal policy parameter θ^* that maximizes the long-horizon accumulated rewards represented by a value function,

$$\theta^* = \arg\max_{\theta} \mathbb{E}_{s\sim d_0(s)}[V_\pi(s)]$$

with

$$V_\pi(s) = \mathbb{E}_\pi\left\{\sum_{t=0}^{\infty} \gamma^t r_t | s_0 = s\right\}, \tag{1}$$

where $d_0(s)$ is the initial state distribution, $V_\pi(s)$ is the value function with respect to policy π_θ.

However, such RL algorithms with high-dimensional observation space will result in inefficient training process. Recently studies have shown the potential of learning a low-dimensional latent dynamic model in such problems. The latent dynamic model aims to extract useful information from the raw observation and represents it with a low-dimensional latent state.

3. Learn to drive with virtual memory

As shown in Fig. 1, there are three alternate steps in a training iteration of LVM: latent dynamic model learning, model-based policy training in latent space, and data collection in the environment. In the first step, the latent dynamic model compresses the high-dimensional inputs into low-dimensional latent states and approximates the environment dynamic in the learned latent space. In the second step, the latent dynamic model is utilized to generate virtual trajectories by imagination. The policy function and value function are then updated by these virtual trajectories. In the third step, new interaction data is sampled from the real environment by the updated policy to enrich the replay buffer.

3.1 Latent dynamic model for autonomous driving

As shown in Fig. 2, the latent dynamic model consists of two basic elements, the encoder and the latent transition model. The encoder maps the observation o into a latent state s. Then, the latent transition model defines the dynamic of the latent state, i.e., $s' = f(s, a)$.

In this paper, the recurrent state space model (RSSM) is used to approximate the environment dynamic effectively. RSSM model represents the real world with two kinds of states, the stochastic state s, and the deterministic state h. The usage of the stochastic state enables the model to represent the uncertainty of the environment. The deterministic state effectively improves the predicting accuracy of the model. Hafner et al. (2018) argued that both stochastic part and deterministic part contribute to the performance of the latent dynamic model.

The RSSM has four main components:

(i) Deterministic state model,

$$h_t = f_\psi(s_{t-1}, a_{t-1}, h_{t-1}),$$

(ii) Stochastic state model,

$$s_t \sim p_\psi(s_t|h_t), \qquad (2)$$

(iii) Reward model,

$$r_t \sim p_\psi(r_t|h_t, s_t),$$

(iv) Observation model,

$$o_t \sim p_\psi(o_t|h_t, s_t),$$

where ψ is the parameter of the RSSM model. The deterministic state model and stochastic model together form the representation model, which maps the high-dimensional observation space into a low-dimensional feature space, which is called the latent space. In the latent space, it is easier to extract useful information for the subsequent policy training. The deterministic state model is usually approximated by recurrent neural networks such as Long Short-Term Memory (LSTM) (Hochreiter and Schmidhuber, 1997) and Gated Recurrent Unit (GRU) (Chung et al., 2014). In such recurrent neural network models, h_t encodes the information of historical observations and actions. The stochastic state model predicts the distribution of the latent state according to the deterministic state. Meanwhile, the above two state models also compose the transition model of the latent dynamics, which can be used to generate virtual trajectories and improve the policy. The policy training process does not interact with the real environment, wherefore, a reward model is essential for the policy improvement and evaluation of RL training. The observation model is an essential part when training the representation model with variational inference method.

The RSSM model is optimized by variational inference method, in which the evidence lower bound (ELBO) is maximized. The objective is defined as

$$J_{\text{RSSM}} = \mathbb{E}_p\left(\sum_t J_o + J_r + J_D\right),$$

with

$$J_o = \ln p_\psi(o_t|s_t)$$
$$J_r = \ln p_\psi(r_t|s_t)$$
$$J_D = -\text{KL}\big(p_\psi(s_t|s_{t-1}, a_{t-1}, o_t)\|q_\psi(s_t|s_{t-1}, a_{t-1})\big). \qquad (3)$$

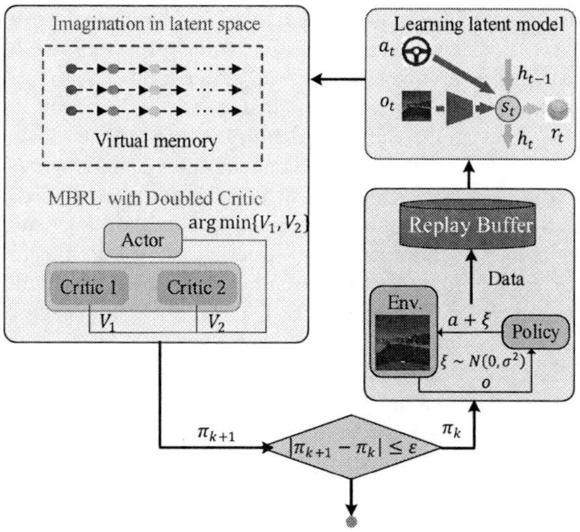

Figure 1. Framework of LVM

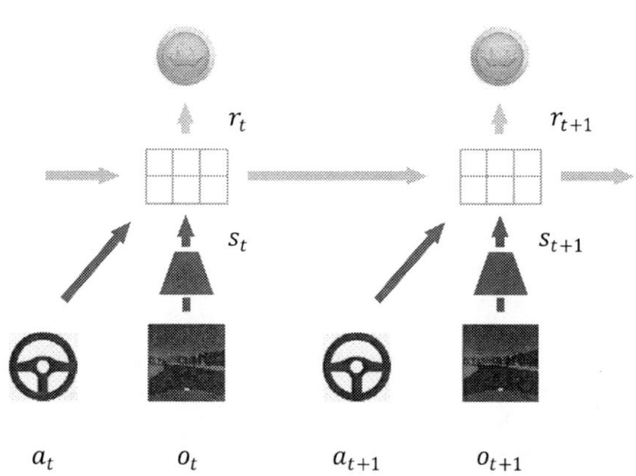

Figure 2. Latent dynamic model

The RSSM loss consists of three parts. J_o and J_r are the log-likelihood of the observation and reward, which can be regarded as a reconstruction loss. In implementation, the distributions of observation and reward are considered as Gaussian distributions, of which the mean is the output of the observation model and reward model, and the variance is a predefined constant. Meanwhile, J_D is the state prediction loss that minimizes the Kullback-Leibler (KL) divergence between the prior distribution and posterior distribution of state transition, which is also a metric for the prediction accuracy of RSSM. Because the agent has no access to the real environment, accurate prediction is necessary for policy training.

3.2 Policy optimization with virtual memory

After obtaining the RSSM model, it can be used to generate trajectories $\hat{\tau}_i$ in virtual memory. At first, several real trajectories are sampled from the replay buffer and serve as the virtual trajectory's starting points. Then, as shown in Fig. 3, the RSSM model is used to generate trajectories of a certain length from these states. Due to the stochastic property of the latent dynamic model, various trajectories are generated from each starting point. These trajectories are used to update the value function and policy function.

Interacting with the learned RSSM model is more efficient than in the real environment. Therefore, a more stable on-policy policy training method is naturally used instead of its off-policy counterparts which are commonly used in model-free RL to increase data efficiency.

In policy optimization, the stochastic state s_t and deterministic state h_t are concatenated to form the policy's real input. The concatenated state not only contains information of the current observation but also contains historical information which is necessary to represent current state. The policy is trained in the actor-critic architecture. The actor outputs the action according to the current state. And the critic approximates the value function of the state.

$$\begin{aligned} a_t &= \pi_\theta(s_t, h_t), \\ V_\phi(s_t, h_t) &= \mathbb{E}_{\pi_\theta}\Big(\sum_{\tau=t}^{H} \gamma^{\tau-t} r_\tau\Big), \end{aligned} \tag{4}$$

where θ, ϕ are the parameters of the approximated policy and value.

The critic is optimized by temporal-difference (TD) updating. Considering that the training data is generated by the learned RSSM model, the variance of value estimation is high because of inevitable model errors. TD(λ) return is used to reduce the variance of policy evaluation. TD(λ) return, $V_\lambda(s_t, h_t)$, is defined as

$$\begin{aligned} V_\lambda(s_t, h_t) = (1-\lambda)\sum_{n=1}^{H-1} \lambda^{n-1} V_N^n(s_t, h_t) \\ + \lambda^{H-1} V_N^H(s_t, h_t) \end{aligned} \tag{5}$$

where

$$\begin{aligned} V_N^k(s_t, h_t) = \mathbb{E}_{\pi_\theta}\Bigg\{\sum_{\tau=t}^{t+k-1} \gamma^{\tau-t} r_\tau \\ + \gamma^k V_\phi(s_{t+k}, h_{t+k})\Bigg\} \end{aligned}$$

Inspired by the Clipped Double Q-learning technique, two value function, $V_\lambda^{(1)}(s, h; \phi_1)$ and $V_\lambda^{(2)}(s, h; \phi_2)$ are learned independently to reduce overestimation and stabilized the training process. When estimating the target, a minimum operation is utilized, i.e.

$$\begin{aligned} &V_{\text{target}}(s_t, h_t) \\ &= \min\left\{V_\lambda^{(1)}(s, h; \phi_1), V_\lambda^{(2)}(s, h; \phi_2)\right\}. \end{aligned} \tag{6}$$

The two value functions are updated independently as

$$\begin{aligned} \min_{\phi_1} J_{V_1} &= \mathbb{E}_{\pi_\theta}\left(\sum_{\tau=t}^{t+H} \frac{1}{2}\left\|V_{\phi_1}(s_\tau, h_\tau) - V_{\text{target}}(s_\tau, h_\tau)\right\|^2\right), \\ \min_{\phi_2} J_{V_2} &= \mathbb{E}_{\pi_\theta}\left(\sum_{\tau=t}^{t+H} \frac{1}{2}\left\|V_{\phi_2}(s_\tau, h_\tau) - V_{\text{target}}(s_\tau, h_\tau)\right\|^2\right). \end{aligned} \tag{7}$$

The actor aims to maximize the expected return of the policy which is estimated by the critic. Thus, its objective is

$$\max_\theta J_\pi = \mathbb{E}_{\pi_\theta}\left(\sum_{\tau=t}^{t+H} V_{\text{target}}(s_\tau, h_\tau)\right). \tag{8}$$

3.3 Data Collection

In each iteration, new trajectories sampled by the current policy are added to the replay buffer. In the sampling process, in order to fully explore the whole environment, Gaussian noise is attached to the policy outputs, i.e.

$$\begin{aligned} a &= \pi_\theta(s_t, h_t) + \xi, \\ &\text{with} \\ \xi &\sim \mathcal{N}(0, \sigma^2) \end{aligned} \tag{9}$$

where σ is a predefined constant. The pseudo-code of LVM is shown in Algorithm 1.

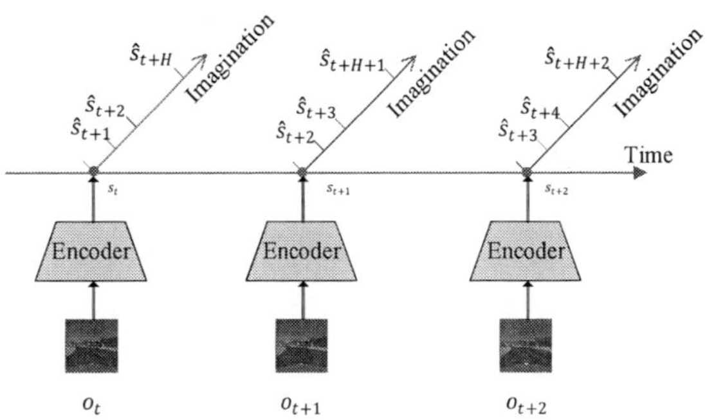

Figure 3. Virtual trajectory imagination

4. Experiments

In this section, LVM is implemented in a lane-keeping task, where the vehicle is expected to drive along the center of the lane. Firstly, the environment setup is introduced. Secondly, the essential details of implementation are described. Finally, the evaluation results are presented and discussed.

Algorithm 1: Learn to drive in virtual memory (LVM)

Initialize replay buffer $\mathcal{D}$ with S episodes with random policy
Initialize neural network parameters $\theta, \phi_1, \phi_2, \psi$
Initialize RSSM learning rate α_m, value learning rate α_v and policy learning rate α_π
for $e = 1 \dots MaxEpoch$ **do**
  for $i = 1 \dots TrainFreq$ **do**
    Sample sequence batch $\{(a_t, o_t, r_t)\}_{t=t:t+L}$ from the replay buffer
    // Dynamics learning
    $\psi \leftarrow \psi + \alpha_m \nabla_\psi J_{\text{RSSM}}$
    // Policy learning
    for $k = 1 \dots TrajNum$ **do**
      Rollout trajectories $\{(s_\tau, a_\tau, r_\tau)\}_{\tau=t:\,t+H}$ from each s_t in the sequence batch
    end
    $V_{\text{target}} \leftarrow \min\left\{V_\lambda^{(1)}, V_\lambda^{(2)}\right\}$
    $\phi_1 \leftarrow \phi_1 - \alpha_V \nabla_{\phi_1} J_{V_1}$
    $\phi_2 \leftarrow \phi_2 - \alpha_V \nabla_{\phi_2} J_{V_2}$
    $\theta \leftarrow \theta + \alpha_\pi \nabla_\theta J_\pi$
  end
  // Data collection
  for $j = 1 \dots DataCollectFreq$ **do**
    $s \leftarrow$ env.reset()
    while not *done* **do**
      $s \leftarrow \pi_\theta(s)$
      $s, r, done =$ env.step(a)
      $\mathcal{D}$.append$(s, a, r, done)$
    end
  end
end

4.1 Environment setup

The high fidelity simulation environment is built in Webots (Michel, 2004), which is a professional mobile robot simulation software package. As shown in Fig. 4, the environment is a two-lane roundabout with various buildings on both sides. Moreover, to make the scene more realistic, we also consider light changes in the environment.

The environment takes the acceleration a_x and steering angle δ as inputs. The feedback is composed of three elements: an image from the camera, a scalar reward, and a bool variable which indicates whether the episode ends. The image has three channels with size 64×64. The underlying vehicle model used in the simulator is a bicycle model in polar coordinates (Rajamani, 2006).

In this task, the vehicle is supposed to drive in the center of the road. Meanwhile, the control variables, the acceleration, and steering angle shall be as small as possible. Therefore, the reward function is designed to penalize the deviation of the vehicle from the center line and large action outputs,

$$r = c_1 y^2 + c_2 \varphi^2 + c_3 \omega^2 + c_4 \beta^2 + c_5 (v - v_0)^2 + c_6 \delta^2 + c_7 a_x^2, \tag{10}$$

where c_i, $i = 1 \dots 7$ are predefined coefficients, y is the distance between vehicle centroid and the road centerline, φ is the error between vehicle heading angle and the road tangent, β is the side-slip angle of the vehicle body, ω is the yaw rate of the vehicle body, v stands for the longitudinal speed of the vehicle, v_0 is the expected longitudinal speed, δ is the steering angle, and a_x is the longitudinal acceleration.

The design of the reward function is critical to the algorithm training. In model-free algorithms, a discrete reward is commonly used as a penalty for the agent's death, i.e., crashing out of the road. However, in model-based training, where a differential reward function is needed for the gradient computation, such a discrete reward would not be as effective as that in model-free training. Therefore, in this work, the reward is designed as a smooth and differential function.

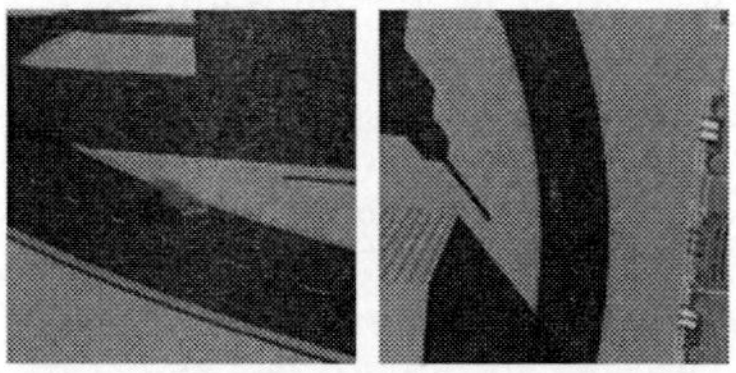

Figure 4. The environment built in Webots

4.2 Implementation details

Before LVM training starts, we pre-train RSSM to build a basic representation of the environment dynamics. Several trajectories are sampled by a random policy, such as Gaussian policy, to build a training dataset. Then the latent dynamic model is pre-trained in the fixed dataset. After pre-training, the main part of LVM, including RSSM training, model-based RL training in latent space, and data collection, is repeated until the algorithm converges.

The detailed structure of the RSSM model is shown in Fig. 5, which consists of two CNNs, three fully connected networks, and a recurrent neural network. The policy and the value are both fully connected networks, each of which has two hidden layers with 256 units. Key hyper-parameters of the proposed algorithm are shown in Table 1.

4.3 Evaluation results

The performance of image reconstruction is demonstrated in Fig. 6. The result shows that the reconstructed image contains all the information used in downstream tasks, such as the lateral position, the curvature of the road, and the lane mark. Therefore, the model-based RL trained with the latent dynamic model is effective in the real environment.

We compared our algorithm with Dreamer (Hafner et al., 2019) and SLAC (Lee et al., 2019). The model-based RL algorithm, Dreamer, solves the pure image-input control task of Mujoco and achieves high final performance and data efficiency. Meanwhile, SLAC combines the latent dynamic model with SAC, a state-of-the-art model-free RL algorithm, and achieves comparative performance compared with Dreamer in Mujoco.

In this work, the average return is utilized as the performance measure. Each algorithm is trained in five different runs with different random seeds. In order to compare the data efficiency of different algorithms, we make different algorithms collect the same amount of data in each training step. Thus, after the same steps of training, the algorithm with higher average rewards has better data efficiency. As shown in Fig. 7, LVM and Dreamer show better data efficiency than SLAC. Furthermore, their final performance apparently outperforms SLAC. LVM and Dreamer use a better representation model, RSSM, which represents the latent states with both deterministic variables and stochastic variables to improve the representing ability of environmental uncertainty and the accuracy of state prediction. Besides, LVM and Dreamer make full use of the learned dynamic models in latent space by directly propagating gradient through the model, which offers better guidance for policy improvement. Meanwhile, LVM has a more stable training process compared with Dreamer. As shown in Fig. 7, after 250k steps of training, the average return of LVM has almost converged. In contrast, the curve of Dreamer still oscillates occasionally. The LVM uses the double critic structure, which reduces the overestimation of the value function by taking minimization of the state values given by the two critics. It provides a more accurate estimation of the policy learning objective, thus improving the stability of algorithm training.

The control performance of LVM and its baselines is demonstrated in Fig. 8. The mean lateral position error is 0.196m for LVM, 0.316m for Dreamer, and 0.875m for SLAC. LVM has the lowest lateral position error compared with its counterparts according to 20 random tests. The average lateral position error of LVM is 38.0% less than that of Dreamer, and 89.0% less than SLAC.

In summary, LVM exhibits the best data efficiency and convergence stability during the training process and superior control performance in the lane-keeping task.

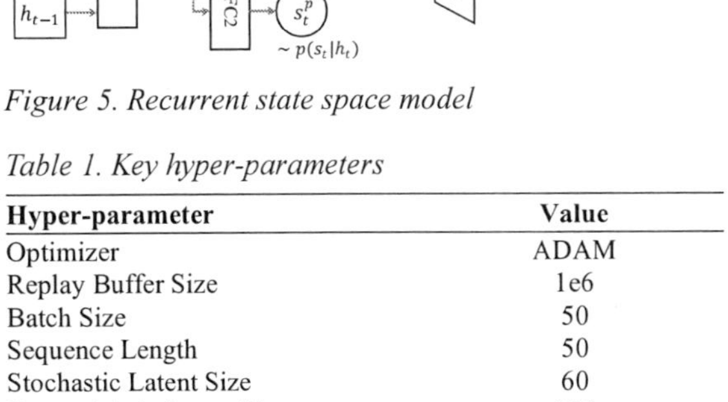

Figure 5. Recurrent state space model

Table 1. Key hyper-parameters

Hyper-parameter	Value
Optimizer	ADAM
Replay Buffer Size	1e6
Batch Size	50
Sequence Length	50
Stochastic Latent Size	60
Deterministic Latent Size	256
Time Step	0.5
Discounted Factor	0.99
Actor Learning Rate	1e-4
Critic Learning Rate	1e-4
RSSM Learning Rate	1e-3

a) Original images

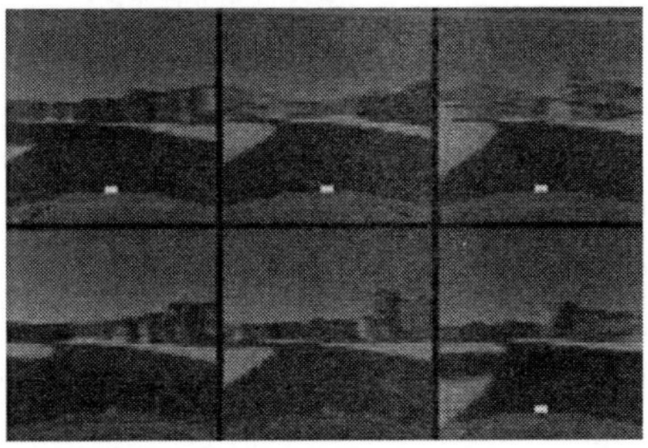

b) Reconstructed images

Figure 6. The reconstruction performance of RSSM

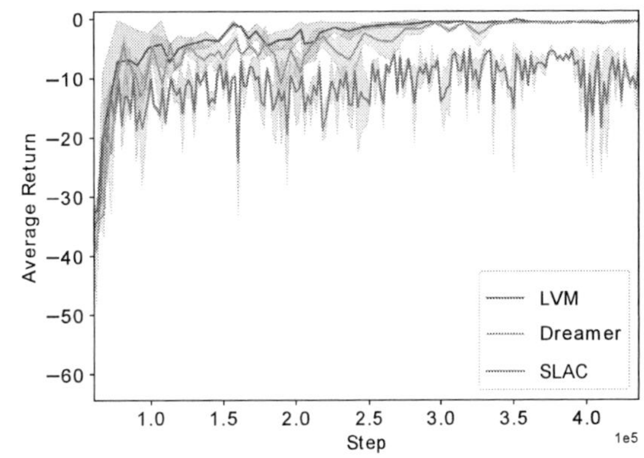

Figure 7. Comparison of average return

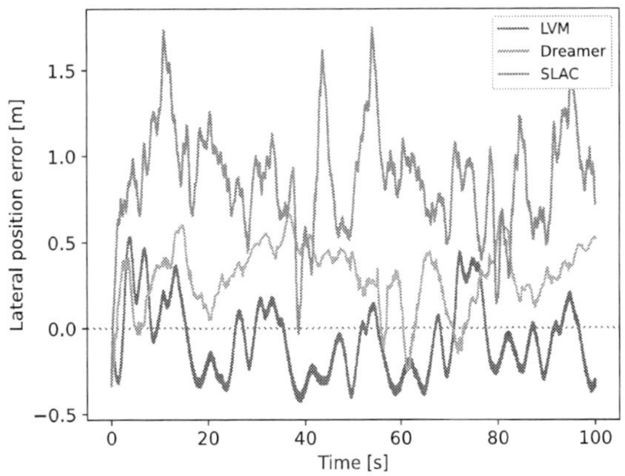

Figure 8. Comparison of control performance

5. Conclusion

This paper proposes an autonomous driving system with a front camera image as the only sensor, which is called Learn to drive with Virtual Memory (LVM). LVM learns a latent dynamic model, which compresses high-dimensional raw camera data into low-dimensional latent states and predicts the environment transition in the latent space. Instead of interacting with the real environment, LVM learns the optimal policy by virtual memory generated by the learned latent dynamic model, which significantly improves the data efficiency. In the policy learning process, the double critic structure is designed to stabilize the algorithm training and reduce the oscillation occurrence. The performance of the algorithms is evaluated in the lane-keeping task in a roundabout. Experiments show that LVM has a more stable training process and better control performance compared with Dreamer and SLAC. We conclude that LVM is a promising method for developing an image-input autonomous driving system.

In the future, the potential of LVM in different driving tasks and more complex scenarios will be investigated. Besides, better latent environment model structure and training method will be explored.

6. Acknowledgements

This study is supported by NSF China with 52072213 and U20A20334. It is also partially supported by Tsinghua University-Tencent Joint Laboratory

7. References

Bojarski, M., D Del Testa, D. Dworakowski, B. Firner, B. Flepp, P. Goyal, L. D. Jackel, M. Monfort, U. Muller, and J. Zhang. 2016. "End to End Learning for Self-Driving Cars. "

Chen, J., S. E. Li, and M. Tomizuka. 2021. "Interpretable End-to-End Urban Autonomous Driving with Latent Deep Reinforcement Learning. "IEEE Transactions on Intelligent Transportation Systems PP (99):1-11.

Chen, S., M. Wang, W. Song, Y. Yang, Y. Li, and M. Fu. 2020. "Stabilization Approaches for Reinforcement Learning-Based End-to-End Autonomous Driving. "IEEE Transactions on Vehicular Technology 69 (5):4740-4750.

Chung, J., C. Gulcehre, K. H. Cho, and Y. Bengio. 2014. "Empirical Evaluation of Gated Recurrent Neural Networks on Sequence Modeling." Eprint Arxiv.

Duan, J., Y. Guan, S. E. Li, Y. Ren, and B. Cheng. 2021. "Distributional Soft Actor-Critic: Off-Policy Reinforcement Learning for Addressing Value Estimation Errors." IEEE Transactions on Neural Networks and Learning Systems PP (99).

Duan, J., S. E. Li, Y. Guan, Q. Sun, and B. Cheng. 2020. "Hierarchical Reinforcement Learning for Self-Driving Decision-Making without Reliance on Labeled Driving Data." IET Intelligent Transport Systems 14 (1).

Fujimoto, S., H Van Hoof, and D. Meger. 2018. "Addressing Function Approximation Error in Actor-Critic Methods."

Ha, David, and Jürgen Schmidhuber. 2018. "World models." arXiv preprint arXiv:1803.10122.

Haarnoja, T., A. Zhou, P. Abbeel, and S. Levine. 2018. "Soft Actor-Critic: Off-Policy Maximum Entropy Deep Reinforcement Learning with a Stochastic Actor."

Hafner, D., T. Lillicrap, J. Ba, and M. Norouzi. 2019. "Dream to Control: Learning Behaviors by Latent Imagination."

Hafner, D., T. Lillicrap, I. Fischer, R. Villegas, D. Ha, H. Lee, and J. Davidson. 2018. "Learning Latent Dynamics for Planning from Pixels."

Hasselt, Hado Van, Arthur Guez, and David Silver. 2015. "Deep Reinforcement Learning with Double Q-learning." Computer Science.

Higgins, I., L. Matthey, A. Pal, C. Burgess, X. Glorot, M. Botvinick, S. Mohamed, and A. Lerchner. 2016. "beta-VAE: Learning Basic Visual Concepts with a Constrained Variational Framework."

Hochreiter, S., and J. Schmidhuber. 1997. "Long Short-Term Memory." Neural Computation 9 (8):1735-1780.

Jaritz, M., R. D. Charette, M. Toromanoff, E. Perot, and F. Nashashibi. 2018. "End-to-End Race Driving with Deep Reinforcement Learning." IEEE.

Kendall, A., J. Hawke, D. Janz, P. Mazur, and A. Shah. 2019. "Learning to Drive in a Day." 2019 International Conference on Robotics and Automation (ICRA).

Keshav, B. 2015. "Autonomous cars: Past, present and future a review of the developments in the last century, the present scenario and the expected future of autonomous vehicle technology." IEEE.

Kingma, D. P., and M. Welling. 2014. "Auto-Encoding Variational Bayes." arXiv.org.

Lee, A. X., A. Nagabandi, P. Abbeel, and S. Levine. 2019. "Stochastic Latent Actor-Critic: Deep Reinforcement Learning with a Latent Variable Model. "

Michel, Olivier. 2004. "Cyberbotics Ltd. Webots™: professional mobile robot simulation. "International Journal of Advanced Robotic Systems 1 (1):5.

Montemerlo, Michael, Jan Becker, Suhrid Bhat, Hendrik Dahlkamp, Dmitri Dolgov, Scott Ettinger, Dirk Haehnel, Tim Hilden, Gabe Hoffmann, and Burkhard Huhnke. 2008. "Junior: The Stanford entry in the urban challenge. "Journal of field Robotics 25 (9):569-597.

Mu, Yao, Baiyu Peng, Ziqing Gu, Shengbo Eben Li, Chang Liu, Bingbing Nie, Jianfeng Zheng, and Bo Zhang. 2020. "Mixed reinforcement learning for efficient policy optimization in stochastic environments. "2020 20th International Conference on Control, Automation and Systems (ICCAS).

Pomerleau, D. A. 1989. "Alvinn: An Autonomous Land Vehicle in a Neural Network. "Morgan Kaufmann Publishers Inc.

Rajamani, R. 2006. Vehicle Dynamics and Control: Vehicle Dynamics and Control.

Shengbo, Eben, Li. 2022. Reinforcement Learning and Control.

Tian, B., B. T. Morris, M. Tang, Y. Liu, and S. Tang. 2016. "Hierarchical and Networked Vehicle Surveillance in ITS-A Survey. "

Urmson, C., J. Anhalt, D. Bagnell, C. Baker, R. Bittner, and E. Al. 2008. "Autonomous driving in urban environments: Boss and the Urban Challenge. "Journal of Field Robotics.

Vinyals, Oriol, Igor Babuschkin, Wojciech M Czarnecki, Michaël Mathieu, Andrew Dudzik, Junyoung Chung, David H Choi, Richard Powell, Timo Ewalds, and Petko Georgiev. 2019. "Grandmaster level in StarCraft II using multi-agent reinforcement learning. "Nature 575 (7782):350-354.

Volodymyr, Mnih, Koray, Kavukcuoglu, David, Silver, Andrei, Rusu, and Joel. 2015. "Human-level control through deep reinforcement learning. "Nature.

Yu, April, Raphael Palefsky-Smith, and Rishi Bedi. 2016. "Deep reinforcement learning for simulated autonomous vehicle control. "Course Project Reports: Winter 2016.

Yurtsever, E., J. Lambert, A. Carballo, and K. Takeda. 2019. "A Survey of Autonomous Driving: Common Practices and Emerging Technologies. "

4782 / Experimental Study of Track-Soil Interactions of the Steering Performance of Tracked Robots over Soft Deformable Terrains

Qiaowen Wang, Wenhui Wang, Shipeng Lyu, and Zhenzhong Jia*
Southern University of Science and Technology (SUSTech), Shenzhen, 518055, China
**Corresponding author: jiazz@sustech.edu.cn*

https://doi.org/10.56884/VGTD2054

Abstract

The research of track-soil interaction modeling over soft deformable terrains is an important direction in terramechanics. Wong proposed a general theory of skidsteering tracked vehicle model by focusing on the trackterrain interaction of the bottom surface of tracks while neglecting the resistance and the bulldozing effect contributed by the laterally accumulated soil to the side of tracked vehicle. The phenomenon becomes nonnegligible and sometimes significant during small radius turn maneuvers, which are quite common in robot motion planning. In order to investigate this quite important and complicated interaction process and establish a complete track-terrain interaction model, we build an instrumented experimental platform. In this platform, the RGB-D information for deformable terrains can be measured and analyzed to obtain several important soil parameters in real-time, including the contours of the soils in contact with the supporting wheels, and the cross-section shapes of the accumulated soils. We apply image segmentation, mapping depth images, point cloud drawing and postprocessing to divide the tracked vehicle and the soil in point cloud images. We reconstruct the morphology of the soil accumulated on the side of vehicles during small radius steering maneuvers, and then obtain aforementioned parameters in order to explain the bulldozing effect.

Keywords: tracked vehicles, skid-steering, deformable terrains, bulldozing effect, experimental study

1. Introduction

In recent years, unmanned ground vehicles and mobile robots have been commonly used in inaccessible, repetitive, and high-risk missions such as mining, agriculture, military, and planetary vehicle explorations. Typical examples include wheeled robots such as NASA Mars rovers, unmanned trucks, and tracked robots such as PackBot. There are some comparative studies between wheeled and tracked platforms, especially on their mobility performance (Guo et al., 2019). Tracked robots tend to have better mobility against wheeled platforms. However, they will both suffer from poor mobility when traveling over loose soft terrains due to slippage and sinkage. In addition, deformable rough terrain also poses great challenge to energy sustainability of these robots (Iagnemma and Dubowsky, 2004). This paper focuses on tracked robots. To solve the above problem, it is important to analyze the soil-track interaction phenomena and establish a suitable track-soil contact model. This will help us to design robots with better mobility, further predict the robot power and energy consumption, and greatly improve the overall performance.

However, due to the large contact area between track and soil and the complex sliding and shearing effects and stress distributions between contact surfaces, it is difficult to model the contact between robots and soft terrains precisely (Guo and Peng, 2013).

Many researchers have conducted relevant theoretical studies. Steeds' models assume that the shear force acting on the track by the ground obeys Coulomb's law of friction, and the friction coefficients in the longitudinal and lateral directions of the track are different (Steeds, 1950). The model assumes that once there is relative movement between the track and the ground, the shear stress immediately reaches its maximum value. However, experimental evidence has shown that the shear stress is dependent on the shear displacement. This means that the shear stress will reach its maximum value only after reaching a certain shear displacement (Wong, 2008). Wong developed a skid-steering model for tracked vehicles with isotropic assumptions of shear stress-shear displacement relationship. It assumes that the robot travels on firm soil, hence, there is neither sinkage nor bulldozing effect caused by the side surface of the tracks.

However, the bulldozing effect cannot be neglected for deformable terrains especially when the turning radius of the track is small. As shown in Fig. 1, the rut depth increases rapidly with the decrease of turning radius. This means that the bulldozing effect will dramatically increase for small-radius turns.

In order to solve the difficulties mentioned above, it is necessary to develop the track-soil interaction model for tracked vehicles under small turning radius in soft terrain by considering the bulldozing effect.

Establishing such model will be quite complicated and challenging. This paper is only the first step to solve this problem. In particular, we solve it from an experimental point of view, observing the track-soil interaction phenomenon when vehicle traveling on soft terrains.

We use RGB-D cameras to record the change of deformation of soil morphology and other soil parameters under small radius condition (Kamari and Ham, 2021), and calculate the regularity of soil volume (Li et al., 2020) changing with the turning radius of tracked vehicle. This will be helpful to the further study of the mechanics model.

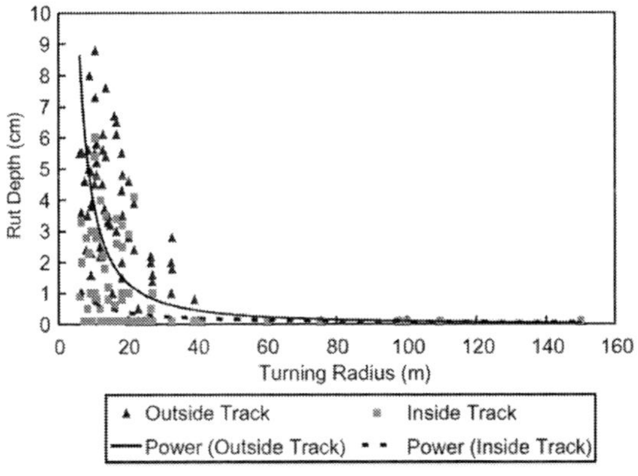

Figure 1. Relationship between Light Armored Vehicle turning radius and rut depth influenced by track locations (Liu et al., 2009)

2. Experimental platform

In order to get the mechanical model of the tracked robot under small-radius steering, we build a testbed and design a customized tracked robot research platform equipped with a variety of sensors, which will be used to measure experimental details such as force, torque, and visual information. The tracked robot and the experimental testbed are shown in Fig. 2.

Figure 2. a) Tracked robot model
b) Tracked robot driving on soft terrain

The robot can be divided into two parts: track module and body. Track module is divided into two parts with central symmetry, including crawler, driving wheel and idler wheels, which are driven by a DC motor. The two track modules and the body are connected by 6-axis F/T sensors, which can measure shear force and other information. The body is equipped with computer, F/T sensor data acquisition card, battery, etc. In order to measure the accumulation of side soil when the tracked robot is driving on soft and deformable soil, especially under the working condition of small-radius turns, a cantilever beam extending from the top of the body to the left and right sides is installed with two RGB-D cameras (model: Intel D435i). We also place an April-tag on top of the robot to track the robot pose. The track width of the machine body is $40mm$, the spacing between tracks is about 190mm, the ground-contact length is about $260mm$, and the robot weight is about $4.7kg$.

This paper mainly introduces the details of the interaction between the side of the crawler and the soil observed by RGB-D cameras and the theoretical analysis of the derived model, which will be covered in the subsequent sections.

3. Track-Soil RGB-D Image Segmentation and Data Processing

To observe the two phenomena (robot sinkage and soil accumulation), a wildly used method is separating the depth information measured by sensors (such as RGB-D camera). Unfortunately, this method may not work in our research condition. In our condition, the robot sinkage is usually small, and the robot track is close to the terrain. So, it is difficult to separate the tracked robot directly by depth information (see Fig. 4b). Some research tries to use the image information to solve this problem (Reina et al., 2006), and it is performance is pretty good in their application. While using their works in our experiment, the performance is affected. The reason is that the light on the robot-soil contact area sometimes could be blocked.

To solve the difficulties above and observe two phenomena, an improved OTSU algorithm is proposed in this paper. By considering the characteristics of the tracked robot (the color of wheels) experimental platform, this method can effectively separate the tracked robot from the soil.

3.1 Image Segmentation based on Improved OTSU method

OTSU is an algorithm to determine the threshold value of binary image segmentation, and it is also called as maximum inter-class variance method. In this paper, an improved OTSU method is used for image segmentation problem, the detailed steps are shown in Fig. 3. Each of following steps is in the same order as in the figure.

1. Firstly, the RGB image of the track-soil interaction area is captured by the on-board camera. The RGB image will be processed by a Gaussian filter to reduce its noise.
2. As the robot wheels are red (special design), We open the R channel to segment those wheels first. Then, we change the wheel's color into black in the image to simplify the next segment process.
3. We use OTSU algorithm to process this image, and some noise will be introduced in the processed image, which could result in incorrect segmentation.
4. Post-processing: use the morphological open operation to eliminate small objects (white noise) in the image, and then perform the dilate operation to shrink the holes in the image.

5. Fill closed holes in the image.
6. We map the color of the wheel and convert its color to black, so that the image is suitable for threshold segmentation.
7. Take operations of OTSU algorithm and hole filling again.
8. Mask of the image to color the soil.

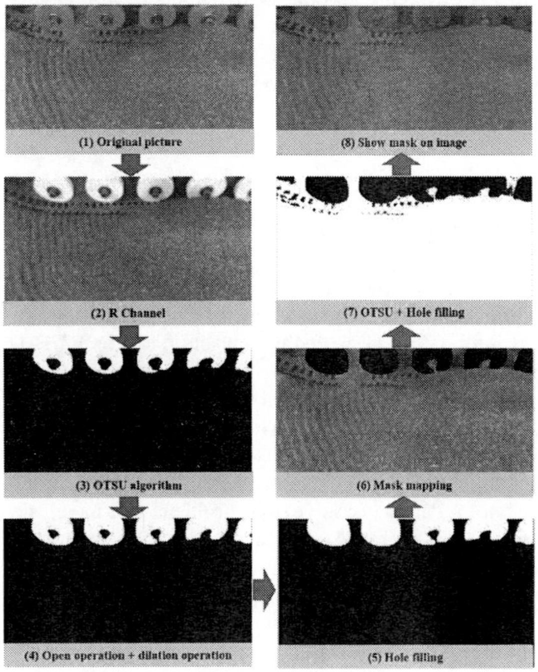

Figure 3. Otsu algorithm processing flow

3.2 Point Cloud Processing

In this part, we use D435i camera to derive depth information of the track-soil interaction area (see in Fig. 4b). As the depth image is set to be aligned to RGB image before, we can map the segmentation results in above image to the point cloud directly (see in Fig. 4d). So, we can derive the point cloud of robot track and soil respectively.

To measure the robot sinkage and the volume of the accumulated sand, we removed the tracked robot and ground from the reconstructed image and place the image in the world coordinate system (see Fig. 5).

As shown in Fig. 5, there are lots of noise points in the point cloud. We choose the statistical filtering method to remove those noise. Assuming that the distance follows a Gaussian distribution, its shape is determined by the mean and standard deviation. These points whose average distance is outside the standard range can be defined as outliers and can be removed from the data. After removing outliers, we need to use the voxel mesh method for down-sampling to reduce the number of point clouds. This method creates a three-dimensional grid through the input point cloud data and uses the center of gravity of all points in the grid to approximately represent other points in the grid.

Due to the similar color of the soil and the screw, a certain amount of incorrect segmentation is caused in the image segmentation process. In other words, the final reconstructed three-dimensional morphology includes some point clouds of the screw. Considering that the camera is rigidly connected to the robot body, the wheel position information is constant in the camera coordinate system, so the misdivided point cloud of the wheel can be removed by a pass filter (Ding et al., 2018). On the negative side of the Y axis, a sufficient width is selected manually so that the intercepted point cloud can include both the undamaged datum and the accumulated soil.

Finally, a voxel down-sampling filter is performed to obtain a clean, sparse, easily observable point cloud image that can be used for volume calculations, as shown in Fig. 6d.

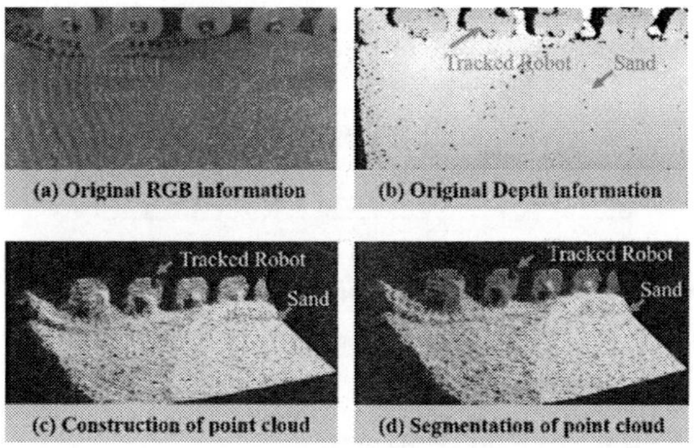

Figure 4. Point cloud reconstruction of track and soil

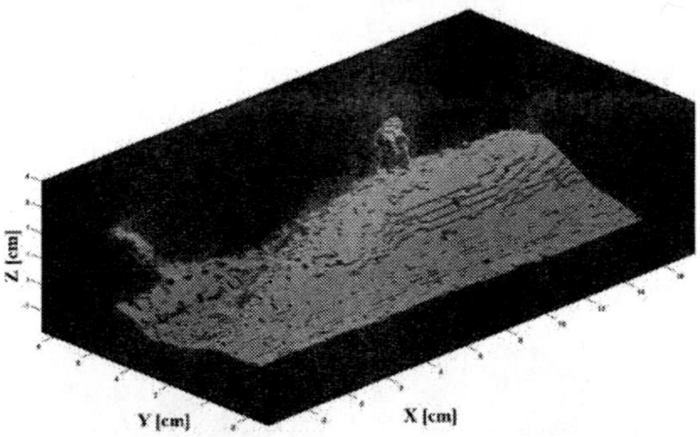

Figure 5. Point cloud after removing track

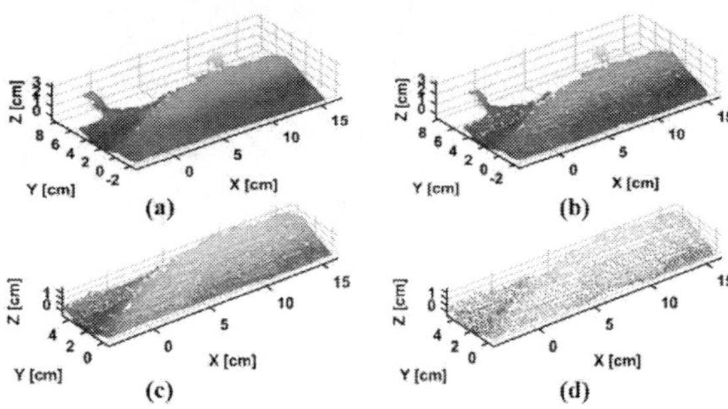

Figure 6. Point cloud processing

In this paper, we use the Delaunay triangulation method to calculate the volume of the accumulated soil (Zhu et al., 2021). The Delaunay triangular mesh is constructed based on the point cloud information, and the three-dimensional spatial coordinates are projected onto the XOY datum plane. The coordinates of the point cloud in Fig. 6(d) projected onto the XOY plane and Delaunay Triangulation result are shown in Fig. 7.

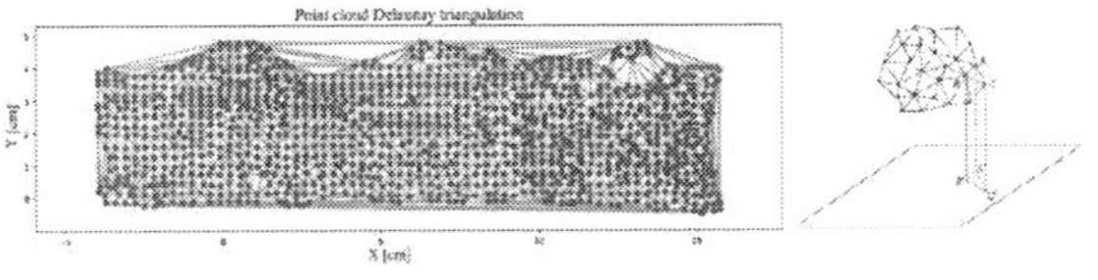

Figure 7. Delaunay triangulation (Liu and Zheng, 2021)

Once the triangular mesh is obtained, the steps for calculating the volume are listed as follows (Guerra, 2021):

1. The sequence of points in 3D space represented by the (x, y, z) triplet is grouped in a mesh grid using Delaunay triangulation in the (x, y) coordinates. This process outputs a collection of points grouped in three sets of 3D coordinates, each representing a triangular plane in 3D space:

$$[A, B, C] = [(x_A, y_A, z_A), (x_B, y_B, z_B), (x_C, y_C, z_C)]$$

2. The plane equation $z = f(x, y)$ is obtained for each group of 3 distinct points (triangles) in 3D space. Given the previous collection of points $[A, B, C]$ in the cartesian system: (1). Vector $\boldsymbol{AB}$ and $\boldsymbol{BC}$ are determined using their coordinates. (2). The cross product $\boldsymbol{AB} \times \boldsymbol{BC}$ generates a perpendicular vector represented by numerical constants (p, q, r). (3). Finally, the corresponding plane equation is given by:

$$p * (x - x_o) + q * (y - y_o) + r * (z - z_o) = 0$$

where (x_0, y_0, z_0) can be any one of the three (A, B, C) points from the plane.

3. Given the plane equation, we can isolate z and obtain a $z = f(x, y)$ function. The volume of each individual triangle is obtained by the sum of 2 double integrals. So, for a triangle with vertices ABC and its plane determined by $z = f(x, y)$ the double integral limits for a single triangular area are determined as follows:

$$vol_{triABC} = \int_{xa}^{xb} \int_{AB}^{AC} z dy dx + \int_{xb}^{xc} \int_{AC}^{BC} z dy dx$$

4. Experiment Results and Discussions

In this suction, we verified the effectiveness and robustness of the proposed method.

4.1 Volume Verification

The soil volume calculation experiments are carried out in this part, and the measurement testbed is shown in Fig. 8. As the shape of the soil in robot-terrain interaction varies while the robot condition changes, the experiments verify the robustness of our soil volume calculation algorithm. Four soil shapes are introduced in the experiments, and they are cones, squares, triangles, and irregular shapes as shown in Fig. 8b-e.

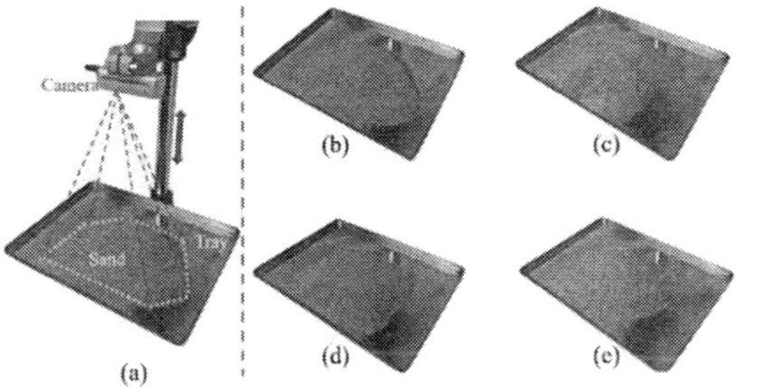

Figure 8. Volume test

The sand used in the experiment is calculated in advance, which is $211 cm^3$. After several measurements of the soil volume in each shape, the measurement error is small which is less than 3.3 %.

4.2 On-board Experiments

In this part, the image segmentation and point cloud processing techniques are used to observe the robot sinkage and soil accumulation phenomena. Some experiments are carried out to measure the soil volume while the robot turning radius varies. During the experiments, the robot velocity is a constant value (0.01618 m/s), and the moving time is the same in each attempt.

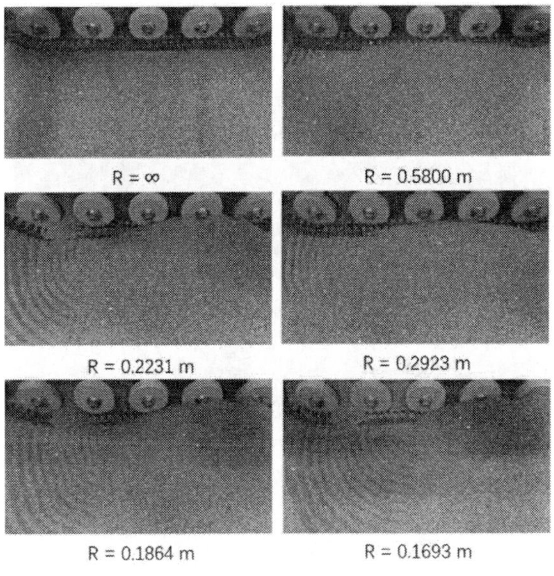

Figure 9. Variation of soil morphology with radius

As shown in Fig. 9, the turning radius of the tracked vehicle is changed by changing the speed ratio of the left and right tracks while keeping the body speed constant. The phenomena are that the soil accumulation increases while the turning radius decreases.

The Fig. 10 shows the relationship between the soil volume and the robot turning radius. When the turning radius is smaller than 0.5 m, the amount of soil volume increases dramatically which can reduce the robot's mobility and increase energy consumption. To overcome these challenges, some interaction model or control algorithm should be introduced, and our work provides a baseline or support for this research.

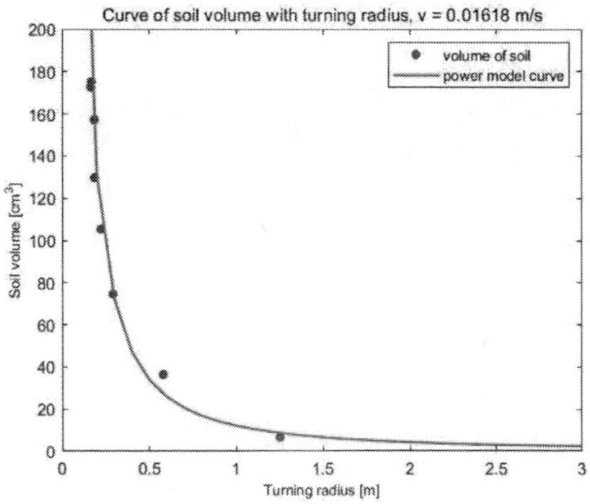

Figure 10. Curve of soil volume with turning radius

The physical interpretation is that in order to ensure a certain speed of the tracked robot, the inner track moves slowly when the turning radius is small, when the side track bulldozes the sand falling on the inner track flows slowly and is easy to accumulate.

4.3 Parameterization

To support future research (such as interaction force modeling), some parameters are introduced to describe the phenomenon of soil accumulation.

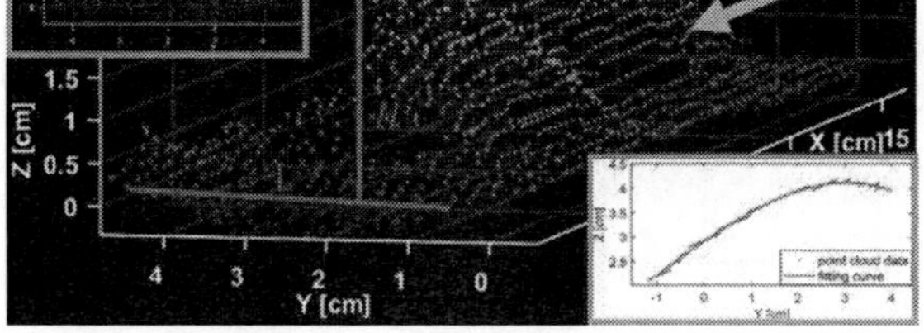

Figure 11. Soil profile and internal friction angle

As shown in Fig. 11, the piled soil is presented by the point cloud. In our experiments, the soil angle α is a constant value when the soil has accumulated to a certain extent. This phenomenon is the same as the robot-terrain interaction theoretical model, and the slope angle α is also called the internal friction angle of the soil in the theoretical model. There are other two parameters called bulldozing width L and bulldozing height H. These two parameters will increase while the robot's turning radius decreased. Based on these three parameters, we can easily describe and reconstruct the area of soil accumulation. What's more, the parameters are conveniently obtained from the point cloud map based on the algorithm in this paper.

4.4 Discussions

In this part, some analyses the accuracy of image segmentation and soil volume calculation are provided.

4.4.1 Image Segmentation

As the traditional method of image segmentation mainly uses the thresholds strategy, this method is heavily influenced by illumination, sand color, etc. For example, when the soil stacking height is consistent with the screw, the soil and screw are divided into a group in the final segmentation result (see Fig. 5, the yellow pattern is the point cloud of the screw). The reason is that the color of the sand and the screw is the same. Although filter method is used to solve the problem in this paper, the filter method needs so much time to adjust the parameter, and its application is constrained. So, we think the deep learning method would be a good method to solve this problem, and it is one of the research projects of our future works.

4.4.2 Soil Volume Calculation

In this paper, we only verify the soil volume calculation algorithm in the measurement testbed (see Fig. 8), and we omit the verification on the real robot-terrain testbed. The reason is that it is too difficult to obtain the actual soil volume in real experiments. To get more accurate data on soil volume, we will introduce the concept of datum plane selection into the Delaunay triangulation algorithm in future work.

5. Conclusions

This paper focuses on the interaction phenomenon between the tracked robot and the soft soil, especially in the small turning radius condition. This paper also provides three simple parameters to describe the interaction phenomenon. To get these parameters, this paper provides an improved OTSU algorithm to segment the image of robot-soil interaction. Then, the segmented image information is projected into the point cloud space to get the soil accumulation information in the robot-soil interaction area. Based on soil accumulation information, we can get these parameters easily, such as the internal friction angle of the soil. This paper also provided a soil volume calculation method based on the Delaunay triangulation algorithm, and the relationship between the soil accumulation and turning radius is analyzed based on experiments. This work in this paper provides solid support for the research of robot-terrain interaction mechanisms.

For future works, we try to explore the relationship between the turning radius of the tracked vehicle and the lateral bulldozing volume. By analyzing the force sensor data carried on the tracked robot, we can understand the relationship between the turning speed, turning radius, soil morphology/volume, and the bulldozing resistance of the tracked vehicle.

6. Acknowledgment

This work was supported in part by the Science, Technology and Innovation Commission of Shenzhen Municipality under Grant No. ZDSYS20200811143601004.

7. References

Ding, S., Zhang, X., Yu, Q., Li, L., & Wang, J. (2018). A Volume Measurement Method for Lunar Soil Collection Based on a Single Monitoring Camera. *Sensors*, *18*(10), 3394.

Guerra, A. (2021). Volume calculations for digital elevation models in python (volpy). https://https://github.com/agu3rra/volpy

Guo, T., Guo, J., Huang, B., & Peng, H. (2019). Power consumption of tracked and wheeled small mobile robots on deformable terrains–model and experimental validation. *Mechanism and Machine Theory*, *133*, 347-364.

Guo, T., & Peng, H. (2013, June). A simplified skid-steering model for torque and power analysis of tracked small unmanned ground vehicles. In *2013 American Control Conference* (pp. 1106-1111). IEEE.

Iagnemma, K., & Dubowsky, S. (2004). *Mobile robots in rough terrain: Estimation, motion planning, and control with application to planetary rovers* (Vol. 12). Springer Science & Business Media.

Kamari, M., & Ham, Y. (2021). Vision-based volumetric measurements via deep learning-based point cloud segmentation for material management in jobsites. *Automation in Construction*, *121*, 103430.

Liu, K., Ayers, P., Howard, H., & Anderson, A. (2009). Influence of turning radius on wheeled military vehicle induced rut formation. *Journal of Terramechanics*, *46*(2), 49-55.

Liu, Y., & Zheng, Y. (2021). Accurate Volume Calculation Driven by Delaunay Triangulation for Coal Measurement. *Scientific Programming*, *2021*.

Li, H., Zou, J., Li, Y., Xu, Y., & Xiao, Z. (2020, December). TOF Camera Array for Package Volume Measurement. In *2020 7th International Conference on Information Science and Control Engineering (ICISCE)* (pp. 2260-2264). IEEE.

Nassar, K., & Jung, Y. H. (2012). Structure-from-motion approach to the reconstruction of surfaces for earthwork planning. *Journal of Construction Engineering and Project Management*, *2*(3), 1-7.

Reina, G., Ojeda, L., Milella, A., & Borenstein, J. (2006). Wheel slippage and sinkage detection for planetary rovers. IEEE/Asme Transactions on Mechatronics, 11(2), 185-195.

Steeds, W. (1950). Tracked vehicles. *Auto. Engr*, April 1950,143–148; May 1950, 187–190; June 1950, 219–222.

Wong, J. Y. (2008). *Theory of ground vehicles*. John Wiley & Sons.

Zhu, W., Xu, H., Huang, Y., & Weng, L. (2021, December). Design of volume calculation algorithm for irregular objects based on reference plane for large scenes. In *International Conference on Computer Vision, Application, and Design (CVAD 2021)* (Vol. 12155, pp. 72-78). SPIE.

4812 / Multi-Fidelity Machine Learning Modeling for Wheeled Locomotion on Soft Soil

***Vladyslav Fediukov*[a,b]*, Felix Dietrich*[b], *and Fabian Buse*[a]**
[a]*Institute of System Dynamics and Control, German Aerospace Center (DLR), Weßling, Germany*
[b]*Institut für Informatik, TU München, Garching b. München, Germany*
**Corresponding author: vladyslav.fediukov@dlr.de*

https://doi.org/10.56884/WGPV6693

Abstract

Wheeled vehicles are the most convenient and widespread locomotion machines for the majority of research, industrial or private tasks. A perceptible share of wheeled vehicles is used on soft soil. Modelling wheel locomotion in these situations is challenging, because of the non-proportional relation between applied shear stress and the soil's deformation. Currently, various conventional simulation approaches are used to describe wheel–soil interaction, ranging from detailed numerical methods with particle-level simulations to simpler empirical models, where a big part of physical formulas are set up a priori, empirically. The ultimate wheel locomotion modelling tool should have high-quality onboard predictions but within a reasonable time. The trade-off is unachievable with the current simulation tools. In this project, we argue that using Machine Learning (ML) we can build a tool with the quality of high-fidelity and speed of lower-fidelity simulations. To fit this requirement, we are combining data from several models with different fidelities, in order to build a multi-fidelity ML model. In the model, forces and torques acting on the wheel are predicted using input data like the wheel's trajectory, surface and soil characteristics. The quality of this model will be validated by Terramechanics Robotics Locomotion Laboratory (TROLL) at Deutsche Zentrum für Luft- und Raumfahrt (DLR), a robotic single-wheel test bed designed to perform wheel–soil interaction experiments automatically. Early results show that, in simplified scenarios, our proposed method can be used to create efficient, multi-fidelity numerical models for locomotion prediction, including uncertainty estimation for the predictions.

Keywords: terramechanics, rover locomotion, machine learning, multi-fidelity

1. Introduction

In the exploration of planetary bodies, the usage of rovers has allowed extending the reach of exploration beyond the immediate vicinity of the landing site. However, in many of those missions, the traversal of soft, sandy areas has been particularly challenging. Extraterrestrial vehicles typically cannot be maintained and repaired during their operation. Any major mistake in movement could damage or incapacitate the rover. Thus, the ability to predict and analyze their locomotion in simulation is key in their operation, and to avoid critical situations like critical sinkage. Models for wheel—soil interaction must be sufficiently accurate and are required to cover many of the potential scenarios which wheel can encounter during extraterrestrial missions. Additionally, models must predict correct forces and torques quickly enough, either for scenario analysis by engineers or even for on-board operation. Providing accurate and fast results in these situations is challenging for current numerical simulation tools. Wheel-ground interaction is highly unpredictable in soft soils, due to their weakly cohesive inner structure and non-proportional relation between applied shear stress and soil deformation. Therefore, it is essential to conduct a large number of simulations, to test different wheel designs in different situations.

The range of simulations extends from lower-fidelity empirical simulations, which use many assumptions that lower approximation capabilities, to higher-fidelity particle simulations, which estimate physical processes by simulating them on a very detailed level. Choosing the proper simulation is usually a trade-off between the accuracy of approximation and time or computation costs. Lower-fidelity, empirical models like TerRA, presented in Barthelmes 2018, provide a simplified approximation of the forces acting on a wheel, in favour of very low computational cost. Medium-fidelity, semi-empirical models like the Soil Contact Model (SCM), presented by Buse 2018, provide better approximation quality, by introducing an explicit model of the soil deformation. However, SCM is already significantly more computationally expensive than TerRA. High-fidelity model Partsival, presented by Lichtenheldt, Ono, and Stubbig 2021, explicitly model the interaction between particles, which rapidly increases computational cost. For comparison, ten seconds of wheel movement in a Partisval simulation require up to several days on current GPUs, whereas the same simulation using SCM will take on the order of minutes, and, with the TerRA model, be completed in seconds.

While useful on their own, wheel-soil simulations can also provide data for Machine Learning (ML) tools. Our objective is to create a grey-box tool, which will be faster than the high-fidelity simulation but preserve the same accuracy level, which enables its efficient usage in extraterrestrial reconnaissance missions. ML has been well proven in terramechanical applications. For example, it was used for terrain classification and slip estimation by Gonzalez and Iagnemma 2018, as well as predicting lateral forces acting on the tire off-road by Dallas et al. 2019, or Stubbig and Lichtenheldt 2021 showed that Bayesian approach can be used to choose a more appropriate wheel shape.

In this paper, we propose a multi-fidelity approach to wheel locomotion prediction. Section 2 contains a description of the multi-fidelity method we use, together with its application in this project. Section 3 describes the theoretical background and current approaches to wheel locomotion simulation. In Section 4, we set up the problem and describe relevant use cases. The data sources we use for our ML methods are briefly discussed in Section 5. Preliminary results are presented in Section 6, and Section 7 contains our conclusions.

2. Methods

Multi-fidelity approaches can be applied if several data sources of different fidelity levels are available and there exists a correlation or non-linear relation between lower- and higher-fidelity levels. If the relation can be approximated accurately enough, predictions generated from lower-fidelity data sources can be mapped to predictions of high fidelity. Given predictions on all fidelity levels separately, for the same input, leads to a classical supervised learning problem for a map from the low- to the high-fidelity predictions. Approximation of this map can be done with various available machine learning methods. If the map between fidelity levels is significantly simpler than the map from input to high fidelity, fewer data points are required to approximate it, compared to a direct approximation of the high-fidelity predictions. Multi-fidelity approaches have already been applied successfully, for example, for predictions of the infection rate of the Plasmodium falciparum in Cutajar et al. 2019. The authors considered older data of malaria distribution as having lower fidelity, and newer distributions as having higher fidelity. They demonstrated that the multi-fidelity approach shows good results, even with a small amount of higher-fidelity training data points.

The following example from Lee et al. 2019 is a simple illustration of why multi-fidelity modelling can be successful. Consider the function $f_h(t) = sin^2(8\pi t)$ to be of "high fidelity," with a "lower-fidelity" approximation of it being $f_l(t) = sin(8\pi t)$. There is a simple but non-linear dependency between two functions, namely $f_h(t) = f_l^2(t)$. Fig. 1 shows that the high-fidelity function can be approximated accurately if the lower-fidelity function values are also used as an input.

This is possible with very few data points because only the simple, quadratic relationship must be learned. Conversely, as shown in Fig. 1, ML methods trained on the dataset $(t, f_h(t))$, i.e., without information on the lower fidelity, show poor results because the amount of training data points is not sufficient to capture the high-frequency oscillations.

In this project, we utilize simulations from several models, but of differing fidelity, as data sources. The simulations provide forces and torques acting on a single wheel. Starting with a lower-fidelity model TerRA, which uses analytical formulas to approximate the movement and is very cheap to simulate, therefore we possess a very large amount of these lower-fidelity simulations. Then medium-fidelity SCM, which uses surface-wheel interaction to compute the force, therefore makes the data-generation procedure longer and costlier, but still reasonable to harvest large datasets. The higher-fidelity simulation tool like Partsival computes forces, considering the effect of each particle on the wheel, which produces very precise outputs, but in extremely small amounts. The real-world testbed provides the most trustable data, but it increases data gathering time even more and restricts, for example, manipulations of gravity. In our view, successful application of multi-fidelity can result in force and torque predictions similar to the ones from high-fidelity models, but with much faster operating time.

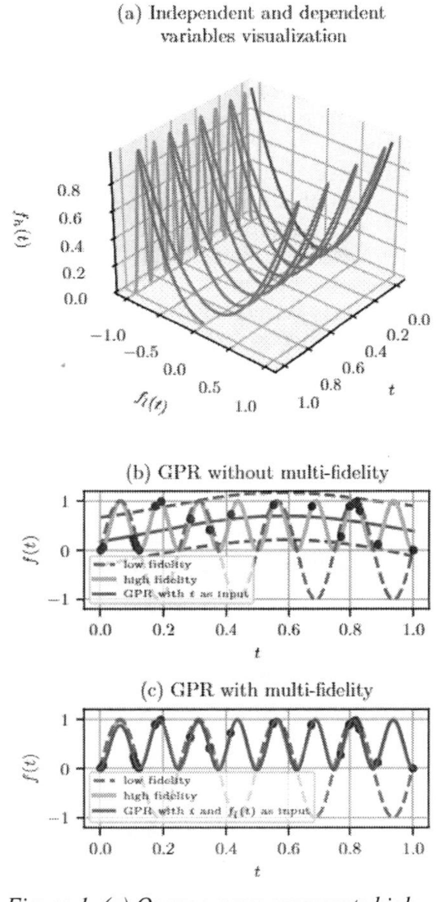

Figure 1. (a) Orange curve represents high-fidelity $sin^2(8\pi t)$ function, while the blue curve shows that f_l and f_h have quadratic dependency. And the green curve shows a two-dimensional manifold, which should be learned by the ML algorithm.
(b) The yellow curve is the high-fidelity data, the blue dots are training samples from the high-fidelity data, the dashed red line is lower-fidelity data and the green line is a prediction of the Gaussian Process Regression (GPR) with only t as an input (dashed green - variation from GPR representing prediction's uncertainty).
(c) When GPR takes as an input both the t and f_l, then prediction and uncertainty are improving considerably. This example is adapted from Lee et al. 2019.

In addition, important part of statistical inference is uncertainties estimation. Hüllermeier and Waegeman 2021 argues that estimation of total uncertainty and its further classification into natural (aleatoric) and knowledge (epistemic) uncertainties should improve understanding and quality of statistical inference because we can reduce later with a better model design. Understanding uncertainties will provide a more nuanced picture for the users of this model.

3. Related Work

Modelling wheel locomotion over soft terrains is a hard task because there is no exact formula for this. To model a robot's locomotion correctly, engineers should simulate shear and normal stress beneath the wheel, sinkage into the ground, slippage and ground resistance. Modelling the locomotion means predicting forces and torques affecting the wheel during its planned route. Predictions should be created with considerations of the soil type, soil surface, wheel type and environmental features (gravity, etc.). Of greatest interest for prediction is the traction force of a wheel or torque in the perpendicular dimension, from which traction force can be derived. As shown in J. Y. Wong 2008, the traction effort of a wheel is produced by the shearing of the terrain. Therefore, the majority of the existing models were approximating traction force with the empirical information about pressure on the soil, soil shear and soil displacement.

First theoretical approaches were introduced by Bekker 1962, where his formula was an exponential dependency of pressure from sinkage, reversed linear from the area of contact and a set of empirically estimated coefficients. From the estimated pressure, applied stress and consequently soil displacement and traction force can be derived. Also, the pressure-sinkage relationship is used to characterize the soil strength and therefore soil behavior during wheel-soil interaction. But this simplicity and several accompanying assumptions, like one that the soil should be perfectly cohesionless, make it not applicable for precise predictions of the traction forces, torques and therefore wheel's locomotion. Another model was introduced by Janosi, Hanamoto, and Galileo Ferraris 1961, where the model expresses shear stress through the normal stress and internal friction angle of the soil. According to Mohr 1914, the Mohr-Coulomb failure criterion for the operatable wheel, shearing stress should be less than a shear strength, to avoid shear failure. Terzaghi and Peck 1948 developed a method for determining bearing capacity, i.e., maximum level of the shearing stress the soil can sustain, which is a function of the internal shearing resistance angle of the soil and friction between the soil and the wheel. Based on two relations, pressure-sinkage and shear stress-shear displacement, J. Y. H. Wong and Reece 1967 has created one of the most popular nowadays wheel-terrain modelling techniques. It is an extension of the Bekker model and it integrates normal and shear stress distribution over the soil-wheel contact area. These models are ubiquitous in off-road wheel locomotion modelling. However, these models are considered lower-fidelity approaches for estimating the traction forces, torques and example, the classical empirical Bekker-Wong-Reece model is invalid once the wheel is deeply embedded into the soil, because resistance force, which is dependent on the wheel's sinkage, became unpredictable because of the bulldozing effect that sinkage provokes and which is unmodeled in this type of simulations. In this empirical model, the contact point creates an analytical plain, and the wheel is always positioned on the plane and cannot go under it. Therefore, the wheel's locomotion could not be predicted for a wheel in an embedded position, which occurs in a substantial number of situations. The highest fidelity level belongs to the numerical methods, which model wheel-soil interaction with different levels of detail. Between exists medium-fidelity levels, semi-empirical models, which are using more computations to estimate more physical processes, but still make a substantial amount of assumptions about underlying physical processes. The large variate of existing models shows that there is no universal approach for this problem. The ultimate goal is to design a tool, which would consider all existing approaches, from each level of fidelity and use for each situation maximum required level of fidelity. Using statistical inference based on data samples from each fidelity level, create the model which would predict wheels traction force and driving torque based on input data from different models.

3.1 Terramechanics for Real-time Application (TerRA)

TerRA is an empirical method, designed to be computationally fast enough for onboard applications. This model omits explicit soil deformation to allow fast calculations and relies on efficient approximations. For example, the effect of grousers is approximated by extending the wheel radius. Force calculation is separated into two parts: normal and tangential forces. The normal forces follow an elasto-plastic behavior with non-linear spring characteristics and modified linear damping to depict effects like slip sinkage. The tangential force is further divided into resistance and traction forces. Resistance is approximated from the passive earth pressure, whereas traction is a function of the shear stress acting on the wheel's surface. The shear stress is approximated from an estimated slip length.

3.2 Soil Contact Model (SCM)

SCM is a semi-empirical method, which provides more accurate information about the forces acting on the wheel, by explicitly modelling soil deformation. The soil is represented by an equidistant grid of nodes, where each node height is modified throughout a simulation to depict soil deformation. Additionally, internal soil states reflect the soil conditions.

To approximate the resulting deformation and soil flow velocity a global flow field is constructed by superpositioning local flow fields, each describing the soil flow introduced from individual points in contact. The local fields themselves are constructed from 2D chords, which describe the direction of the flow, the depth of the flow, and the propagation angle in a 2D cross-section outwards from a contact point. From the resulting soil flow velocity, a shear length describing the soil condition is derived. The normal and shear stress acting on the wheel is then derived from this shear length and the present geometry.

SCM provides a better approximation of the forces and torques affecting the wheel during the movement and does not increase computation time drastically, but still has some drawbacks.

3.3 Partsival

The Discrete Element Method (DEM) is a particle-based method, for example, presented by Knuth et al. 2012. It can express large plastic deformations, soil fracture and soil flow around wheel's grousers.

As DEM is based on fewer assumptions and follows a more physical approach higher quality prediction can be expected. The soil is represented as a set of particles that interacts using contact laws. For this project, Partsival is used, DLR's software tool which implements DEM. Partsival implements three contact models: normal (which uses Hertzian contact force), tangential (which implements stick-slip friction) and rolling (rolling resistance, which is based on two previous forces). High detail of the wheel-soil interaction is achieved by using pseudo-particles. Smaller sizes of pseudo-particles correlate to a more detailed depiction of the soil flow. However, this significantly increases computational costs.

Models like Partsival simulation can provide insights into the potential behavior in low-gravity applications as their physical approach makes these extrapolations feasible. In these applications, cohesion forces based on Van-der-Waals, electrostatic and self-gravitation forces play a larger role.

4. Problem Setting

At DLR we are interested in using wheeled rovers in space missions. Very limited maintenance could be provided on an extraterrestrial object. Therefore, it is very important to ensure the predictability of the rover wheels' movements in all potential situations, on the most popular types and forms of the surface, to avoid critical situations. A list of relevant scenarios for wheel locomotion could be found in the Table 1. For example, Scenario III mentioned in the Table 1 could be seen on the Fig. 2c. Opportunity rover is going uphill on the sandy Martian dune and its wheel is embedded up to 80% of its diameter into the sand. This situation is potentially critical; therefore researchers and engineers should be able to model them before the mission, but also during the actual mission. The models should predict the drawbar pull traction force and forces in other directions, using input information about the wheel speed, angular wheel speed, soil surface and soil types. In this way, we can model the wheel's behavior and predict possible harmful situations, like excessive slippage or sinkage. The ability to predict the forces and torques acting on the wheel is important both for the wheels' design and for the behavior predictions of the rover. Precise and relatively cheap predictions of the forces could be used both for day-to-day predictions or online movements and for the strategic decision of the rover's movements.

Maintenance and retraining resulting ML pipeline is easy, because each time a small dataset of new higher-fidelity data points will be introduced, we would need to only retrain a small part of the entire pipeline.

Table 1. Overview of the selected scenarios

No.	Scenario	Description
I	Horizontal movement	Straight traverse of mostly flat terrain at a constant velocity. Most popular movement, plain terrain depicted on Fig. 2b
II	Steering	Steering both with an overlayed horizontal velocity and while stationary
III	Up-/Downhill	Longitudinal traverse of inclined terrain. For example, on Fig. 2a
IV	Tilting	Lateral traverse of inclined terrain. Example is on Fig. 2d
V	Acceleration	Acceleration or deceleration on flat terrain
VI	Obstacle Interaction	Increased resistance due to some interaction at another part of the rover
VII	Impacts	Vertical impacts into soil, for example after driving over an obstacle

5. Data Sources

As a source for training data as well as for use in the multi-fidelity setup the terramechanical models TerRA, SCM, and Partsival are used, testbed data is not yet incorporated in this work.

To use those models in the proposed machine learning scheme it is required to define a representation of the in- and outputs that can be applied to each model and it is sufficiently comprehensive to fully describe the interaction. Further, it is desired that the inputs in this representation are conceivable as data that could be provided by the onboard sensors of a rover.

Fig. 3 shows a 2D representation of the selected in- and outputs. Table 2 provides a description of the variables. In and outputs to the ML model as well as the data captured when using one of the terramechanical models are all described in the wheel ($_w$) coordinates system. This system is positioned in the center of the wheel, its y-axis is aligned with the rotational axis of the wheel and its x-axis is constructed so that it is parallel to the wheel reference plane. This plane is the best-fit plane in a circular area with a radius equal to the wheel radius. The origin of the surface reference system ($_s$) is the intersection of the wheels system z-axis and the plane.

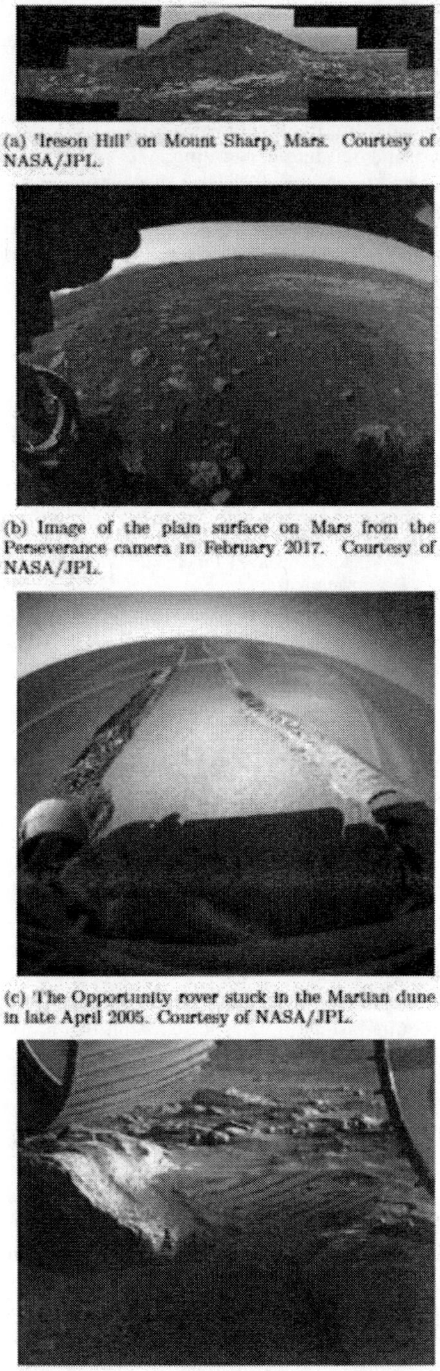

(a) 'Ireson Hill' on Mount Sharp, Mars. Courtesy of NASA/JPL.

(b) Image of the plain surface on Mars from the Perseverance camera in February 2017. Courtesy of NASA/JPL.

(c) The Opportunity rover stuck in the Martian dune in late April 2005. Courtesy of NASA/JPL.

(d) Perseverance tilts itself, to get rid of the interfering pebbles, January 2022. Courtesy of NASA/JPL.

Figure 2. Illustrations of the Martian terrain to consider as potential encounters

The selected inputs are the position of the wheel $\vec{r}_w = \vec{o}_w - \vec{o}_s$ with respect to the surface system. Further inputs are the velocity $\vec{v}_w$ and angular velocity $\vec{w}_w$ of the wheel. To provide information on the orientation of the wheel the gravity direction is provided by $\vec{n}_g$. To extend information above the surface geometry beyond simple sinkage information, which could be extracted from $\vec{r}_w$, a 64 x 64 pixel height-map $\boldsymbol{H}_s$ of the surrounding surface is provided. Each pixel describes the distance between the *x-y* plane of the wheels system to the surface along the wheel system *z*-axis. In all cases, the heightmap is based on the un-deformed surface as this mimics the information that could be extracted from 3D reconstructions of the surface provided by the rover's navigation cameras. Fig. 4 shows the generated heightmap and corresponding simulation.

In total the inputs to describe the interaction are:

$$\boldsymbol{X} = \{\vec{r}_w, \vec{v}_w, \vec{w}_w, \vec{n}_g, \mathbf{H}_s\} \quad (1)$$

The outputs of each model are the forces and torques acting on the wheel, however, depending on the used model they are represented differently.

Forces $\vec{F}_w$ and torques $\vec{T}_w$ applied to the wheel represented in the wheel system are used as an output:

$$\boldsymbol{Y} = \{\vec{F}_w, \vec{T}_w\} \quad (2)$$

To generate useful data as a foundation for the machine learning approach the input data has to be created on randomized terrains. Two aspects of the terrain need to be treated separately, the local small-scale unevenness as well as the large-scale trend. The first needs to reflect the ubiquitous surface roughness present on natural terrain. The second reflects the scenarios introduced in Table 1.

To generate the local unevenness Perlin noise with an amplitude on the cm scale is used. The large-scale surface geometry is defined by randomly generated surface profiles. Fig. 5 shows two examples with the generated terrain. With this setup, it is possible to generate large data sets with well-defined inputs.

Referencing back to the scenarios introduced in Table 1 it is to note that not all simulation models are equally well suited to cover all situations. In general, the more dynamic soil deformation is involved in a scenario the more sophisticated the model is required to model it adequately.

Table 2 shows a categorization of the different tools concerning the different scenarios.

Three categories are used "-" indicating that a model is unsuited to simulate the given scenario, "+" indicating a sufficient depiction of the given situation and "++" indicating a good depiction. Additionally, cases marked with an "!" indicate that the applications of this model in this situation are deemed excessive and wasteful in terms of computation resources.

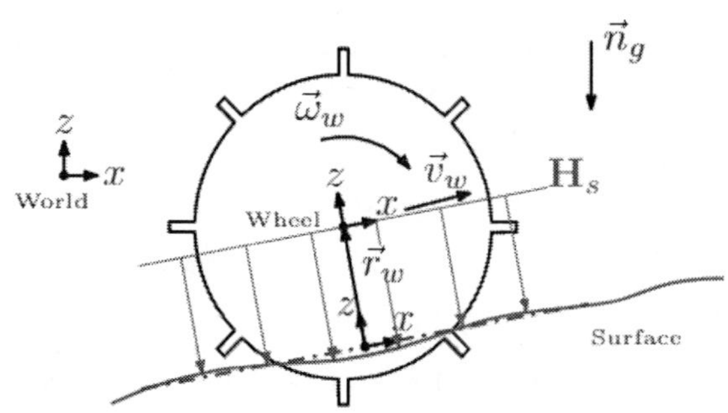

Figure 3. Representation of the input data used for in the ML model

To note is the split evaluation of TerRA in the steering scenario (II), as TerRA is deemed viable when the steering is overlayed with a horizontal movement but not when the steering is performed while stationary. The missing depiction of soil deformation, as shown in Fig. 6, leads to an underestimation of the steering resistance predicted by TerRA (Fig. 6a). Both SCM (Fig. 6b) and Partsival (Fig. 6c) show a buildup of material which subsequently leads to increased resistance. The impact scenario (VII) is only well captured by the DEM model, as all other models do not explicitly cover dynamic events like impacts. The data generation concept is to use the tool with the highest rating that is not deemed excessive as a primary data source for these situations.

To use data from different sources in a single setup it is required that the sources have some kind of relationship for identical inputs. To demonstrate this, we used linear Pearson correlation between TerRA and SCM the results, based on 20 simulation runs.

$$\rho = \frac{E((F_{TerRA}-\mu_{TerRA})-(F_{SCM}-\mu_{SCM}))}{\sigma_{TerRA}\sigma_{SCM}} \qquad (3)$$

where μ is forces' mean, σ is standard deviation and E is mathematical expectation. A correlation of 1 would indicate perfect correlation and 0 shows no correlation. Resulting correlation $\rho = 0.76$, indicates a perceptible linear correlation. Pearson correlation is a simplest way to roughly estimate features significance for the prediction process, however it doesn't show exact significance, while we are not using linear models.

This type of correlation doesn't capture the entire non-linear relationship between the two simulations, but it demonstrates they are correlated even on such a basic level. The results are shown in Fig. 7, the top plot shows the corresponding simulated traction force of SCM plotted against the traction force simulated by TerRA. Points on the diagonal indicate an identical simulation. The lower plot shows one example run. To note is the higher correlation of forces with lower magnitude. This variation is the result of the missing explicit depiction of grousers in TerRA. In simulations performed with SCM, the grousers lead to an oscillating behaviuor which is averaged out in TerRA. With increasing forces, correlation decreases. In a high-dynamic zone, grousers in SCM simulations introduce fluctuation, while TerRA simulation in the same situation shows steadily increasing force (x-axis).

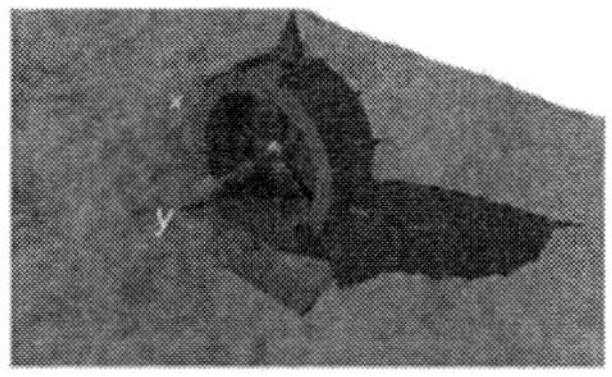

(a) Surface during simulation

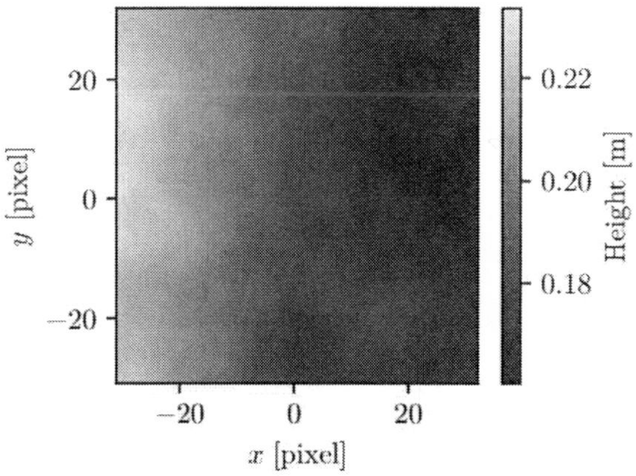

(b) Height map visualization

Figure 4. Simulated terrain and corresponding generated 64×64-pixel height map $\mathbf{H}_s$. *As in the shown situation the wheel is position on a tilted inclination, the values at high x and y values are the lowest*

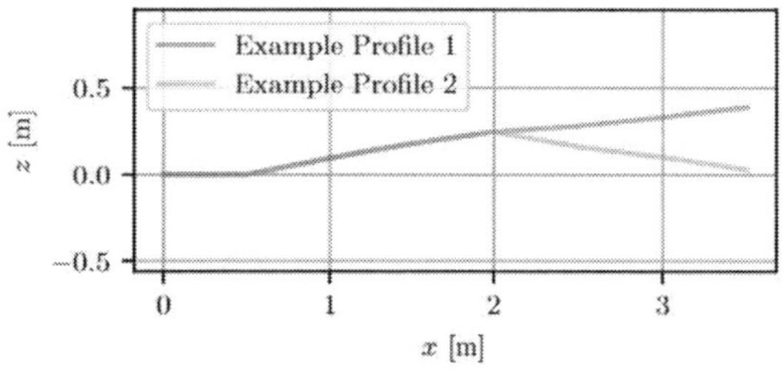

(a) Surface profile definition

(b) Generated terrain corresponding to profile 1

(c) Generated terrain corresponding to profile 2

Figure 5. With these randomly generated surface profiles, we tried to create diverse space of possible wheel movements, for the combination of lower- and medium-fidelity simulations

Table 2. Variables description

Variable	Specification	Description
$\vec{r}_w$	Input	Wheel position w.r.t the surface
$\vec{v}_w$	Input	Wheel velocity
$\vec{w}_w$	Input	Angular velocity of the wheel
$\vec{n}_g$	Input	Gravity direction
$\boldsymbol{H}_s$	Input	Height-map of the surroundings
$\vec{F}_w$	Output	Forces acting on the wheel
$\vec{T}_w$	Output	Torques acting on the wheel

Table 3. List of potential situations and respective simulations. The table assesses the applicability of each model to the introduced scenarios

Scenario	TerRA	SCM	Partsival
I	+	++	++!
II	-/+	+	++
III	-	+	++
IV	-	++	++!
V	-	+	++
VI	-	++	++!
VII	-	-	++

(a) Steering simulation performed with TerRA, no visible surface defromation.

(b) Steering simulation performed with SCM, surface defromation leads to successfive increase in resistance.

(c) Steering simulation performed with Partisval, surface defromation leads to successfive increase in resistance.

Figure 6. Comparison of the steering maneuver in different simulations.

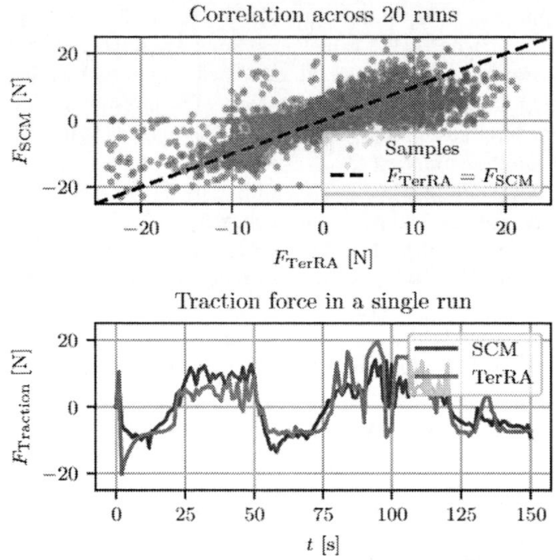

Figure 7. Top plot shows the correlation between TerRA and SCM simulation results. Each sample shows TerRA and SCM traction forces with identical inputs. The bottom plot shows one of the 20 runs

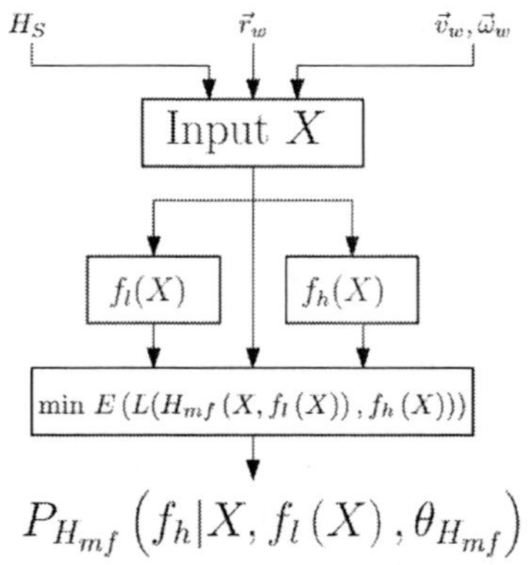

Figure 8: Training procedure

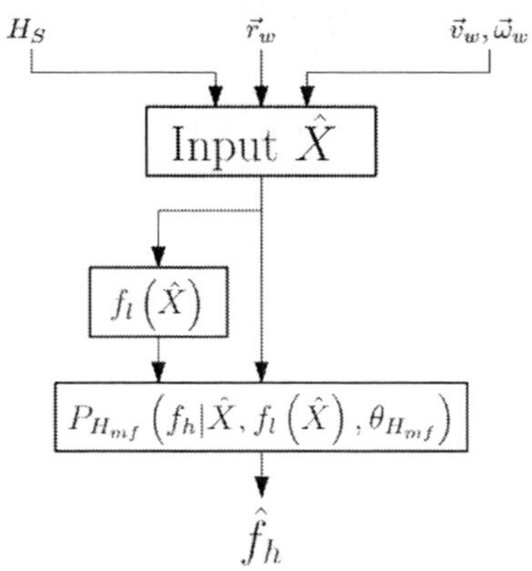

Figure 9. Application procedure

6. Preliminary results

1000 data points from the lower-fidelity TerRA simulation and 100 data points from the medium-fidelity SCM simulation were used to train the model depicted on Fig. 10. All points were randomly chosen from the dataset created on the randomly generated surfaces. The amount of data points was selected empirically, as more points increase costs without a noticeable increase in accuracy. As a reference, a one-wheel SCM simulation was run. Its resulting trajectory was then used as a constrained movement for TerRA, to compute its outputs on exact same inputs.

The training pipeline is depicted on Fig. 8. Given predetermined inputs, like soil characteristics, trajectories and velocities of the wheel, we can create input data $\boldsymbol{X}$ and outputs of the lower- and higher-fidelity simulations, $f_l(\boldsymbol{X})$ and $f_h(\boldsymbol{X})$ respectively, which are simulated forces and torques, $\boldsymbol{Y}$. Then we create an MF ML model, which represents our hypothesis, i.e., belief that the true underlying relationship is described by the chosen model and is denoted by H_{mf}. In this case, the MF model was trained by Non-Linear Auto-Regressive Gaussian Process, presented in Lee et al. 2019, where the nonlinear auto-regressive part comes from the GP's kernel structure.

It represents auto-regression by kernel multiplication, where the first kernel represents an approximation of lower-fidelity function and the second takes the first kernel as prior and represents the mapping between lower- and medium-fidelity outputs. MF ML model takes $\boldsymbol{X}$ and $f_l(\boldsymbol{X})$ as inputs. Then the loss function L is determined. This depends on the requirements, but for this project, we have used Mean Squared Error. Then, using optimization techniques, the expectation E of loss functions is minimized. After this minimization we achieve set of model's parameters $\theta_{H_{mf}}$ and distribution $P_{H_{mf}}$, which predicts values of f_h based on our hypothesis. After the model is trained, we are again determining desired inputs, running them through the lower-fidelity model. Then using distribution from the trained MF ML model H_{mf}, we are making prediction $\hat{f}_h$ which should approximate real higher-fidelity simulations of real data f_h, see Fig. 9. Preliminary results show that multi-fidelity can achieve reasonable results, which could be seen on Fig. 10, where the MF ML model approximates traction force (F_x from $\boldsymbol{Y}$). Using this model, we can achieve comparable results with the actual SCM simulation, but with limited usage of SCM itself. For example, for this MF prediction, only 100 SCM points (green line on Fig. 10) were used, while running SCM simulation (blue line on Fig. 10) over this entire run requires 300 SCM calls. The shown result of the MF model is a better approximation of SCM than TerRA. A threefold decrease in calls of the higher-fidelity function makes a big advantage, especially if we want to approximate DEM simulations like Partsival, which is very costly. On the other hand, training the ML model on the 100 medium-fidelity training points (black line on Fig. 10) will not bring desired performance. Adding multi-fidelity to the Gaussian process, which tries to approximate SCM forces, decreases MSE considerably, from 91.6 to 21.7. Also on the plots the uncertainty estimates of each method can be seen, where light-green and -grey shaded areas are 95%-confidence intervals, showing ranges, where 95% of all potential forces values could be, according to the trained model. With the more accurate mean prediction, multi-fidelity improves the certainty of the model. The MF method is completely developed and applied for reference models of two different fidelity levels, namely TerRA and SCM. The same procedure can be repeated and applied to the combination of SCM with higher-fidelity data, DEM-simulation like Partsival or testbed experiments. Then, the MF ML model is lifted to the next fidelity level. Development of the sustainable and informative higher-fidelity dataset, from both Partsival and TROLL, is in progress.

Taking into conclusion all the above, MF modelling seems relevant for this problem, where semi-empirical and analytical methods can give us a solid basis for the simplest movements of the wheel and small amounts of higher-fidelity numerical simulations or real experiments will provide missing complexity, but its final evaluation is work in progress.

7. Conclusion

This paper provides an overview of the current tasks in the field of terramechanics, a more specific look at the available data in our laboratory, and about potential utilization of this data. A wide variety of data covering a diverse set of locomotion problems creates opportunities to apply ML algorithms. We have shown that wheel behavior in lower-fidelity simulations has a visible relationship with the wheel's behavior in higher-fidelity simulations. We also classify different scenarios to emphasize the variability of challenging problems in wheel-soil interaction. We argue that using simulation data from many different scenarios, optimizing computation efficiency and approximation precision, and multi-fidelity modelling can lead to a combined approach that can be both fast and precise. For example, there is no need to use high-precision simulations for simple tasks, like driving forward on plain terrain. In these scenarios, we thus need only a few high-fidelity simulations for enhancing and refinement of a large amount of low- and medium-fidelity data. For rare and extreme scenarios, a larger amount of expensive, high-fidelity particle simulations will be necessary. The multi-fidelity approach was successfully used in other domains, so we argue that with the right task design, it can also be employed to solve terramechanical tasks. Successful implementation of multi-fidelity modelling, based on different machine learning and physical models, creates a robust model that produces accurate and fast predictions for a wide variety of off-road wheeled vehicle scenarios. It will also be helpful for the operation of extraterrestrial vehicles, where manual maintenance is unavailable and therefore engineers can rely only on simulations.

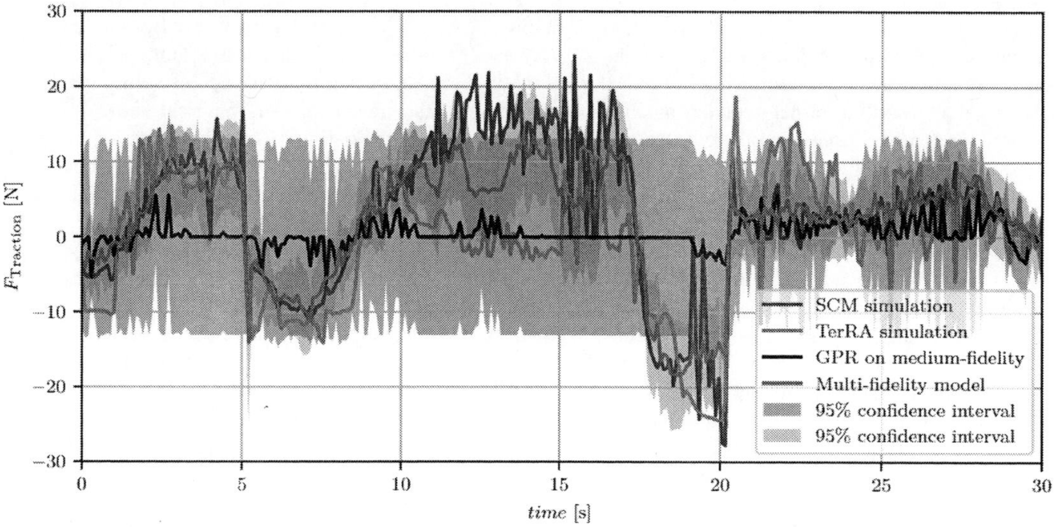

Figure 10. Example of multi-fidelity and single-fidelity ML approximations of a SCM simulation in comparison

8. Acknowledgements

This research is supported by the Helmholtz Association under the joint research school "Munich School for Data Science–MUDS."

The co-author of this paper, Felix Dietrich, is funded by the Deutsche Forschungsgemeinschaft (DFG, German Research Foundation)—468830823.

9. References

Barthelmes, Stefan (June 2018). "TerRA: Terramechanics for Real-time Application." In: *5th Joint International Conference on Multibody System Dynamics*. url: https://elib.dlr.de/121815/.

Bekker, Mieczysław G. (1962). "Theory of Land Locomotion: The Mechanics of Vehicle Mobility." In. Buse, Fabian (Oct. 2018). "Using superposition of local soil flow fields to improve soil deformation in the DLR Soil Contact Model-SCM." In: *5th Joint International Conference on Multibody System Dynamics*. url: https://elib.dlr.de/121794/.

Cutajar, Kurt et al. (2019). *Deep Gaussian Processes for Multi-fidelity Modeling*. arXiv: 1903.07320 [stat.ML].

Dallas, James et al. (2019). Neural network based terramechanics modeling and estimation for deformable terrains. Tech. rep. US ARMY CCDC GVSC.

Gonzalez, Ramon and Karl Iagnemma (2018). "DeepTerramechanics: Terrain Classification and Slip Estimation for Ground Robots via Deep Learning." In: *arXiv e-prints*, arXiv–1806.

Hüllermeier, Eyke and Willem Waegeman (2021). "Aleatoric and epistemic uncertainty in machine learning: An introduction to concepts and methods." In: *Machine Learning* 110.3, pp. 457–506.

Janosi, Zoltan J., Ben. Hanamoto, and Istituto elettrotecnico nazionale Galileo Ferraris (1961). "The analytical determination of drawbar pull as a function of slip for tracked vehicles in deformable soils." In. Knuth, Margaret A., et al. (2012). "Discrete element modeling of a Mars Exploration Rover wheel in granular material." In: *Journal of Terramechanics* 49, pp. 27–36.

Lee, Seungjoon, et al. (2019). "Linking Gaussian process regression with data-driven manifold embeddings for nonlinear data fusion." In: *Interface focus* 9.3, p. 20180083.

Lichtenheldt, Roy, Shoko Ono, and Leon Stubbig (2021). "Large scale discrete element simulation campaigns—Simulating extraterrestrial soils in PartsiVal." In: *14th World Congress on Computational Mechanics (WCCM)*. url: https://elib.dlr.de/141296/.

Mohr, Otto (1914). *Abhandlungen aus dem Gebiete der technischen Mechanik.* W. Ernst & Sohn.

Stubbig, Leon, and Roy Lichtenheldt (2021). "Optimizing the Shape of Planetary Rover Wheels using the Discrete Element Method and Bayesian Optimization." In: *VII International Conference on Particle-Based Methods* (PARTICLES 2021). Ed. by Peter Wriggers et al. International Center for Numerical Methods in Engineering (CIMNE). url: https://elib.dlr.de/148275/.

Terzaghi, Karl, and Ralph Peck (1948). "Soil Mechanics." In: *Engineering Practice. John Wiley and Sons, Inc., New York.*

Wong, Jo Yung (2008). *Theory of ground vehicles.* John Wiley & Sons.

Wong, John Y. H., and A. R. Reece (1967). "Prediction of rigid wheel performance based on the analysis of soil-wheel stresses part I. Performance of driven rigid wheels." In: *Journal of Terramechanics* 4, pp. 81–98. Paper 4812 | Page 10

4827 / Introducing Polibot: A High Mobility Tracked Robot with Innovative Passive Suspensions

Rocco Galati, Andrea Grazioso, Angelo Ugenti, Giacomo Mantriota, and Giulio Reina*
Department of Mechanics, Mathematics and Management, Politecnico di Bari, Bari, Italy
**Corresponding author: giulio.reina@poliba.it*

https://doi.org/10.56884/QYUY9071

Abstract

In the context of autonomous off-road mobility, tracked robots feature larger contact areas compared to equivalent wheeled vehicles, which turn into improved traction especially on soft terrains. This paper discusses a novel tracked robot, named Polibot, that is outfitted with passive articulated suspensions. The multibody model is developed within the MSC Adams environment leveraging on the Adams Tracked Vehicle (ATV) Toolkit. The different subsystems composing the vehicle assembly are outlined, with specific attention to the modeling strategies of the swing arms constituting the suspension system, as well as the characterization of the track belt and inertial properties. Efforts in validating the multibody twin against the real prototype are also presented, showing a good agreement in experiments performed on asphalt.

Keywords: multibody modeling and analysis, rough-terrain mobile robots, tracked robot, passive articulated suspension

1. Introduction

Tracks are known to outperform wheels in terms of improved traction over both steep hard surfaces and deformable terrains, and better obstacle negotiation. Notable examples proposed in the literature to further improve mobility in tracked robots are, for example, PackBot (Yamauchi, 2004), Quince (Yoshida et al., 2014), and Mothership (Huff et al., 2012). New architectures need to be investigated to push forward the next generation of tracked vehicles (Galati et al., 2021). In this context, Polibot (contraction of POLItecnico of Bari's roBOT) represents a novel contribution. Highlights of the proposed vehicle refer to the adoption of an articulated suspension that ensures high load capacity, vibration reduction, and negotiation of highly challenging terrains. Polibot that is shown in Fig. 1 has a footprint of 1.5 × 1 m and a weight of about 70 kg with a payload up to 40 kg. The control and acquisition system has been implemented under ROS (Robot Operating System). The robot is outfitted with a rich suite of sensors, including sensors to measure the electrical currents drawn by the drive motors, a GNSS navigation system supporting RTK cm-accuracy, a range finder and various monitoring cameras (Milella and Reina, 2014).

This paper focuses on the development of the multibody (MB) model of Polibot. Leveraging on the increasing development of computational resources, numerical simulations have become an important tool to predict the behavior of complex mechanical systems (Balena et al., 2021). Tracked vehicles are examples of flexible multibody systems with challenges like frictional contacts, track-terrain interaction, large deformations of the track. In this research, the Polibot MB model is developed within MSC Adams using the toolkit Adams Tracked Vehicle (ATV) that is under testing and development at the Politecnico of Bari's Robotic Mobility Lab.

Figure 1. Polibot: a tracked robot with passive articulated suspensions

2. Virtual multibody twin

The multibody model of the Polibot prototype is developed by using the plugin ATV of MSC Adams, which offers an environment specifically dedicated to the mechanical simulation of tracked vehicles. Following the template-based design methodology proper to the used software, the rover is divided into six major subsystems as shown in Fig. 2: the hull, the powertrain, the front, central and rear suspension unit together with their road wheels and the track belt. These subsystems are modelled independently from each other in such a way as to combine correctly to form the assembly to be simulated. Despite the set of elementary geometries available within the ATV toolkit, particular care is taken in importing into MSC Adams the Parasolid files of the real CAD geometries and assigning them to the corresponding parts of the different subsystems.

The powertrain comprises, on each side of the robot, a gearbox whose input shaft is driven by an electric motor, while the output shaft carries the sprocket that engages with the track belt. The angular speeds of the two shafts are constrained to each other to reflect the gearbox transmission ratio; either a prescribed motion or torque can be imparted to the input shaft, the sprocket being consequently made to rotate. To allow for the transmission of motion, contact force instances are created to simulate the insertion of the thirteen sprocket teeth into the holes of the track belt.

All the four ground wheels of the rover are suspended with respect to the hull by means of trailing arm suspensions. In particular, the front and the central suspension units, shown in Fig. 3, are specifically modelled to reflect some features that are not present in the standard library of the ATV shared database. The front suspension arm shown in Fig. 3a is composed of two parts: the first one, hinged to the hull via a revolute joint, is free to rotate about the attachment point; the second rigid body is connected to the first by means of a translational joint that allows it to slide back and forth, thus varying the overall length of the arm itself. The sliding motion due to such a constraint is necessary to define the tensioner entity of the ATV toolkit between the two parts of the front arm, resembling the screw-nut tensioning system of the real prototype. The possibility to adjust the track belt tension is crucial to keep the track segments in contact with the inner running gear of the track system. The two front road wheels are carried by the sliding part of the suspension arm, whose rotation is opposed by a linear spring-damper force that represents the shock absorber attached to the hull.

The central road wheels represented in Fig. 3b are carried by two swing arms hinged to the hull at the same point by means of revolute joints. A linear spring-damper force simulate the shock absorber acting between the two arms, such that their rotations are not independent: when one wheel is lifted, the other is pushed against the ground, thus providing a system capable of better adapting to the unevenness of the terrain being traversed and redistributing the pressure over the track contact area. The stiffness and damping coefficients of the linear spring-damper forces that represent the shock absorbers of the front, central and rear suspension units are set equal to 12 N/mm and 0.5 Ns/mm respectively. The front and the central springs are preloaded, the former with 120 N and the latter with 60 N, while the rear one is not preloaded.

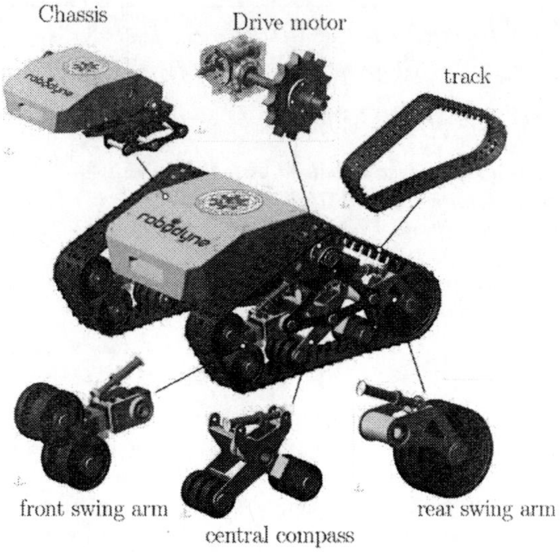

Figure 2. Multibody model of Polibot and its subsystems

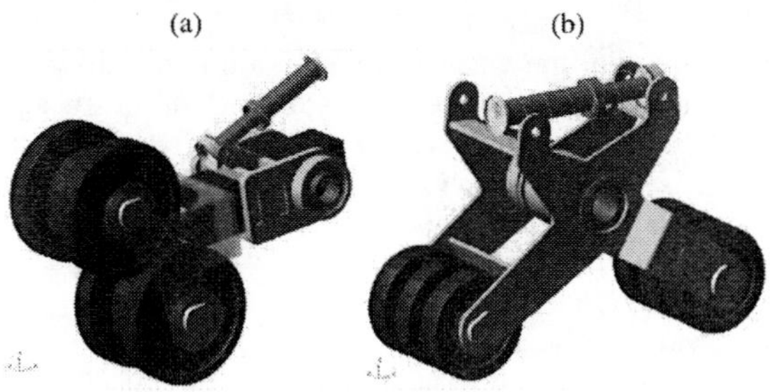

Figure 3. Front a) and central b) suspension units and wheels subsystems

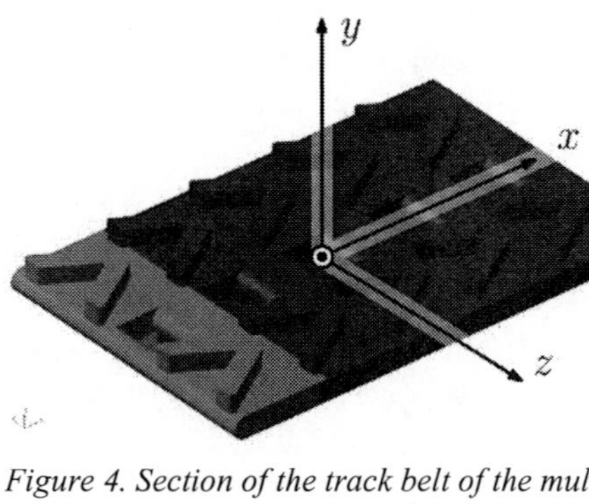

Figure 4. Section of the track belt of the multibody model of Polibot and coordinate system for the representation of the connection forces between the segments

The Polibot prototype is equipped with rubber made tracks reinforced with steel cable souls. Within the ATV toolkit the tracks are discretized into a finite number of rigid segments linked to one another through compliant connections that are created during the automatic wrapping of the track system around the rolling wheels (see Fig. 4). When creating the track segments, it is possible to choose between two different options: steel type and rubber type. If the steel track type is selected, as done in (Grazioso et al., 2022), the segments parts are connected to each other by pin connections and a limitation of the relative motion between adjacent elements is posed through a torque impact. In the present paper the rubber type is chosen, as it is more suitable for modelling the tracks of Polibot: in this case the segments parts are connected to each other with a 6 × 6 force field element whose stiffness matrix is that of the Timoshenko beam, the damping matrix being calculated as a fraction of the stiffness matrix. The elastic and damping forces and torques exchanged by two consecutive segments depend on the relative displacement and velocity of the frames associated with the frames, as well as on the translational, rotational, and cross coefficients in the stiffness and damping matrices. Each side of the multibody model of Polibot presents a track made up of 56 segments, each of which is 18 mm long, 120 mm wide and 9 mm high. Based on such dimensions and assuming the E-Modulus and G-Modulus of the rubber to be equal to 9 MPa and 3 MPa respectively, the rubber field properties are calculated (see Table 1).

Table 1. Properties of the rubber field used for the track segment connections

	x	y	z
Trans. stiffness [N/mm]	5.3E+04	69.4	147.4
Trans. damping [Ns/mm]	534.0	0.7	1.5
Rot. stiffness [Nmm/deg]	3.7E+03	1.1E+06	163.3
Rot. damping [Nmms]	37.5	1.1E+04	1.6

3. Inertial properties

One of the first requirements for the modeling of a mobile robot such as Polibot is to define its inertial properties. This section describes the procedure followed to identify the position of the center of gravity with respect to a reference system located at the center point of an imaginary line connecting the centers of the two sprockets, as indicated in Figs. 5 and 6.

A first test is designed to estimate x_G, that represents the x coordinate of the center of gravity, using the suspension and ground reaction method. Note that the procedure here described is usually applied to wheeled vehicles, while being controversial for tracked vehicles, for which the ISO Standard 789-6 prescribes the use of decking and knife edges. The objective is to suspend the robot in such a way that only the four front and rear wheels are exchanging forces with the ground, to measure F_F and F_R, that indicate the fractions of the rover weight that loads the front and rear wheels respectively, as shown in Fig. 5.

To achieve this result, after locking the suspension to avoid internal configuration variations, the rover is lifted onto two bars, with the front (left and right) wheels resting on the first bar and the rear wheels on the second one, to emulate the front and rear axle of a vehicle. Each bar is placed on a scale, and once F_F and F_R are known, x_G can be estimated as follows:

$$x_G = b' - \frac{F_F \cdot l}{W} = b' - \frac{F_F \cdot l}{F_F + F_R} \tag{1}$$

where W is the robot weight, b' and l are geometric parameters as indicated in Fig. 5.

A similar procedure is implemented for the estimation of the y coordinate of the center of gravity, y_G. The robot is settled onto two bars, but in this case the bars are placed parallel and underneath each track, to measure F_{left} and F_{right}, that represent the fractions of the rover weight that loads the left and right tracks respectively, as shown in Fig. 6. Each bar is placed on a scale, and once F_{left} and F_{right} are known, y_G can be estimated as follows:

$$y_G = \left(\frac{1}{2} - \frac{F_{left}}{F_{left} + F_{right}}\right) t \tag{2}$$

where t is the distance between the tracks, as shown in Fig. 6.

A final procedure is designed to estimate z_G, that represents the z coordinate of the center of gravity. With the suspension motion locked up, the robot front axle is jacked up so that the lower parts of the tracks are at an angle, α, with the horizontal. The rear wheels are positioned on scales, while the front axle is raised as shown in Fig. 7.

$$z_G = R_F + \frac{\frac{F_R l_1}{\cos\alpha} - Wa}{W \tan\alpha} - h_O \tag{3}$$

where W is the rover weight previously measured, F_R is the weight on the rear wheels with front elevated, h_O is the distance between the center of the sprocket and the bottom part of the track, while a and l_1 are geometric parameters, as indicated in Fig. 7.

The procedures described in this section have been applied to the prototype of Polibot, leading to the results shown in the second column of Table 2. Also, the overall robot weight is 97 kg. The resulting y coordinate is not equal to zero, implying a slight asymmetry of the system with respect to the x-z plane. However, 8 millimeters correspond to only 1.3% of the distance between the tracks, and it could be justified with manufacturing tolerances and inaccuracy of the measurements of the scales.

Similar tests are now performed on the multibody model of Polibot. Note that each subsystem has been designed to closely match the prototype counterpart in terms of geometry and materials. This modelling effort led to the results shown in the third column of Table 2. As expected, the simulated y coordinate of the center of gravity is zero, as the model is perfectly symmetric. The x and y coordinates differ by 7% and 1% respectively between empirical and simulated values. The overall mass of the modeled rover is 97 kg, which exactly matches the actual weight of the prototype.

To complete the set of inertial properties, the multibody model is used to compute the mass inertia tensor. These values are not validated against experimental estimations, because of the difficulty of measuring moments of inertia. However, since the results of the model in terms of mass and position of center of gravity were close enough to the empirical values, it was assumed that the simulated moments of inertia are accurate, and they are shown in Table 2.

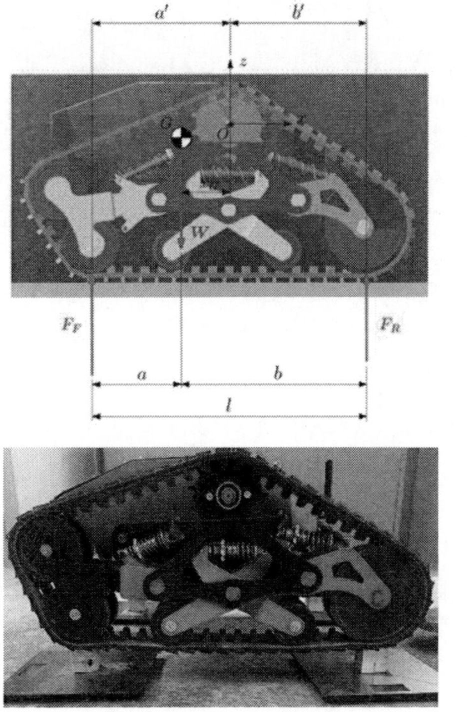

Figure 5. Procedure for the determination of the x coordinate of the center of gravity

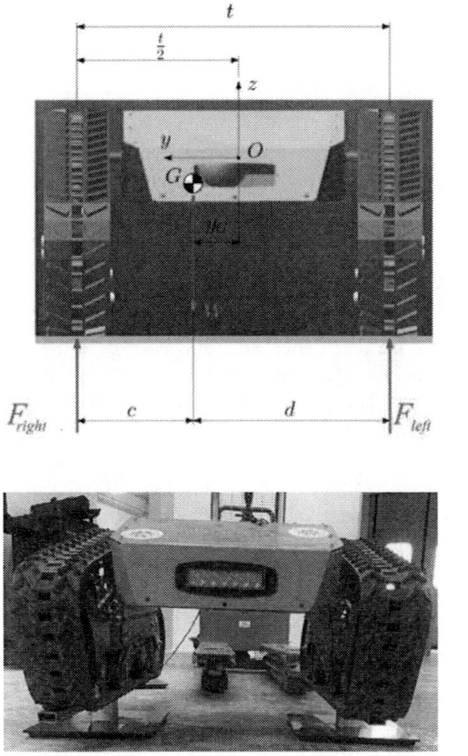

Figure 6. Procedure for the determination of the y coordinate of the center of gravity

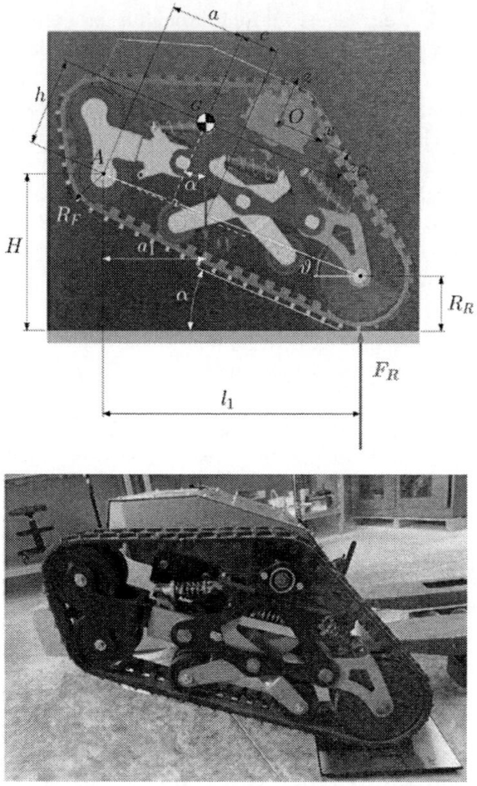

Figure 7. Procedure for the determination of the z coordinate of the center of gravity

Table 2. Empirically estimated and simulated position of the center of gravity of Polibot

Coordinate	Empirical value [mm]	Simulated value [mm]	Error [%]
x_G	-52	-56	7
y_G	8	0	-
z_G	-100	-99	1

Table 3. Moments of Inertia as obtained from the MD model of Polibot.

Moment of inertia	Value [kg m^2]
I_{xx}	8.1
I_{yy}	6.3
I_{zz}	10.1

4. Model validation

Field tests are conducted to validate the multibody model of Polibot. The prototype was fitted with a sensor frame that mounted GPS receivers, an Inertial Measurement Unit (IMU) and stereocameras, as shown in Fig. 8. The data gathered by the sensor is logged and stored by the robot. During the field tests, the robot was teleoperated to move across different type of surfaces such as asphalt, concrete, grass, and clay. It is commanded by means of PWM signals that modulate the torque delivered by the two electric motor (one per each sprocket). The actual rotational speed of the motors is then measured during the tests with the use of encoders. The measured rotational speeds are finally provided as input to the multibody model for the simulation, while the rest of the data gathered is used for validation purposes.

For the validation, the multibody model is used to simulate Polibot while moving on asphalt. Due to the nature of the traversed terrain, it was chosen to run the simulation of the model on a flat hard soil: the normal force exchanged between each track segment and the ground is calculated according to a nonlinear viscoelastic model with stiffness 2.0E+05 N/m, damping 2.0E+03 Ns/m, exponent 2.0 and penetration depth 1 mm. A modified continuous Coulomb friction model (Vasileiou et al., 2021) is used to predict the friction force developed over the contact area between tracks and ground: the static and dynamic friction coefficients are equal to 0.9 and 0.7 respectively, the stiction transition velocity is set to 0.1 m/s, while the friction transition velocity is assigned a value of 0.5 m/s.

The results of the simulation are compared against the correspondent field test and discussed in this section. Figure shows the angular velocities of the motors measured by the encoders while Polibot was moving on asphalt. The corresponding trajectory, as measured by the GPS, is shown by the red line in Fig. 10a and it is compared against the simulated trajectory, in blue. The initial heading is measured by the IMU and it is also provided as input to the model.

Figure 10b compares the measured and simulated rover travel velocity. In the field tests, the robot velocity is measured by the GPS. Figure 10c shows a similar comparison in terms of yaw rate, where the simulation results are plotted against the data measured by the IMU.

The overall distance covered by the robot during the considered test, taken from the origin of the frame of Fig. 10a, is $d_{mea} = 22.58$ m. The percentage deviation of the simulation from the measured data, in terms of final position, is calculated as:

$$E = 100\left(1 - \frac{d_{mea}-d}{d_{mea}}\right) \tag{4}$$

where d is the distance between the final points of the real and the simulated path. Similarly, the following two metrics are adopted to evaluate the local error between the real and the simulated robot position and velocity:

$$E_p = \sqrt{\frac{1}{n_p}\sum_{k=1}^{n_p} |\boldsymbol{r}_{mea} - \boldsymbol{r}_{sim}|^2} \tag{5}$$

$$E_v = \sqrt{\frac{1}{n_v}\sum_{k=1}^{n_v} |\boldsymbol{v}_{mea} - \boldsymbol{v}_{sim}|^2} \tag{6}$$

where $\boldsymbol{r}_{mea}$ is the measured position vector, $\boldsymbol{r}_{sim}$ is the simulated position vector, $\boldsymbol{v}_{mea}$ is the measured velocity vector, $\boldsymbol{v}_{sim}$ is the simulated velocity vector, n_p and n_v are the sizes of position and velocity vectors respectively. The results obtained from Eq. (4), (5) and (6), are summarized in Table 4.

Figure 8. Polibot during field tests, with sensor frame mounted

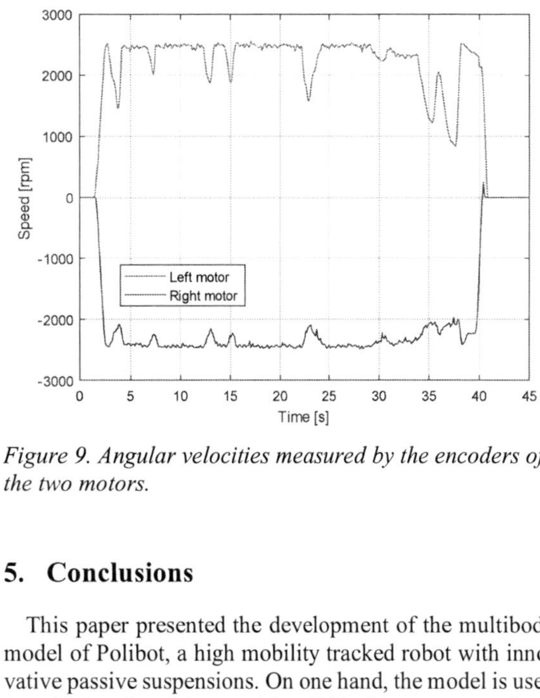

Figure 9. Angular velocities measured by the encoders of the two motors.

5. Conclusions

This paper presented the development of the multibody model of Polibot, a high mobility tracked robot with innovative passive suspensions. On one hand, the model is used to calculate inertial properties difficult to measure (e.g., moments of inertia). On the other hand, it is used to simulate the rover while moving on asphalt, and the results are validated against experimental data gathered during field tests performed with the prototype of Polibot.

The comparison between measured and simulated quantities (Fig. 10 and Table) show good agreement, proving how the modelling efforts spent in designing an accurate virtual twin of Polibot led to results that almost match the experimental data.

As future work, the model developed in this work and the MSC Adams ATV toolkit will be used to model more challenging deformable terrains, and the results will be validated against field tests performed on clay. In addition, different maneuvers will be considered, such as rotations on the spot and closed loops.

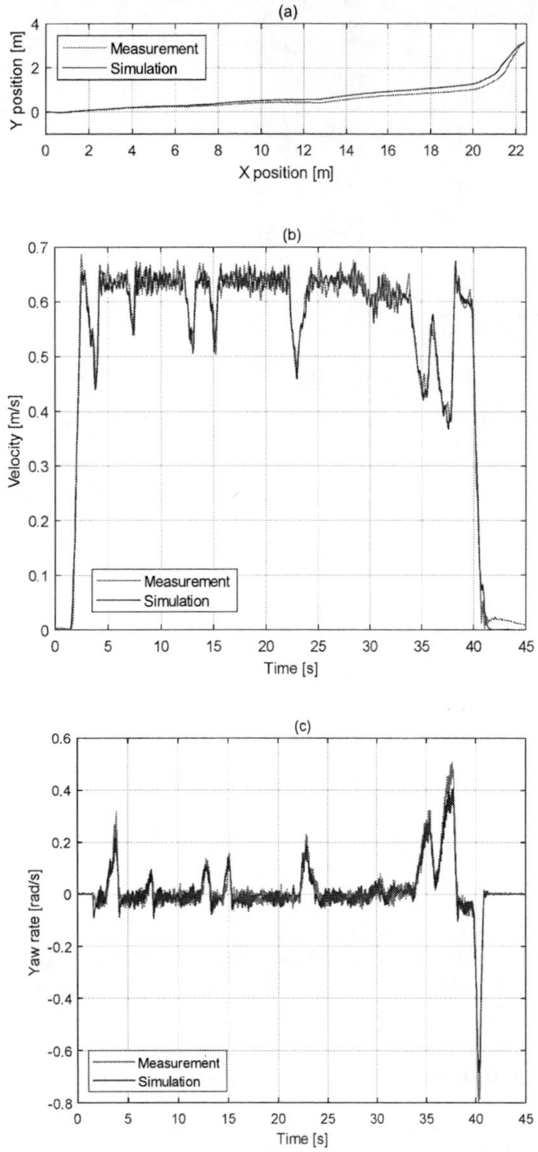

Figure 10. Comparison between measured and simulated data: a) path followed by the robot, b) velocity, and c) yaw rate

Table 4. Summary of errors of the multibody simulation in comparison to experimental data

Error	Value
E [%]	0.23
E_P [m]	0.18
E_V [m/s]	0.04

6. Acknowledgements

The financial support of the project Agricultural inTeroperabiLity and Analysis System (ATLAS), H2020 (Grant No. 857125) is gratefully acknowledged.

7. References

Balena, M., Mantriota, G., Reina, G., 2021. Dynamic Handling Characterization and Set-Up Optimization for a Formula SAE Race Car via Multi-Body Simulation. *Machines*, 9(126).

Galati, R., Mantriota, G., Reina, G., 2021. Design and development of a tracked robot to increase bulk density of flax fibers. *Journal of Mechanisms and Robotics*. 13(5): 050903.

Grazioso, A., Galati, R., Mantriota, G., Reina, G., 2022 Multibody simulation of a novel tracked robot with innovative passive suspension. In: Advances in Italian Mechanism Science. Mechanisms and Machine Science, IFToMM Italy, 1-8.

Huff, J., Conyers, S., Voyles, R., 2012. Mothership—a serpentine tread/limb hybrid marsupial robot for USAR. *IEEE Intl. Symp. on Safety, Security, and Rescue Robotics (SSRR)*. 1–7.

Milella, A., Reina, G., 2014. 3D reconstruction and classification of natural environments by an autonomous vehicle using multi-baseline stereo. Intel Serv Robotics 7, 79–92.

Vasileiou, C., Smyrli, A., Drogosis, A., Papadopoulos, E., 2021. Development of a passive biped robot digital twin using analysis, experiments, and a multibody simulation environment. *Mechanism and Machine Theory*. 163, 104346.

Yamauchi, B. M., 2004. PackBot: a versatile platform for military robotics. *Unmanned ground vehicle technology VI*. 5422.

Yoshida, T., Nagatani, K., Tadokoro, S., Nishimura, T., Koyanagi, E., 2014. Improvements to the rescue robot Quince towards future indoor surveillance missions in the Fukushima Daiichi nuclear power plant. *Springer Tracts in Advanced Robotics*. 92, 19–32.

5060 / Bionic Quadruped Robot for Mars Surface Exploration

Long Qiao[a], Guangming Chen[a]*, Lutz Richter[b], and Aihong Ji[a]

[a]Lab of Locomotion Bioinspiration and Intelligent Robots, College of Mechanical and Electrical Engineering, Nanjing University of Aeronautics and Astronautics, 29 Yudao Street,Nanjing, 210016, China.

[b]Large Space Structures GmbH, Hauptstr. 1e, D-85386 Eching, Germany

**Corresponding author: guangming2017@nuaa.edu.cn*

https://doi.org/10.56884/WFTS1248

Abstract

Mars surface exploration has attracted significant attention of scientists for exploiting new resources and space. To perform explorations on Mars surface, various structures of planetary rovers have been proposed. The Mars surface contains loose granular materials and various sizes of rocks. Traditional wheeled, crawler and legged structures of Mars rovers are mainly designed to walk on granular materials terrain, which are incapable of adapting to rocky surfaces. To improve the adaptations for both granular and rocky surfaces, this paper introduces a quadruped legged robot inspired by the locomotion of desert animal lizard that can walk on granular and rocky surfaces. The main feature is that the structure of the proposed robot possesses bionic multi-toe foot and flexible active spine. To verify the robot locomotion, kinematics on foot, leg and spine of the quadruped robot are anal••yzed. Furthermore, robot motions are analytically predicted with respect to two types of gaits. Combining control framework for adapting to both granular and rocky surfaces, a prototype of Mars robot has been manufactured. Experimental tests demonstrated that the bionic robot can walk on granular surfaces, and can also climb on rocky surface using the multi-joint toe with claw. Therefore, this bionic quadruped robot can have higher adaptability for Mars surface environment.

Keywords: planetary rover, mobile robot, Mars terrain, flexible spine, bionic design

1. Introduction

In order to increase the knowledge and discover new resources beyond the earth, planet exploration has become a new fascinating field. Over the last decade, the U.S. launched the rover 'Perseverance' to Mars, and developed 'VIPER' for the polar regions of the Moon (Stukes et al., 2021). The European Union is ready to launch 'Exomars' Rover to Mars (Allender et al., 2020). China has sent 'YuTu' probe to the moon (Chunlai Li et al., 2019) and 'ZhuRong' to Mars (Yang et al., 2021). The United Arab Emirates is prepared to send the Rashid rover to the moon (Almehisni, Garg, & Wali, 2021). Planetary surface exploration mainly includes surface observation, sample acquisition, instrument deployment and infrastructure construction (Nishida & Wakabayashi, 2012). For conducting these tasks, a series of planetary surface rovers have to be proposed. The planetary surfaces such as for lunar and Mars are distributed with granular soils and rocks of different sizesh (Medina et al., 2021), researchers mainly proposed five types of rover structures to adapt such environment, namely, wheeled (Yang et al., 2021) (Ellery, 2016), crawler (Chunri et al., 2013), legged (Görner, Wimböck, & Hirzinger, 2009) (Takuma, Ikeda, & Masuda, 2010), wheel-legged (Meng et al., 2016; Sunspiral et al., 2012), and gripper-foot (Parness et al., 2017; Parness et al., 2013).

The wheeled robot is the most common structure for planetary rovers due to high flexibility, fast speed and simple structure. Nonetheless, wheeled robot is incapable of walking on very soft soils. For instance, the Martian robot 'Spirit' suffered from stuck in soft sand (Arvidson et al., 2010). In comparison, the crawler robots can produce large friction such that they achieve superior adaptability on soft terrain. However, the crawler robots result in insufficient maneuverability in cornering and obstacle avoidance and consume large energy. The legged and leg-wheeled robot of NASA's Athlete (Sunspiral et al., 2012) are based on the mechanism and structures of animal legs. The legged robots prohibit higher maneuverability in cornering and avoiding obstacles (Mooney & Johnson, 2014) (Bidgoly et al., 2010; Pouya et al., 2012), which are regarded as one main direction for designing planetary robots. However, current legged robot foots are composed of flat or spherical pads, which is not capable of stabilizing in soft terrains (Zhang, Y., Huang, J., Han, 2021). This issue also applies for the leg-wheeled robot. Parness et al. (Zhao et al., 2021) (Parness et al., 2013) designed rock-climbing robots which utilized grippers on the foot toes. It can climb on vertical, and inverted rock surfaces, which is thus applicable to Mars rocky environment. Nevertheless, the trunks of such legged robots are rigid, which is adverse to adjusting the stride for accelerating and turning.

Stable movements of planetary robot requires that the foot can adapt to both granular and rocky terrains. Bionic foot structures can adopt the mechanisms and functions of animals, such as bionic ostrich foot (R. Zhang et al., 2020; Roderick, Cutkosky, & Lentink, 2021), bionic goat foot (Tong Yan, Bo Su, Wei Xu, 2021), hook claw foot (Yanwei et al., 2019). Many desert animals utilize multiply toes and claws for grasping sand and rock surfaces, so that they can enhance stability for the granular and rocky terrains. To facilitate the adaptability of the legged robots for planetary surfaces, the multi-toe foot structures of desert lizard are mimicked. In addition, the flexible truck of lizard is referred to increase the stride which is beneficial to efficient walking and turning.

In this work, a bionic quadruped robot with multi-toe foot and flexible spine is proposed to adapt to planetary surface environment. The kinematics analysis of the bionic foot and spine are carried out. The coordination motions between robot feet, legs and spine are numerically modelled. Furthermore, field tests of walking and turning over granular and rocks were conducted. This work demonstrates that the bionic quadruped robot is adaptive to granular and rocky surface environments, which is promoting to be used for Mars surface exploration.

2. Structure design

To ensure high adaptability on planetary surfaces that are covered of granular materials and rocks, this section elaborates on the structure design of multi-toe foot and flexible trunk the bionic quadruped robot. Moreover, the gait plan for the robot walking straight and turning is presented.

2.1 Biological function analysis

The desert lizard is a quadruped animal that can efficiently move in granular sands and rocks, as shown in *Fig. 1*. Its foot pad contains outer main toes and separated inner auxiliary toes. The foot and toe pads can increase the contact area with respect to soil to prevent subsidence on the soft sand (Phan et al., 2017). All toes connect to the ankle of the foot, and each toe has a claw at the tip, which can efficiently grasp the sand to solid it with toes and pads. During walking, the shins of the leg swing synchronizing with thighs shins movements. Meanwhile, the flexible spine swings to adapt to the change of strides to assist movements. For climbing on rocks, the claw can grasp tightly on the surfaces and move freely in combination with the flexible spine.

Figure 1. Desert lizard

2.2 Bionic structure

The bionic structure of the quadruped robot is composed of foot, leg and spine (Fig. 2). The foot pad connects three main toes and one auxiliary toe. Each toe consists of three joints and a claw at the toe tip. The claw connects the joints of the toe through rotors. A thin spring is adopted between the claw and the pad. At the bottom of the toe, a rope is applied to connect the claw and the pad for each toe (Kitano et al., 2016). A steering gear is placed between the ankle joint and the foot pad (Roderick, Cutkosky, & Lentink, 2021). When the toe grasps soft surface, the ropes are dragged by the steering gear. Then the claws and toe joints bend inward to grasp the surface. Meanwhile, the spring is dragged and bend downward. When the gear motion is turned off, the toes bend up and the spring retrieves. The ankle is a passively rotational and is composed of ball joint and spring. Through the ankle, the foot pad connects the leg. The leg design adopts hinged structure, which has the advantages of simple structure and high adaptability (Park, Park, & Kim, 2015). The leg is composed of motors (marked by green color) and different shapes of brackets (H. Zhang et al., 2014; Chen Li, Hsieh, & Goldman, 2012). The steering gears work with the bracket to produce swing motion and forward motion for the leg.

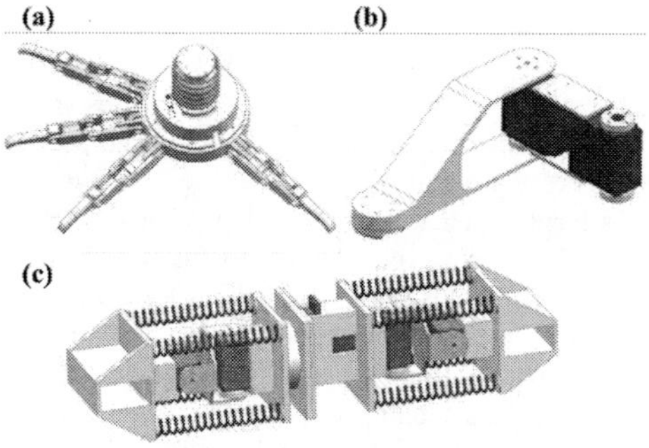

Figure 2. Bionic structures: a) foot b) leg c) spine

For the structure of flexible symmetrical spine, eight springs are used to actively produce deformation in in synchronization with leg movements (Bidgoly et al., 2010; Pouya et al., 2012; Takuma, Ikeda, & Masuda, 2010). Both ends of the spine are respectively connected steering gears that connects leg parts. The middle steering gear connects a multi-functional bracket, which forms two degrees of freedom with up-down and left-right rotation. The middle segment of the joint utilizes single-axis steering gear, a bearing and container of structures. The steering gear is used for producing torsion degree of freedom. The bearing is added in the middle of the flange rod to connect the front and back of spine to for the torsion movement of the robot spine. The spine totally can produce five degrees of freedom. Therefore, the spine can actively bend and twist under the control of the steering gears during movements. Based on the analysis of the foot, leg and spine, the design of the quadruped robot structural model is shown in the Fig. 3.

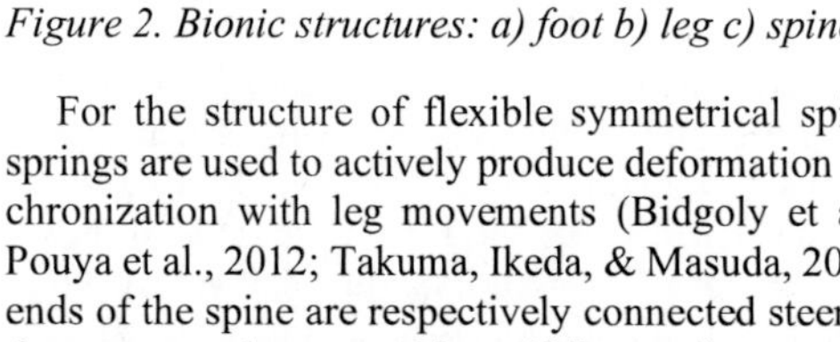

Figure 3. Structure of quadruped robot inspired by desert lizard

3. Theoretical analysis

Kinematics analysis can be used to predict quadruped robot performances in accordance with the predetermined gaits and trajectory. Through analysis, the defects of the robot in various aspects during movement can be predicted, thus necessary adjustments for the structures can be made timely. This section presents kinematics of the quadruped robot. Moreover, the gait plan for walking straight and turning is illustrated.

3.1 Kinematics analysis

The foot motion is divided into toe movement and ankle movement. Toe movement consists of grasping motion that the foot touches the ground and releasing motion that the foot leaves the ground, as shown in Fig. 4. When grasping downward, the rotation angle of toe and claw is related to the hardness of actual contact surface. The maximum rotation angle (θ) of the toes is calculated at 96° according to the structure and geometric sizes. The ankle generates passive rotational motions subjected to irregularities of contact surfaces.

According to the leg model diagram, D-H parameter method (Yuan, Zhou, & Luo, 2019) is used to analyze the forward motion of a single leg of a quadruped robot. A schematic diagram of the coordinate system corresponding to the leg of the robot is established as shown in Fig. 5. The establishment of each coordinate system adopts the rules of the rod and the corkscrew rule. The D-H parameter table of the leg of the robot can be established, as given in Table 1.

According to the designed robot structure, the length of each part of robot leg is measured as $L_1 = 40mm$, $L_2 = 76mm$, $L_3 = 119mm$. The Monte Carlo method (Liu et al., 2013) is used to simulate the mapping of joint space to workspace. Using the forward kinematics solution, a series of random variables within the range of joint angles and a set of joint space vectors are randomly generated. The distribution of the workspace of the foot is obtained. Based on the above modeling analysis, it can be seen that the angles of the joints of the robot's single leg are respectively $-pi/3 \leq \theta 1 \leq pi/3$, $-pi/2 \leq \theta 2 \leq pi/2$ and $-5pi/6 \leq \theta 3 \leq 5pi/6$. The foot reachable space can be solved by using robotics-Toolbox 10.3, as shown in Fig. 6. As can be seen from the figure, maximum reachable space of the leg for X, Y and Z direction are respectively (-43,235), (-203,203), and (-195,195).

If the position changes of the robot's foot in each axis are known, the angle changes of each joint can be solved, which is used as the input signal to drive the robot's leg to move in accordance with the established position. In this manner, the trajectory of the robot's foot can be planned. Using initial coordinate of the foot as the coordinate center, the planning trajectory is assumed to be a semicircular curve. The foot moves forward 40mm each time and the trajectory curve is expressed as:

$$(z - 20)^2 + y^2 = 400 \ (z > 0) \tag{6}$$

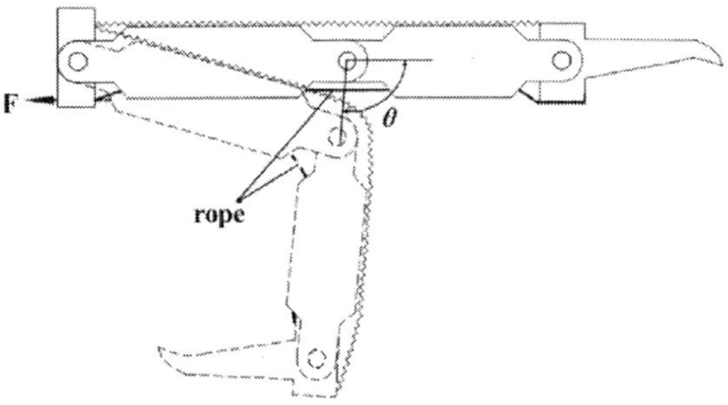

Figure 4. Toe motion diagram angle (θ)

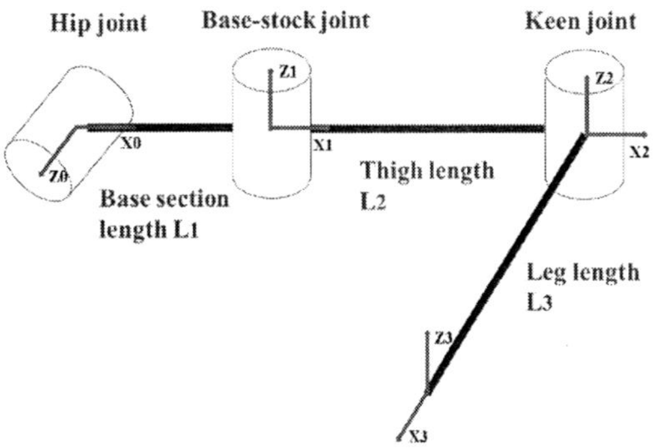

Figure 5. Schematics of D-H coordinate system of robot leg

Table 1 D-H parameters

Link	Angle	Torsion	Distance	Length
n	θ_n	α_n	d_n	a_n
1	θ_1	90°	0	L1
2	θ_2	0°	0	L2
3	θ_3	0°	0	L3

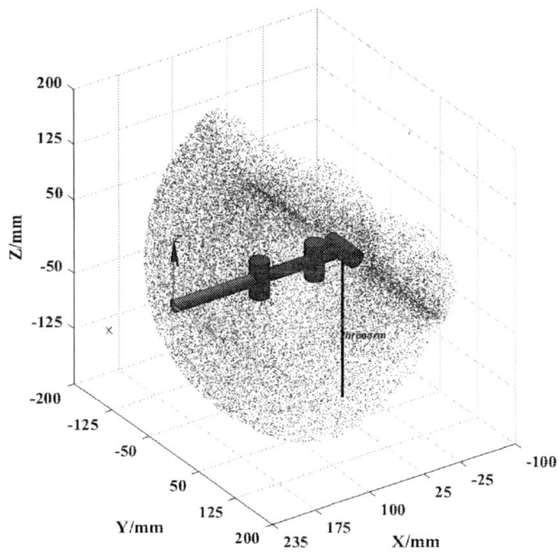

Figure 6. Workspace diagram of leg swing

Convert this into an expression for time:

$$\begin{cases} X = 116 \\ Z = 20 - 20cos(t)\ (0 \le t \le 2 * pi) \\ Y = 20sin(t)(0 \le t \le pi), 0\ (pi \le t \le 2 * pi) \end{cases} \quad (7)$$

Using equation (6), the angle changes of each joint relative to time can be obtained, as shown in Fig. 7. The initial and eventual values of the three joints were all zero. The Hip joint first increases, and after reaching a peak, it decreases gradually to zero as completing one period. This means that the hip joint firstly turns forward. At the peak the velocity decreases to the zero. After that, it reverses and the velocity increases again. The Base-stock joint initially increases approximately by a sinusoid curve and reaches peak at time Pi, and overall process is symmetrical. The Keen joint experiences a converse curve compared to Base-stock joint and its rotation range of the base joint is slightly greater.

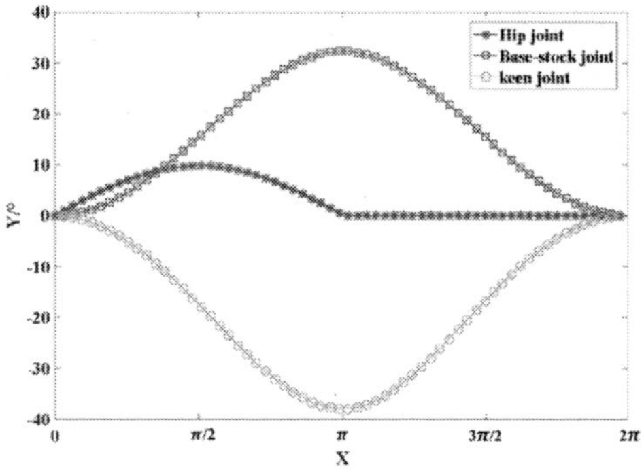

Figure 7. Joint angle position diagram

When the robot moves on the flat ground, the spine swings or twists with the feet, enabling the robot to move stably. Therefore, the movement of the robot's spine must be coordinated with the movement of its legs. The test shows that when the swing range of the spine is ±5°, the robot's center of mass is located within the range of the front and hind diagonal of the foot, indicating that the robot's static standing is relatively stable. When the robot is in motion, the maximum swing of the robot’s spine is 30°, swing angle changes with time as :

$$\theta = 0.2618 * \sin(t/2) \quad (t > 0) \quad (8)$$

3.2 Gait plan

After kinematic analysis on the movements of the foot, leg and spine, the robot motion can be planned. Fig. 8 illustrates the tripod gait for a motion period, in which the arrow designates the “forward” direction; the black square represents the support phase; the white square represents the swing phase; the white point represents the revolute joint of the robot; dashed triangle represents the stable area, and the black point *M* represents the center of mass. In the step 1, the spine bends left while the right hind foot strides forward, and other legs support the body. Then the spine returns to normal and simultaneously move the right front foot (Step 2). In Step 3 & 4, robot carries out contrary motions and accomplishes a full period.

Fig. 9 presents the turning gait for a full period consisting of four postures. The curve arrow denotes the “turning” direction. At first, the left hind leg steps back and constraining inward (Step 1). Then the right hind leg lifts up and simultaneously the rear half of the spine bends to the right (Step 2). Subsequently, the left hind leg returns to the initial angle. Next, the right front leg moves forward and constrains inward (Step 3). The left front leg raises while the front half of the spine bends to the left. Finally, the right front leg retrieves to the initial angle (Step 4), so that the robot turns at an angle and then forms a turning cycle. Under this strategy, this robot can turn to any direction and the turning radius is about half that of the body size.

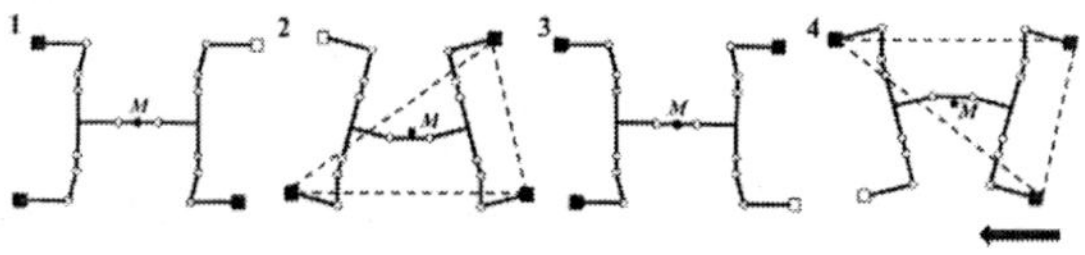

Figure 8. Schematics of tripod gait

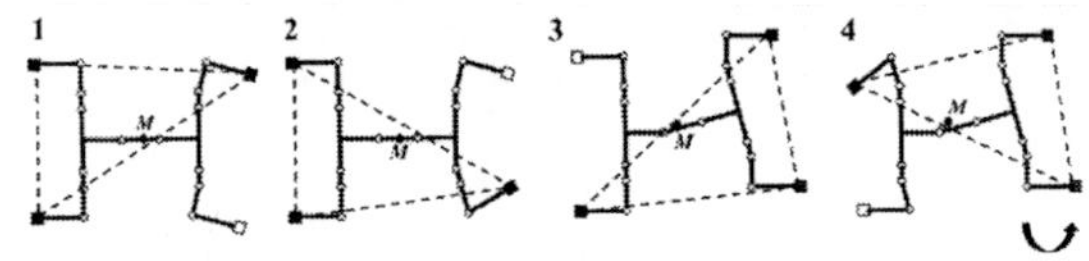

Figure 9. Schematics of turning gait

4. Motion simulation

The above theoretical analysis illustrates the motions of the robot based on the movements of foot, leg, spine, and the gaits. In this section, the software Adams is used to predict the movements based on the theoretical results. Specifically, the foot trajectories, the walking and turning performances of the robot are modelled.

4.1 Foot movement simulation

The foot movements are critical to functions for the robot to move on sands or rocks. The foot movements can be studied from the toe and ankle trajectories. In the simulation, theoretical analysis is applied to model the variation of the rotational angle of each joint. The foot structures are printed using resin materials. The spring material is manganese steel and the rope is carbon fiber. Their values of materials for simulation are referred to the database in the Adams manual. Fig. 10a shows the trajectories of the toes bending down to grab inward. Unidirectional force with the same size and direction is applied at the connection between the toe and the driving rope. When the force is exerted, the toe grabs downward. When the force is canceled, the toe recovers to the initial state under the action of springs. The bending angle is also larger than 90° which is consistent with theoretical analysis.

The swing phase simulation of a single leg was carried out to verify the theoretical analysis for solving the results of each joint angle, as shown in Fig. 10b. The angle changes of each joint (Eq. 3, 4, 5) were adopted into the simulation model. The red trajectory was the predicted swing phase trajectory of the foot obtained. The trajectory is represented in the base coordinate system (O_o). X-axis is the displacement perpendicular to the forward direction, which is 0.22mm. Y-axis represents the leg lift height, which is 20.36mm by measuring the leg lift height, Z-axis is the span in the forward direction, the measured span is 41.04mm. Compared with the theoretical analysis results, the error range in X-axis is 0.22mm, Y-axis is 0.36mm, Z-axis is 1.04mm. These errors may be caused by the measurement error of the leg size or the setting position error of the drive. Nonetheless, the overall error is small, which verified the correctness of the theoretical analysis.

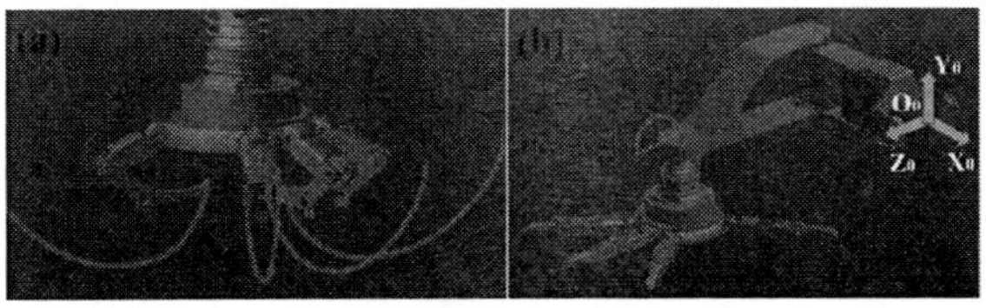

Figure 10. a) Toe trajectory. b) Ankle trajectory

4.2 Gait simulation

The gait plan is adopted to predict the walking and turning performances of the robot. These simulations involve the contact between robot feet and ground surface. For conveniently examine the walking and turning performance by the adopted tripod gait, the ground material is set to steel. The contact parameters can be determined by referring to Adams manual, which are given in Table 2. The integral solver uses GSTIFF, integral format is I3, the maximum number of iterations is 25, surface tolerance is 200, and the simulation step size is 0.01s.

Table 2. The contact parameters in simulation

Stiffness [N/mm]	Force index	Damping [Ns/mm]	Penetration depth[mm]
1150	2	0.68	0.1
Static friction coefficient	**Dynamic friction coefficient**	**Static friction velocity [mm/s]**	**Dynamic friction velocity [mm/s]**
0.15	0.1	0.1	10

Fig. 11 shows the simulations of robot walking straight based on the tripod gait plan. Firstly, the spine bends to the left, and the right hind leg swing forward. Then the spine bends to the left to adapt to the stride. Meanwhile, the left hind leg and left front leg swing forward. For next half cycle, robot carries out contrary motions. It demonstrates that the legs and spine of the robot can cooperate with each other and move normally, which shows the feasibility of theoretical linear gait planning and the results are in consistent with the theoretical tripod gait movement.

It reveals that the center of mass change of the robot in all directions when it moves forward by simulation measurement. The X-axis is the forward direction, which can move forward 58.7mm per cycle. The Y-axis is the swinging direction of the robot's spine, which can swing within $\pm$56.1mm. The initial offset of the center of mass on the Y-axis is 2.4mm in each cycle. Z-axis denotes the movement of the robot center of mass in the vertical direction. The deviation of the center of mass in the z-axis direction is almost 0. These variation ranges can be considered small and can still result in high stability.

Fig. 12 shows the simulations of the robot turning in accordance with the turning gait. It demonstrates that through the coordination of legs and spine, the later part of the robot moves to the right first, and then the first part of the robot moves to the left. In this manner, a complete cycle of turning can be made, which shows the feasibility of the theoretically turning gait.

After a quarter of a turn, the displacements of the center of mass for X, Y, and Z axis are respectively 164.6mm, 125.2mm and 9.3mm. The simulation results show that the average rotation is 10° after each cycle, and the radius of turning circle is about 146mm. The total length of the robot's spine is about 380mm, so the change range of the center of mass is small, indicating that the robot has a small turning radius.

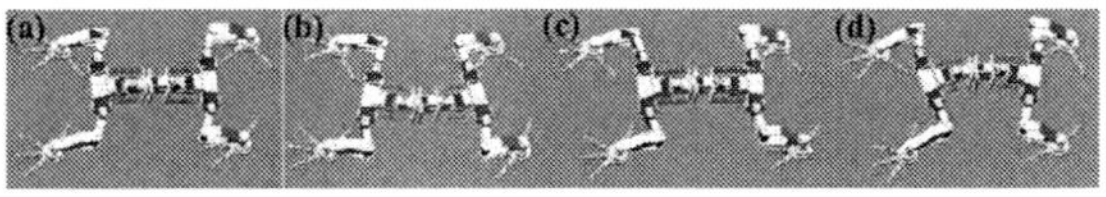

Figure 11. a-d) Simulation motions of tripod gait test

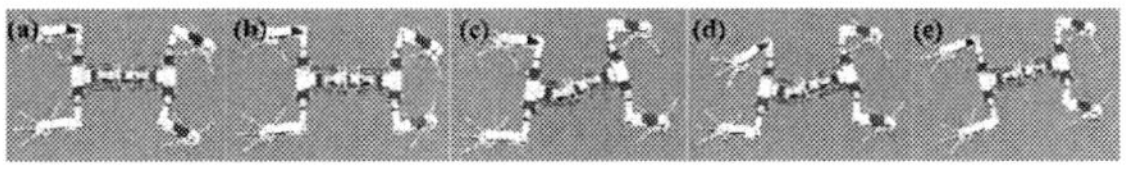

Figure 12. a-e) Simulation motions of turning gait test

5. Experimental tests

To validate the performances of the bionic robot, this robot is manufactured. Besides, the walking gait and turning gait is experimentally evaluated with respect to granular surfaces. Toward application of the quadruped robot in Mars surface environment, the discussions for this work are presented.

5.1 Manufacture

The quadruped robot mainly consists of mechanical part and electronical part. The mechanical part contains the structures as presented in Fig. 3. The basic components for composing foot, leg and spine are fabricated by 3D printing using resin materials. The springs and steering gears are commercially purchased. The steering gear control panel will receive the signals from the control center. Using high accuracy potentiometer, the deviations of rotation angles are transferred to the steering gear. Then the steering gear adjusts the positions and voltages to obtain prescribed rotation angles. At the same time, it adopts asynchronous serial bus communication. The port can send data on one line and receive data on the other line. Therefore, simple and long-distance communication can be realized, which greatly simplifies the complexity of wiring.

Quadruped robot electronical system hardware mainly includes the control center, driving part and the power, as is shown in Fig. 13. The 17 steering gears are LX-224 serial bus. The steering gear control board is utilized to adjust central interface and power supply pressure. Raspberry PI 4B is selected as the control center, which can use wireless technology for remote control. Through the selection of the above parts and electrical appliances. The power supply voltage is 6V-12V.

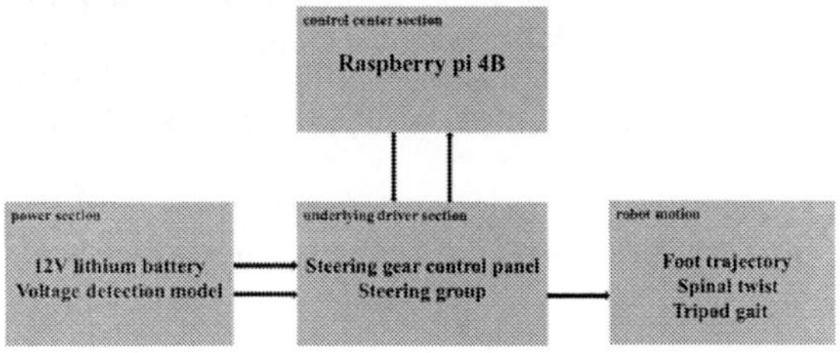

Figure 13. Hardware control system diagram

5.2 Adaptability tests

To validate the motions of the robot, the walking and turning tests are carried out on the sand surfaces. In addition, the climbing tests of the robot are conducted with respect to rocky surface. For evaluating the adaptability of the robot on sand surface, the effectiveness of grasping ability is tested, which is shown in Fig. 14. The sand surface is mostly composed of fine coarse particles with sizes less than 1 mm. The bulk density is around 1700kg/m^3. In the initial state, the toe is relaxed. When the motor drives the rope, the toe can bend down to grasp sands. The maximum grasping angle is higher than 90° as expected.

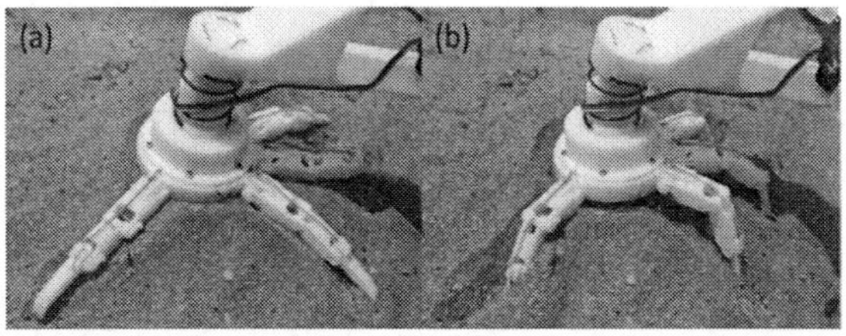

Figure 14. Foot grab test: a) relaxing state b) toes bend inward

Figs 15a-d shows the four postures of the robot when it walks in a straight line on sand surface. It was observed that its gait was consistent with the tripod gait in the theoretical analysis, and it was relatively stable during the walking process. The feet could keep its stability on the sandy land without subsidence. The test shows that it can move forward according to the tripod gait made by theoretical analysis.

Figs 16a-d shows the robot's turning gait. It is found that the actual robot turns at an average angle of 10° each time, which is slightly smaller than the angle obtained by theoretical analysis. The results showed that robot turning gait stable and consistent with the theoretical analysis. After in situ turn a circle the center of mass in the lower right produces small range of deviation. The average rotation is about 10° per cycle, which is consistent with the results of the simulation analysis.

Fig. 17 shows preliminary test for the robot climbing rocky surfaces which consists of distinct irregular and rough asperities. The joint roughness coefficient is averagely greater than 20. Results demonstrate that the robot can attach the rocks by toe grasping using interlocking between the claws and the surface asperities. In addition to the interlocking, the passive deformations of the toes and ankle, and the active flexible spine contributes significantly to the attachment of the whole robot body.

Figure 15. a-d) Four motions in tripod gait test

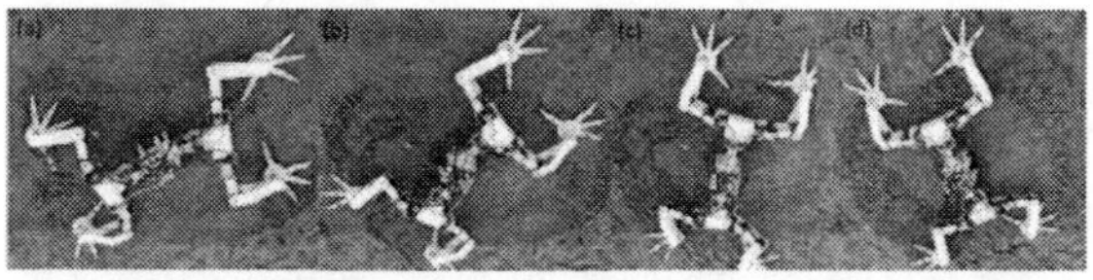

Figure 16. a-d) Four motions in turning gait test

Figure 17. Rock climbing test

6. Discussion

In order to propose a quadruped robot for Mars surface exploration, the inspirations were obtained from the desert animal Lizard that can efficiently move on soft sands and climb on irregular surfaces of rocks. The bionic structures have adopted the characteristics of biological foot, leg, and spine. Thus, the bionic robot model has potentials on the adaptability to granular and rocky environments.

Kinematics of robot toe, foot, and spine were theoretically assessed. The gait plan for robot walking straight and turning were designated. The theoretical results were employed by simulations to predict the motions. The simulations demonstrated the actions of the toe grasping and adaptability of foot. In addition, the motions for walking and turning were performed, which confirms the effectiveness of the theoretical analysis.

Finally, the validation tests of the prototype were carried out with respect to both granular and rocky surfaces. It demonstrated the robot can efficiently grasp the sand to prevent sinking and can move effectively on the sand surfaces. In addition, the robot can attach the irregular surfaces of rocks by the coordination's of claw, toes and spine. Though the experimental tests were conducted under the earth's surface environment, the gravity effects on the adaptability of the robot are not significant.

The adaptability for the rocky surfaces is not fully demonstrated from this research, which will be the further work for the improvement of this robot. Moreover, the feedback control for the adaptability of the robot over varied terrains should be applied. Meanwhile, the contact model between robot foot and different terrains should be justified for aching effective control. In addition, the structure materials of the robot such the resin material must be replaced by metal to resist the thermal effect. Last not least, the power supply system should be taken into consideration.

7. Conclusions

To adapt to Mars surface terrain, this paper proposed the bionic structure of the quadruped robot based on the desert lizard. To verify the robot locomotion, kinematics on foot, leg and spine of the quadruped robot are analyzed. Furthermore, robot motions are analytically predicted by applying two types of gaits. Experimental tests of a prototype of Mars robot demonstrated that the bionic robot can walk on granular surfaces and can also attach on rocky surface. This work demonstrates that the bionic quadruped robot is adaptive to granular and rocky environments, which is highly potential to be used for Mars surface exploration.

8. Nomenclature

θ	*Angle of toe joint*	*[rad]*
θ_1	*Angle of hip joint*	*[rad]*
θ_2	*Angle of base-stock joint*	*[rad]*
θ_3	*Angle of keen joint*	*[rad]*
a_n	*Link length*	*[m]*
d_n	*Link distance*	*[m]*
α_n	*Link angle*	*[rad]*
${}_{n-1}^{n}T$	*Pose matrix between the two links*	*[-]*
P_x	*Position coordinates of the foot on the X axis under the base system*	*[-]*
P_y	*Position coordinates of the foot on the Y axis under the base system*	*[-]*
P_z	*The position coordinates of the foot on the Z axis under the base system*	*[-]*

9. Acknowledgement

The authors gratefully acknowledge financial support from the Foundation Research Project of Jiangsu Province Natural Science Fund (No. BK20190415) and International Cooperation Fund for Foreign Experts Working in NUAA (G2021181006L).

10. References

Allender, E. J., Cousins, C. R., Gunn, M. D., & Caudill, C. M., 2020. Multiscale and Multispectral Characterization of Mineralogy with the ExoMars 2020 Rover Remote Sensing Payload. Earth Sp. Sci. 7.

Almehisni, R., Garg, P., & Wali, M., 2021. Thermal Management System Design and Analysis of Rashid Rover—Emirates Lunar Mission. Proc. Int. Astronaut. Congr. IAC C2.

Arvidson, R. E., Bell, J. F., Bellutta, P., et al., 2010. Spirit Mars Rover Mission: Overview and selected results from the northern Home Plate Winter Haven to the side of Scamander crater. J. Geophys. Res. E Planets 115, 1–19.

Bidgoly, H. J., Vafaei, A., Sadeghi, A., & Ahmadabadi, M. N., 2010. Learning approach to study effect of flexible spine on running behavior of a quadruped robot. Emerg. Trends Mob. Robot. Proc. 13th Int. Conf. Climbing Walk. Robot. Support Technol. Mob. Mach. CLAWAR 2010 1195–1201.

Chunri, P., Guozheng, Y., Zhiwu, W., Hua, L., & Pingping, J., 2013. Design of a tracked robot and analysis of its obstacle- climbing 4–8.

Ellery, A., 2016. Planetary Rovers Robotic Exploration of the Solar System. Astronautical Engineering.

Görner, M., Wimböck, T., & Hirzinger, G., 2009. The DLR Crawler: Evaluation of gaits and control of an actively compliant six-legged walking robot. Ind. Rob. 36, 344–351.

Kitano, S., Hirose, S., Horigome, A., & Endo, G., 2016. TITAN—XIII : sprawling-type quadruped robot with ability of fast and energy - efficient walking. RO-BOMECH J. 1–16.

Li, Chen, Hsieh, S. T., & Goldman, D. I., 2012. Multi-functional foot use during running in the zebra-tailed lizard (Callisaurus draconoides). J. Exp. Biol. 215, 3293–3308.

Li, Chunlai, Liu, D., Liu, B., Ren, X., et al., 2019. Chang'E-4 initial spectroscopic identification of lunar far-side mantle-derived materials. Nature 569, 378–382.

Liu, Z., Liu, H., Luo, Z., & Zhang, X., 2013. Improvement on Monte Carlo method for robot workspace de-termination. Nongye Jixie Xuebao/Transactions Chinese Soc. Agric. Mach. 44, 230–235.

Medina, F. J., Manzano, A., Villacampa, A., Ciska, M., & Herranz, R., 2021. Understanding Reduced Gravity Effects on Early Plant Development Before Attempting Life-Support Farming in the Moon and Mars. Front. Astron. Sp. Sci. 8, 1–7.

Meng, X., Wang, S., Cao, Z., & Zhang, L., 2016. A review of quadruped robots and environment perception. Chinese Control Conf. CCC 2016-Augus, 6350–6356.

Mooney, J. G., & Johnson, E. N., 2014. A Comparison of Automatic Nap-of-the-earth Guidance Strategies for Helicopters. J. F. Robot. 33, 1–17.

Nishida, S. I., & Wakabayashi, S., 2012. Lunar surface exploration using mobile robots. Cent. Eur. J. Eng. 2, 156–163.

Park, H., Park, S., & Kim, S., 2015. Variable-speed Quadrupedal Bounding Using Impulse Planning: Un-tethered High-speed 3D Running of MIT Cheetah 2 5163–5170.

Parness, A., Abcouwer, N., Fuller, C., Wiltsie, N., Nash, J., & Kennedy, B., 2017. LEMUR 3: A limbed climbing robot for extreme terrain adaptability in space. Proc.—IEEE Int. Conf. Robot. Autom. 5467–5473.

Parness, A., Frost, M., Thatte, N., King, J. P., Witkoe, K., Nevarez, M., Garrett, M., Aghazarian, H., & Kennedy, B., 2013. Gravity-independent rock-climbing robot and a sample acquisition tool with microspine grippers. J. F. Robot. 30, 897–915.

Phan, L. T., Lee, Y. H., Lee, Y. H., Lee, H., Kang, H., & Choi, H. R., 2017. Effects of spinal joint on quadruped-al bounding. 2017 14th Int. Conf. Ubiquitous Robot. Ambient Intell. URAI 2017 688–689.

Pouya, S., Khodabakhsh, M., Moeckel, R., & Ijspeert, A. J., 2012. Role of Spine Compliance and Actuation in the Bounding Performance of Quadruped Robots. In Dyn. Walk. Conf.

Roderick, W. R. T., Cutkosky, M. R., & Lentink, D., 2021. Bird-inspired dynamic grasping and perching in arboreal environments. Sci. Robot. 6.

Stukes, S., Allan, M., Bajjalieh, G., Deans, M., Fong, T., Hihn, J., & Utz, H., 2021. An Innovative Approach to Modeling VIPER Rover Software Life Cycle Cost. IEEE Aerosp. Conf. Proc. 2021 March, 1–16.

Sunspiral, V., Wheeler, D. W., Chavez-Clemente, D., & Mittman, D., 2012. Development and field testing of the FootFall planning system for the ATHLETE robots. J. F. Robot. 29, 483–505.

Takuma, T., Ikeda, M., & Masuda, T., 2010. Facilitating multi-modal locomotion in a quadruped robot utilizing passive oscillation of the spine structure. IEEE/RSJ 2010 Int. Conf. Intell. Robot. Syst. IROS 2010 - Conf. Proc. 4940–4945.

Tong Yan, Bo Su, Wei Xu, T. Q., 2021. Research on Bionic Foot Design and Climbing Strategy of Quadruped Robot. China North Veh. Res. Inst.

Yang, J., Zhang, T., Tian, H., & Ji, L., 2021. Experimental verification of Zhurong Mars rover. Exp. Tech-nol. Manag. 38, 1–5.

Yanwei, L., Xiang, H., Limeng, W., Tao, M., & Yan, L., 2019. Design and Analysis of a Bio-inspired Biped Wall-climbing Robot with Spines. Mech. Sci. Technol. Aerosp. Eng. 38, 1185–1190.

Yuan, S., Zhou, Y., & Luo, C., 2019. Crawling Gait Planning Based on Foot Trajectory Optimization for Quadruped Robot. 2019 IEEE Int. Conf. Mechatronics Autom. 1490–1495.

Zhang, Y., Huang, J., Han, L., 2021. Research status of planetary surface mobile exploration robots: Review. Hangkong Xuebao/Acta Aeronaut. Astronaut. Sin. 42(1), 523–909.

Zhang, H., Liu, Y., Zhao, J., Chen, J., & Yan, J., 2014. Development of a Bionic Hexapod Robot for Walking on Unstructured Terrain. J. Bionic Eng. 11, 176–187.

Zhang, R., Pang, H., Wan, H., Han, D., Li, G., & Wen, L., 2020. Design and Analysis of the Bionic Mechanical Foot with High Trafficability on Sand. Appl. Bionics Biomech. 2020.

Zhao, J., Han, T., Wang, S., Liu, C., Fang, J., & Liu, S., 2021. Design and Research of All-Terrain Wheel-Legged Robot. Sensors 1–18.

5408 / Ride Comfort Comparison Between Suspension Modes: Input Towards Designing Difference Threshold Experiments During Driving

Cor-Jacques Kat[a]*, Kylian Praet[b], Miguel Dhaens[b], and Schalk Els[a]

[a]Vehicle Dynamics Group, Department of Mechanical and Aeronautical Engineering, University of Pretoria, Pretoria, South Africa

[b]Tenneco Automotive Europe–Monroe® METC, Sint-Truiden, Limburg, Belgium

**Corresponding author: cor-jacques.kat@up.ac.za*

https://doi.org/10.56884/MGKW8503

Abstract

Ride comfort is an important topic for on- and off-road suspension design. Difference thresholds of whole-body vibration is important to determine perceptibility of changes in a vehicle's dynamics. Difference thresholds can be used to guide ride comfort improvements. Difference thresholds have been estimated for vertical and multi-axial seat vibration in laboratory settings. In order to determine the applicability of these laboratory difference thresholds and/or to estimate difference thresholds during driving, it is required that changes can be made in the vehicle's vibration that is transmitted to the occupants i.e., the stimulus. Ride comfort is quantified by the weighted vertical seat pad vibration and compared between four suspension modes of a vehicle over three roads from ten repeat runs. Significant differences in the median weighted vertical seat pad vibration were found between Mode 1 and the other three modes over Road 1 and Road 2. No significant differences were found over Road 3. The significant differences over Road 1 are in the range of the median relative difference threshold reported in literature. Over Road 2 the differences are below the reported 25th percentile relative difference thresholds. Some combinations of the suspension modes and roads result in ride comfort differences. The suspension mode and road combinations could be used to verify the applicability of available difference thresholds during driving.

Keywords: ride comfort, vehicle vibration, difference thresholds

1. Introduction

The difference threshold for vehicle vibration on a seat is defined by Mansfield and Griffin (2000) as: "...the minimum change in the magnitude of the whole-body vibration required for the seat occupant to perceive the change in magnitude." The difference threshold is also referred to as the just noticeable difference. The relative difference threshold is obtained by taking the percentage of the Weber fraction which is the ratio between the just noticeable change in stimulus magnitude and the reference stimulus magnitude.

Knowledge of the difference threshold is useful in determining whether changes in a vehicle's vibration, for example due to different suspension characteristics, would be perceivable by vehicle occupants.

Studies have estimated difference thresholds for whole-body vibration for participants seated on a rigid surface exposed to vertical sinusoidal vibration (Morioka and Griffin, 2000, Matsumoto et al., 2002, Forta et al., 2009), for participants seated on a car seat exposed to vertical random vibration (Mansfield and Griffin, 2000, Pielemeier et al., 1997), and for participants in a vehicle on a 4-poster subjected to multi-axis random vibration (Gräbe et al., 2020). The studies using random vibration considered stimulus changes in magnitude resulting from changes over the entire spectrum. All these studies were conducted in laboratory conditions. The study with conditions closest to driving is that of Gräbe et al. (2020). Participants were seated in a vehicle that was excited on a 4-poster test rig in a semi-anechoic chamber, creating an environment with limited aural and visual inputs.

In order to determine the applicability of these laboratory difference thresholds, or to estimate difference thresholds during driving, it is required that changes can be made in the vehicle's vibration that is transmitted to the occupants i.e. the stimulus.

The aim of this paper is to determine the difference in ride comfort, as quantified by the vertical seat pad vibration between four suspension modes of a vehicle over three roads. Significant differences between modes are compared to difference thresholds in literature.

2. Materials

The vehicle is fitted with a developmental Monroe Intelligent Suspension CVSA2/Kinetic® suspension system that can be set to different modes. Four different modes were defined for this study. The CVSA2/Kinetic® system was deactivated in one mode and activated in different development settings for the other three modes. Acceleration was measured on the driver seat surface in the vertical direction below the ischial tuberosities of the driver using a seat pad accelerometer (356B40, PCB Piezotronics). Data was sampled at 10 kHz using IPETRONIK (Mx-SENS2, IPETRONIK) with a 1250 Hz anti-aliasing filter. For ride comfort measurements, a lower sampling rate would be sufficient (e.g., sample rate of 400 Hz and anti-aliasing filter at 100Hz (Paddan and Griffin, 2002)). The selected sample rate was with respect to additional measures that were also recorded and used for other aspects. Details about the road profile and suspension modes are not provided as this information is proprietary.

3. Method

The vehicle was driven over three different roads in each of the four suspension modes. Ten repeat runs of each suspension mode over each road were completed. The cruise control of the vehicle was set to the desired speed for each road.

The effect of whole-body vibration on comfort was evaluated according to BS 6841 (1987) in order to compare to the relative difference thresholds obtained by Gräbe et al. (2020). They estimated difference thresholds using the BS6841 weighted vertical seat pad acceleration and the combined point ride value. No significant difference was found between the medians of the relative difference threshold of the two ride values. Therefore, the current study will consider the weighted vertical seat pad acceleration. The vertical seat vibration is measured by the seat pad accelerometer. The vertical seat acceleration is weighted in the frequency domain during post processing using the applicable weighting function and multiplication factor as defined in BS 6841 (1987). All frequency content below 0.5 Hz and above 80 Hz is discarded. This results in the vertical component ride values (CRVz). The crest factors for all suspension modes over Road 1 and 3 are below six. The crest factors for the suspension modes over road 2 range between 5.8 to 6.8. The root-mean-square (r.m.s.) is used to represent the magnitude of the acceleration since the crest factor is below six for most of the modes and roads. The severity of high peaks over Road 2 may be underestimated by using r.m.s. (BS 6841, 1987).

Statistical analyses, including hypothesis testing, were done with MATLAB® version 2020b (Mathworks, Natick, MA, USA). Because of the small sample size, non-parametric hypothesis tests were used. Friedman's test and Dunn's multiple comparison tests were used to determine if the vertical component ride value differ significantly between the four suspension modes.

The percentage relative difference in the median vertical component ride value between suspension modes are calculated using Eq. 1. The convention used with respect to the reference mode is that the mode with the lower mode number in a pair is chosen as the reference mode.

$$perc.rel.diff. = \frac{mode - reference\ mode}{reference\ mode} \times 100 \qquad (1)$$

4. Results

The distributions of the vertical component ride value are shown in the box plots in Figure 1, Figure 2 and Figure 3 for Road 1, Road 2 and Road 3, respectively.

Friedman's test did indicate a significant difference between at least two median vertical component ride values across the four suspension modes on Road 1 (p-value < 0.001) and Road 2 (p-value < 0.001). Friedman's test did not indicate a significant difference between any of the median vertical component ride values across the four suspension modes on Road 3 (p-value = 0.266). Table 1 and Table 2 present the adjusted p-values (Bonferroni correction) for Dunn's multiple comparison tests over Road 1 and Road 2, respectively.

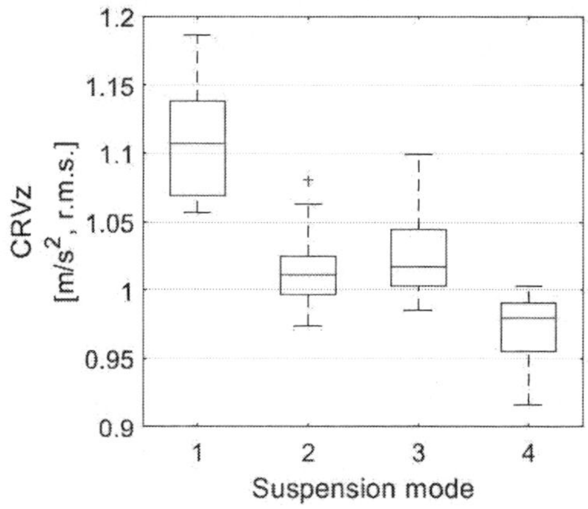

Figure 1. Box plots showing the distribution of the vertical component ride value (CRVz) over Road 1

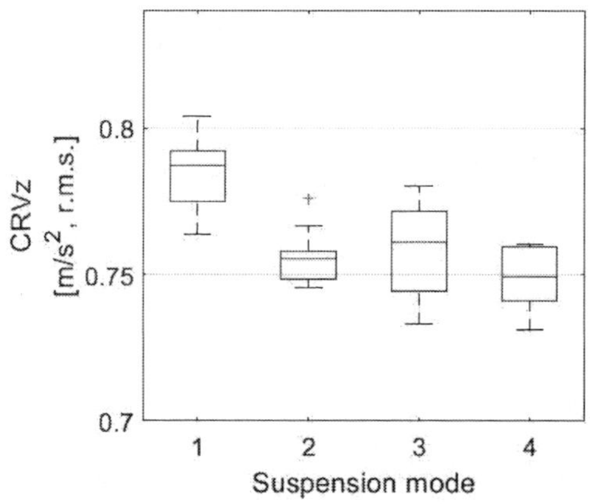

Figure 2. Box plots showing the distribution of the vertical component ride value (CRVz) over Road 2

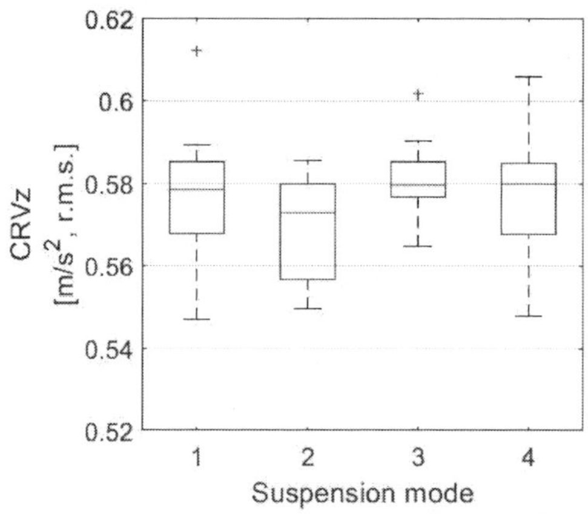

Figure 3. Box plots showing the distribution of the vertical component ride value (CRVz) over Road 3

Significant differences were found over Road 1 between suspension Mode 1 and Mode 2 (p-value = 0.006) and between Mode 1 and Mode 4 (p-value = 0.000). Between Mode 1 and Mode 3 a marginally significant difference was found (p-value = 0.056). Over Road 2, significant differences were found between suspension Mode 1 and the other 3 modes (Mode 1 vs Mode 2, p-value = 0.019; Mode 1 vs Mode 3, p-value = 0.034; Mode 1 vs Mode 4, p-value = 0.000).

The percentage relative difference in the median component ride values between Mode 1 and the other three modes over Road 1 are -8.7%, -8.1% and -11.5%. Over Road 2, the percentage relative difference between Mode 1 and the other three modes are -4.0%, -3.3% and -4.8%, respectively.

Table 1. Adjusted p-values (Bonferroni correction) for Dunn's multiple comparison tests over Road 1
⁺p-value 0.05-0.1: marginally significant;
**p-value < 0.05: significant difference at 5% level*

	Mode 1	Mode 2	Mode 3	Mode 4
Mode 1	-	0.006*	0.056⁺	0.000*
Mode 2		-	1.000	1.000
Mode 3			-	0.341
Mode 4				-

Table 2. Adjusted p-values (Bonferroni correction) for Dunn's multiple comparison tests over Road 2
**p-value < 0.05: significant difference at 5% level*

	Mode 1	Mode 2	Mode 3	Mode 4
Mode 1	-	0.019*	0.034*	0.000*
Mode 2		-	1.000	0.500
Mode 3			-	0.341
Mode 4				-

5. Discussion

Mansfield and Griffin (2000) estimated difference thresholds for 10 males seated on an automobile seat exposed to vertical vibration recorded in a vehicle. Four reference stimuli were considered i.e. tarmac stimuli at weighted magnitudes of 0.2, 0.4 and 0.8 m/s^2, r.m.s. and a paved stimulus with a weighted magnitude of 0.4 m/s^2, r.m.s. Over the four stimuli, the median relative difference thresholds range between 11.8 % and 14.1 %. They exposed participants to seated whole-body vibration in the vertical direction only. Gräbe et al. (2020) estimated difference thresholds for 10 males seated in a vehicle subjected to multi-axis vibration resulting from the vehicles response to two road profiles on a 4-poster test rig. Over the two roads, considering the vertical component ride value, the median relative difference thresholds were 8.58 % and 10.99 %. Both Mansfield and Griffin (2000) and Gräbe et al. (2020) determined relative difference thresholds using the same psychophysical method and level of detection probability (i.e., 79.4 %).

Considering the suspension pairs which were found to have significant differences over Road 1, the percentage relative differences in median vertical component ride value are in the range of the median relative difference thresholds reported in Mansfield and Griffin (2000) and Gräbe et al. (2020). This means that 50% of occupants would be able to correctly identify the suspension mode with the larger vibration magnitude 79.4% of the time. Over Road 2, significant differences were found between Mode 1 and the other three. These relative differences are however below the median relative difference thresholds reported in Mansfield and Griffin (2000) and Gräbe et al. (2020). The relative differences are not only below the median relative difference thresholds but also below the 25th percentile reported. This would imply that only a few occupants may be able to identify the suspension mode with the larger vibration magnitude. No significant differences were found in the vertical component ride values between the suspension modes on road 3. The percentage relative differences were close to zero, well below the median relative difference thresholds. This would imply that almost none of the occupants will be able to identify the suspension mode with the larger vibration magnitude.

No formal subjective evaluation was conducted during this study but the driver and passenger (being two of the authors) did make some observations. Both the driver and passenger were able to perceive changes between suspension modes and identify the suspension mode with the larger vibration magnitude over all three roads. Over Road 1 and 2 differences in suspension modes were most noticeable, with differences noticeable on Road 3 but more difficult to identify between some of the modes.

The subjective observations seem to agree with the objective findings over Road 1 and 2. However, over Road 3 it seems that there is a disagreement. Further investigation is needed into the applicability of the available difference thresholds to real world driving situations.

6. Conclusions

The combination of the suspension modes and roads result in ride comfort differences, as quantified by the BS681 weighted vertical seat pad vibration, that is in the range and below the available difference thresholds in literature. The suspension mode and road combinations could be used to verify the applicability of available difference thresholds during driving.

7. Nomenclature

CRVz Vertical component ride value [m/s^2, r.m.s]

8. Acknowledgements

This research was funded from the European Union Horizon 2020 Framework Program, Marie Skłodowska-Curie actions, under grant agreement no. 872907

9. References

British Standards Institution, BS 6841 (1987)—British Standard Guide to measurement and evaluation of human exposure to whole-body mechanical vibration and repeated shock, London.

Forta, N.G., Morioka, M., Griffin, M.J., (2009), Difference thresholds for the perception of whole-body vertical vibration: dependence on the frequency and magnitude of vibration, Ergonomics, vol. 52, no. 10, pp. 1305-1310, doi: 10.1080/00140130903023709.

Gräbe, R.P., Kat. C-J, van Staden, P.J. and Els, P.S, (2020), Difference thresholds for a vehicle on a 4-poster test rig, Applied Ergonomics, vol. 87, 103115.

Griffin, M.J. (1990), Handbook of Human Vibration, Academic Press Limited, London.

Mansfield, N.J., Griffin, M.J. (2000), Difference thresholds for automobile seat vibration, Applied Ergonomics, vol. 31, no. 3, pp. 255-261.

Matsumoto, Y., Maeda, S., Oji, Y. (2002), Influence of frequency on difference thresholds for magnitude of vertical sinusoidal whole-body vibration, Industrial Health, vol. 40, no. 4, pp. 313-319.

Morioka, M., Griffin, M.J. (2000), Difference thresholds for intensity perception of whole-body vertical vibration: Effect of frequency and magnitude, J. Acoustical Society Am., vol. 107, no. 1, pp. 620-624.

Paddan, G.S., Griffin, M.J. (2002), Evaluation of whole-body vibration in vehicle, Journal of Sound and Vibration, vol. 253, no.1, pp. 195-213.

Pielemeier, W.J., Jeyabalan, V., Meier, R.C., Otto, N.C. (1997), Just noticeable differences in vertical vibration for subjects on an automobile seat, Proceedings of the 32nd United Kingdom Group Meeting on Human Response to Vibration, Southampton, England, pp. 333-344.

5800 / Interaction Modeling and Dynamic Control Strategy for C-shaped Legs with Sandy Terrain in Terradynamics

Chuanxiao Yang[a], Zhiyue Xin[a], Xiong Hu[a], Shibin Sun[a], Liang Ding[b]*, and Dewei Tang[b]

[a]Shanghai Maritime University, Shanghai, China

[b]Harbin Institute of Technology, Harbin, Heilongjiang, China

**Corresponding author: liangding@hit.edu.cn*

https://doi.org/10.56884/KBPT8232

Abstract

Abstract: High trafficability and robustness on terrain surfaces composed of granular substrates can be obtained through configurations of C-shaped legs. Recently, C-shaped configuration has been widely used in the locomotion mechanism design of the legged mobile robots on harsh grounds, especially for sandy terrains. In case of rotational gait, based on the stress-macro deformation relationship according to the Resistive Force Theory (RFT) in terradynamics, an interaction model between a C-shaped leg and the sandy terrain is established in this paper. Considering the influence of velocity field of C- leg particle systems on Fourier linear coefficient in fitting precondition, the inversion characters of mechanical parameters through integral derivation and linearized expression of the model are established. Then, the experimental data samples are employed to identify the mechanical parameters of the sandy terrain, followed by checking its validity through setting a limit for the tolerance between predicted results and experimental data. Furthermore, mechanical behaviors under swing gaits are analyzed, for which some basic dynamic control strategies are suggested in this study. In this work, the outputs of driving torque within the range of a feasible region experienced at the joint are explored. It is concluded that the mechanical properties dominated by the sandy terrain and leg geometry can be revealed by the identified parameters, some driving outputs can be evaluated through interaction models in terradynamics.

Keywords: interaction model, C-shaped legs, resistive force theory, parameters identification, driving torques

1. Introduction

It has been proved that a better trade-off can be achieved by combining the robust legs with fast-wheeled locomotion when locomotors walk on harsh, rugged or deformable terrain types,[1] even constituted by granular substrates.[2] Due to this principle, *C*-shaped or *C*-typed legs (simplified as *C*-legs in the following text) offer a typical prototype with a strong mobility, which is able to provide sufficient dragging and supporting power on soft terrains [3]. Distinguished from pure paddling or pushing patterns triggered by a flat leg, two gaits categorized as "digging" and "intruding rolling" can be generated by *C*-legs.[4] Interaction modeling between *C*-legs and deformable terrains is essential to the leg design, terrain classification, as well as their advanced applications, such as robotic kinematic simulation, trajectory planning and intelligent control, etc. Recently, several approaches based on traditional Resistive Force Theory (abbreviated as RFT) in terradynamics have been developed to settle the problem of modeling between legs and substrates, since the influences of motion velocity and shape factors on properties of the stress field has been described.[5] Such models are semi-empirical formulas corresponding to fitting conditions, a linear coefficient dominated by physical properties of the terrains and geometrical features of the legs are required to scale the stress variations. In traditional terramechanics, the entry and leaving angle have been employed to predict the mechanical behavior of wheeled locomotion.[6] Similarly, the intrusion and attack angle are primary factors that affect the contact properties of terradynamics, which provides a sufficient description when legs intrude into granular substrates.[5] The enclosed configuration of the free end of locomotors is often designed as a convex hull, which can be achieved by rotating the curved edge of *C*-legs around a fixed axis. Through superimposing projected forces, RFT fitting models can also be expanded to spatial occasions.[7] However, sufficient experimental data groups are required to scale the benchmark stress coefficient for the designative particle types.

In order to make an intelligent perceptual decision, the properties of locomotors with a C-leg configuration have to be identified and the type of terrain has to be classified. These methods including the accelerated iteration of Newton, genetic algorithm (GA), support vector machine (SVM) and recursive least square, which are often employed to perform the identified tasks.[8-10] Usually, more accurate identification results are led by massive computations, such as the extended Kalman filter and composite Simpson-Newton methods. Wu has successfully classified the terrain types depending on the data tracked from capacitive tactile sensors, which are mounted on a C-leg that has a relatively small size.[11] Concentrated on a hexapod robot walking on outdoor environments, the SVM algorithm can be employed to classify the terrain substrates governed by the data extracted through experiments.[12]

Modeling based on RFT and its verification of the interaction of *C*-legs interaction with the sandy terrains are involved in this paper. Furthermore, these mechanical parameters indicated by Fourier form are identified through sufficient experimental studies and recursive least square method. Set as the boundary condition, a mechanical model is employed to generate the output of control strategy in dynamic occasion in this work.

2. Modeling the interaction for C-legs

According to Resistive Force Theory (RFT), there is a linear stress Fourier function with variables including attack angle, intrusion angle and subsidence. The schematic principles of side views have been illustrated in Fig. 1. Since the integral bounds of a stress model are governed by the rotation state, two phases are categorized into the consideration of the interaction period, including the incompletely submerged state and submerged state. The incompletely submerged state begins from the moment that the initial contact occurs in the threshold state when the free end of a C-leg is the same height with terrain surface, as is shown in Fig. 1.

2.1 Modeling Based on Terradynamics

The strip element along the width direction represented by the Point P is an infinitesimal component of the C-leg, with its tangent plane indicated by l_{TP}, which is perpendicular to Line O_2P.

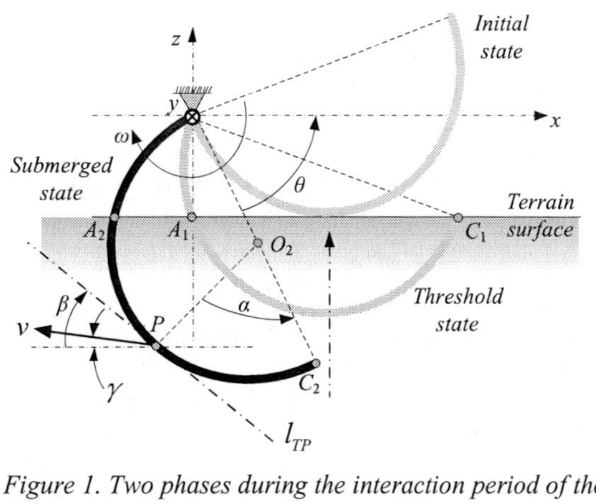

Figure 1. Two phases during the interaction period of the C-leg with sandy terrain

The Cartesian coordinate is set in Fig. 1 and the C-leg is driven clockwise around the stationary Y-shaft. The interaction stresses predicted on the basis of RFT can be calculated as follows:

$$\begin{aligned} &\sigma_{z,x} = \alpha_{z,x} \cdot \operatorname{sgn}(\Delta z) \cdot \Delta z \\ &\alpha_{z,x} = Fourier\{\beta, \gamma\} \\ &\beta = \beta(\theta, \alpha) \\ &\gamma = \gamma(\theta, \alpha) \end{aligned} \tag{1}$$

where $\sigma_{z,x}$ denotes the contact stresses respectively experienced on the element P along the axis z and x respectively, attack angle β refers to the inclination of tangent plane l_{TP}, while intrusion angle γ indicates the inclination of velocity vector, Δz is the sinking displacement of the arbitrary element P and θ are the rotary angles that the C-leg passes from the initial state, α is the parameter revealing the location of the element P at the edge of the C-leg, which acts as the only integral variable in the following net force and driving torque functions.

$$F_{z,x} = \int \sigma_{z,x}(\theta,\alpha) \mathrm{d} A_s = \int [\sigma_{z,x}(\theta,\alpha) \cdot \frac{\partial A_s}{\partial \alpha}] \mathrm{d}\alpha \tag{2a}$$

$$T_y = \int [\sigma_z(\theta,\alpha) x_p(\theta,\alpha) + \sigma_x(\theta,\alpha) z_p(\theta,\alpha)] \frac{\partial A_s}{\partial \alpha} \mathrm{d}\alpha \tag{2b}$$

where A_s is the distributed area of stresses. A high plasticity is performed by the granular medium of sandy soil. Hence, sand particle systems present creepage under external loads. The spatial area of deformation in sand will spread slowly to adjacent regions. Therefore, there is no need in considering the magnitude of interaction velocity between C-leg and sandy terrain. The linear Fourier coefficients $\alpha_{z,x}$ are expressed as Eq. 2 and 3.

$$\alpha_z = \sum_{i=0}^{1} A_i \cos(2i\beta) + \sum_{j=0}^{1} B_j \sin(2j\beta + \gamma) \tag{3}$$

$$\alpha_x = \sum_{m=0}^{1} C_m \cos(2m\beta + \gamma) + D_1 \sin(2\beta) \tag{4}$$

where A_i, B_j and C_m denote the fitting coefficient in the *Fourier* form. Considering the full range of variations of parameters θ and α, a unique mapping relation of them with the attack and intrusion angle can be obtained as:

$$\{\beta,\gamma\} = \left\{ -(\theta + \alpha + \frac{\pi}{2}), \theta + \frac{\alpha}{2} + \frac{\pi}{2} \right\} \tag{5}$$

Then the integral boundary conditions for Eqs. (2a) and (2b) are quantitatively determined by the following expressions, which have been categorized into two phases:

At the incompletely submerged stage:

$$\theta \in \left\{ -\arcsin\frac{h}{2R}, \arcsin\frac{R-h}{R} \right\} \tag{6a}$$

$$\text{s.t. } \alpha \in [\alpha_l, \alpha_r]$$

$$\begin{cases} \alpha_l = \arcsin(\frac{h}{R} + \sin\theta) - \theta - \pi \\ \alpha_r = \arcsin(\frac{h}{R} + \sin\theta) + \theta \end{cases} \tag{6b}$$

At the submerged stage:

$$\theta \in \left\{ \arcsin\frac{h}{2R} - \pi, -\arcsin\frac{h}{2R} \right\} \tag{7a}$$

$$\text{s.t. } \alpha \in [\alpha_l, \alpha_r] = \left[\arcsin(\frac{h}{R} + \sin\theta) - \theta - \pi, 0 \right] \tag{7b}$$

where α_l and α_r are the left and right boundary conditions, h is the raising height from terrain surface to the rotary axis Y, and R is the radius of curvature of the C-leg. Through early derivations, it has been demonstrated that analytical solutions to Eq. 2a performing a linear form (Eq. 8) can be achieved through the transformation of these angles and the partial integration of these functions. The analytical results have been shown in Table.1, in which the boundary conditions are governed by Eq. 6 and 7.

$$[F_z, F_x] = \left[\sum_{i=0}^{1} A_i T_{Ai} + \sum_{j=0}^{1} B_j T_{Bj}, \sum_{m=0}^{1} C_m T_{Cm} + D_1 T_{D1} \right] \quad (8)$$

Table 1. Analytical solutions of Eq. 2 in the linear form

Linear term	Analytical results
T_{A0}	$(\sin\theta + \frac{h}{R})\alpha\Big\|_{\alpha_l}^{\alpha_r} - \cos(\theta+\alpha)\Big\|_{\alpha_l}^{\alpha_r}$
T_{A1}	$(\frac{R}{2}\sin\theta + \frac{h}{2})\sin(2\theta+2\alpha)\Big\|_{\alpha_l}^{\alpha_r} + R\cos(\theta+\alpha)\Big\|_{\alpha_l}^{\alpha_r} - \frac{2R}{3}\cos^3(\theta+\alpha)\Big\|_{\alpha_l}^{\alpha_r}$
T_{B0}	$\frac{2}{3}(R\sin\theta + h)\sin(\theta + \frac{3}{2}\alpha)\Big\|_{\alpha_l}^{\alpha_r} + \frac{2}{5}R\cos\theta\sin\theta\sin(\frac{5}{2}\alpha)\Big\|_{\alpha_l}^{\alpha_r} + 2R\cos^2\theta\cos(\frac{\alpha}{2})\Big\|_{\alpha_l}^{\alpha_r} + 2R\left[2\cos^3(\frac{\alpha}{2}) - \frac{8}{5}\cos^5(\frac{\alpha}{2}) - \cos(\frac{\alpha}{2})\right]\Big\|_{\alpha_l}^{\alpha_r}$
T_{B1}	$2(R\sin\theta + h)\sin(\theta + \frac{\alpha}{2})\Big\|_{\alpha_l}^{\alpha_r} + \frac{2}{3}\sin\theta\cos\theta\sin(\frac{3\alpha}{2})\Big\|_{\alpha_l}^{\alpha_r} - \frac{2}{3}\cos^2\theta\cos(\frac{3\alpha}{2})\Big\|_{\alpha_l}^{\alpha_r} + \left[\frac{4}{3}\cos^3(\frac{\alpha}{2}) - 2\cos(\frac{\alpha}{2})\right]\Big\|_{\alpha_l}^{\alpha_r}$
T_{C0}	$(\frac{R}{2}\sin\theta + \frac{h}{2})\cos(2\theta+2\alpha)\Big\|_{\alpha_l}^{\alpha_r} + \frac{2R}{3}\sin^3(\theta+\alpha_r)\Big\|_{\alpha_l}^{\alpha_r}$
T_{C1}	$\frac{2}{3}(R\sin\theta + h)\cos(\theta + \frac{3}{2}\alpha)\Big\|_{\alpha_l}^{\alpha_r} + \frac{2}{5}R\cos\theta\sin\theta\cos(\frac{5}{2}\alpha)\Big\|_{\alpha_l}^{\alpha_r} + \frac{2}{5}R\sin^2\theta\sin(\frac{5}{2}\alpha)\Big\|_{\alpha_l}^{\alpha_r} + 4R\sin^3(\frac{\alpha}{2})\Big\|_{\alpha_l}^{\alpha_r} - \frac{16R}{5}\sin^5(\frac{\alpha}{2})\Big\|_{\alpha_l}^{\alpha_r}$
T_{D1}	$2(R\sin\theta + h)\cos(\theta + \frac{\alpha}{2})\Big\|_{\alpha_l}^{\alpha_r} + \frac{2}{3}\sin\theta\cos\theta\cos(\frac{3\alpha}{2})\Big\|_{\alpha_l}^{\alpha_r} - \frac{2}{3}\cos^2\theta\sin(\frac{3\alpha}{2})\Big\|_{\alpha_l}^{\alpha_r} + 2\sin(\frac{\alpha}{2})\Big\|_{\alpha_l}^{\alpha_r} - \frac{4}{3}\sin^3(\frac{\alpha}{2})\Big\|_{\alpha_l}^{\alpha_r}$

Besides, the driving torque about y-axis is:

$$T_y = \int_{\alpha_l}^{\alpha_r} [C_0 \sin(\theta + \frac{\alpha}{2}) + C_1 \sin(\theta + \frac{3\alpha}{2}) - D_1 \sin(2\theta + 2\alpha)] \cdot [\sin\theta + \sin(\theta+\alpha) + \frac{h}{R}][\sin\theta + \sin(\theta+\alpha)]\,\mathrm{d}a +$$

$$\int_{\alpha_l}^{\alpha_r} [A_0 - A_1\cos(2\theta+2\alpha) - B_0\cos(\theta + \frac{3\alpha}{2}) + B_1\cos(\theta + \frac{\alpha}{2})] \cdot [\sin\theta + \sin(\theta+\alpha) + \frac{h}{R}][\cos\theta + \cos(\theta+\alpha)]\,\mathrm{d}a \quad (9)$$

2.2 Modeling of Dynamic System

Taking the interaction force as the physical constraint, gravity, inertial force, driving torque, supporting force, and braking force are contained in the components of the entire C-leg system. The location of center of mass (*COM*) of the C-leg will change as the rotary angle θ increases, as the function of time.

$$x_{com} = \frac{R^2}{m_C}\int_0^{\pi} \rho\left[\cos\theta + \cos(\theta+\alpha)\right]\mathrm{d}\alpha \quad (10a)$$

$$z_{com} = \frac{R^2}{m_C}\int_0^{\pi} \rho\left[\sin\theta + \sin(\theta+\alpha)\right]\mathrm{d}\alpha \quad (10b)$$

where x_{com} and z_{com} denote the position of *COM*, ρ and m_C are the linear density and mass of the C-leg respectively. The forward direction of moment is set to be anticlockwise, and the dynamic equilibrium equations can be obtained as:

$$\sum F_x = F_{\mathrm{D}x} + F_x - m_C \frac{\mathrm{d}^2 x_{com}}{\mathrm{d}t^2} = 0 \quad (11a)$$

$$\sum F_z = F_{\mathrm{D}z} + F_z - m_C g - m_C \frac{\mathrm{d}^2 z_{com}}{\mathrm{d}t^2} = 0 \quad (11b)$$

$$\sum T_y = T_{\mathrm{D}} + T_y - m_C g x_{com} - I_y \frac{\mathrm{d}^2\theta}{\mathrm{d}t^2} = 0 \quad (11c)$$

where $F_{\mathrm{D}x}$ and $F_{\mathrm{D}z}$ denote the driving forces from the motor, T_{D} is the driving torque; I_y is the moment of inertia of axis y for the C-leg. If the thickness of the C-leg is ignored, the density can also be derived from including its mass and dimensions of leg.

$$I_y = \rho\int_0^{\pi} R_P^2(\theta,\alpha)\,\mathrm{d}\alpha = \frac{m_C}{\pi R}\int_0^{\pi} R_P^2(\theta,\alpha)\,\mathrm{d}\alpha \quad (12a)$$

$$R_P^2 = \left[\cos\theta + \cos(\theta+\alpha)\right]^2 + \left[\sin\theta + \sin(\theta+\alpha)\right]^2 \quad (12b)$$

Substitute the locations of COM in Eq. 10a, 10b and the inertial moment in 12a, 12b, into functions 11a and (11b), the dynamic functions of the force are transformed to:

$$\left(\frac{2\sin\theta}{\pi} - \cos\theta\right)\left(\frac{\mathrm{d}\theta}{\mathrm{d}t}\right)^2 - \left(\sin\theta + \frac{2\cos\theta}{\pi}\right)\frac{\mathrm{d}^2\theta}{\mathrm{d}t^2} = \frac{1}{m_C R}\left\{\sum_{m=0}^{1} C_m T_{Cm}(\theta) + D_1 T_{D1}(\theta) + F_{\mathrm{D}x}\right\} \quad (13a)$$

$$\left(\cos\theta - \frac{2\sin\theta}{\pi}\right)\frac{\mathrm{d}^2\theta}{\mathrm{d}t^2} - \left(\sin\theta + \frac{2\cos\theta}{\pi}\right)\left(\frac{\mathrm{d}\theta}{\mathrm{d}t}\right)^2 = \frac{1}{m_C R}\left\{\sum_{i=0}^{1} A_i T_{Ai}(\theta) + \sum_{j=0}^{1} B_j T_{Bj}(\theta) + F_{\mathrm{D}z} - m_C g\right\} \quad (13b)$$

Similarly, the dynamic driving torque is written as:

$$T_{\mathrm{D}} + T_y(\theta) = \frac{m_C g R}{\pi}(\pi\cos\theta - 2\sin\theta) + \frac{2m_c}{R}\frac{\mathrm{d}^2\theta}{\mathrm{d}t^2} \quad (14)$$

The driving torque T_y is governed by the rotary angle θ as well as its matched boundary conditions α_l and α_r, as is shown in Eq. 9. It is difficult to obtain analytical solutions to driving torques of the angle θ. Instead, numerical solutions can be alternative means to solve the problem. In previous studies on C-legs, we found that the linear coefficients B_0 and A_1 are contributed more to the vertical force, while C_0 is more influential for horizontal force.

3. Experimental Study and Establishment

An experimental platform is carried out in this work, developed in Harbin Institute of Technology. The joint height from the terrain surface is adjusted through balanced conditions including the lifting jack, counterweights, and the dead-weight of the C-leg. A conveyor belt is employed to realize the horizontal locomotion of the C-leg. IPC is used to drive the servo motor to realize the rotation around the fixed axis. The net force torques as well as their corresponding altitude angles are measured through a six- dimensional sensor. The schematic principles of the experiment apparatus are illustrated in Fig. 2.

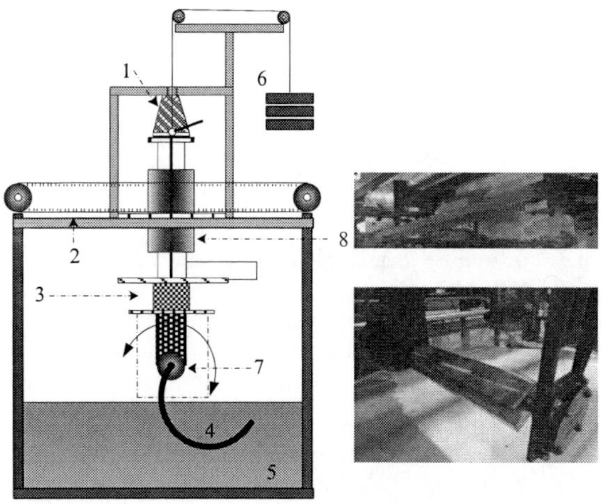

Figure 2. Experimental apparatus of interaction mechanics between rotary C-leg and sandy terrain

It has been found that the maximum interaction force would be increased by only 20 percent if the rotary angular velocity raises from 0.2 to 1 rad/s. Therefore, a relatively low angular velocity is chosen in this work to achieve a steady "paddling" which makes it more precise when being tracked by industrial cameras. The ratio of the peak upper bearing load to sand height should also be considered during experimental conduction. Before each experiment, calibration through encoders and repeated flattening are required. Automatic filtering is adopted to contribute to the oscillation descending of data in a signal processing system, and all configured parameters are presented in Table 2.

Table 2. Configured parameters in the experiment

Parameters	Numerical values
Width of C-leg: b (mm)	100, 150, 200
Radius of C-leg: R (mm)	250
Rotary joint height: h (mm)	150
Angular velocity: ω (rad/s)	0.45
Sandy layer height: (mm)	1000
Internal friction angle: (rad)	0.437
Bulk density: (kg/m^3)	1.61×10^3

Based on the parameters listed above, the net force is estimated through Eq. 8 in advance, and the magnitude of the maximum interaction force is not larger than 300N. In order to measure and handle the sensor and design errors, structural steel is chosen to build the C-leg, of which the total deformation under distributed loading conditions is predicted through finite element analysis software, as is shown in Fig. 3. It can be found that when the thickness of the C-leg is set as 5mm, the maximum deformation in simulation will not exceed 2.2mm. Compared with dimensional sizes of the C-leg, this deformation is not large and can be neglected.

The experimental results in width groups of 100 mm and 150 mm are illustrated in Fig. 4. It can be concluded that the peak force occurs earlier than the middle location revealed by 1.57 rad. Through comparisons, it has also been induced that both the maximum vertical and horizontal force will increase as the width increases. However, the predicted force almost remains unchanged though the width continues rising after the threshold of 200mm. Hence, the magnitude of force tends to rise disproportionately as the width is increased. while the width of the leg rises from 100 mm to 150 mm, the peak value of vertical force increases by 52.0%, and that of horizontal force increases by 42.0%. However, only 15.8% of increment is generated from the vertical force if the width is increases to 200 mm, which is lower than 26.6% of the horizontal force. When the magnitude of width is changed to 250mm, obvious augments of forces can hardly be observed in the experiments. In Fig. 4, only typical width groups including b=100 mm and b=150mm are presented.

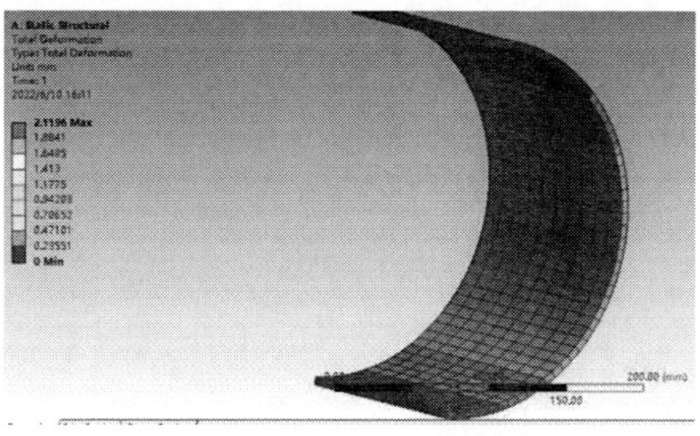

Figure 3. Maximum deformation at static bearing condition

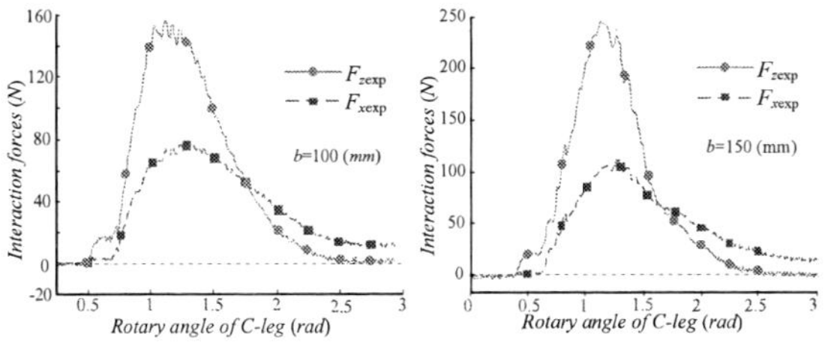

Figure 4. Experimental results of interaction forces experienced on C-leg

4. Parameters Identification Based on Interaction Model

The value of linear parameters A_0, A_1, B_0, B_1, C_0, C_1 and D_1 is obtained through recursive least square (RLS) method, the iteration of vertical force term by term has been presented as.

$$\left[{}^{n+1}A_0, {}^{n+1}A_1, {}^{n+1}B_0, {}^{n+1}B_1\right]$$

$$=\left\{\begin{bmatrix} {}^{n}A_0 \\ {}^{n}A_1 \\ {}^{n}B_0 \\ {}^{n}B_1 \end{bmatrix}^{\mathrm{T}} \boldsymbol{T}_{4n}(\theta_1,\ldots,\theta_n)+F_{\exp z(n+1)}\begin{bmatrix} T_{A0}(\theta=\theta_{n+1}) \\ T_{A1}(\theta=\theta_{n+1}) \\ T_{B0}(\theta=\theta_{n+1}) \\ T_{B1}(\theta=\theta_{n+1}) \end{bmatrix}^{\mathrm{T}}\right\}\cdot$$

$$\left\{\boldsymbol{T}_{4n}(\theta_1,\ldots,\theta_n)+\begin{bmatrix} T_{A0}(\theta=\theta_{n+1}) \\ T_{A1}(\theta=\theta_{n+1}) \\ T_{B0}(\theta=\theta_{n+1}) \\ T_{B1}(\theta=\theta_{n+1}) \end{bmatrix}\begin{bmatrix} T_{A0}(\theta=\theta_{n+1}) \\ T_{A1}(\theta=\theta_{n+1}) \\ T_{B0}(\theta=\theta_{n+1}) \\ T_{B1}(\theta=\theta_{n+1}) \end{bmatrix}^{\mathrm{T}}\right\}^{-1} \quad (15)$$

where $\boldsymbol{T}_{4n}$ is a 4×4 symmetric matrix governed by the Eq. 16; the functions of T_{A0}, T_{A1}, T_{B0} and T_{B0} have been expressed in Table.2; $F_{\exp n}$ indicate an array of experimental data ($1\times n$) of vertical forces; n+1 denotes the serial number to be identified.

$$\boldsymbol{T}_{4n}=\begin{bmatrix} \sum_{t=1}^{n} T_{A0}^2(\theta_t) & T_{A0A1} & T_{A0B0} & T_{A0B1} \\ T_{A1A0}=T_{A0A1} & \sum_{t=1}^{n} T_{A1}^2(\theta_t) & T_{A1B0} & T_{A1B1} \\ T_{B0A0}=T_{A0B0} & T_{B0A1}=T_{A1B0} & \sum_{t=1}^{n} T_{B0}^2(\theta_t) & T_{B0B1} \\ T_{B1A0}=T_{A0B1} & T_{B1A1}=T_{A1B1} & T_{B1B0}=T_{B0B1} & \sum_{t=1}^{n} T_{B1}^2(\theta_t) \end{bmatrix} \quad (16)$$

The elements T_{MiNj} (M=A, B; N=A, B; i=0,1; j=0,1) are expressed as:

$$T_{MiNj}=\sum_{t=1}^{n} T_{Mi}(\theta_t)T_{Nj}(\theta_t) \quad (17)$$

Concerning Eq. 16, the initial value of these parameters is achieved by four pieces of experimental data corresponding to the initial four timing points. An iteration form similar to Eq. 16 can be obtained for horizontal force, in which the experimental data $F_{\exp x(n+1)}$ should be obtained in advanced. Based on the above analysis, the ultimate value of these parameters has been identified and presented in Table 3. Considering all experimental width groups, it has been found the tolerance between the results of prediction and from experiments won't exceed 5% in previous studies. To have an intuitive understanding of these errors, the presentation of identified results of vertical and horizontal forces for the width configuration b=100 mm has been illustrated in Fig. 5.

Table 3. Configured parameters in the experiment

Parameters	Identified values based on ***RLS*** method		
	b=100 mm	b=150 mm	b=200 mm
A_0	102.64	151.83	162.46
A_1	210.18	290.05	298.75
B_0	-603.30	-862.72	-857.67
B_1	72.18	90.95	103.68
C_0	-186.00	-230.64	-271.44
C_1	132.11	172.67	212.38
D_1	4.66	5.89	7.01

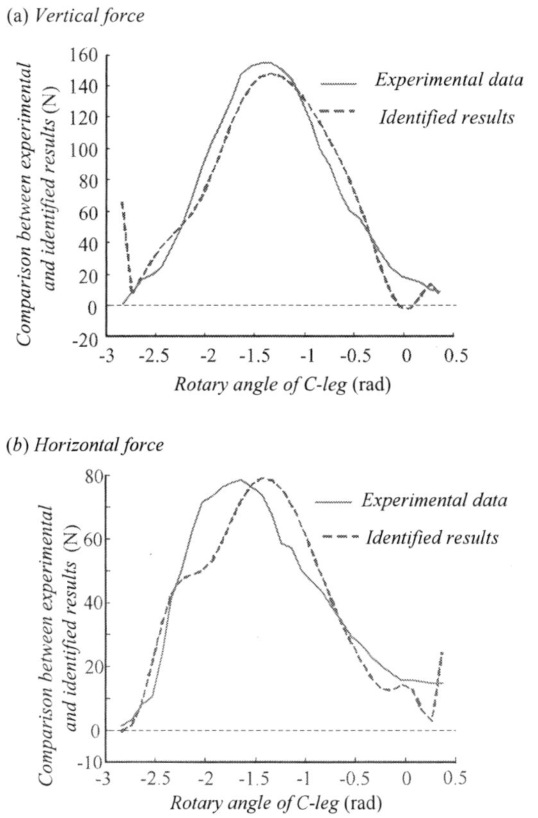

Figure 5. Comparisons between experimental data and identified results

5. Torques and Forces in Uniform Speed Situation

In order to predict the driving torques, the Newton-Cotes method is used to solve the Eq. 9. Four intervals are separated into in the range of (0, π), and the Cotes solution to Eq. 9 can be obtained as:

$$T_y = \frac{\pi}{90}\{7[g(\alpha=\alpha_l)+g(\alpha=\alpha_r)]+32[g(\alpha=\frac{3\alpha_l+\alpha_r}{4})+ g(\alpha=\frac{3\alpha_r+\alpha_l}{4})]+12g(\alpha=\frac{\alpha_l+\alpha_r}{2})\} \quad (18)$$

where $g(\alpha, \theta)$ denotes the integrand in Eq. 9. Hence, depending on identified parameters, the outcome of driving torques using Cotes solutions are presented in Fig. 6. The maximum driving torque that can be used to complete a full rotation into the sandy terrain with a fixed axis is about 1767.92N· m.

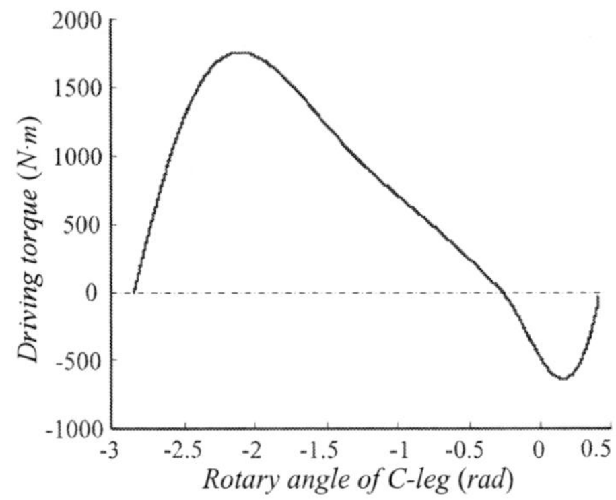

Figure 6. Driving torque based on Eq. 15 in the Cotes form

If the C-leg rotates in a relatively uniform angular velocity, the angular acceleration in Eq. 13 and 14 can be neglected. First of all, it is assumed that the mass m_C and radius R of the C-leg are constants, and control velocity is set as ω, then the outputs of F_{Dx}, F_{Dz}, T_D can be expressed as:

$$F_{Dx} = m_C R\omega^2\left(\frac{2\sin(\omega t)}{\pi} - \cos(\omega t)\right) - F_x(\omega t) \quad (19a)$$

$$F_{Dz} = -m_C R\omega^2\left(\sin(\omega t) + \frac{2\cos(\omega t)}{\pi}\right) - F_z(\omega t) + m_c g \quad (19b)$$

$$T_D = \frac{m_C g R}{\pi}[\pi\cos(\omega t) - 2\sin(\omega t)] - T_y(\omega t) \quad (19c)$$

The angular velocity ω can be replaced by the equation $\omega=\Delta\theta/\Delta t$. Thus, the functions predicting forces in Eq. 19 *a~c* dominated by the rotary angle θ can be obtained, as is illustrated in Fig. 6. The values designed for the mass and radius are m_C=10kg and R=250mm respectively. The initial value of the angular velocity ω=0.2rad/s, and it has been found that F_{Dx}, F_{Dz} as well as T_D almost remains unchanged even if ω=2rad/s. This can be a further proof in dynamics that rotary velocity has little impact on driving force. However, if the mass of the C-leg increases, the image of vertical force will move upwards accordingly. Through observation of Fig. 7, it can be found that the horizontal and vertical force can be approached by trigonometric functions.

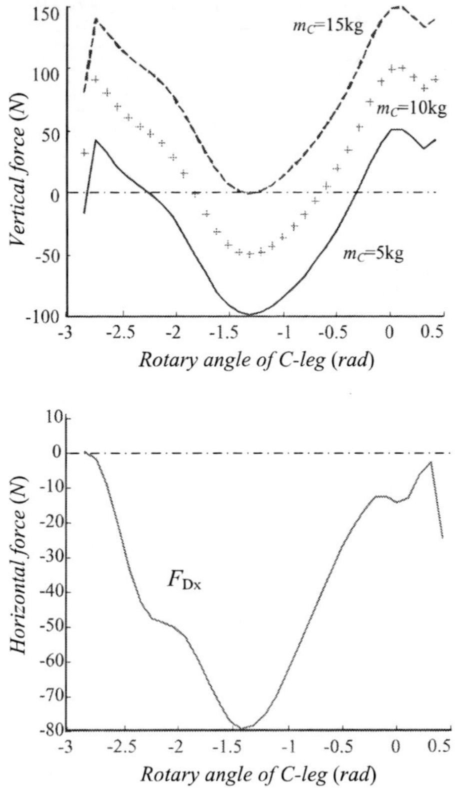

Figure 7. Control forces in the uniform speed situation

6. Conclusions

In this paper, we focus on the modeling of interaction mechanics between a C-leg and a sandy terrain based on terradynamics as well as its preliminary application to the identification and outputs of driving forces/torques.

Through integration by parts, an analytical solution to the interaction force can be achieved in the Fourier form, of which both the motion and configuration parameters are involved in the systematic models. Then through an experimental study, it has been found that limited reinforcements in resisting sinkage and slides are generated as the width of the C-leg increases. Using the RLS method, we have successfully identified the linear coefficients of interaction models with an acceptable tolerance. Depending on these identified parameters and models, the required time-dependent driving torques are emulated numerically in the four-level Newton-Cotes form. It is concluded that the time when the maximum driving torque occurs is earlier than that of the maximum force. Finally, the input passive force governing a relatively steadily rotation is presented in the simulation. The magnitude of rotary angular velocity almost has no effect on interaction mechanics, avoiding too heavy or too light mass design of the C-leg is beneficial to the steady supporting and rotation control.

7. Acknowledgements

This paper was partly supported by the National Key R&D Program of China (91948202).

8. References

[1] Chao L., Xicen T., Yanan Y., Zhiyang F., 2022. Design and analysis of a novel deformable wheel-legged robot. *Journal of Mechanical Engineering*. 58(3), 65-73.

[2] Shrivastava S., Karsai A., Aydin Y.O., et al., 2020. Material remodeling and unconventional gaits facilitate locomotion of a robophysical rover over granular terrain. *Science Robotics*, 5(42), 10.1126/scirobotics.aba3499.

[3] Vina A., Barrientos A., 2021. C-legged hexapod robot design guidelines based on energy analysis. *Applied Sciences*, 11(6), 2513.

[4] Bai G., Thomaszewski B., 2020. A C-Legged Monopedal Robot and its Transition from Multiple Locomotion Modes. *Journal of Mechanisms and Robotics*, 12(4):1-24.

[5] Chen L., Tingnan Z., Daniel I G., 2013. A terradynamics of legged locomotion on granular media. *Science*, 339(6126), 1408-1412.

[6] Junlong G., Tianyou G., Ming Z., et al., 2020. In-situ evaluation of terrain mechanical parameters and wheel-terrain interactions using wheel-terrain contact mechanics for wheeled planetary rovers. *Mechanism and Machine Theory*, 145, 103696.

[7] Treers L., Cao C., Stuart H., 2021. Granular resistive force theory implementation for three-dimensional trajectories. *IEEE Robotics and Automation Letters*, 6(2), 1887-1894.

[8] Dianlei H., Rui Z., Qingqiu Cao., et al., 2021. Research in mechanical model of bionic foot intruding into sands with different physical characteristics[J]. *Journal of Terramechanics*. 98: 25-33.

[9] Arreguin A.L., Montenegro S., Dilger E., 2021. Towards in-situ characterization of regolith strength by inverse terramechanics and machine learning: A survey and applications to planetary rovers. *Planetary and Space Science*, 204, 105271.

[10] Riya Z., Yiting K., Jue Y., et al., 2021. An integrated terrain identification framework for mobile robots: system development, analysis, and verification. *IEEE-ASME Transactions on Mechatronics*, 26(3), 1581-1590.

[11] X.Alice W., Tae Myung H., Rudranarayan M., et al., 2016. Integrated ground reaction force sensing and terrain classification for small legged robots. *IEEE Robotics & Automation Letters*, 1125-1132.

[12] Yue Z., Feng G., Qiao S., Yunpeng Y., 2021. Terrain classification and adaptive locomotion for a hexapod robot Qingzhui. *Frontiers of Mechanical Engineering*, 16(2): 271-284.

5979 / Research on Drag Reduction Performance of Sliding Plate of Rice Direct Seeding Machine Based on Non-Smooth Structure of Loach Surface

Hongchang Wang[a,b], Zhen Jiang[a,b], Kaiquan Ding[a,b], Guozhong Zhang*[a,b], Abouelnadar Salem[a,b,c], and Yuan Gao[a,b]

[a]College of Engineering, Huazhong Agricultural University, Wuhan, Hubei, China
[b]Agricultural Equipment Engineering Technology Research Center, Wuhan, Hubei, China
[c]Soil Conservation Department, Desert Research Center, Cairo, Egypt
**Corresponding author: zhanggz@mail.hzau.edu.cn*

https://doi.org/10.56884/KUAL5071

Abstract

Sliding plate was the key soil-engaging component of rice direct seeding and planting machinery. It has the problems of large sliding resistance and serious soil adhesion. which has a serious negative impact on the operation efficiency and quality of rice planting machinery. Reducing the soil adhesion and resistance between sliding plate and soil can significantly improve the operation effect of the whole machine and reduce power consumption. Loach moves freely and flexibly in the mud and has highly efficient lubrication and drag reduction effects. The movement state of sliding plate was similar to that of loach, as well as the working environment and conditions. Therefore, the sliding plate of rice direct seeding machine was selected as the research object and the loach as the bionic prototype. The macroscopic and microscopic structure characteristics of the scales were observed, the results showed that the body surface of the loach was covered by scales, and the scales had a ridged non-smooth structure. The simulation analysis of the drag reduction performance of the non-smooth structure based on Fluent was carried out, the results show that the maximum drag reduction rate was 2.55% at the speed of 1m/s. The bionic sliding plate of rice direct-seeding machine was constructed based on the non-smooth structure of loach body surface, and its working performance was simulated and analyzed. The single factor test results show that at the speed of 1m/s, the drag reduction rate of ribbed height in the range of 3.5mm to 4.5mm was relatively high, the drag reduction rate of ribbed width in the range of 4mm-5mm was relatively high; the drag reduction rate was relatively high when the distance between ribs was in the range of 4mm5mm. The results of orthogonal test show that the order of primary and secondary factors of bionic structure parameters affecting drag reduction rate was ribbed spacing > ribbed width > ribbed height. The optimal parameter combination was ribbing height 4mm, ribbing width 4.5mm, ribbing spacing, and the optimal drag reduction rate was 4.21%. The results of this study can provide theoretical support for bionic design of soil engaging components of rice planting machinery in wet and soft paddy field.

Keywords: loach, non-smooth, sliding plate, drag reduction, paddy field

1. Introduction

Paddy field was an important part of cultivated land in China, accounting for about 24.5% of the cultivated land area, and rice was generally cultivated. Rice was one of the three main grains in China, with an annual planting area of about 30 million hectares and more than 60% of the domestic population as the main grain. Therefore, the production of rice plays a crucial role in China's food security. At present, the key working components of rice planting machinery, such as rice transplanter seedling transplanting mechanism (Xu et al., 2020), pot seedling transplanting mechanism of pot seedling transplanting machine (Wang et al., 2021), seeder of precision seed-planter (Cai et al., 2020; Xing et al., 2021), which can basically meet the agronomic requirements of rice planting.

The soil-engaging components of sliding plates such as furrow sliding plates and seeding sliding plates are the key auxiliary working components of rice planting machinery. In the working process, the soil-engaging components of sliding plates directly and continuously contact with disturbed saturated paddy soil, support the weight, also have the function of the open ditch and flat land, can produce extrusion, shear on paddy soil, resulting in slurry paddy field to spill out of the soil, the resistance to flow and indicates, on the one hand will affect seedling, bowl, and seeds, such as lodging, buried, affect the growth of rice seedlings and the water seed germination. On the other hand, soil adhesion to the surface of the soil-engaging components leads to increased resistance, increased power consumption, and reduced performance (Cao et al., 2022; Zhang et al., 2016). The average slip rate of the tractor in paddy field was up to 34%, the maximum traction efficiency was only about 46%, and the average fuel consumption was up to 544g/(kw.h). When the boat tractor works in paddy field, the sliding resistance accounts for about 20%~40% of the total resistance. Due to the decrease of soil bearing capacity and soil adhesion, it was difficult to sink and move the operating tools, which leads to the decrease of the working efficiency and performance of the whole rice planting machine.

In order to improve the working performance and operation effect, scientists have carried out optimization research on sliding plate type soil-engaging components. In order to reduce the soil pressure on paddy field and reduce the sludge, side plates and auxiliary plates and other mechanisms were attached to the sliding plate or boat plate (Liu et al., 1994; Li et al., 1997). Zhu et al. (1985, 1989) put forward a mathematical model of resistance and drag reduction rate under the condition of air curtain, calculated the area of inflatable layer, resistance of slide plate and drag reduction rate, and carried out experimental verification. Zhang et al. (2017) improved the sludge and reduced sliding resistance by changing the structural parameters such as rake angle and radius of curvature of the sliding plate (Wang, 2016; Jiang et al., 2021). A bionic boat model of rice transplanter was designed based on the microscopic structure of fish scales, and the contact between the bionic model and paddy soil was simulated by using discrete element method. The research results showed that the bionic model had a good drag reduction effect, the maximum total resistance was reduced by 49.84% (Wu et al., 2020, Jiao et al., 2017). The simulation and experimental research on hull with non-smooth surface of concave and convex hull were conducted, and the results show that the hull with non-smooth surface microstructure has obvious drag reduction effect (Zhou et al., 2019; Zhou et al., 2021). The resistance of the ship hull, at low speed, based on the non-smooth surface structure of the crocodile abdomen was studied, and the results showed that the resistance of the hull with a non-smooth surface can be reduced by 6.3% (Yan et al., 2020). Although the existing studies have achieved certain results in reducing mud blocking and sliding resistance, it still fails to solve the bottleneck problems of mud blocking, large sliding resistance and excessive flow of mud and water. In particular, there was a lack of research on the drag reduction performance of the sliding plate of rice direct planting machine.

Loach is a typical aquatic animal. It moves freely and flexibly in the mud. Its body skin was covered with scales and mucus deep in the skin, which can play a protective role and reduce resistance, and had efficient drag reduction and lubrication effect (Sun, 2019). The surface microstructural characteristics of *Paramisgurnus dabryanus* were extracted and studied, the results showed that the drag reduction rate of the bionic sample made of rib-shaped structure on the surface of *Paramisgurnus dabryanus* could reach 23% (Wu et al, 2020, 2021). The body surface mucus of loach was studied, and the results showed that the body surface mucus of loach had rheological properties and had good effects on drag reduction and interface lubrication (Seo et al., 2020).

The movement state of the sliding plate of the rice direct-seeding machine was similar to that of the loach, as well as the working environment and conditions. Therefore, the sliding plate of the rice direct seeding machine was taken as the research object in this study, and the loach as the bionic prototype. The microstructure of the loach surface was using observed by using the ultra depth of field microscope and scanning electron microscopy (SEM), and the drag reduction performance of non-smooth surface structure of the loach surface was simulated and analyzed by using the FLUENT software. The bionic slide plate of rice direct seeding machine was constructed based on the results, and the interaction with paddy soil and working performance were studied.

2. Materials and Methods

This study absolutely conforms to the animal protection law of China and has passed the ethical review of animal experiments of Huazhong Agricultural University. Loach used in this experiment were bought from fish market of Wuhan.

2.1 Microstructure Characteristics of Loach Body Surface

2.1.1 Ultra depth of field microscope test

Before observation, the loach needed to be pretreated including the loach was anesthetized with ether and killed. The body surface of the loach was washed with distilled water to remove impurities on the body surface. A 5% sodium hydroxide solution was prepared, and the cleaned loach was soaked in the solution for 30 minutes to remove the mucus on the body surface. After removal from the solution, rinse with distilled water and 75% alcohol solution to thoroughly remove mucus and impurities. Samples with the size of 10mm*10mm were taken from different parts of the body surface of the loach, cleaned with distilled water and placed on the slides. The scales on the body surface are removed and cleaned with an ultrasonic cleaner to remove surface impurities. The samples and scales of the body surface were observed with a digital ultra depth of field microscope (VHX-6000) (Fig. 1 and Fig. 2).

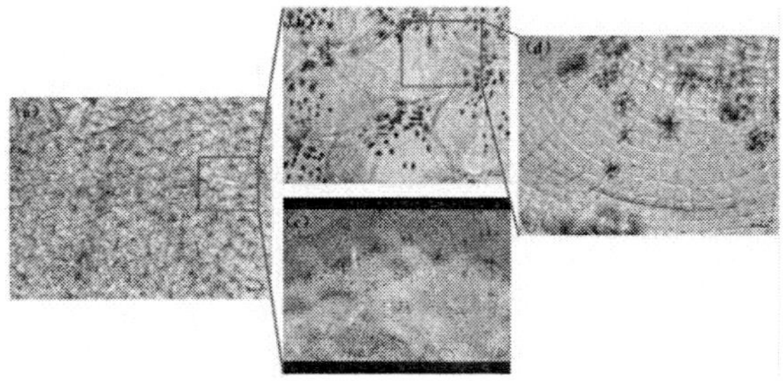

Figure 1. Characteristics of body surface of loach

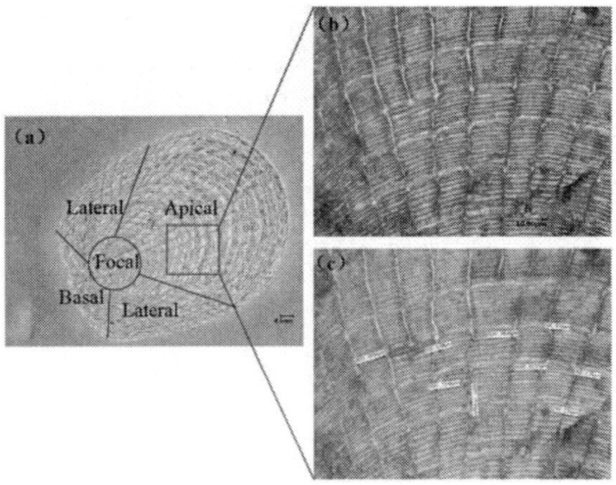

Figure 2. Characteristics of body surface scales of loach

2.1.2 SEM test

The scales were obtained according to the method of ultra depth of field microscope test. The treated scales were put into 2.5% glutaraldehyde fixed solution and dried and fixed at low temperature. After drying and fixing, spray gold for 30 minutes. The morphological characteristics of scales were observed at 1000, 1700 and 3000 magnification, respectively (Fig. 3).

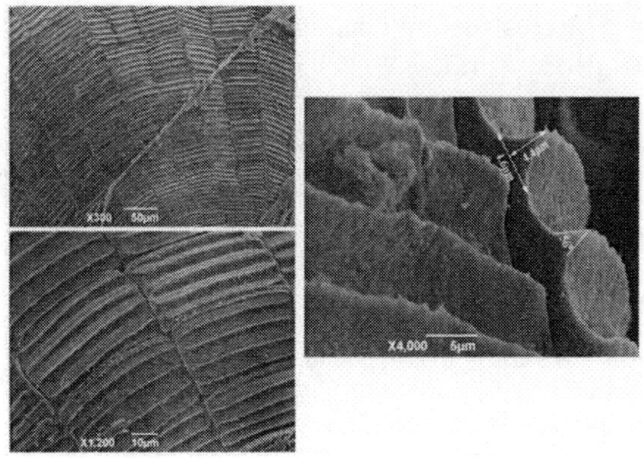

Figure 3. Microstructural characteristics of scales

2.2 Simulation Model Construction

2.2.1 Establishment of paddy soil model

The paddy field soil was in a disturbed saturation state after treatment, and its rheological properties were significantly enhanced. The contact process between the sliding plate and paddy field soil was relative sliding. In this study, the rigid body of the sliding plate was treated as fixed, and the paddy field soil was treated as flow domain, to simulate the mutual movement of the sliding plate and paddy soil. According to the soil parameters of paddy soil of Table 1, the paddy soil watershed was established in FLUENT software.

Table 1 Physical characteristics of paddy soil

W %	Density kg/m-3	M g/mol	λ K-1	δ s Pa	PV Pa S
30	1792	35	0.0002	1220.3	215

2.2.2 Simulation model of sliding plate

In order to investigate the drag reduction performance of microstructure of loach scale, a micro-scale sliding plate model was constructed by using Solidworks software (Fig. 4), and the length, width and height of the model were 50μm, 10μm and 15.6μm, respectively. The constructed sliding plate model was imported into the fluent software and the paddy field soil watershed with the length, width and height of 60μm, 20μm and 20μm, respectively. Then, the basin surface was defined as inlet and outlet. The surface to be measured was defined as wall-drag, the flow surface was defined as SYM-moving, and the symmetry surface was defined as SYM, and the grid was meshed. The flow rate at the inlet of paddy field soil basin was set as 1m/s, and the gravity acceleration was set as -9.8m/s^2 on z-axis.

On the basis of micro-scale simulation analysis, the existing rice direct-seeding slide plate was taken as the prototype, and a bionic sliding plate of rice direct-seeding was constructed according to the characteristics size of scale microstructure through the similarity principle (Fig. 5). The corresponding paddy field soil watershed model was constructed (Fig. 6), and the watershed parameter settings were shown in Table 1.

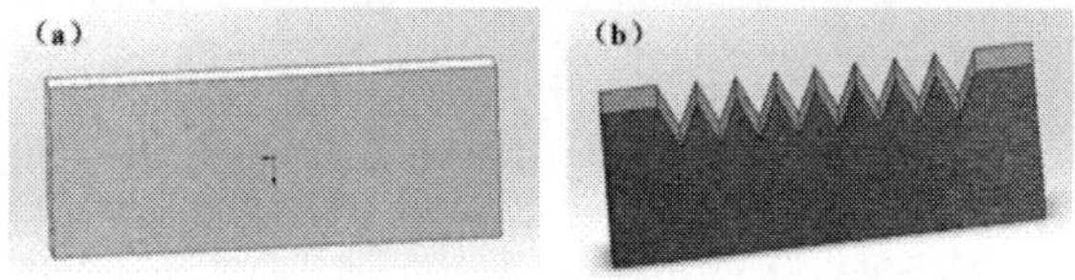

Figure 4. Micro scale sliding plate model:
a) smooth sliding plate, B) bionic sliding plate

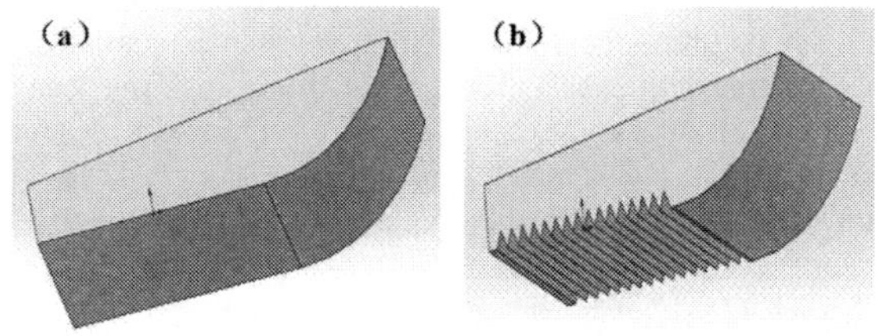

Figure 5. Sliding plate of rice direct seeding machine:
a) traditional sliding plate, b) bionic sliding plate

Figure 6. Paddy soil watershed model

2.2.3 Simulation Parameters

The total resistance of sliding plate components in mud basin was divided into differential pressure resistance and viscosity resistance. Differential pressure resistance refers to the formation of pressure difference between front and rear parts caused by eddy current when mud flows through characteristic structural surface, resulting in resistance. Viscosity resistance refers to the friction generated on the surface of the characteristic structure, due to the viscosity of the mud when the mud medium flowing through the surface of the characteristic structure, resulting in resistance. The total resistance of sliding plate components was calculated as shown in *Eq.1*.

$$F = F_1 + F_2 \tag{1}$$

Where, F was the total resistance, F_1 was the differential pressure resistance, and F_2 was the viscosity resistance.

The simulation analysis in this study mainly investigated the drag reduction characteristics of non-smooth structure characteristics under different velocity conditions in the basin, so the medium velocity in the basin was selected in the range of 1m/s-4m/s. The calculation *Eq.* of drag reduction rate of slide plate was as follows:

$$\gamma = (F_3 - F_4)/F_3 \times 100\% \qquad (2)$$

Where, γ was the drag reduction rate, F_3 was the resistance of the smooth sliding plate, F_4 was the resistance of the bionic sliding plate.

In the interaction between the sliding plate of rice direct-seeding machine and paddy field soil, the influence of velocity and non-smooth structure characteristic parameters on the drag reduction performance of the sliding plate was simulated and analyzed. According to the actual working conditions of the rice direct seeding machine, the speed range was set as 0.5m/s-1.5m/s. The parameters of non-smooth feature (rib) include rib height, rib width and rib space (Table 2).

Table 2. Influence parameters of drag reduction performance

Velocity (m/s)	Rib height (mm)	Rib width (mm)	Rib space (mm)
0.5	3.0	2.0	2.0
0.75	3.5	2.5	2.5
1	4.0	3.0	3.0
1.25	4.5	3.5	3.5
1.5	5.0	4.0	4.0
	5.5	4.5	4.5
	6.0	5.0	5.0

3. Results and Discussion

3.1 Geometric structure characteristics of body surface of loach

It can be found that superdepth of the body surface of the loach were covered with layers of overlapping scales by the observation of ultra-depth field microscopy and bios-canning electron microscopy. The single scale was oval, and the scale can be divided into four areas: Basal, Lateral, Focal and Apical (Wu, 2019). The Basal and Focal areas were embedded in the skin to fix the scale, while the Basal and Focal and Lateral blocked by other scales. The Apical area was exposed to external mucus. The scale had a wavy and regularly arranged rib geometric structure (as shown in Fig. 3) from the Focal area to the surrounding. The measured parameters of the scale microtopography were as follows: the rib height was 4.4μm, the rib width was 4.6μm, and the rib vertex angle was 56°.

3.2 Drag reduction performance of non-smooth structure characteristics of loach scales

According to the Fluent software simulation in the micro scale condition, the velocity of medium in the basin was in the range of 1m/s-4m/s. The differential pressure resistance, viscosity resistance, total resistance and drag reduction rate of the smooth sliding plate and bionic sliding plate, were measured (Fig. 7 and Fig. 8).

Fig. 7 showed that the differential pressure resistance, viscosity resistance and total resistance of smooth sliding plate and increase with the increase of mud flow rate. The viscosity resistance and total resistance of the bionic sliding palte were less than that of smooth sliding plate, while the differential pressure resistance of the bionic sliding plate was greater than the smooth sliding plate. Because the bottom surface of the bionic sliding plate was a non-smooth surface, which made the fluid generate eddy current at the rib structure, and a low-pressure area was generated between the ribs, resulting in the surrounding high-pressure area fluid flowing into the area between the ribs, forming a local water circulation, and the non-smooth structure forms a thicker water film than the smooth surface. Therefore, the pressure difference become larger, the differential pressure resistance increases, the adhesion between the sliding plate and the watershed decreases, and the viscosity resistance decreases. It can be seen from the Fig. 7 that the decrease of the viscosity resistance was greater than the increase of the differential pressure resistance, and the total resistance decreases.

It can be seen from Fig. 8, the drag reduction rate decreases with the incease of mud flow velocity. When the velocity of mud was 1 m/s, there was the maximum drag reduction rate, which was 2.55%. Combined with the differential pressure resistance curve in Fig. 7, with the increase of speed, the eddy current effect between non-smooth structure increases, the pressure difference increases, and the differential pressure resistance increases, resulting in the reduction of total resistance, and reduction of the drag reduction rate.

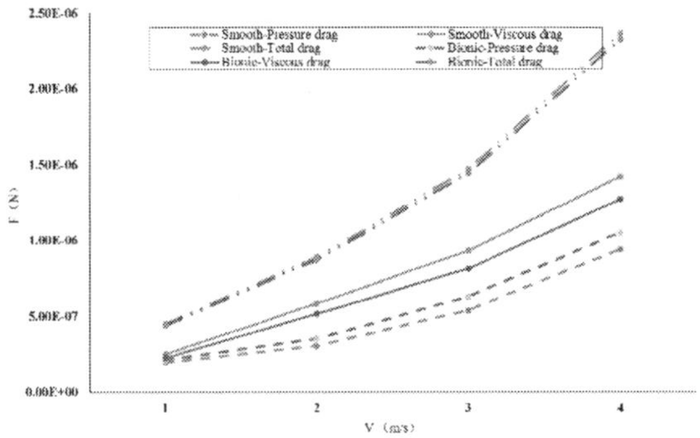

Figure 7. Variation law of resistance with forward speed

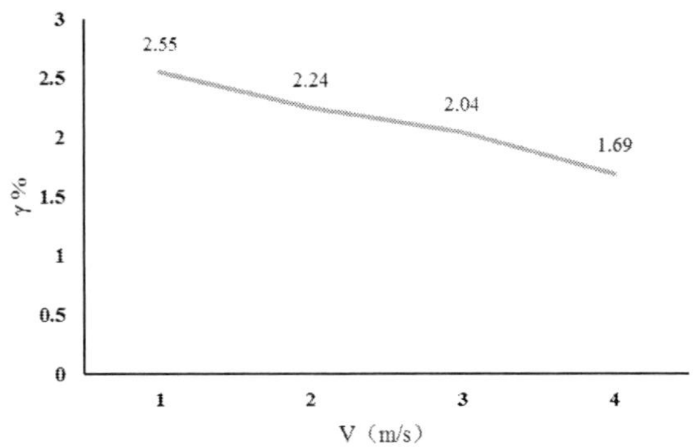

Figure 8. Variation law of drag reduction rate with forward speed

3.3 Analysis of drag reduction of sliding plate of rice direct seeding machine

In order to investigate the drag reduction performance and effect of the rid non-smooth structure of loach scale on macro scale, the geometric dimensions of rid were applied to the underside of the sliding plate of the rice direct seeding machine according to the similarity principle, and the bionic sliding plate with non-smooth surface was obtained, and its drag reduction performance was simulated.

3.3.1 Influence of velocity on drag reduction

According to the working speed of the sliding plate of the rice direct seeding machine, the medium flow rate of in the watershed was set at 0.5m/s-1.5m/s. The interaction between the sliding plate and the paddy field soil watershed was simulated. The differential pressure resistance, viscosity resistance and total resistance of the smooth sliding plate and the bionic sliding plate under different speeds were obtained respectively (Fig. 9), and also the drag reduction rate (Fig. 10).

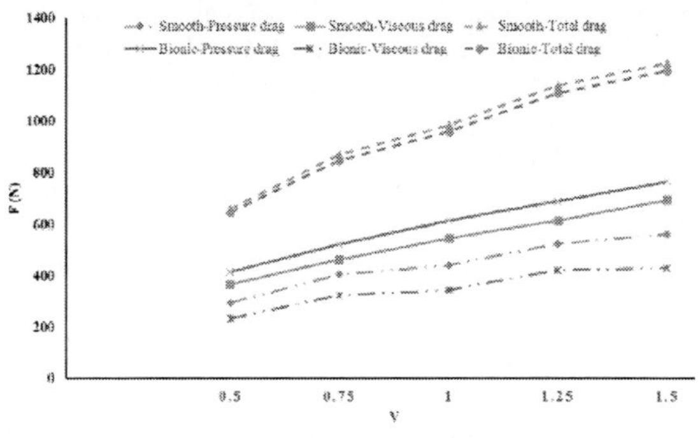

Figure 9. Variation law of resistance with forward speed

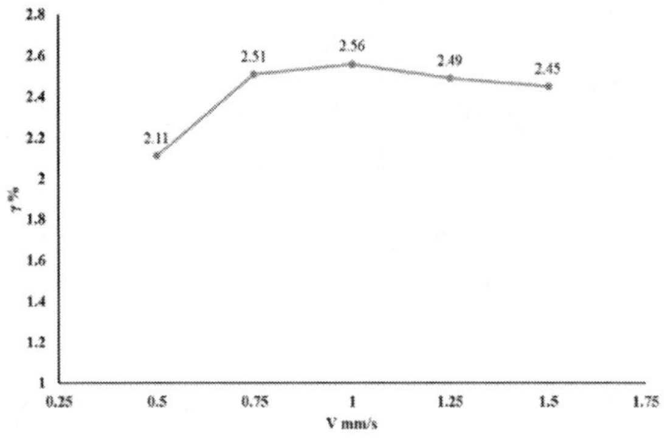

Figure 10. Variation law of drag reduction rate with forward speed

Fig. 9 showed that after macro the rib non-smooth structure, the trend of resistance of the sliding plate was consistent, that was, namely the differential pressure resistance of bionic sliding plate was greater than the smooth sliding plate, viscosity resistance and total resistance of the bionic sliding plate were less than the smooth sliding plate. With the increase of speed, differential pressure resistance, viscosity resistance and total resistance also increase. Fig. 10 showed that the variation law of drag reduction rate was consistent with that of micro-scale, that is, with the increase of velocity, drag reduction increases first and then decreases, and the maximum drag reduction rate was 2.56% at 1m/s. Comparing the drag reduction performance of the micro-scale non-smooth geometric features, the rib structure of loach scales had the effect of drag reduction.

3.3.2 Influence of rib geometric characteristic parameters

In order to further investigate influence of the geometric characteristics parameters of the rib on the drag reduction of sliding plate, the single factor simulation was carried out for three factors with the rib height, rib width and rib space as variables, respectively, to study the effect of every factors on the drag reduction effect. According to the foregoing research, the optimal drag reduction effect was achieved when the medium speed was 1m/s, so the speed was set as 1m/s in the simulation analysis in this section. The values of rib height, width and space ranged from 3mm-6mm, 2mm-5.5mm and 2mm-5.5mm, respectively. Other parameters were the same as those in the previous study. The total resistance with different rib height, width and space were obtained (Fig. 11).

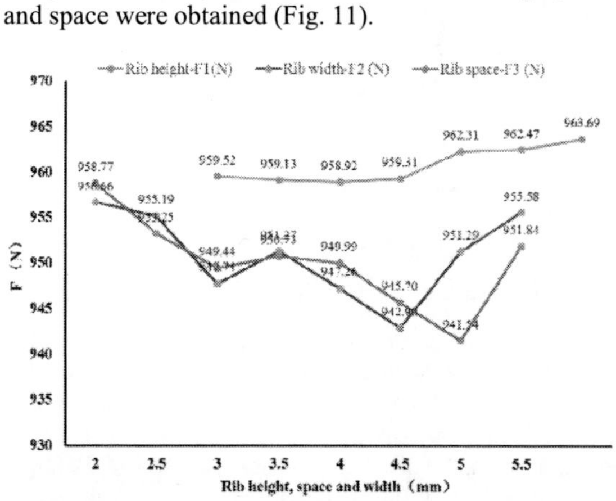

Figure 11. Influence of rib geometric characteristic parameters

It indicated that the total resistance of the bionic sliding plate increases with the increase of the rib height. When the rib height was 3.5mm-4.5mm, the total resistance of the bionic sliding plate was small. When the rib height was 4mm, the total resistance had the minimum value, which was 958.92N. With the increase of rib space, the total resistance of the bionic sliding plate decrease first and then increases. When the rib space was 5mm, the total resistance was the lowest, which was 941.54N. When the spacing was small, the space between rib was limited, the eddy current effect was weak, and the drag reduction effect was not obvious. With the increase of rib space, the space between rib also increases, the eddy current effect was enhanced, the water film thickens, and the drag reduction effect was obvious. When the rib space was too large, the space between rib was larger, the eddy current effect was weakened, the water film thickness was reduced, and the drag reduction effect was reduced. As can be seen from Fig. 12, when the spaces was small, the resistance was large and uneven, mainly concentrated in the middle and front end of the sliding plate.

With the increase of rib width, the total resistance of bionic sliding plate decreases first and then increases. When the rib width was 4.5mm, the total resistance had the lowest value, which was 942.90N. The variation trend was consistent with the rib space, because the change of rib width was the change of the bottom width, which was essentially the change of the space between ribs and had a similar effect with the change of rib space.

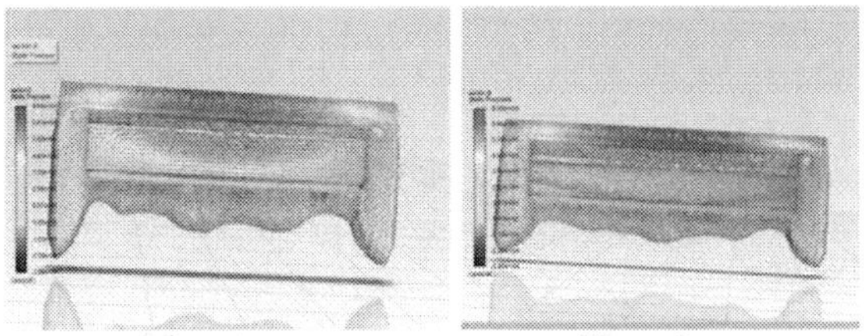

Figure 12. Resistance cloud of different rib space

In order to further understand the influence degree and priority of the three parameters of rib height (A), rib width (B) and rib space (C) on the resistance, L9(34) orthogonal table was selected for orthogonal experimental analysis on the basis of the above results (Table 3). The results of orthogonal experimental were shown in Table 4.

Table 3. Rib parameter factor land evel

Level	(A)	(B)	(C)
1	3.5	4	4
2	4	4.5	4.5
3	4.5	5	5

Table 4 Analysis of orthogonal experimental results

Test No.	A	B	C	D	F (N)	γ %
1	1	1	1	1	952.43	3.24
2	1	2	2	2	946.60	3.83
3	1	3	3	3	945.47	3.95
4	2	1	2	3	949.28	3.56
5	2	2	3	1	942.90	4.21
6	2	3	1	2	951.74	3.31
7	3	1	3	2	947.60	3.72
8	3	2	1	3	950.90	3.40
9	3	3	2	1	951.69	3.61
K1	11.02	10.52	9.95	11.06		
K2	11.08	11.44	11.00	10.86		
K3	10.73	10.87	11.88	10.91		
k1	3.67	3.51	3.32	3.69		
k2	3.69	3.81	3.67	3.62		
k3	3.58	3.62	3.96	3.64		
R	0.12	0.31	0.64	0.07		
The order of factors primary → secondary					C>B>A	
The optimal combination					$A_2B_2C_3$	

It indicated that the order of influencing factors on drag reduction rate of the bionic sliding plate of rice direct-seeding machine was C>B>A, that was, the rib space had the greatest influence on drag reduction rate, followed by rib width, and rib height has the least influence. The optimal parameter combination was $A_2B_2C_3$, that was, the rib height was 4mm, the rib width was 4.5mm, and the rib space was 5mm. This parameter combination can be found in the analysis table of orthogonal test results, which was Test No. 5, with the total resistance was 942.9N, and the drag reduction rate was 4.21%. Therefore, when the rib non-smooth structure was 4mm in height, 4.5mm in width and 5mm in space, the optimal drag reduction effect was achieved, and the drag reduction rate was 4.21%.

4. Conclusions

In this study, the microscopic morphology and structural characteristics of the scales of loach body surface were observed and analyzed by using ultra depth of field microscope and bioscanning electron microscope. It was found that the surface of loach was covered with overlapping scales, which could be divided into Basal, Lateral, Focal and Apical areas. The Basal and Focal areas were embedded in the skin to fix the scales, while the basic, focal, and later areas were blocked by other scales. The Apical area was exposed to external mucus. The scale had a wavy and regularly arranged rib geometric structure from the focal area to the surrounding. The rib structures can store mucus and reduce movement resistance. The drag reduction mechanism of rib structure was revealed by fluent simulation analysis. The existence of rib structure increases the differential pressure. Although it increases the differential pressure resistance, due to the increase of the thickness of the surface water film, the viscosity resistance was greatly reduced, and the total resistance was reduced.

Based on the microstructural characteristics of loach scales, a bionic sliding plate of rice direct seeding machine was designed, and the interaction between the sliding plate and paddy soil was simulated. The variation of sliding plate resistance with speed and rib structural parameters was obtained, and the optimal rib parameter combination and the maximum drag reduction rate were obtained.

5. Nomenclature

M	*Molar mass*	*[g/mol]*
PV	*Plastic viscosity*	*[Pa·S]*
W	*Soil moisture content*	*[%]*
λ	*Thermal conductivity*	*[k^{-1}]*
δ_s	*Yield stress*	*[Pa]*

6. Acknowledgements

This study was supported by the National Natural Science Foundation of China (No. 51775220).

7. References

Xu, C., Shan Y., Xin, L., Xie, J., Li, Z., Zhao, Y., 2020. Design and Experiment of High-speed Rice Transplanter with Extensible Mulch Cutting Mechanism in Mulching Cultivation System. Transactions of the Chinese Society for Agricultural Machinery. 51, 79-87.

Wang, L., Sun, L., Xu, Y., Yu, G., Zhang, W., Zheng, J., 2021. Design Method of Transplanting Mechanism of Planetary Gear Train Based on Spatial Trajectory. Transactions of the Chinese Society for Agricultural Machinery. 52, 51-9.

Cai, J., Liu, M., Xiao, L., Lin, J., Ye, Y., Yang, H., 2020. Design and Experiment of Transplanting Device with Variable Row-spacing of Rice Potted-seedling Transplanter. Transactions of the Chinese Society for Agricultural Machinery. 51, 50-9.

Xing, H., Wang, Z., Luo, X., Zang, Y., He, S., Xu, P., Liu, S., 2021. Design and experimental analysis of rice pneumatic seeder with adjustable seeding rate. Int J Agr Biol Eng. 14, 113-22.

Cao, M., Long, Z., Wang, Y., Pan, Y., Wang, J., 2022. Impact of soil physical properties on the driving performance of a tracked tractor on paddy soils in the plastic state. Journal of Zhejiang University (Agriculture and Life Sciences). 48, 125-34.

Zhang, X., Xia, J., Zhang, J., He, X., Liang, S., Zhang, S., Wu, H., Wan, S., 2016. Working performance experiment of combination blade roller for straw returning in paddy field and dry land. Transactions of the CSAE. 32, 9-15.

Liu, Q., Li, B., Huang, J., 1994. Research and experiment on the mechanism of preventing mud mound in the motorized rice transplanter. Modern Agricultural Equipment. 11-5.

Li, Zh., Li, Q., Li, J., Chen, H., 1997. The study on the theory and measure of waveproof of high-speed rice transplanter. Transactions of the Chinese Society for Agricultural Machinery. 34-40.

Zhu, G., Qiao, X., 1985. Experimental research on parameters of the body of bttm. Transactions of the Chinese Society for Agricultural Machinery. 3, 17-28.

Zhu, G., Jiang, C., 1989. The development performance test and structure analysis of boat- tractor. Transactions of the Chinese Society for Agricultural Machinery. 2, 46-52.

Wang, C., 2016. Design and experiment of precision rice drop drilling machine with simple structure. Guang Zhou.

Zhang, M., Luo, X., Wang, Z., Wang, B., Xue, Zh., 2017. Optimization design and experiment of profiling and slide board mechanism of precision rice hill-drop drilling machine. Transactions of the CSAE. 33, 18-26.

Jiang, Q., He, Z., Wang, Y., Wang, J., 2021. Optimizing the working performance of a boat-type tractor using central composite rotatable design and response surface method. Comput electron agr. 181, 105944.

Wu, L., Wang, H., Song, Y., Zhang, B., Xu, Y., Liu, C., Yan, Y., 2020. Drag reduction mechanism of Paramisgurnus dabryanus loach with self-lubricating and flexible micro-morphology. Sci Rep-Uk. 10.

Wu, L., Wang J., Luo G., Wang S., Qu J., Fan X., Liu C., 2021. Study on the Drag Reduction Characteristics of the Surface Morphology of Paramisgurnus dabryanus Loach. Coatings. 11.

Jiao, Zh., 2017. Research on drag reduction of transplanter float with bionic microstructure. Shen Yang.

Zhou, M., Chen, L., Liu, Y., Guo, F., 2019. Research on drag reduction mechanism of non-smooth surface for bottom plate of the boat-tractor. Journal of Agricultural Mechanization Research. 41, 258-63.

Zhou, J., Zou, X., 2021. The analysis of drag reduction performance of the boat-type tractor. Journal of Agricultural Mechanization Research. 43, 227-33.

Yan, G., Tian, P., Mo, J., Xie, H., Wei, H., 2020. Low-velocity resistance distortion and bionic drag reduction for ship-type paddy field machinery. Int J Agr Biol Eng. 13, 7-14.

Sun, P., 2019. Study on extraction, isolation, purification and characterization of lectin from loach skin mucus. Tian Jin.

Seo, E., Yoon, G.Y., Kim, H.N., Lim, J.H., Kim, S., Kim, B., Kim, K.H., Lee, S.J., 2020. Morphological features of mucous secretory organ and mucous secretion of loach Misgurnus anguillicaudatus skin for friction drag reduction. J Fish Biol. 96, 83-91.

6174 / Factors Affecting Bevameter Soil Characterization

Schalk Els*, Herman Hamersma, Ray Kruger
Department of Mechanical and Aeronautical Engineering, University of Pretoria, South Africa
**Corresponding author: schalk.els@up.ac.za*

https://doi.org/10.56884/IKRL4697

Abstract

The Bekker-Wong soil-wheel interaction model has been widely adopted in the terramechanics community. This approach requires data measured with a Bevameter. The Bevameter test includes a pressure-sinkage test and a shear strength test. Although the original intention of the Bevameter method is to closely replicate the interaction between the wheel/track and the soil, tests are often done at lower vertical loads with smaller contact areas to keep tests within practical experimental limits. The traditional Bevameter shear strength test, which uses a rotational shear ring with grousers, tends to overpredict the drawbar pull of vehicles by 30 to 40% [(Chang and Baker, 1973, Shoop, 1993)]. Translational shear arguably more closely resembles the shear mechanism that occurs between a wheel/track and the soil. The literature indicates that terramechanics investigations are significantly influenced by soil condition and preparation. This study investigates (a) the influence of soil preparation on the pressure-sinkage and shear-displacement relationship of sandy soil, (b) the influence of the shearing mechanism on shear strength, and (c) the influence of shear contact area on shear strength. Experimental results indicate that soil preparation has a substantial influence on the pressure-sinkage relationship. The shearing mechanism was investigated by comparing traditional rotational shear with translational shear. The findings indicate that the shearing mechanism may explain the overprediction of shear strength typical of rotational shear tests. The shear contact area also exhibits a significant influence on the measured shear strength. Further investigation into the shear mechanism and size and shape of the shear contact area is recommended.

Keywords: Bevameter, pressure-sinkage, soil shear strength, shear mechanism, contact area

Editor's note: full-width tables and figures located at end of paper.

1. Introduction

Humans have been involved in off-road locomotion since the invention of the wheel in circa 3500 B.C. Scientific method has only been applied to off-road transport in the last century, prior methods relying mostly on empiricism and trial-and-error (Wong, 1984). Scientific interest was stimulated in this field during the 1950s and 1960s when Dr M.G. Bekker published his *Theory of land locomotion* (Bekker, 1956). Bekker laid the foundations of what is today known as terramechanics.

Wong (1984) broadly defines terramechanics as the study of the performance of a machine in relation to its operational environment. This may include the machine's mobility and its ability to perform some task such as earth moving or tilling. The field of terramechanics remains relevant today, with global industries relying on machinery that are capable of working off-road. Industries such as agriculture and forestry, earthmoving, construction, and mining rely heavily on off-road locomotion. A proper understanding of the terrain-vehicle interaction and terrain-implement interaction may result in economic activity in these industries becoming more efficient, more environmentally friendly, more economically feasible and more sustainable.

Fundamental to terramechanics is understanding the terrain-vehicle interaction. As a result, a large body of work is available in the literature that discusses several modelling approaches to characterise this interaction. Crucial to the modelling of the terrain-vehicle interaction is experimental measurement. Three prevailing measurement techniques exist in the literature to determine the soil strength (Wong, 1984), the first utilising the cone penetrometer and the second utilising the Bevameter. The third characterisation technique is based on civil engineering soil mechanics, but these techniques are not particularly suited to the study of mobility and locomotion and has seen little application in the literature.

The cone penetrometer was developed during the Second World War with the intention to provide military vehicle operators with a simple field device to assess terrain trafficability on a 'GO'/'NO GO' basis (Wong, 1984). The mechanical properties of the terrain are lumped into a single parameter known as the cone index. The cone index reflects the combined shear and compressive characteristics of the terrain. In practice, differentiation between the contribution of each factor is very difficult (Wong, 1984).

The Bevameter technique (Bevameter is a contraction of Bekker Value Meter) is an analytical approach that attempts to determine the soil behaviour in shear and compression (Golob, 1981). A moving vehicle exerts both normal and shear loads to the terrain and the Bevameter technique attempts to replicate these loads with a plate penetration test and a shear test (Wong, 1984).

The Bevameter soil characterization technique is a popular and widely employed terrain characterization method in the terramechanics community (Wong, 2009). The most prominent terrain-tyre interaction model employed today is the Bekker-Wong model (Lyasko, 2010, Harnisch et al., 2005, Gallina et al., 2014, Taheri et al., 2015, Apfelbeck et al., 2011). This model is based on, and traditionally requires, Bevameter soil characterization (Bekker, 1969, Gallina et al., 2014, Padmanabhan et al., 2018). According to the recently revised ISTVS standards, the Bevameter is used to determine the following parameters (He et al., 2020):

- cohesion,
  angle of internal friction,
- sinkage moduli (those associated with Bekker's sinkage model), and
  sinkage exponent.

Shoop (1993) classifies the Bekker values into two sets of empirical terrain model parameters: one set describes the in-situ pressure-sinkage (bearing capacity) relationship (n, k_c, k_ϕ) and the other the in-situ shear-stress displacement behavior (c, ϕ, K).

Although Bevameter soil characterization is widely employed, outdated practices exist for terrain characterisation with this method (He et al., 2020). There is no fixed test equipment specification, and test equipment is typically tailored to the need of the researcher (He et al., 2020).

It has long been recognised that the test equipment and experimental test setup influence the soil properties obtained using the Bevameter technique (Reece, 1964, Bekker, 1969). Although decades of research exist on the topic, a number of factors have been identified that have not been thoroughly investigated. One prominent but seldom investigated factor identified is how the in-situ shear mechanism affects the identified mechanical soil properties. Specifically, how the in-situ longitudinal translational shear apparatus differs from in-situ torsional shear apparatus. Another factor includes the influence of the shear tool contact area. He et al. (2020) recommends that tests should be conducted at contact areas and loads comparable to the intended running gear of the target vehicle. However, tests are often done at smaller loads and contact areas to keep tests within practical experimental limits.

Finally, investigating the mentioned test setup factors itself raises technical challenges. Typical Bevameters found in the literature are tailored for specific use cases and not to the study of a wide range of experimental setup factors. As part of this study, it is required to develop a device to enable fundamental research to be conducted on a wide range of factors that affect Bevameter soil characterisation.

This study investigates (a) the influence of soil preparation on the pressure-sinkage relationship of soil, (b) the influence of the shearing mechanism on shear strength, and (c) the influence of shear contact area on shear strength. Tests are performed on a sandy soil.

2. Methodology

2.1 Bevameter description and specifications

A new Bevameter was designed and commissioned for this study. The developed Bevameter aimed to address the shortcomings identified of existing equipment encountered in the literature. The most prominent shortcoming identified was that existing equipment are tailored to either performing translational or torsional shear tests, and not capable of both. Furthermore, small scale equipment is typically employed that is not capable of executing Bevameter tests at the full tyre contact patch areas of full-scale off-road vehicles. The developed Bevameter, depicted in Figure 1, addressed these shortcomings and present a fully self-contained in-situ soil testing unit with onboard power sources, instrumentation, and data logging equipment. For this study, the Bevameter is deployed on a dedicated outdoor soil bin for laboratory-style testing. However, field application is also possible. The Bevameter can perform three types of in-situ soil characterization tests. They are: 1) pressure-sinkage tests, 2) translational (longitudinal) shear tests, and 3) torsional shear tests. For pressure-sinkage tests, the Bevameter acts in displacement control mode. For translational and torsional shear tests, force control is applied through the use of dead weights. The translational and torsional shear motions are through controlled displacements/rotations. Various ground engagement tools can be fitted to the Bevameter to study the effect of contact area and the inclusion of grousers.

The Bevameter technical specifications are indicated in Table 1.

Table 1. Bevameter technical specifications

Specification	Value
Normal force (displacement control) [kN]	15
Normal force (force control) [kN]	6.5
Shear force [kN]	3.5
Torque [Nm]	630
Vertical velocity [mm/s]	16
Horizontal velocity [mm/s]	20
Torsional shear velocity [mm/s, RPM]	600, 50
Vertical travel [mm]	400
Horizontal travel [mm]	850
Overall Dimensions (h, w, l) [mm]	1200, 1480, 2230
Total gross mass [kg]	1020

2.2 Soil selection and preparation

Considering the application of this study, it would be ideal to conduct the experimental investigation using a soil with texture classification that is representative of agricultural soil. It is generally agreed upon that loam soil(50% sand, 20% silt, 30% clay) is the best agricultural soil (Abdulazeez, 2017). Finding a source of a large homogenous quantity of soil of this composition is challenging. The difficulties include finding an excavation site, matching the desired composition, finding a method to sieve approximately 8000 kg of excavated soil from larger particles and ensuring homogeneity. An alternative option is to engineer an artificial soil using professionally graded materials to have the desired composition and characteristics. However, the logistics of sourcing and mixing large quantities of graded sand, silt, and clay were not practical for a large outdoor soil bin. Finally, the inclusion of clay also makes it hard to control the moisture content because of the low water permeability.

For the reasons mentioned above, a sandy soil was selected for this study. The selection of sandy soils is common practice for laboratory Bevameter testing (Bekker,

1969, Van et al., 2008). Van et al. (2008) state the reason for this is that sandy soil produces repeatable soil conditions for testing in laboratory conditions. This is corroborated by Bekker (1969), who states that dry cohesionless soils exhibit lower variance in laboratory testing than other soil types. Sandy soil is also said to be the easiest soil to prepare for repeated tests (Bekker, 1969).

The specific soil selected is a fine sand known as Cullinan sand. This particular sand is selected because it is readily available to this study. During preparation, the sand was washed, baked, and graded by the supplier.

It was decided to test the soil in the partially saturated state. It is expected that the inclusion of moisture is more representative of natural soils encountered by vehicles in the field than completely dry soil. It is also postulated that the inclusion of moisture will exhibit a shear rate effect through activation of pore pressure. Furthermore, the inclusion of moisture has other benefits, such as facilitating compaction (Craig, 2013). However, a downside of the inclusion of moisture is that it adds complexity to the study as it introduces an additional variable that needs to be measured and controlled. The target moisture content was selected based on the SANS 3001-GR30:2014 (South African Bureau of Standards, 2014) modified AASHTO test for optimal compaction density.

A thorough soil characterisation investigation was performed using standard test procedures. The selected soil's engineering characteristics and other important metrics essential to the study of mechanical soil strength are indicated in Table 2.

Five different soil preparation methodologies that lead to five different soil densities were investigated. Figure 3 indicates the general methodology for soil preparation.

The soil preparation procedure consists of five steps as indicated in Figure 3. Photos of the five standard steps (including one additional optional compaction step) are provided in Figure 4.

Step 1: Agitation

The agitation procedure was kept constant throughout the soil preparation investigation. It consisted of the following two steps.

i. Machine raking with a rake attachment to a depth of 200mm as indicated in Figure 4a. The 200mm working depth is constrained by the 400mm of Bevameter vertical travel. It consists of one forward pass and one reverse pass at a rate of 15mm/s.
ii. Hand raking was performed, as indicated in Figure 4b, to roughly level the surface and further agitate and loosen clumps that are present after the deep agitation step. A procedure of three forward and three reverse passes was followed. The rake has a soil penetration depth of 50 mm.

Step 2 & 4: Levelling

A levelling tool is attached to the Bevameter, and levelling is performed as indicated in Figures 4c and 4f. This allows the surface to be levelled relative to the Bevameter ground engaging tool. In the final levelling step, 8 mm of the compacted layer is removed. This is not ideal, as it preshears and disturbs the surface from the compaction step; however, it is a necessary requirement to ensure a level surface.

Step 3: Compaction

In this study, different compaction methods are investigated. Figure 4d indicates the primary compaction method, which consisted of impacting a base plate with a tamping rod. Optional further compaction is indicated in Figure 4e. Although not indicated, the final method utilized a vibrating plate compactor, which consisted of three passes at walking pace in 100mm layers as filling of soil bin progressed.

Two methods are used to quantify the soil compaction state. The first method is the relative compaction R_c,

$$R_c = \frac{\rho_d}{\rho_{dmax}} \times 100 \tag{1}$$

Where ρ_d is the in-situ dry density and ρ_{dmax} is the maximum density from the Mod. AASHTO tests from Table 2. ρ_d was calculated using a photogrammetry method as described by Berney et al. (2018). The second method is the relative density,

$$D_d = \frac{\rho_{dmax}(\rho_d - \rho_{dmin})}{\rho_d(\rho_{dmax} - \rho_{dmin})} \times 100 \tag{2}$$

Where ρ_{dmax} is based on the ASTM D4253-92-1B vibrating table test and ρ_{dmin} is from the ASTM D4254-91-1A inverted cylinder test from Table 2.

Details of the achieved soil density and compaction state are indicated in Table 3. The results from Table 3 are in line with what is expected, an increase in compaction effort resulted in an increase in measured density. Note, method three exhibits negative relative density. This is explained by the inclusion of moisture which causes the soil particles to stick or clump together, as observed in Figures 4a-c, leading to the formation of macroscopic and microscopic pockets of air which decrease soil density. The ASTM D4254-91-1A (ASTM International, 2016) minimum density tests utilised completely dry soil. Although no density data was collected for methods one and two, these methods are expected to have even larger negative relative density. On the other hand, method five exhibits a high compaction state and is classified as dense soil. Considering the large increase in compaction energy compared to method four, a larger increase in density was expected. It may indicate the soil is approaching maximum compaction achievable in field conditions. Overall, Table 3 gives an indication of the compaction state and field densities relative to two standard laboratory methods.

2.3 Experimental procedure

Three types of experimental tests are performed, namely: 1) in-situ pressure-sinkage tests, 2) in-situ rotational shear tests and 3) in-situ translational shear tests. An example of each is indicated in Figure 5.

For the pressure-sinkage test, the Bevameter acts in displacement control mode in the vertical direction. First, the end effector is suspended approximately 10 mm above the soil. Data recording is then started. In post-processing, this acts as the load offset removal for the six load cell channels. The penetrating plate is brought to the surface until it just touches the surface as indicated in Figure 5 and held for a period of time. The point of "just touching" is determined based on combination of visual inspection and monitoring the vertical load cell output. This point is used in post-processing as a reference point for zero sinkage. The test is then continued until the depth of 80 mm is reached.

The rotational shear test starts by removing the gearbox backlash by loading the effector opposite to the direction of motion. Data recording is then started with the Bevameter end effector suspended approximately 10mm above the soil. The effector is then slowly lowered in displacement control mode until the grouser is fully embedded in the soil, and the plate underside just touches, as indicated by Figure 5b. This forms the zero position for the shear sinkage measurement in post-processing. By continually monitoring the vertical force from the calibrated readout, the end effector is slowly lowered further until the soil has taken the full weight of the Z-carriage. The Z-carriage is then decoupled and the soil engagement tool acts in force control mode by dead weight. The rotational shear test is started, as indicated in Figure 5b, and the test continues until a shear sinkage of 80 mm is reached.

The procedure is exactly the same for translational shear test, except for two additional considerations. For the translational shear test, it is critically important that the load cell is perfectly perpendicular to the direction of motion. Load cell alignment is provided by markings on the shaft indexed in 2-degree increments. The shaft is then locked to prevent rotation during translational shear tests using a friction shaft clamp mechanism. The test is continued until shear sinkage of 80 mm is reached or 250 mm shear displacement is reached. For the translational shear test, the latter is often the case.

3. Results

3.1 The effect of soil preparation

Figure 6 indicates the influence of the soil preparation method and density on soil bearing capacity. The soil preparation remained the same for all steps in Figure 3, except for the compaction step; therefore, the soil preparation methods from Table 3 primarily reflect a difference in soil density.

It is observed in Figure 6 that the soil preparation method and associated density have a significant influence on the soil pressure-sinkage response. Method (1) represents the minimum soil density in field conditions (no compaction), and Method (5) represents a dense soil near the maximum field density achievable using standard compaction methods (86% Mod. AASHTO). Method (1) exhibits the exponentially increasing trend typical for Bevameter tests found in the literature. Method (5) exhibits an initial linear increase that gradually transitions into a plastic flow. At intermediate densities, the soil exhibits a piecewise linear response with a sharp transition point between gradients. Overall, Figure 6 indicates the soil's bearing capacity is highly sensitive to density, with small changes in density resulting in large changes in the bearing capacity.

Table 2 indicates the differences between the relative soil bearing capacity for the different soil preparation methods.

Table 2. Relative soil bearing capacity for the different soil preparation methods

Soil preparation method	Bearing Capacity at 80mm [kPa]	Increase Relative to Method 1
1	64	-
2	418	653%
3	445	695%
4	611	954%
5	671	1048%

Figure 7 indicates the influence of soil preparation methods and density on translational shear stress. It is observed that the soil preparation has a less significant effect on shear stress compared to the influence it had on pressure-sinkage response. For Method (5), the soil exhibits a clear maximum shear stress. However, at lower soil densities associated with Methods 2-4, the data exhibits no clear maximum shear stress with an increasing non-asymptotic trend, rather than the gradual decrease in gradient at higher displacements. Note that the very low bearing capacity observed for Method 1 (see Figure 6), rendered the shear tests for this method impractical, because the sinkage at 93 kPa normal pressure exceeds 80 mm.

3.2 The effect of shearing mechanism on shear strength

Before quantitatively comparing translational and torsional shear, it is beneficial to compare the physical processes side by side. Figure 8 indicates the shear evolution at equivalent linear displacement (equivalent linear displacement indicated by j).

From Figure 8, it is observed that the physical shear mechanism is significantly different for linear and torsional shear. For translational shear, the soil tends to form a passive Rankine failure zone that extends in front of the shear tool in the direction of motion. The shear tool is observed to plough underneath the soil and experience significant bulldozing of the first grouser. At 70mm displacement, the centre shaft starts to engage the soil, arguably invalidating further shear measurements. Figure 8 indicates translational shear exhibits a 3D soil failure pattern with significant soil disturbance ($j \geq 30$mm) lateral to the tool direction of motion.

Figure 8 indicates the torsional shear mechanism tends to push the soil radially out and upward transverse to the direction of the annular ring displacement. The soil failure beneath the torsional shear ring is complex and differs significantly from the translational shear mechanism. Unfortunately, the failure pattern beneath the soil surface is not observable. No evidence indicates torsional shear is invalidated at larger shear displacements, as is the case for translational shear.

Figure 9 compares translational and torsional shear stress at nominal vehicle ground pressure and a significant difference between linear and torsional results is observed. Torsional shear exhibits larger shear across all shear displacements with a maximum difference of 16.7%. This is contrary to the expected result. Translational shear includes a bulldozing force due to the leading edge of the first grouser not present for torsional shear. Furthermore, the centre shaft of the translational shear mechanism is in contact with the soil, artificially increasing the shear stress. Despite these factors, rotational shear still exhibits larger shear stress. One hypothesis for the observed difference is the larger perimeter, and therefore lateral grouser area, of torsional shear at equivalent areas. Another hypothesis is that translational shear forms a plastic Rankine failure zone in front of the tool in the direction of motion, as indicated by the plastic failure far in front of the tool in Figure 8 at $j = 10$ mm, weakening the soil in the direction of motion.

3.3 The effect of shear contact area on shear strength

Figure 10 indicates the influence of contact area on shear at nominal vehicle ground pressure. Note the 583 cm^2 tests in Figure 10 were terminated early by the automatic load cell overload protection circuit. The load cell was designed for an expected shear stress ratio of 0.54 at 100 kPa and 583 cm^2; however, as observed in Figure 10, the realized stress ratio was significantly higher (0.65-0.92).

Figure 10 indicates the shear contact area has a significant influence on shear strength. The smallest area exhibits the largest shear stress, followed by the intermediate area and lastly, the largest area. A maximum difference of 15% is observed between the largest and smallest area. The difference is expected to be significantly larger (≈35%) had the tests not been terminated early. The result is contrary to the expected result. The large shear ring was expected to exhibit higher shear stress due to its larger perimeter. One hypothesis for the observed behaviour is that the grouser effects are proportionally larger for the small area since the grouser height was fixed at 10 mm. Similarly, soil fall-back and other drag shear effects may be more significant at the smaller absolute torque values. Another hypothesis is that the soil bearing capacity is higher for large area, causing less slip sinkage and therefore lower shear stress.

3.4 Soil parameter estimation

In order to model the pressure-sinkage relationship the Bernstein (1913) sinkage equation is implemented,

$$p = kz^n \tag{3}$$

In this application, the popular Bekker (1969) sinkage model is not advantageous since only a single plate size was utilized for the preliminary soil preparation study. For modelling shear stress, the Mohr-Coulomb failure criterion employed for maximum shear stress,

$$\tau_{max} = c + \sigma tan\phi \tag{4}$$

For shear stress-displacement modelling the Janosi-Hanemoto equation (Janosi and Hanamoto, 1961) is selected:

$$\tau = \tau_{max}(1 - e^{\frac{-j}{K}}) \tag{5}$$

This model is specifically recommended by the SAE J939 standard for data of increasing-non asymptotic form. Since linear shear is invalidated at larger shear displacement, a 50 mm cut-off is selected for all shear tests to enable fair comparisons. The parameters of equations (3)-(5) are determined using the nonlinear least squares estimation and uncertainty quantification technique laid out by Apfelbeck et al. (2011) for Bevameter testing. The results from parameter estimation are indicated in Table 5. For reference, the range of observed soil moisture content for all tests in this study are also indicated.

4. Discussion

The results presented in this study are preliminary, but some key insights highlighting the importance of accurate testing and reporting during soil characterization are immediately evident.

As expected, diligent soil preparation is of utmost importance for repeatable laboratory testing. The soil preparation method and resulting compaction level may influence the results by orders of magnitude. The repeatability during soil preparation is also crucial; this is not an easy task, as the moisture content is at best observable and almost impossible to control.

The shearing mechanism also influences the measured shear strength. Rotational shear tests are very prevalent in the literature, but the literature indicates that the results obtained with this method often over-predict the drawbar-pull performance of wheeled vehicles. It is postulated that the shearing mechanism used during rotational shear tests may be responsible for this discrepancy. This is a point that should be investigated in more detail, however, that falls outside of the scope of this paper.

The effect of shear stress area was also significant, although not as much as that of the shear mechanism. Nevertheless, using representative contact areas during shear strength characterisation is crucial to accurately predict the tractive capabilities of off-road vehicles.

5. Conclusion and recommendations

The aim of the study was to investigate (a) the influence of soil preparation on the pressure-sinkage relationship of sandy soil, (b) the influence of the shearing mechanism on shear strength, and (c) the influence of shear contact area on shear strength

The investigation found soil density to have a significant influence with an order of magnitude change in pressure-sinkage response due to changes in soil density. Shear tests exhibited less sensitivity to soil preparation.

A significant effect was observed when comparing linear and torsional shearing mechanisms, with torsional shear tests exhibiting an increase of up to 17.6% over translational shear tests. Considering that the literature indicates torsional Bevameter soil characterization often over-predicts drawbar pull performance, translational shear tests are believed to be more representative for predicting vehicle traction.

The shear contact area also showed a noteworthy effect on shear strength, with the smallest contact area indicating a 20% increase in shear over the full contact patch area. Furthermore, the small area exhibited a significantly higher variance compared to intermediate and large areas.

Although these results are the initial findings of this study, several recommendations can already be made:

- The difference between the shear mechanism of an actual wheel and that from Bevameter tests should be investigated. A half-round tyre or soil engagement tool can be mounted to the Bevameter and translational shear tests performed. It is hypothesised that the shear stress-displacement curve will differ significantly from standard Bevameter tests due to the different physical shear mechanisms.

Investigate if numerical and analytical methods can replicate the experimental results from this study. The numerical and analytical methods can be validated on Bevameter data before moving on to full wheel-soil interaction.

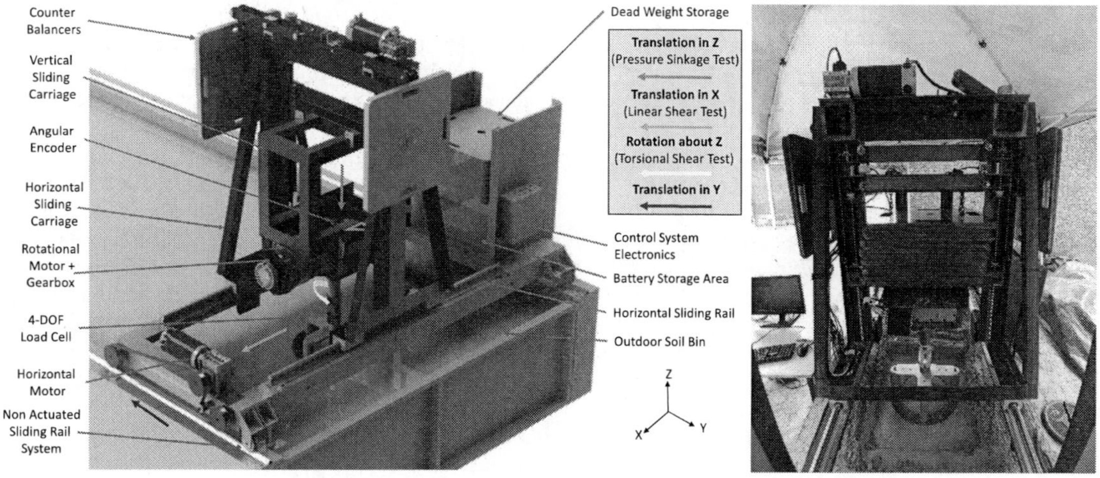

Figure 1. Isometric view of the Bevameter (left) and physical Bevameter on soil bin (right)

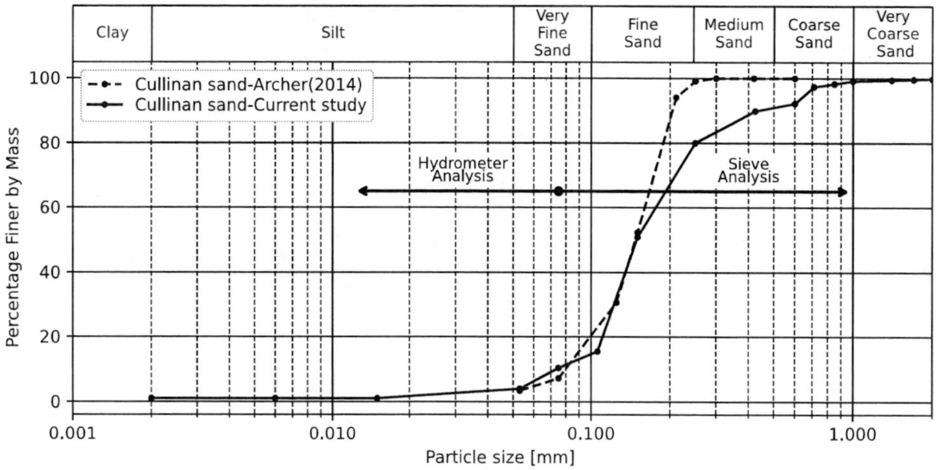

Figure 2. Soil particle size distribution. Particle size classes from USDA soil classification system (United States Department of Agriculture, 1987)

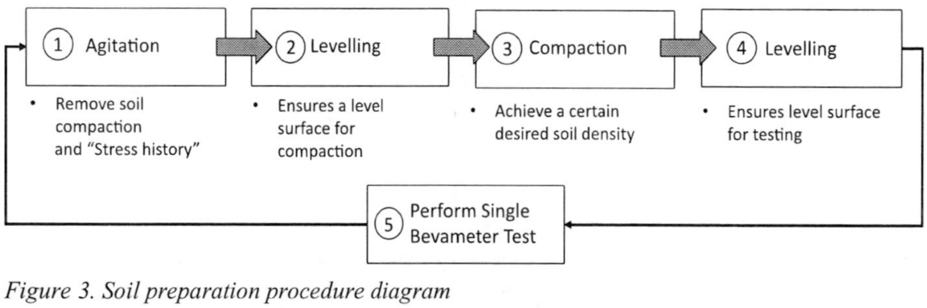

Figure 3. Soil preparation procedure diagram

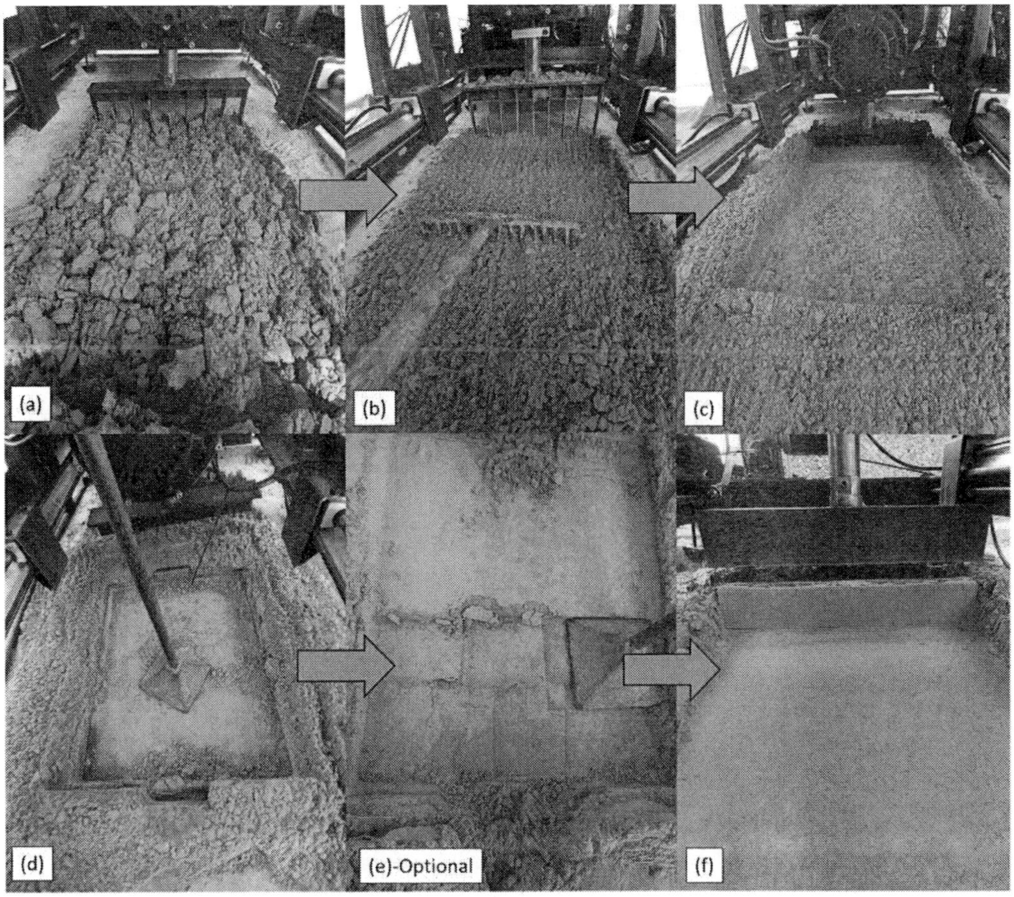

Figure 4. Soil preparation procedure. a) Deep machine raking. b) Hand raking. c) First levelling. d) Indirect tamping. e) Optional direct tamping. f) Final levelling

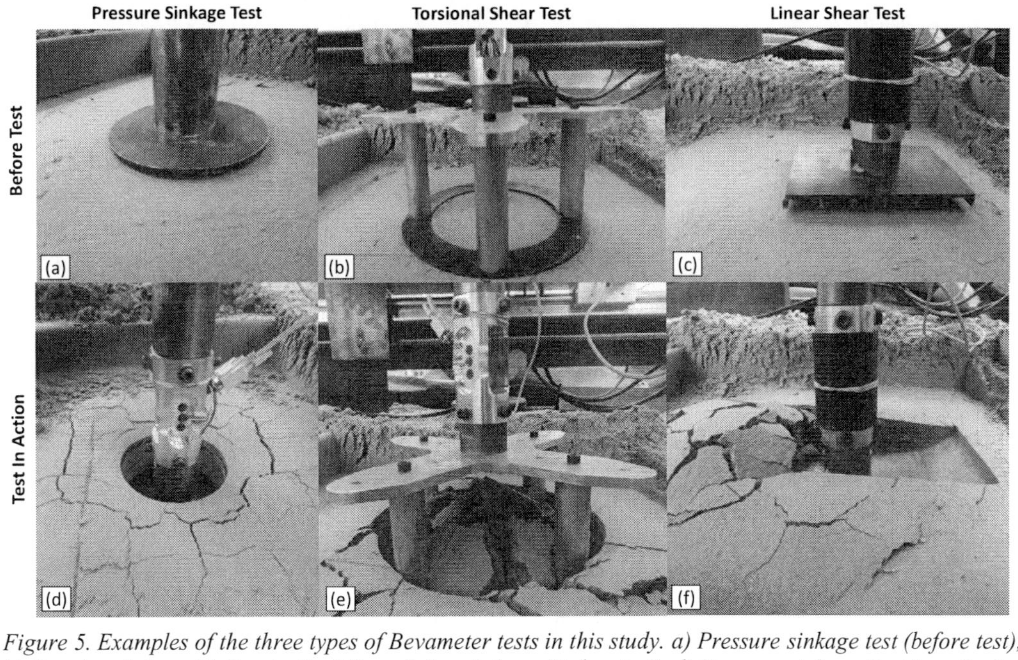

Figure 5. Examples of the three types of Bevameter tests in this study. a) Pressure sinkage test (before test), b) Rotational shear (Before test), c) Translational shear (Before test), d) Pressure sinkage (test in action), e) Rotational shear (test in action), f) Translational shear (test in action)

Table 2. Soil properties

Soil property	Value	Standard/Reference
Specific gravity λ [kg/m^3.]	2651	SANS 5844:2014
Maximum dry density ρ_{dmax}[Kg/m^3]	1580.7	ASTM D4253-92-1B
Minimum dry density ρ_{dmin}[Kg/m^3]	1429.6	ASTM D4254-91-1A
Modified Procter (Mod. AASHTO) maximum dry density ρ_{dmax} $[kg/m^3]$	1784	SANS 3001-GR30:2015
Optimum moisture content (dry basis) [%]	6.5	
$D_{50}[\mu m]$	149	ASTM D6913-17[2]
$D_{60}[\mu m]$	181.5	SANS 3001-GR3:2014[3]
$D_{10}[\mu m]$	73.5	
Coefficient of uniformity C_u	2.46	ASTM D2487-06
Uniformity classification	Uniformly graded	
Sand [%]	96.2	USDA Soil Classification (United States
Silt [%]	2.8	Department of Agriculture, 1987, He et al., 2020)
Clay [%]	<1	
Textural classification	Fine sand	
Particle shape[1]	Sub Angular	
Effective friction angle ϕ' [Deg.]	36.1	ASTM 3080-72[4]
Effective cohesion c' [kPa]	0	
Target moisture content w (dry basis) [%]	6.5	ASTM D4643-93
Maximum void ratio e_{max}	0.86	Craig (2013)
Minimum void ratio e_{min}	0.48	
Degree of saturation at max density S_{rmax} [%]	35.4	
Degree of saturation at min density S_{rmin} [%]	20.2	

[1]Particle shape from electron microscope imaging on Cullinan sand by Archer (2014).
[2]Sieve analysis.
[3]Hydrometer analysis.
[4]Direct shear box test under consolidated drained conditions, consolidated to 1784 kg/m^3 dry density, 6.5% moisture content.

Table 3. Soil compaction procedure and achieved densities

Method	Compaction method	Compaction energy $[kJ/m^2]$	Dry density $[kg/m^3]$	% Mod. AASHTO [1]	Relative density[2] [%]	Density classification[3]
1	None	0	No data collected	-	-	-
2	20 indirect blows with tamper rod on base plate	1.1	No data collected	-	-	-
3	40 indirect blows with tamper rod on base plate	2.2	1390	77.4	-29	Very loose
4	20 indirect blows tamper rod on base plate + 20 direct blows with the tamper to soil	2.2	1511	84.7	56.4	Medium Dense
5	Vibrating plate compactor - 3 passes in 100mm layers	48	1538	86.2	73.1	Dense

[1] Based on the SANS 30001-GR30: 2015 Modified AASHTO Procter test.
[2] Based on the maximum density from ASTM D4254-91-1A vibrating table test and minimum density from ASTM D4253-92-1B inverted cylinder test from Table 2.
[3] Soil density classification from Kaniraj (1988) based on relative density.

6. References

Abdulazeez, A. 2017. Effects of soil texture on vegetative and root growth of senna obtusifolia seedlings indigenous to bichi, Sudan savannah of Northern Nigeria, in green house conditions. IOSR Journal of Agriculture and Veterinary Science (IOSR-JAVS), 10, 70-74.

Apfelbeck, M., Kuß, S., Rebele, B. & Schäfer, B. 2011. A systematic approach to reliably characterize soils based on Bevameter testing. Journal of Terramechanics, 48, 360-371.

Archer, A. 2014. Using small-strain stiffness to predict the settlement of shallow foundations on sand. University of Pretoria.

ASTM International 2016. ASTM D4254-16 Standard Test Methods for Minimum Index Density and Unit Weight of Soils and Calculation of Relative Density. West Conshohocken, PA.

Bekker, M. G. 1956. Theory of land locomotion, Ann Arnor, MI, University of Michigan Press.

Bekker, M. G. 1969. Introduction to terrain-vehicle systems, Ann Arbor, University of Michigan Press.

Berney, E. I., Ganesh, N. B. & Pratt, T. C. 2018. A photogrammetric method for obtaining soil density. Engineer Research and Development Center, Vicksburg, Mississippi, United States.

Bernstein, R. 1913. Probleme zur experimentellen Motorpflugmechanik. Der Motorwagen, 16, 199-206.

Chang, B.-S. & Baker, W. 1973. Soil parameters to predict the performance of off-road vehicles. Journal of Terramechanics, 9, 13-31.

Craig, R. F. 2013. Soil mechanics, Springer.

Gallina, A., Krenn, R., Scharringhausen, M., Uhl, T. & Schäfer, B. 2014. Parameter identification of a planetary rover wheel–soil contact model via a Bayesian approach. Journal of Field Robotics, 31, 161-175.

Golob, T. B. 1981. Development of a terrain strength measuring system. Journal of Terramechanics, Vol. 18, pp. 109-118.

Harnisch, C., Lach, B., Jakobs, R., Troulis, M. & Nehls, O. 2005. A new tyre–soil interaction model for vehicle simulation on deformable ground. Vehicle system dynamics, 43, 384-394.

He, R., Sandu, C., Mousavi, H., Shenvi, M. N., Braun, K., Kruger, R. & Els, P. S. 2020. Updated standards of the international society for terrain-vehicle systems. Journal of Terramechanics, 91, 185-231.

Janosi, Z. & Hanamoto, B. 1961. An analysis of the drawbar pull vs slip relationship for track laying vehicles. Army Tank-Automotive Center, Warren, Michigan.

Kaniraj, S. R. 1988. Design aids in soil mechanics and foundation engineering, Tata McGraw-Hill.

Lyasko, M. I. 2010. How to calculate the effect of soil conditions on tractive performance. Journal of Terramechanics, 47, 423-445.

Padmanabhan, C., Gupta, S. & Mylswamy, A. 2018. Estimation of terramechanics parameters of wheel-soil interaction model using particle filtering. Journal of Terramechanics, 79, 79-95.

Reece, A. R. 1964. Problems of soil vehicle mechanics. Army Tank-Automotive Center, Warren, Michigan.

Shoop, S. 1993. Thawing soil strength measurements for predicting vehicle performance. Journal of terramechanics, 30, 405-418.

South African Bureau of Standards 2014. SANS 5844:2014 Edition 2.2 Particle and relative densities of aggregates. Pretoria, South Africa: South African Bureau of Standards.

Taheri, S., Sandu, C., Taheri, S., Pinto, E. & Gorsich, D. 2015. A technical survey on Terramechanics models for tire–terrain interaction used in modeling and simulation of wheeled vehicles. Journal of Terramechanics, 57, 1-22.

United States Department of Agriculture (1987) Soil Mechanics, Level 1, Module 3, USDA Soil Textural Classification Study Guide. USDA Soil Conservation Service, Washington DC.

Van, N. N., Matsuo, T., Koumoto, T. & Inaba, S. 2008. Experimental device for measuring sandy soil sinkage parameters. Bulletin of the Faculty of Agriculture Saga University, 93, 91-99.

Wong, J. Y. 1984. An introduction to terramechanics. Journal of Terramechanics, Vol. 21, pp. 5-17.

Wong, J. Y. 2009. Terramechanics and off-road vehicle engineering: terrain behaviour, off-road vehicle performance and design, Butterworth-Heinemann.

6316 / Perceptive Locomotion of Legged Robot Coupling Model Predictive Control and Terrain Mapping

Boyang Xing[a]*, Bo Su[a], Lei Jiang[ab]*, Yufei Liu[a], Zhirui Wang[a], Jianxin Zhao[a], and Tianqi Qiu[a]
[a]Unmanned Center, China North Vehicle Research Institute, Beijing
[b]College of Computer Science and Technology, Zhejiang University, Hangzhou
**Corresponding author: golaced@163.com*

https://doi.org/10.56884/KPGL5403

Abstract

Legged robots promise an advantage over traditional wheeled systems; however, most legged robots are still confined to structured and flat environments. In this paper, we present a motion planner for the perceptive rough-terrain locomotion with quadruped robots. One of the main reasons for this is the difficulty in planning complex whole-body motions while taking into account the terrain conditions. This problem is very high-dimensional as it considers the robots dynamics together with the terrain model in a suitable problem formulation. In this work, we propose a novel trajectory and foothold optimization method that plans dynamically both foothold locations and motions (coupled planning). It jointly optimizes body motion, step duration and foothold selection, considering the terrain topology. Our model predictive controller tracks compliantly trunk motions while avoiding slippage. We test our method and comparative evaluations over a set of terrains of progressively increasing difficulty. To this end, we present a novel pose optimization approach that enables the robot to climb over significant obstacles. We experimentally validate our approach with the quadrupedal robot Panda5 autonomously traversing obstacles such steps, inclines, and stairs. The locomotion planner re-plans the motion at every step to cope with disturbances and dynamic environments.

Keywords: quadruped robots, perceptive locomotion, model predictive control, geographic terrain

1. Introduction

At present, MPC controller has been the popular motion control algorithm of quadruped robot. Compared with the traditional VMC control algorithm (Jiaping Cheng, 2002). It considers the trajectory of the robot in the future, and then using the online optimal control method to realize the trajectory tracking. In paper (Bledt, G, 2019) a simplified single rigid body model is established, and several gait locomotion such as trot, bound and pace are realized by the time-trigger constraints. In paper (QuanN, 2019) an off-line trajectory planning method is proposed, which can realize forward flip and back flip. Therefore, MPC controller shows excellent performance in the motion control of legged robot. To realize the maneuvering under complex terrain, the robot not only needs to the stable control of support, but also the planning of swing behavior is equally important. The selection of foothold points can ensure the stable movement of the robot in the discrete support area. The above problem of motion planning and balance control combined with perception is called as locomotion perception (Bellicoso, 2017). Researchers have adopted a variety of methods to solve this problem. In Xingdong (2002), the author establishes the elevation map of the terrain and selecting the foothold position that is not at the terrain edge. However, the global optimization of the algorithm cannot be guaranteed because the motion state of the robot is not considered. In paper (Shon D, 2022), an end-to-end deep learning method is proposed. The depth map is sent into the neural network, the future multi-step foothold position can be generated by the training network. However, it required a lot of training data and time.

Traditional MPC control models often use fixed time-trigger constraints or open-loop phase sequence. However, in complex terrain, the constraint conditions are inconsistent with the contact conditions due to falling or slip. To solve the above problems, this paper proposes a pr behavior planning framework as shown in the figure, which modifies the MPC model constraints Online based on the touch TD event trigger, improving the stability control performance on rough terrain, At the same time, digital terrain GIS analysis technology is used for online processing of local elevation map for perceptual landing planning, and multi-dimensional constraints such as slip model and robot ZMP stability are further integrated into the traditional landing point selection algorithm to select the global optimal landing point by online optimization iteration. The proposed algorithm is tested in typical steps and plum blossom pile terrain built in webots simulation, and it is verified that the proposed algorithm has good performance in obstacle surmounting and motion planning.

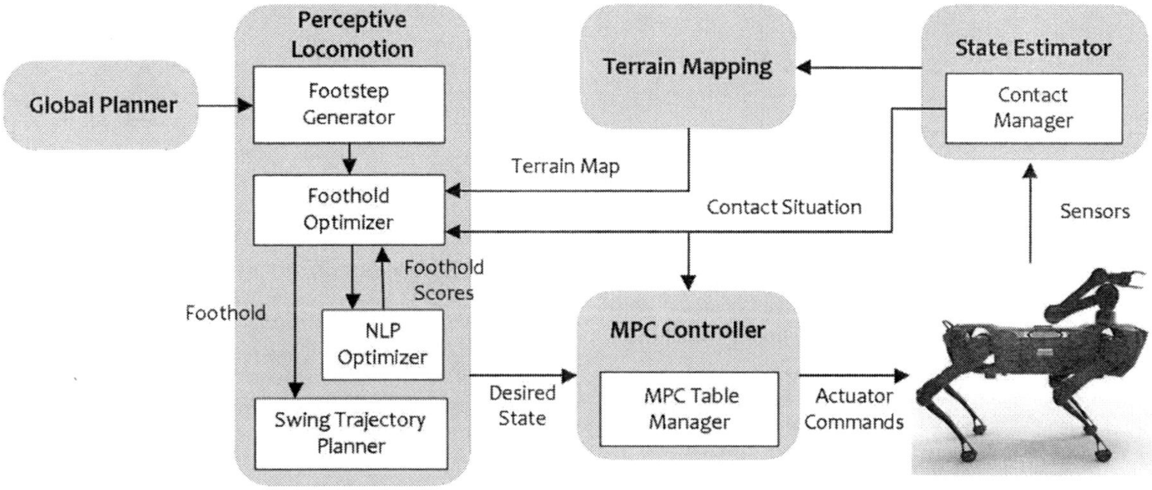

Figure 1. Proposed Perception Locomotion Framework

2. Event-Trigger Based MPC controller

2.1 Convex MPC Controller

In [3], an MPC control model based on a simplified model of single rigid body with multiple support points is proposed, as shown in the figure. In the figure, $f_i \in R^3$ is ground reaction force, $p \in R^3$ is the robot's centroidal position, $\boldsymbol{I}$ is centroidal rotational inertia, $\omega \in R^3$ is the body's attitude angle, $r_i \in R^3$ is the direction vector from foot to the robot's centroid.

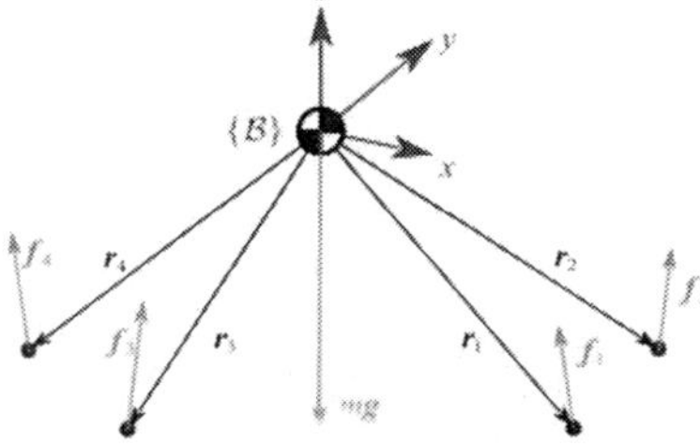

Figure 2. Simplified Rigid Body Model

In the world coordinate system, the single rigid body dynamic model is modeled as follows:

$$\ddot{\mathrm{p}} = \frac{\sum_{i=1}^{n} f_i}{m} - \mathbf{g} \quad (1)$$

$$\frac{d}{dt}(I\omega) = \sum_{i=1}^{n} f_i \times f_i \quad (2)$$

$$\dot{R} = [\omega]_{\times} R \quad (3)$$

Where $[\omega]_{\times} \in R^3$ is the antisymmetric matrix of ω, attitude angle ω and rotation matrix $\boldsymbol{R}$ satisfy $[\omega]_{\times} = \dot{R} \cdot R^T$.

So far, the concentrated mass modeling of quadruped robot pose changes caused by ground reaction force on system input is completed, the state-space representation is:

$$\dot{x}(k+1) = A_k(\psi)x(k) + B_k(r_1, \cdots, r_n, \psi)u(t) \quad (4)$$

Based on the above discrete model, the original complex nonlinear system can be successfully simplified into a linear time-varying discrete system suitable for convex optimization, and the corresponding MPC control problem can be transformed into the following convex optimization problem with constraints:

$$\min_{x,u} \sum_{i=0}^{k-1} \left\| x_{i+1} - x_{i+1,ref} \right\|_{Q_i} + \left\| u_i \right\|_{R_i} \quad (5)$$

$$x_{i+1} = A_i x_i + B_i u_i,\ i = 0 \cdots k-1$$

$$\underline{c}_i \le C_i u_i \le \overline{c}_i,\ i = 0 \cdots k-1$$

$$D_i u_i = 0, i = 0 \cdots k-1$$

A rolling optimization problem is described in the above formula, which considers the error between the model and the expected trajectory in the future K steps, and finally solves the optimal reaction force for the stable control of the robot through optimization method.

2.2 Event-Trigger Based Constraint

The MPC controller realizes prediction of system state in the future N prediction windows and calculates the optimal control command. For N prediction windows, N*3 constraint equations can be constructed, which contain the the foot force amplitude and the constraints of the friction cone:

$$C_i \Rightarrow \begin{cases} G_i \cdot f_{\min} \le f_z \le G_i \cdot f_{\max} \\ -\mu f_z \le f_x \le \mu f_z \\ -\mu f_z \le f_y \le \mu f_z \end{cases} \quad (7)$$

Where G_i represents the stance situation, which can realize the limitation of the maximum force in the Z direction of the foot, and then the force in the X and Y directions is also constrained to be 0 by the friction cone. G_i can be used to describe the stance or swing of each leg in one gait cycle or multiple gait cycles in the future. The sequence described by G_i can realize the adjustment of MPC constraints, thereby affecting the solved optimal ground reaction force:

$$\underline{c}_i \le C_i u_i \le \overline{c}_i \text{ , } i = 0 \cdots N-1$$

$$C = \begin{bmatrix} C_0 & & & 0 \\ & C_1 & & \vdots \\ & & \ddots & \\ 0 & \cdots & & C_i \end{bmatrix} \quad (8)$$

In paper [3], an open-loop sequence is used to describe the stance and swing of each leg, which is consistent with the expected gait phase sequence. This method can achieve stable movement of the robot on flat terrain, but the position error of footstep location and swing leg on rough terrain will cause the inconsistency between the touchdown moment and the open-loop phase sequence. However, the constraints set by the MPC based on this method will cause errors in the results of the optimization force. Aiming at the above problems, this paper proposes a dynamic constraint matrix adjustment method based on an event-triggered mechanism.

The Event-MPC method uses the phase sequence diagram prediction to fine-grained correction of the open-loop phase sequence of each leg. It calculates the corresponding event error based on the touchdown TD event of the swing state machine, and further combines the swing time parameter to correct the G_i sequence and realizes the synchronous update of the G_i phase sequence by designing an independent timer. The flow chart of the method is shown in Figure 3.

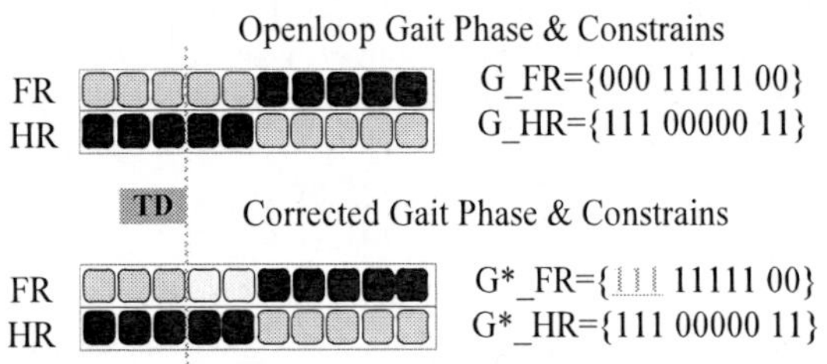

Figure 3. Gait Phase Sequence Constraint Correction

As shown in the figure above, the phase sequence diagram of the right front (RF) leg and right hind (RH) leg in the Trot gait is given. At the dotted line, the RF leg touches the ground in advance to generate a TD event. For the original open-loop phase sequence, the leg has not yet entered the stance phase. If this constraint matrix is used to solve the MPC model, the correct force cannot be solved. Based on the Event-MPC algorithm correction, after the TD event, the corresponding prediction period is solved based on the remaining swing time and the original open-loop phase sequence is corrected, which realizes the consistency between the stance phase and the phase sequence constraint matrix, thus ensuring the correctness of the MPC solution.

3. Perception Locomotion Based on GIS

In Section 2.2, Online correction of MPC problem constraint matrix by constructing event trigger signal through feedback of foot force estimation to achieve stable control of robot support phase. To improve the complex terrain trafficability, it is necessary not only to accomplish tracking control of the desired trajectory by the center of mass, but also to ensure landing on a safe terrain area by visual landing. For the vision-based motion control problem, this paper proposes an optimization algorithm to select footholds based on GIS digital map analysis (Cheng J, 2010) technology, and its main process is shown in the following figures.

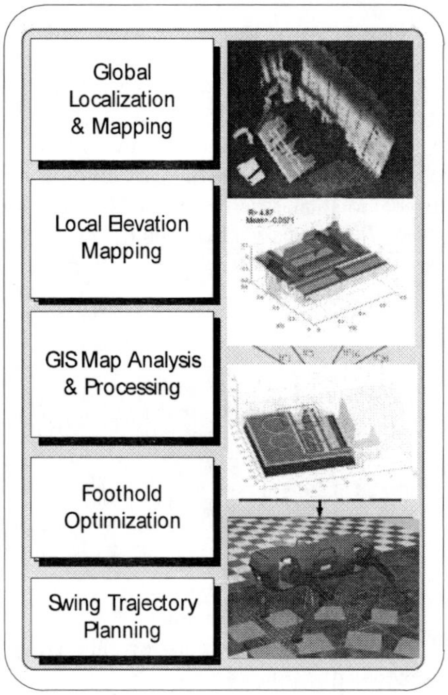

Figure 4. Proposed Perception Locomotion Method

As shown in Figure 4, the proposed method first uses depth camera or 3D radar to build a global depth map based on the SLAM algorithm and obtains local elevation map by combining robot pose data while performing pre-treatment such as filtering and normalization. Then, through GIS digital map analysis technology, the macro parameters of the local elevation map such as ruggedness and terrain slope are calculated as the reference trajectory constraint of MPC controller, and the elevation information of the safe landing area is established by eliminating the landing location such as step edge. Finally, the global cost function is constructed by considering the multidimensional constraints of the center-of-mass trajectory, SLIP model nominal foothold and ZMP through the foothold optimizer to filter the optimal foothold.

3.1 Foothold Search Space

The foothold selection of the quadruped robot has an impact on the static acceleration of the support phase, and the nominal foothold Ps calculated by the SLIP model is able to stabilize the body speed. Thus, the search space S is constructed as shown below, which contains M discrete subsequent footholds in a rectangular region centered on the current nominal foothold.

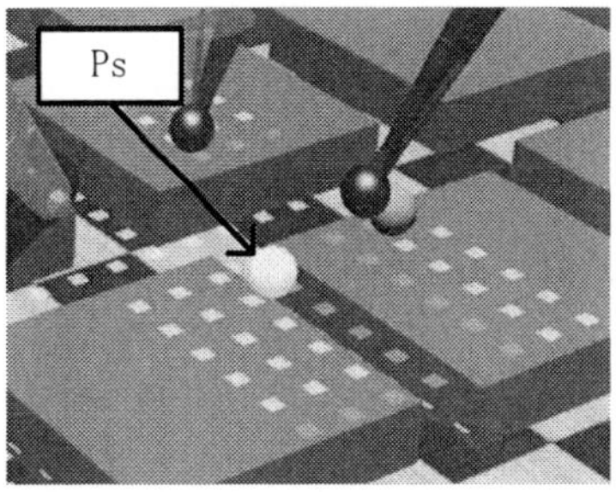

Figure 5. Search Space S and SLIP Model Nominal Foothold

3.2 Global Cost Function

The global cost function constructed based on the elevation constraint, SLIP velocity constraint, and motion constraint in the search space S, and the optimal foothold p_i^* is obtained by repeated iterations. Based on the cost function and the elevation weight of the terrain, the optimal foothold with the largest weight is calculated by iterating through the candidate footholds in the search area S. To avoid jittering back and forth of foothold in complex terrain, a dead zone $s_{\max}$ is proposed to ensure that the interval between the selection of two footholds is not too large:

$$p_i^* \Rightarrow \sum s_f(p_i^*, p_i) > s_{\max} \qquad (9)$$

3.3 The Weights of Optimization Constraints

A. The weight of velocity error: The foothold calculated by the SLIP model can achieve optimal control of velocity, therefore, the cost function is defined to minimize the error between the foothold and the SLIP nominal foothold:

$$\min_{p_i \in S} \left\| \hat{p}_{slip} - l_{norm} - p_i \right\|_2 \qquad (10)$$

Where l_{norm} is the maximum leg movement space, SLIP foothold calculates by Raibert's simplified model.

$$\hat{p}_{hip,i} = \frac{T_{st}}{2}\hat{v} + k_v(\hat{v} - v_d) \qquad (11)$$

B. The weights of foothold elevation: In order to achieve safe landing, the foothold needs to avoid both steep slopes and terrain edges within a certain elevation range. Therefore, the GIS digital terrain calculation model is used to traverse the elevation map $h_{map,S}$ in the search area to calculate the weights of each candidate foothold:

$$s_f(h_{map,S}) = \lambda_1 \underset{Rough}{R} + \lambda_2 \underset{Edge}{D} + \lambda_3 \underset{Slope}{P} \qquad (12)$$

Where $\lambda_1 + \lambda_2 + \lambda_3 = 1$, The above terrain factors unevenness R, edge D and slope P can be calculated by GIS model.

C. The weights of support area: To ensure the stability of the platform, the foothold should increase the support area. The cost function is as follows:

$$\min_{p_i \in S} \sum_{j \neq i} \left\| \hat{p}_j - p_i \right\|_2^{-1} \qquad (13)$$

D. The collision avoidance of landing: To avoid collision with other limbs during the swinging movement, the motion planning should avoid the foothold of the current support foot. The cost function is as follows:

$$\left\| p_j^{st} - p_i \right\|_2 > d_{\min} \qquad (14)$$

E. Minimum step length: The cost function constrains the shortest step when crossing the barrier to reduce the energy consumption:

$$\min_{p_i \in S} \left\| p_{i,LF} - p_i \right\|_2 + \left\| p_{i,TD} - p_i \right\|_2 \qquad (15)$$

4. Simulation and Experiment

In order to verify the proposed algorithm in the paper, we built the system based on c++ in the webots environment and carried out experiments on perception locomotion and stability control under two typical complex terrains: steps and piles. In the experiment, Panda5 quadruped is used as the platform, relevant parameters of robot and simulation environment are shown in Table 1.

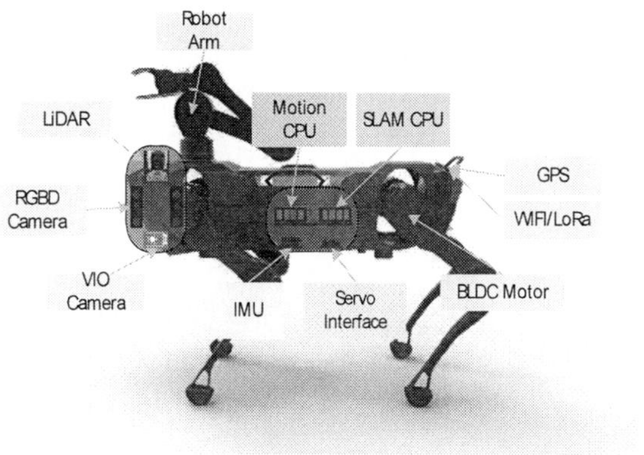

Figure 6. Panda5 Quadruped Robot

Table 1. The parameters of the Panda5 Robot

Contents	Value
Mess of body	*30 kg*
Mess of swing leg	*7 kg*
Mess of knee leg	*6 kg*
Length of swing joint	*530 mm*
Length of knee joint	*580 mm*
Max torque of joints	*130 N.m*
Ground friction	*0.45*

4.1 Experiment of Event-MPC Controller

In order to verify the proposed event MPC control algorithm, the robot walk on the rugged terrain with a height difference of 5cm and compared with the traditional MPC strategy. The experimental results are shown in the Figure 7.

As shown in Figure 7a, the experimental results of solving MPC with open-loop phase sequence constraint are shown. As shown in the figure, after the early touchdown, because the open-loop phase sequence is still swinging, it does not calculate the GRF, resulting in the fluctuation of the robot height; Figure 7b shows the proposed algorithm. It can be seen that when a touchdown event occurs, the MPC controller plans the corresponding GRF immediately. Compared with the traditional algorithm, the height and attitude control of the robot is more stable.

4.2 Experiment of Perception Locomotion

The proposed GIS based elevation map method is further tested. Firstly, discrete elevation terrain is constructed in MATLAB, and the terrain safety landing area is selected through the proposed algorithm.

4.2.1 GIS elevation map analysis and processing

The following figure shows the GIS elevation map processing results, where Figure 8a is the input original elevation map, Figure 8b is the processed elevation map, where red and blue are the terrain edge and variation area, R is the terrain roughness, Mean is the elevation variation coefficient.

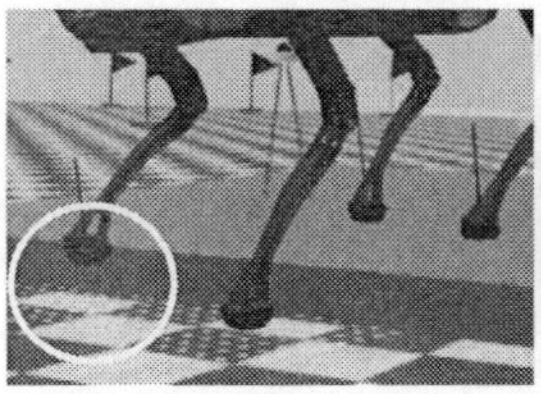

a) Time-triggered

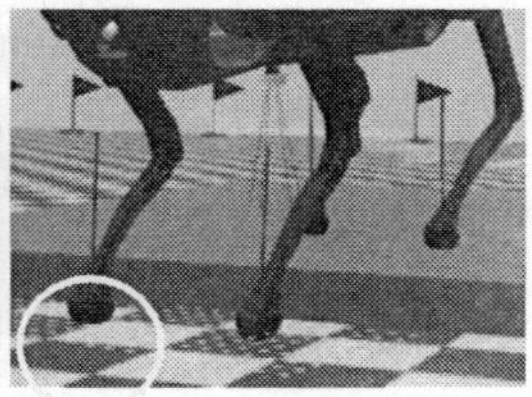

b) Event-triggered

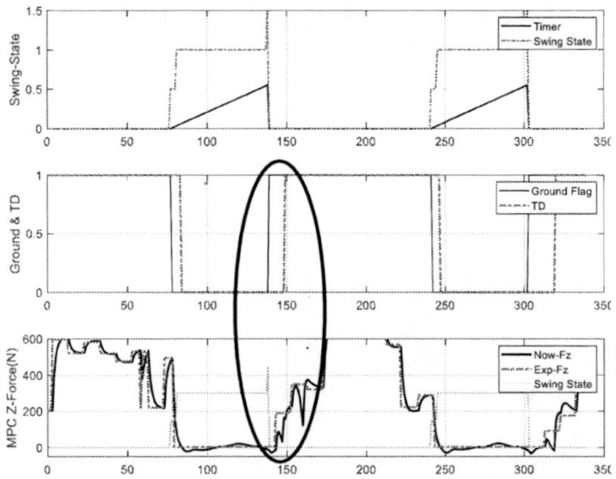

c) Time-triggered

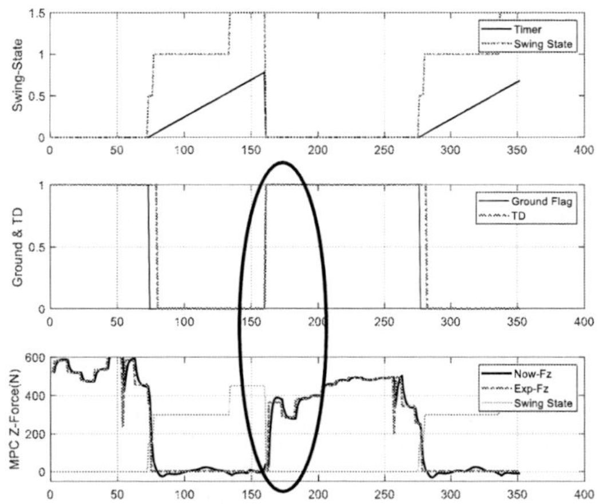

(d) Event-triggered gait

Figure 7 Time-triggered and Event-triggered MPC

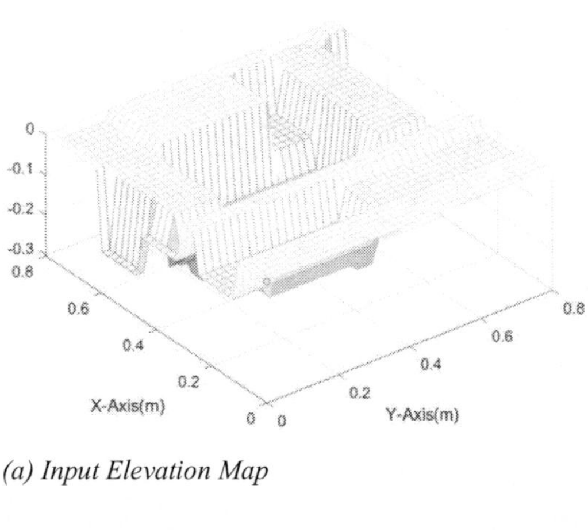

(a) Input Elevation Map

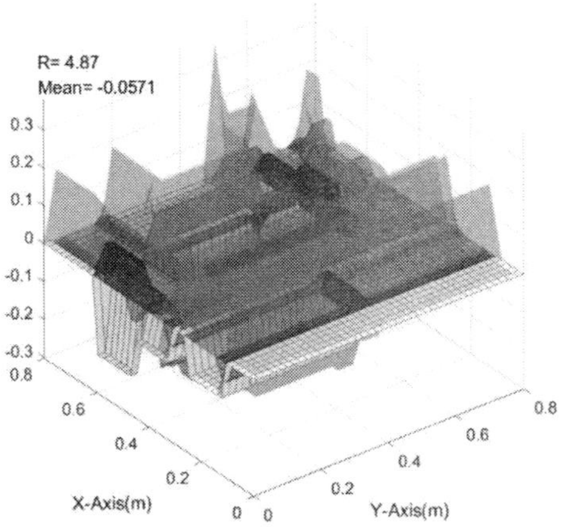

(b) GIS processed map
Figure 8. GIS Processed Analysis and Processing

4.2.2 Iterative optimal selection of foothold

The digital processing results based on GIS further verify the proposed iterative optimization algorithm for foothold. The optimal foothold is selected under ZMP and other constraints by combining the robot motion state and the nominal slip foothold. The experimental results are shown in the Figure.9.

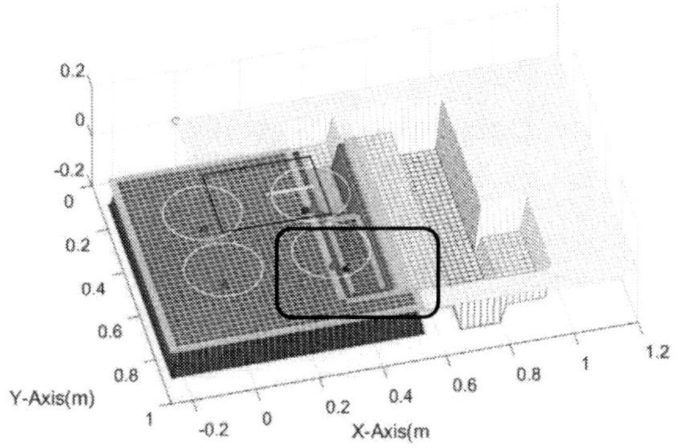

a) Front-leg foothold selection

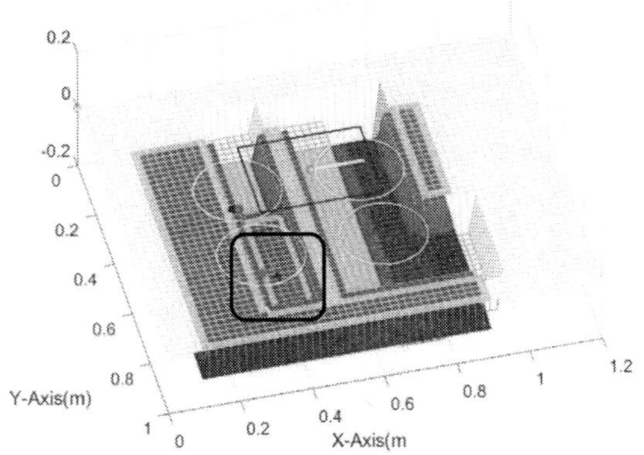

b) Hind-leg foothold selection
Figure 9. Optimal Selection of Foothold

As shown in Figure 9a, the red and pink points are slip footholds, and the blue is the optimal foothold. As shown in the figure, the optimal foothold is selected on the safe area.

4.3 Terrain Experiment

Finally, in order to verify the proposed framework, we set two typical complex terrain trench, stairs and piles.

4.3.1 Trench and Stair Terrain

Firstly, three stairs with different widths, and a maximum width of 15cm is set up. As we can that the proposed control framework realizes the safe foothold of the robot. Further, when crossing the stairs, avoid the edge of stairs based on the proposed algorithm.

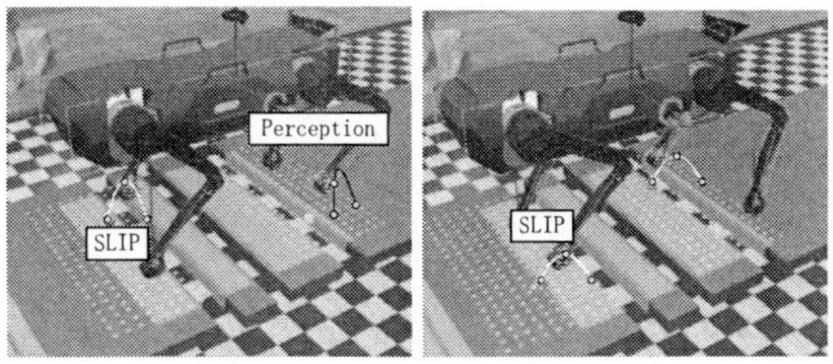

a) Trench terrain

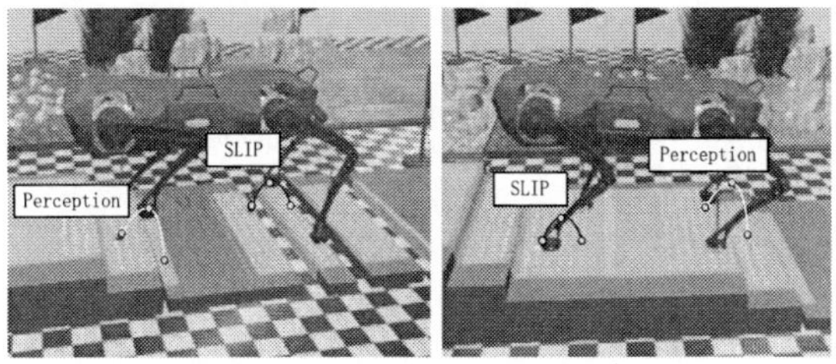

b) Stair terrain
Figure 10. Complex terrain experiment

4.3.2 Step Terrain

Further carry out obstacle surmounting on the terrain of piles. In the experiment, the size of the piles is 25*25cm, and the spacing is smaller than 20cm. As shown in Figure 11a, the robot completes the global trajectory planning based on (Piazzi A, 2000) to ensure movement along the central axis of the piles. The experimental results based on the proposed method are shown in Figure 11b.

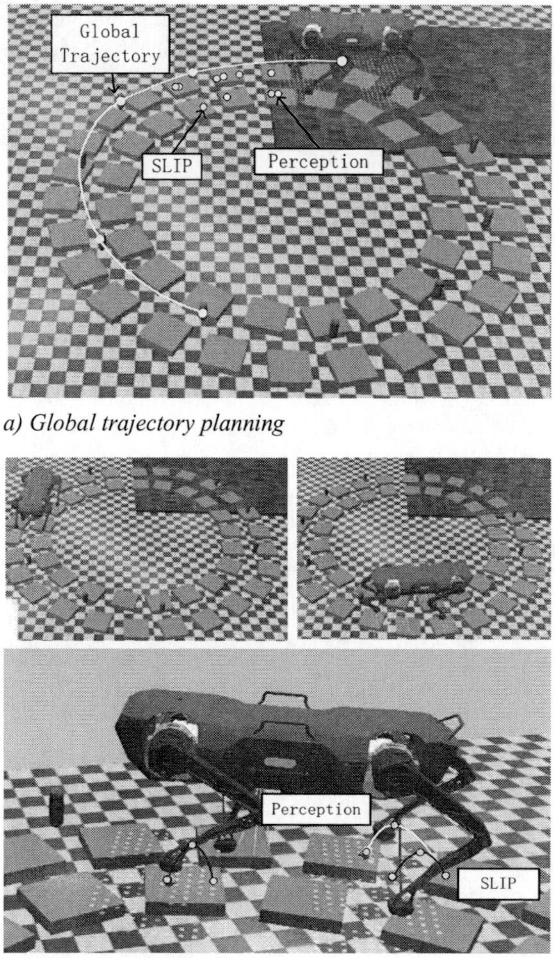

a) Global trajectory planning

b) Foothold optimization
Figure 11. Circular Square Piles Experiments

5. Conclusion

In this work, we propose a foothold optimization method that plans dynamically both foothold locations and motions (coupled planning). It jointly optimizes body motion, step duration and foothold selection, considering the terrain topology. Our MPC tracks compliantly trunk motions while avoiding slippage. We test our method and comparative evaluations over a set of terrains of progressively increasing difficulty. To this end, we present a novel pose optimization approach that enables the robot to climb over significant obstacles. We experimentally validate our approach with the quadrupedal robot Panda5 autonomously traversing obstacles such steps, inclines, and stairs. The locomotion planner replans the motion at every step to cope with disturbances and dynamic environments.

6. Acknowledgements

This work is supported by the State Administration of Science, Technology and Industry for National Defense, PRC under Grant JCKY2019208B024.

7. References

Jiaping Cheng, Xiyu Geng. Intuitive control of a quadruped walking robot based on virtual model [J]. Journal of shanghai jiaotong university, 2002, 36(8):1150-1154.

Bledt, G., Powell, M. J., Katz, B., Carlo, J. D., Wensing, P. M., & Kim, S. (2019). MIT Cheetah 3: Design and Control of a Robust, Dynamic Quadruped Robot. 2018 IEEE/RSJ International Conference on Intelligent Robots and Systems (IROS). IEEE.

Quan, N., Powell, M. J., Katz, B., Carlo, J. D., & Kim, S. (2019). Optimized Jumping on the MIT Cheetah 3 Robot. 2019 International Conference on Robotics and Automation (ICRA). IEEE.

Bellicoso, C. D., Gehring, C., Hwangbo, J., Fankhauser, P., & Hutter, M. (2017). Perception-less terrain adaptation through whole body control and hierarchical optimization. IEEE-RAS International Conference on Humanoid Robots. IEEE.

Xingdong, Hewei, Gao, Fusheng, Zha, & Jian, et al. (2019). Learning the cost function for foothold selection in a quadruped robot. Sensors (Basel, Switzerland).

Shon, D., Noh, B., & Byun, N. (2022). Identifying the exterior image of buildings on a 3d map and extracting elevation information using deep learning and digital image processing.

Cheng, J., & Wang, S. (2010). Research on Analysis of Site Selection of the University Supermarket Based on SuperMap GIS. IEEE.

Piazzi, A., & Visioli, A. (2000). Global minimum-jerk trajectory planning of robot manipulators. IEEE Transactions on Industrial Electronics, 47(1), 140-149.

6718 / Research on Vehicle Running Performance on Paved Roads Covered with Falling Volcanic Ash

Junya Yamakawa[a]*, Ryosuke Eto[a], Yasuhiro Ichikado[a], Mitsuhiro Yoshimoto[b], Tatsuji Nishizawa[b], Tomohiro Kubo[b], and Hiroyuki Yamada[a]
[a] *National Defense Academy, Yokosuka, Japan*
[b] *Mount Fuji Research Institute, Yamanashi, Japan*
**Corresponding author: yamakawa@nda.ac.jp*

https://doi.org/10.56884/YTYF8036

Abstract

Japan has many active volcanoes, and a large eruption can cause ash fall over a wide area. The accumulation of volcanic ash on paved roads affects the driving of vehicles. Therefore, we collected data by driving over volcanic ash spread on paved roads with a vehicle equipped with instruments that can measure the force applied to the running tires. Three types of volcanic ash with different grain sizes and thicknesses of 1 cm, 5 cm, and 10 cm were used. Experiments were conducted at constant speeds of 10 to 40 km/h and at rapid acceleration and deceleration. Data were collected not only on flat straight roads but also on curved roads with a radius of 30 m. In addition to the tire's longitudinal and side forces and the normal load, the camber angle, rotational speed, and ground speed were measured. The slip ratio can be obtained from the difference between the tire rotation speed and the ground speed. Further, since the ground speed of the tire is measured in two directions, the longitudinal and lateral directions of the tire, the sideslip angle of the tire can be calculated. From these data, the characteristics of rolling resistance, driving force, and side force on an ash-fall road surface can be obtained.

Keywords: volcanic ash, asphalt road, rolling resistance, side force, sideslip angle, gaussian process regression.

1. Introduction

Among the many active volcanoes in the Japanese archipelago, Mt. Fuji is the highest active volcano in Japan at 3776 m above sea level with a beautiful conical shape. Mt. Fuji has erupted repeatedly in the past in cycles of about 100 years, but more than 300 years have already passed since the last eruption. Although there are currently no signs of eruption, it is undeniable that an eruption will occur in the near future. Depending on the magnitude of the eruption and weather conditions, ash fall of 2 to 10 cm is expected in the Tokyo metropolitan area about 100 km away, and 50 cm in areas closer to Mt. Fuji. For this reason, Yamanashi Prefecture, where Mt. Fuji is located, provides education on how to prepare for eruptions as part of its disaster countermeasures. In October 2021, an event was held in an asphalt paved parking lot with an area of approximately 20,000 m^2, where driving paths was created by spreading volcanic ash to allow citizens to experience how vehicle driving would be affected by the ash fall caused by the eruption. Flat roads and inclined simulated ashfall roads covered with volcanic ash of different grain sizes and thicknesses were prepared. In addition to straight roads, curved roads were also prepared.

Previously, we had been conducting laboratory experiments to determine the effects of volcanic ash on vehicle driving (Yamakawa et al., 2016). This time, more data was collected by driving an instrumented vehicle on the site of this event. This paper describes the simulated ash-fall asphalt surfaces that the instrumented vehicle run and the results obtained from the analysis of the collected data.

2. Experiment

2.1 Experiment fields

This section describes the experimental field in which the measurement vehicle ran. A volcanic ash course sponsored by Yamanashi Prefecture was created in the parking lot of Fuji-Hokuroku Park in Fujiyoshida City, at the foot of Mt. The experiment was conducted with the cooperation of the Mount Fuji Research Institute. Figure 1 shows a plan view of the fields, and Table 1 lists the conditions for each course. The volcanic ash spread on the asphalt road surface is of three types: coarse-grained, medium-grained, and fine-grained, as shown in Figure 2. Course A is a flat, straight path smoothly connected with 15 m paths each of 1 cm, 5 cm, and 10 cm volcanic ash thickness with different grain sizes. Course B is curved roads with a 30m radius constant flat and a gradient of 2.5%. The thickness of the volcanic ash is 3 cm in both cases. Course C is straight slopes, coarse and fine grained, 3 and 5 cm thick on 5% slopes. The harsh environment includes course D with a volcanic ash depth of 30 cm. The coarse and fine grains are excavated volcanic ash deposited by past eruptions, while the medium grains are volcanic ash from Sakurajima. Because the experimental site was outdoors, it was affected by rainfall, but the weather was generally sunny throughout the experiment, and data were collected under mostly dry conditions. The experimental path (A4) in wet conditions was watered.

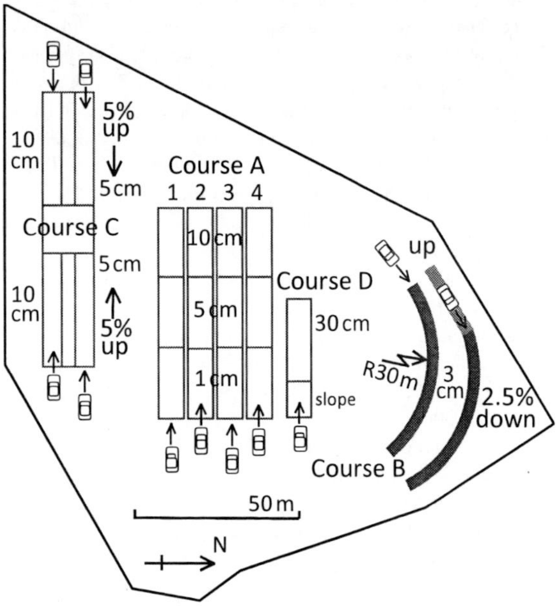

Figure 1. Plan of the experimental site

Table 1. Conditions for each course

Course ID	Course	Grain Size	Condition	Thickness (cm)
A1	Flat	coarse	Dry	1, 5, 10
A2	Flat	medium	Dry	1, 5, 10
A3	Flat	fine	Dry	1, 5, 10
A4	Flat	fine	Wet	1, 5, 10
B1	Flat Curve	fine	Dry	3
B2	Down Curve	fine	Dry	3
C1	Up	coarse	Dry	5
C2	Up	fine	Dry	5
C3	Up	fine	Dry	5
C4	Up	fine	Dry	3
D1	Flat	coarse	Dry	30

Figure 2. Photographs of volcanic ash (coarse, medium, and fine grains)

2.2 Instrumented vehicle

Figure 3 shows the vehicle used in the driving experiment, a Mitsubishi Motors Type 73 light-duty truck, whose main specifications are shown in Table 2. It is 4-wheel drive but can be switched to 2-wheel drive on the rear wheels. In the experiment, they were switched according to the purpose. The tires were YOKOHAMA GEOLANDAR M/T+ for off-road use, and the front two wheels are equipped with measuring instruments. They can measure the sextant force applied to the front tires, the ground speed by laser Doppler velocimetry, and the attitude angle of the tires. In addition to the ground speed of the tire, the rotational angular velocity of the tire can be measured, and the tire slip ratio can be calculated by comparing these values. Furthermore, ground speed is measured in two directions of the plane of movement, so the travel direction of the tires can be determined. From this, the tire's angle of sideslip angle can be determined. The laser displacement sensors, which basically measure the effective radius of the tires based on the distance from the tire rotation axis to the road surface, can be used to determine the amount of tire sinkage into volcanic ash. The tire pressure was set to 180 kPa, the standard pressure for size 218/85R 18LT.

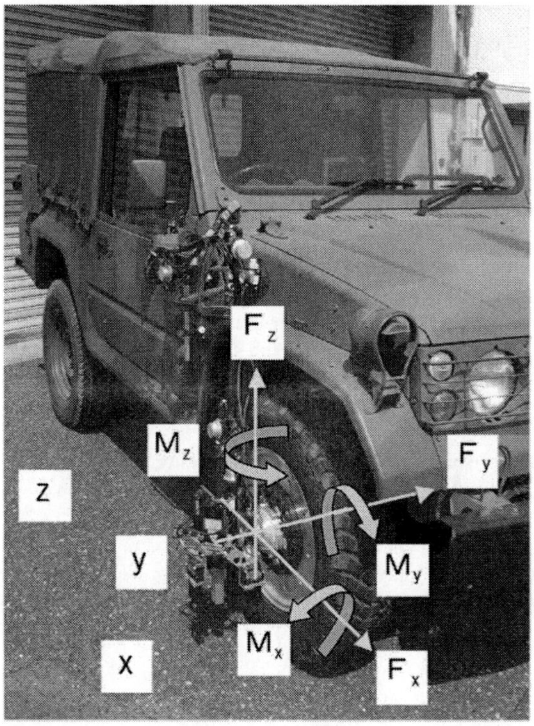

Figure 3. The instrumented vehicle and coordinates in Measurement

Table 2. Main specification of the instrumented vehicle

Length / Width / Height	4140 mm / 1765 mm /1970mm
Wheelbase	2420 mm
Tread (front/rear)	1485 mm / 1500 mm
Minimum Ground Height	245 mm
Vehicle Weight	1940 kg
Engine	4M40 diesel, supercharged
Cylinder / Displacement	In-line 4-cylinder/2835 cc
Maximum Power	125 PS (4000 rpm)
Maximum Torque	30.0 kgm (2000 rpm)

2.3 Conditions for experiment

There were four main driving patterns in the experiment:

a) Passing through at a constant speed of 10-40 km/h. Rolling resistance was measured by passing through a volcanic ash road at a constant speed. Trends by volcanic ash type, thickness, and velocity were investigated. Since rolling resistance is measured in a rolling condition, the vehicle was driven on the rear wheels. Rolling resistance is equal to the longitudinal force on free rolling tires.

b) Acceleration and sudden stop experiments. This was done to determine the relationship between tire slip ratio and longitudinal forces (driving force and braking force). The experiments were conducted with four-wheel drive by generating driving force at the front wheels.

c) Trafficability experiment at low speed. In conditions where driving is difficult, such as a volcanic ash thickness of 30 cm, the vehicle was driven at low speeds to determine if passage was possible. 2-wheel drive and 4-wheel drive were used.

d) Driving experiment on the curved roads. To clarify the relationship between the tire's side force and sideslip angle, the vehicle was driven in two directions on the curved road while maintaining speed at 10 to 30 km/h to obtain data on right and left turns.

3. Experimental results

From the data obtained by the instrumented vehicle, it is possible to determine the characteristics of rolling resistance, tractive force, and side force on the ash fall road surface. Each of these items is described below.

3.1 Rolling resistance

Figure 4 shows the experimental results obtained on course A, showing the relationship between tire sinkage and rolling resistance coefficient. A1 is coarse grain, A2 is medium grain, and A3 is fine grain. A4 is fine grained but watered and moistened. The rolling resistance coefficient is a dimensionless quantity dividing the longitudinal force by the normal load on the tire. Sinkage is calculated as the difference between the distance from the tire rotation axis to the volcanic ash surface. The rolling resistance increases as the sinkage into the volcanic ash increases. The sinkage differs depending on the grain size of the volcanic ash. The rolling resistance coefficient μ_r is approximately linear with the sinkage z and is shown by the lines in Figure 4. The coefficients are shown in Table 3.

$$\mu_r = a_r z + a_0 \qquad (1)$$

Wet, fine-grained volcanic ash showed less sinkage and less rolling resistance. For asphalt roads without volcanic ash, the value was about 0.018.

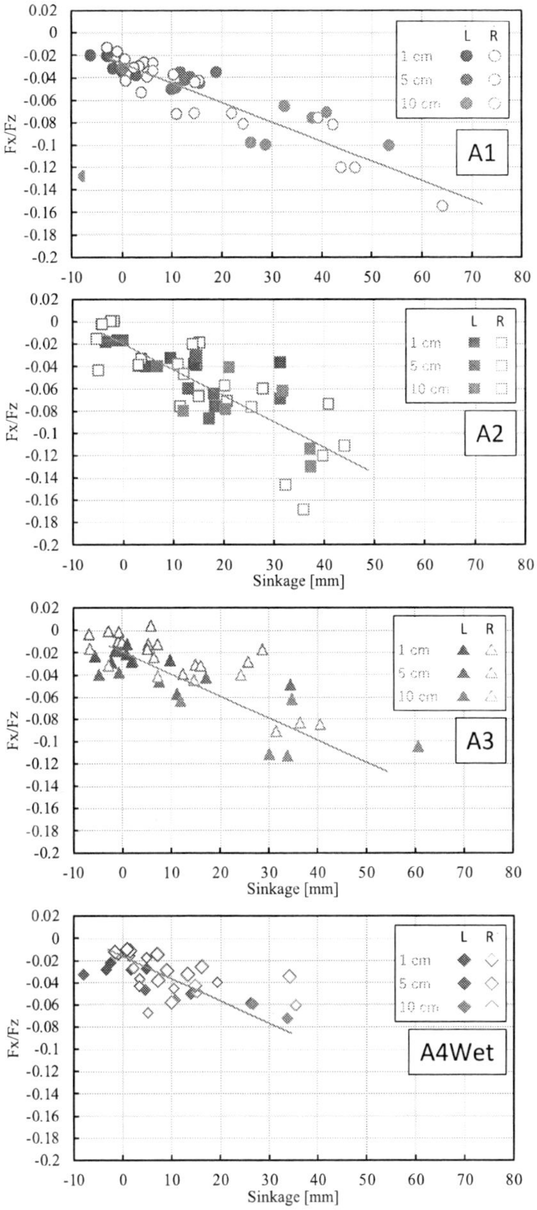

Figure 4. Relationship between rolling resistance and sinkage

Table 3. Rolling resistance coefficients a_r and a_0

Course	a_r [/cm]	a_0 [-]
A1	0.017	0.027
A2	0.022	0.022
A3	0.02	0.022
A4 (wet)	0.02	0.018

Figure 5 shows the relationship between vehicle velocity and sinkage. The increased thickness of the volcanic ash increases rolling resistance. For the same volcanic ash thickness, sinkage tends to decrease with increasing velocity. This is likely due to the velocity effect, where the volcanic ash is less deformed by the increase in velocity, even when the same load is applied.

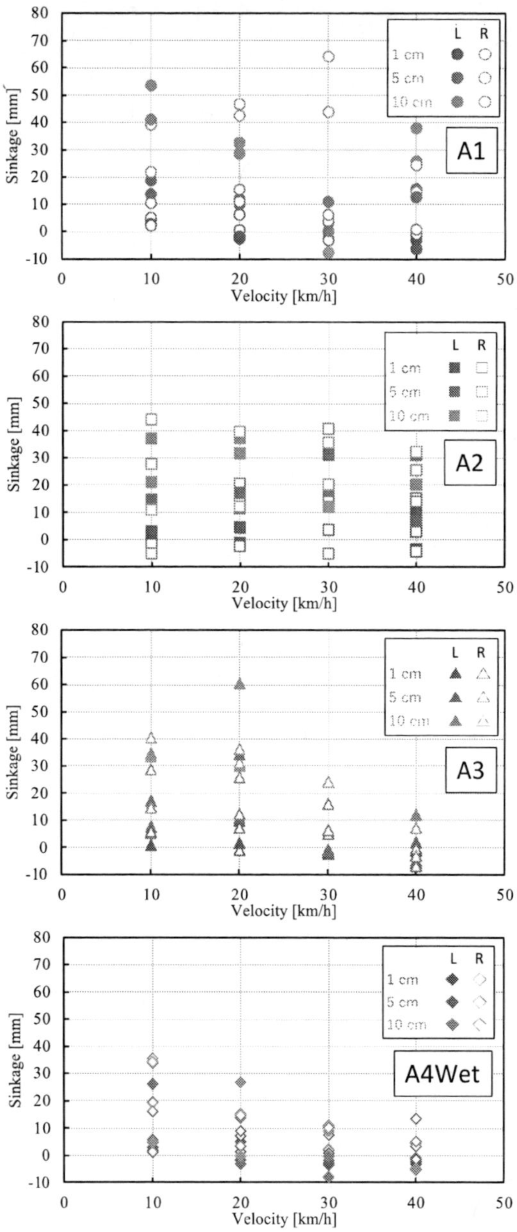

Figure 5. Relationship between vehicle speed and tire sinkage

3.2 Slip ratio and tractive force

The slip ratio is calculated from the difference between the rotational speed of the tire and the ground speed. The effective radius of the tire was determined on a firm surface (on asphalt) as a constant so that the travel speed obtained from the ground and the tire rotation speed coincide when the tire is in rolling condition without driving force applied. The effective radius of the tire r_0 is expressed by the ground velocity v_{t0} and the angular velocity ω_0,

$$r_0 = v_{t0}/\omega_0. \tag{2}$$

The slip ratio s is expressed in terms of the measured ground speed v_t and the angular velocity of the tire rotation ω,

$$s = \frac{r_0\omega - v_t}{r_0\omega} \quad \textit{for driving,} \tag{3}$$

$$s = \frac{r_0\omega - v_t}{v_t} \quad \textit{for braking.} \tag{4}$$

Figure 6 shows the relationship between slip ratio and longitudinal force measured in each of Course A. The longitudinal force is dimensionless, divided by the normal load. The lines shown in each figure are due to data obtained in one run. 2-3 measurements were taken under the same conditions, and the same trend was observed, so a representative right or left wheel result is shown. The measured longitudinal force represents the tractive force, which is the driving or braking force of the tires minus the resistant force. In each graph, the right side is for acceleration and the left side is for deceleration. For comparison, Fig. A1 also includes an asphalt result (Yamakawa et al., 2017). The thicker the volcanic ash, the smaller the slope of the traction force near the slip ratio of 0, indicating that it is more slippery. The slope tends to be larger for smaller volcanic ash particles. In driving force, due to the limits of engine power, it is not always measured to saturation, but in braking force, the tires lock up and some are measured to the maximum value. Measurements in wet conditions are for fine grains, which tend to be somewhat slippery in thicker volcanic ash than in dry conditions.

3.3 Thickly accumulated volcanic ash

Under loose volcanic ash with a thickness of 30 cm, the vehicle in 4-wheel drive condition were able to drive. The two-wheel drive could pass the 10-meter section if it could maintain some speed. However, once the speed was reduced, tire slippage increased, resulting in getting stuck. Escape was possible by switching to 4-wheel drive. Running on 30 cm thickness ash in loosely packed conditions resulted in about 10-12 cm of sinkage.

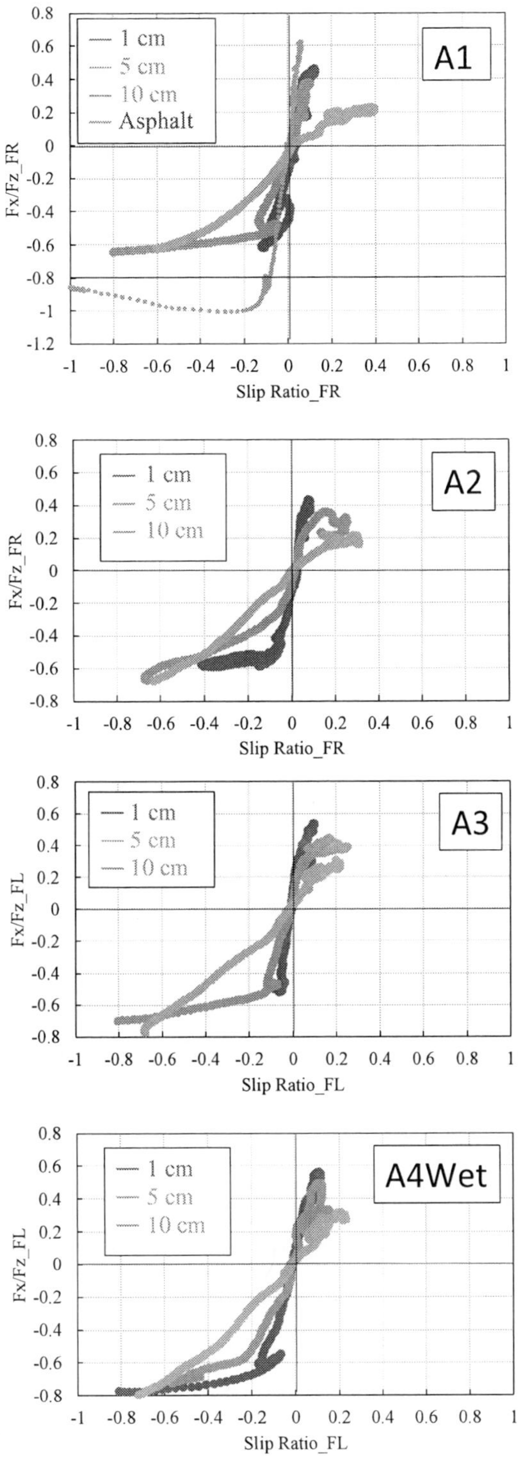

Figure 6. Relationship between tractive and braking forces and slip ratio

3.4 Sideslip and side force

The relationship between side force and sideslip angle was clarified from data obtained during curve driving. Since the ground speed in the two directions near the wheel can be measured, the sideslip angle β, which is the angle between the tire direction and the movement direction, can be expressed by the following equation,

$$\tan\beta = V_y/V_x \qquad (5)$$

where, (V_x, V_y) is the ground speed near the tire in the x and y directions, respectively. During turn, not only the side force due to the sideslip angle but also the side force called camber thrust due to the camber angle are generated, and the measured side force is the sum of these. Since the centrifugal force increases the normal load on the outer wheels, this effect must be considered. Here, Gaussian process regression for multiple variables was used to obtain the relationship between the sideslip angle and side force (Yamakawa et al., 2021). Although a detailed description of Gaussian process regression (Bishop, 2006) is omitted, we used the following equation as the kernel function.

$$k(\boldsymbol{x}_m, \boldsymbol{x}_n) = \eta_0 exp\left\{-\frac{1}{2}(\boldsymbol{x}_m - \boldsymbol{x}_n)^{\mathrm{T}}\Sigma_1(\boldsymbol{x}_m - \boldsymbol{x}_n)\right\} + \boldsymbol{x}_m^{\mathrm{T}}\Sigma_2\boldsymbol{x}_n \qquad (6)$$

$$\Sigma_1 = \begin{bmatrix}\eta_1 & 0 & 0\\ 0 & \eta_2 & 0\\ 0 & 0 & \eta_3\end{bmatrix};\ \Sigma_2 = \begin{bmatrix}\eta_4 & 0 & 0\\ 0 & \eta_5 & 0\\ 0 & 0 & \eta_6\end{bmatrix} \qquad (7)$$

where, $\boldsymbol{x}_m^{\mathrm{T}}$ is the mth experimental data (β_m, γ_m, ΔF_{zm}), consisting of a combination of three variables: the sideslip angleβ, the camber angle γ, and the ground normal load change ΔF_z from the standard load. In addition, η_0–η_6 are hyperparameters. Data were combined from the left and right wheels obtained in the two-run experiments. The entire data set was randomly divided into four sets, one set of which was the training data and another set of which was the validation data. The number of data in one set is N = 1469.

The combination of hyperparameters that maximizes the likelihood of the regression model for the experimental data was obtained using a combination of the gradient and Newton-Raphson methods. To ensure that there was no over-fitting, validation data was used to confirm that the mean squared error with the experimental values was almost the same as that of the training data.

Figure 7 shows the relationship between the sideslip angle and side force based on all left-right wheel experimental data obtained for one run each up and down the curved road, indicated by circles. Since the normal load changes between the outside and the inside during a turn and the camber angle also changes, side force values are obtained under different conditions. Changes in normal load, which has the greatest effect on side force, are indicated by the color shade. The normal load in straight run was used as the standard, and its change was indicated by the color shade. A large increase in normal load is close to black and a large decrease is close to white. The solid line in the figure is the result obtained using Gaussian process regression and shows the relationship between the sideslip angle and side force when the effect of camber angle is excluded. The results are shown for changes in the normal load. The regression curve obtained by Gaussian process is expressed as a probability distribution. The solid lines are their means, while the dashed lines show the mean ± standard deviation. The larger the number of data points obtained, the smaller the standard deviation, indicating that the regression model is more accurate. The figure below compares the data with those in an asphalt road (Yamakawa et al., 2021). The slope near the zero-sideslip angle is not affected much by the normal load for asphalt road, but for volcanic ash, the slope tends to have a greater effect on the change in normal load. However, the maximum side force is sufficiently greater for asphalt.

4. Discussion

As the volcanic ash thickness increased, the amount of sinkage increased, and the rolling resistance increased. As the velocity increased, the phenomenon of reduced sinkage was observed because of the rate effect on the deformation of the volcanic ash. This means that higher speeds tend to decrease rolling resistance.

As for the effect of volcanic ash grain size, rolling resistance was about the same for coarse and fine grains, and rolling resistance tended to be slightly higher for medium-grained volcanic ash. Rolling resistance was reduced in wet fine grains due to less sinking. This was due to the moderately moist condition, and a moister condition would have produced different results. Volcanic ash, which has an even smaller grain size and a high clay content, has a significant impact on driving when it becomes muddy.

The graphs of slip ratio and driving force or braking force show that thicker volcanic ash tended to slip more easily for both driving and braking. At a volcanic ash thickness of 10 cm, the maximum coefficient of friction (ratio of the longitudinal force to the normal load) for driving was about 0.2 for coarse and medium grains, and about 0.4 at a thickness of 5 cm. In comparison, the value for fine grains was slightly higher, around 0.3 at a thickness of 10 cm. When wet in fine grains, they tended to be slightly slippery than when dry.

At a volcanic ash thickness of 30 cm, the 4-wheel drive vehicle was able to drive without problems, but the 2-wheel drive vehicle was sometimes unable to drive. If some speed was maintained and sinkage was controlled, passage was possible even with two-wheel drive. Even if the vehicle was disabled in two-wheel drive, escape was possible by switching to four-wheel drive.

Experiments on the curves were performed with a volcanic ash thickness of about 3 cm. The results showed that cornering power, the coefficient of side force generation near a zero-sideslip angle (slope, the increment of side force with respect to the sideslip angle), varied significantly with normal load. The greater the normal load, the greater the slope. The slope at a zero-sideslip angle was the same as that on the asphalt surface at the standard normal load, but the cornering power was about 1.5 times greater when the normal load was increased by 2350 N. For runs in volcanic ash, saturation occurred at relatively small lateral slip angles of 3-5°.

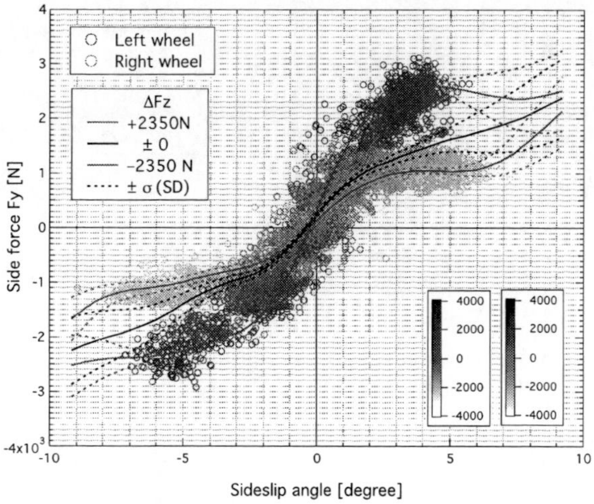

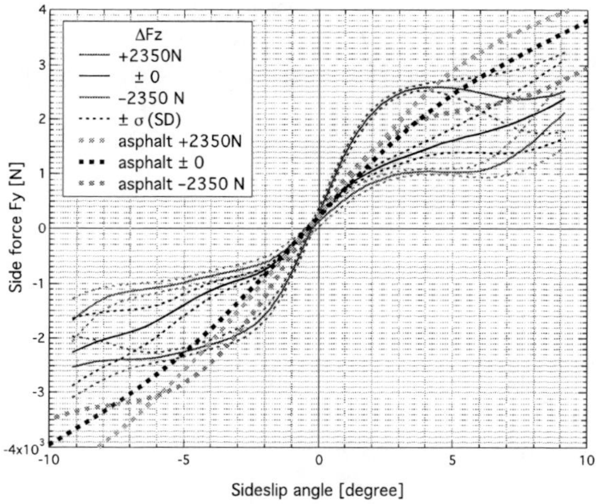

Figure 7. Relationship between sideslip angle and side force (when camber angle is set to 0)

5. Conclusions

In order to clarify the impact of a volcanic eruption disaster on vehicle driving, data was obtained by driving the instrumented vehicle on asphalt roads covered with volcanic ash. Although the data is for one type of tire mounted on an off-road vehicle, it confirms the impact of tires on driving. The following is a summary of the data.

Rolling resistance increases with the amount of sinkage due to the thickness of the volcanic ash. With increasing speed, sinkage and rolling resistance decrease. Rolling resistance was similar for coarse and fine grains and tended to be slightly greater for medium grains. Rolling resistance was reduced in wet fine grains.

Thick volcanic ash tended to make vehicles slippery. The maximum coefficient of friction in driving was about 0.2-0.4 at a volcanic ash thickness of 10 cm.

At a volcanic ash thickness of 30 cm, 4-wheel drive vehicles can drive without problems, but 2-wheel drive vehicles may not be able to drive.

The cornering power of the side force at a sideslip angle near 0 degree increased with increasing normal load. Side forces saturated at relatively small sideslip angles of 3 to 5°.

6. Acknowledgements

This work was supported by JSPS KAKENHI Grant Number JP20K04357.

7. References

Bishop, C. M., 2006., *Pattern Precognition and Machine Learning,* Springer Science+Business Media, LLC.

Yamakawa, J., Aota, K., Eto, R., 2021. Acquisition of interaction characteristics between tire and ground by Gaussian process regression. *Proceedings of the ISTVS 20th International Conference, online.*

Yamakawa, J., Enari, A., Eto, R., Soe, P.T., 2017. Measurement of tire characteristics traveling in actual ground environment with an instrumented vehicle. *Proceedings of the 19th International & 14th European-African Regional Conference of the ISTVS, Budapest.*

Yamakawa, J., Yamashita, Y., Eto, R., 2016. Measurement of vehicle performance traveling on asphalt-paved roads covered with volcanic ash. *Proceedings of the 8th ISTVS Americas Regional Conference, Troy, MI.*

6796 / Nonparametric Terrain Estimation Based on the Interaction Simulation Between Planetary Penetrator and Soil

Xintao Yang[a], Han Huang[a]*, Zhixin Xiang[a], Qinghao Yan[a], Haozhe Wang[a], and Shucai Xu[b]
[a]Nanjing University of Aeronautics and Astronautics, Jiangsu, Nanjing, China
[b]State Key Laboratory of Automotive Safety and Energy, Tsinghua University, Beijing, China
**Corresponding author: huanghan@nuaa.edu.cn*

https://doi.org/10.56884/MBEM9373

Abstract

Penetration detection is an important way to method for in-situ scientific exploration of planets, which is used to indirectly detect the mechanical properties of the planetary subsurface soil. In this paper, four ovoid-nosed penetrators with different radius penetrating different compactness of planetary soil under different impact velocity were simulated using nonlinear finite element (FE) method through Hypermesh/Ls-dyna. A nonparametric estimation method for estimating the mechanical properties of planetary soil is presented. Three main characteristic parameters in the process of penetrator penetration were adopted in the method, including maximum acceleration am, penetration depth z, and maximum deflection angle θm. Twenty identification parameters for identifying planetary soils properties were derived based on these three characteristic parameters. The terrain consisted of simulant soil classified nonparametrically and artificially defined as three states (low, medium, and high compactness) based on different harnesses, and five identification criteria (slack, ideal, partially strict, relatively strict and strict) were put forward. Four superior identification parameters were derived through the analysis of the evaluation indexes (recognition rate, accurate rate, and conservative rate) under different identification criteria. and are selected as the optimal identification parameters and verified by simulation test, the accurate success rate and conservative success rate of the final estimation are 58.3% and 91.7%, respectively, which indicated that the nonparametric estimation method in this work can be used to evaluate the mechanical properties of planetary soil effectively.

Keywords: planetary penetrator, terramechanics, estimation method, simulation analysis, mechanical properties of planetary soil

Editor's note: full-width tables and figures located at end of paper.

1. Introduction

With the development of deep space exploration, the preliminary study on the surface soil of the planet has gradually become a hot topic for scientists. Compared with the disadvantages of difficult landing technology and high cost of lander detection adopted in current deep space exploration (Sun et al. 2010), penetrator impact detection has some advantages, such as low cost, relatively simple technology and accurate detection, which has been the attention of domestic and foreign scholar in the field of deep space exploration.

In 1977, Ames of NASA envisioned releasing a 37 kg penetrator from a planetary probe at a speed of 60 m/s, and finally achieving a penetration speed of 135-165 m/s through a reducer (Lorenz et al. 2011). The penetrator could withstand a load of 1500-2000 g for the exploration of the Martian surface. In 1990s, ESA proposed the MARSNET (Scoon et al. 1993), which envisaged using 3-4 penetrators weighing about 65 kg to realize the network exploration of Mars surface at a speed of about 80 m/s. In 1996, Russia launched the penetrator on the Mars-96 (Harvey 2007), which finally realized the penetrating exploration of the Martian surface at the speed of 60-80 m/s with decelerating by parachute. The maximum penetrating depth was about 5 m, and the maximum acceleration was about 500 g. In 2002, the Lunar-A exploration mission of the Japanese ISAS and JAXA released three penetrators on the probe and impacted the Lunar surface at a speed of nearly 300 m/s to explore of the Lunar soil (Shiraishi et al. 2008). The penetration depth of the penetrators was about 1-3m. The maximum overload during penetration was about 10000 g.

Iagnemma et al. (2004) presented a nonparametric online terrain estimation method based on the wheel-terrain interaction model, and then Kang (2003) improved this method so that it can accurately measure the terrain characteristics of many different terrain types. The process of penetrating planetary soil with penetrator is affected by many factors, such as penetration velocity (Xu et al. 2020; Zhou et al. 2020), penetration angle (Ren et al. 2018), penetrator head shape (Xie et al. 2019) and compactness of soil.

In this paper, the ovoid-nosed penetrator with different velocity and radius penetrating different compactness of planetary soil were simulated in FE simulation by Hypermesh/Ls-dyna. By analyzing the simulation results, the maximum acceleration a_m, penetration depth z and maximum deflection angle θ_m generated by penetrator penetrating different compactness of soil were summarized. A nonparametric estimation method was proposed to estimate the properties of planetary soil. The appropriate identification parameters and critical values were determined by combining simulation data and evaluation criteria, and the specific technical route was shown in Fig. 1.

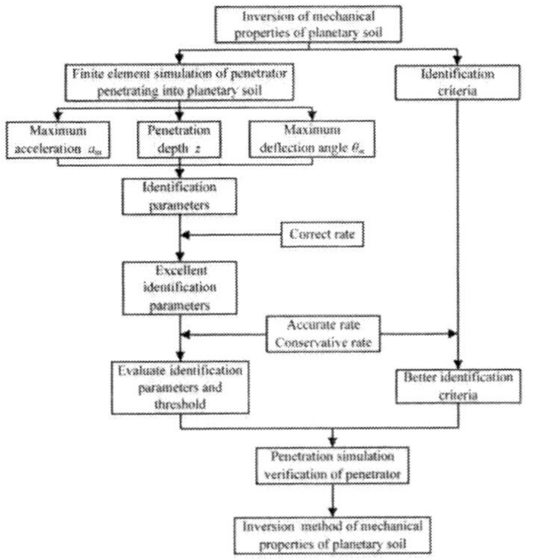

Figure 1. Inversion of mechanical properties of simulated planetary soil

2. Modeling for Numerical Simulation

2.1 FE Model of Penetrator

Considering the overall size of the planetary soil is large, in order to reduce the number of FE unit division and save calculation time, the 1/2 symmetric model of the penetrator and the planetary soil was used for modeling (Wang et al. 2019). The total size of the satellite soil model was set as 3200 mm (diameter) ×6000 mm (depth). The FE mesh was relatively dense in the region with a radius of 550 mm in the center of the planetary soil, and relatively sparse outside the region (Chen et al. 2022). The number of mesh elements of planetary soil model was 3931200. As shown in Fig. 2, there are four penetrators with the same length of 500 mm, the same oval head radius of 125 mm and different structure radius of 50 mm, 60 mm, 70 mm and 80 mm, which were respectively marked as P1, P2, P3 and P4. In order to avoid prolongation of calculation time due to large size difference, the mesh size of the penetrator was the same as that of the concentrated region of the planetary soil.

The FE mesh of the planetary soil and the penetrator used 3D SOLID 164 (Pan et al. 2011), the contact between the penetrator and the planetary soil used CONTACT_ERODING_SURFACE_TO_SURFACE (Li et al. 2008). Six degrees of freedom were constrained to the bottom nodes of the planetary soil FE model, and the upper surface of the model was set as free surface boundary (Nie et al. 2021). The planetary soil model's outer surface used no reflection boundary condition in order to eliminate the influence of expansion wave and shear wave (Meng et al. 2019).

2.2 Material Parameters of FE Model

In this paper, the FE simulation material model's system of units used t-mm-s. The penetrator structure was set as *MAT_RIGID and used LC9 super-rigid aluminum material widely used in the aerospace field. The material density ρ was set as 2.804×10^3 kg/m^3, the elasticity modulus E was set as 72610 MPa and the poison's ratio μ was set as 0.3.

Planetary soils used the *MAT_SOIL_AND_FOAM model, which was mainly used to represent soil, concrete and crushable foam and was proposed by Krieg in 1972. In this paper, three kinds of soil with different compactness were defined as low, medium, and high compactness, and respectively labeled as L, M and H. The specific material parameters are shown in Table 1.

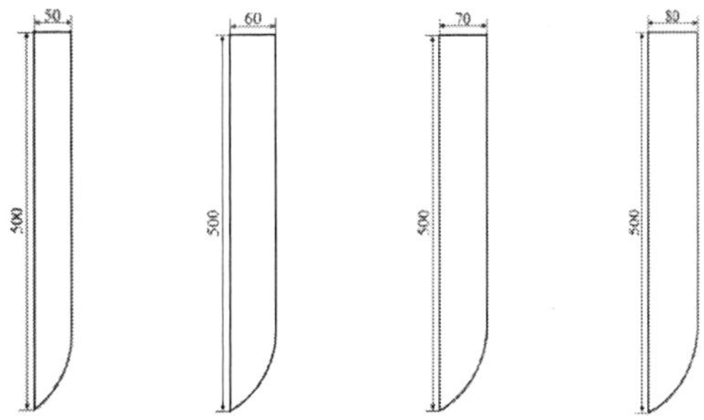

a) b) c) d)

Figure 2. Schematic diagram of penetrator structure

Table 1. Material parameters of planetary soils.

Soil compactness	ρ [kg/m^3]	G [MPa]	γ [MPa]
Low	*1.80×10^3*	*16*	*132.8*
Medium	*2.05×10^3*	*59.6*	*138*
High	*2.10×10^3*	*386*	*630*

2.3 Simulation Results

The different radius penetrators with different velocity penetrating different compactness of planetary soil were simulated in FE simulation by Hypermesh/Ls-dyna, in order to explore the process of penetrating planetary soil. The stress nephogram of the process of vertical penetration of medium compactness planetary soil by a 50 cm radius penetrator at a speed of 200 m/s was shown in Fig. 3. Because of the high kinetic energy of the penetrator in the initial stage of soil penetration, biggish stress is generated in the planetary soil near the head of the penetrator. Meanwhile, with the continuous penetration of the penetrator, the area of the biggish stress area near the head of the penetrator is constantly shrinking, and the stress is constantly spreading around in the form of stress waves until the penetrator completely slows down and stops.

The relationship between penetration velocity, maximum acceleration and soil compactness is shown in Fig 4. Through the analysis of the simulation results, when the same penetrator penetrates soil with different compactness at the same speed, the soil with higher compactness will be penetrated with higher maximum acceleration. At the same time, when the velocity of the penetrator is small, the maximum acceleration produced during the penetration has little difference. On the contrary, when the velocity of the penetrator is large, the maximum acceleration produced during the penetration has great difference.

The relationship between penetration velocity, penetration depth and soil compactness is shown in Fig. 5. As can be seen from the figure, the higher the penetration velocity of the penetrator, the deeper the penetration depth. For the same velocity, the relationship between soil compactness and penetration depth is negatively correlated, that is, the higher soil compactness, the smaller penetration depth.

The relationship between penetration velocity, maximum deflection angle and soil compactness is shown in Fig. 6. In the process of vertical penetration of soil, the deflection angle of the penetrator has a certain randomness, but the general rule can still be analyzed from the figure. In the range of penetration velocity higher than 200 m/s, maximum deflection angle of penetrator in medium and high compactness soil is smaller than that in low compactness soil, and most of the penetrators show a trend of the smaller the maximum deflection angle in the higher compactness soil. When the penetration velocity is less than 200 m/s, the maximum deflection angle of the penetrator is small in the penetration process, and the soil compactness has little influence on the maximum deflection angle of the penetrator.

3. Methods for Evaluating Mechanical Properties of Planetary Soil

3.1 Identification Parameters

According to the analysis of the simulation results of penetrator penetrating soil in Section 2.3, the relationship between unitary identification parameter and the compactness of planetary soil can be obtained. In the process of penetration, the higher the maximum acceleration a_m, the higher the soil compactness, which can be defined as positive correlation. The bigger maximum deflection angle θ_m during vertical penetration of the penetrator, the lower compactness of planetary soil, which can be defined as a negative correlation. The deeper the penetration depth z, the lower compactness of planetary soil, which can be defined as negative correlation. At the same time, considering that the simulation process may be affected by some uncontrollable factors, the selection of a single identification parameter cannot be considered. Therefore, equation (1) is defined to carry out a comprehensive non-parametric evaluation, and 20 binary and ternary identification parameters can be concluded according to the equation, as shown in Table 2.

$$A_i = \frac{a_m^{(k_1)}}{\theta_m^{(k_2)} \cdot z^{(k_3)}} \qquad (1)$$

where k_1, k_2 and k_3 are index coefficients of identification parameter, respectively, the corresponding value is 0, 1 or 2, and at least two index coefficients are not 0 at the same time.

3.2 Identification Criteria

In the process of mechanical property recognition of planetary soil by using identification parameters, different mechanical property of planetary soil will lead to the distribution of identification parameters as shown in Fig. 7: separation, partial overlap and significant overlap. Because of the discreteness of simulation data, the distribution of mechanical property identification parameters may be intertwined. Therefore, the key of soil mechanical property identification is to seek the identification parameter with large differences as far as possible, so as to avoid the overlapping multi-solution problems. By averaging the identification parameters of each group, the phenomenon of introducing random error due to the large discreteness of identification parameters can be effectively avoided. The average values of low, medium, and high compactness soil identification parameters are defined as A_L, A_M and A_H respectively. The soil identification parameters in the transition zone between low and medium compactness are $\overline{A_{LM}}$, which is the average of A_L and A_M, and the soil identification parameters in the transition zone between medium and high compactness are $\overline{A_{MH}}$, which is the average of A_M and A_H.

According to actual requirements, five identification criteria are defined in Table 3.

In addition to the identification criteria of strict in the table, the following evaluation identification criteria should be analyzed by designing the following indicators:

1) Identification rate: $\eta_j = \sum \frac{N_a}{N}$, the probability of identifying the structure.

2) Accurate rate: $\eta_a = \frac{1}{3}\sum \eta_p$, the probability that the recognition result conforms to the actual state, such as low compactness is judged as low compactness, medium compactness is judged as medium compactness, and high compactness is judged as high compactness.

3) Conservative rate: $\eta_c = \frac{1}{3}\sum \eta_q$, the identification results tend to be strict, so as to improve the safety factor. In addition to accurate identification, for example, the medium compactness is judged as low compactness, the high compactness is judged as low compactness, and the high compactness is judged as medium compactness.

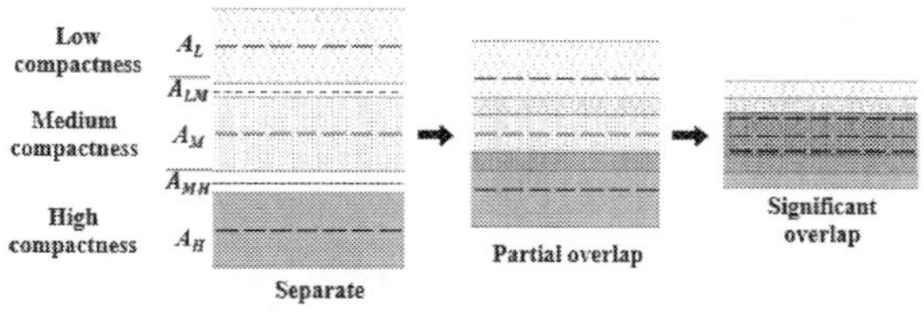

Figure 7. Schematic diagram of planetary soil's mechanical properties identification parameters distribution

4. Identification Results of Planetary Soil Mechanical Properties

4.1 Identification Rate Analysis

A total of 60 groups of data were obtained from the simulation test of penetrator penetration into planetary soil. All test data are processed with the above identification parameters. According to identification parameters results under different identification criteria, the excellent identification parameter is defined as having at least two identification rates greater than or equal to 0.8. All 9 excellent identification parameters selected from the above 20 identification parameters are shown in Table 4, in which L-x, M-x and H-x in the identification column are respectively expressed as low compactness, medium compactness or high compactness is identified as x compactness (x = L, M, H).

It can be seen from the analysis of Table 4, as the identification criteria gradually change from slack to strict, the identification rate of L-L increases, most of them reach 0.80 or more in the partially strict criteria, and all of them reach 0.85 or more in the relatively strict criteria. In the slack criteria, the identification rate of M-M is obviously higher than that of the other three criteria, while in the ideal criteria, the identification rate of H-H is the highest among all criteria. For all the excellent identification parameters, the identification rate of ternary identification parameters is better than that of binary identification parameters. Among them, the identification parameters of $\frac{a_m}{\theta_m^2 \cdot z}$, $\frac{a_m}{\theta_m \cdot z^2}$, $\frac{a_m^2}{\theta_m^2 \cdot z}$, $\frac{a_m}{\theta_m^2 \cdot z^2}$ and $\frac{a_m^2}{\theta_m^2 \cdot z^2}$ contains three or more identification rates higher than 0.8.

4.2 Evaluation Index Analysis of Identification Parameters

According to the identification rate results of the identification parameters in Section 4.1, the accurate rate and conservative rate of the excellent identification parameters under different identification criteria is calculated as shown in Table 5. The average criteria is the average value of the accurate rate and conservative rate under the slack, ideal, partially strict and relatively strict criteria, and all the accurate rate and conservative rate in the strict criteria are taken as 0.33 and 1 respectively.

According to the analysis of Table 5, under the slack criteria, the accurate rate and conservative rate of $\frac{a_m}{\theta_m^2 \cdot z}$ are the highest, which are 0.62 and 0.75 respectively. Under ideal and relatively strict criteria, the accurate rate of $\frac{a_m}{\theta_m \cdot z^2}$ is the highest, reaching 0.67 and 0.57 respectively, and the conservative rate of $\frac{a_m^2}{\theta_m^2 \cdot z^2}$ is the highest, reaching 0.83 and 0.92 respectively. When the judgment criterion changes from slack to relatively strict, the conservative rate shows a gradual increase trend, while the accurate rate reaches the maximum in the middle part (ideal criteria and partially strict criteria). For the above 9 identification parameters, the maximum values of average accurate rate is taken at $\frac{a_m}{\theta_m \cdot z^2}$, which is 0.61, and the maximum average conservative rate is taken at $\frac{a_m^2}{\theta_m^2 \cdot z^2}$, which is 0.84.

By observing the above identification parameters, it is found that all 9 identification parameters contain items a_m and θ_m. Therefore, it is analyzed and compared according to the number of quadratic terms contained in a_m and θ_m in the identification parameter. The optimal identification parameter and the identification criteria are selected by horizontally comparing the best accurate rate and conservative rate in each identification criteria and vertically comparing the number of optimal judgment results in each criterion.

According to Table 5, among the identification parameters containing zero quadratic terms in a_m and θ_m, with the increase of the number of times of penetration depth z in the denominator, the accurate rate and conservative rate show a gradual increasing trend in each criterion. Therefore, $\frac{a_m}{\theta_m \cdot z^2}$ is selected as one of the identification parameters to evaluate the mechanical properties of planetary soil. Among the identification parameters containing one quadratic term in a_m and θ_m, the accurate rate of $\frac{a_m}{\theta_m^2 \cdot z}$ is better than $\frac{a_m^2}{\theta_m \cdot z^2}$ and $\frac{a_m}{\theta_m^2 \cdot z^2}$, which is 0.6, and the conservative rate of $\frac{a_m}{\theta_m^2 \cdot z^2}$ is also better than $\frac{a_m^2}{\theta_m \cdot z^2}$ and $\frac{a_m}{\theta_m^2 \cdot z^2}$, which is 0.81. So, it is considered to select $\frac{a_m}{\theta_m^2 \cdot z}$ and $\frac{a_m}{\theta_m^2 \cdot z^2}$ as the evaluation identification parameters. Both a_m and θ_m are square terms in the identification parameters, the accurate rate and conservative rate of $\frac{a_m^2}{\theta_m^2 \cdot z^2}$ is obviously better than $\frac{a_m^2}{\theta_m^2}$ and $\frac{a_m^2}{\theta_m^2 \cdot z}$, so $\frac{a_m^2}{\theta_m^2 \cdot z^2}$ is selected as one of the identification parameters to evaluate the mechanical properties of planetary soil.

By comparing the number of optimal judgment results under each criterion, the ideal criteria is better than other criterions, so it is more reasonable to choose ideal criteria.

Through the above analysis, the final conclusion is that the superior identification parameters used for the identification of the mechanical property of planetary soil are $\frac{a_m}{\theta_m^2 \cdot z}$, $\frac{a_m}{\theta_m \cdot z^2}$, $\frac{a_m}{\theta_m^2 \cdot z^2}$ and $\frac{a_m^2}{\theta_m^2 \cdot z^2}$, the better criteria is ideal.

Finally, the ideal critical value in the actual assessment is determined, as shown in Table 6. The number used for assessment is calculated according to the simulation test data. If it is less than the lower critical value, the soil is determined as a low compactness, if it is higher than the upper critical value, it is determined as a high compactness, otherwise it is determined as a medium compactness.

Table 1. Superior identification parameters and critical value

Number	Identification parameter	Lower critical value	Upper critical value
15	$\frac{a_m}{\theta_m^2 \cdot z}$	0.10	0.23
16	$\frac{a_m}{\theta_m \cdot z^2}$	0.19	0.36
19	$\frac{a_m}{\theta_m^2 \cdot z^2}$	0.03	0.07
20	$\frac{a_m^2}{\theta_m^2 \cdot z^2}$	0.65	1.83

4.3 Validation of Evaluation Methods

According to the identification parameters and its critical value summarized above, the FE simulation verification of penetrator penetration is carried out. In the FE simulation verification test, the penetrator structure with radius R = 65 mm (the other parameters are the same as those mentioned in Section 2.1) was used to penetrate low, medium and high compactness soil at four different speeds (75 m/s, 125 m/s, 225 m/s and 275 m/s), and a total of 12 groups of simulation validation test were carried out. Table shows the mechanical property evaluation data and results of planetary soil. In accurate evaluation, L-L, M-M and H-H are marked as 1, and rest of the evaluation results are marked 0. In conservative evaluation, L-L, M-y (y = L, M) and H-z (z = L, M, H) are marked as 1, and rest of the evaluation results are marked 0.

The verification results show that the accurate success rate of the four identification parameters is higher than 50%, among which the accurate success rate of $\frac{a_m}{\theta_m^2 \cdot z}$ and $\frac{a_m}{\theta_m \cdot z^2}$ are the highest, 58.3%. At the same time, the conservative success rate of four identification parameters are higher than 80%, among which the conservative success rate of $\frac{a_m}{\theta_m^2 \cdot z}$, $\frac{a_m}{\theta_m \cdot z^2}$ and $\frac{a_m}{\theta_m^2 \cdot z^2}$ is the highest, 91.7%. Therefore, if conditions permit, it can choose the identification parameters $\frac{a_m}{\theta_m^2 \cdot z}$ and $\frac{a_m}{\theta_m \cdot z^2}$, and the ideal criteria for estimating the mechanical properties of planetary soil, so as to ensure the success rate of results.

5. Conclusions

The relationship between the three main characteristic parameters was summarized by analyzing the simulation process of penetrator penetrating into planetary soil. The maximum acceleration of penetrator was positive correlation with the soil compactness. The higher soil compactness, the higher maximum acceleration produced by penetrator in the process of penetrating. The maximum deflection angle and penetration depth were negatively correlated with the soil compactness. The higher soil compactness, the smaller maximum deflection angle and penetration depth of the penetrator in the penetration process. A nonparametric estimation method was proposed to estimate the properties of planetary soil based on the analysis of the penetration process. It can list 20 binary and ternary identification parameters. By analyzing the accurate rate and conservative rate of the identification parameters to the mechanical properties of planetary soil, four superior identification parameters were finally selected, which were $\frac{a_m}{\theta_m^2 \cdot z}$, $\frac{a_m}{\theta_m \cdot z^2}$, $\frac{a_m}{\theta_m^2 \cdot z^2}$ and $\frac{a_m^2}{\theta_m^2 \cdot z^2}$, and the identification criteria was determined to be the ideal criteria. Through the verification and evaluation of the selected superior identification parameters under the ideal criteria, the two optimal identification parameters were finally obtained as $\frac{a_m}{\theta_m^2 \cdot z}$ and $\frac{a_m}{\theta_m \cdot z^2}$, and the final prediction accurate success rate and conservative success rate were 58.3% and 91.7% respectively.

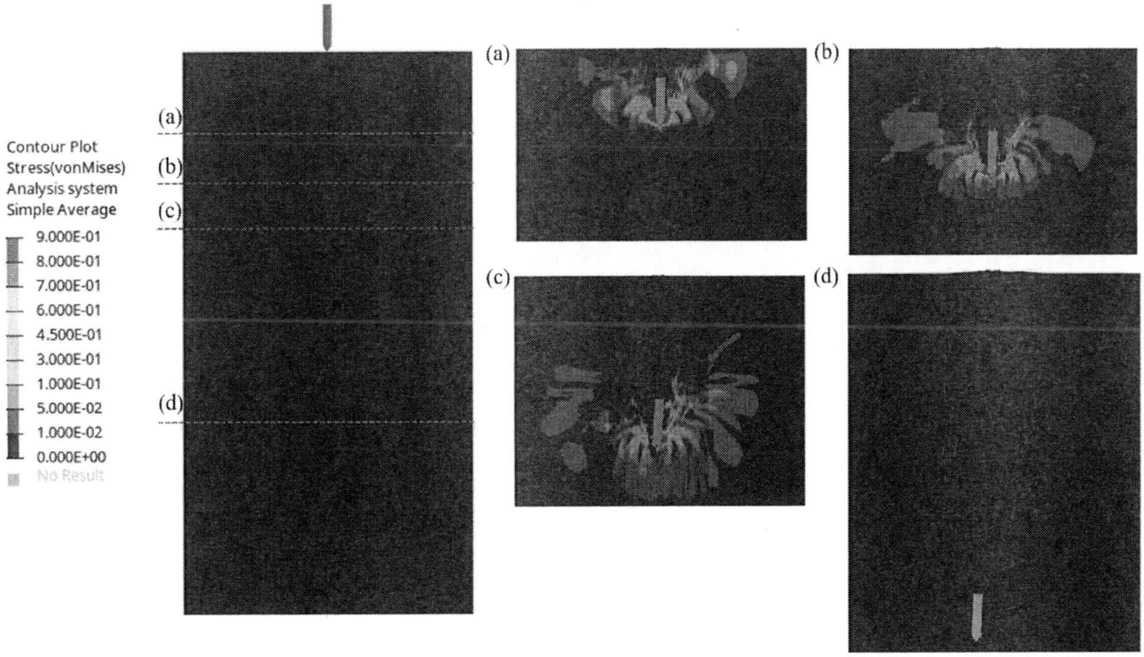

Figure 3. The stress nephogram of soil during the penetration process.
a) 0.005 s, b) 0.010 s, c) 0.015 s, d) 0.100 s

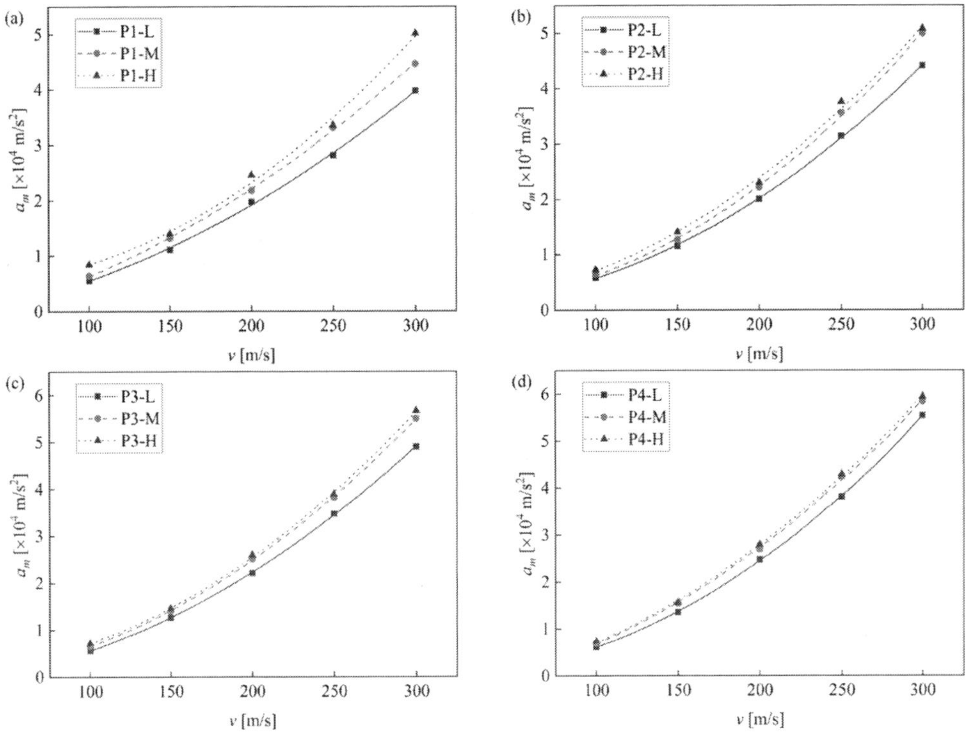

Figure 4. The maximum acceleration under different velocity and soil compactness conditions.
(a) P1, (b) P2, (c) P3, (d) P4

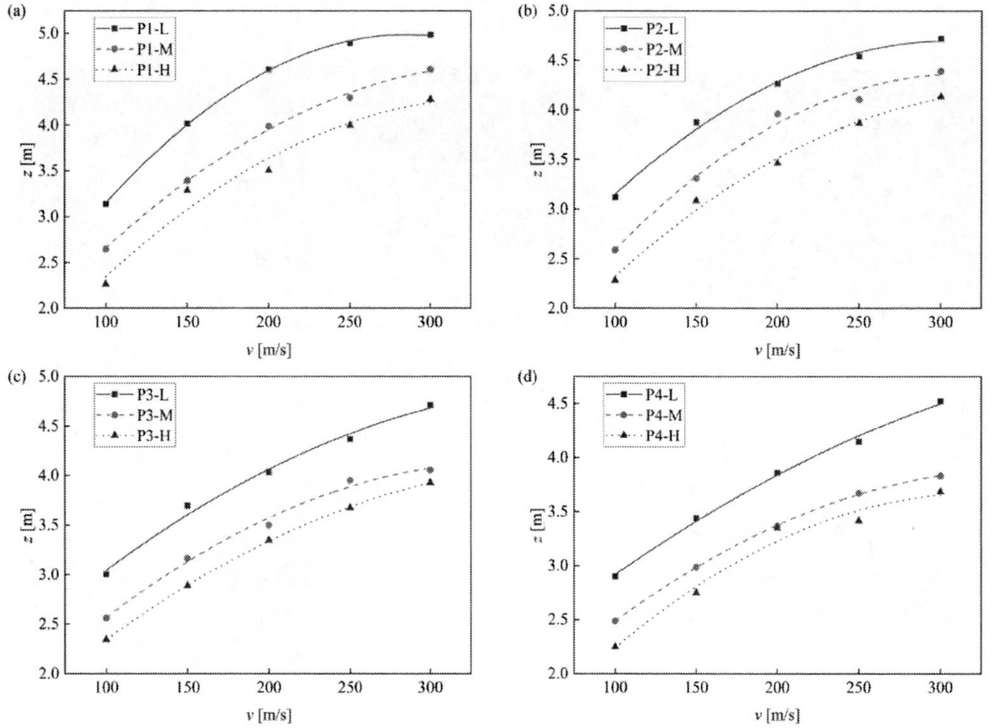

Figure 5. The penetration depth under different velocity and soil compactness conditions
a) P1, b) P2, c) P3, d) P4

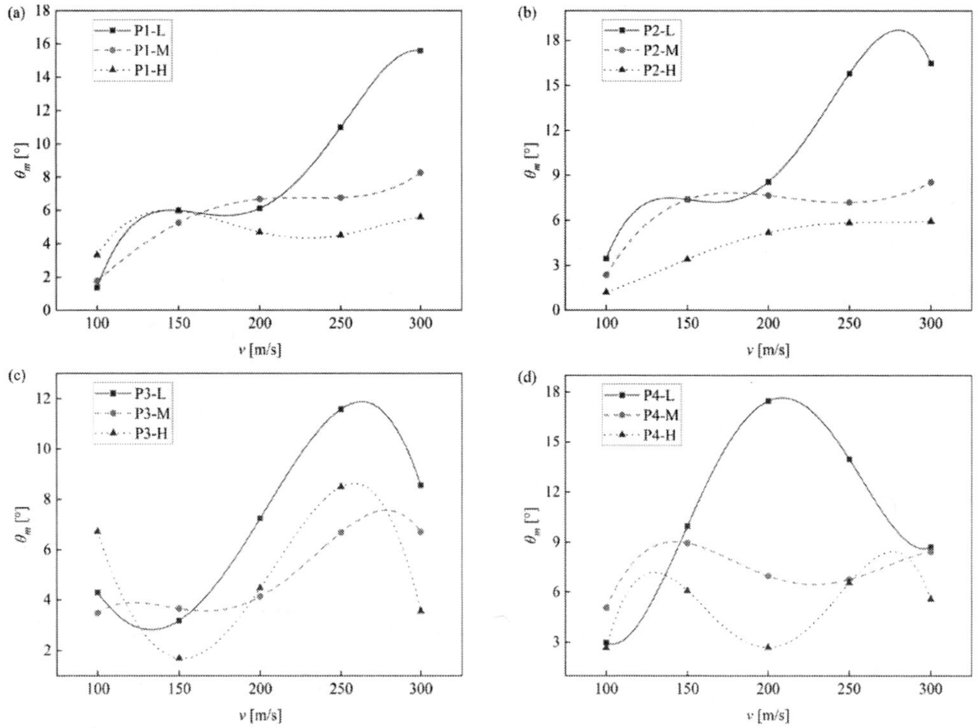

Figure 6. The maximum declivity angle under different velocity and soil compactness conditions.
a) P1, b) P2, c) P3, d) P4

Table 2. Identification parameters.

Number	Identification parameter	Number	Identification parameter	Number	Identification parameter	N
1	$\frac{a_m}{\theta_m}$	2	$\frac{a_m^2}{\theta_m}$	3	$\frac{a_m}{\theta_m^2}$	
5	$\frac{a_m}{z}$	6	$\frac{a_m^2}{z}$	7	$\frac{a_m}{z^2}$	
9	$\frac{1}{\theta_m \cdot z}$	10	$\frac{1}{\theta_m^2 \cdot z}$	11	$\frac{1}{\theta_m \cdot z^2}$	
13	$\frac{a_m}{\theta_m \cdot z}$	14	$\frac{a_m^2}{\theta_m \cdot z}$	15	$\frac{a_m}{\theta_m^2 \cdot z}$	
17	$\frac{a_m^2}{\theta_m^2 \cdot z}$	18	$\frac{a_m^2}{\theta_m \cdot z^2}$	19	$\frac{a_m}{\theta_m^2 \cdot z^2}$	

Table 3. Identification criteria for mechanical properties of planetary soil

Identification criteria	Identification basis		
	Low compactness	Medium compactness	High compactness
Slack	$A_i < A_L$	$A_L \le A_i \le A_H$	$A_i > A_H$
Ideal	$A_i < \overline{A_{LM}}$	$\overline{A_{LM}} \le A_i \le A_{Z2}$	$A_i > \overline{A_{MH}}$
Partially strict	$A_i < A_M$	$A_M \le A_i \le A_H$	$A_i > A_H$
Relatively strict	$A_i < \overline{A_{MH}}$	$\overline{A_{MH}} \le A_i \le A_H$	$A_i > A_H$
Strict	-	-	$A_i > 0$

Table 4. The identification rate of excellent identification parameters

Identification criteria	Identification result	$\frac{a_m}{\theta_m}$	$\frac{a_m^2}{\theta_m^2}$	$\frac{a_m}{\theta_m \cdot z}$	$\frac{a_m}{\theta_m^2 \cdot z}$	$\frac{a_m}{\theta_m \cdot z^2}$	$\frac{a_m^2}{\theta_m^2 \cdot z}$	$\frac{a_m^2}{\theta_m \cdot z^2}$	$\frac{a_m}{\theta_m^2 \cdot z^2}$	$\frac{a_m^2}{\theta_m^2 \cdot z^2}$
Slack	L-L	0.40	0.40	0.35	0.45	0.40	0.35	0.40	0.40	0.40
	L-M	0.50	0.50	0.65	0.45	0.60	0.60	0.50	0.55	0.60
	L-H	0.10	0.10	0.00	0.10	0.00	0.05	0.10	0.05	0.00
	M-L	0.20	0.20	0.05	0.00	0.00	0.15	0.15	0.00	0.05
	M-M	0.35	0.45	0.75	**0.80**	0.75	0.65	0.40	**0.80**	0.75
	M-H	0.45	0.35	0.20	0.20	0.25	0.20	0.45	0.20	0.20
	H-L	0.05	0.05	0.05	0.00	0.00	0.05	0.05	0.00	0.05
	H-M	0.35	0.40	0.40	0.40	0.40	0.40	0.35	0.50	0.40
	H-H	0.60	0.55	0.55	0.60	0.60	0.55	0.60	0.50	0.55
Ideal	L-L	0.60	0.65	0.75	0.55	0.75	**0.80**	0.75	0.60	**0.80**
	L-M	0.30	0.25	0.15	0.30	0.20	0.10	0.15	0.25	0.15
	L-H	0.10	0.10	0.10	0.15	0.05	0.10	0.10	0.15	0.05
	M-L	0.30	0.35	0.20	0.15	0.15	0.30	0.40	0.15	0.30
	M-M	0.25	0.20	0.45	0.55	0.45	0.35	0.10	0.50	0.40
	M-H	0.45	0.45	0.35	0.30	0.40	0.35	0.50	0.35	0.30
	H-L	0.20	0.25	0.10	0.05	0.05	0.10	0.20	0.00	0.10
	H-M	0.05	0.10	0.15	0.15	0.15	0.30	0.10	0.20	0.30
	H-H	0.75	0.65	0.75	**0.80**	**0.80**	0.60	0.70	**0.80**	0.60
Partially strict	L-L	**0.80**	**0.80**	**0.80**	0.70	**0.85**	**0.80**	**0.85**	**0.80**	**0.80**
	L-M	0.10	0.10	0.20	0.20	0.15	0.15	0.05	0.15	0.20
	L-H	0.10	0.10	0.00	0.10	0.00	0.05	0.10	0.05	0.00
	M-L	0.40	0.45	0.35	0.25	0.35	0.40	0.50	0.30	0.35
	M-M	0.15	0.20	0.45	0.55	0.40	0.40	0.05	0.50	0.45
	M-H	0.45	0.35	0.20	0.20	0.25	0.20	0.45	0.20	0.20
	H-L	0.25	0.25	0.15	0.10	0.10	0.25	0.25	0.05	0.15
	H-M	0.15	0.20	0.30	0.30	0.30	0.20	0.15	0.45	0.30
	H-H	0.60	0.55	0.55	0.60	0.60	0.55	0.60	0.50	0.55
Relatively strict	L-L	**0.90**	**0.90**	**0.90**	**0.85**	**0.95**	**0.90**	**0.90**	**0.85**	**0.95**
	L-M	0.00	0.00	0.10	0.05	0.05	0.05	0.00	0.10	0.05
	L-H	0.10	0.10	0.00	0.10	0.00	0.05	0.10	0.05	0.00
	M-L	0.55	0.55	0.65	0.70	0.60	0.65	0.50	0.65	0.70
	M-M	0.00	0.10	0.15	0.10	0.15	0.15	0.05	0.15	0.10
	M-H	0.45	0.35	0.20	0.20	0.25	0.20	0.45	0.20	0.20
	H-L	0.25	0.35	0.25	0.20	0.20	0.40	0.30	0.20	0.40
	H-M	0.15	0.10	0.20	0.20	0.20	0.05	0.10	0.30	0.05
	H-H	0.60	0.55	0.55	0.60	0.60	0.55	0.60	0.50	0.55

Table 5. The accurate rate and conservative rate of the excellent identification parameters

Identification criteria	**Evaluation criteria**	$\frac{a_m}{\theta_m}$	$\frac{a_m^2}{\theta_m^2}$	$\frac{a_m}{\theta_m \cdot z}$	$\frac{a_m}{\theta_m^2 \cdot z}$	$\frac{a_m}{\theta_m \cdot z^2}$	$\frac{a_m^2}{\theta_m^2 \cdot z}$	$\frac{a_m^2}{\theta_m \cdot z^2}$	$\frac{a_m}{\theta_m^2 \cdot z^2}$	$\frac{a_m^2}{\theta_m^2 \cdot z^2}$
Slack	η_a	0.45	0.47	0.55	0.62	0.58	0.52	0.47	0.57	0.57
	η_c	0.65	0.68	0.72	0.75	0.72	0.72	0.65	0.73	0.73
Ideal	η_a	**0.53**	0.50	**0.65**	**0.63**	**0.67**	0.58	**0.52**	**0.63**	0.60
	η_c	0.72	0.73	0.80	0.75	0.78	0.82	0.75	0.75	0.83
Partially strict	η_a	0.52	0.52	0.60	0.62	0.62	**0.58**	0.50	0.60	**0.60**
	η_c	0.78	0.82	0.87	0.83	0.87	0.87	0.80	0.87	0.87
Relatively strict	η_a	0.50	**0.52**	0.53	0.52	0.57	0.53	0.52	0.50	0.53
	η_c	0.82	0.85	0.90	0.88	0.90	0.90	0.82	0.88	0.92
Strict	η_a	0.33	0.33	0.33	0.33	0.33	0.33	0.33	0.33	0.33
	η_c	1.00	1.00	1.00	1.00	1.00	1.00	1.00	1.00	1.00
Average	η_a	0.50	0.50	0.58	0.60	**0.61**	0.55	0.50	0.58	0.58
	η_c	0.74	0.77	0.82	0.80	0.82	0.83	0.75	0.81	**0.84**

Table 7. Evaluation data and results of planetary soil mechanical properties

	Number	**L**	**L**	**L**	**L**	**M**	**M**	**M**	**M**	**H**	**H**	**H**	**H**	
Identification parameter	15	0.03	0.05	0.04	0.71	0.20	0.03	0.12	0.06	0.07	0.06	0.25	0.26	
	16	0.11	0.09	0.36	0.12	0.25	0.10	0.13	0.33	0.24	0.19	0.38	0.38	
	19	0.26	0.01	0.01	0.01	0.02	0.05	0.03	0.01	0.04	0.02	0.07	0.06	
	20	0.93	0.10	0.24	0.28	0.09	0.10	0.92	1.99	0.21	0.26	2.03	2.74	
Evaluation results	15	L	L	L	H	M	L	M	L	L	L	H	H	
	16	L	L	H	L	M	L	L	M	M	L	H	H	
	19	H	L	L	L	L	M	M	L	M	L	H	M	
	20	M	L	L	L	L	L	M	H	L	L	H	H	Success rate
Accurate evaluation number	15	1	1	1	0	1	0	1	0	0	0	1	1	58.3%
	16	1	1	0	1	1	0	0	1	0	0	1	1	58.3%
	19	0	1	1	1	0	1	1	0	0	0	1	0	50%
	20	0	1	1	1	0	0	1	0	0	0	1	1	50%
Conservative evaluation number	15	1	1	1	0	1	1	1	1	1	1	1	1	91.7%
	16	1	1	0	1	1	1	1	1	1	1	1	1	91.7%
	19	0	1	1	1	1	1	1	1	1	1	1	1	91.7%
	20	0	1	1	1	1	1	1	0	1	1	1	1	83.3%

6. Nomenclature

γ	*Bulk modulus*	*[MPa]*
E	*Elasticity modulus*	*[MPa]*
G	*Shear modulus*	*[MPa]*
μ	*Poisson's ratio*	
ρ	*Density*	*[kg/m^3]*

7. Acknowledgements

This present work was supported by National Natural Science Foundation of China (Grant No. 51305223), Post-doctoral Science Foundation of China (Grant No. 2018M641338) and funded by Scientific Research Foundation of Nanjing University of Aeronautics and Astronautics, PR China (Grant No. 1011-YAH20001).

8. References

Chen, Y., C. Tan, Y. Guo, 2022. Comparative study of numerical simulations of projectile penetration into metal targets. *Explos. Shock Waves*, 42(04):111-127. http://dx.chinadoi.cn/10.11883/bzycj-2021-0125

Harvey, B., 2007. *Russian Planetary Exploration: History, Development, Legacy, and Prospects*. Springer-Praxis, Berlin.

Iagnemma, K., S. Kang, H. Shibly, and S. Dubowsky, 2004. Online terrain parameter estimation for wheeled mobile robots with application to planetary rovers. *IEEE Trans. Rob.*, 20 (5): 921–927. https://doi.org/10.1109/TRO.2004.829462

Kang, S. 2003. Terrain parameter estimation and traversability assessment for mobile robots. Ph.D. dissertation, Dept. of Mechanical Engineering, Seoul National Univ.

Li, X., L. Jiang, Z. Zhao, D. Liu, Y. Ou, 2008. Numerical study on penetration of a high-speed-rotating bullet into the moving sheet-metal plate. *Explos. Shock Waves*, 2008(01):57-61. http://dx.chinadoi.cn/10.3321/j.issn:1001-1455.2008.01.010

Lorenz R. D., 2011. Planetary penetrators: Their origins, history and future. *Adv. Space Res.*, 48(3):403-431. https://doi.org/10.1016/j.asr.2011.03.033

Meng, X., G. Liu, L. Fu, 2019. Simulation analysis of impact and penetration for small celestial body penetrator. *J. Northeast. Univ., Nat. Sci.*,40(02):251-255. https://doi.org/10.12068/j.issn.1005-3026.2019.02.019

Nie, Z., Y. Peng, R. Chen, Z. Cai, F. Li, 2021. Sensitivity analysis of RHT model parameters for rock materials under penetrating condition. *J. Vib. Shock*, 40(14):108-116. https://doi.org/10.13465/j.cnki.jvs.2021.14.015.

Pan, X., Y. He, Y. He, X. Zhao, 2011. Simulation and experimental research on mechanics of static rotary penetration to concrete. *J. Nanjing Univ. Sci. Technol.*, 35(03):355-358+375. https://doi.org/10.14177/j.cnki.32-1397n.2011.03.001.

Ren, B., G. Tao, 2018. Experimental investigation and numerical simulation of ballistic when projectile penetrating natural soil. *Sci. Technol. Eng.*,18(01):28-33. http://dx.chinadoi.cn/10.3969/j.issn.1671-1815.2018.01.006

Scoon, G. E. N., G. P. Whitcomb, 1993. MARSNET: A precursor to the surface exploration of Mars. *Acta Astronaut.,* 29(10-11), 823-831. https://doi.org/10.1016/0094-5765(93)90164-R

Shiraishi, H., S. Tanaka, A. Fujimura, H. Hayakawa, 2008. The present status of the Japanese penetrator mission: LUNAR-A. *Adv. Space Res.*, 2008, 42(2):386-393. https://doi.org/10.1016/j.asr.2007.08.022

Sun, Z., H. Zhang, X. Wu, J. Ma, 2010. Mission analysis of a lunar soft lander. *Spacecr. Eng.*, 19(05):12-16. http://dx.chinadoi.cn/10.3969/j.issn.1673-8748.2010.05.003

Wang, Y., L. Xu, C. Yu, L. Cao, 2019. Structural design and penetration numerical simulation of a deep space impactor. *Mach. Des. Res.*, 35(06):188-191+196. https://doi.org/10.13952/j.cnki.jofmdr.2019.0342.

Xie, G., F. Chen, S. Gong, W. Lu, 2019. Numerical simulation on the projectile with different nose shapes penetrating moving target based on material point method. *Chin. J. Appl. Mech.*, 36(03):573-579+758. http://dx.chinadoi.cn/10.11776/cjam.36.03.D019

Xu, L., W. Wang, Y. Peng, Y. Wang, C. Zhou, 2020. Structural design and experiment of deep space impactor. *J. Mach. Des.*, 37(09): 67-70. https://doi.org/10.13841/j.cnki.jxsj.2020.09.011.

Zhou, C, Y. Yuan, C. Yu, Y. Wang, L. Gan, 2020. Experimental study on penetration of deep space impactor. *J. Nanjing Univ. Sci. Technol.*, 44(03):354-358+378. https://doi.org/10.14177/j.cnki.32-1397n.2020.44.03.014.

7018 / A Review of Modeling and Validation Techniques for Tire-Deformable Soil Interactions

Varsha Swamy*, Rashna Pandit, Alba Yerro-Colom, Corina Sandu, Denise Rizzo, and Katherine Sebeck
[a]*Terramechanics, Multibody, and Vehicle Systems (TMVS) Laboratory, Department of Mechanical Engineering, Virginia Tech, Blacksburg, Virginia, United States*
[b]*Department of Civil and Environmental Engineering, Virginia Tech, Blacksburg, Virginia, United States*
[c]*U.S. Army CCDC Ground Vehicle Systems Center, 6501 E. 11 Mile Road, Warren, MI 48397. DISTRIBUTION A. Approved for public release, distribution unlimited. OPSEC #6709*
**Corresponding author: varshasswamy@vt.edu*

https://doi.org/10.56884/IKFA4933

Abstract

The mobility of vehicles in off-road environments is critical for many applications. Understanding and predicting the tire-soil interaction is a challenge, especially when the terrain is largely deformable and plastic, and non-linear behavior is expected. This paper presents an overview of the bibliographic references on tires–deformable soil interactions and identifies the gaps present in the literature. The capabilities and challenges of different modeling methods used to assess the traction performance of a vehicle on deformable soil (i.e., analytical, semi-empirical, and numerical are discussed. Also, since the performance of the vehicle highly depends on the soil characteristics, a summary of the different material models used for non-cohesive and cohesive soil and the methods that have been used to characterize the soil is presented. The soil behavior and strength also depend on the water content and permeability along with the soil type. To highlight this, the available works on tire-wet soil interactions are discussed. Lastly, the different tire models and the validation methods that have been used for the soil, tire, and tire-deformable soil interaction are summarized. To conclude, this bibliographic review highlights that there is very limited work in the physics-based study of trafficability in wet soils, in particular in cohesive soil.

Keywords: deformable soil, terrain mobility, tire-soil interaction, wet soils, soil modeling

1. Introduction

Off-roading vehicles face several types of challenging terrains. The mobility of these vehicles, such as military trucks, earth-moving machines, and tractors, is essential in adverse environments. Over the years, modeling tire-terrain interactions has been a major focus in the Terramechanics community. Different technological advancements and recent innovations have helped bridge the gap between the results from simulations and reality, to support the modeling of virtual proving grounds.

The modeling of tire-terrain interactions is challenging. First, the non-linear behavior of the deformable terrain, which is affected by many factors (e.g., cohesive or non-cohesive, dry or wet, and strain-rate effects), needs to be characterized (i.e., the terrain model must be built). Next, the tire (i.e., tire model) is made of a deformable, non-linear material with complex geometry; the treads and lugs affect the overall dynamics of the system. Finally, the description of the tire-terrain interface (i.e., contact model) is also important because it influences the sinkage and the final traction obtained in the tire. The calibration and validation of all the models require input data which needs to be measured from feasible experiments that can capture the key characteristics of the problem.

Below, a summary of the literature review concerning modeling approaches to address tire–deformable soil interactions is provided. Special emphasis is given to numerical solutions based on continuum mechanics and the validation techniques that have been used.

2. Modeling Approaches

The existing models for tires on deformable terrain are empirical, semi-empirical, or based on numerical techniques. Empirical and semi-empirical techniques are very useful when real-time applications are concerned. However, they struggle to accurately capture the physics of the problem in different scenarios. Contrarily, solutions from physics-based numerical techniques require more computational time but can capture the response and the dynamics of the problem more accurately. Combinations of these modeling techniques, for example, a numerical tire model with an empirical terrain model have also been carried out (Chen, 1988).

When modeling numerically, a common approach for the tire is the Finite Element Method (FEM) which is based on continuum mechanics. The terrain model, however, can be based on different techniques from continuum mechanics or discrete mechanics. The Discrete Element Method (DEM), formulated from discrete mechanics, has been used to capture the micro-mechanics with a grain-level depiction of particles (*Fig. 1*). Granular soils have been successfully modeled using DEM. They are often coupled with Computational Fluid Dynamics (CFD) to simulate soil-water multiphase interactions. However, they take a huge amount of computational time and often fail to represent the real shapes of the soil particles and the interactions between them. Parameterizations of such models are extremely difficult.

The continuum mechanics-based approaches used in the literature to simulate the terrain are the Finite Element Method (FEM), Arbitrary Lagrangian-Eulerian (ALE), Smoothen Particle Hydrodynamics (SPH), and Material Point Method (MPM). These continuum-based approaches are macro-scale and require a material model (material constitutive law) to uniquely describe the stress-strain behavior of the material. FEM is a Lagrangian mesh-based approach that can handle the modeling of complex geometries and material constitutive laws (*Fig. 2*). However, they fail at capturing large deformations. ALE can be useful in such cases due to its adaptive meshing scheme. It is an intermediate method between Lagrangian and Eulerian and the mesh is not fixed to the material or in space. Larger deformations can be handled by SPH which is a Lagrangian mesh-less technique (*Fig. 3*). SPH is a particle-based method where each particle represents an aggregate of soil grains (and not soil grains themselves like DEM). It can be used for multiphase modeling too. However, a big disadvantage of SPH is its struggle in applying boundary conditions. This can be straightforwardly dealt with by MPM, a fairly new method that implements the Lagrange approach for the material points and a background Eulerian mesh (*Fig. 4*). This hybrid technique solves the governing equations at the mesh nodes and interpolates the results back to the material points. The different modeling approaches for soil discussed above have been shown in *Fig. 1, 2, 3, and 4* from state-of-the-art work in each technique.

Different modeling tools, such as those presented above, should be used depending on the application. Below, a discussion is provided on how the different approaches have been used to model tire-terrain interactions.

3. Modeling of Deformable Terrain

A critical role in characterizing the terrain is held by the soil properties. This study concentrates on the deformable soil type of terrains only, from now on. Deformable soils can be modeled using empirical and semi-empirical equations or numerical techniques.

3.1 Empirical and semi-empirical based models for deformable soils

Traditionally, the soil has been modeled using empirical and semi-empirical equations due to the complex nature of the task. These equations are usually in the form of pressure-sinkage equations and shear stress-shear displacement relations. The most commonly used relations are the Bekker (1969) pressure-sinkage relations and the Janosi and Hanamoto (Janosi, 1961) shear stress-shear displacement relations. Many corrections and updates on these relations have been proposed over the years. These corrections and the parameterization testing methodologies are available in the ISTVS standards (He et al., 2018).

Although the above method has been used over the years to describe the soils, they come with all the disadvantages related to empirical models. With the advancement of technology, we are at a stage where physics-based models are necessary for further enhancement of technology. Modeling the terrain, using physics-based continuum mechanics approaches requires soil material models.

Figure 1. DEM tire-soil interaction: Vertical displacement counter (Zhao and Zang, 2017)

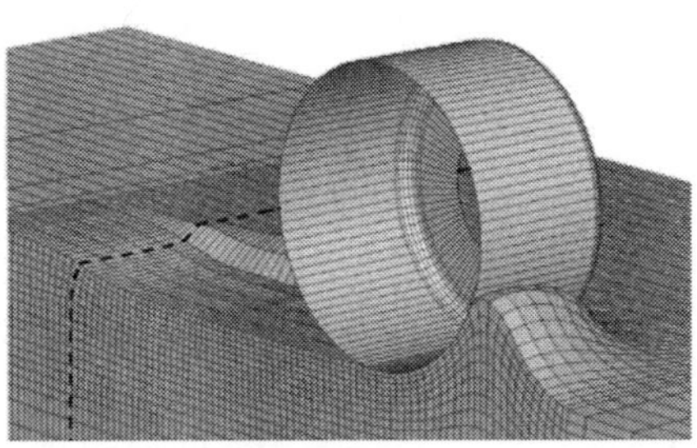

Figure 2. FEM tire-soil interaction: Final configuration of the rolling wheel (Hambleton and Drescher, 2009)

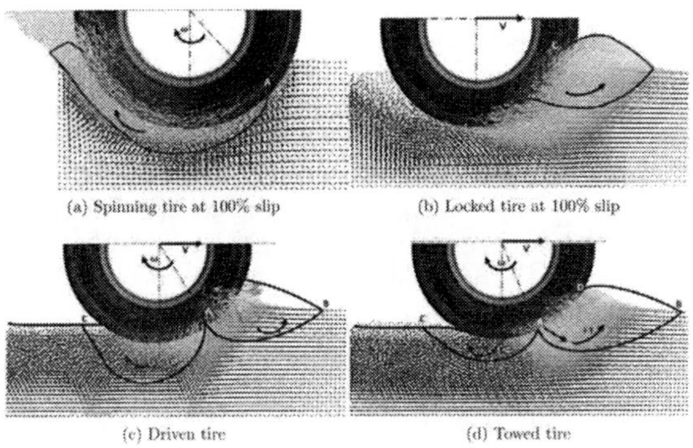

Figure 3. SPH tire-soil interaction: Flow patterns for different driving conditions (El-Sayegh, 2020)

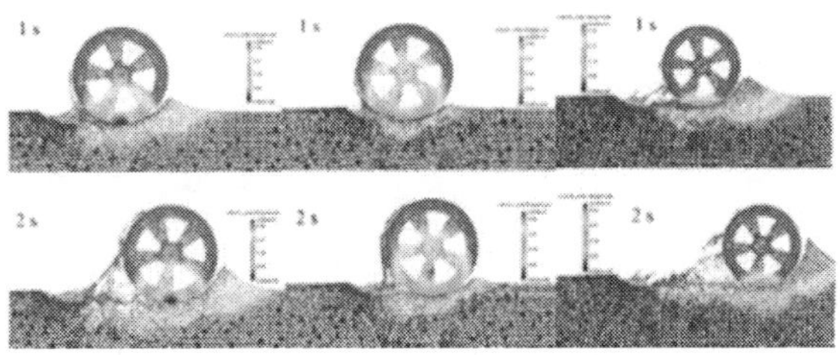

Figure 4. MPM tire-soil interaction: Velocity distribution for different slip ratios (Xu et al., 2021)

3.2 Physics-based material models for deformable soils

There are many material models available to describe the behavior of soils, each with different capabilities. No single material model can universally be used for all types of soils to model every scenario (Contreras et al., 2013). The users must select the material model based on the abilities they want to model and the abilities the material model has to offer, keeping in mind the disadvantages. A few abilities to consider while modeling are yield surface shape, hardening and softening behavior, strain rate dependency, plastic flow rule (associative or non-associative), multi-phase modeling and pore water pressure prediction, hysteresis on cyclic loading and reloading, number of invariants considered, and viscoplastic behavior. The flow rule defines the direction of the plastic deformation. Associative models prescribe plastic potential to be equal to the yield surface and hence the plastic strain increment to be normal to the yield surface. A non-associative flow rule requires the user to separately describe the plastic potential.

More advanced models may have added capabilities that can accurately capture the soil for more complex conditions. However, they require more computational time, input data, and effort. It must be noted that additional inputs may necessitate more experimentation to be conducted and calibration processes based on trial and error to fit the data. It is essential to select the correct material model based on the material type and also obtain the correct material parameters for the same. The yield surfaces of different common soil material models are presented in *Fig. 5*.

3.2.1 Non-cohesive soils

Modeling of non-cohesive soils, like sand, has been approached by many researchers. A linearly elastic perfectly-plastic material model like the Mohr-Coulomb law with the non-associative flow and zero dilation angle has been used (Saliba, 1990). The zero-dilation angle signifies that the plastic potential is horizontal and there is no plastic volumetric deformation while plastic shearing. Another elastic perfectly-plastic material model is the Drucker-Prager model which has been used with non-associative behavior (Hambleton and Drescher, 2008; Ozaki et al., 2015) for sandy soils and soft soil (Moslem and Hossein, 2014). Along with the elastic perfectly-plastic behavior, modeling strain-rate effects which scale up or down the yield surface based on the strain rate have been implemented. Wright (2012) used this approach to model sandy loam.

Alternatively, the "elastic-plastic material model" available in PAM-CRASH has been employed to model dry sand (Allen II et al., 2007; Dhillon, 2013), sandy loam (Slade, 2009), and LETE sand (Dhillon, 2013; Sharifi, 2020). A more advanced elastic-plastic model is the Drucker-Prager cap model. The cap is a yield surface that couples the volumetric deformations with the shear deformations. It has been used to model McCormick Ranch sand which is a clayey sand (Sally A. Shoop, 2001) and soft soil (Yamashita et al., 2018)

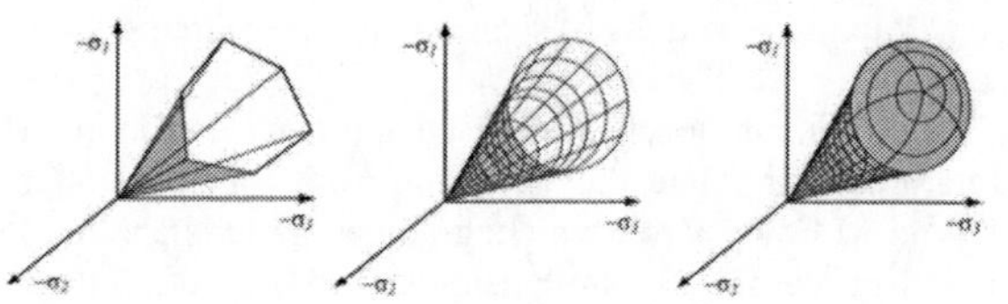

Figure 5. Yield surface of different soil models; Mohr-Coulomb, Drucker-Prager, and Strain Hardening Cap Model (Contreras et al., 2013)

3.2.2 Cohesive soils

The most basic model to use for cohesive soils is the Von-Mises model with a friction angle set to zero (Hambleton and Drescher, 2008). The Drucker-Prager model has been employed to model soft clay (Ozaki et al., 2015), loamy clay texture (Cueto et al., 2016), and clayey loam soil with the associative flow (Farhadia et al., 2019). Similarly, the elastic-plastic material model on PAM-CRASH has been used to model heavy clay (Dhillon, 2013), clayey soil (Dhillon, 2013; Ragheb et al., 2013), and clayey loam (Sharifi, 2020). The modified Cam-Clay model based on critical state soil mechanics theories has been used to accurately model Ultisol at different moisture percentages (Silva et al., 2018). The Drucker-Prager cap model has also been used to effectively model cohesive soils; a loose loam with high cohesion (Fervers, 2004) and clayey loam soil (Jafari, 2015; Maria et al., 2021) have been modeled. Soils with different moisture contents have been modeled using total stress approaches without separating the soil and the water properties.

3.3 Material properties for material models for deformable soils

Suitable material properties for the soil material model are essential to capture the dynamics of tire-soil interactions. Each soil material model has different inputs. There are two methods researchers have used over the years. The first method involves conducting geotechnical tests to obtain the parameters and then validating them. Depending on the material model's requirements, often mentioned by the developers, different geotechnical tests are required to parameterize the models. Common geotechnical tests required for the parameterization include the triaxial test to obtain the strength parameters and hence the yield surface, uniaxial compression tests to obtain the stiffness parameters like the Young's modulus, or even the direct shear tests or ring shear tests to obtain the secant modulus. Additionally, data is required if strain dependency is to be captured in the model. The testing conditions must be for similar conditions we observe in the field, for low confinement pressures. Also, in-situ tests can be conducted like the cone penetrometer test which was developed by the U.S. Army Corps of Engineering to obtain material parameters like the friction angle and cohesion angle based on correlations present in literature. By testing various depths, we can obtain the Young's Modulus for each depth and use it to accurately model the terrain (Maria et al., 2021).

In the second method, the material parameters are calibrated based on tests conducted by the modeler using trial and error or optimization techniques. A common practice is to parameterize the soil model by calibrating the material input parameters to the empirical correlations obtained from pressure sinkage (Bekker, 1969; Wong, 2001) and direct shear tests (Janosi, 1961). Wheel sinkage tests have also been used to parameterize the soil material model (Papamichael and Vrettos, 2021).

4. Modeling of Tires for Deformable Soil Navigation

Tires are complex anisotropic structures that are challenging to model. Simple models assume the tire to be rigid or a combination of spring and dampers. More advanced models are valid for different loads, inflation pressure, and camber angle of the tire.

4.1 Semi-empirical tire models for deformable soil interactions

A few semi-empirical models include the Nepean wheeled vehicle performance model NWVPM (Wong and Asnani, 2008), B. Chan and C. Sandu model (Chan and Sandu, 2014), Vehicle Dynamic Analysis Non-linear (VDANL) model (Liang et al., 2004), AESCO Soft Soil Tire Model (AS2TM) (Harnisch et al., 2005), Soil Contact Model (SCM) (Krenn and Gibbesch, 2011), Rigid ring model (Chae, 2006; Slade, 2009; Zegelaar and Pacejka, 1997), Distributed lumped mass-spring model (Wakui and Terumichi, 2011), radial-inter radial spring-damper tire model (Madsen et al., 2012), Flexible Ring Tire Model (FTire) (Gipser, 2007), HSSTM (Sandu et al., 2020, 2019b, 2019a), and Modified Magic Formula tire model (He et al., 2019a, 2019b). Models like the AS2TM, FTire, and HSSTM allow the users to input their user-defined models for the tire and the terrain.

The terrain models for these models are usually empirical formulations. The tire model for the SCM, FTire, Distributed lumped mass-spring model, radial-inter radial spring-damper tire model, and the HSSTM are 3D, while the rest of the models described are 2D models. The parameterization of these models usually involves data, either tested by the user or available already in the literature. Further reviews of these models are present in the literature (Sharifi, 2020; Taheri et al., 2015).

4.2 Physics-based FEM tire models for deformable soil interactions

Numerical tire modeling using FEM has also been used increasingly over the past years for tire-terrain interactions. It is important to capture the tire deformation accurately as it seems to affect the sinkage (Hambleton and Drescher, 2008) and the value of motion resistance, especially on firm soils (Maria et al., 2021). Based on the application and the resources available, researchers have used both simple models and advanced models in FEM. Initially, 2D models of the tire had been used for modeling. It is only in recent time's 3D models for tires to account for the depth effect has been used. It is mainly the carcass that is affected by the strain in the 3rd direction (Fervers, 2004). Comparing the results obtained from 2D and 3D analysis it was concluded that the 3D effect is larger in sands than in clays (Hambleton and Drescher, 2009, 2008). When the terrain is highly deformable, simpler material models for the tire like the rigid wheel or elastic tire of one material and rigid rims have been used. Rubinstein et al. (2018) calibrated the tire material model from static vertical loading experiments on rigid surfaces. Similarly, Farhadia et al., (2019) did the same for different combinations of inflation pressure and vertical load for their tire material model.

A more detailed tire model considers the various parts of a tire and its construction. *Figure 6* provides a good view of the cross-section of a FEM tire model with the different parts. The construction of a tire depends on whether the tire is bias-ply or radial-ply. More advanced tire models employ modeling the different parts of the tire including the tread, undertread, carcass and belt plies, shoulder, sidewall, beads, and rim using the correct geometric dimensions and material models. The modeling of tire treads and grooves is important when validating the contact area patch. When data from manufacturers are not available, researchers have also cut the tire and traced the tire cross-section. The rubber material present in the tread, undertread, bead fillers or shoulder has been usually modeled as a hyperelastic Mooney-Rivlin material is seen in many works (Allen II et al., 2007; Dhillon, 2013; Gheshlaghi et al., 2021; Mohsenimanesh et al., 2009b; Slade, 2009). The belt and carcass plies are multilayered and have been (Xia, 2011; Xia and Yang, 2012) modeled as a multilayered membrane with isotropic rubber to model the carcass and belt and radial plies layered strategically at specific angles (Chae, 2006; Dhillon, 2013; El-Sayegh et al., 2019; Ragheb et al., 2013). To simplify this, researchers (Hambleton and Drescher, 2008; Moslem and Hossein, 2014) have also resorted to modeling the carcass as linear elastic. The rim is usually rigid shell elements and the beads are beams (El-Sayegh et al., 2019). Another different method to model tires was by Yamashita et al. 2018 by simplifying the fiber-reinforced rubber tire model as composite shell elements. The Darnell-Shoop model was used by Shoop, (2001) to model the tire. Xia (Xia, 2011; Xia and Yang, 2012) used a finite strain hyperelasticity user-defined material model to accurately model the rubber parts of the tire.

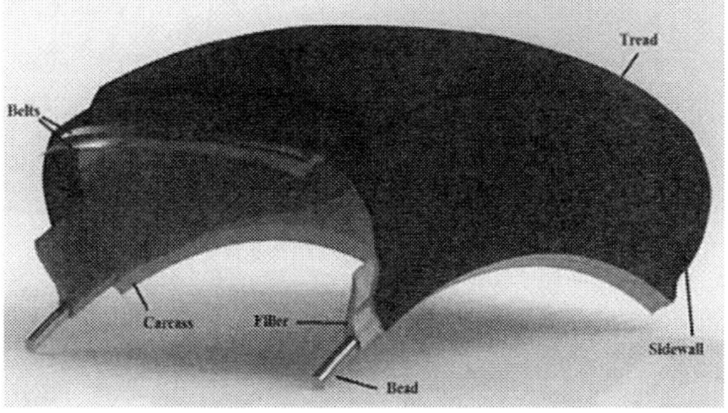

Figure 6. FEM Tire cross-section (Sh. Taheri, 2014)

5. Modeling of Tire-Soil Interface

Modeling the tire soil interface is crucial and is an area that still requires additional research to accurately simulate the interaction. Contact definition is key to understanding how two bodies are normally interacting and transmitting forces between them. Surface to surface contacts are very commonly used when dealing with FEM and nodes to surface contact is used when using SPH. The solver will require the user to define the master and the slave surface. Generally, the master surface is one with a more coarse mesh and it is made sure that the slave is not allowed to penetrate the master.

Contact mechanisms are usually penalty-based for this application. The penalty-based formulations resist the penetration of the slave into the master surface by introducing linear springs in between the surfaces. The smaller the distance of penetration between them, the higher the spring force. It must be kept in mind that this method is meant for pseudo-static analysis as the inertia forces are not considered (Sally A. Shoop, 2001). The kinematic contacts on the other hand consider the inertial forces and are more suitable for dynamic simulations.

The tangential shear forces in the model are captured by the friction law and must be accurate to precisely predict the traction. Friction models present in the literature are usually empirical or semi-empirical. Most of the models for friction implemented by researchers are based on the Mohr-Coulomb friction model. It was seen that while using the Mohr-Coulomb model, the lateral force and the friction coefficient, are related almost quadratically (Slade, 2009; Xia, 2011). A higher frictional coefficient would provide better traction. Slade suggests correct testing of the tire in the soil to find the static and dynamic coefficient of friction is hence required. Experimental tests for traction and braking can be used to find friction (Sally A. Shoop, 2001).

The Mohr-Coulomb model is very simple. It is unable to capture any rate-based frictional contact behavior (Xia, 2011). Hence, other models have also been used to model friction. Liu and Wong, (1996) used a unique nonlinear model for friction and defined friction as a function of relative velocity from their traction and wheel slip data. In 2016, Ozaki and Kondo, (2016) executed a user-defined contact to model anisotropic friction based on elastoplastic models which implement penalty-based formulation. The friction cone is orthotropic anisotropic with rotational hardening instead of the isotropic-sliding surface that is usually used from Mohr-Coulomb friction. Hence, now, the frictional property will vary depending on the sliding direction. The paper further investigated the effect of the anisotropy of the contact on the drawbar pull and the side force for different slip angles and slip ratios. Yamashita et al., (2018) successfully used the LuGre friction model to carry out tire-soil simulations. Later, Papamichael and Vrettos, (2021) conducted direct shear tests between the tire and the soil to model the soil-terrain interaction. It was experimentally found that the shear strength between the tire and soil is much lower than the soil-soil interaction. These interference shear properties between the tire and soil were used for the soil model.

Shoop, (2001) mentioned that the friction models present in literature (as of the time of publication of her study) are not very precise and will also not be for the next immediate future. There still is a need for further research and accurate modeling of friction.

6. Validation Methods

Numerical techniques for modeling tire-terrain interactions require validation at several levels to ensure the correct inputs have been used. It is proposed that every tire-deformable soil study should have three levels of validation: validation of the deformable soil, validation of the tire model, and validation of the tire and deformable soil together (static and dynamic). Different studies have accomplished different levels of validation and have been presented below.

6.1 Deformable soil validation

Validation of the soil is achieved either through comparing the experimental results obtained through quasi-static and dynamic triaxial testing and direct shear test (Cueto et al., 2013; Farhadia et al., 2019; Yamashita et al., 2018), through pressure sinkage tests, or relations available in the Wong (2001) textbook (El-sayegh, 2020; Ragheb et al., 2013; Sharifi, 2020; Slade, 2009), shear displacement and shear stress relations in Wong (El-sayegh, 2020; Gheshlaghi et al., 2021; Sharifi, 2020) or cone penetrometer results obtained (Rubinstein et al., 2018).

6.2 Tire Validation

Different static and dynamic tests have been used to validate the tire model. Vertical static deflection tests on a rigid surface to obtain the tire's stiffness are the most common test and have been used (Chae, 2006; El-sayegh, 2020; Fervers, 2004; Gheshlaghi et al., 2021; Slade, 2009; Yamashita et al., 2018) also validates the lateral stiffness and longitudinal stiffness along with the vertical stiffness. Additionally, the static footprint test can be used to evaluate the contact area on a rigid surface (Mohsenimanesh et al., 2009a, 2009b; Slade, 2009) and even obtain the contact stress distribution (Li and Schindler, 2013a, 2013b; Sally A Shoop, 2001). A dynamic test used for validation is the drum-cleat test used by (Chae, 2006; El-Sayegh et al., 2019, 2018; El-Sayegh and El-Gindy, 2018; Sharifi, 2020) to validate the harmonics of the first mode of vibration. A tire is made to roll over a cleat and the vertical forces at the spindle are converted into the frequency domain from the time domain using an FFT. It is used to then calculate the sidewall damping (Chae, 2006; El-sayegh, 2020). Another dynamic test is to obtain the dynamic rolling radius for different loads and inflation pressure on a rigid surface as done by Fervers, 2004; Ragheb et al., 2013 used cornering characteristics data on a rigid surface to validate the tire model.

6.3 Tire—Deformable soil validation

The tire with deformable soil validation is necessary to ensure the interaction or contact model and all the different components are involved with each other suitably to generate the right results. A basic test is to statically load the tire vertically on the deformable soil and record the sinkage and contact area for different inflation pressure and vertical loads (Cueto et al., 2016; Mohsenimanesh et al., 2009b). Another simple technique is to validate the rolling resistance from the virtual rolling tests with the data already present in the literature (Dhillon, 2013).

Full validation in detail requires the use of large-scale repeatable testing. Soil bin testing facilities indoors (He et al., 2019a) or single-wheel vehicle testing outdoors (Li and Schindler, 2013a) have been used over the years for such large-scale validation. Indoor soil bin validation techniques are usually preferred to easily control the surrounding environment and have homogeneous soil. An indoor testing facility (Sandu et al., 2008) is illustrated in *Fig. 7*.

The soil bin testing facility indoor usually consists of a soil bin, soil conditioning equipment to prepare the terrain (leveling and compaction), a motion control system to roll the tire forward and also allow for braking at different slip ratios, attachments to change the camber and the toe of the tire if required, a normal load controller, a data acquisition system and analysis to process and then save the data, power sources to drive the whole system, and additional small equipment like shovels and lifters to support the experiment. Usual tests conducted with the tires on such devices include free-
rolling tests, traction tests at different slip ratios, braking, and cornering tests at different slip angles. Indoor scaled models have also been used in the past to validate numerical models as it is easier to regulate and perform. In their study, Hambleton and Drescher (2008) build an apparatus with a test bed that moves forward on a linear bearing rail on which the aluminum wheels connected to a load cell operate. Efforts must be taken to ensure that the scaled model is a good representation of the actual phenomenon. Further review and survey on the different soil bin facilities and what they were employed for can be found by Ani et al., (2018).

These full-scale testing mentioned above are used to record data and are later processed and compared to results in the simulations. The dynamics rolling tests are performed to record the soil's plastic deformation and sinkage (Maria et al., 2021) and ruts profile through 3D scanning (Li and Schindler, 2013a). Farhadi et al., (2018) additionally obtained the contact area and volume by creating plaster mold from the tire footprint and 3D scanning the same, as shown in *Fig. 8*. Ozaki et al., (2015) used a novel Particle Image Velocimetry (PIV) analysis to qualitatively validate the ripple effect found under rigid wheels on soft soils. Further, the rolling resistance can be experimentally obtained for different velocities, normal loads, and inflation pressures and used for validation (Jafari, 2015; Maria et al., 2021; Yamashita et al., 2018). Cueto et al., (2016) conducted static indentation tests on the soil to measure the vertical stress in the soil at different depths and water content. Sharifi, (2020) carried out experimentation with load cells to obtain the vertical stress in tire rolling conditions for various speeds on two types of soils: clayey loam and lete sand. Similarly, Gheshlaghi (Gheshlaghi et al., 2021; Gheshlaghi and Mardani, 2021) performed the free-rolling tests on clayey loam with different moisture contents and validated it with the SPH predictions. Another parameter obtained from the dynamic rolling tests which can be used to validate the model is the gross traction (Rubinstein et al., 2018).

Figure 7. Indoor testing facility at the TMVS Laboratory in Virginia Tech with soil-bin (He et al., 2019a)

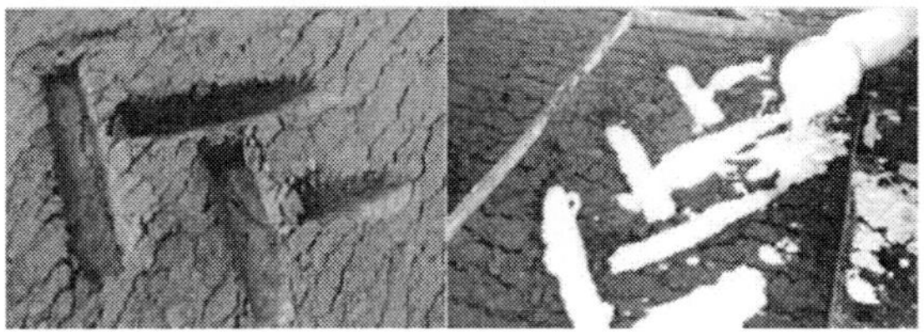

Figure 8. Tire footprint formation during testing and plaster mold creation to measure contact area and contact volume (Farhadi et al., 2018)

7. Final Discussion

A summary of modeling and validation approaches for a few of the prominent works in this field has been provided in Tables 1, 2, and 3 for different modeling approaches based on continuum mechanics.

From the literature review, we see a shift from empirical models to more physics-based models to model tire-deformable soil interactions. FEM-based tire models have been advanced over the years to accurately capture the dynamics of tires. Simulations have been able to obtain precise soil deformations and the rut. Yet, we still need to investigate the effect of the tire tread width. Moreover, there is still a need to develop improved soil models and interface friction models. From Tables 1, 2, and 3, we see that the soil models that have been explored are limited to the Mohr-Coulomb, Drucker-Prager and Drucker-Prager Capped models. Although granular and non-cohesive soiled terrains have been investigated, more plastic materials which show history-dependent behavior are yet to be approached in detail. Wet soils have been approached in the last few years. Most of the models are based on total stress analysis with no indication of the pore water pressure behavior. The interface contact models are still based on Mohr-Coulomb which is a fairly simple model. Many works assume friction coefficient without proper testing. Modeling of soil adhesion to the tire and stone picking are areas not ventured into in detail yet.

Validation techniques have grown significantly in importance recently. Validation of the tire with vertical static loading to obtain the vertical stiffness and the contact area has been conducted by many researchers. Tire-soil interactions have been usually validated by acquiring the rolling resistance and soil sinkage. However, many works are incompletely validated or validated for only one scenario for a particular set of test conditions. Full-scale validation to different levels is required. Many tire-soil models still require validation for drawbar pull, lateral forces, combine slip, aligning moment, multipass, and bulldozing force.

8. Conclusions

An extensive literature survey was conducted to identify the different types of modeling and validation approaches present for tire-deformable soil interactions. Although empirical and semi-empirical methods have aided the development of this field in the past, they seldom capture the dynamics of a range of scenarios. Physics-based numerical approaches have gained importance for modeling virtual proving grounds. Discussions on the different material models used for modeling deformable soils and tires are provided. Additionally, the calibration processes for these material models when extensive material testing cannot be conducted are examined. Recent developments in friction modeling between the tire and the deformable soil and the testing procedures that have been used are presented.

A major part of this paper deals with the validation techniques. It is necessary to have different levels of validation; validation of deformable soil model, validation of tire model, and validation of tire-deformable soil interaction model. Different techniques used by researchers are investigated in detail. The paper is concluded by summarizing the major contributions to this field of tire-deformable soil interactions through summary tables with important details on modeling and validation. It is identified that the major gaps in the literature are on the modeling of plastic, wet cohesive soils, contact modeling of adhesion at the interface, studies on cornering and maneuverability, multi-pass studies, and full-scale validation of the different models.

9. Acknowledgments

This study has been partially supported by the Excellence in Research Award received by Ms. Varsha Swamy during the 2021 summer and by the ARC Award received under the Cooperative Agreement W56HZV-19-2-0001 U.S. Army DEVCOM Ground Vehicle Systems Center, as well as by the TMVS Lab and Geotechnical Engineering Lab at Virginia Tech.

10. References

Agarwal, S., Senatore, C., Zhang, T., Kingsbury, M., Iagnemma, K., Goldman, D.I., Kamrin, K., 2019. Modeling of the interaction of rigid wheels with dry granular media. J. Terramechanics 85, 1–14. https://doi.org/10.1016/j.jterra.2019.06.001

Allen II, J., Shoffner, B., Öijer, F., El-Gindy, M., Trivedi, M., Johansson, I., 2007. Predictions of tire-terrain interaction using finite element analysis models, in: Proceedings of the ASME 2007 International Design Engineering Technical Conferences & Computers and Information in Engineering Conference. pp. 995–1000.

Ani, O.A., Uzoejinwa, B.B., Ezeama, A.O., Onwualu, A.P., Ugwu, S.N., Ohagwu, C.J., 2018. Overview of soil-machine interaction studies in soil bins. Soil Tillage Res. 175, 13–27. https://doi.org/10.1016/j.still.2017.08.002

Bekker, M.G., 1969. Introduction to Terrain-Vehicle Systems. Part I: The Terrain. Part II: The Vehicle.

Chae, S., 2006. Nonlinear finite element modeling and analysis of a truck tire.

Chan, B.J., Sandu, C., 2014. Development of a 3-D quasi-steady state tyre model for on-road and off-road vehicle dynamics simulations: Part III-off-road flexible tyre model. Int. J. Veh. Syst. Model. Test. 9, 151–176. https://doi.org/10.1504/IJVSMT.2014.061456

Chen, S., 1988. Study of tire-terrain interaction using the Finite Element Method.

Contreras, U., Li, G., Foster, C.D., Shabana, A.A., Jayakumar, P., Letherwood, M.D., 2013. Soil models and vehicle system dynamics. ASME. Appl. Mech. Rev. 65. https://doi.org/10.1115/1.4024759

Cueto, O.G., Coronel, C.E.I., Bravo, E.L., Morfa, C.A.R., Suarez, M.H., 2016. Modelling in FEM the soil pressures distribution caused by a tyre on a Rhodic Ferralsol soil. J. Terramechanics 63, 61–67.

Cueto, O.G., Coronel, C.E.I., Morfa, C.A.R., Sosa, G.U., Gómez, L.H.H., Calderón, G.U., Suárez, M.H., 2013. Three dimensional finite element model of soil compaction caused by agricultural tire traffic. J. Terramechanics 99, 146–152.

Dhillon, R.S., 2013. Development of truck tire-terrain finite element analysis models. https://doi.org/10.1190/segam2013-0137.1

El-Gindy, M., Lescoe, R., Öijer, F., Johansson, I., Trivedi, M., 2011. Soil modeling using FEA and SPH techniques for a tire-soil interaction, in: Proceedings of the ASME 2011 Design Engineering Technical Conferences and Information in Engineering Conference. pp. 793–802. https://doi.org/10.1115/DETC2011-47104

El-sayegh, Z., 2020. Modeling and Analysis of Truck Tire-Terrain Interaction.

El-Sayegh, Z., El-Gindy, M., 2018. Cornering characteristics of a truck tire on wet surface using finite element analysis and smoothed-particle hydrodynamics. Int. J. Dyn. Control 6, 1567–1576. https://doi.org/10.1007/s40435-018-0403-5

El-Sayegh, Z., El-Gindy, M., Johansson, I., Öijer, F., 2018. Improved tire-soil interaction model using FEA-SPH simulation. J. Terramechanics 78, 53–62. https://doi.org/10.1016/j.jterra.2018.05.001

El-Sayegh, Z., Sharifi, M., Gheshlaghi, F., Mardani, A., 2019. Development of an HLFS agricultural tire model using FEA technique. SN Appl. Sci. 1. https://doi.org/10.1007/s42452-019-1524-y

Farhadi, P., Golmohammadi, A., Malvajerdi, A.S., Shahgholi, G., 2020. Tire and soil effects on power loss: Measurement and comparison with finite element model results. J. Terramechanics 92, 13–22.

Farhadi, P., Golmohammadi, A., Sharifi, A., Shahgholi, G., 2018. Potential of three-dimensional footprint mold in investigating the effect of tractor tire contact volume changes on rolling resistance. J. Terramechanics 78, 63–72.

Farhadia, P., Golmohammadia, A., Malvajerdib, A.S., Shahgholi, G., 2019. Finite element modeling of the interaction of a treaded tire with clay-loam soil. Comput. Electron. Agric. 162, 793–806.

Fervers, C.W., 2004. Improved FEM simulation model for tire-soil interaction. J. Terramechanics 41, 87–100. https://doi.org/10.1016/j.jterra.2004.02.012

Gheshlaghi, F., El-Sayegh, Z., El-Gindy, M., Oijer, F., Johansson, I., 2021. Advanced Analytical Truck Tires-Terrain Interaction Model, in: SAE Technical Papers. https://doi.org/10.4271/2021-01-0329

Gheshlaghi, F., Mardani, A., 2021. Prediction of soil vertical stress under off-road tire using smoothed-particle hydrodynamics Fatemeh. J. Terramechanics 95, 7–14.

Gipser, M., 2007. FTire—The tire simulation model for all applications related to vehicle dynamics. Veh. Syst. Dyn. 45, 139–151. https://doi.org/10.1080/00423110801899960

Hambleton, J.P., Drescher, A., 2009. On modeling a rolling wheel in the presence of plastic deformation as a three- or two-dimensional process. J. Terramechanics 51, 846–855. https://doi.org/10.1016/j.ijmecsci.2009.09.024

Hambleton, J.P., Drescher, A., 2008. Development of Improved Test Rolling Methods for Roadway Embankment Construction, Minnesota Department of Transportation. MN/RC 2008-08.

Harnisch, C., Lach, B., Jakobs, R., Troulis, M., Nehls, O., 2005. A new tyre-soil interaction model for vehicle simulation on deformable ground. Veh. Syst. Dyn. 43, 384–394. https://doi.org/10.1080/00423110500139981

He, R., Sandu, C., Khan, A.K., Guthrie, A.G., Schalk Els, P., Hamersma, H.A., 2018. Review of terramechanics models and their applicability to real-time applications. J. Terramechanics 81, 3–22. https://doi.org/10.1016/j.jterra.2018.04.003

He, R., Sandu, C., Osorio, J.E., 2019a. Systematic tests for study of tire tractive performance on soft soil: Part I—Experimental data collection. J. Terramechanics 85, 59–76. https://doi.org/10.1016/j.jterra.2019.07.004

He, R., Sandu, C., Osorio, J.E., 2019b. Systematic tests for study of tire tractive performance on soft soil: Part II—Parameterization of terramechanics model and tire model. J. Terramechanics 85, 77–88. https://doi.org/10.1016/j.jterra.2019.07.003

Jafari, H., 2015. 3D finite element model to simulate the rolling resistance of a radial ply tire validated by experiments in soil bin testing facility. J. Adv. Veh. Eng. 1, 37–45.

Janosi, Z., 1961. The analytical determination of drawbar-pull as a function of slip for tracked vehicles in deformation soils., in: Proc. of the 1st Int. Conf. of Terrain_Vehicle Systems, Torino, Italy.

Krenn, R., Gibbesch, A., 2011. Soft soil contact modeling technique for multi-body system simulation, in: Trends in Computational Contact Mechanics. pp. 135–155. https://doi.org/10.1007/978-3-642-22167-5_8

Lescoe, R., El-Gindy, M., Koudela, K., Öijer, F., Volvo, M.T., Johansson, I., 2010. Tire-soil modeling using finite element analysis and smooth particle hydrodynamics techniques, in: Proceedings of the ASME 2010 International Design Engineering Technical Conferences and Computers and Information in Engineering Conference.

Li, H., Schindler, C., 2013a. Analysis of soil compaction and tire mobility with finite element method, in: Proceedings of the Institution of Mechanical Engineers, Part K: Journal of Multi-Body Dynamics. pp. 275–291. https://doi.org/10.1177/1464419313486627

Li, H., Schindler, C., 2013b. Investigation of tire-soil interaction with analytical and finite element method. Mech. Based Des. Struct. Mach. 41, 293–315. https://doi.org/10.1080/15397734.2012.744677

Liang, C.Y., Allen, R.W., Rosenthal, T.J., Chrstos, J.P., Nunez, P., 2004. Tire modeling for off-road vehicle simulation. SAE Tech. Pap. 113, 1063–1073. https://doi.org/10.4271/2004-01-2058

Liu, C.H., Wong, J.Y., 1996. Numerical simulation of tire-soil interation based on critical state soil mechanics. J. Terramechanics 33, 209–221.

Madsen, J., Negrut, D., Reid, A., Seidl, A., Ayers, P., Bozdech, G., Freeman, J., O'Kins, J., 2012. A Physics-Based Vehicle/Terrain Interaction Model for Soft Soil Off-Road Vehicle Simulations. SAE Int. J. Commer. Veh. 5, 280–290. https://doi.org/10.4271/2012-01-0767

Maria, E. di, Reina, G., Ishii, K., Giannoccaro, N.I., 2021. Rolling resistance and sinkage analysis by comparing FEM and experimental data for a grape transporting vehicle. J. Terramechanics 97, 59–70.

Mohsenimanesh, A., Ward, S.M., Gilchrist, M.D., 2009a. Stress analysis of a multi-laminated tractor tyre using non-linear 3D finite element analysis. Mater. Des. 30, 1124–1132. https://doi.org/10.1016/j.matdes.2008.06.040

Mohsenimanesh, A., Warda, S.M., Owendea, P.O.M., Javadi, A., 2009b. Modelling of pneumatic tractor tyre interaction with multi-layered soil. Biosyst. Eng.

Moslem, N., Hossein, G., 2014. Numerical simulation of tire/soil interaction using a verified 3D finite element model. J. Cent. South Univ. 21, 817–821.
https://doi.org/10.1007/s11771-014-2005-5

Ozaki, S., Hinata, K., Senatore, C., Iagnemma, K., 2015. Finite element analysis of periodic ripple formation under rigid wheels. J. Terramechanics 61, 11–22.

Ozaki, S., Kondo, W., 2016. Finite element analysis of tire traveling performance using anisotropic frictional interaction model. J. Terramechanics 64, 1–9.

Papamichael, S., Vrettos, C., 2021. Indentation tests and rolling simulations of a compliant wheel on soil at different consistencies. J. Terramechanics 94, 39–48.

Ragheb, H., El-Gindy, M., Kishawy, H., 2013. Multi-Wheeled Combat Vehicle Tire Modeling on Rigid and Soft Terrain, in: 15th International Conference on Aerospace Sciences and Aviation Technology.
https://doi.org/10.21608/asat.2013.22236

Rubinstein, D., Shmulevich, I., Technion, N.F., 2018. Use of explicit finite-element formulation to predict the rolling radius and slip of an agricultural tire during travel over loose soil. J. Terramechanics 80, 1–9.

Saliba, J.E., 1990. Elastic-viscoplastic finite-element program for modeling tire/soil interaction. J. Aircr. 27, 350–357. https://doi.org/10.2514/3.25279

Sandu, C., Taylor, B., Biggans, J., Ahmadian, M., 2008, Building Infrastructure for Indoor Terramechanics Studies: The Development of a Terramechanics Rig at Virginia Tech, 16th Int. Conf. of ISTVS, Turin, Italy

Sandu, C., Taheri, Sh, Taheri, S., Els, S., Jimenez, E., 2020. Hybrid soft soil tire model (HSSTM). Part III: Model parameterization and validation. J. Terramechanics 88, 1–15. https://doi.org/10.1016/j.jterra.2019.11.002

Sandu, C., Taheri, Sh, Taheri, S., Gorsich, D., 2019a. Hybrid Soft Soil Tire Model (HSSTM). Part II: Tire-terrain interaction. J. Terramechanics 86, 15–29.
https://doi.org/10.1016/j.jterra.2019.08.004

Sandu, C., Taheri, Shahyar, Taheri, Saied, Gorsich, D., 2019b. Hybrid Soft Soil Tire Model (HSSTM). Part I: Tire material and structure modeling. J. Terramechanics 86, 1–13. https://doi.org/10.1016/j.jterra.2019.08.002

Sharifi, M., 2020. Modelling and Validation of Agricultural Tire-Terrain Interaction.

Shoop, Sally A., 2001. Finite Element Modeling of Tire-Terrain Interaction.

Shoop, Sally A, 2001. Finite Element Modeling of Tire—Terrain Interaction Cold Regions Research and Engineering Laboratory.

Silva, R.P., Rolim, M.M., Gomes, I.F., Pedrosa, E.M.R., Tavares, U.E., Santos, A.N., 2018. Numerical modeling of soil compaction in a sugarcane crop using the finite element method. Soil Tillage Res. 181, 1–10.
https://doi.org/10.1016/j.still.2018.03.019

Slade, J., 2009. Development of A New Off-Road Rigid Ring Model For Truck Tires Using Finite Element Analysis Techniques.

Taheri, S., 2014. Finite Element modeling of tire-terrain dynamics interaction for full vehicle simulation application.

Taheri, Sh., Sandu, C., Taheri, S., Pinto, E., Gorsich, D., 2015. Terramechanics models for tire—terrain interaction used in modeling and simulation. J. Terramechanics 57, 1–22.

Wakui, F., Terumichi, Y., 2011. Three-dimensional numerical simulation of tire behavior on soft ground. Nihon Kikai Gakkai Ronbunshu, C Hen/Transactions Japan Soc. Mech. Eng. Part C 77, 3264–3277.
https://doi.org/10.1299/kikaic.77.3264

Wong, J.Y., 2001. Theory of Ground Vehicles.

Wong, J.Y., Asnani, V.M., 2008. Study of the correlation between the performances of lunar vehicle wheels predicted by the Nepean wheeled vehicle performance model and test data. Proc. Inst. Mech. Eng. Part D J. Automob. Eng. 222, 1939–1954.
https://doi.org/10.1243/09544070JAUTO811

Wright, A., 2012. Tire / soil interaction modelling within a virtual proving ground environment.

Xia, K., 2011. Finite element modeling of tire/terrain interaction: Application to predicting soil compaction and tire mobility. J. Terramechanics 48, 113–123.
https://doi.org/10.1016/j.jterra.2010.05.001

Xia, K., Yang, Y., 2012. Three-dimensional finite element modeling of tire/ground interaction. Int. J. Numer. Anal. Methods Geomech. 36, 498–516.
https://doi.org/10.1002/nag

Xu, X., Yerro, A., Soga, K., Li, M., Jin, F., 2021. Numerical Modelling and Simulation of the Wheel Rotation Problems by the Material Point Method, in: International Conference of the International Association for Computer Methods and Advances in Geomechanics. pp. 810–818.
https://doi.org/10.1007/978-3-030-64514-4_87

Yamashita, H., Jayakumar, P., Alsaleh, M., Sugiyama, H., 2018. Physics-based deformable tire–soil interaction model for off-road mobility simulation and experimental validation. J. Comput. Nonlinear Dyn. 13.
https://doi.org/10.1115/1.4037994

Zegelaar, P.W.A., Pacejka, H.B., 1997. Dynamic tyre responses to brake torque variations. Veh. Syst. Dyn. Int. J. Veh. Mech. Mobil. 27, 65–79.
https://doi.org/10.1080/00423119708969645

7092 / A Time Domain Passivity Controller for Teleoperation of Four Wheeled Differential Mobile Robot on Slippery Surface

Hui Xi, Yanjing Li, Pengyu Sun, Weihua Li*, and Jianfeng Wang
Harbin Institute of Technology (Weihai), Weihai, Shandong, China
**Corresponding author: liweihua@hit.edu.cn*

https://doi.org/10.56884/BSME8486

Abstract

Four wheeled differential mobile robots have been widely used in many fields. However, on the slippery surfaces (e.g., marble floor), the tire-terrain contact surfaces may induce phenomena of slipping especially when it turns, which brings big challenges for its teleoperation. Aiming at this difficulty, a time-domain passive controller (TDPC) is proposed in this study to compensate for the active energy generated by the environment termination induced by the tire-terrain interaction, and then a stable teleoperator is designed under the coordination of master robot's position and slave robot's velocity. Experiments with a commercial four wheeled mobile robots on the marble floor are conducted to validate the proposed approach.

Keywords: wheeled mobile robot, tire-terrain interaction, teleoperation

1. Introduction

Under the development of artificial intelligence and Internet of Things technology, driverless technology has attracted much attention from academia and industry. However, in some hazard environments, the vehicle's automation is hard to be fully implemented. Under this case, the teleopetaion technology becomes an effective approach to address its control issue.

The four wheeled differential mobile robot (FWDMR) is a common kind of the driverless vehicle, which is widely used in the field of transportation, exploration and other tasks [1-3]. However, due to the fact that the road surface may be rough and slippery (e.g., marble floor) at the wheel-terrain contact area, the ideal pure rolling and nonholonomic constraints can no longer be assumed for each wheel, which will lead to longitudinal slipping and lateral slipping [4,5], inducing that the command-tracking performance between the remote site and the human operator becomes poor. Therefore, under this environment, the FWDMR cannot maintain its passive characteristics, which then brings new challenges to the stability of its control system.

In order to address the control issues induced by the wheel longitudinal slipping, some works have been conducted in [6-8]. The required rotational speed of each driving wheel was regulated by the online estimated longitudinal slippage in [6]. And a new approach based on the wheel-terrain interaction forces was proposed in [7], corresponding to wheel-terrain interaction states. Similarly, an adaptive controller for trajectory tracking was also proposed in [8]. In general, these approaches address the velocity deviation induced by the longitudinal slipping through the online estimated wheel-terrain interaction information.

Teleoperation is an effective approach when a robot works in hazardous or remote environments, and an appropriately designed haptic force can further improve the teleoperation performance [9,10]. Compared with the teleoperation of non-mobile robot, the workspace of the FWDMR is relatively larger than the master robot's workspace, so that the position-velocity coordination between the master site and the slave site is usually adopted [11-13]. In addition, to tele-drive the FWDMR with the non-holonomic constraint under the position-position coordination, a semi-autonomous approach was proposed in [14] by using a weighting matrix.

Although the research on teleoperation of the FWDMR is relatively mature [9-14], there are rare works considering the influence of the longitudinal slipping during its teleoperation process [15]. In general, the teleoperation controllers are mostly at the kinematic velocity level. In the presence of wheel longitudinal slipping, this kind of kinematic control will generate an avoidable error between the desired speed and the actual speed, which poses new challenges for the teleoperation of the remote FWDMR. Therefore, this paper also employs an acceleration level controller based on a time-domain passive controller (TDPC) to compensate for the influence induced by wheel slipping.

The remainder of this paper is organized as follows. In Section 2, an augmented kinematic model of the FWDMR that considers the longitudinal slipping under the acceleration-level controller is developed. In Section 3, a bilateral teleoperator for the FWDMR with a designed TDPC to compensate for the environment's non-passivity is proposed. In Section 4, experiments of the proposed approach are reported to validate the improved teleoperation system's performance. Finally, the conclusions are given in Section 5.

2. Kinematic Modeling of FWDMR with Longitudinal Slip

Due to the characteristics of the four-wheeled driving, the slave robot FWDMR occurs the phenomenon of wheel longitudinal slipping during the driving process on slippery surface. In this case, the slave environment termination will generate a certain degree of activity, which then causes the system to become unstable and unable to accurately complete the control task. On the basis of the related work of [1, 2], this paper firstly proposes an augmented kinematic model of a four-wheel differential mobile robot, and its kinematic parameters are shown in Figure 1.

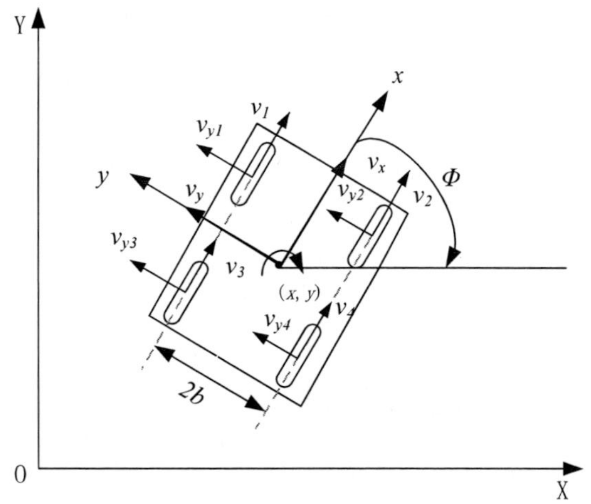

Figure 1. Unmanned vehicle motion structure

For the described FWDMR in Fig. 1, its kinematics model can be described as

$$\begin{bmatrix} v_x \\ v_y \\ \omega \end{bmatrix} = \begin{bmatrix} \cos(\phi) & \sin(\phi) & 0 \\ \sin(\phi) & -\cos(\phi) & 0 \\ 0 & 0 & 1 \end{bmatrix} \begin{bmatrix} \dot{x} \\ \dot{y} \\ \dot{\phi} \end{bmatrix} \tag{1}$$

where x, y and ϕ are the position and yaw angle of the robot base, v_x is the linear velocity of the FWDMR in the forward direction (m/s), v_y is the lateral velocity in the lateral direction (m/s), and ω is the base's angular velocity (rad/s).

Under ideal non-holonomic conditions, there is no lateral sliding of the FWDMR, so we have

$$v_y = \dot{x}\sin(\phi) - \dot{y}\cos(\phi) = 0 \tag{2}$$

Therefore, the inverse kinematics model of the slave FWDMR can be written as

$$\begin{bmatrix} v_x \\ \omega \end{bmatrix} = \begin{bmatrix} \cos(\phi) & \sin(\phi) & 0 \\ 0 & 0 & 1 \end{bmatrix} \begin{bmatrix} \dot{x} \\ \dot{y} \\ \dot{\phi} \end{bmatrix} \tag{3}$$

When the wheel is purely rolling, its forward velocity is strictly equal to the input of the motor for the FWDMR:

$$v_{xd} = r\omega_i \tag{4}$$

where r is the radius of wheel (m), and ω_i is the wheel rotation speed (rad/s) , and i=1, 2, 3, 4.

Therefore, the following slip ratio can be used to describe the degree of wheel slipping

$$S_i = \begin{cases} (r\omega_i - v_i)/v_i & (v_i \neq 0) \\ 0 & (v_i = 0) \end{cases} \tag{5}$$

where S_i is the wheel slip ratio.

Under the influence of wheel slip, the kinematic model of the slave unmanned vehicle is corrected as:

$$\begin{aligned} \begin{bmatrix} v_x \\ \omega \end{bmatrix} &= E\begin{bmatrix} v_1 \\ v_2 \\ v_3 \\ v_4 \end{bmatrix} = E\begin{bmatrix} r\omega_1 \\ r\omega_2 \\ r\omega_3 \\ r\omega_4 \end{bmatrix} = E\begin{bmatrix} r\omega_1 - S_1v_1 \\ r\omega_2 - S_2v_2 \\ r\omega_3 - S_3v_3 \\ r\omega_4 - S_4v_4 \end{bmatrix} \\ &= E\begin{bmatrix} r\omega_1 \\ r\omega_2 \\ r\omega_3 \\ r\omega_4 \end{bmatrix} - E\begin{bmatrix} S_1v_1 \\ S_2v_2 \\ S_3v_3 \\ S_4v_4 \end{bmatrix} = \begin{bmatrix} v_{xd} \\ \omega_d \end{bmatrix} - E\begin{bmatrix} S_1v_1 \\ S_2v_2 \\ S_3v_3 \\ S_4v_4 \end{bmatrix} \end{aligned} \tag{6}$$

where v_i is the wheel linear velocity (m/s), and

$$E = \begin{bmatrix} \frac{1}{4} & \frac{1}{4} & \frac{1}{4} & \frac{1}{4} \\ -\frac{1}{4b} & \frac{1}{4b} & -\frac{1}{4b} & \frac{1}{4b} \end{bmatrix}.$$

The kinematic model in Eq. (6) shows that the FWDMR does not have good command tracking performance, so that the speed of the FWDMR at the slave site is adjusted by the acceleration level controller as [15] did. Here, only the longitudinal slip of the unmanned vehicle is considered for analysis. The lateral slipping on flat terrains under a low speed can be ignored due to the small lateral load. First, the schematic diagram of the acceleration level controller for the FWDMR is given as shown in Fig. 2.

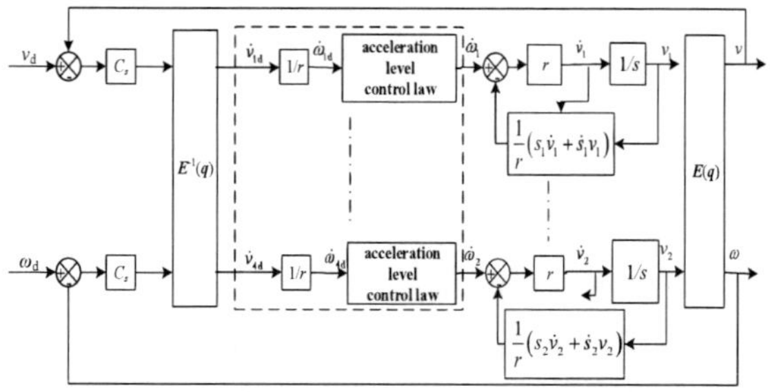

Figure 2. Acceleration level control for FWDMR.

Assuming that the transfer matrix from $\dot{\omega}_{1d}$ to $\dot{\omega}_1$ is an identity matrix, by differentiating it can be obtained:

$$\begin{cases} r\left\{\dot{\omega}_1 - \frac{1}{r}(S_1\dot{v}_1 + \dot{S}_1 v_1)\right\} = \dot{v}_1 \\ r\left\{\dot{\omega}_2 - \frac{1}{r}(S_2\dot{v}_2 + \dot{S}_2 v_2)\right\} = \dot{v}_2 \\ r\left\{\dot{\omega}_3 - \frac{1}{r}(S_3\dot{v}_3 + \dot{S}_3 v_3)\right\} = \dot{v}_3 \\ r\left\{\dot{\omega}_4 - \frac{1}{r}(S_4\dot{v}_4 + \dot{S}_4 v_4)\right\} = \dot{v}_4 \end{cases} \tag{7}$$

Then, combining Eq. (6) and Eq. (7), the kinematic model of the FWDMR can be augmented as

$$\begin{bmatrix} \dot{v}_x \\ \dot{\omega} \end{bmatrix} = E \begin{bmatrix} \dot{v}_1 \\ \dot{v}_2 \\ \dot{v}_3 \\ \dot{v}_4 \end{bmatrix} = u_s - \delta_{x\varepsilon} \tag{8}$$

where u_s is the desired acceleration of the FWDMR at the slave site, and $\delta_{x\varepsilon}$ is the environment termination (ET), which can be written in the following form:

$$\delta_{x\varepsilon} = -\begin{bmatrix} F_{e1} \\ F_{e2} \end{bmatrix} = \underbrace{E \begin{bmatrix} \dot{S}_1 & 0 & 0 & 0 \\ 0 & \dot{S}_2 & 0 & 0 \\ 0 & 0 & \dot{S}_3 & 0 \\ 0 & 0 & 0 & \dot{S}_4 \end{bmatrix}}_{K_1} E^{-1} \begin{bmatrix} v_x \\ \omega \end{bmatrix} + \underbrace{E \begin{bmatrix} S_1 & 0 & 0 & 0 \\ 0 & S_2 & 0 & 0 \\ 0 & 0 & S_3 & 0 \\ 0 & 0 & 0 & S_4 \end{bmatrix}}_{K_2} E^{-1} \begin{bmatrix} \dot{v}_x \\ \dot{\omega} \end{bmatrix} \tag{9}$$

In the above equation, the matrix K_1 and K_2 cannot be guaranteed to be positive definite, so the passivity of this environment termination cannot be guaranteed, which may lead to the teleoperation system to be potentially unstable. Therefore, in the teleoperation process, this phenomenon should be specially cared to guarantee the system passivity.

3. Bilateral Teleoperation of FWDMR

3.1 Master robot

Herein, a joint robot with two degrees of freedom (DOF) is employed as the master robot, which is modeled as

$$M_m \ddot{X}_m + C_m \dot{X}_m = F_m + F_h, \tag{10}$$

where M_m and C_m are the mass and Coriolis/centrifugal matrices, respectively; X_m ([X_{m1}, X_{m2}]) is the joint position; and F_m ([F_{m1}, F_{m2}]) and F_h ([F_{h1}, F_{h2}]) are the forces generated by robot actuators and human operators, respectively.

In this study, the coordination between the master robot's position (X_m) and the slave FWDMR's velocities (v_s, ω_s) is also employed similar to that in [17]. In addition, a new variable $r_m = \lambda X_m + \dot{X}_m$ ($0<\lambda<1$) is also introduced with a modified damping coefficient B_{mv}, and then the master robot can be described as

$$M_m \ddot{X}_m + (C_m + B_{mv}) \dot{X}_m = \bar{F}_m + F_h, \tag{11}$$

where $B_{mv} + C_m \geq M_m / \lambda$. According to the results in [17], the sub-master-system (11) is passive with the introduced r_m. Therefore, the coordination ([X_{m1}, X_{m2}], [v_s, ω_s]) is regulated as ([r_{m1}, r_{m2}], [v_s, ω_s]).

3.2 Nonpassivity Compensation with TDPC

Based on the above analysis, due to the existence of external forces at the wheel-terrain contact surface, the longitudinal and lateral slip induces the ET become an active subsystem, which injects additional energy into the teleoperation system. In order to address this control issue and guarantee the system stability, a TDPC approach is proposed herein to compensate for the ET's activity. A time-varying damping coefficient is designed and added to the linear velocity control and angular velocity control of the FWDMR.

Based on the kinematic model of the FWDMR in Eq. (8), the interaction between the slave robot and the environment can be obtained, mainly including the force in the forward direction of the unmanned vehicle and the force in the rotation process, respectively:

$$F_{e1} = u_{s1} - \dot{v}_s \tag{12}$$

$$F_{e2} = u_{s2} - \dot{\omega}_s \tag{13}$$

When the TDPC is not employed here, the kinetic energy of the ET includes the energies generated by the interaction relevant with the linear velocity and the angular velocity of the robot base, respectively:

$$e_{e1}(n) = e_{e1}(n-1) + F_{e1} v_s \Delta T \tag{14}$$

$$e_{e2}(n) = e_{e2}(n-1) + F_{e1} \omega_s \Delta T \tag{15}$$

In each sampling time, the above kinetic energy generated by the interaction between the robot and the environment is irrelevant to the teleoperator. It is mainly caused by the longitudinal slippage of the FWDMR during the teleoperation process, which generates additional energy injection into the teleoperation system, making the system potentially unstable.

Therefore, from the perspective of the system energy, this paper designs a passive controller (PC) to consume the energy generated by the environment, that is, an activity compensation algorithm for the environment termination based on the time-domain passivity. With the PC, the teleoperator for the FWDMR teleoperation can be expressed in a two-port network as:

$$\begin{bmatrix} F_h \\ F_e \end{bmatrix} + \begin{bmatrix} 0 \\ PC_e \end{bmatrix} = \begin{bmatrix} z_{11} & z_{12} \\ z_{21} & z_{22} \end{bmatrix} \begin{bmatrix} X_m \\ -v_s \end{bmatrix} \tag{16}$$

The PC_e is the TDPC controller designed to compensate for the active energy at the ET. Therefore, a passivity observer (PO) is also required to monitor the energy generated in the environment sub-system in real time. The bilateral teleoperation system using TDPC is shown in Fig. 3.

The PO_e mainly observes the energy generated in each sampling time of the system, that is, the energy generated in n-1 to n steps, the energy input of the ET during the sampling time, and the energy consumed by the PC_e. The passivity controller is implemented by a damping coefficient α_e that varies with time to offset the active energy generated by the environment. The kinetic energy of the ET after using TDPC is expressed as:

$$E_{e1}(n) = E_{e1}(n-1) + (F_{e1}(n)v_s(n) + \alpha_{e1}(n-1)v_s^2(n-1))\Delta T \tag{17}$$

$$E_{e2}(n) = E_{e2}(n-1) + (F_{e2}(n)\omega_s(n) + \alpha_{e2}(n-1)\omega_s^2(n-1))\Delta T \tag{18}$$

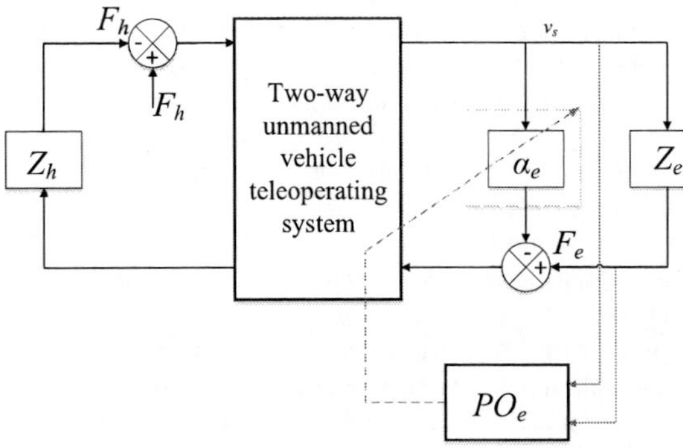

Figure 3. FWDMR teleoperation system with TDPC

When the PO_e observes that the energy at the ET of the system is negative, it means that the interaction force between the FWDMR and the environment is active at this moment, resulting in energy injection. At this moment, the system is active, the stability is reduced, and it is difficult to control for the slave sub-system. Therefore, it is necessary to design a damping coefficient α_e to consume the injected energy to maintain the passivity nature of the slave system. However, when the PO_e observes that the energy is positive or equal to 0, it means that there is no active energy generated at this moment, and the system is still stable. Therefore, there is no need to adjust the damping coefficient, and α_e is set to be 0. For the energy generated by the linear velocity control and the angular velocity control, the damping coefficient α_e is designed to be adjusted respectively, and the design of α_e is as follows.

$$\alpha_{e1}(n) = \begin{cases} -\dfrac{E_{e1}(n)}{v_s^2(n)\Delta T} & (E_{e1}(n) < 0) \\ 0 & (E_{e1}(n) \geq 0) \end{cases} \tag{19}$$

$$\alpha_{e2}(n) = \begin{cases} -\dfrac{E_{e2}(n)}{\omega_s^2(n)\Delta T} & (E_{e2}(n) < 0) \\ 0 & (E_{e2}(n) \geq 0) \end{cases} \tag{20}$$

When the PO_e observes that the system energy is negative, the PC_e begins to work, and the corresponding slave controller is also regulated. In addition to the command transmission of the master, the controller input of the slave robot also plays the role of the PC_e.

The following equations prove that the activity brought by the tire-terrain interaction is effectively compensated, and then the stability of the slave robot subsystem can be guaranteed.

$$\begin{aligned} \widetilde{E}_{e1}(n) &= E_{e1}(n) + \alpha_{e1}(n)v_s^2(n)\Delta T \\ &\geq E_{e1}(n) - \frac{E_{e1}(n)}{v_s^2(n)\Delta T}v_s^2(n)\Delta T \geq 0 \end{aligned} \tag{21}$$

$$\begin{aligned} \widetilde{E}_{e2}(n) &= E_{e2}(n) + \alpha_{e2}(n)\omega_s^2(n)\Delta T \\ &\geq E_{e2}(n) - \frac{E_{e2}(n)}{\omega_s^2(n)\Delta T}\omega_s^2(n)\Delta T \geq 0 \end{aligned} \tag{22}$$

3.3 Slave teleoperator

In this paper, the slave robot is the above described FWDMR in Eq. (8), and a haptic joint robot is employed as the master robot in Eq. (11). Here, the slave robot only owns two degrees of mobility (linear velocity v_s and angular velocity ω_s of the base), and the acceleration level control mode in Fig. 2 is used for improving the performance of command-tracking. The first joint angle X_{m1} of the master robot is used to map the linear velocity v_s of the slave robot, and the second joint angle X_{m2} of the master robot is used to map the linear velocity ω_s of the slave robot. Based on the position-velocity coordination errors, the controllers of the slave robot for its linear velocity and angular velocity are respectively designed as

$$u_{s1} = k_p(\gamma_1 r_{m1} - v_s) \tag{23}$$

$$u_{s2} = k_p(\gamma_2 r_{m2} - \omega_s) \tag{24}$$

The controllers for the master robot are designed as:

$$\tau_{m1} = k_m(v_s - \gamma_1 r_{m1}) \tag{25}$$

$$\tau_{m2} = k_m(\omega_s - \gamma_2 r_{m2}) \tag{26}$$

where k_p and k_m are controller gains; γ_1 and γ_2 are the scaling factors of the coordination.

With the above controller, the motion coordination between the master robot and the slave robot can be implemented, and the stability of the teleoperation system is then analyzed to determine the controller parameters by using the Llewellyn's criterion (see appendix).

The control structure of the FWDMR teleoperation system is based on the position-velocity coordination. After adding the PC controller from the end to adjust the environmental energy, the control structure of the system is shown in Fig. 4.

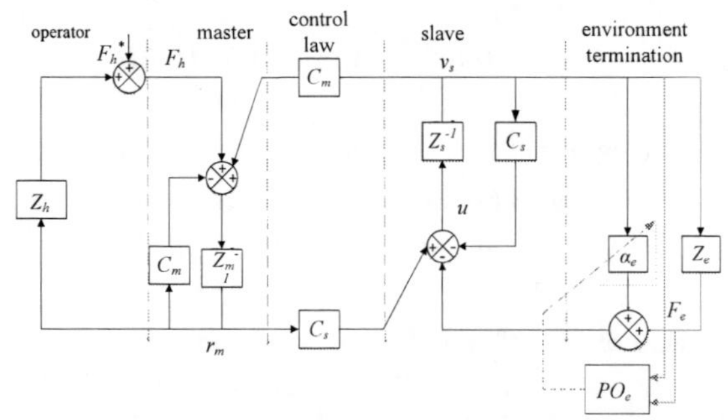

Figure 4. FWDMR bilateral teleoperation system

Substitute the PC controller into Eq. (16) to obtain a two-port network model. Firstly, the stability for the linear velocity control channel of the slave FWDMR is analyzed. The two-port network model can be described as:

$$\begin{bmatrix} F_{h1} \\ F_{e1} \end{bmatrix} + \begin{bmatrix} 0 \\ \alpha_{e1} v_s \end{bmatrix} = \underbrace{\begin{bmatrix} z_{11} & z_{12} \\ z_{21} & z_{22} \end{bmatrix}}_{Z} \begin{bmatrix} r_{m1} \\ -v_s \end{bmatrix} \tag{27}$$

The impedance matrix Z in Eq. (28) is:

$$Z = \begin{pmatrix} M_{m1}s + \gamma_1 k_m & k_m \\ \gamma_1 k_p & k_p + s \end{pmatrix} \tag{28}$$

According to the Llewellyn's criterion, when the system stability conditions are met, the impedance matrix should satisfy certain conditions. And the value range of the control parameters is shown in Eq. (29):

$$\begin{cases} k_m \geq 0 \\ k_p \geq 0 \end{cases} \tag{29}$$

Then the stability for the angular velocity control channel is analyzed. Herein, its two-port network model can be described as:

$$\begin{bmatrix} F_{h2} \\ F_{e2} \end{bmatrix} + \begin{bmatrix} 0 \\ \alpha_{e2} \omega_s \end{bmatrix} = \underbrace{\begin{bmatrix} z_{11} & z_{12} \\ z_{21} & z_{22} \end{bmatrix}}_{Z} \begin{bmatrix} X_{m0} \\ -\dot{X}_{s2} \end{bmatrix} \tag{30}$$

The impedance matrix Z in *Eq.* (31) is:

$$Z = \begin{pmatrix} M_{m2}s + \gamma_2 k_m & k_m \\ \gamma_2 k_p & k_p + s \end{pmatrix} \tag{31}$$

When the system stability condition is satisfied, the impedance matrix should also satisfy the following conditions.

$$\begin{cases} k_m \geq 0 \\ k_p \geq 0 \end{cases} \tag{32}$$

In the actual experiment, as long as the control parameters meet the above conditions in Eqs. (29) and (32) with the designed TDPC in Eqs. (19) and (20), the phenomenon of active energy will no longer occur at the ET, and the teleoperation system will be stable.

4. Experiment

In order to verify the FWDMR teleoperator proposed in this paper, a teleoperation control platform is built, which is mainly composed of a master robot, a slave robot and a communication channel. The master robot is a force feedback handle device (Geomagic Touch in Fig. 5, 3D Systems Inc., Cary, NC, USA), and the slave robot is a FWDMR (SCOUT in Fig. 6, AgileX Inc., Dongguan, China), and the communication channel is implemented by a LAN (local area network). The control system on the FWDMR is implemented by the Robot Operating System (ROS) and is positioned by the SLAM technology based on Lidar after mapping the test environment. In addition, the FWDMR runs on the smooth marble floor, which can induce obvious wheel slipping.

In the following experiments, the teleoperator parameters in Eqs. (23) ~ (26) all meet the relevant conditions based on the Llewellyn's criterion, which are set to be

$k_p = k_m = 2$, $\gamma_1 = \gamma_2 = 1$.

Herein, two sets of tests are conducted to validate the proposed teleoperator in this paper. First, a test considering the designed master-slave controllers without TDPC compensation is done as shown in Fig. 7; second, a test with TDPC to compensate for the activity of the slave environment is done as shown in Fig. 8. In the developed experimental system, the controller frequency for the TDPC is set to be 20 Hz, and the maximum value of both damping coefficients in the designed PC_e is set to be 3.

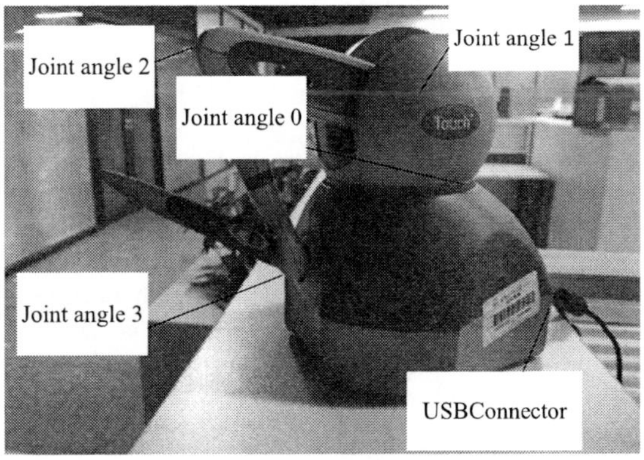

Figure 5. Geomagic Touch handle

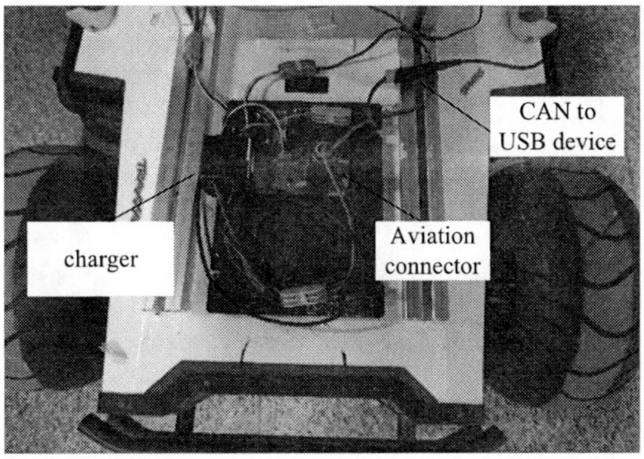

Figure 6. FWDMR

In the experiments, when the velocity of the slave robot is less than the position of the master robot, the operator can feel an obvious pushing-back force; when the velocity of the slave robot is greater than the position of the master robot, the operator can feel an obvious pulling-forward force, which means that there exists active energy in the teleoperation system, and the system is potentially unstable.

In the results without TDPC in Fig. 7, it can be seen that the command-tracking error is large, and the linear velocity of the unmanned vehicle cannot be well tracked in Fig 7a. From 20s to 30s, the linear velocity of the slave robot exceeds the desired values from the master site, and the operator can feel a pulling-forward force, which means that the system is potentially unstable. According to the change trend of the energy generated by the linear velocity channel in (c), it can be seen that in the 20s-30s, the energy changes suddenly to be negative, indicating that the ET generates active energy, and the system is no longer stable, resulting in poor command-tracking performance of the slave robot. In addition, the relative stable errors from 40s to 80s in Figs. 7a and 7b are mainly induced by the wheel slipping under the constrained maximum velocities (0.2 m/s for linear velocity and 0.2 rad/s for angular velocity).

After adopting the TDPC, as shown in Fig. 8, the tracking performance of the slave robot is obviously improved. The linear velocity and the angular velocity of the slave robot can well track the position of the master robot, resulting in good command-tracking performance. In most cases, the energy generated at the channels of the linear velocity and the angular velocity by using TDPC is greater than or equal to 0, as shown in 8c and 8d. From 10s to 20s, although the linear and angular velocity of the slave robot exceeds the desired values from the master site, the ET's energy is still close to 0, meaning that smaller active forces are generated. This is caused by the constrained damping coefficient (set to be 3 as shown in 8e and 8f), but most of the active energy has been consumed by the designed TDPC, which further guarantees the passivity of the teleoperation system and improves the command-tracking performance of the slave robot.

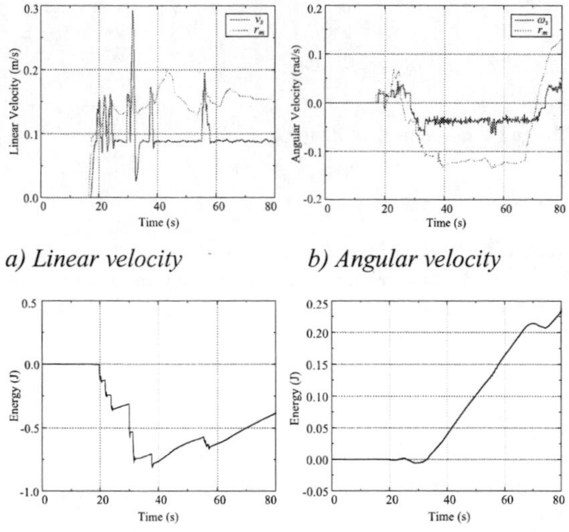

a) Linear velocity *b) Angular velocity*

(c) Kinetic energy for v_s *d) Kinetic energy for ω_s*

Figure 7. Teleoperation test of FWDMR without TDPC

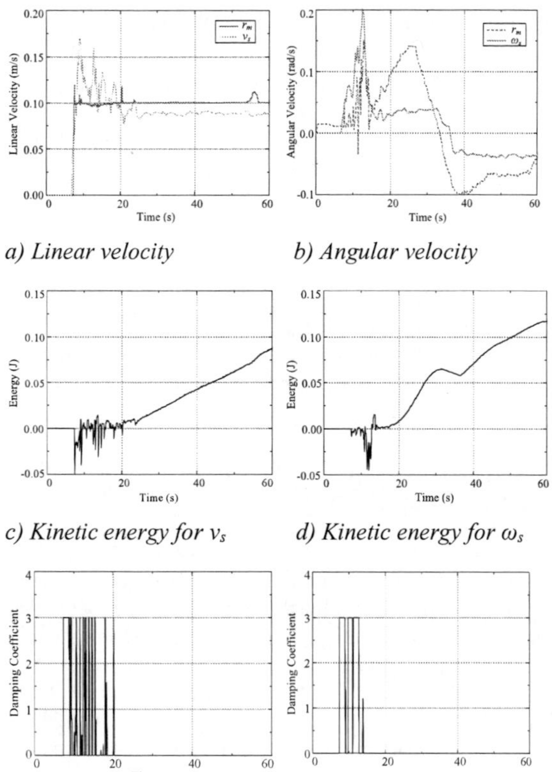

a) Linear velocity *b) Angular velocity*

c) Kinetic energy for v_s *d) Kinetic energy for ω_s*

e) Damping coefficient for v f) Damping coefficient for ω_s

Figure 8. Teleoperation test of FWDMR with TDPC

5. Conclusion

This paper presents a master-slave bilateral teleoperator for the FWDRM considering the wheel longitudinal slipping. In the proposed teleoperation system, the position of the master robot is employed to coordinate with the velocity of the FWDMR. And the structure based on position-velocity coordination error and the two-port network model are developed for the bilateral teleoperation of the FWDMR. The non-passivity of the ET is analyzed based on its augmented kinematic model, and then a TDPC controller is designed to compensate for the negative influence induced by the wheel slipping based on the passivity theory. Finally, the stability of the proposed teleoperator is constrained by the Llewellyn's criterion. Two set of comparison experiments are conducted and validate that the proposed teleoperation controller owns good teleoperation for the FWDRM under wheel slipping.

6. Acknowledgment

This work was supported by the National Natural Science Foundation of China (52175007/52175012 /51705096), the Self-Planned Task (NO. SKLRS202005B) of State Key Laboratory of Robotics and System (HIT), the China Postdoctoral Science Foundation (2018M630348/ 2019T120259), and the Scientific Research Foundation of Harbin Institute of Technology (Weihai).

7. Appendix

Lemma 1 (Llewellyn's criterion [16]) The two-port network in (27) is absolutely stable (i.e., the overall system is bounded-input/bounded output stable assuming the passivity of both terminations) if and only if

(1) $Z_{11}(s)$ and $Z_{22}(s)$ have no poles in the right half plane.

(2) Any poles of $Z_{11}(s)$ and $Z_{22}(s)$ on the imaginary axis are simple with real and positive residues.

(3) For $s = j\omega$ and all real values of ω:

$$\mathrm{Re}(Z_{11}) \geq 0$$

$$\mathrm{Re}(Z_{22}) \geq 0$$

$$2\,\mathrm{Re}(Z_{11})\mathrm{Re}(Z_{22}) - \mathrm{Re}(Z_{12}Z_{21}) - |Z_{12}Z_{21}| \geq 0 .$$

8. References

[1] Moudoud B, Aissaoui H, Diany M, 2020. Robust trajectory tracking control based on sliding mode of Differential Driving Four-Wheeled Mobile Robot. *IEEE 6th International Conference on Optimization and Applications (ICOA)*.

[2] Xie Y, Zhang X, Meng W, et al., 2020. Coupled fractional-order sliding mode control and obstacle avoidance of a four-wheeled steerable mobile robot. *ISA Transactions*, 108(4), 282-294.

[3] The Rover Team, 1997. Characterization of the Martian Surface Deposits by the Mars Pathfinder Rover, Sojourner. *Science*, 278(5344), 1765-1768.

[4] Liang Z., Zhao J., Dong Z., et al., 2020. Torque vectoring and rear-wheel-steering control for vehicle's uncertain slips on soft and slope terrain using sliding mode algorithm. *IEEE Trans. Veh. Technol.*, 69(4), 3805-3815.

[5] Li W., Li Z., Liu Y., et al., 2019. Semi-autonomous bilateral teleoperation of six-wheeled mobile robot on soft terrains. *Mech. Syst. Signal Process.*, 133, 106234.

[6] Ishigami G., Miwa A., Nagatani K., et al., 2007. Terramechanics-based model for steering maneuver of planetary exploration rovers on loose soil. *Journal of Field Robotics*, 24(3), 233-250.

[7] Kim D., Min K., Kim H., et al., 2020. Vehicle sideslip angle estimation using deep ensemble-based adaptive Kalman filter. *Mech. Syst. Signal Process.*, 144, 106862.

[8] Cui M., Huang R., Liu H., et al., 2014. Adaptive tracking control of wheeled mobile robots with unknown longitudinal and lateral slipping parameters. *Nonlinear Dynamics*, 78, 1811-1826.

[9] Farkhatdinov I., Ryu J.H., 2007. Hybrid position-position and position-speed command strategy for the bilateral teleoperation of a mobile robot. *International Conference on Control, Automation and Systems*, 2442-2447

[10] Luo J., Lin Z., Li Y., Yang C., 2020. A Teleoperation Framework for Mobile Robots Based on Shared Control. *IEEE Robotics and Automation Letters*, 5(2), 377-384.

[11] Lee D., Palafox O.M., Spong M.W., 2006. Bilateral Teleoperation of a Wheeled Mobile Robot over Delayed Communication Network. *Proc. IEEE Int. Conf. Robot. Autom.*, 3298-3303.

[12] Santos J., Conceição A., Santos T., Araújo H., 2018. Remote control of an omnidirectional mobile robot with time-varying delay and noise attenuation, *Mechatronics*, 52, 7-21.

[13] Slawinski E., Mut V.A., Fiorini P., Salinas L.R., 2012. Quantitative Absolute Transparency for Bilateral Teleoperation of Mobile Robots. *IEEE Transactions on Systems, Man, and Cybernetics—Part A: Systems and Humans*, 42(2), 430-442.

[14] Malysz P., Sirouspour S., 2011. A Task-space Weighting Matrix Approach to Semi-autonomous Teleoperation Control. *Proc. IEEE/RSJ Int. Conf. Intell. Robots Syst.*, 645-652.

[15] Li W., Liu Y., Ding L., et al., 2022. Teleoperation of wheeled mobile robots subject to longitudinal slipping and lateral sliding by time-domain passivity controller. *Mechatronics*, 81, 102705.

[16] Llewellyn F.B., 1952. Some Fundamental Properties of Transmission Systems. Proceedings of the IRE, 40(3): 271–283.

[17] Li W., Liang D., Zhen L., et al. Kinematic Bilateral Tele-Driving of Wheeled Mobile Robots Coupled with Slippage[J]. IEEE Transactions on Industrial Electronics, 2017, 64(3):2147-2157.

7199 / Vehicle Dynamic Factor Characterized by Actual Velocity and Combined Influence of the Transmission and Driveline System

Vladimir Vantsevich[a], David Gorsich[b], Jesse Paldan[c]*, Jordan Whitson[c], Brian Butrico[b], and Oleg Sapunkov[b]
[a] *Worcester Polytechnic Institute, Worcester, Massachusetts, United States*
[b] *U.S. Army DEVCOM Ground Vehicle Systems Center, Warren, Michigan, United States*
[c] *University of Alabama at Birmingham, Birmingham, Alabama, United States*
**Corresponding author: jpaldan@uab.edu*

https://doi.org/10.56884/ARAO9883

Abstract

Four wheeled differential mobile robots have been widely used in many fields. However, on the slippery surfaces (e.g., marble floor), the tire-terrain contact surfaces may induce phenomena of slipping especially when it turns, which brings big challenges for its teleoperation. Aiming at this difficulty, a time-domain passive controller (TDPC) is proposed in this study to compensate for the active energy generated by the environment termination induced by the tire-terrain interaction, and then a stable teleoperator is designed under the coordination of master robot's position and slave robot's velocity. Experiments with a commercial four wheeled mobile robots on the marble floor are conducted to validate the proposed approach.

Keywords: terrain mobility, dynamic factor, vehicle performance

1. Introduction

The traction and acceleration performance are operational properties important for maintaining the mobility of off-road terrain vehicles. Operational properties for mobility should be enhanced during the conceptual design phase of manned and autonomous vehicles by correlating vehicle dynamic characteristics to mobility indices related to the tire-terrain interactions. A common mobility index based on terramechanics and vehicle dynamics is the dynamic factor, used for characterizing the traction and acceleration performance. The dynamic factor has been in use for many decades (Chudakov, 1950; Genta and Morello, 2009; Grishkevich, 1986; Litvinov and Farobin, 1989; Zimelev, 1957).

The expression for the dynamic factor is derived from the traction balance equation of vehicle moving in a straight-line motion; the traction balance of a vehicle with n-number of drive axles and m-number of total axles is given in Eq. (1).

$$F_{x\Sigma} = \sum_{i=1}^{n}(F'_{xi} + F''_{xi}) = \sum_{i=1}^{m}(R'_{xi} + R''_{xi}) \pm W_a sin\theta_n + \frac{W_a}{g}\delta_r a_a + D_{ad} \qquad (1)$$

Here, $F_{x\Sigma}$ is the sum of all circumferential forces of the left (F'_{xi}) and right (F''_{xi}) wheels of $i = 1, n$ axles. These forces are balanced by the vehicle's acceleration and motion resistance forces: this includes wheel rolling resistances $R'^{('')}_{xi}$, the longitudinal component of the vehicle weight W_a on a slope with angle θ_n, aerodynamic drag D_{ad}, and the inertia force at longitudinal acceleration a_a, where δ_r is a rotating mass factor. At an assigned speed, the total circumferential force is determined from the powertrain and the aerodynamic force from the vehicle design. The difference $(F_{x\Sigma} - D_{ad})$ thus gives the potential remaining circumferential force that can be provided by the powertrain to overcome grade resistance, rolling resistance, and accelerate the vehicle. Normalized by the gross weight of the vehicle W_a, this margin is called the dynamic factor D_a.

$$D_a = \frac{F_{x\Sigma} - D_{ad}}{W_a} = \frac{\sum_{i=1}^{n}(F'_{xi} + F''_{xi}) - D_{ad}}{W_a} = \frac{\sum_{i=1}^{m}(R'_{xi} + R''_{xi})}{W_a} \pm sin\theta_n + \frac{\delta_r}{g}a_a \qquad (2)$$

Eq. (2) is the general form of the dynamic factor. Specific cases include steady motion on a flat surface (Eq. (3)), steady motion on a slope (Eq. (4)), and acceleration on a flat surface (Eq. (5)).

$$D_a = \frac{F_{x\Sigma} - D_{ad}}{W_a} = \frac{\sum_{i=1}^{n}(F'_{xi} + F''_{xi}) - D_{ad}}{W_a} = \frac{\sum_{i=1}^{m}(R'_{xi} + R''_{xi})}{W_a} \qquad (3)$$

$$D_a = \frac{F_{x\Sigma} - D_{ad}}{W_a} = \frac{\sum_{i=1}^{n}(F'_{xi} + F''_{xi}) - D_{ad}}{W_a} = \frac{\sum_{i=1}^{m}(R'_{xi} + R''_{xi})}{W_a} \pm sin\theta_n \qquad (4)$$

$$D_a = \frac{F_{x\Sigma} - D_{ad}}{W_a} = \frac{\sum_{i=1}^{n}(F'_{xi} + F''_{xi}) - D_{ad}}{W_a} = \frac{\sum_{i=1}^{m}(R'_{xi} + R''_{xi})}{W_a} + \frac{\delta_r}{g}a_a \qquad (5)$$

Uses of the dynamic factor include determining rolling resistance that can be overcome without switching from the highest transmission gear, determining the maximum rolling resistance that can be overcome on the lowest gear, determining maximum potential acceleration, and determining the range of velocities that correspond to stable work of the internal combustion engine at full throttle on the highest gear. The normalized form of the dynamic factor allows comparing the traction and acceleration capabilities of various vehicles of different weight.

Dynamic factor is presented in most vehicle dynamics studies as a function of the theoretical vehicle linear velocity V_a computed from the characteristics of the powertrain at full throttle only without including the speed reduction from tire slippage. This dynamic factor D_a is referred to as the "theoretical" dynamic factor in this paper with subscript a, also used for the acceleration a_a in Eqs. (1-5). A nonlinear function between the circumferential force and tire slippage was used in (Grishkevich, 1986) to analyze the dynamic factor of a 4x2 car on a snowy road as a function of the actual velocity. In its non-normalized form, the dynamic factor is the relationship between the vehicle's tractive effort and theoretical speed. This relationship is used in predicting the vehicle's speed-made-good, the effective maximum possible vehicle speed from one location to another along a straight-line under steady-state conditions, proposed as a mobility metric for next generation mobility models (Wong et. al., 2020). The speed-made-good is predicted from the theoretical velocity using Eq. (6) (Wong et. al., 2020)

$$V_x = V_a(1 - s_\delta) \qquad (6)$$

where V_x is the vehicle's actual velocity or speed-made-good, V_a is the theoretical velocity, and s_δ is the slippage determined using the tractive effort-slip relationship. The tire slippage can considerably impact the velocity for off-road vehicles that require large traction in most operational conditions. For this reason, the actual velocity V_x should be computed using the tire slippage.

Furthermore, for vehicles with four or more driving wheels, the wheels may develop different slippages due to different terrains and operating conditions. The tire slippages and therefore the actual velocity of multi-wheel drive vehicles significantly depends on the configuration of the driveline system, which plays a key role in the engine power distribution among the driving wheels. Conventional methods do not include driveline characteristics in the dynamic factor analysis.

2. Driveline Effect on Dynamic Factor and Actual Velocity

This paper proposes a new method for analysis of the dynamic factor which takes into account terrain and slippage effects which limit the dynamic factor and actual velocity in a way that includes the contribution of the driveline configuration on the traction of individual wheels. The actual velocity is computed using a system of generalized parameters which allow determination of the driveline's effect on the wheel power distribution. The generalized slippage of the vehicle, $s_{\delta a}$, is used as the measure of the slippage from Eq. (6) to characterize the speed reduction from theoretical velocity V_a to actual velocity V_x.

$$V_x = V_a(1 - s_{\delta a}) \qquad (7)$$

$s_{\delta a}$ is a generalized parameter of the vehicle computed from individual tire slippages $s_{\delta i}^{\prime(\prime\prime)}$ based on the number and configuration of power dividing units in the driveline.

In this paper, the analysis of the dynamic factor is derived for the FED Alpha vehicle with open and symmetric differentials in the transfer case. The full vehicle parameters are available at (KRC, 2021). An open differential always provides the same torques to its two output shafts. The friction in the differential is small and can be ignored. The wheel's rolling radius in the driven mode, r_w^0, is used for determining the circumferential force F_x from the torque T_w (Andreev et. al., 2010).

$$F_x = \frac{T_w}{r_w^0} \qquad (8)$$

r_w^0 is the rolling radius of a wheel in the driven mode when the torque is zero; it can be determined experimentally or estimated from wheel load and tire characteristics and is considered known in this paper. Thus, applying Equation (7) and the equality of the output torques for three open differentials establishes the following relationships between the wheel circumferential forces:

$$\frac{F''_{xi} r^{0''}_{wi}}{F'_{xi} r^{0'}_{wi}} = 1, \; i = 1{,}2 \qquad (9)$$

$$\frac{F''_{x2} r^{0''}_{w2} + F'_{x2} r^{0'}_{w2}}{F''_{x1} r^{0''}_{w1} + F'_{x1} r^{0'}_{w1}} = 1 \qquad (10)$$

Circumferential forces are a function of tire slippage; the exponential equation given by Eq. (11) is used in this paper (Andreev et. al., 2010).

$$F_{xi}^{\prime(\prime\prime)} = \mu_{pxi}^{\prime(\prime\prime)} R_{zi}^{\prime(\prime\prime)} \left\{1 - \frac{s_{\delta ci}^{\prime(\prime\prime)}}{2 s_{\delta i}^{\prime(\prime\prime)}} \left[1 - exp\left(-\frac{2 s_{\delta i}^{\prime(\prime\prime)}}{s_{\delta ci}^{\prime(\prime\prime)}}\right)\right]\right\} \qquad (11)$$

Here, $\mu_{pxi}^{\prime(\prime\prime)}$ is the peak friction coefficient, $R_{zi}^{\prime(\prime\prime)}$ is the normal reaction of a wheel, $s_{\delta i}^{\prime(\prime\prime)}$ is the tire slippage, and $s_{\delta ci}^{\prime(\prime\prime)}$ is the tire characteristic slippage. Parameters $\mu_{pxi}^{\prime(\prime\prime)}$ and $s_{\delta ci}^{\prime(\prime\prime)}$ are terrain characteristics which are considered known/given.

The theoretical velocity of the vehicle (before the velocity reduction from slippage) is calculated from the speed of the engine:

$$V_a = \omega_0 r_a^0 = \frac{\omega_e}{u_{tr}} r_a^0 \qquad (12)$$

Here, ω_0 is the angular velocity of the input shaft of the transfer case, ω_e is the angular velocity of the crankshaft, and u_{tr} is the transmission gear ratio. r_a^0 is the generalized rolling radius of the vehicle in the driven mode, which can be computed using Eq. (13) and Eq. (14) derived for open differentials (Andreev et. al., 2010):

$$r_{ai}^0 = \frac{2 r^{0''}_{wi} r^{0'}_{wi}}{(r^{0''}_{wi} + r^{0'}_{wi})}, \; i = 1{,}2 \qquad (13)$$

$$r_a^0 = \frac{r_{a1}^0 r_{a2}^0 (1 + u_d)}{r_{a2}^0 u_1 + r_{a1}^0 u_2 u_d} \qquad (14)$$

r_{ai}^0 is the generalized rolling radius of an axle. u_d is the internal gear ratio of the differential (equal to 1 for the FED Alpha's symmetric differential).

The actual velocity is re-written as follows using Eq. (7) and Eq. (12) (Andreev et. al., 2010).

$$V_x = \omega_0 r_a^0(1 - s_{\delta a}) = \frac{\omega_e}{u_{tr}} r_a^0(1 - s_{\delta a}) \tag{15}$$

The generalized slippage, $s_{\delta a}$, is mathematically linked to the tire slippages through Eq. (16) and Eq. (17) derived for open differentials. The generalized slippage of the axles $s_{\delta ai}$ is computed from the individual wheel slippages and rolling radii in the driven mode (Andreev et. al., 2010).

$$s_{\delta ai} = 1 - \frac{\left(r_{wi}^{0''} + r_{wi}^{0'}\right)(1 - s''_{\delta i})(1 - s'_{\delta i})}{r_{wi}^{0'}(1 - s'_{\delta i}) + r_{wi}^{0''}(1 - s''_{\delta i})}, \; i = 1,2 \tag{16}$$

The generalized slippage of the vehicle $s_{\delta a}$ depends on the axle generalized slippages and rolling radii (Andreev et. al., 2010).

$$s_{\delta a} = 1 - \frac{(r_{a1}^0 u_2 u_d + r_{a2}^0 u_1)(1 - s_{\delta a1})(1 - s_{\delta a2})}{r_{a1}^0 u_2 u_d (1 - s_{\delta a1}) + r_{a2}^0 u_1 (1 - s_{\delta a2})} \tag{17}$$

The generalized rolling radius in the driven mode from Eq. (14) and generalized vehicle slippage from Eq. (17) allow calculation of the actual velocity from Eq. (15). When applying the actual velocity and acceleration, the traction balance of the vehicle derived from Eq. (1) is

$$\sum_{i=1}^{m}(R'_{xi} + R''_{xi}) + W_a sin\theta_n + \frac{W_a}{g}\delta_r a_x + k_w A_f V_x^2 + F_d - \sum_{i=1}^{n}(F'_{xi} + F''_{xi}) = 0 \tag{18}$$

Here, F_d is the drawbar pull force, A_f is the frontal area projection of the vehicle, and k_w is the aerodynamic drag factor computed from Eq. (19) widely utilized in conventional vehicle dynamics

$$k_w = 0.5\rho C_x \tag{19}$$

where and ρ the air density and C_x is the drag coefficient. The total circumferential force in Eq. (18) comes from the engine torque T_e split among all wheels

$$\frac{\sum_{i=1}^{2}(F''_{xi} r_{wi}^{0''} + F'_{xi} r_{wi}^{0'})}{u_i u_{tc} u_{tr}} = T_e \tag{20}$$

where u_{tc} is the gear ratio of the transfer case and u_i is the gear ratio of the driveline system from a wheel to the transfer case including the gear ratio of the final drive. A lookup table at wide-open throttle is used to determine engine torque at an assigned engine speed.

By replacing F_x forces in Eq. (9, 10, 18, 20) with their components from the right side of Eq. (11), the circumferential forces are expressed in terms of the unknown slippages. Finally, by solving a system of equations consisting of Eqs. (9, 10, 15, 18, 20) and then computing the dynamic factor with one of Eq. (2-5), the dynamic factor can be established as a function of the vehicle actual velocity.

3. Dynamic Factor of FED Alpha

In this section, the dynamic factor of the FED Alpha is computed using the conventional and proposed approaches. The terrain condition is a meadow having a loam soil sublayer with moisture content varying from 8% to 12% and grass cover of 0.05m to 0.10m. The vehicle moves at constant velocity on a flat surface (case 1 of the dynamic factor from Eq. (3)).

Figure 1 portrays the dynamic factor according to the conventional approach. Dynamic factor D_a is computed using the powertrain characteristics only and plotted against the vehicle theoretical velocity V_a. The first 6 curves display the dynamic factor for the FED Alpha's 6 transmission gears across the range of its engine speed at full throttle.

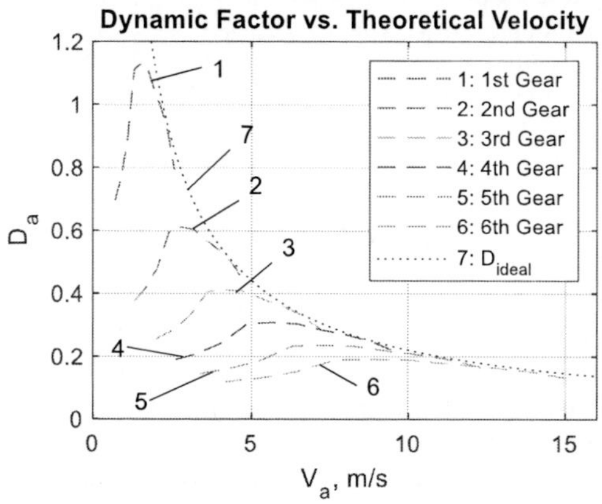

Figure 1. Dynamic factor of FED Alpha, potential values not limited by terrain grip or slippage, D_a

Curve 7 is the ideal dynamic factor computed from Eq. (21)

$$D_{ideal} = \frac{N_{e\,max}\eta_m}{W_a V_a} \tag{21}$$

where $N_{e\,max}$ is the engine max power and η_m is the mechanical efficiency of the transmission and driveline (assumed constant). D_{ideal} is the best dynamic performance possible if the vehicle had a continuously variable transmission and could completely convert the maximum engine power into traction. The dynamic factor of mechanical gear transmissions will underutilize the engine power to some extent and therefore curves 1-6 are enveloped by curve 7. A higher number of gears will bring the theoretical dynamic factors closer to the ideal curve to increase the power utilization. At higher gears, if the resistance to motion increases and velocity drops, the gear should be switched to a lower gear at the maximum point of the dynamic factor to increase the dynamic factor further and continue stable movement.

In Fig. 1, the dynamic factor is computed without any influence of the terrain. However, the maximum traction of the vehicle is limited by the terrain gripping conditions as well as the vehicle powertrain. As seen from the form of Eq. (11), the circumferential force is a function of slippage. Therefore, the circumferential force has a maximum value characterized by μ_{px}, R_z, and $s_{\delta c}$ at which no further increase in F_x is possible and mobility is lost. The grip-limited dynamic factor, labeled D_{ag}, is plotted in Fig. 2. In Fig. 2, it can be seen that while potential at gears 3-6 is unaffected because the forces do not reach the terrain limit, the maximum potential of gears 1-2 is limited compared to in Fig. 1. This is because the terrain grip limit is lower than the potential provided by the powertrain and it cannot be fully taken advantage of on this terrain. Furthermore, curve

8 shows the normalized total rolling resistance for the vehicle on the terrain computed from $\frac{\sum_{i=1}^{m}(R'_{xi}+R''_{xi})}{W_a}$. The dynamic factor needs to be higher than this line for the vehicle to have the potential for steady motion at a given velocity and gear. As the dynamic factor on gear 6, with the lowest dynamic factor, does not dip below curve 8, the vehicle has the potential for steady motion at all gears and velocities shown in Fig. 2.

While Fig. 2 includes the terrain grip, it does not account for tire slippage, as it is still plotted using the vehicle's theoretical velocity. In Fig. 3, the dynamic factor is plotted using the proposed method for computing actual velocity using the vehicle generalized slippage. The actual dynamic factor (also limited by terrain grip) is labeled D_{xg}. Since the dynamic factor is computed at steady motion, acceleration a_x in Eq. (18) is assigned zero. To balance both sides of Eq. (18) for F_x computed at full throttle, the drawbar pull force F_d in Eq. (18) is varied to ensure the balance of forces. This has an effect on the load distribution which determines the front/rear split of the normal reactions R_{zi}.

Figure 4 is a comparison of the theoretical and actual dynamic factors from Figs. 2-3.

As shown, the difference between the dynamic factors D_{ag} and D_{xg} becomes greater when the vehicle is operating at higher traction loads on lower transmission gears. The vehicle could not operate on this terrain at full throttle on gear 1 without losing mobility and the maximum potential on gear 2 is further reduced from D_{ag} to D_{xg}. Therefore, for off-road vehicles that mostly operate at higher circumferential forces and at lower gears in transmission, the impact of driveline characteristics and slippage should be taken into account when analyzing the dynamic potential.

3.1 Dynamic Factor and Acceleration Limits

It is possible to obtain limiting characteristics of the dynamic potential from Fig. 4, including the max dynamic factor and its velocity on the highest gear, the max velocity and dynamic factor on the highest gear, and the max dynamic factor on the lowest gear. These values are provided Table 1. While the grip does not limit the dynamic factor at the highest gear, on the lowest gear the dynamic factor drops 49% due to the grip limits and another 22% due to the tire slippages.

Another limiting characteristic is the theoretical and actual acceleration, which are computed by solving Eq. (18) for acceleration at full engine throttle with zero drawbar pull and slope angle.

$$a_a = [D_a - \frac{\sum_{i=1}^{m}(R'_{xi}+R''_{xi})}{W_a}]\frac{g}{\delta_r} \tag{22}$$

$$a_x = [D_x - \frac{\sum_{i=1}^{m}(R'_{xi}+R''_{xi})}{W_a}]\frac{g}{\delta_r} \tag{23}$$

Accelerations are shown in Fig. 5, which follow a pattern similar to the dynamic factor.

The limits for acceleration are provided in Table 2 for all gears of the transmission.

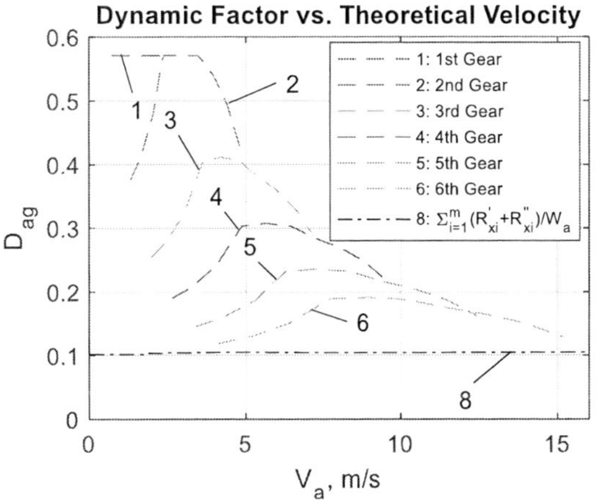

Figure . Dynamic factor of FED Alpha, Potential values limited by grip, D_{ag}

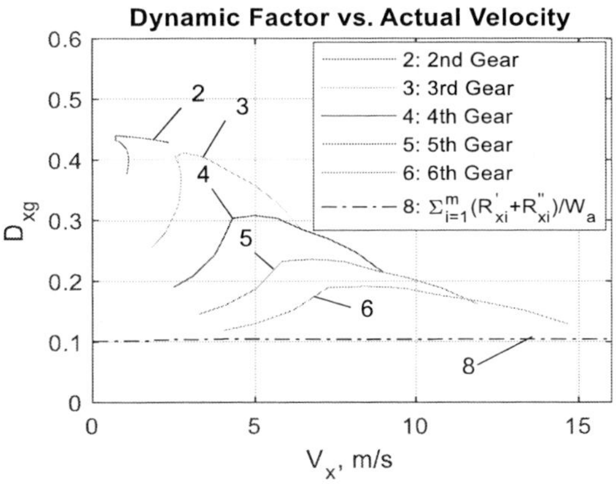

Figure 3. Dynamic factor of FED Alpha, Actual values limited by grip D_{xg}

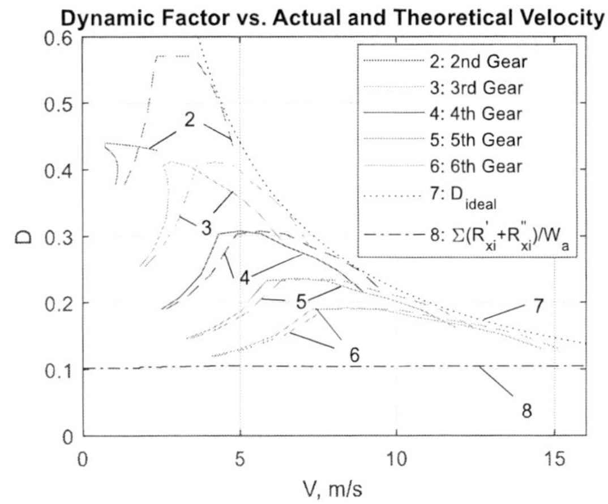

Figure 4. FED Alpha: Theoretical dynamic factor D_{ag} (dash curves) and actual dynamic factor D_{xg} (solid curves)

Table 1. FED Alpha Maximum Dynamic Factor and Velocity

Dynamic factor computation	Highest Gear				Lowest Gear
	Max Dynamic Factor	Velocity at max dynamic factor [m/s]	Max velocity [m/s]	Dynamic factor at max velocity	Max dynamic factor
Theoretical dynamic factor (not grip limited)	0.1914	8.9240	15.1970	0.1297	1.1287
Theoretical dynamic factor (grip limited)	0.1914	8.9240	15.1970	0.1297	0.5710
Actual dynamic factor (grip limited)	0.1916	8.4216	14.6510	0.1301	0.4471

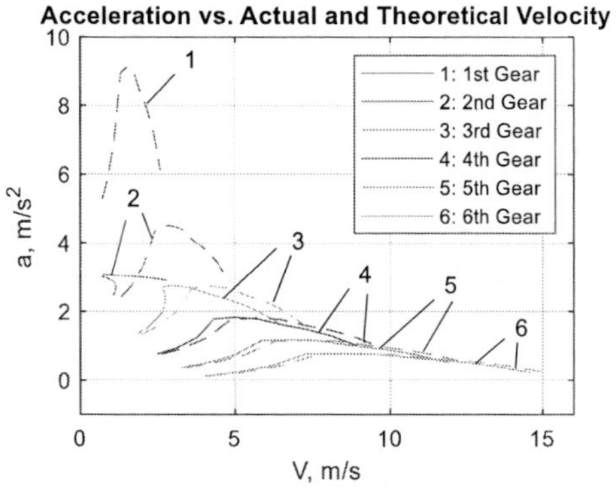

Figure 5. Max acceleration of FED Alpha, Theoretical (dash lines) vs. actual (solid lines) values

Table 2. FED Alpha Acceleration Limits

Gear	Max acceleration [m/s²]	
	Theoretical, a_a	Actual, a_x
1	9.128	-
2	4.510	3.078
3	2.732	2.760
4	1.805	1.824
5	1.164	1.177

4. Integral Dynamic Factor

An approach to characterize numerically the effects of the engine, transmission, and driveline system on the actual dynamic factor and, thus, on dynamic ability of vehicles to accelerate from zero speed to a specified velocity, the following approach is considered in this section. The area under the dynamic factor curves is computed and plotted in Fig. 4. This area is the integral of the $D(V)$-function and characterizes the level of engine power utilization for vehicle acceleration performance through converting the power into vehicle traction and speed. The area under the hyperbolic curve of the ideal dynamic factor, A_{ideal}, is the largest, showing the ideal conversion of engine power into traction and speed. Area A_a, the area under the theoretical dynamic factor curve, is limited by the transmission gears and is smaller. A further decrease is shown by the terrain grip limitation, A_{ag}. Finally, the lowest value is seen with the inclusion of limitations of the transmission gears, terrain grip, and driveline configuration, A_{xg}. Ratios of the areas given in Eq. (24) show the drop in vehicle performance from the ideal dynamic factor to the actual dynamic factor.

$$A_{ideal} : A_a : A_{ag} : A_{xg} =$$

$$1 : 0.81 : 0.47 : 0.35 \quad (24)$$

The integral dynamic factor, given by Eq. (25), can be used to assess the acceleration performance of a vehicle to accelerate to a specific velocity.

$$D_{\int} = \int D dV \quad (25)$$

Integral dynamic factors are plotted in Fig. 7 for the four cases of the ideal dynamic factor, theoretical dynamic factor, and actual dynamic factor adding limitations of the transmission gears, terrain grip, and driveline system. Fig. **Error! Reference source not found.** shows the drop in potential of the power utilization at any velocity within the entire range of velocities.

5. Design Factors Based on the Integral Dynamic Factors

The integral dynamic factors characterize the completeness of the power conversion into traction and acceleration performance, which is a consideration for mobility performance of terrain vehicles. As an element of vehicle mobility, assessments of the transmission and driveline system are proposed based on the dynamic factors. Equation (26) is the drivetrain design factor, C_{dt}.

$$C_{dt} = D_{\int a}/D_{\int ideal} \quad (26)$$

Factor C_{dt}, the ratio of the theoretical and ideal dynamic factors, characterizes the theoretical ability of a conceptual transmission to utilize the engine power for vehicle acceleration performance.

Equation (27) gives the driveline design factor, C_{dl}.

$$C_{dl} = D_{\int x}/D_{\int a} \quad (27)$$

C_{dl} is the ratio of the actual and ideal dynamic factors. This factor characterizes the influence of the driveline system on the wheel power distribution. Because $D_{\int a}$ is a function of the theoretical velocity V_a and $D_{\int x}$ is a function of the actual velocity V_x, these velocities should be correlated to each other through Eq. (7) when computing C_{dl}.

The powertrain design factor, C_{pt}, combines the previous two design factors to integrate the influence of the transmission and driveline system.

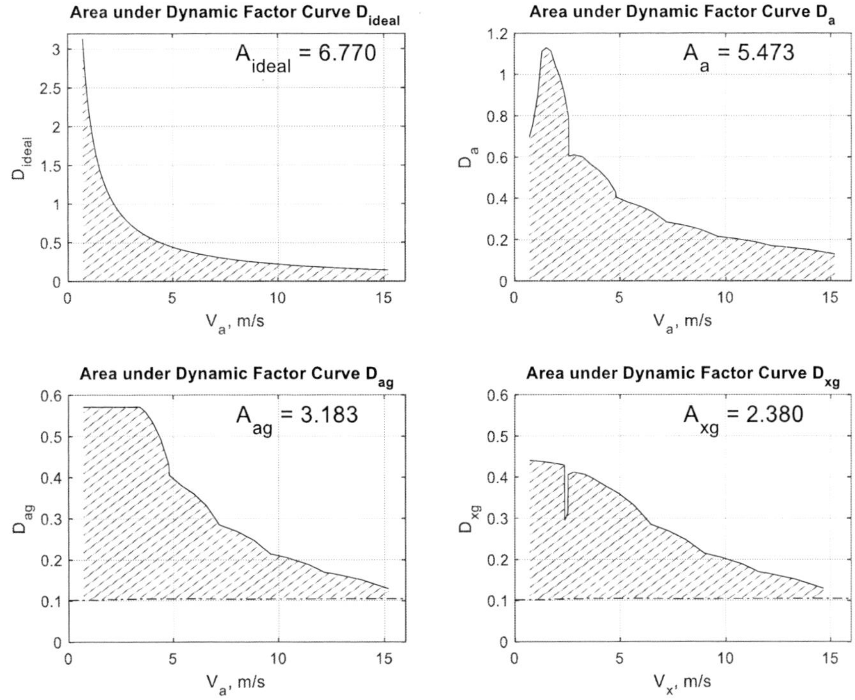

Figure 6. Dynamic factor of FED Alpha: a) Potential ideal values A_{ideal}*,*
b) Potential values limited by transmission A_a*, c) Potential values limited by transmission and grip* A_{ag}*,*
d) Actual values limited by transmission, grip, and driveline A_{xg}

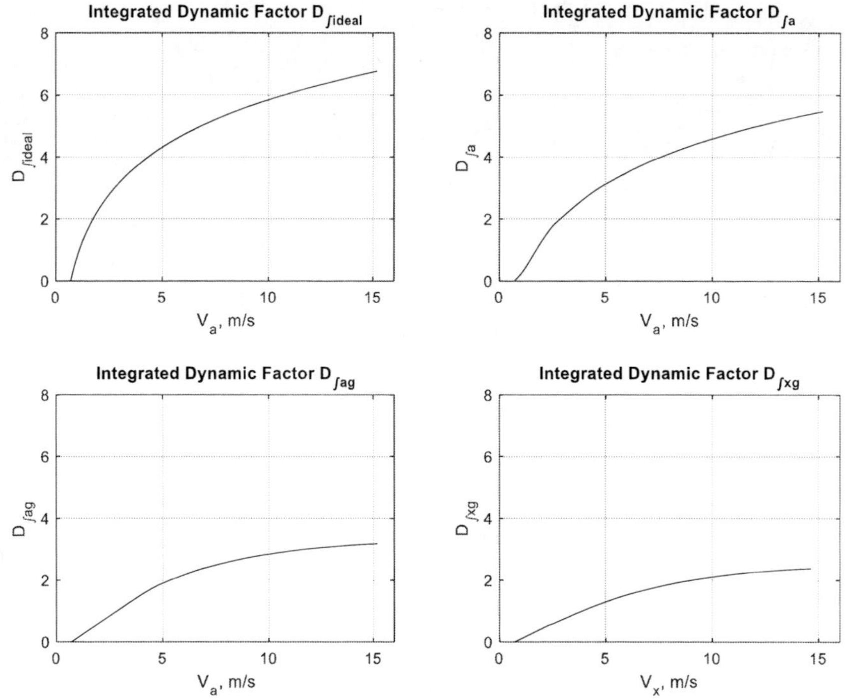

Figure 7. Integral dynamic factor of FED Alpha: a) Theoretical ideal $D_{\int ideal}$, b) Theoretical limited by transmission $D_{\int a}$, c) Potential limited by transmission and grip $D_{\int ag}$, d) Actual limited by transmission, grip, and driveline system $D_{\int xg}$

$$C_{pt} = C_{dt}C_{dl} = \left(\frac{D_{\int a}}{D_{\int ideal}}\right)\left(\frac{D_{\int x}}{D_{\int a}}\right) = \frac{D_{\int x}}{D_{\int ideal}} = f(V_x) \quad (28)$$

Figures **Error! Reference source not found.-Error! Reference source not found.** plot the proposed design factors for the FED Alpha.

The drivetrain design factor does not show a significant drop in the integral dynamic factor at mid and high velocities because the limits of the terrain grip and tire slippages are not included in C_{dt}. Compared with C_{dt}, the driveline design factor C_{dl} shown in Fig. 9 is about 40% lower at the mid and high velocities once the impact of tire slippages on the vehicle velocity is included. The powertrain design factor C_{pt} shown in Fig. 10 shows the overall reduction in performance is over 60%. These dimensionless factors facilitate comparison between vehicles with different transmissions and drivelines and for assessing these systems in design.

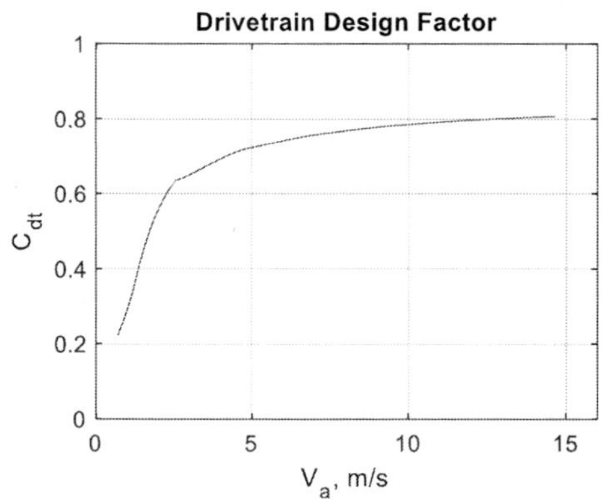

Figure 8. FED Alpha Drivetrain design factor, C_{dt}

6. Conclusions

The dynamic factor is a dimensionless characteristic that characterizes the potential traction normalized by the vehicle weight. For off-road vehicles with high traction loads, the influence of tire slippage on the vehicle's mobility is considerable. The vehicle's driveline configuration, which determines the split of power from the engine to the wheels, is an important factor in the distribution of tire slippages and traction not typically considered in analysis of the dynamic factor. In this paper, a method was proposed to include the driveline effect in the calculation of the vehicle's actual velocity based on a system of generalized vehicle parameters. The method was demonstrated in a computation of the dynamic factor of the FED Alpha 4x4 vehicle in a simulated terrain condition. A set of vehicle design factors was proposed based on the integral dynamic factor. Being dimensionless, the design factors allow for comparing and ranking different vehicles with different transmissions and drivelines and, also, for assessing and selecting better transmission and driveline concepts for a vehicle under design.

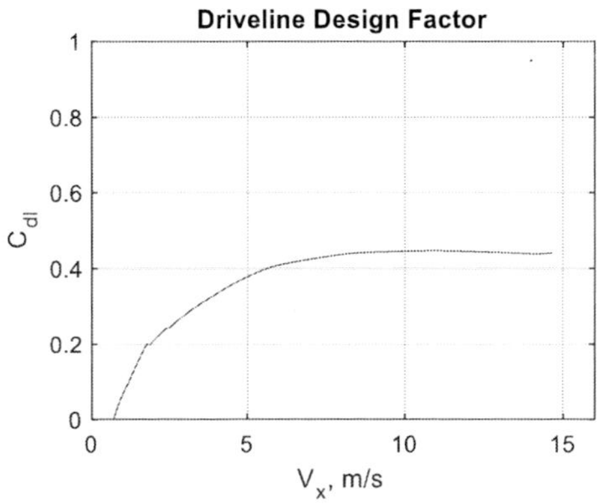

Figure 9. FED Alpha Driveline design factor, C_{dl}

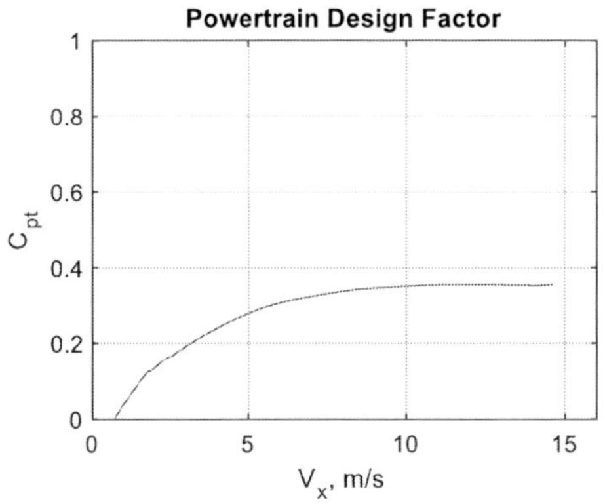

Figure 10. FED Alpha Powertrain factor, C_{pt}

7. Nomenclature

a_a	*Theoretical linear acceleration*	*[m/s²]*
a_x	*Actual linear acceleration*	*[m/s²]*
C_{dl}	*Driveline Design Factor*	
C_{dt}	*Drivetrain Design Factor*	
C_{pt}	*Powertrain Design Factor*	
C_x	*Drag coefficient*	
D_a	*Theoretical dynamic factor*	
D_{ad}	*Aerodynamic drag*	*[kN]*
D_{ag}	*Theoretical (grip limited) dynamic factor*	
D_{ideal}	*Ideal dynamic factor*	
D_{xg}	*Actual (grip limited) dynamic factor*	
F_a	*Linear inertia force*	*[kN]*
F_d	*Drawbar Pull*	*[kN]*
F_x	*Wheel circumferential force*	*[kN]*
$F_{x\Sigma}$	*Sum of circumferential forces*	*[kN]*
g	*Acceleration of gravity*	*[m/s²]*
i	*Index of an axle*	
k_w	*Aerodynamic drag factor*	*[kg/m³]*
m_a	*Vehicle mass*	*[kg]*
n	*Number of axles*	
$N_{e\,max}$	*Engine max power*	*[kW]*
r_a^0	*Generalized rolling radius (driven mode)*	*[m]*
r_w^0	*Rolling radius (driven mode)*	*[m]*
R_x	*Rolling resistance*	*[kN]*
R_z	*Normal reaction*	*[kN]*
s_δ	*Tire slippage*	
$s_{\delta a}$	*Generalized slippage*	
$s_{\delta c}$	*Characteristic slippage*	
T_e	*Engine torque*	*[Nm]*
T_w	*Wheel torque*	*[Nm]*
u_d	*Open differential internal gear ratio*	
u_i	*Final drive gear ratio*	
u_{tc}	*Transfer case gear ratio*	
u_{tr}	*Transmission gear ratio*	
V_a	*Theoretical linear velocity*	*[m/s]*
V_x	*Actual linear velocity*	*[m/s]*
W_a	*Vehicle weight*	*[kN]*
η_m	*Efficiency of transmission and driveline*	
θ_n	*Ground slope angle*	*[rad]*
δ_r	*Rotating mass inertia factor*	
μ_{px}	*Peak friction coefficient*	
ρ	*Air density*	*[kg/m³]*
ω_e	*Engine speed*	*[rad/s]*
ω_0	*Transfer case input shaft speed*	*[rad/s]*
$'$	*Left wheels*	
$''$	*Right wheels*	
$'('')$	*Includes both left and right wheels*	

8. Acknowledgements

This research is supported by Automotive Research Center (ARC) under federal award No. W56HZV-19-2-0001 and ARC subaward No. SUBK00014000.

Disclaimer: Reference herein to any specific commercial company, product, process, or service by trade name, trademark, manufacturer, or otherwise, does not necessarily constitute or imply its endorsement, recommendation, or favoring by the United States Government or the Department of the Army (DoA). The opinions of the authors expressed herein do not necessarily state or reflect those of the United States Government or the DoA and shall not be used for advertising or product endorsement purposes.

DISTRIBUTION A. Approved for public release; distribution unlimited. OPSEC # 6691

9. References

Andreev, A. F., Kabanau, V., Vantsevich, V. V., 2010. *Driveline systems of ground vehicles: theory and design*, CRC Press.

Chudakov, E. A., 1950. *Theory of Automobile*, State Publishing House of Machine-Building Literature, Moscow, Russia (in Russian).

Genta, G., Morello, L., 2009. *The Automotive Chassis, Volume 2: System Design*, Springer.

Grishkevich, A. I., 1986. *Automobile. Theory*, Vysheishaia Shkola Publishing House, Minsk (in Russian).

Litvinov, A. S., Farobin, Y. E., 1989. *Automobile: Theory of Operational Properties*, Mashinostroenie Publishing House, Moscow, Russia (in Russian).

Michigan Technological University Keweenaw Research Center (KRC), 2021. NATO Applied Vehicle Technology (AVT) Panel Cooperative Demonstration of Technology CDT-308, Next-Generation NATO Reference Mobility Model (NG-NRMM) Demonstration. https://www.mtu.edu/cdt/ (accessed 2021).

Wong, J. Y., Jayakumar, P., Toma, E., Preston-Thomas, J., 2020. A review of Mobility Metrics for Next Generation Vehicle Mobility models. *J. Terramechanics* 87, 11-20.

Zimelev, G. V., 1957. *Theory of Automobile*, second ed., Military Publishing House, Moscow, Russia (in Russian).

7233 / Study of Passive Steering Mechanism for Mars Surface Exploration Rovers

Asahi Oe, Shin-Ichiro Nishida*, and Shintaro Nakatani
Tottori University, Tottori, Japan
**Corresponding author: s-nishida@tottori-u.ac.jp*

https://doi.org/10.56884/RDKV9886

Abstract

In planetary exploration, rovers are used to move across the surface to obtain high-resolution topographic elevation maps, as well as to conduct mining and rock sampling. However, because of the harsh planetary environment, surface exploration poses many technical challenges. In this study, we assume an unmanned exploration rover that is small enough to be carried on a Japanese launch vehicle, and its weight is limited by the payload capacity of the launch vehicle. It is also important that the rover is not prone to failure. The surface of Mars is covered with fine sand called regolith, and this sand can get between the mechanisms and cause them to fail. Therefore, a rover with lightweight and simple mechanisms is needed. In addition, Mars has a large slope, and there are not only rocky terrain but also uneven terrain such as depressions and craters. For smooth exploration, the rover must travel a safe and efficient path to avoid tipping over or getting stuck. However, wheel slippage on the regolith makes it difficult to follow the target path precisely and can result in getting stuck. If an active steering mechanism is omitted in order to configure a simple and lightweight traveling system, and if the path following is attempted to be performed by the difference in rotation speed between left and right traveling mechanisms such as wheels and crawlers when traveling on a curved path, the ground of the traveling system is forced to skid, which places a severe load on the traveling system. Therefore, this study proposes a lightweight, simple, and passive mechanism that reduces skidding and provides good path-following performance. The proposed mechanism uses a passive Ackermann mechanism without actuators, and the steering is performed by the difference in rotational speed between the left and right wheels. Since there is no actuator for steering, the mechanism is lightweight and simple. In order to suppress steering in unintended directions, a weak spring is added in the direction that returns the steering angle to a straight line. This paper describes the results of a prototype wheeled rover mechanism incorporating the proposed mechanism in the size assumed and confirming its ability to follow a target path by running experiments on soil simulating the surface of Mars. The experimental results show that the proposed mechanism has a higher ability to follow the target path than a rover without a steering mechanism. In addition, the rover's ability to follow the target path was improved by correcting the steering angle with respect to the rotational speeds of the left and right wheels.

Keywords: Mars exploration rovers, passive steering mechanism, path following

1. Introduction

In planetary exploration, rovers travel on the surface of a planet to obtain detailed topographic data and to mine and collect rocks. However, the harsh environment of a planet poses many technical challenges.

Because the rover is transported by rocket, the weight of the rover is limited by the payload capacity of the rocket. In addition, it is difficult to repair the rover manually in the event of a failure, so a rover with a low risk of failure is required. The surface of Mars is covered with fine sand called regolith. If regolith penetrates between mechanisms, it can cause failures, so dustproofing is necessary, but it is difficult to do so for complex mechanisms [4]. Because of the severe temperature changes and strong ultraviolet rays on Mars, rubber tires like those used on Earth will deteriorate and become unusable. In fact, NASA's Mars exploration rovers use metal wheels [3].

Mars has rocky terrain, plains covered with regolith, and terrain with steep and uneven slopes. The rover needs to travel along a safe and efficient path for smooth exploration. However, if the traveling mechanism slips or gets stuck due to regolith, it causes wear, energy loss, and the inability to follow the target path accurately. In addition, the accuracy and efficiency of the path is reduced when the traveling mechanism skids on a curved path. Therefore, there is a need for a driving/steering mechanism and control law that can reduce slippage and path errors and follow the target path.

In this study, we proposed a passive steering mechanism using springs as a lightweight and simple mechanism, built an experimental rover, and conducted driving experiments. In addition, a running experiment was conducted to compensate for the deviation from the target path. The results of these experiments are reported.

2. Mechanism consideration

2.1 Traveling mechanism

Currently, there are two types of traveling mechanisms being studied: wheel-type and crawler-type. In this study, the wheel type was adopted because of its light weight and simplicity. In addition, although wheeled models with six wheels are widely used, a four-wheel mechanism was used in this study in order to reduce weight and size.

2.2 Steering mechanism

In a rover without a steering mechanism, the direction of the wheels differs from the direction of travel when turning, as shown in Figure 1a, causing wheels to skid. Therefore, Uemura (2018) proposed a passive steering mechanism with a rotating link at the center of the front wheels as shown in Figure 1b to reduce wheel skidding without using an actuator for steering, and a spring is added to allow the steering angle to be changed. This mechanism does not use an actuator for steering, and the steering angle is changed by the difference in rotational speed between the left and right wheels, and the steering angle is maintained by balancing the tension of the spring [5]. However, this mechanism tends to cause the wheels to interfere with the vehicle body during steering, which places restrictions on the mechanism configuration. Therefore, the distance between the wheel and the axis of rotation of the steering wheel is shortened to make the steering mechanism less prone to wheel interference with the vehicle body.

Typical steering mechanisms for 4-wheeled vehicles are the Ackermann mechanism, parallel steering, and 4-wheel steering (in-phase and inverse phase). A rover without a steering mechanism and parallel steering at low speeds have the problem of skidding when turning. Four-wheel steering requires four actuators for steering, which makes the mechanism complex. Since an exploration rover requires a lightweight and simple mechanism with minimal wheel slippage, the Ackermann mechanism is used in this study because the rover runs at low speeds.

The Ackermann mechanism can change the angle of the left and right wheels to match the center of rotation of each wheel. On Earth, the Ackermann mechanism is mainly used, which is approximated by a linkage such that the knuckle arm extensions intersect on the rear wheel axis.

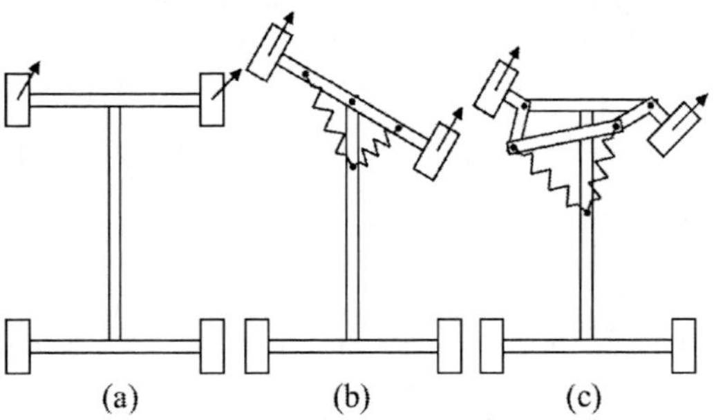

Figure 1. Wheel direction and direction of travel when turning

3. Consideration of the proposed mechanism

When traveling on rough terrain or overcoming rocks, complex external forces are applied to the wheels, which may cause the wheels to steer in unintended directions. Therefore, we propose a mechanism in which a spring-loaded restoring force is added to the steering mechanism to suppress unintended steering. In the future, the addition of a viscous friction element is considered necessary to cope with running on uneven terrain such as rocky terrain. However, in order to simplify the problem, analysis, experiments, and evaluation were conducted using only spring elements.

We propose the mechanism shown in Figure 2. When turning, as shown in Figure 3, the force received from the ground changes by changing the rotation speed of the left and right wheels, and the steering is turned off. At this time, the spring extends and exerts a force in the direction of returning the steering wheel, and the steering wheel is maintained at a point where the force of the wheel is balanced with the force received from the ground.

The force of the wheel in balance with the spring is obtained. The forces acting on the tie rods are shown in Figure . The Eq. 1-4 show the balance of forces applied to the tie rods. The symbols used here are as shown in Table 1. The forces that the wheels receive from the ground are assumed to be equal in size and opposite in direction, and forces perpendicular to the direction of the wheels are not considered.

$$F_c - F_{br}\cos\alpha_4 - F_{kr}\cos\theta_2 - \frac{l_o}{l_b}F_{wr}\cos\left(\alpha_4 - \frac{\pi}{2}\right) = 0 \quad (1)$$

$$F_{br}\sin\alpha_4 - F_{kr}\sin\theta_2 + \frac{l_o}{l_b}F_{wr}\sin\left(\alpha_4 - \frac{\pi}{2}\right) = 0 \quad (2)$$

$$-F_c + F_{bl}\cos\alpha_3 + F_{kl}\cos\theta_1 + \frac{l_o}{l_b}F_{wl}\cos\left(\alpha_3 + \frac{\pi}{2}\right) = 0 \quad (3)$$

$$F_{bl}\sin\alpha_3 - F_{kl}\sin\theta_1 + \frac{l_o}{l_b}F_{wl}\sin\left(\alpha_3 + \frac{\pi}{2}\right) = 0 \quad (4)$$

From Eq. 1-4, the wheel force balanced by the spring force is given by Eq. 5.

$$F_{wr} = \frac{l_b}{l_o}\frac{F_{kl}\sin(\theta_1+\alpha_3)\sin\alpha_4 - F_{kr}\sin(\theta_2+\alpha_4)\sin\alpha_3}{\sin\alpha_3 + \sin\alpha_4} \quad (5)$$

The stronger the spring force, the more resistant the wheels are to disturbance from forces applied to the wheels on uneven terrain, but the greater the force exerted by the wheels on the ground when turning, the more likely the soft ground is to shear and fracture. This steering mechanism does not use an actuator for steering, so it has a low risk of failure, is lightweight and simple, and can change the wheel angle.

Table 1. Symbol description

Symbol	**Description**
F_{bl}, F_{br}, F_c	Axial force of each link
F_{kl}, F_{kr}	Left and right spring force
F_{wl}, F_{wr}	Force received from the ground by the left and right wheels
α_1, α_2, α_3, α_4	Angle of Ackermann link
θ_1, θ_2	Angle between left and right springs and tie rods
l_b	Knuckle arm length
l_o	Distance between kingpin and wheel

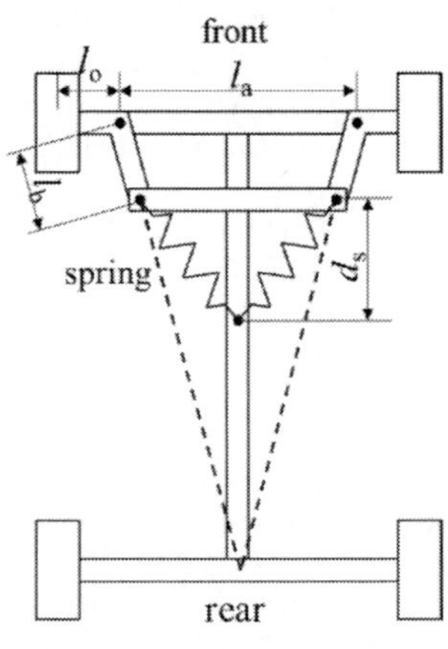

Figure 2. Proposed mechanism

This study assumes a small rover that can be placed on a Japanese rocket (H-IIB rocket). The mass of the rover is assumed to be 30 kg out of the payload capacity of the H-IIB rocket. The size of the experimental rover is assumed to be the same as that of the assumed Mars exploration rover, with a wheelbase of 500 mm and a tread base of 500 mm. Since the surface gravity of Mars is 0.38 times that of the Earth, the mass of the rover should be about 12 kg, considering the difference in gravity [2]. The rover configuration is summarized in Table 2.

l_a and l_b were considered as the lengths of the Ackermann links. The combination with the smallest turning radius at the maximum steering angle was calculated and decided so that the wheel skid angle would be small.

For the spring, the spring constant, natural length, and mounting position d_s were considered. The spring conditions were calculated and decided based on the balance between the force the wheel receives from the ground and the force of the spring.

4. Running experiments

4.1 Purpose of experiment

The purpose of the experiment is to confirm the effectiveness of the proposed method by driving in a circular arc on flat sand. In addition, to improve the tracking performance to the target path by compensating for the deviation of the steering angle.

4.2 Experimental equipment and environment

The experimental rover is shown in Figure 5. The mass was 11.9 kg. Rigid and elastic wheels are used. The elastic wheel deforms to increase the ground contact area to lower and equalize the ground contact pressure, thereby reducing wheel slippage due to shear failure of the ground. The experiment uses an elastic wheel with annular springs proposed by Nishio and Nishida (2019) in Figure 6. The inner spring distributes the load, and the outer spring keeps the ground pressure low and equal. When this elastic wheel was driven straight on sand on flat ground, the slip rate was less than 1% [1]. The radius and width of the wheel are 108 mm and 60 mm, respectively. Figure 7 shows an elastic wheel attached to the experimental rover, which flexed by about 12 mm under 1 G on Toyoura sand, resulting in a ground contact length of about 95 mm. The maximum steering angle of the outer wheel when turning was set at 30 deg, and a stopper was provided to prevent the wheel from turning beyond this angle. The steering can be fixed by passing a pin through the center of the tie rod, allowing the rover to be used as a rover without a steering mechanism.

The experimental field is covered with Toyoura sand to simulate regolith.

The rover's position information is obtained by detection of the rover's marker with a camera mounted above the experimental field.

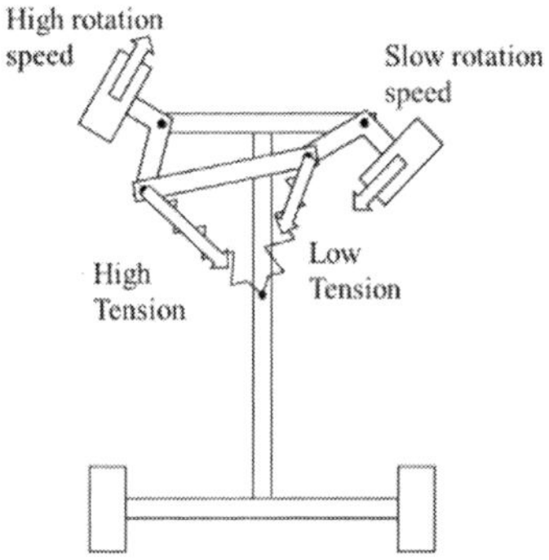

Figure 3. Steering during turning

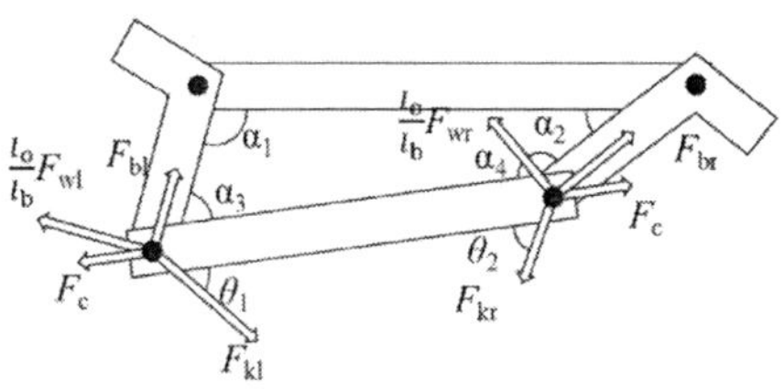

Figure 4. Balance of the force applied to the axis of the tie rod

Table 2. Configuration of the rover

Item	Specifications
Traveling mechanism	Wheel type 4 wheels
Steering mechanism	Ackermann mechanism, passive steering
Wheelbase	500 mm
Tread base	500 mm
Mass	12 kg

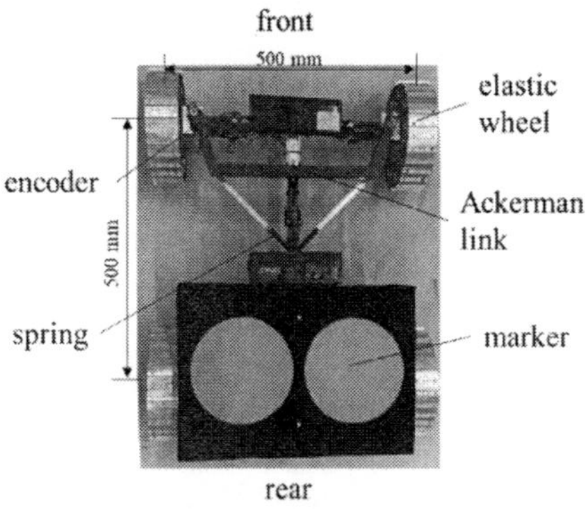

Figure 5. Experimental rover

Figure 6. Elastic wheel

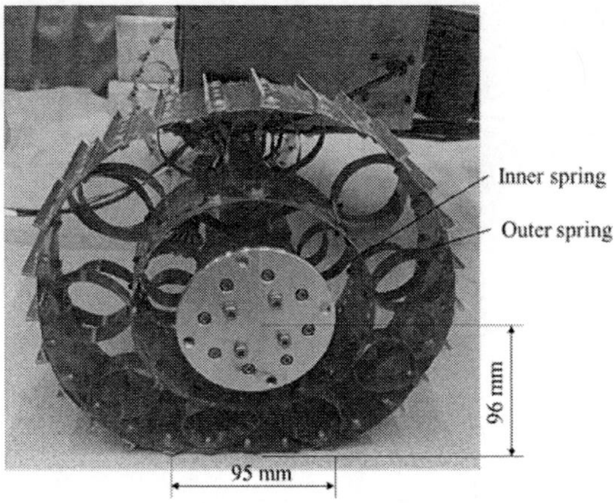

Figure 7. Deformed elastic wheel

4.3 Experimental method

This is an experiment in which an arc-shaped path is run as the target path. A rover is driven in an arc from a straight steering position. A rover is driven in an arc on flat sand by varying the spring, wheels, target path curvature, and target turning speed. Three types of springs with spring constants k = 0.09, 0.18, and 0.27 N/mm were used, as well as without springs. If the spring constant is greater than this, the ground is more prone to shear failure and the target steering angle will not be achieved. The rovers were also compared with a fixed steering wheel as a rover without a steering mechanism. The wheel rotation speed is calculated as the ratio of the distance between the center of the turn and each wheel so that the center speed of the rover becomes the target turning speed. Two target turning speeds (v_{ref}=10, 20 mm/s) and five target path curvatures (0.2, 0.4, 0.6, 0.8, 1 1/m) are used.

The radius of the circle is calculated from the rover's travel path detected from the marker by a camera mounted above the experimental field using the least-squares method and compared to the target by the curvature of the path. In this case, the traveling path until the steering angle becomes constant is not used to calculate the radius of the circle.

Corrective driving feeds back the steering angle to change and correct the difference in rotation speed between the left and right wheels. In this steering angle compensation, the deviation e input to the PI control is added to the rotation speed deviation as shown in Eq. 6, with the outer wheel having the steering angle deviation added and the inner wheel having the steering angle deviation subtracted.

$$e = \begin{cases} \omega_{\mathrm{ref}} - \omega + k_{\theta}(\theta_{\mathrm{oref}} - \theta_{\mathrm{o}}) \text{ (outer front wheel)} \\ \omega_{\mathrm{ref}} - \omega - k_{\theta}(\theta_{\mathrm{oref}} - \theta_{\mathrm{o}}) \text{ (Inside front wheel)} \end{cases} \quad (6)$$

A block diagram of the control is shown in Figure 8. The rotation speed and steering angle of each wheel are PI-controlled, and the steering angle is changed by the difference in rotation speed between the left and right wheels by turning the motor of each wheel.

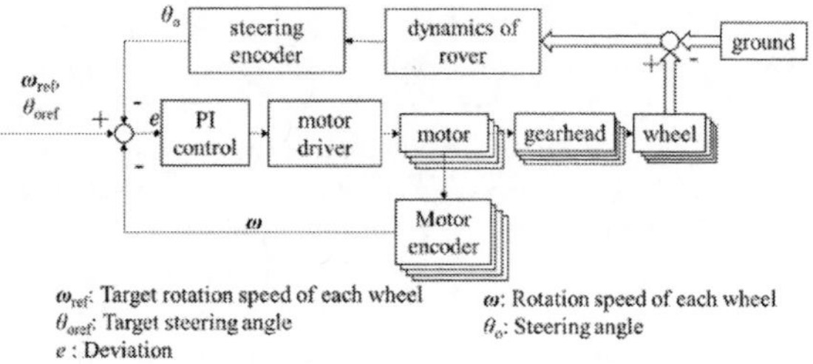

Figure 8. Block diagram of control

4.4 Experimental results

Figure 9 shows a graph of the path curvature error for the experiment with no correction, rigid wheels, and a turning speed of 10 mm/s. It was confirmed that the smaller the spring constant, the smaller the path curvature error. The larger the target path curvature, the larger the path curvature error.

Figure 10 shows the path curvature error graphs for the experiments with correction, rigid wheels, and a turning speed of 10 mm/s. It can be seen that the path curvature error is reduced by the correction.

4.5 Considerations

In the case without the steering mechanism, the wheels do not change direction, resulting in skidding when turning, which causes an error in path curvature compared to the target path curvature. In the case of no correction, the path curvature error of the proposed mechanism is smaller than that without steering, confirming that the proposed mechanism is effective in driving in a circular arc. This is considered to be due to the steering mechanism's ability to reduce wheel skidding. The cause of the error from the target path curvature is thought to be that the steering angle became smaller than the target value due to wheel slippage.

When corrected by the steering angle, the error was reduced for all target path curvatures. When the target path curvature was 1 1/m, the error was 0.48 to 0.61 times that of the uncorrected method. This result confirms the effectiveness of the proposed correction method. The reason for this is thought to be that the correction reduced skidding as the steering angle approached the target value, and the difference in wheel speed between the front wheels increased.

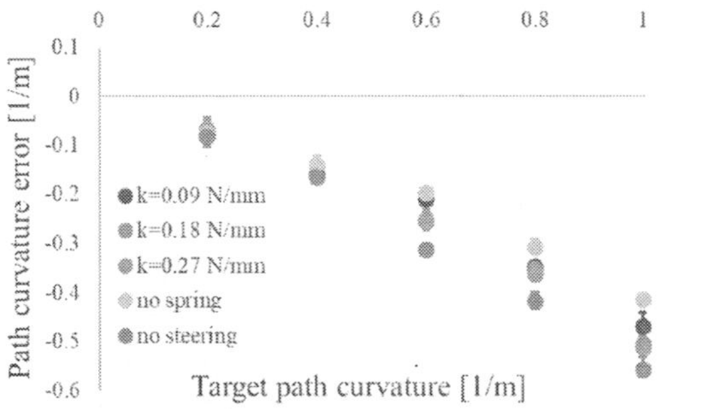

Figure 9. Target path curvature and path curvature error for each spring without correction (rigid wheel, turning speed 10mm/s)

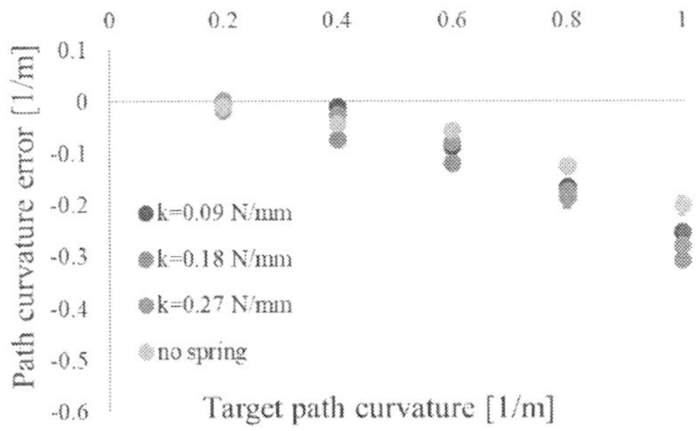

Figure 10. Target path curvature and path curvature error for each spring in steering angle correction (rigid wheel, turning speed 10mm/s)

5. Conclusions

In this study, a passive steering mechanism using springs was proposed as a lightweight and simple mechanism for reducing wheel skidding on a Mars rover. The length of the Ackermann link was determined from the turning radius at the maximum steering angle and the wheel skid angle, and the spring conditions were calculated to balance the wheel forces. Driving experiments were conducted with the lightweight and simple mechanism of the experimental rover, and it was confirmed that the proposed mechanism is more effective in driving a circular arc than a rover without a steering mechanism. The rover was able to approach the curvature of the target path by compensating the rotational speed of each wheel based on the steering angle feedback.

Future tasks include improving the correction method to reduce the path curvature error and running on uneven terrain where complex external forces are applied to the wheels.

6. Acknowledgements

This work was supported by JSPS KAKENHI Grant Number JP17K06119.

7. References

[1] Nishio N, Nishida S. Study of flexible wheel for Mars Exploration rover. 28th Space Engineering Conference [SEC'19]. No.19 – 344.

[2] Nishio N, Nishida S, Nakatani S. Study of a Planetary Exploration Rover with a Hinge-type Center-of-gravity Shift Mechanism. Trans. JSASS Aerospace Tech. Japan 2021;19(4):469-476.

[3] Rankin A, Maimone M, Biesiadecki J, Patel N, Levine D,Toupet O. Mars Curiosity Rover Mobility Trends During the First 7 Years. Journal of Field Robotics 2021;38(5):759-800.

[4] Shigehara M, Nishida S. Rover Design for Lunar and Planetary Exploration. Journal of Robotics Society of Japan 2003; 21(5):472-476.

[5] Uemura T, Nishida S. A study on arc path following of Mars exploration rover. ROBOMECH2018; 2A2-B08.

7399 / Tire-Soil Tangential Force Reinforcement Learning Modeling

Yingchun Qi[a]*, Jiaqi Zhao[a], Ye Zhuang[a], Weiguang Fan[b], Hui Ye[a], and Jianzhong Zhu[a]
[a] *Jilin University, Changchun, Jilin, China*
[b] *China North Vehicle Research Institute, Beijing, China*
**Corresponding author: qiyc@jlu.edu.cn*

https://doi.org/10.56884/RHBE9228

Abstract

Abstract: Tire-soil tangential interaction involves complex terramechanics. When modeling the tire-soil tangential forces, a great amount of experiments is needed to identify the parameters in the currently available models. The efficiency and accuracy of the current models is still time and cost consuming. The control of the terrain vehicle is introduced gradually to the off-road vehicles. Such application requires more accurate and real-time tire-soil force models. Therefore, the machine learning technique, the reinforcement learning algorithm, is introduced to the tire-soil tangential force modelling. First, the tire-soil rolling experiment is carried out under longitudinal and lateral slip condition with the tire-soil test facility. The tire-soil forces vs slip ratios test data is obtained on the sand and mud road surfaces. The reinforcement learning model, which including the physical interpretation and the uncertainty (with Gaussian Process), is proposed. The model parameters is identified through the supervised learning (training) by the model from the acquired experimental data. The model accuracy could be improved gradually with the iterative offline learning (training). The trained model could calculate the force-vs-slip ratio relationship with high accuracy and efficiency. The proposed model could also be updated with the new data learning.

Keywords: off-road vehicle, reinforcement learning, tire-soil tangential force modelling

1. Introduction

As the only component connecting the vehicle to the ground, the tire-soil interaction characteristics are one of the important factors affecting the vehicle dynamic performance. Various off-road vehicles, such as agricultural machinery, military vehicles, desert vehicles and forestry vehicles, are mostly driven on soil roads (Wong 2009). The control of the terrain vehicle is introduced gradually to the off-road vehicles. Such application requires more accurate and real-time tire-soil force models.

In recent years, the exploration of the moon and Mars has become a hot topic. Studying the relationship between the vehicle and the planet soil has also become a necessary research content for the development of the planet rover. The establishment of the tire-soil force model requires an in-depth study of soil mechanics and terramechanics. A large amount of experimental data is needed to identify the model parameters, and the model cannot meet the real-time performance for control requirements (Ding et al. 2011, Chhaniyara et al. 2012).

A Gaussian process-based reinforcement learning model is proposed. The model parameters are identified through the supervised learning (training) by the model from the acquired experimental data (Williams and Rasmussen 2006). The trained model could calculate the force-vs-slip ratio relationship with high accuracy and efficiency. With the further enrichment of experimental data, the reinforcement learning algorithm will further improve the accuracy of the model.

2. Tire-soil tangential force experiment

The training data acquisition for the tire-soil tangential force model will be first described. The data used to learn the Gaussian Process model was acquired from the tire-soil tangential force experiment. Fig. 1 shows the experiment plant, and Fig. 2 shows the test tire used. The plant is mainly composed of motor, main frame, connecting rod, trailer, guide rail, soil trough, sensor, traction device, computer, etc. The load on the wheel can be adjusted by changing the counterweight on the frame, and the wheel can move freely along the soil groove. The tension sensor on the trailer is used to measure the traction force of the hook, which also is the tire-soil tangential force. The horizontal displacement sensor is used to measure the horizontal displacement of the wheel, then the velocity could be calculated.

The data obtained through the experiment include: the displacement and rotational velocity of the wheel, motor driving torque. The value of sinkage is measured by a handheld laser range finder (Jester and Klik 2005, Eitel et al. 2011).

The test wheel and the test trolley are driven separately in the experiment. The angular velocity of the wheel ω was given. Due to soil resistance and tire deformation, the velocity of the trolley V_t is different from the driving velocity of the wheel ωR. The ratio between ωR and the difference of V_t and ωR is the wheel slip rate s.

$$s = \frac{\omega R - V_1}{\omega R}$$

Different wheel slip ratios can be obtained by adjusting the test trolley and wheel driving velocity, and then the longitudinal friction coefficient can be calculated. The test condition of experiment is illustrated as Table 1.

The soil used in the experiment is the lunar soil simulant of JLU-M-1, and the mechanical parameters are shown in Table 2 (Gromov 1999, Li et al. 2012).

The measurement repeatability was considered fully in the experiment plan, and each test was repeated 10 times. Considering the disturbance of sandy and muddy soils, the test ground was reconditioned after each test.

The quantitative relationship between the tangential force of the tire and the slip ratio when the load on the wheel is fixed, the tangential force-slip ratio curve is shown as Fig. 3 The tangential force increases with slip rate.

Figure 1. Tire-soil force experiment plant

Figure 2. The tire in the force experiment

Table 1. Test condition of tire-soil force experiment

Load *W* (N)	Velocity V_1 (*m/s*)	Slip Ratio *s*
3750		
4250	0.12	3%-70%
4750		

Table 2. JLU-M-1 lunar soil mechanical parameters

Parameters	Lunar soil	JLU-M-1
exponent of sinkage n	0.8-1.2	1.03
cohesion modulus k_c(kN/m^{n+1})	1.4	22.14
friction modulus k_φ(kN/m^{n+2})	820	1438.5
cohesion c(kPa)	0-0.59	0.32
friction angle ϕ(°)	30-50	34.6
shear modulus K(cm)	1.78	1.21
density γ(g/cm^3)	0.83-2.29	1.87

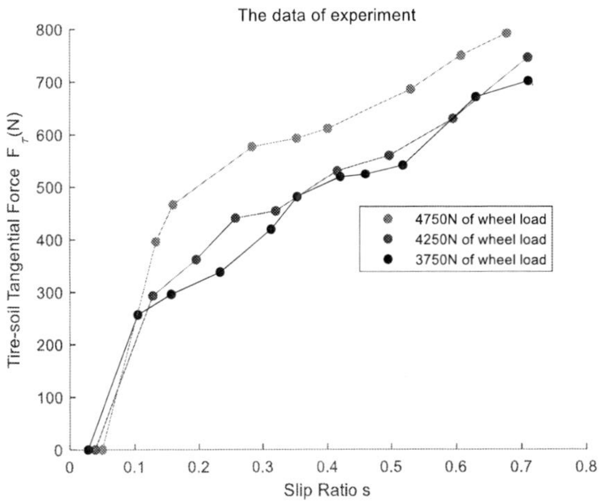

Figure 3. Tire-soil force with different wheel load

3. Gaussian Process modeling for tire-soil tangential force

As the experimental data illustrate, the tire tangential force is related to the load of the wheel *W* and slip ratio *s*. During the process of experiment, there is a certain measurement deviation of data, so the following assumption was shown:

$$\begin{cases} F_\tau = F_t(x_k) + \varepsilon, \quad F_t: \mathbb{R}^{n_x} \to \mathbb{R} \\ \varepsilon \sim \mathcal{N}(0, \sigma_n^2) \end{cases}$$

Considering the uncertainty of the force measurement, it is assumed that the tire tangential force is a function of wheel load and slip ratio. Where, $x_k = [W_k \quad s_k]$is correlation feature. The measurement deviation follows a normal distribution, the variance is σ_n^2 and mean is 0, F_t is a tire-soil tangential force function which we want to model.

According to the theory of Gaussian Process modeling, the function of F_t is a GP (Gaussian Process):

$$F_t \sim \mathcal{GP}$$

A GP is defined by its mean function $m(x)$ and covariance function, $k(x, x')$ then:

$$E(F_t) = m(x)$$

$$\mathrm{Cov}(F_t(x), F_t(x')) = E[\{F_t(x) - m(x)\}\{F_t(x') - m(x')\}]$$
$$= k(x, x')$$

Given the prior assumption and training data set $\mathcal{D}$ which can be obtained via experiment, the training point is:

$$X = [W_k \ \ s_k] = [x_1, x_2, \dots, x_m]^T \in \mathbb{R}^{m\times 2}.$$
$$F_\tau = [F_{\tau 1}, F_{\tau 2}, \dots, F_{\tau m}]^T \in \mathbb{R}^{m\times 1}$$

Then the posterior distribution of F_t^* is:

$$F_t^* | X^*, X, F_\tau \sim \mathcal{N}(\mu^*, \Sigma^*)$$

where, $X^*$is the predicted point, F_t^* is the corresponding output, then:

$$\mu^* = \mathbb{E}[F_t^* | X, Y, X^*]$$
$$= K(X^*, X)[K(X, X) + \sigma_n^2 I]^{-1} F_\tau$$
$$\Sigma^* = K(X^*, X^*)$$
$$-K(X^*, X)[K(X, X) + \sigma_n^2 I]^{-1} K(X, X^*).$$

In the tire-soil tangential force model, input-output characteristics are smooth and continuous, so the ardSE function is chosen as kernel function here (Alvarez et al. 2012):

$$k(x, x') = \sigma_f^2 \exp(-\frac{1}{2}(x - x')^T M^{-1}(x - x'))$$

Where,

$$M = \begin{bmatrix} l_1 & 0 \\ 0 & l_2 \end{bmatrix}$$

The ardSE kernel function uses different kernel hyperparameters $l_i (i = 1,2)$ which are determined by the relative importance of different input features. During the process of training model, a Bayesian Optimization was used to optimize the hyperparameter in the GP model, which is introduced in chapter 4.

4. Bayesian Optimization of GP Model

In the GP model established, some hyperparameters need to be optimized to achieve better model performance. However, the relationship between hyperparameters and loss functions is generally not clear. The Bayesian optimization method could find optimization directions in the global scope of the hyperparameters search domain and help us to gain suitable GP model hyperparameters (Garnett 2022).

The error between the true value y and the predicted value $\hat{y}$ was defined as the loss of model. Where, the minimum squared error (Minimum Squared Error, MSE) is used as the loss function L:

$$L = \sum_j^n (\hat{y} - y)^2$$

It is assumed that the hyperparameters θ in the GP model have a certain mapping relationship with the loss function L:

$$g : \theta \to L$$

Then, the process of optimizing hyperparameters θ is transformed into an optimization problem that, which θ minimize the loss function L (Mockus 2012).

The map g from hyperparameters θ to the loss function L is difficult to evaluate. $g(\theta)$ could be observed and its derivatives are not evaluated. So objective function is defined to substitute the map g .The objective function, $h(\theta)$, is formed by the posterior distribution in Bayesian Optimization (Bergstra et al. 2011). Here, the objective function $h(\theta)$ was treated as the probability of a GP:

$$F_\tau | \theta, (x, y) \sim \mathcal{N}(0, k\{(x, y), (x, y)\} + \sigma_n^2 I)$$
$$h(\theta) = p(F_\tau | \theta, (x, y))$$

The acquisition function AF is a function which determines what areas in the hyperparameters domain are worth exploiting and what areas are worth exploring. The value of AF is equal to the performance of trained model proximately. The AF was chosen as follow:

$$AF(\theta) = \log p(F_\tau | \theta, (x, y))$$

Where, (x, y) is the points of training data. When the OB find the maximum value of $AF(\theta)$, the optimal hypermeters θ^* that is equal to the hyperparameters θ, are found. The process of OB was shown as Table 3.

In this paper, hyperparameters can be regarded as two cases: model parameters (basis function, initial value of the coefficients for the basis matrix, initial value for the noise standard deviation of the GP model) and training parameters (the amount of data for sub-training (batch size), etc.)

Table 3. The process of Bayesian Optimization

Algorithm 1 Bayesian Optimization

1. **INPUT**: GP model, training data (x, y) , L , epochs,
   acquisition function (AF),
   hyperparameters θ
2. Optimization begins:
3. $\theta \leftarrow$ Initial (θ) in search domain
4. **for** $i = 1, \dots,$ epochs **do**:
   GP model $\leftarrow$ training data (x, y)
   $P(L | \theta, (x, y)) \leftarrow fitmodel($ GP model$|_\theta, (x, y))$
   $\theta' \leftarrow \arg\max_{\theta'}(AF(\theta'))$
   $\theta \leftarrow \theta'$
5. **end for**
6. **OUTPUT**: θ^* ▷ optimized hyperparameters

5. The result of trained GP model

The purpose of training model is to select the most suitable model. For supervised learning, the selected model is hoped to have a strong generalization ability to unknown data, so the process of model validation is required to reflect the prediction performance of different models on unknown data.

At first, all training data was tested to measure the performance of the model, but this method will lead to model overfitting for the trained data. Then, to solve this problem, all training data was divided into training set and test set. The training set was used for model training, and the predictive performance of the trained model is measured on the test set. This measurement method is called at Test

Accuracy (TA), which can effectively avoid model overfitting. However, there is a disadvantage of TA, that is the test accuracy is a high variance estimate and depend on the test sets used (Riley et al. 2015). The model maybe got different test accuracy from different test sets.

K-fold cross-validation is proposed in this paper. The training data is divided into K parts, after that, K-1 parts of data are taken out to train the model, one part of data is used for the trained model prediction. The prediction error of this part of the data is calculated, then another one part of data was chosen for testing the model. and their sum of squares record. The error will be recorded until all parts of data have been used for testing once and only once. The sum of all errors is substituted for the test accuracy to measure the model predictive performance (Trippa et al. 2015). In this paper, 5-fold cross-validation is selected.

The process of training GP model is running on MATLAB environment. The value of epochs is 35, and the search domain of the optimization hyperparameter is shown as Table 4.

As Fig. 4 shows, the MSE of GP model converge to the minimum at 18th iteration. The hyperparameters at this iteration are selected as optimal hyperparameters. Fig. 5 illustrate the predictive curve of slip ratio and tire-soil tangential force from the learned GP model with different wheel load. The value of R-Squared is equal to 0.98 in the validation of the learned model.

In Fig. 6-8, the green line represents the mean of GP model predictive value with various slip ratio, and the dark areas represent the 95% prediction intervals which indicates the uncertainty of the model. Because the noise of the tire-soil model was considered, the GP model could predict the tangential force in a small interval. Due to there is no experiment data in large slip ratio, the uncertainty area above 0.8 slip ratio become open out.

As Fig. 9 shows, the predictive result in various wheel loads with a 0.35 slip ratio is presented. Because the slip ratio cannot be absolutely fixed during the experiment, there is only data in 0.32 of slip ratio and 3750N of wheel load. Excepted for the red point which is in 0.35 slip ratio, the blue point which is in 0.32 slip ratio is also plotted in the figure.

Fig. 10 is the predictive value of the soil-tire tangential force from the GP model in two input features (slip ratio and wheel load). The red points are experiment data. The GP model could predict the force accurately in huge wheel load. The GP model learned the positive relationships between slip ratio and tangential force, wheel load and tangential force respectively.

Table 4. Optimization hyperparameter domain

Hyperparameter	Search Domain
σ_n	[1,2000]
σ_f	[1,2000]
l_1	[1,1000]
l_2	[1,1000]
Basis Function	[Zero, Constant, Linear]

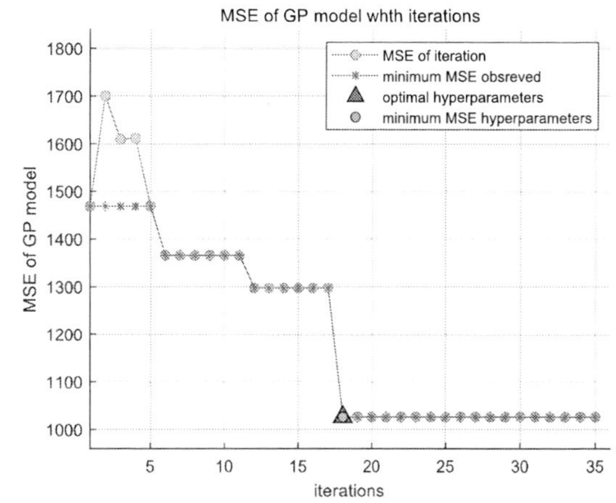

Figure 4. MSE of GP model during training process

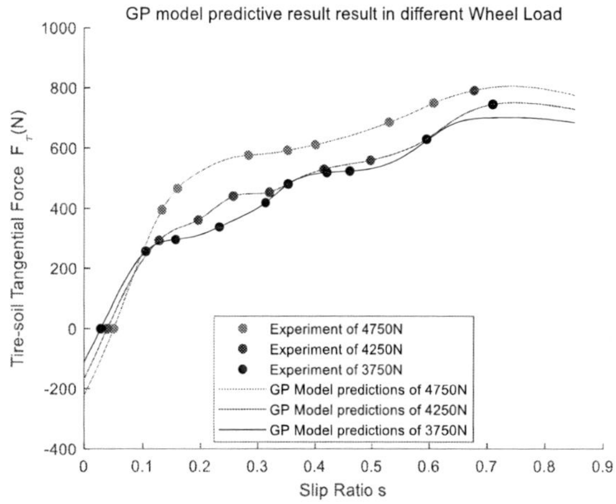

Figure 5. Soil-tire tangential force vs slip ratio in different wheel load

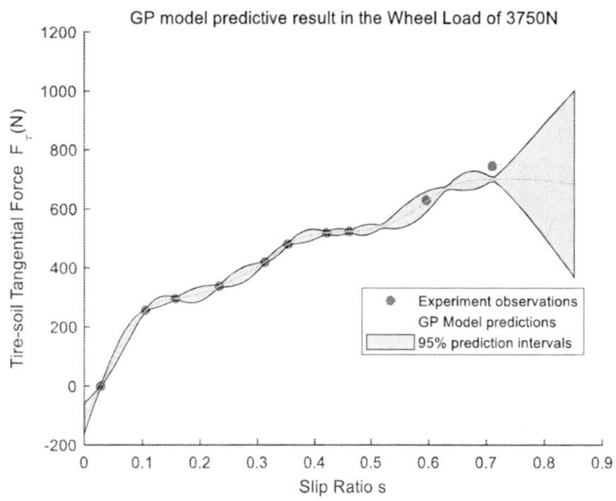

Figure 6. The predictive result with uncertainty from GP model in the wheel load of 3750N

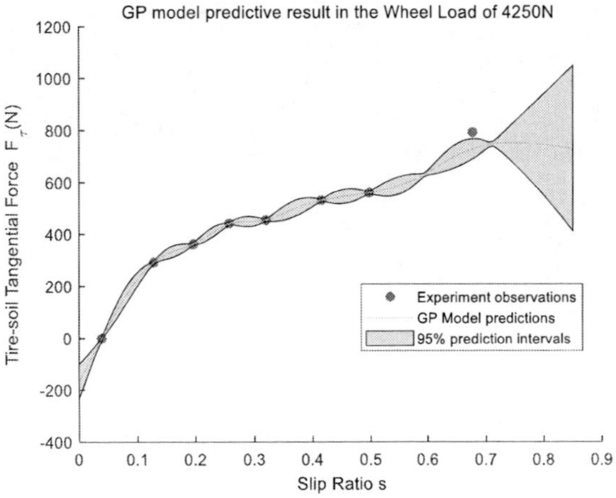

Figure 7. The predictive result with uncertainty from GP model in the wheel load of 4250N

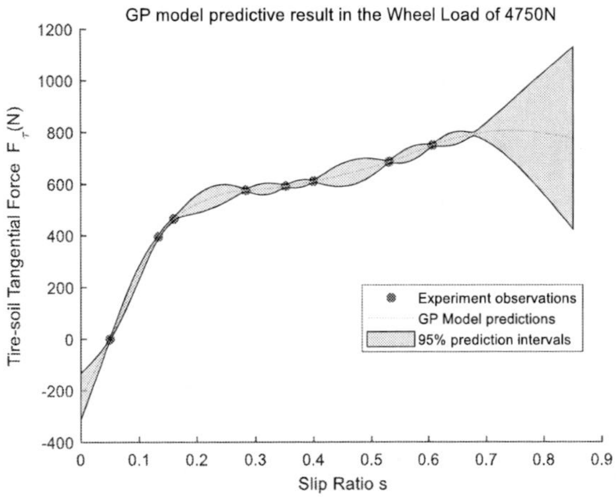

Figure 8. The predictive result with uncertainty from GP model in the wheel load of 4750N

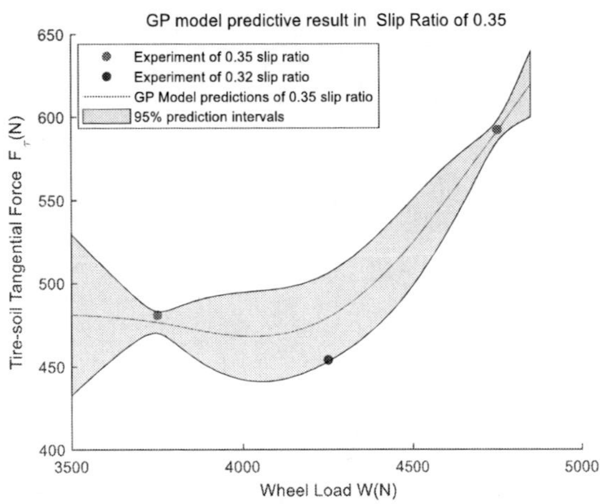

Figure 9. The predictive result with uncertainty from GP model in slip ratio of 0.35

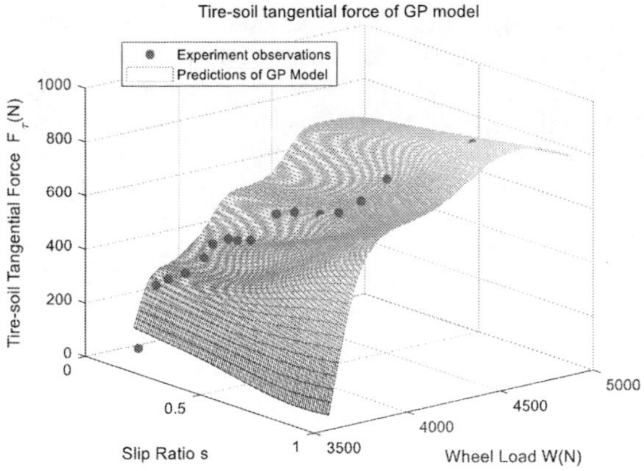

Figure 10. The two-feature predictive result of GP model

6. Conclusion

Since the theory model of terramechanics is complex, this paper explores a novel tire-soil modeling method based on Gaussian Process and Machine Learning.

Taking into account the uncertainty of the observed experiment data and the tire-soil system, a GP model was established. The soil-tire tangential force data of the wheel on the simulated lunar soil JLU-M-1 is tested by the experimental plant, and the data is used to learn the model. In the learning process, an advanced Bayesian Optimization in machine learning is applied to optimize the hyperparameters to accelerate the learning process. The learning model converged at the 18th iteration, and the value of R-Squared was 0.98 in the validation for experimental data. The prediction results of the model are in good agreement with the experimental data points, and the model can also predict some intervals which means the uncertainty mentioned above. The tire-soil tangential force model can be learned by the data-driven method without understanding the properties and mechanisms of the environmental soil, and the model has the ability to iteratively improve when the amount of experimental data increases. The proposed modeling method can contribute to the control and modeling of off-road vehicle.

7. References

Alvarez, M. A., L. Rosasco N. D. Lawrence, 2012. “Kernels for vector-valued functions: A review.” *Foundations and Trends in Machine Learning* 4(3): 195-266.

Bergstra, J., R. Bardenet, Y. Bengio B. Kégl, 2011. “Algorithms for hyper-parameter optimization.” *Advances in neural information processing systems* 24.

Chhaniyara, S., C. Brunskill, B. Yeomans, M. Matthews, C. Saaj, S. Ransom L. Richter, 2012. “Terrain trafficability analysis and soil mechanical property identification for planetary rovers: A survey.” *Journal of Terramechanics* 49(2): 115-128.

Ding, L., H. Gao, Z. Deng, K. Nagatani K. Yoshida, 2011. "Experimental study and analysis on driving wheels' performance for planetary exploration rovers moving in deformable soil." *Journal of Terramechanics* 48(1): 27-45.

Eitel, J. U., C. J. Williams, L. A. Vierling, O. Z. Al-Hamdan F. B. Pierson, 2011. "Suitability of terrestrial laser scanning for studying surface roughness effects on concentrated flow erosion processes in rangelands." *Catena* 87(3): 398-407.

Garnett, R., 2022. *Bayesian Optimization,* Cambridge University Press Cambridge, UK.

Gromov, V., 1999. *Physical and mechanical properties of lunar and planetary soils,* Springer.

Jester, W. A. Klik, 2005. "Soil surface roughness measurement—methods, applicability, and surface representation." *Catena* 64(2-3): 174-192.

Li, J. Q., H. Li, D. W. Jin, L. He M. Zou, 2012. Study on the Pressure-Sinkage Relationship of Lunar Soil Simulant. Advanced Materials Research, *Trans Tech Publ.*

Mockus, J., 2012. *Bayesian approach to global optimization: theory and applications,* Springer Science & Business Media.

Riley, R. D., I. Ahmed, T. P. Debray, B. H. Willis, J. P. Noordzij, J. P. Higgins J. J. Deeks, 2015. "Summarising and validating test accuracy results across multiple studies for use in clinical practice." *Statistics in Medicine* 34(13): 2081-2103.

Trippa, L., L. Waldron, C. Huttenhower G. Parmigiani, 2015. "Bayesian nonparametric cross-study validation of prediction methods." *The Annals of Applied Statistics* 9(1): 402-428.

Williams, C. K. C. E. Rasmussen, 2006. *Gaussian processes for machine learning,* MIT Press Cambridge, MA.

Wong, J. Y., 2009. *Terramechanics and off-road vehicle engineering: terrain behaviour, off-road vehicle performance and design,* Butterworth-Heinemann.

7878 / A Method for Fast Obtaining of Soil Shear Strength Index Based on DEM Free-Fall Cone Penetration Simulation

Jincheng Diao*, Jingwei Gao, Xiaobo Song, and Baojun Di

National University of Defense Technology, Changsha, Hunan, China
**Corresponding author: 1130201222@qq.com*

https://doi.org/10.56884/RHBE9228

Abstract

Abstract: Tire-soil tangential interaction involves complex terramechanics. When modeling the tire-soil tangential forces, a great number of experiments is needed to identify the parameters in the currently available models. The efficiency and accuracy of the current models is still time and cost consuming. The control of the terrain vehicle is introduced gradually to the off-road vehicles. Such application requires more accurate and real-time tire-soil force models. Therefore, the machine learning technique, the reinforcement learning algorithm, is introduced to the tire-soil tangential force modelling. First, the tire-soil rolling experiment is carried out under longitudinal and lateral slip condition with the tire-soil test facility. The tire-soil forces vs slip ratios test data is obtained on the sand and mud road surfaces. The reinforcement learning model, which including the physical interpretation and the uncertainty (with Gaussian Process), is proposed. The model parameters is identified through the supervised learning (training) by the model from the acquired experimental data. The model accuracy could be improved gradually with the iterative offline learning (training). The trained model could calculate the force-vs-slip ratio relationship with high accuracy and efficiency. The proposed model could also be updated with the new data learning.

Keywords: discrete element method, free-fall cone penetration, soil shear strength index, triaxial compression test

1. Introduction

The soil shear strength index includes internal friction angle (φ) and internal cohesion (c). The current methods for obtaining soil shear strength mainly include empirical methods through the cone index (*CI*) (Zhang, 2002) and experimental methods through the triaxial compression test. Aiming at the shortcomings of the current methods, this paper tries to explore a method for fast obtaintion of soil mechanical properties based on DEM cone penetration simulation. DEM was first proposed by Cundall (Cundall and Strack, 1971), and its theoretical development and application fields are expanding.

Hu et al. (2021) established an off-road tire steering model on sandy soil by using DEM, and the parameters of the sandy soil model are calibrated based on the triaxial compression test and inclined plate experiment. Nakanishi et al. (2020) analyzed the traction performance of the lugged wheel by using the open source discrete element software OSS. Zhang (Zhang and Hao, 2015) found that the penetration depth of torpedo anchor was negatively correlated with φ & c of seabed soil through numerical simulation. Tanaka et al. (2000) proposed a method for determining the time interval used in EDEM software simulation on the stability of the algorithm during the calculation process. Asaf et al. (2007) optimized the Nelder-Mead algorithm and established a DEM soil mechanics model. Konishi et al. (2005) carried out DEM simulations for different materials and got conclusion of the penetration resistance-depth curve. Alvaro Janda (Janda and Ooi, 2016) studied the scaling rules of the contact model parameters during the modeling process of EDEM cohesive soils. Lin et al. (2018) qualitatively revealed the flow of lunar soil particles and the development of resistance during the static cone penetration process with FEM-DEM coupled simulation.

The present research on the DEM penetration simulation has the shortcoming that it is difficult to control the constant speed of the penetration process. Therefore, based on the 3D particle EDEM software soil modeling, this paper set the gravity environment in RecurDyn to make the penetrating rod fall freely, and realized the EDEM-RecurDyn coupled simulation to explore the corresponding relationship between the penetration depth and the soil shear strength index. The research process of this paper is shown in Fig. 1.

2. Test Research

This paper conducts three kinds of test to explore the inner relationship between height of free-fall, moisture content, and soil shear strength index.

2.1 Free-fall cone penetration test

Compared with the method of controlling the speed of the penetration rod in the traditional cone penetration test, this paper adopts the method of free-fall penetration, trying to obtain the corresponding relationship between penetration depth and soil mechanical properties, which is easy to operate and requires less equipment. The test site in this paper is near Jingbian County, Yulin City, Shaanxi Province, which is a typical loose soil landform. In order to ensure the universality and generality of samples, 12 soil sample points are randomly selected around here, and their specific coordinates are shown in Table 1.

The free-fall cone penetration test was carried out using the patented digital soil cone penetrometer applied by our laboratory to measure the penetration depth of different soil samples with different free-fall heights (H). Its structure is shown in Fig. 2.

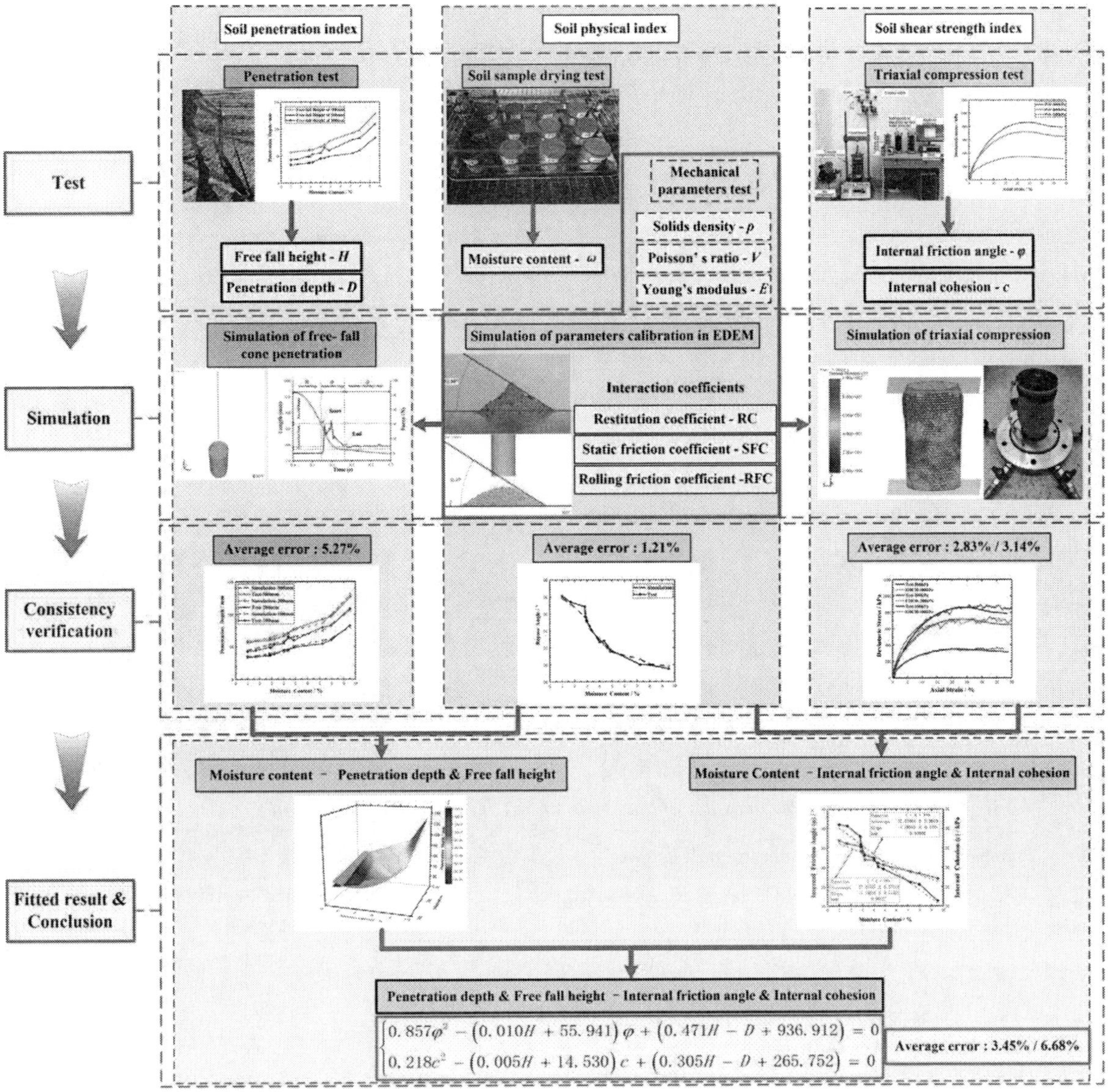

Figure 1. The research process of this paper

12 sample points were selected, and the penetration depths were measured by laser rangefinders. In order to reduce the measurement error, a laser rangefinder was installed on both sides of the penetration rod, and the average value of the two was taken as the measurement value. At each free-fall height at each point, repeat the free-fall process three times and take the average value of the penetration depth as the penetration depth test value at that point. The test results are shown in *Table 1*. Another concern is that as long as the test is carried out near the same sampling point, the test is repeatable given that the soil quality is almost unchanged. It can be seen from the test results Table 1 that the penetration depth of each sample point only increases with the increase of the height of the falling cone, but there is no obvious rule to follow between the sample points. Therefore, it is necessary to further explore the corresponding relationship between soil samples in different states and penetration depths by testing the differences between soil samples at different points. Next, it is necessary to carry out the moisture content test.

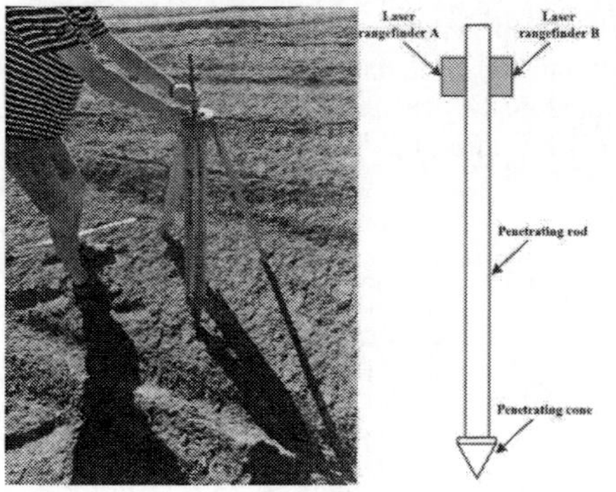

Figure 2. The structure of cone penetrometer

2.2 Drying test

Moisture content (ω) is another important parameter of soil composition and it greatly affects the mechanical properties of soil such as pressure and shear. The drying test refers to measuring the weight difference of the sample before and after drying to determine the percentage of the moisture to the dry soil mass, which is the moisture content.

First, the boxes containing 12 groups of soil samples were labeled and weighed, and the soil samples were respectively loaded and then weighed to obtain the total weight of the box and the wet soil. After that, the drying test was carried out, the drying time was 8 hours, and the drying temperature was 107°C. The soil samples before and after drying are shown in Fig. 3. Through the above process, the weight of dry soil and wet soil can be obtained, and the moisture content is calculated according to Eq. 1. Three sets of moisture content data are obtained, as shown in Table 2.

$$\omega = \frac{m - m_d}{m_d} \times 100\% \qquad (1)$$

Where ω is moisture content in natural state, m is the mass of wet soil, m_d is the mass of dry soil.

The moisture content of the 12 groups of samples from the test data are rearranged according to the order of low to high, as shown in *Table 3*.

Table 3. Moisture content of 12 sample points (%)

P12	P4	P7	P11	P10	P1	P3	P8	P2	P5	P6	P9
0.99	1.66	2.81	2.86	2.93	3.20	4.01	4.37	4.85	7.24	7.96	9.54

According to *Table 1* and *Table 3*, at the same sampling point, that is, under the condition of the same moisture content, the penetration depth increases with the increase of the free-fall height. At the same free-fall height, the penetration depth increases with the continuous increase of the moisture content, which is a positive correlation increase.

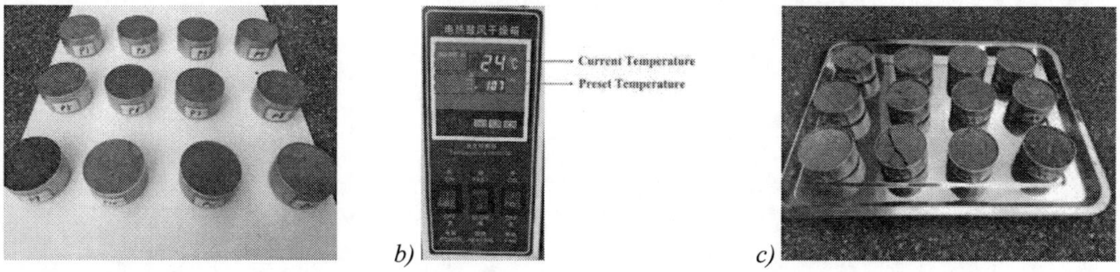

Figure 3. a) Wet soil b) Temperature controller of the electric blast drying oven c) Dried soil

Table 1. Penetration depths in different free-fall height

Sample Points	Longitude	Latitude	Altitude (*m*)	Penetration Depths (*mm*)		
				100*mm*	200*mm*	300*mm*
P1	117°52′29.81″	38°47′16.30″	1293	38.0	53.0	61.5
P2	117°52′32.72″	38°47′17.42″	1293	49.0	60.0	77.5
P3	117°52′01.32″	38°47′48.42″	1290	45.0	55.0	66.5
P4	117°52′00.78″	38°47′46.27″	1288	34.5	44.0	58.5
P5	117°52′09.88″	38°47′40.09″	1290	54.5	80.0	92.5
P6	117°52′13.26″	38°47′41.19″	1290	60.0	82.5	97.0
P7	117°52′20.02″	38°47′30.36″	1292	36.5	49.0	60.0
P8	117°52′20.59″	38°47′26.75″	1293	45.5	65.5	70.0
P9	117°50′31.47″	38°47′09.34″	1286	83.0	108.5	127.0
P10	117°50′43.94″	38°47′03.46″	1287	37.5	48.5	61.0
P11	117°44′44.14″	38°37′14.85″	1339	37.0	48.5	62.0
P12	117°44′45.36″	38°37′15.07″	1339	33.5	42.5	56.0

Table 2. Data during the drying process

Sample Points	Box (g)	Box+Wet soil (g)	Box+Dried soil (g)	Wet soil (g)	Dried soil (g)	Water (g)	Moisture content (%)
P1	18.5	189.1	183.8	170.6	165.3	5.3	3.20
P2	17.1	179.2	171.7	162.1	154.6	7.5	4.85
P3	18.1	168.5	162.7	150.4	144.6	5.8	4.01
P4	16.4	168.9	166.4	152.5	150.0	2.5	1.66
P5	17.8	158.5	149.0	140.7	131.2	9.5	7.24
P6	17.1	163.5	152.7	146.4	135.6	10.8	7.96
P7	17.8	167.8	163.7	150.0	145.9	4.1	2.81
P8	17.9	153.8	148.1	135.9	130.2	5.7	4.37
P9	16.6	134.9	124.6	118.3	108.0	10.3	9.54
P10	18.2	183.3	178.6	165.1	160.4	4.7	2.93
P11	18.1	176.3	171.9	158.2	153.8	4.4	2.86
P12	17.3	190.7	189.0	173.4	171.7	1.7	0.99

2.3 Triaxial compression test

Instrument used in triaxial compression test is the 28-WF4005 static strain semi-automatic triaxial compression instrument produced by Wykeham Farrance, Italy, which is shown in *Fig. 4*.

Three groups confining pressure of 100kPa, 200kPa, and 300kPa were set for each soil sample. Three tests were carried out under different confining pressure, and the test would be terminated automatically when the force-time curve reached the peak value and began to drop. After the test, take sample P10 as an example, the shear failure curve of the soil sample was drawn, as shown in *Fig. 5*.

Through the above-mentioned triaxial compression test, three Mohr's stress circles (Zhang and Liu, 2010) of each group of samples under different confining pressure conditions were obtained, and the optimal common tangent equation of these three Mohr's stress circles was obtained through MATLAB software programming. According to the Mohr-Coulomb strength theory (Huang et al., 2017), the longitudinal intercept of the line of the Mohr failure envelope is c, and the horizontal inclination is φ. The test results are shown in *Table 4*. Since the φ and c obtained by the test are the true and accurate values, they are represented by φ_0 and c_0 respectively.

It can be seen from *Table 4* that as the moisture content increases, the internal friction angle decreases from 36.91° to 27.15°, and the internal cohesion decreases from 41.05kPa to 21.36kPa, both of which are close to linear and negatively correlated with moisture content decrease.

3. Numerical Simulation Research

The software used in the numerical simulation process are EDEM2018 and RecurDyn. Since the simulation process is the same as the above test process, the cone penetration in the simulation environment is set as a free-fall in the gravity environment.

3.1 Preliminary calibration of parameters using repose angle simulation

Zou et al. (2020) optimized the contact model between particles in the discrete element method, so that the shear parameters in the simulation process have more accurate results.

Repose angle test was used to preliminary calibration the contact coefficients of the particles, including recovery coefficient (RC), rolling friction coefficient (RFC) and static friction coefficient (SFC). Take sample P10 as an example, when RC=0.25, SFC=0.33, RFC=0.47, the simulated repose angle was 32.37°, and the error was only 1.31%. Therefore, this set of contact parameters could represent the actual situation of the sample P10 well, as shown in *Fig. 6*.

Since φ & c of the soil were measured under the static friction of soil particles, the main factor that affected the mechanical properties of soil particles was the SFC. That is, φ & c has almost no correspondence with RC and RFC, but has a close correspondence with SFC.

According to this principle, the RC and RFC were kept constant, and the SFC was constantly adjusted so that the simulation results of the repose angle could be consistent with the test results. Finally, the parameter calibration results of 12 groups of soil samples are shown in *Table 5*. The error between the simulation results and the test results of the repose angle is also reflected in *Table 5*.

According to the results of *Table 5*, **the average error of the preliminary calibration for the parameters is only 1.21%**. And the preliminary calibration of the parameters are accurate and effective. The set soil particles can well represent the actual situation of the soil.

Combined with the moisture content obtained by the test in *Table 3*, it can be seen from *Table 5* that the higher the moisture content, the smaller the SFC and the smaller the φ. SFC is negatively correlated with moisture content, and positively correlated with φ.

a)

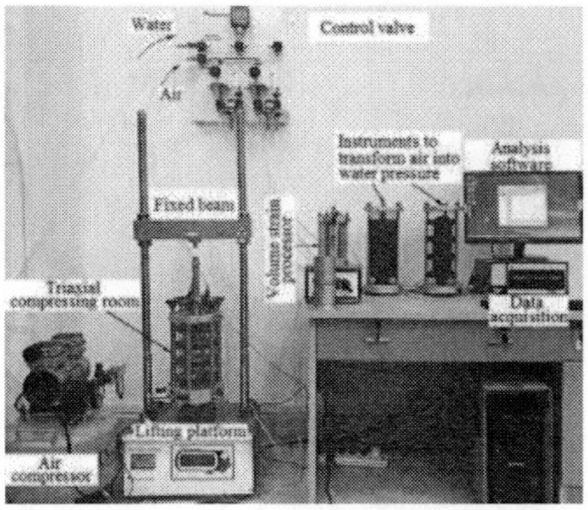

b)

Figure 4. a) Instruments of triaxial compression test
b) Sample before and after compression

a)

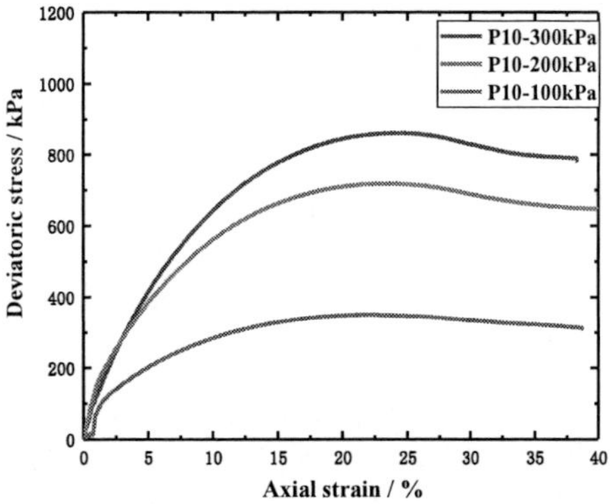

b)

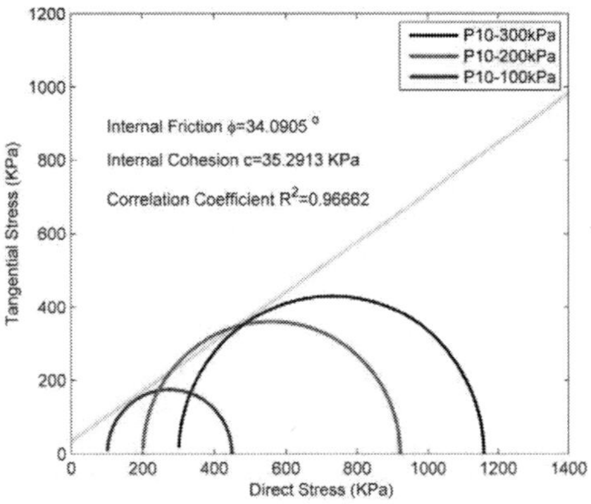

Figure 5. a) The stress-strain curve
b) Mohr's stress circle of sample P10 in the test

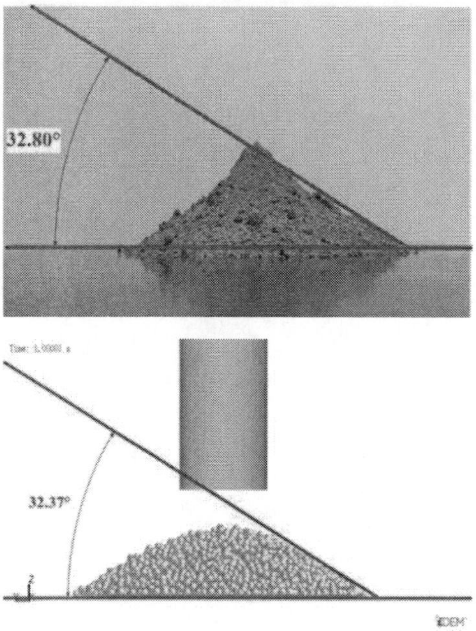

Figure 6. Test and simulation results of sample P10 repose angle

Table 4. ϕ & c of the 12 sample points in the test

Sample Points	φ_0 (°)	c_0 (*kPa*)
P12	36.91	41.05
P4	36.10	40.68
P7	35.47	37.95
P11	34.13	36.04
P10	34.09	35.29
P1	31.94	33.19
P3	31.86	32.43
P8	31.09	30.19
P2	30.64	29.59
P5	28.90	26.24
P6	28.50	25.66
P9	27.15	21.36

Table 5. Repose angle simulation results

Sample Points	SFC	Test Result (°)	Simulation Result (°)	Error (%)
P1	0.30	34.57	34.29	0.81
P2	0.20	31.77	31.54	0.72
P3	0.26	32.65	33.01	1.10
P4	0.48	37.40	37.21	0.56
P5	0.16	30.02	30.65	2.10
P6	0.14	29.98	30.07	0.30
P7	0.33	36.86	35.69	3.17
P8	0.24	35.16	35.69	1.51
P9	0.10	29.50	29.82	1.08
P10	0.33	32.80	32.37	1.31
P11	0.33	36.11	35.69	1.16
P12	0.71	38.10	37.83	0.71
			Average Error (%)	**1.21**

3.2 Further calibration of parameters using triaxial compression simulation

The soil model was generated according to the contact parameters obtained by preliminary calibration, and the triaxial compression simulation was used to further verify whether the mechanical properties were consistent. Still taking sample P10 as an example, RC=0.25, SFC=0.33, RFC=0.47 were set to the particle parameters of the triaxial simulation. Confining pressure of 100kPa, 200kPa, and 300kPa were set and the stress-strain curve & the Mohr's Stress Circle were obtained by simulation, as shown in *Fig. 7*.

Similarly, the triaxial compression simulation results of the 12 samples are shown in *Table 6* by using the Mohr's Stress Circle common tangent equation. Combined with the actual soil shear strength index obtained by the test in *Table 4*, the error between the simulation results and the test results of the triaxial compression is also reflected in *Table 6*.

According to the results of *Table 6*, **the average errors of φ & c of the triaxial compression simulation parameter further calibration are only 2.83% and 3.14% respectively**, indicating that the further calibration of the parameter fits well, and the set particles can indeed represent the actual force in the soil.

3.3 EDEM-RecurDyn coupled simulation of free-fall cone penetration

RecurDyn simulation software was used to build the model of cone penetrometer, and EDEM simulation software was used to build soil model. In DEM, the particle shape and physical parameters have a great influence on the simulation results (Sweijen et al., 2020). All the parameters are shown in *Table 7*.

The initial position of penetration and the simulation process are shown in *Fig. 8*.

The contact parameters and free-fall height of different samples were set, and the EDEM-RecurDyn coupled simulation were performed. The penetration depth of the simulation result is shown in *Table 8*. Combined with the actual penetration depth obtained by the test in *Table 1*, the error between the simulation results and the test results of the penetration depth can be obtained, which is also reflected in *Table 8*.

According to the results of *Table 8*, the calculation indicates that **the average error of the penetration simulation is 5.27%**. It shows that the simulation process can replace the actual process of penetration to some extent.

The penetration depths comparison between the simulation and the test is shown in *Fig. 9*. It can be seen from *Fig. 9* that under the same moisture content, the penetration depth increases with the increase of the free-fall height. At the same free-fall height, as the moisture content increases, the penetration depth also increases in a positive correlation.

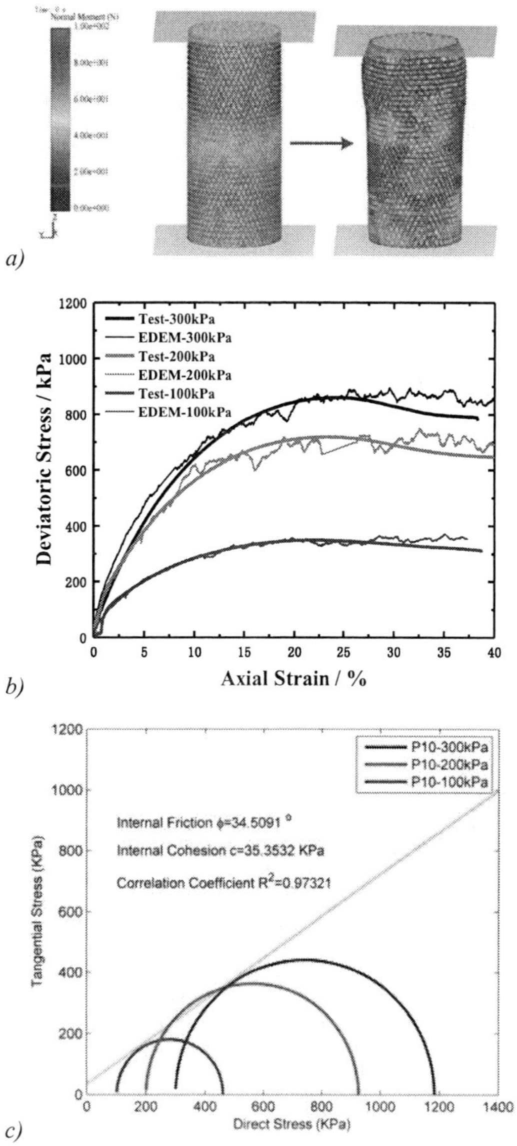

Figure 7. a) Triaxial compression simulation by EDEM b) The stress-strain curve in the simulation c) The Mohr's stress circle in the simulation

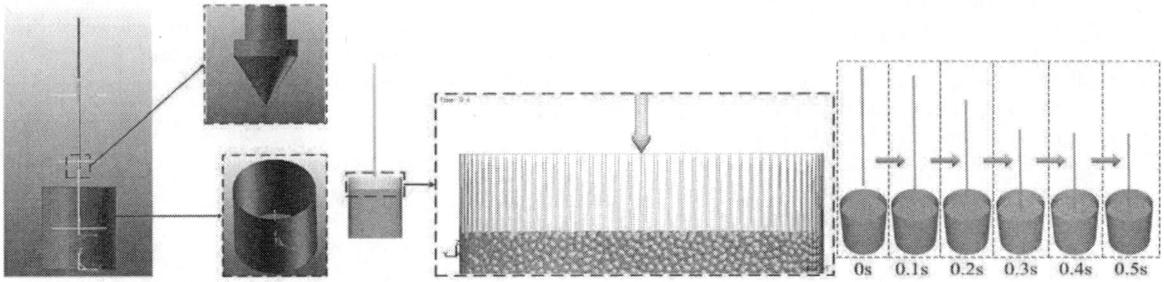

Figure 8. EDEM-RecurDyn coupled simulation

Table 6. ϕ and c of the 12 sample points in the triaxial compression simulation

Sample Points	φ (°)	φ_0 (°)	Error (%)	c (kPa)	c_0 (kPa)	Error (%)
P1	33.65	31.94	5.35	34.16	33.19	2.92
P2	31.23	30.64	1.93	30.55	29.59	3.24
P3	30.11	31.86	5.49	30.97	32.43	4.50
P4	35.89	36.10	0.58	39.40	40.86	3.57
P5	29.76	28.90	2.98	26.10	26.24	0.53
P6	30.02	28.50	5.33	25.94	25.66	1.09
P7	34.51	35.43	2.60	35.35	37.95	6.85
P8	31.55	31.09	1.48	30.88	30.19	2.29
P9	28.22	27.15	3.94	23.47	21.36	9.88
P10	34.51	34.09	1.23	35.35	35.29	0.17
P11	34.51	34.13	1.11	35.35	36.04	1.91
P12	36.18	36.91	1.98	41.34	41.05	0.71
Average Error (%)			**2.83**			**3.14**

Table 7. Parameters of the cone penetrometer model and mechanical parameters of the soil

Cone Parameter		**Soil Parameter**	
Density	7800$kg \cdot m^{-3}$	Density	2650 $kg \cdot m^{-3}$
Poisson's ratio	0.30	Poisson's ratio	0.25
Young's modulus	2.08×105MPa	Young's modulus	75MPa
Rod length	910mm	RC	0.25
Rod diameter	13mm	SFC	0.10
Column height above cone	5.68mm	RFC	0.47
Column diameter above cone	20mm	Number of particles	52794
Cone height	17.32mm	Radius of particles	5mm
Cone diameter	20mm		
Cone angle	60mm		
Cylinder diameter	400mm		
Cylinder height	500mm		

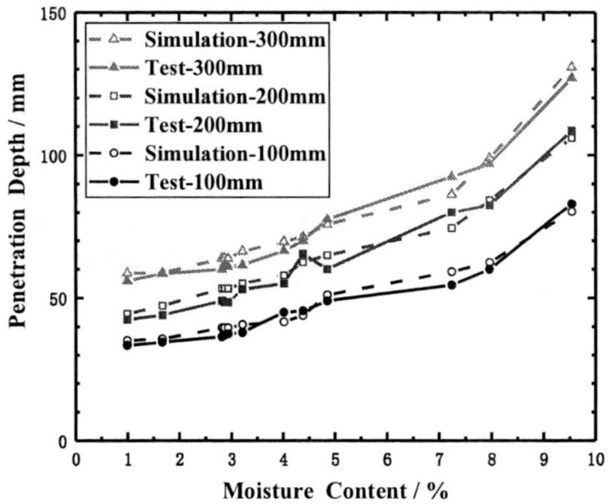

Figure 9. Penetration depths comparison between simulation and test

Table 8. Penetration depth of the simulation

Free-Fall Height		Penetration Depth (*mm*)												Average Error (%)
		P1	P2	P3	P4	P5	P6	P7	P8	P9	P10	P11	P12	
100 *mm*	**Test$_1$**	38.0	49.0	45.0	34.5	54.5	60.0	36.5	45.5	83.0	37.5	37.0	33.5	
	Simulation$_1$	40.7	51.0	41.7	35.7	59.2	62.4	39.6	43.9	80.2	39.6	39.6	35.2	
	Error (%)	7.11	4.08	7.33	3.48	8.62	4.00	8.49	3.52	3.37	5.60	7.03	5.05	**5.64**
200 *mm*	**Test$_2$**	53.0	60.0	55.0	44.0	80.0	82.5	49.0	65.5	108.5	48.5	48.5	42.5	
	Simulation$_2$	55.1	64.9	57.9	47.3	74.4	84.3	53.3	62.5	106.0	53.3	53.5	44.5	
	Error (%)	3.96	8.17	5.27	7.50	7.00	2.18	8.78	4.58	2.30	9.90	9.90	4.71	**6.19**
300 *mm*	**Test$_3$**	61.5	77.5	66.5	58.5	92.5	97.0	60.0	70.0	127.0	61.0	62.0	56.0	
	Simulation$_3$	66.3	75.7	69.7	58.7	86.3	99.0	63.8	71.4	130.7	63.8	63.8	58.7	
	Error (%)	7.80	2.32	4.81	0.34	6.70	2.06	6.33	2.00	2.91	4.59	2.90	4.82	**3.97**

4. Equation Fitting and Verification

4.1 Nomenclature

c	*Internal cohesion*	*[KPa]*
D	*Penetration depth*	*[mm]*
H	*Free-fall height*	*[mm]*
φ	*Internal friction angle*	*[°]*
ω	*Moisture content*	*[%]*

4.2 Equation fitting

Combining the data of triaxial compression test, the corresponding relationship between the moisture content and the shear strength index of the 12 groups of sample points is obtained, as shown in *Fig. 10*. It can be seen from *Fig. 10* that both φ and c have a negative correlation with the moisture content, which is close to a linear decrease relationship. Fitting the corresponding relationship between ω and φ, c in the figure to obtain the *Eq. 2*, the fitting degrees R^2 are 0.93609 and 0.90127, respectively. Fitting the corresponding relationship between the simulation results of D and ω, H in *Fig. 10* into a continuous function to obtain *Eq. 3*, the degree of fitting R^2 is 0.93561.

$$\begin{cases} \varphi = 37.255 - 1.150\omega \\ c = 42.440 - 2.280\omega \end{cases} \tag{2}$$

$$D = 41.899 - 9.045\omega + 0.082H + 1.132\omega^2 + 0.012\omega H \tag{3}$$

4.3 Verification

Equation 2 and *Eq. 3* are combined to get a quadratic *Eq. 4*, which is about the relationship between D, H and φ, c. Therefore, the mechanical properties of the soil can be inferred through the simulation results of the penetration process. It should be noted here that when there is no solution, the value of the unknown is –b/2a as the closest solution.

$$\begin{cases} 0.857\varphi^2 - (0.010H + 55.941)\varphi + (0.471H - D + 936.912) = 0 \\ 0.218c^2 - (0.005H + 14.530)c + (0.305H - D + 265.752) = 0 \end{cases} \tag{4}$$

In order to verify whether *Eq. 4* was accurate, the H, D of the 36 sets of simulation data were substituted into the equation to compare the simulation results with the actual test results. The specific verification process is shown in *Table 9*. The error between the equation derived from simulation and the test results of the soil shear strength index is also reflected in *Table 9*.

Overall, the average value of error φ & c in the fitted equation are 3.45% and 6.68%, respectively. It shows that *Eq. 3* derived from the simulation results is highly accurate, and the DEM free-fall cone penetration simulation can replace the triaxial compression test to obtain soil mechanical properties with very low error.

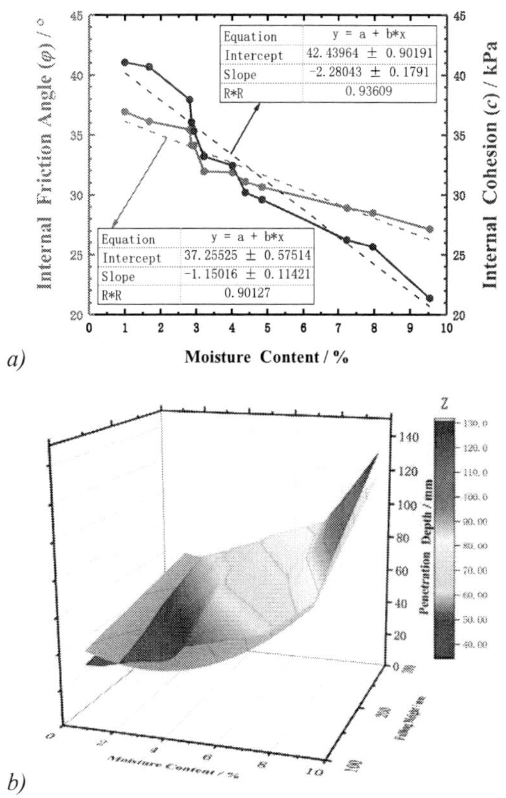

Figure 10. (a) Fitted relationship of φ, c-ω
(b) Fitted relationship of D-ω, H

Table 9. Equation consistency verification

Free-Fall Height		P1	P2	P3	P4	P5	P6	P7	P8	P9	P10	P11	P12	Average Error (%)
	φ_1 (°)	33.22	29.67	30.40	33.22	28.86	28.18	33.22	30.30	25.99	33.22	33.22	33.22	
	φ_0 (°)	31.94	30.64	31.86	36.10	28.90	28.50	35.43	31.09	27.15	34.09	34.13	36.91	
100	**Error (%)**	4.01	3.17	4.58	7.97	0.14	1.14	6.23	2.54	4.27	2.55	2.66	9.99	**4.10**
mm	c_1 (*KPa*)	32.55	27.11	28.49	34.47	25.56	24.24	34.47	28.30	19.98	33.28	34.47	34.47	
	c_0 (*KPa*)	33.19	29.59	32.43	40.68	26.24	25.66	37.95	30.19	21.36	35.29	36.04	41.05	
	Error (%)	1.94	8.37	12.16	15.26	2.58	5.52	9.16	6.26	6.48	5.69	4.35	16.02	**7.82**
	φ_2 (°)	32.61	30.71	31.87	33.80	28.07	27.82	33.80	29.80	25.67	33.80	33.80	33.80	
	φ_0 (°)	31.94	30.64	31.86	36.10	28.90	28.50	35.43	31.09	27.15	34.09	34.13	36.91	
200	**Error (%)**	2.10	0.22	0.02	6.36	2.89	2.40	4.59	4.14	5.45	0.84	0.95	8.41	**3.20**
mm	c_2 (*KPa*)	32.02	28.90	30.91	35.62	23.92	23.44	35.62	27.23	19.26	35.62	35.62	35.62	
	c_0 (*KPa*)	33.19	29.59	32.43	40.68	26.24	25.66	37.95	30.19	21.36	35.29	36.04	41.05	
	Error(%)	3.54	2.32	4.68	12.44	8.84	8.65	6.14	9.79	9.82	0.93	1.17	13.23	**6.80**
	φ_3 (°)	34.39	30.54	32.97	34.39	28.70	28.26	34.39	31.92	25.87	34.39	34.39	34.39	
	φ_0 (°)	31.94	30.64	31.86	36.10	28.90	28.50	35.43	31.09	27.15	34.09	34.13	36.91	
300	**Error (%)**	7.66	0.34	3.49	4.74	0.69	0.85	2.94	2.67	4.72	0.87	0.76	6.83	**3.05**
mm	c_3 (*KPa*)	36.77	28.49	32.52	36.77	25.05	24.20	36.77	30.93	19.57	36.77	36.77	36.77	
	c_0 (*KPa*)	33.19	29.59	32.43	40.68	26.24	25.66	37.95	30.19	21.36	35.29	36.04	41.05	
	Error (%)	10.77	3.71	0.28	9.62	4.54	5.69	3.12	2.45	8.36	4.18	2.01	10.44	**5.43**

5. Conclusions

1) Penetration depth and moisture content is positively correlated, and φ, c and moisture content are linearly negatively correlated.

2) The EDEM-RecurDyn simulation fits the penetration process of the test well.

3) The DEM free-fall cone penetration simulation can replace triaxial compression test to obtain soil mechanical properties φ & c with extremely low error, which are only 3.45% and 6.68%. A soil model database can be constructed on this basis to replace actual tests.

4) A new method to calculate the bearing capacity of soil, evaluate the stability of soil foundation, and analyze the earth pressure is proposed.

6. Acknowledgements

Sincere thanks to Professor Gao for his guidance on the idea of this paper, Xiaobo Song and Baojun Di for their assistance in the test, and Cheng Hu for his great help in the numerical simulation.

7. References

Asaf, Z., Rubinstein, D., Shmulevich, I., 2007. Determination of discrete element model parameters required for soil tillage. *Soil and Tillage Reserch*. 92, 227-242.

Cundall, P. A., Strack, O. D., 1971. A discrete numerical model for granular assemblies. *Géotechnique*. 29, 47-65.

Hu, C., Gao, J.W., Diao, J.C., Song, X.B., 2021. Numerical simulation of tire steering on sandy soil based on discrete element method. *AIP Advances*. 11, 015015.

Huang, Q., Jia, G.Y., Zhang, Y. J., 2017. *Soil Mechanics*. Science Press, Beijing, pp. 139-141.

Janda, A., Ooi, J. Y., 2016. DEM modeling of cone penetration and unconfined compression in cohesive solids. *Powder Technology*. 293, 60-68.

Konishi, T., Oida, A., Nakashima, H., Kawase, Y., 2005. Cone penetration resistance of a mesoscopic soil model. *Proceedings of the 15th International Conference of ISTVS (Hayama, Japan)*. pp. 72-79.

Lin, C. X., Tu, F. B., Ling, D. S., Hu, C. B., 2018. FEM-DEM coupled modeling of cone penetration tests in lunar soil. *Journal of Central South University*. 25, 392-405.

Nakanishi, R., Nakashima, H., Miyasaka, J., Ohdoi, K., 2020. Tractive performance analysis of a lugged wheel by open-source 3D DEM software. *Journal of Terramechanics*. 92, 51-65.

Sweijen, T, Hassanizadeh, S. M., Aslannejad, H., Leszczynski, S., 2020. The effect of particle shape on porosity of swelling granular materials: Discrete element method and the multi-sphere approximation. *Powder Technology*. 360, 1295-1304.

Tanaka, H., Momozu, M., Oida, A., Yamazaki, M., 2000. Simulation of soil deformation and resistance at bar penetration by the Distinct Element Method. *Journal of Terramechanics*. 37, 41-56.

Zhang, J. L., Hao, J. W., 2015. Numerical Analysis of Torpedo Anchor Penetration in Soil. *Journal of Water Resources and Architectural Engineering*. 13, 33-39.

Zhang, K. G., Liu, S. Y., 2010. *Soil Mechanics*. China Architecture & Building Press, Beijing, pp. 189.

Zhang, K. J., 2002. *Vehicle Terramechanics*. National Defense Industry Press, Beijing, pp. 201.

Zou, M., Xue, L., Gai, H., 2020. Identification of the shear parameters for lunar regolith based on a GA-BP neural network. *Journal of Terramechanics*. 89, 21-29.

8131 / Parameters Calibration of Red Clay Soil in Hilly Area of Southwest China for Discrete Element Simulation Based on Repose Angle Test

Le Yang, Qinghui Lai, Liangliang Zhao, Peihang Li, Zhihong Zhang, and Zhaoyang Chen*
Kunming University of Science and Technology, Kunming, China
**Corresponding author: zzh_0822@hotmail.com*
https://doi.org/10.56884/BOCW9283

Abstract

Its southwest's rugged and mountainous terrain has thick soil, which causes high resistance, poor efficiency, and sometimes even the impossibility to operate agricultural equipment. Using discrete element simulation, it is possible to optimize the agricultural machinery elements that come into contact with the soil in order to reduce drag and increase efficiency. Although there are presently no precise and trustworthy discrete element modeling parameters for red clay, the physical characteristics of red clay in the hilly and mountainous regions of the southwest are unique. As a result, in this study, the soil moisture content was 12.5% and 18.3%, respectively, for the actual working environment of the soil moisture content of 10%–20% in the hilly and mountainous areas of southwest China, and the experiment's measurement of the accumulation angle was 38.54°. This subject of the study was clay. To calibrate the appropriate model's physical characteristics, use the Hertz-Mindlin with JKR contact model in the EDEM simulation software. Prior to simulating the accumulation angle of soil particles, the intrinsic physical parameter values of the red loam soil are first obtained through actual experiments. The range of soil contact mechanical parameters in the GEMM database is then used to determine the optimal value interval of the contact parameters determined by the steepest slope test. To determine the regression model of the soil accumulation angle, the quadratic regression rotation orthogonal combination test is used to obtain the second-order regression model of the accumulation angle and the significant parameters. The significant parameters are then optimized using the actual accumulation angle as the target. In the end, it was found that the following contact mechanics characteristics worked well together in the EDEM simulation test: JKR surface energy 8 J/m2, restitution coefficient 0.35, dynamic friction coefficient 0,13, and static friction coefficient 0.56. The relative inaccuracy determined by the actual physical test is 1.80%, and the stacking angle is 39.24°. The study's findings demonstrate that the method has a high degree of calibration accuracy and is both reasonable and useful for calibrating soil discrete element simulation parameters. The pertinent calibration parameters can serve as a technological foundation for investigating machine-soil interaction and machine-tool optimization research in southwest China's mountainous regions.

Keywords: red clay, discrete element method, repose angle, parameter calibration, response surface methodology

1. Introduction

Hilly and mountainous regions in southwest of China plays a significant role in the cultivation of priceless Chinese medicines like ginseng and Panax notoginseng. It is ideal for the cultivation of some crops and Chinese medicinal ingredients because this region has an extended sunny period. One of the crucial farming steps in the southwest's hilly and mountainous regions is the farming operation before seeding agricultural crops. To decrease the adhesion resistance brought on by soil adhesion and boost production efficiency, it is crucial that high-passage crawler equipment be developed that can operate in the southwest's hilly and mountainous locations [1]. Studying the process of interaction between red clay and soil-contacting parts is the main challenge in the development of high-efficiency operating machinery. The relative motion between soil particles cannot be accomplished using the current finite element model while conducting field tests or soil tank experiments. Currently present models are based on continuous motion between media to replicate the overall behavior of soil degradation in addition to the dynamic process while engaging with the soil-contacting portions [2-3]. In recent years, discrete element method modeling has been gradually applied to agriculture. Modeling the material mechanics of minute particles is appropriate for this approach. Through this process, a single particle is divided into a number of smaller particles, each with a certain mass and form. The contact problem between various particles and their boundaries is frequently solved using it. so that its inner workings are seen [4-5]. There are numerous potential applications for the contact model between soil-soil and soil-contacting parts [6–8]. In discrete element simulations of other crops including rice, potatoes, and wheat, several contact models are typically employed [9–13]. Due to alteration in the field environment, while agricultural machinery is operating, the interactions between the soil and the contact components will also clearly differ [14]. To effectively calibrate the parameters of the chosen model and offer accurate simulation parameters for discrete element simulation experiments, different contact models are used when employing discrete elements for modeling in accordance with the real soil condition [15–18].

This research investigates the inter-particle cohesion using the "Hertz-Mindlin with JKR" contact model in the EDEM software to address the issue of the lack of precise and trustworthy discrete element simulation model parameters for the interaction mechanism between agricultural machinery and red clay in hilly and mountainous areas.

2. Parameter Calibration of Soil Particle Contact Model

2.1 Selection of Contact Models

The water content of red sticks employed in physical tests in hilly and mountainous places, respectively, was 12.5%±1% and 18.3%±1% in the study, which is based on the concept of combining physical test and EDEM simulation test.

The choice of the contact model will directly impact the accuracy of the findings since the red clay in hilly and mountainous places exhibits a hard plastic condition and the thickness of the soil layer varies substantially. The classic Hertz-Mindlin contact model cannot mimic the adhesion between soils, just its elastic deformation, making it unsuitable for red clay with adhesive forces. The Hertz-Mindlin with JKR Cohesiveness model is a cohesion contact model that is based on the Hertz-Mindlin contact model mixed with the JKR theory, although it is only relevant to the occurrence of static electricity, moisture content, and other external factors between the material's particles [19-20]. Adhesion or agglomeration that is obviously present, such as with damp soil, agricultural products, etc. Red clay has a high water content, which increases the adhesion force between soil and soil or between soil and farming equipment. The parameters in this study were calibrated based on the Hertz-Mindlin with JKR Cohesion model, which was chosen to represent the red clay in hilly and mountainous environments. (*Equation 1 shows calculates the JKR normal elastic contact force whereas Eq. 2 defines the amount of normal overlap of particles in contact with each other*)

The model calculates the JKR normal elastic contact force F_1 to describe the cohesiveness between particles.

$$F_1 = \frac{4E^*}{3R^*} - 4\alpha^{\frac{3}{2}}\sqrt{\pi\gamma E^*} \qquad (1)$$

$$\delta = \frac{\alpha^2}{R^*} - \sqrt{\frac{4\pi\gamma\alpha}{E^*}} \qquad (2)$$

where: F_1 is mean JKR normal elastic contact force; γ is mean surface tension; E^* is mean equivalent modulus of elasticity; α is mean the amount of tangential overlap of particles in contact with each other; R^* is mean equivalent contact radius.

The cohesion force created by the mutual attraction of the two particles also be calculated to approximate the model when the two particles are not in direct contact [22]. (*Eq. 3 defines the maximum clearance in the normal direction when the cohesion between particles is not zero; Eq.4 defines When the cohesion between particles is not zero, the maximum gap in the tangent direction*)

$$\delta_c = \frac{\alpha_c^2}{R^*} - \sqrt{\frac{4\pi\gamma\alpha_c}{E^*}} \qquad (3)$$

$$\alpha_c = \left[\frac{9\pi\gamma R^{*2}}{2E^*}\left(\frac{3}{4} - \frac{1}{\sqrt{2}}\right)\right]^{\frac{1}{3}} \qquad (4)$$

The intrinsic parameters of the research object, its contact parameters, and the contact model parameters are how the model parameters are primarily broken down. By reading the relevant literature, you may find out the intrinsic parameters and contact model parameters of the study item. The cohesion between particles or the cohesion between particles and the surface of soil-contacting components should be stressed when evaluating the contact model for laterite with heavy adhesion in hilly and mountainous environments. Using the EDEM software, the JKR model may be utilized to describe the cohesive force between particles. The major task of this article is to calibrate the red clay contact model experimentally using two moisture contents.

2.2 Experiment Items and Simulation Parameters

With time and climate, clay moisture content does fluctuate dramatically. In order to establish a theoretical framework for investigating the interaction between implements and soil as well as the anti-sticking and drag reduction of soil-contacting portions, two typical moisture contents of clays12.5%±1% and 18.3%±1% were chosen as the research objects in this paper. This experiment suggested in this study can also be used to conduct calibration research for other clays with a high water content. The scientific foundation for the calibration of soil characteristics with various water contents is also provided by this study. Furthermore the simulated angle of repose of soil particles is employed as the response value, and the physical parameters serve as the calibration objects. The regression model of the angle of repose is created based on the Box-Behnken response surface optimization approach, and the regression model is optimized to get the optimization of model parameters for two water contents.

Red clay with moisture contents of 12.5%±1% and 18.3%±1% is used as the research object to calibrate the discrete element parameters of the soil particle contact model and conduct the angle of repose test. The soil type is typical hilly and mountainous red clay, with a bulk density of 1156 kg/m^3. Table 1 displays the intrinsic characteristics of red clay and related minerals.

Table 1. EDEM simulation parameters of soil repose angle

Material	Parameter	Value	Source
Soil	density/kg·m^{-3}	1156.4	measured
	Poisson	0.4	measured
	Shearing model/MPa	1	literature [22]
UHMW-PE	density/kg·m^{-3}	964	literature [23]
	Poisson	0.38	literature [23]
	Shearing model/MPa	8.4×10^9	literature [24]
65Mn	density/kg·m^{-3}	7861	literature [25]
	Poisson	0.29	literature [26]
	Shearing model/MPa	7.9×10^9	literature [25-26]
PTFE	density/kg·m^{-3}	2.1×10^3	literature [27]
	Poisson	0.4	literature [28]
	Shearing model/MPa	100	literature [27-28]
245 Boron Steel	density/kg·m^{-3}	7850	literature [29]
	Poisson	0.25	literature [30]
	Shearing model/MPa	2.06×10^8	literature [29-30]

The red clay contact model parameters may be derived by importing the soil model parameters from the pre-experiment into the aforementioned material library, which is based on the pertinent properties of soil particles in the EDEM general granular material database [21]: JKR surface energy is 4~12 J/m^2, the coefficient of restitution is 0.15~0.35, the rolling coefficient is 0.05~0.2, and the static friction coefficient is 0.2~0.92.

2.3 Angle of Rest Experiment

2.3.1 Soil Moisture Determination

Experiments were conducted in the laboratory of modern equipment engineering, Faculty of Modern Agricultural Enginerring, Kunming University of Science and Technology. The collected dirt is screened by a vibrating pendulum sand screener to reduce the soil particle size range to less than 1 mm after contaminants such big clods of soil and stones are removed. To equally spread the moisture inside the soil, wet, agitate, and wrap it in plastic wrap for a week. Then, use a moisture content measuring equipment (YF-100F, Puyan Trading Co., Ltd., Dongguan) to take multiple measurements. The soil then returns to its natural state of accumulation after a random measurement to ensure water distribution is uniform.

The experiment took place in a largely enclosed internal setting. The error of the measured moisture content was allowed to be 1% based on the soil moisture determination method (GB 7172-87), and the drying method was chosen to determine the soil moisture content parameters. A cutting ring with a specific volume was also used at the same time for sampling. In a state of natural accumulation, the earth is gathered into the ring knife and fills it.

Equipment are as follows:

1. Drying box (HG101-2A): Electric heating drying box should be able to regulate the temperature between 105 and 110 degrees Celsius.
2. Electronic balance: weighing more than 100g. Sensitivity 0.01g.
3. Others: Vernier calipers, crucible tongs and cutting ring, etc.

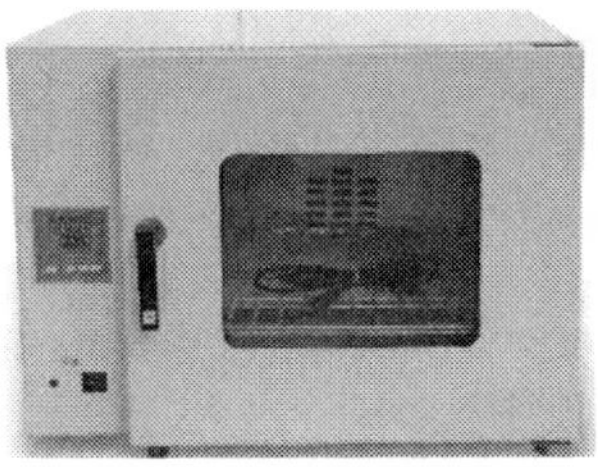

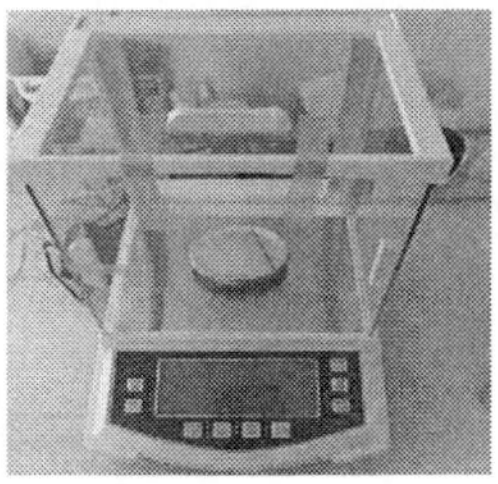

a) HG101-2A Drying box *b) Electronic balance*
Fig.1 Main test equipment

After cleaning, heating, and drying the cutting ring, weigh it using an electronic scale and note the mass as M_1. Using a cutting ring to collect soil samples; the soil fills the ring knife naturally while it is being accumulated, and minimal pressure is applied to it during the sampling process. Re-weigh the dirt sample and the ring knife using the electronic balance, and then note their combined mass as M_2. In the drying box, place the cutting ring and soil sample, set the temperature to 105 °C, and run the drying process for at least 24 hours. The weighing process is carried out when the drying is finished, the cutting ring is removed every hour, and it is repeated until the number of electronic days does not change for two consecutive times. According to the formula, the total weight M3 of the sample, the ring cutter, and the soil moisture content may all be calculated (5).

$$\omega = \frac{M_2 - M_3}{M_3 - M_1} \times 100\% \quad (5)$$

Where : ω indicates soil moisture content; M_1 indicates the quality of the cutting ring; M_2 indicates the total mass of the ring knife and soil before drying, M_3 indicates the total mass of the ring knife and soil after drying

2.3.2 Physical test of angle of repose

When soil is supposed to settle freely into a conical funnel, it assumes the shape of cones. The calibrated angle of repose is the interior angle created between the surface of the earth and the horizontal tabletop. In order to describe the flow, friction, and other properties of the material, the size of the angle of repose is frequently utilized. Such an angle of repose is connected to the density, surface area, and friction coefficient of the material to be calibrated. So, the majority of granular materials are expressed by the angle of repose in the discrete element modeling of granular materials. For measuring the angle of repose, follow these simple steps: Prepare heavy red clay with a moisture level of 12.5%±1%, and 18.3%±1% respectively. The physical experiment involves the test bench (Fig. 1), measuring the soil's accumulation angle with a protractor, repeating the process 10 times, and taking the red clay's average result, which comes out to 38.21°.

a) Moisture content 12.5%

b) Moisture content 18.3%
Figure 2. Physical test of repose angle of red clay soil

2.4 Simulation Investigation of Response Angle

The response surface optimization approach was applied, with the soil angle of repose being utilized as the response value, based on the reference value range of the soil model discussed above. Instead, Box-Behnken technique was employed to carry out the calibration test based on the Design-Expert software. Table 2 shows each measurement factor's level range.

In the simulation test, the angle of repose established using the same parameters was repeated five times. Images of the simulation model were intercepted in the X and Y directions of the accumulation angle simulation model produced by EDEM and imported into AutoCAD [21]. Table 3 displays the measurements' findings.

Table 2. Horizontal factor coding for simulation tests with angle of rest

Level	JKR surface energy x_1	Restitution coefficient x_2	Friction coefficient x_3	Rolling friction coefficient x_4
Low Level	4	0.15-	0.2	0.05
Center Level	8	0.35	0.56	0.125
High Level	12	0.35	0.92	0.2

Table 3. Experimental design and results of soil rest angle simulation

Test No.	x_1	x2	x3	x4	Response angle /(°)
1	4	0.35	0.56	0.2	30.45
2	8	0.35	0.56	0.125	38.52
3	4	0.35	0.56	0.05	16.35
4	12	0.55	0.56	0.125	26.82
5	8	0.55	0.92	0.125	24.56
6	12	0.35	0.56	0.2	54.48
7	8	0.15	0.56	0.05	43.86
8	12	0.15	0.56	0.125	42.56
9	8	0.35	0.56	0.125	37.45
10	8	0.35	0.56	0.125	39.51
11	4	0.15	0.56	0.125	46.21
12	4	0.35	0.2	0.125	25.48
13	8	0.55	0.56	0.05	12.25
14	4	0.35	0.92	0.125	25.16
15	8	0.35	0.92	0.05	33.84
16	12	0.35	0.92	0.125	51.47
17	8	0.35	0.92	0.2	40.23
18	8	0.55	0.2	0.125	16.45
19	8	0.35	0.2	0.05	31.20
20	12	0.35	0.2	0.05	40.26
21	8	0.35	0.56	0.05	47.25
22	4	0.55	0.56	0.125	11.78
23	12	0.35	0.2	0.125	48.89
24	8	0.55	0.56	0.2	25.53
25	8	0.15	0.56	0.2	46.72
26	8	0.35	0.56	0.125	36.21
27	8	0.15	0.2	0.125	46.56
28	8	0.35	0.56	0.125	35.48
29	8	0.15	0.92	0.125	43.14

3. Calibration of Simulation Parameters

Based on the above test values of the angle of repose, the slope rolling, the static friction, and the collision recovery experiments were used to acquire the reliable and precise simulation parameters of the red clay in the hilly and mountainous areas of the southwest, to maximize the drag reduction performance of agricultural machinery, and to give matching parameters for the design of bionic materials. Tests to confirm the range of parameters discovered using the test for the angle of repose. For 245 boron steel and three other drag-reducing materials, (UHMW-PE, PTFE, 65Mn) the soil-contacting material for agricultural machinery, a static friction test was conducted to determine the typical soil-contacting surface materials of red clay in hilly and mountainous places under four moisture contents. According to the results of the measured static friction coefficient, the slope rolling test was used to calibrate the discrete element parameters of red clay under high moisture content (18.3%±1%) on the surface of different soil-contacting materials.

3.1 Slope Experiments for Rolling Coefficient of Friction

As increasing water contents, the soil adhesion at various water contents (water contents below the soil liquid limit) steadily improves [31]. Using the physical test rig depicted in Figure 2, the static friction coefficients of two types of typical earth-contacting component materials under high moisture content were calculated. With a 10 mm radius, the 3D-printed spherical model was utilized to represent soil particles. A ruler is used to measure the ball's rolling distance S_2 in the horizontal direction as it rolls down from its static position at the top of the inclined plane until it stops. This process is repeated five times, and the average value is taken as the rolling distance in the horizontal direction. Simultaneously, its simulation experiment was done in the EDEM software, and the parameter values match those of the actual test. Key parameters of the slope created by the experimental tests are provided in Table 4 under the results of the physical test.

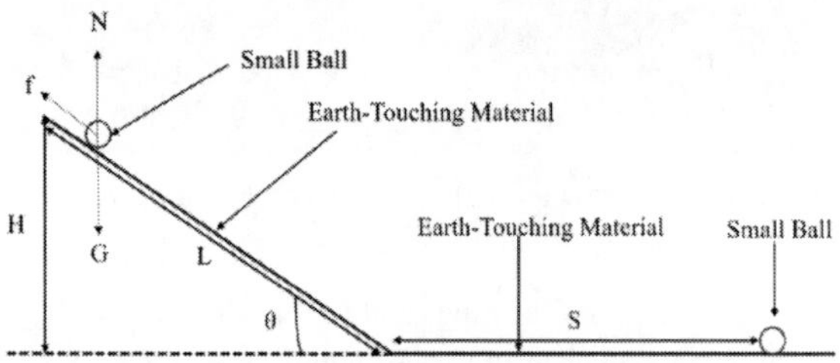

a) Schematic of the rolling coefficient of friction

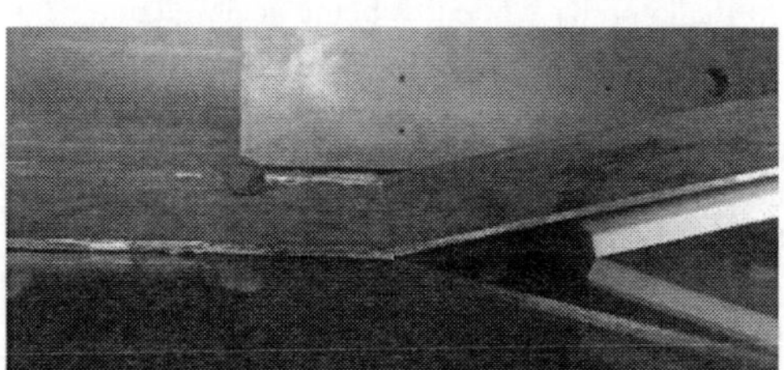

b) Physical tests

Figure 3. Measure the coefficient of friction of rolling

Table 4. Parameters related to inclined test benches

Material	Bevel length [mm]	Bevel height [mm]
65Mn	220	25
PTFE	180	25
UHMW-PE	140	25
245 Boron Steel	100	25

The parameters in the experiment are designed for the calibration test using the Box-Behnken method in the Design-expert software, referring to the relevant parameters of plastic and steel in the EDEM material database, corresponding to the relevant literature and the results of the actual test, and exporting the soil. Table 5 shows the slope simulation test factors and levels between the 245 boron steel and the PTFE plate.

Figure 4. Simulation test of soil and different soil-touching materials

The parameters in the experiment are designed for the calibration test using the Box-Behnken method in the Design-expert software, referring to the relevant parameters of plastic and steel in the EDEM material database, corresponding to the relevant literature and the results of the actual test, and exporting the soil. Table 5 shows the slope simulation test factors and levels between the 245 boron steel and the PTFE plate

4. Experimental Results and Analysis

4.1 Angle of Rest Regression Model

The analysis of variance function of the Design-Expert software is used to examine the data in Table 2, and the results are shown in Table 3. Information in Table 6 is then subjected to multivariate regression analysis. The red clay regression model for the southwest hilly mountain region is shown below.

The variance analysis of the response values R_1 is shown in Table 6. According to the data in the table, we can know the significance of factors and their interaction effect. P-value for R_1 equation model is less than 0.0001. The determination coefficients R_1 is 0.9964, and the adjusted determination coefficient $Radj^2$ is 0.980, respectively. These results indicate a highly significant ($P < 0.01$) relationship between the model response value (angle of repose) and the independent variables. These data illustrate that the model has good reliability and accuracy. Therefore, Eq. 5 could truly and reliably reflect the variation law of the angle of repose was used to predict the accumulation angle under different water contents.

$$R_1 = 37.43 + 11.99x_1 - 13.20x_2 + 5.61x_4 + 4.67x_1x_2 + 2.89x_2x_3 + 2.6x_2x_4 - 2.42x_3x_4 - 5.4x_2^2 - 9.14x_1x_2^2 \quad (6)$$

4.1.1 Optimization and verification of red clay rest angle parameters with different moisture content

A group with an angle of repose that is most comparable to the angle of repose obtained in the physical test is obtained using the parameter optimization function in the Design-expert software, utilizing the angle of repose of 38.21° formed by the red clay with a moisture content of 12.5%±1% in the EDEM simulation test as the optimization target. The data, which are the size of the static friction coefficient, collision recovery coefficient, JKR surface energy, and sliding friction coefficient of the red clay particles, are 0.711, 0.301, 8.156 J/m^2, and 0.113, respectively, and the angle of repose at this time is 38.982°; the error with the actual test results is 2.02 percent. Figure 3 displays the ratio of the actual test pile type to the optimized simulation test of the angle of repose. It is clear from comparing the simulation test and the actual measurement test that they are very similar. The collision restitution coefficient, followed by the kinetic friction coefficient and surface energy, is the most important of the four variables used to calibrate angle of repose.

The difference between the physical experimental data and EDEM simulation test results for the red clay used in the experiment, which had a moisture content of 12.5%±1%, is within 3%, and the parameter values of the simulation test's results within the error range are consistent with the results of the physical test, namely JKR. Moreover, static friction coefficient is 0.6~0.8, collision recovery coefficient is 0.2~0.5, and rolling friction coefficient is 0.125~0.19; surface energy is 7~10 J/m2. Similar to the previous example, for the red clay with a moisture content of 18.3%±1%, applying angle of repose 42.21° as the optimization target of the simulation test, the JKR surface energy between soils can be calculated to be 8.964 J/m2, static friction coefficient is 0.889, and soil particle recovery can be achieved. The figure below, which is similar in angle and accumulation model. Also, the coefficient is 0.348, and the coefficient of rolling friction is 0.162. Similar to the simulation test parameters, the red clay's moisture content ranges from 18.3% ± 1%, which means that the JKR surface energy is 8–11 J/m2, the static friction coefficient is 0.6–0.8, the collision recovery coefficient is 0.2–0.5, and the rolling friction coefficient is 0.135–0.19.

Table 5. Factors and levels of slope simulation test

Level	Material	JKR surface energy x_1	Restitution coefficient x_2	Friction coefficient x_3	Rolling friction coefficient x_4
Low Level	245 boron steel	2.00	0.20	0.20	0.03
	PTFE	1.00	0.10	0.10	0.03
Center Level	245 boron steel	7.00	0.45	0.45	0.055
	PTFE	5.50	0.40	0.40	0.055
High Level	245 boron steel	12.00	0.70	0.70	0.08
	PTFE	10.00	0.70	0.70	0.08

Table 6. Analysis of variance for soil angle of rest regression model

Source	Squares	Degree of freedom	F Value	P Value
Model	3910.82	22	69.21	< 0.0001**
x_1	574.56	1	223.78	< 0.0001**
x_2	696.96	1	271.45	< 0.0001**
x_3	4.80	1	1.87	0.2207
x_4	125.89	1	49.03	0.0004**
x_1x_2	87.33	1	34.01	0.0011**
x_1x_3	2.10	1	0.8189	0.4004
x_1x_4	0.0036	1	0.0014	0.9713
x_2x_3	33.47	1	13.03	0.0112 *
x_2x_4	27.14	1	10.57	0.0174 *
x_3x_4	23.33	1	9.09	0.0236 *
x_1^2	6.21	1	1.92	0.2149
x_2^2	122.93	1	64.71	0.0002**
x_3^2	6.71	1	2.10	0.1971
x_4^2	0.8586	1	0.7885	0.4087
$x_1^2x^2$	0.8646	1	0.3367	0.5828
$x_1^2x^3$	5.51	1	2.15	0.1932
$x_1^2x_4$	4.32	1	1.68	0.2421
$x_1x_2^2$	166.99	1	65.04	0.0002**
$x_1x_3^2$	0.3961	1	0.1543	0.7081
$x_2^2x_3$	10.19	1	3.97	0.0934
$x_2^2x_4$	4.96	1	1.93	0.2139
$x_2x_3^2$	2.15	1	0.8385	0.3951
$x_1^2x_2^2$	1.40	1	0.5015	0.5015
Residual	14	6		
Lack of Fit	3.20	2	0.8513	0.4920
Pure Error	10.81	4		
Total	3924.83	28		

*Note: ** indicates extremely significant impact (P<0.01), and * indicates significant impact (0.01<P<0.05)*

4.2 Regression Mode for Angle of Rest

Figure 4 displays the results of the static friction measurement test, which was used to determine the static friction coefficients for various water contents. The figure shows that the static friction coefficient is positively correlated with the soil moisture content for the material of typical soil-contacting parts in the plastic range of the soil, and the increase range is fairly obvious. Its coefficient of static friction and the rise in water content do not appear to be correlated in a linear fashion for plastic earth-contact parts. Red clay is still able to slide freely even when the water content is above the range that makes it plastic. Indeed the friction coefficient also does not significantly increase.

Figure 5 demonstrates how the vertical (90°) 245 boron steel adheres to the red clay with a moisture content of 25%. Figure 9 shows that the soil has completely adhered when the moisture content is 25% (less than the red clay's liquid limit). As well the soil ball remains stationary on the 245 boron steel plate even after the physical test bench's angle is increased by more than 90 degrees, proving that the soil ball's gravity is significantly lower than the force that holds it to the plate. For typical plastic soil-contacting parts with low surface energy, when the moisture content is 25%, they can still slide freely within a certain lifting angle range, indicating that tires or tracks with smaller surface energy have a certain effect of reducing adhesion and desorption. It can also provide some ideas for the design of biomimetic viscosity reduction and desorption.

a) Moisture content 12.5%

b) Moisture content 18.3%
Figure 5. The moisture content is 12.5% and 18.3% soil rest angle

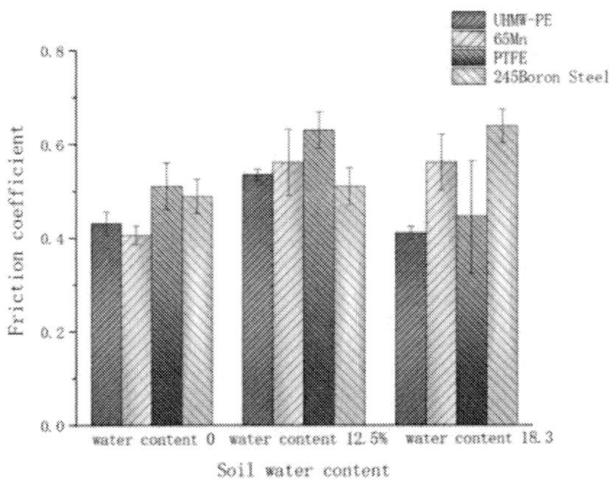

a) The coefficient of static friction when the moisture content does not exceed 25%

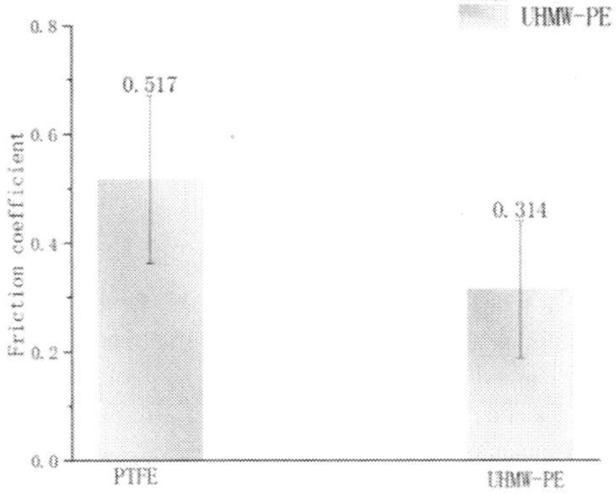

b) Moisture content 25% ±1%
Figure 6. Static friction coefficient between red clay with different moisture content and different soil-touching materials

Figure 7. Red clay with a moisture content of 25% adheres to a vertical 245 boron steel plate

5. Conclusions

The Hertz-Mindlin with JKR Cohesion contact model in EDEM was used in this study to calibrate the parameters of the red clay under the two moisture contents based on the actual operating soil moisture content of agricultural machinery in the hilly and mountainous parts of southwest China. Basic contact model parameters of the red clay beneath the two water contents were derived using the response surface optimization approach, and the regression equation of the angle of repose was obtained for comparison:12.5%±1%, the JKR surface energy was 8.156 J/m^2, the rolling friction coefficient between soil particles was 0.113, the coefficient of restitution was 0.301, and the coefficient of static friction was 0.711; With a moisture content of 18.3%±1%, soil particles' JKR surface energy is 8.964 J/m^2, their coefficient of restitution is 0.348, their coefficient of static friction is 0.889, and their rolling friction with other soils is 0.964 J/m^2,its value is 0.162.The rolling friction coefficient between the red clay particles (or between the particles and the material of the soil-contacting parts) and the particles (or the contact between the particles) is determined through the regression model variance analysis of the angle of repose of the red clay in the southwest hilly and mountainous areas under two moisture contents. According to the physical test phenomenon, there is a strong correlation between the surface energy parameters of the soil component material and the contact model. It can be seen that the calibration results under the two different water contents are consistent with the actual situation. Therefore, for the red clay with a moisture content of 10% to 20%, when the relevant discrete element simulation test is carried out, the corresponding parameter values can take values between the upper and lower limits.

Under various moisture contents, the static friction coefficients of four typical soil-contacting part materials were measured. For 245 boron steel, which is used in more earth-contacting parts, the increase in static friction coefficient with increased moisture content is more noticeable. The adhesion between the 245-boron steel and the soil is already greater than its own gravity when the soil moisture content is higher than 25%, and the soil ball will not slide relatively. For two common types of plastic soil-contacting parts, the static friction coefficient gradually rises as soil moisture content rises, but the rise is gradual. When the soil moisture content is greater than 25%, the soil can still slide freely and its friction coefficient does not increase significantly. For PTFE sheet, due to its extremely low surface energy, it has good debonding and debonding properties. The adhesion reduction and desorption effect of low surface energy contact parts is also better.

6. Acknowledgements

This study was financially supported by a program of China's National Natural Science Foundation Project (Grant No. 52065031, 51605210). The authors declare no conflict of interest.

7. References

[1] Yan X L, Wang C, Li M, et al. 2019, Design and experiment on the chisel-shape deep-shovel in red loam. J. EI. Commun., 41(4):135-142.

[2] Mustafa U, John M Fielke. 2014, 3D DEM tillage simulation: Validation of a hysteretic spring (plastic) contact model for a sweep tool operating in a cohesionless soil. J. Sci. Commun., 220-227.

[3] Zhang R, Li J Q, Zhou C H, et al. 2007, Simulation of dynamic behavior of soil ahead of the bulldozing plates with different surface configurations by discrete element method. J. EI. Commun. 13-19.

[4] Mustafa Ucgul, John M Fielke, Chris Saunders. 2015, Three-dimensional discrete element modeling DEM of tillage accounting for soil cohesion and adhesion. J. SCI. Commun: 298-306.

[5] Wang X L, Hu H, Wang Q J, et al. 2017, Calibration method of soil contact characteristic parameters based on DEM theory. J. EI. Commun 48(12): 78-85.

[6] Ma Y H, Ma S S, Jia H L, et al. 2014, Measurement and analysis on reducing adhesion and resistance of bionic ripple opener. J. EI. Commun 30(5): 36-41.

[7] Zheng K, He J, Li H W, et al. 2016, Research on polyline soil-breaking blade subsoiler based on subsoiling soil model using discrete element method. J. EI. Commun, 47(9): 62-72.

[8] Deng J Y, Hu J, Li Q D, et al. 2016, Simulation and experimental study on the subsoiler based on EDEM discrete element method. J. EI. Commun, 37(4): 14 - 18.

[9] Lu F Y, Ma X, Tan S Y, et al.2017, Simulative calibration and experiment on main contact parameters of discrete elements for rice bud seeds. J. EI. Commun, 49(2): 94-99.

[10] Wang Y X, Liang Z J, Zhang D X, et al. 2016, Calibration method of contact characteristic parameters for corn seeds based on EDEM. J. EI. Commun, 32(22): 36-42.

[11] Liu F Y, Zhang J, Li B, et al. 2016, Calibration of parameters of wheat required in discrete element method simulation based on repose angle of particle heap. J. EI. Commun, 32(12): 247-253.

[12] Liu F Y Zhang J, Chen J. 2018, Construction of visco-elasto-plasticity contact model of vibratory screening and its parameters calibration for wheat. J. EI. Commun, 34(15): 37 -43.

[13] Peng F, Wang H Y, Fang F, et al. 2018, Calibration of discrete element model parameters for pellet feed based on injected section method. J. EI. Commun, 49(4): 140-147.

[14] Zhang T, Liu F, Zhao M Q, et al. 2017, Measurement of physical parameters of contact between soybean seed and metering device and discrete element simulation calibration. J. EI. Commun, 22(9): 86-92.

[15] Xing J J, Zhang R, Wu P, etc. 2020, Parameter calibration of discrete element simulation model of brick red soil particles in Hainan thermal zone. J. EI. Commun, 36(05):158-166.

[16] Yu J Q, Fu H, Li H, et al. 2005, Application of discrete element method to research and design of working parts of agricultural machines. J. EI. Commun, 21(5): 1-6.

[17] Santos K G, Campos A V P, Oliveira O S, et al. 2015, Dem simulations of dynamic angle of repose of acerola residue: a parametric study using a response surface technique J. SCI. Commun, 1(2):11326-11333.

[18] Mohammad Almagzoub, Mohammad,Carlo Menon. 2010, DEM Numerical Simulation of Abrasive Wear Characteristics of a Bioinspired Ridged Surface J. SCI. Commun, 7(02):175-181.

[19] Shi L R, Zhao W Y, Sun W. 2017, Soil particle contact model and parameter calibration of farmland in arid areas of Northwest China based on discrete elements. J. EI. Commun, 33(21):181-187

[20] Wang X L, Zhong X K, et al. 2021, Calibration of no-till soil parameters based on discrete element nonlinear elastoplastic contact model. J. EI. Commun, 37(23):100-107.

[21] Li J W, Tong J, Hu B, et al. 2019. Discrete element simulation parameter calibration of interaction between clay-weight black soil and contact parts with different moisture content. J. EI. Commun, 35(06):130-140

[22] Wu T, Huang W F, Chen X S, et al. 2017, Calibration of discrete element model parameters for cohesive soil considering the cohesion between particles. J. EI. Commun, 38(3): 93-98.

[23] DEM 2.5 theory reference guide[R/OL]. 2014-12-05. [2015-06-07]. http://www.docin.com/p-980174717. html.

[24] Xiong P Y, Yang Z, Sun Z Q, et al. 2018, Simulation analysis and experiment for three-axis working resistances of rotary blade based on discrete element method. J. EI. Commun, 34(18): 113-121.

[25] Shi M L. 2021. Preparation and structural properties of UHMWPE composites[D]. J. EI. Commun,

[26] Mustafa Ucgul, Chris Saunders, John M Fielke. 2017.Discrete element modelling of tillage forces and soil

movement of a one-third scale mouldboard plough. J. SCI. Commun, (155): 44-54.

[27] J Y Sun, Y M Wang, Y H Ma, et al. 2018. DEM simulation of bionic subsoilers (tillage depth >40 cm) with drag reduction and lower soil disturbance characteristics. J. SCI. Commun, (119): 30-37.

[28] Teng B. Study on Face-on-Face Torsional Tribological Interfacial Behaviors[D]. J. EI. Commun, 2014.

[29] Hu C M, He H L, Hu S S. 2003. Study on the dynamic mechanical properties of No. 45 steel. J. EI. Commun t, (02):188-192.

[30] Ren L Q et.al.2011, Soil adhesion mechanics[M]. Beijing.

[31] Li H J. 2007. Application research on mechanized broad bean on-demand sowing technology[J]. J. EI. Commun (09):187-188.

8349 / The Effect of Integrating a Bio-Inspired Convex Structure with a Low-Surface Energy Polymer on Soil Adhesion and Friction

Abouelnadar Salem[a,b], Guozhong Zhang[a]*, and Hongchang Wang[a]

[a]*College of Engineering, Huazhong Agricultural University, Wuhan, 430070, China*

[b]*Soil Conservation Department, Desert Research Center, Cairo 11753, Egypt*

**Corresponding author: zhanggz@mail.hzau.edu.cn*

https://doi.org/10.56884/GULN6927

Abstract

The capacity of soil-burrowing animals to move freely in sticky soil is a motivational trait for developing soil engaging tools with high operational efficiency. The hydrophobicity and morphological profiles of soil animals' skin were reported to be the key pillars in producing their anti-adhesive mechanisms. Ultra-high molecular weight polyethylene (UHMW-PE) possess outstanding corrosion resistance, hydrophobicity, and chemical stability, which qualify it as a potential choice in soil adhesion reduction. Hence, this study aimed to investigate the feasibility of integrating a domed surface inspired by the micro-convex structure of the dung beetle skin with the UHMW-PE as a surface coating for soil engaging components in terms of soil adhesion reduction. The sliding resistance of three sliding plates (flat plate of carbon steel, flat plate of UHMW-PE, and domed plate of UHMW-PE), entirely identical in the projected area, was evaluated in two soil textures of silty clay and sandy clay loam, at four moisture levels of 18, 23, 28, and 33% and four drag speeds of 0.15, 0.2, 0.25, and 0.3 m s-1 using a completely randomized design. The dimensions of the embossed domes on the tested plate surface and their distribution pattern were established based on the previously published structural optimization of the bioinspired convex surface. In each treatment, the tested plate was dragged for 0.7 m of the soil bin length, and the sliding resistance was recorded continuously using the distributed stress and strain test and analysis system (DH3820 N). The coefficients of adhesion and friction were calculated according to the Mohr-Coulomb failure criterion. The variance analysis revealed that all investigated parameters significantly affected coefficients of adhesion and friction. In addition, as compared to flat steel plates, UHMW-PE coated plates exhibited much lower adhesion in all treatments, paving the way for practical applications in soil adhesion reduction and soil engaging component optimization.

Keywords: soil adhesion, external friction coefficient, ultra-high molecular weight polyethylene, bio-inspired surface

Editor's note: full-width tables and figures located at end of paper.

1. Introduction

Adhesion is considered a major issue when mechanizing agriculture production systems, especially in wet soils, where soil particles cling to soil-engaging components, increasing moving resistance. This naturally occurring phenomenon has several negative repercussions for farming operations, including reduced germination rates (Goshtasb, Fielke 2009), up to a 30% increase in energy consumption (Ren, Tong 2001), and a restriction on no-tillage planting in sticky soils (Khosravani, Desbiolles 2011). Another effect of soil adhesion is a 20-30% reduction in the operating efficiency of loading and excavation equipment (Ren, Han 2006).

The largest segment of the energy consumed in mechanizing agricultural production systems is squandered trying to overcome soil adhesion and friction at the soil-tool interface. Therefore, several approaches were proposed to address soil adhesion, including the continuous air or water injection into the contact interface (Li, Kou 2010, Zhang, Chen 2016), thermal desorption (Azadegan and Massah 2012), mechanical vibration (Liu, Xia 2022), mechanical detachment (Liu, Zhao 2018), electro-osmosis (Heuser, Spagnoli 2012), soil-engaging component coating (Barzegar, Hashemi 2016, Marani, Shahgholi 2019), and biomimetic concept of similarity (Guan, Fu 2022, Soni, Salokhe 2007).

In most soil-engaging components of farming machinery, soil slides tangentially over the tool surface. The shear stress at the soil-tool contact interface is a function of soil-tool friction, adhesion, and normal pressure applied to the contact interface (Barzegar, Hashemi 2016, Marani, Shahgholi 2019, Salem, Wang 2021). During dynamic equilibrium, the adhesion and external friction angle can be calculated simultaneously based on the numerical values of the tangent and vertical forces at the soil-tool contact interface using equation 1. The effect of surface structural optimization on soil-tool adhesion has been extensively studied (Guan, Fu 2022, Salem, Zhang 2022, Soni, Salokhe 2007, Zhang, Huang 2013). They concluded that adequately designed bio-inspired surfaces could significantly reduce the sliding resistance at soil-tool contact interface compared with ordinary surfaces.

$$\tau = C_a + \mu \sigma \qquad (1)$$

Here, τ is the shear stress at the soil-solid contact interface, N/m^2, C_a is the coefficient of adhesion, N/m^2, μ is the soil-solid coefficient friction, dimensionless, and σ is the normal stress on the soil-solid contact interface, N/m^2.

Several studies have been reported on employing the low-surface energy polymers as a surface coating to combat soil adhesion (Barzegar, Hashemi 2016, Marani, Shahgholi 2019, Nalbant and Palali 2011). However, the practical utility of organic polymers as a surface coating for soil-engaging components to reduce soil adhesion remains limited due to their poor abrasion resistance against soil, particularly in harsh farming environments. Unlike many other organic polymers, ultra-high molecular weight polyethylene (UHMW-PE) has proved its operating efficiency and sustainability in many practical applications due to its intrinsic characteristics such as corrosion resistance, hydrophobicity, biocompatibility, and chemical stability. These exceptional properties qualified UHMW-PE as a potential choice in many engineering applications, including soil adhesion reduction and medicine as an artificial substitute for human joints (Chirende, Li 2010, McGloughlin and Kavanagh 2000).

Barzegar et al. examined the draught force of two thin tines that were identical in shape and size, one covered with a thick layer of UHMW-PE and the other made of polished steel. The performance evaluation was conducted in a soil bin under different moisture content and compaction conditions. The results revealed that the UHMWPE coated tine significantly reduced the draught force in all treatments. In the same context, Marani et al. used a simple slider system to evaluate the sliding resistance of nanocoated plates against the St37 steel plate under different operating and soil conditions. The results showed that, compared to St37 steel plates, the nanocoated plates reduced adhesion coefficient by up to 9 times while reducing external friction angle by up to 2.5 times.

Salem et al. conducted a structural optimization study on a bio-inspired domed surface inspired by the microconvex structure of the dung beetle skin. They reported that a surface coverage ratio of 60%, dome height of 5 mm, and dome base diameter of 20 mm were the best bio-inspired domed surface geometries for minimizing shear stress at the soil-sample interface under paddy field conditions. This study aimed to validate the feasibility of combining the optimized geometries reported by Salem et al. with ultra-high molecular weight polyethylene as a surface coating for soil-engaging components. For this purpose, three sliding plates, equal in the projected area, were evaluated in terms of their coefficients of adhesion and friction under different operating and soil conditions.

2. Materials and methods

2.1 Experiment design and fabrication of tested samples

Three sliding plates, equal in the projected area of 0.0805 m^2 (*Fig. 1*), were evaluated in two different soil textures (Table 1), at four different moisture levels of 18, 23, 28, and 33% and four drag speeds of 0.15, 0.2, 0.25 and 0.3 m s^{-1}. The domes were distributed on the plate surface using a parallelogram pattern. The dimensions of the domes on the tested plate surface were established based on the previously published structural optimization study of the bio-inspired domed surface (Salem, Wang 2021). A completely randomized design with three replications was adopted to conduct the current experiments, producing 288 data sets.

a

b

c

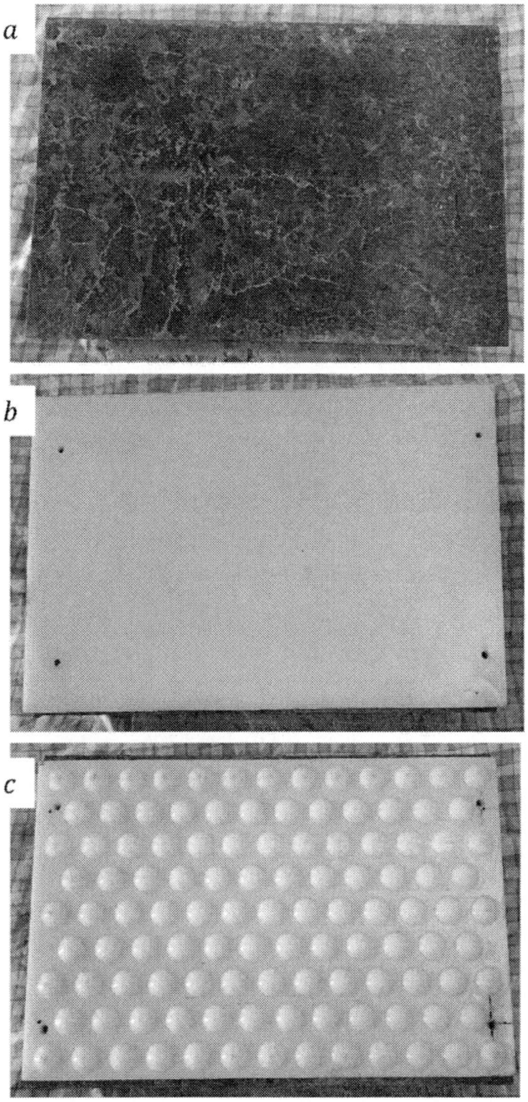

Figure 1. The tested samples: (a) flat plate of carbon steel; (b) flat plate of UHMW-PE; (c) domed plate of UHMW-PE

Table 1. Experimental soil consistency limits (%)

Soil texture	Liquid limit	Plastic limit
Silty clay	42	17
Sandy clay loam	33	18

Soil liquid and plastic limits were determined using the liquid and plastic limits tester (LP-100D) following GB/T50123-2019 geotechnical test method standard. The test concept is based on the relationship between cone penetration and soil shear strength (Hansbo 1957). The liquid limit of a particular soil is the moisture content at which the undrained shear strength of that soil is 1.7 KPa, whereas the plastic limit is the moisture content corresponding to the undrained shear strength of 170 KPa (Wroth and Wood 1978).

$$C_u = KQ/h^2 \quad (2)$$

Here, C_u is the undrained shear strength, Q is the cone's weight, h is the cone's penetration depth into the soil, and K is constant, which is wholly influenced by the cone angle. This study used the 0.59 N-60° cone assigned to soft soils, with a K value of 0.3 for remolded soil. The liquid and plastic limits were determined as the moisture content values corresponding to a 10.2 mm and 1 mm penetration distance on the best-fitted line, respectively (*Fig. 2*) for more details see (Sharma and Sridharan 2018).

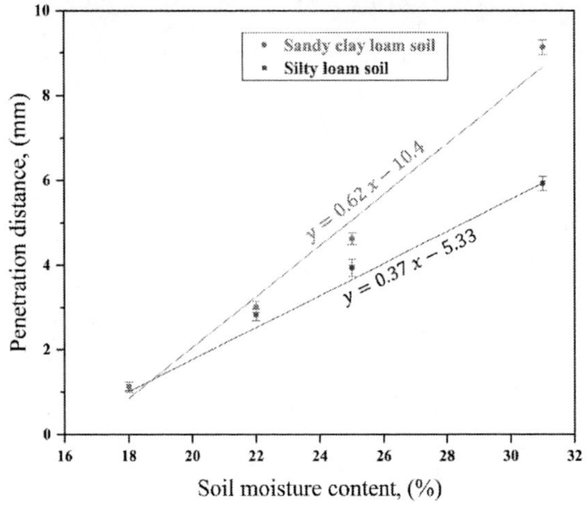

Figure 2. The investigation of soil plastic and liquid limits

2.2 Measurement method and procedure

Each tested plate was placed in contact with the top surface of the soil paste and subjected to a specified normal load (*Fig. 3*). The tested plate was dragged at the specified speed for the desired displacement. This procedure is repeated using four normal loads of 65, 95, 125, and 155 N (this load includes the plate weight). At ultimate dynamic equilibrium, the corresponding traction force of each applied normal load is recorded.

The numerical values of soil adhesion and external friction coefficients were calculated according to the Mohr-Coulomb failure criterion (equation 1), which shows the relationship between tangential and normal stresses during ultimate dynamic equilibrium. According to the Mohr-Coulomb failure criterion, the adhesion value is determined by intersecting the failure line created by linear fitting of the obtained points with the y-axis, and the slope of the failure line represents the external friction angle (*Fig. 4*). Many researchers have utilized the above technique to measure soil adhesion and friction with other materials (Abbaspour-Gilandeh, Hasankhani-Ghavam 2018, Marani, Shahgholi 2019).

2.3 Soil preparation and implementation of the experiment

Initially, the tested soil was soaked in water for two days, and then the moisture content of the soil was measured and tracked twice daily until it reached the highest desired moisture content in the current study (33% d.b.). The relevant tests were then carried out, and so on until the lowest studied moisture content (18% d.b.) was achieved. A precise test bench was used to assess soil adhesion and friction coefficients at various combinations of the investigated parameters (*Fig. 5*).

An S-shaped load cell (Kyowa, 50 kg) was used to measure the sliding resistance of the different sliding plates. Each sliding plate was tested individually using a 0.4 kW electric motor with an inverter controlling the motor speed. The plate was pulled for 0.7 m, and the sliding resistance was continuously acquired by the distributed stress and strain test and analysis system (DH3820 N) (see the supplementary figure).

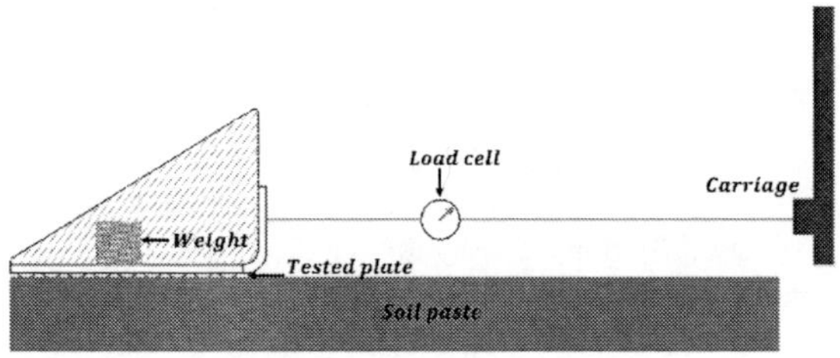

Figure 3. A simple slider technique for measuring soil adhesion and friction coefficients

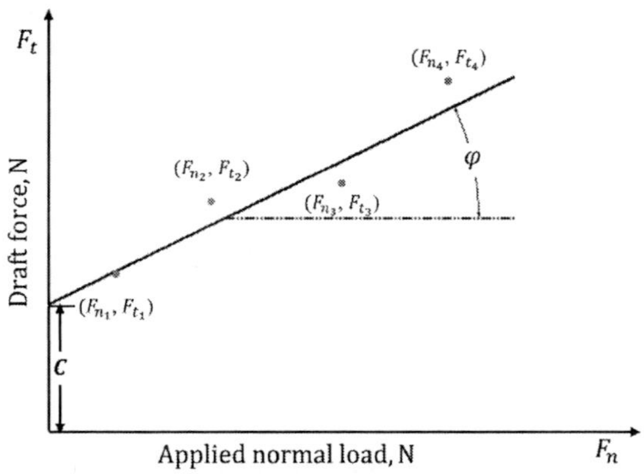

Figure 4. Schematic representation of a failure line at the soil-plate contact interface

3. Results and discussion

The average of the calculated coefficients of adhesion and friction for all combinations of investigated parameter levels are presented in a spreadsheet (see the supplementary tables). The standard deviations of replicate measurements were computed and shown on charts in the form of error bars to evaluate the variation of repeated data from mean values.

The IBM SPSS 25 statistical software package was used to analyze the data based on a complete randomized design. According to the results of the variance analysis (Table 2), all investigated parameters and binary interaction effects were statistically significant at the 5% probability level, except for the binary interaction effect between soil moisture content and sliding speed on the coefficient of friction.

Figure 5. The experimental setup for determining soil adhesion and friction coefficients.

3.1 The interactive impacts of soil texture, test-plate material, and moisture content on soil adhesion

As shown in *Fig. 6*, an increase in moisture content in silty clay soil increased soil adhesion for all tested plates. In contrast, soil adhesion values in sandy clay loam soil were increased by increasing its moisture content up to 28%. When the moisture content in the sandy clay loam soil was exceeded 28%, the values of adhesion coefficients rapidly decreased for all tested plates, which may be attributed to the low water-holding capabilities of light soils. As a result, when the moisture level of the sandy clay loam soil exceeds 28%, it behaves like a fluid; similar behaviour has been observed by other researchers in light soils with a high sand percentage (Burbaum and Sass 2017, Hasankhani-Ghavam, Abbaspour-Gilandeh 2015, Manuwa 2012).

Due to its high surface energy, the flat steel plate had the highest adhesion coefficient in both soil textures and all moisture contents. In comparison, the UHMW-PE plate with domed protrusions exhibited the lowest adhesion coefficient due to its lower wettability (Nalbant and Palali 2011). Furthermore, the capacity of an adequately designed non-smooth surface to interrupt the continuity of the interfacial water film enhances soil adhesion reduction (Chirende, Li 2010, Wang, Xue 2019). In general, the hydrophobic effect of UHMW-PE plates is due to the high water contact angle of Ultra-high molecular weight polyethylene (94°), which hinders a continuous interfacial water film from developing at the contacting interface, thereby reducing the attraction forces between soil particles and the plate surface (Tong, Ren 1999).

As observed from *Fig. 6*, silty clay soil is more adhesive due to fine particles with a large surface area (clay and fine silt). Furthermore, increasing the moisture content value increased adhesion, implying that a high moisture content creates a suitable environment for ionic bond formation. Consequently, increased adhesion is predicted in clay and moisture-rich soils (Chen, Tong 1994). The maximum soil-plate adhesion value for steel plate was 1199.55 N/m^2 in silty clay soil with a moisture content of 33% and a sliding speed of 0.3 m s^{-1}, while the maximum soil-plate adhesion value for sandy clay loam soil was 842.29 N/m^2 with a moisture content of 28% at the same sliding speed of 0.3 m s^{-1}. Generally, UHMWPE-coated plates (i.e., the flat UHMWPE plate and domed UHMWPE plate) exhibited much lower adhesive force than the flat steel plate.

Another interesting aspect to notice is that the soil adhesion reduction of the plates with a surface coating of UHMWPE polymer increased compared to the steel plate at the high moisture content levels. This result demonstrates that increasing soil moisture content increased soil adhesion to steel plates more effectively than to plates with a UHMWPE polymer surface coating. It was concluded that the plates with a surface coating of UHMWPE polymer preserved their ability to minimize soil adhesion in moist and sticky soils due to their inherent hydrophobic properties. Other researchers have reported similar behavior for nano-coated and polymer-coated materials in other experiments (Barzegar, Hashemi 2016, Marani, Shahgholi 2019, Ren, Deng 2004, Salokhe, Chuenpakaranant 1999, Salokhe, Gee-Clough 1990).

It can be observed that the optimized bio-inspired surface generated a lower adhesion coefficient and a higher friction coefficient compared to a flat surface made of the same material. However, because the resultant decrease in adhesion coefficient was more pronounced than the resultant increase in friction coefficient, the overall impact was a reduction in the optimized bio-inspired surface sliding resistance compared to flat surface sliding resistance. This demonstrates the viability of coupling the optimized bio-inspired surface structure with a low surface energy polymer, which opens up many opportunities for developing anti-adhesive soil-contacting tools.

3.2 The tendency of the test-plate material to adhere to soil

The soil tendency for adhering to the test-plate material was evaluated by collecting, drying, and weighing the mass of soil clinging to each plate after sliding across the surface of silty clay soil with a moisture content of 33% for 0.7 m under an applied load of 155 N (Fig. 10). Although this simple measure doesn't fully reflect the complicated process of soil adhesion during paddock preparation, it may be used to compare and identify how different materials impact soil adhesion (Zumsteg and Puzrin 2012). After three replications, the average dried mass of soil clinging to the ultra-high molecular weight polyethylene plate was 427 gm with a standard deviation of 95 gm, compared to 767 gm with a standard deviation of 217 gm for the carbon steel plate. These results indicate that the ultra-high molecular weight polyethylene polymer has a significantly lower tendency to adhere to the soil than carbon steel plate.

Figure 9. The effect of test-plate material on the adhered soil mass to each plate under identical experimental settings

3.3 The interactive impacts of soil texture, test-plate material, and moisture content on external soil friction

Figure 7 displays the variation in soil external friction angle coefficient for the investigated plates at all studied moisture contents in both soil textures. Among all treatments, the largest friction coefficient values were obtained at a moisture content of 28% in sandy clay loam soil at a sliding speed of 0.3 m s^{-1}, with values of 1.03, 0.71, and 0.78 for flat steel plate, flat UHMWPE plate, and domed UHMWPE plate, respectively. In contrast, the lowest values in the sandy clay loam soil were obtained at a moisture content of 33% and a sliding speed of 0.15 m s^{-1} with values of 0.27, 0.21, and 0.21 for flat steel plate, flat UHMWPE plate, and domed UHMWPE plate, respectively. Gill and Vanden Berg stated that there is a direct relationship between soil sand percentage and friction coefficient; hence increasing the proportion of coarse particles in sandy clay loam soil compared to silty clay soil might be one of the causes (Gill and Vanden Berg 1967).

In sandy clay loam soil, the friction coefficients for all studied plates decreased rapidly when the soil moisture content exceeded 28%. This reduction may be attributed to the lubricating effect provided by forming a thick interfacial water layer at the contacting surface, which is related to the low water-holding capacity of such light soil (Ren, Han 2006). The results revealed that the plates with a surface coating of UHMWPE polymer had a lower coefficient of friction than steel plates. It may be inferred that plates with a surface coating of UHMWPE polymer due to their inherent properties produce more favorable conditions for the soil particles to slide over the plate surface.

3.4 The interactive impacts of soil texture, test-plate material, and sliding speed on coefficients of adhesion and friction

As can be seen in *Fig. 8* in both types of soil texture and for all sliding plates, the external friction coefficient, and soil adhesion increased with increasing sliding speed. Generally, the increased adhesion and external friction coefficients of all studied plates with increasing sliding speed may be attributed to the increased shear rate at the contacting interface (Swick and Perumpral 1988, Yang, Tong 2019). Among the studied plates engaged with two soil textures at all investigated sliding speeds, the maximum adhesion was 1199.55 N/m^2, related to a flat steel plate in silty clay soil at a speed of 0.3 m s^{-1}. The flat UHMWPE plate and domed UHMWPE plate had corresponding values of 815.37 and 719.32 N/m^2, respectively. While the smallest adhesion values were obtained in sandy clay loam soil at a speed of 0.15 m s^{-1} for flat steel plate, flat UHMWPE plate, and domed UHMWPE plate with values of 405.22, 362.48, and 289.5 N/m^2, respectively.

4. Conclusion

The effects of test-plate material, soil moisture content, soil texture, and sliding speed on coefficients of adhesion and friction were investigated. It was observed that the binary interaction effect of soil texture and soil moisture content on coefficients of adhesion and friction is highly significant. Furthermore, the results revealed a significant difference in soil adhesion between light and heavy soil at the same moisture content. Combining a bio-inspired convex shape with a surface coating of UHMWPE polymer significantly reduced the adhesion coefficients compared to carbon steel, especially in moist and sticky soils. The effect of tested-plate material on soil adhesion levels was investigated by collecting, drying, weighing, and comparing the amount of soil sticking to each plate. According to the results, ultra-high molecular weight polyethylene polymer has a significantly lower tendency to stick to soil than carbon steel. The lower adhesion coefficients of bio-inspired polymer-coated surfaces pave the way for practical applications in soil adhesion reduction and soil-engaging component optimization.

Table 2. The effects of the studied parameters on external friction and soil adhesion

Source of variation	df	Mean squares		F-value	
		Adhesion	Friction	Adhesion	Friction
Soil texture (A)	1	2898909.34	0.646	1482.95*	413.92*
Material and structure (B)	3	2201778.56	4.82	1126.33*	3088.56*
Soil moisture content (C)	3	519582.47	1.237	265.8*	792.65*
Sliding speed (D)	3	992319.95	0.681	507.63*	436.22*
A × B	3	244593.88	0.02	125.12*	12.55*
A × C	3	663398.89	1.773	339.37*	1136.13*
A × D	3	10732.32	0.058	5.49*	37.15*
B × C	9	74768.28	0.055	38.25*	35.12*
B × D	9	28123.4	0.01	14.39*	6.52*
C × D	9	4885.46	0.003	2.5*	1.73^{ns}
Error	256	1954.83	0.002		
Total	383				

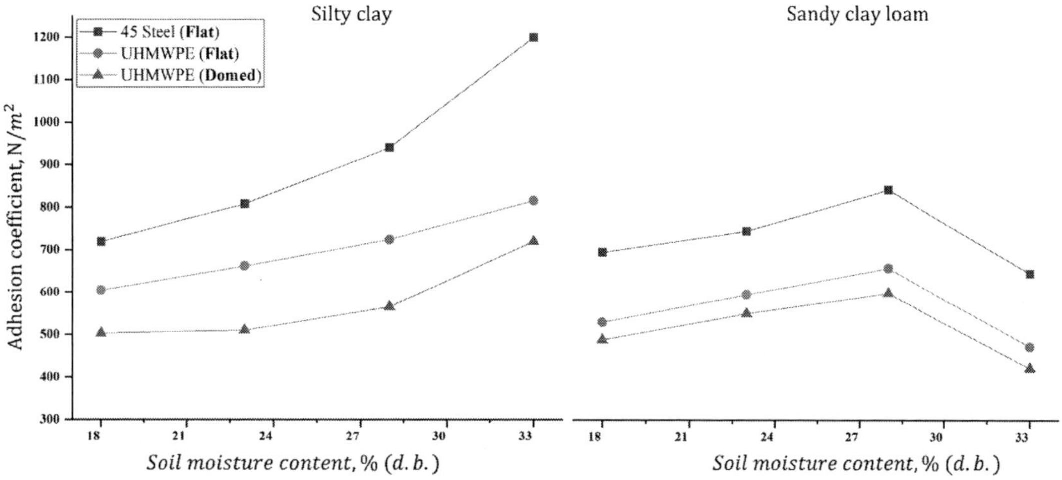

Figure 6. The variation in adhesion coefficients between both soil textures according to test-plate material and soil moisture content

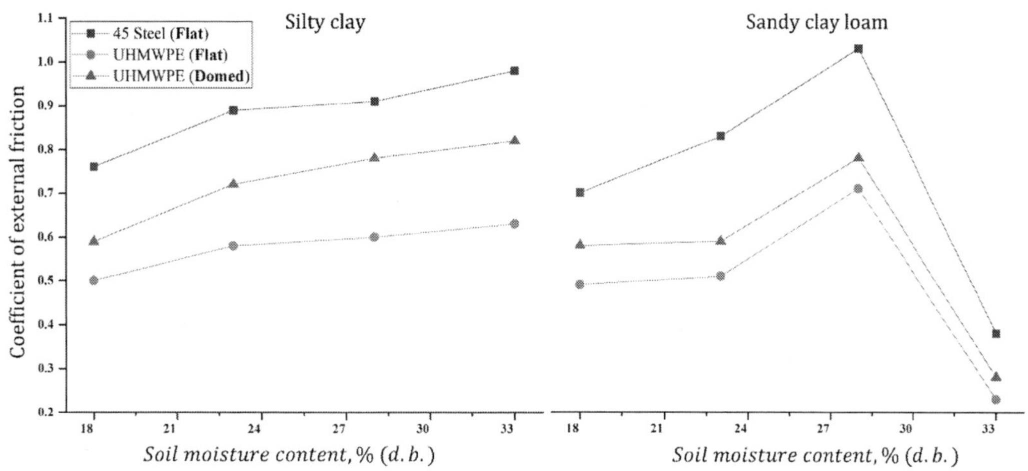

Figure 7. The variation in external friction coefficients between both soil textures according to test-plate material and soil moisture content

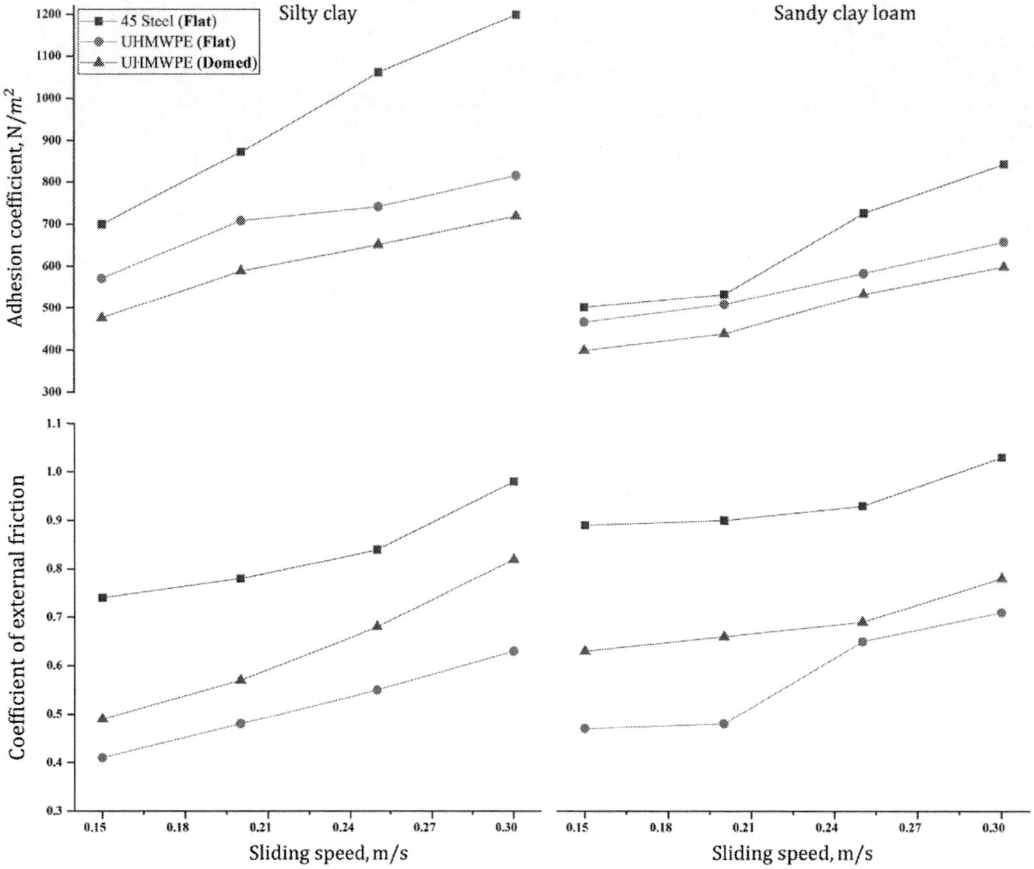

Figure 8. The variation in adhesion and external friction coefficients between both soil textures according to test-plate material and sliding speed

5. Acknowledgements

This study was financially supported by the National Natural Science Foundation of China; (Grant No. 51775220) titled "Contract–Rheology-Resistance coupling mechanism between the flat plate type soil working parts and disturbed paddy soil". The authors would like to express their gratitude to the Huazhong Agricultural University soil physics laboratory staff for their assistance with the soil bin experiments.

6. References

Abbaspour-Gilandeh, Y., Hasankhani-Ghavam, F., Shahgoli, G., Shrabian, V. R., Abbaspour-Gilandeh, M., 2018. Investigation of the effect of soil moisture content, contact surface material and soil texture on soil friction and soil adhesion coefficients. Acta technologica agriculturae. 21, 44-50

Azadegan, B., Massah, J., 2012. Effect of temperature on adhesion of clay soil to steel. Agronomic Research in Moldavia. 45, 21–7

Barzegar, M., Hashemi, S., Nazokdast, H., Karimi, R., 2016. Evaluating the draft force and soil-tool adhesion of a UHMW-PE coated furrower. Soil and Tillage Research. 163, 160-7

Burbaum, U., Sass, I., 2017. Physics of adhesion of soils to solid surfaces. Bulletin of Engineering Geology and the Environment. 76, 1097-105

Chen, Y., Tong, J., Ren, L., Chen, B., Ge, L., 1994. Abrasive wear behavior of several soil adhesion-reducing materials. Transactions of the Chinese Society of Agricultural Engineering. 10, 91-5

Chirende, B., Li, J., Wen, L., Simalenga, T. E., 2010. Effects of bionic non-smooth surface on reducing soil resistance to disc ploughing. Science China Technological Sciences. 53, 2960-5

Gill, W. R., Vanden Berg, G. E. Soil Dynamics in Tillage and Traction US Department of Agriculture: Agricultural Research Service; 1967.

Goshtasb, A. K., Fielke, J., Desbiolles, J. A review of soil/tool adhesion principles and approaches to reducing limitations of disc seeders. In: Climate ATIaC, editor. The 2009 CIGR International Symposium of the Australian

Society for Engineering in Agriculture. Brisbane, Queensland: Engineers Australia; 2009. p. 208-15.

Guan, C., Fu, J., Xu, L., Jiang, X., Wang, S., Cui, Z., 2022. Study on the reduction of soil adhesion and tillage force of bionic cutter teeth in secondary soil crushing. Biosystems Engineering. 213, 133-47. https://doi.org/https://doi.org/10.1016/j.biosystemseng.2021.11.018

Hansbo, S. A new approach to the determination of the shear strength of clay by the fall-cone test. Stockholm, Sweden: Royal Swedish Geotechnical Institute 1957.

Hasankhani-Ghavam, F., Abbaspour-Gilandeh, Y., Shahgoli, G., Rahmanzadeh-bahram, H., 2015. Design, manufacture and evaluation of the new instrument to measure the friction coefficient of soil. International Commission of Agricultural Engineering. 17, 101–9

Heuser, M., Spagnoli, G., Leroy, P., Klitzsch, N., Stanjek, H., 2012. Electro-osmotic flow in clays and its potential for reducing clogging in mechanical tunnel driving. Bulletin of Engineering Geology and the Environment. 71, 721-33

Khosravani, A., Desbiolles, J., Fielke, J. Opportunities for improving the performance of single disc seeders in sticky soil conditions. In: Hegarty R, Banhazi T, Saunders C, editors. 2011 Society for Engineering in Agriculture Conference: Diverse Challenges, Innovative Solutions. Barton, A.C.T.: Engineers Australia; 2011. p. 256-65.

Li, J., Kou, B., Liu, G., Fan, W., Liu, L., 2010. Resistance reduction by bionic coupling of the earthworm lubrication function. Science China Technological Sciences. 53, 2989-95

Liu, G., Xia, J., Zheng, K., Cheng, J., Wang, K., Zeng, R., et al., 2022. Effects of vibration parameters on the interfacial adhesion system between soil and metal surface. Soil and Tillage Research. 218, 105294-304

Liu, H., Zhao, S., Tan, H., Yang, Y., Zhang, X., 2018. Investigation on press device in reducing adhesion and resistance based on scrape and vibration principle. Transactions of the Chinese Society of Agricultural Machinery. 49, 86-92

Manuwa, S., 2012. Evaluation of soil/material interface friction and adhesion of Akure sandy clay loam soils in southwestern Nigeria. Advances in Natural Science. 5, 41-6

Marani, S. M., Shahgholi, G., Moinfar, A., 2019. Effect of nano coating materials on reduction of soil adhesion and external friction. Soil and Tillage Research. 193, 42-9

McGloughlin, T. M., Kavanagh, A., 2000. Wear of ultrahigh molecular weight polyethylene (UHMWPE) in total knee prostheses: a review of key influences. Journal of Engineering in Medicine. 214, 349-59

Nalbant, M., Palali, A. T., 2011. Effects of different material coatings on the wearing of plowshares in soil tillage. Turkish journal of agriculture and forestry. 35, 215-23

Ren, L.-Q., Tong, J., Li, J.-Q., Chen, B.-C., 2001. SW—soil and water: soil adhesion and biomimetics of soil-engaging components: a review. Journal of Agricultural Engineering Research. 79, 239-63

Ren, L., Deng, S., Wang, J., Han, Z., 2004. Design principles of the non-smooth surface of bionic plow moldboard. Journal of Bionic Engineering. 1, 9-19

Ren, L., Han, Z., Li, J., Tong, J., 2006. Experimental investigation of bionic rough curved soil cutting blade surface to reduce soil adhesion and friction. Soil and Tillage Research. 85, 1-12

Salem, A. E., Wang, H., Gao, Y., Zha, X., Abdeen, M. A., Zhang, G., 2021. Effect of Biomimetic Surface Geometry, Soil Texture, and Soil Moisture Content on the Drag Force of Soil-Touching Parts. Applied Sciences. 11, 8927-38. https://doi.org/10.3390/app11198927

Salem, A. E., Zhang, G. Z., Abdeen, M. A. M., Wang, H., Gao, Y., 2022. Optimizing the adhesion of soil-touching parts based on biomimetic concepts using the Taguchi method. International Journal of Agricultural and Biological Engineering. 15, 147-54

Salokhe, V., Chuenpakaranant, W., Niyampa, T., 1999. Effect of enamel coating on the performance of a tractor drawn rotavator. Journal of terramechanics. 36, 127-38

Salokhe, V., Gee-Clough, D., Manzoor, S., Singh, K., 1990. Improvement of the tractive performance of cage wheel lugs by enamel coating. Journal of Agricultural Engineering Research. 45, 209-24

Sharma, B., Sridharan, A., 2018. Liquid and plastic limits of clays by cone method. International Journal of Geo-Engineering. 9, 22. https://doi.org/10.1186/s40703-018-0092-0

Soni, P., Salokhe, V., Nakashima, H., 2007. Modification of a mouldboard plough surface using arrays of polyethylene protuberances. Journal of Terramechanics. 44, 411-22

Swick, W., Perumpral, J., 1988. A model for predicting soil-tool interaction. Journal of Terramechanics. 25, 43-56

Tong, J., Ren, L.-Q., Yan, J.-L., Ma, Y.-H., Chen, B. C., 1999. Adhesion and abrasion of several materials against soil. International Agricultural Engineering Journal. 8, 1-22

Wang, Y., Xue, W., Ma, Y., Tong, J., Liu, X., Sun, J., 2019. DEM and soil bin study on a biomimetic disc furrow opener. Computers and Electronics in Agriculture. 156, 209-16

Wroth, C. P., Wood, D. M., 1978. The correlation of index properties with some basic engineering properties of soils. Canadian Geotechnical Journal. 15, 137-45. https://doi.org/10.1139/t78-014

Yang, Y., Tong, J., Ma, Y., Jiang, X., Li, J., 2019. Design and experiment of biomimetic rotary tillage blade based on multiple claws characteristics of mole rats. Transactions of the Chinese Society for Agricultural Engineering. 35, 37-45 (in Chinese)

Zhang, D., Chen, Y., Ma, Y., Guo, L., Sun, J., Tong, J., 2016. Earthworm epidermal mucus: Rheological behavior reveals drag-reducing characteristics in soil. Soil and Tillage Research. 158, 57-66

Zhang, Y., Huang, H., Ren, L., 2013. Experiment and drag reduction mechanism of bionic excavator tooth during soil cutting. Nongye Jixie Xuebao= Transactions of the Chinese Society for Agricultural Machinery. 44, 258-61

Zumsteg, R., Puzrin, A. M., 2012. Stickiness and adhesion of conditioned clay pastes. Tunnelling and Underground Space Technology. 31, 86-96

8654 / Construction of a Soil Clods Recognition Bench-Scale Experiment for Discrete Element Method Modeling of Tilling Phenomena

Shuto Ishii[a]*, Isami Suto[a], Hiroaki Tabe[a], Keisuke Nagato[a]*, Moju Zhao[a]*, Yoshifumi Ueshige[b], Takashi Iritani[b], and Masayuki Nakao[a]

[a]Department of Mechanical Engineering, The University of Tokyo, Bunkyo-ku, Tokyo, Japan 113-8656

[b]Analysis Center, Technology Innovation Research and Development Unit, Research and Development Headquarters, Kubota Corporation, Sakai-City, Osaka, Japan 590-0823

**Corresponding author: ishii.s@hnl.t.u-tokyo.ac.jp*

https://doi.org/10.56884/VBXY3767

Abstract

The development of rotary claws in tiller machines has been evaluated using actual equipment and analysis based on the discrete element method (DEM), an effective method for modeling the movement of granular materials. The required functions of the rotary claw are to break the soil into small pieces and plow the soil flat. However, evaluation by testing using actual equipment is difficult for precise measurements, and a considerable amount of time is required for one test. Additionally, DEM simulations have difficulty reproducing the soil behavior owing to the large number of parameters that determine their accuracy. Therefore, a comparison with the experimental results is necessary when determining DEM parameters. This research aims to develop a soil clod recognition method and an evaluation system for evaluating tractor performance. The evaluation system was designed to measure the soil shape before and after tilling with a 1/4-scale tillage claw. To capture the changes in the soil shape and distribution and size of the soil clods, we developed an image processing method that uses point cloud data from a depth camera to calculate the difference before and after tilling. The experimental parameters were the soil moisture content and claw rotation speed. Their effects on the formation and decomposition of the soil clods were evaluated. To confirm the feasibility of the results obtained from the proposed evaluation system, DEM simulations were performed under identical conditions to compare the distribution and size of the soil clods and changes in soil shape. The experimental results showed that the location, number, and size of scattered soil clods varied with the moisture content of the soil and rotational speed of the claw. Based on the experimental results, a comparison was made with the results of the DEM analysis to clarify the differences. The developed model experiment system for soil clods and the image processing method made it possible to quantitatively compare soil dispersal between the experiment and the DEM, which may accelerate the search for parameters for the DEM to reproduce the tilling.

Keywords: discrete element method (dem), recognition of soil clods, tillage claw, moisture content ratio, rotation speed

1. Introduction

In recent years, design evaluation of rotary development has been conducted through tests using actual tractor. Tests have been conducted in many fields. Usaborisut et al. (2020) conducted field tillage trials using a rotary tiller and power harrow with and without pre-applied subsoil and compared the characteristics of both. In these tests, the soil quality and moisture content ratio were controlled. Therefore, a single experimental trial takes a significant amount of time to complete. It is also difficult to collect accurate data from large-scale environments.

In previous studies, an experimental-scale model was created to reduce the time and effort required to use a machine. It also evaluated the shape of the claw with greater precision. For example, Chertkiattipol and Niyamapa (2010) analyzed the torque characteristics of a single claw with different shapes. They found that the position of the cutting edge and shape of the rotating blade affected the torque characteristics. Matin et al. (2015, 2016) analyzed the motion of four blades using a high-speed camera. They discovered that the blade penetration depth increased with increasing rotational speed, and peak torque was generated. Matin et al. (2021) conducted experiments with various blades, tilling widths, geometries, and tilling depths. They found that straight blades produced the most optimal soil clods and uniform grooves.

In addition, many studies have used the DEM as a method other than testing in actual equipment for design evaluation of tillage claw. DEM is an effective method for modeling the flow of granular materials. Cheng et al. (2021) conducted a simulation test based on DEM to analyze the wet clay soil adhering to the rotary-tillage during rotary-tillage in paddy fields. Li et al. (2018) conducted a soil-cutting simulation to optimize the rotary blade and save energy for the tilling operation. Zhao et al. (2020) analyzed the torque applied to claws and soil dispersal when the shape of claw was changed. However, the mechanism is complex and difficult to test in practice since the adhesive forces between soil particles vary with changes in conditions, such as moisture content. In addition, the number of parameters that determine the accuracy of the model is large and difficult to identify. This makes it difficult to perfectly reproduce the rotary-tillage process.

To improve the accuracy of the DEM simulations, it is necessary to compare them with real phenomena. Studies to improve the accuracy of DEM simulations include comparisons of DEM simulations with experiments on actual equipment and comparisons of DEM simulations with experiments using bench-scale test equipment. Hirasawa et al. (2013, 2014) compared the experimental and DEM simulation results for soil leveling, soil-particle dispersion velocity, and torque applied to the claw.

In this study, we evaluated the important functions of a tractor as a novel subject for a comparison between experiments and DEM. The purposes of tillage are to level the soil for efficient seeding after tilling and break the soil into small pieces to aerate the soil for plant growth. For both purposes we focused on changes in soil conditions before and after tillage.

Vision and force sense are important tools in measuring soil conditions. In this system, we focused on visual methods of measurement to focus on the number, size, and location of soil clods when observing their decomposition before and after tilling. As a future prospect, we plan to combine sensing other than visual information, for example, with information on torque applied to the claws during tilling. To recognize soil conditions, Grundy et al. (2020) used a depth camera to measure soil-surface roughness and introduced it as an alternative to existing measurement techniques. Although it recognizes the soil shape, it cannot detect soil clods on the top of the soil. Therefore, this study aimed to develop a system that recognizes soil conditions before and after tillage and detects soil clods. The developed system was verified in a 1/4-scale bench experiment. The size and position data obtained from the recognition results were used to confirm the dispersal of soil clods.

2. Method

2.1 Experimental equipment and systems

In this study, a scale model of a rotary experiment was developed. Figure 1a shows a diagram of the experimental apparatus. The experimental apparatus consisted of a motor and 1/4-scale tilling claw. The motor drives the tilling claw, which is attached to the shaft for rotation. Rotation of the claw cultivated the soil bed. An area around the soil bed was used for the detection.

Figures 1b and 1c show the apparatus used in this experiment. A scale model was created using a 1/4-scale tilling claw.

An Intel RealSense LiDAR Camera L515 was used to capture the changes in soil shape and distribution and size of soil clods before and after tilling. This camera is equipped with both an RGB camera and LiDAR and can acquire colored point cloud data with a single device. Figure 1(d) shows the location of the depth camera. A depth camera was mounted directly above the soil bed. The minimum depth distance of the depth camera was 0.25 m. In addition, the accuracy of the depth measurements was less than 2 cm at 1 m.

Figure 1e shows the soil bed used in this study. Soil beds and floor surfaces were created to observe soil dispersal. The soil bed had the smallest size that would be unaffected by the wall surface.

The bench-scale experiment was conducted using the following procedure. First, the dry soil was weighed and mixed with a calculated weight of water to reach the moisture content suitable for the experiment. The soil was then compacted into a container to achieve an appropriate soil volume fraction under experimental conditions. The soil bed was recorded using a depth camera as the pre-tillage data. The motor was then driven to rotate the tilling claw. A tilling test was conducted. Finally, the data were recorded as post-tillage data.

2.2 Experimental conditions

The experimental conditions are presented in Table 1. To investigate the mechanism of soil clod breakdown, experiments were conducted by varying two parameters: moisture content ratio of the soil and rotational speed of the claw. Three conditions were used for moisture contents of 8 %, 10 %, and 12 %. These moisture contents were common during tilling and is most likely to generate clods. If the moisture content is too small (less than 4%), clods are not produced because adhesion between soil particles does not occur easily. Conversely, if the moisture content is too high (more than 16%), the soil is clayey, and in the case of a one-jaw, one-rotation cultivation, soil clod breakdown does not occur, and a single soil clod is generated according to the trajectory of the rotating claw. Therefore, the observation of soil breakdown in these cases is considered unnecessary. Therefore, we set an intermediate moisture content ratio that facilitates the observation of the decomposition of soil clods. The moisture content ratio (w) can be derived from Equation (1) using the weight of soil (M_s) and weight of water (M_w).

$$w = \frac{M_w}{M_s} \times 100 \quad (1)$$

The rotation speeds during the experiment were 100, 200, and 300 rpm, which are common rotation speeds for tractor tilling.

Next, the fixed parameters are described. The volume fraction of soil was set to 44 %. It ensured that the soil was sufficiently soft, and the decomposition of the soil clod was easily observed. The soil volume fraction of 44 % used in this study corresponded to approximately 1200 kg/m^3 in terms of soil bulk density. The soil volume fraction (P) can be derived from Equation (2) using the volume of air (V_a), water (V_w), and soil (V_s). The tilling depth was set to 35 mm, which is approximately 1/4 times that of the actual machine.

$$P = \frac{V_s}{V_a+V_w+V_s} \times 100 \quad (2)$$

The soil properties used in the experiments are listed in Table 2. The experiment was repeated five times for each condition.

Table 1. Operating conditions during the experiments

Variable parameters	Rotational speed [rpm]	100, 200, 300
	Moisture content ratio [%]	8, 10, 12
Fixed parameters	Soil volume fraction [%]	44
	Tilling depth [mm]	35

Table 2. Soil properties in the experiments

	Particle size distribution mass percentage [%]		
	Sand	Silt	Clay
Soil type	2–0.42 mm	0.42–0.002 mm	< 0.002 mm
Clay Loam (CL)	46.0	38.4	15.6

2.3 Soil clod recognition method

In this section, we describe a method for recognizing the soil clods. To recognize the soil clods after tilling, it is necessary to recognize randomly distributed clods in the soil. Therefore, methods such as ordinary RGB image recognition cannot distinguish between clods and soil, and the recognition is unsuccessful.

To classify clods and mountains, we used the geometrical characteristics of the soil clods in the recognition method. If there is a soil clod, it is assumed that the height will increase rapidly compared to the height of the surrounding area. In contrast, the mountainous shape varied continuously with the surrounding height. Therefore, when photographed from above, there was an occlusion area at the bottom of the soil clods.

Figure 2a shows a conceptual diagram of soil clod detection. First, a contour was obtained for each clod and mountain height. The areas of the contours were then calculated. The difference between the two is that the area reduction rate is small for clods and large for mountains.

Figure 2b shows the details of the proposed algorithm. First, a depth image is prepared before and after tilling. Each pixel contains a depth value expressed as a value from 0 to 255.

A difference in the depth image before and after tilling was created. The differences in images were binarized by varying the threshold values. The difference between images implies the difference in depth values for each pixel. Furthermore, by changing the threshold value and binarizing it, an image that resembles a surface cut can be created at a certain height.

Contours were extracted for each image. In conventional contour extraction, contours are merged when clods are adjacent to each other. Therefore, we used a watershed algorithm (Beucher, 1992). This algorithm is a classical region segmentation method widely used for the image segmentation of rock particles (Guo et al., 2022). It separates and recognizes objects in contact with each other. This method can distinguish between adjacent clods and extract them as separate contours.

Next, each contour from the first different image was registered as a group. The first difference image appeared to be cut at the lowest height.

Then, by gradually changing the threshold value, the contours of the image with a slightly higher cut surface were obtained. For each contour, the position of the geometric center was compared to that of the contour group previously registered to determine whether it was included in the contour group. If included, the contour is added to the contour group. Otherwise, a new contour group was created.

For each contour group, we determined the rate of decrease in the contour area. If the rate of decrease in the contour area was large, the contour group was considered to be a mountain. If the rate of decrease was small, the contour group was considered to be a clod. Based on these classifications, the contour groups were labeled as clods or mountains.

The largest contour in a group of contours labeled as a clod is the clod contour. The area and center of geometry of the largest contour can be determined.

2.4 DEM simulation

To compare the experimental results obtained using the soil clod recognition system proposed in this study with the simulation results of the DEM, the DEM was used to simulate tilling. Figure 3 shows a schematic of the simulation model. A 1/4-scale model of an actual machine was used for the simulation.

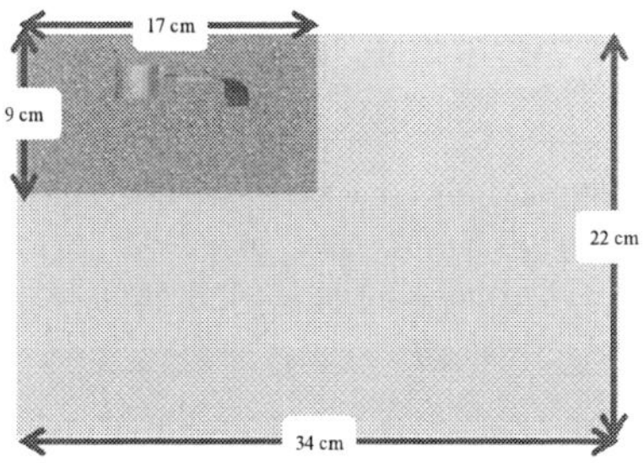

Figure 3. Schematic of the simulation model

The simulation was performed using the following procedure: First, the particle size distribution for the particles to be filled was created based on the particle size distribution of the particles used in the experiment. A container simulating the field was filled with particles. The pressure was applied from the top of the container to simulate the compacted soil. The simulation was performed by rotating the tilling claw at the same rotational speed as that in the experiment. Finally, the behavior of the particles during tilling and their distribution after tilling were compared.

Table 3 shows the properties of the soil in the DEM simulation. Additional parameters called Edinburgh-Elasto-Plastic-Adhesion (EEPA) were used in this analysis to analyze the behavior of the soil clods. The EEPA parameters are the parameters of tensile force generation and nonlinear force-displacement behavior. This parameter makes it possible to represent particle adhesion. Therefore, it was considered to be a necessary parameter for representing the soil clods in this study.

Table 3. Properties of the soil in DEM simulation

Parameter Type	Value
Constant pull-off force $[N]$	0
Surface Energy $[J/m^2]$	10
Constant Plasticity Ratio	0.9
Slope Exp.	1.5
Tensile Exp.	20
Tangential Stiff Multiplier	0.28571
Coefficient of Restitution	0.5
Coefficient of Static Friction	0.5
Coefficient of Rolling Friction	0.01
Poisson's Ratio	0.25
Solids Density $[kg/m^3]$	2500

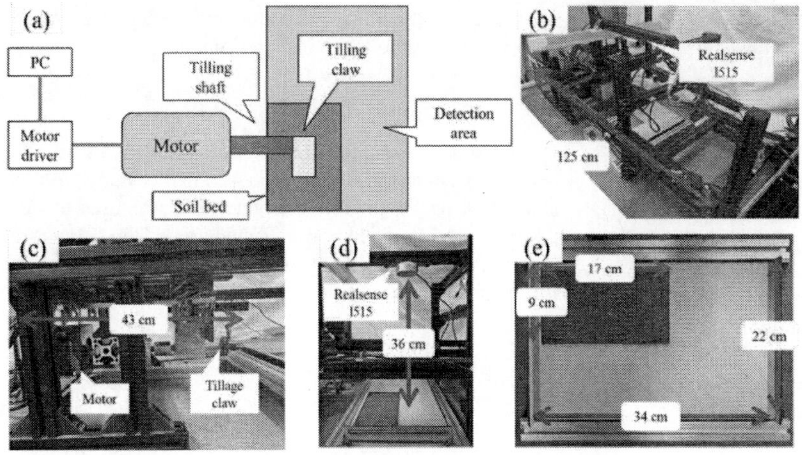

Figure 1. Experimental apparatus: (top-left view) a) System diagram, (top-right view) b) Overall appearance, (bottom-left view) c) Appearance around tilling shaft, (bottom-center view) d) View of soil profile recording, and (bottom-right view) e) Soil bed for the experiment

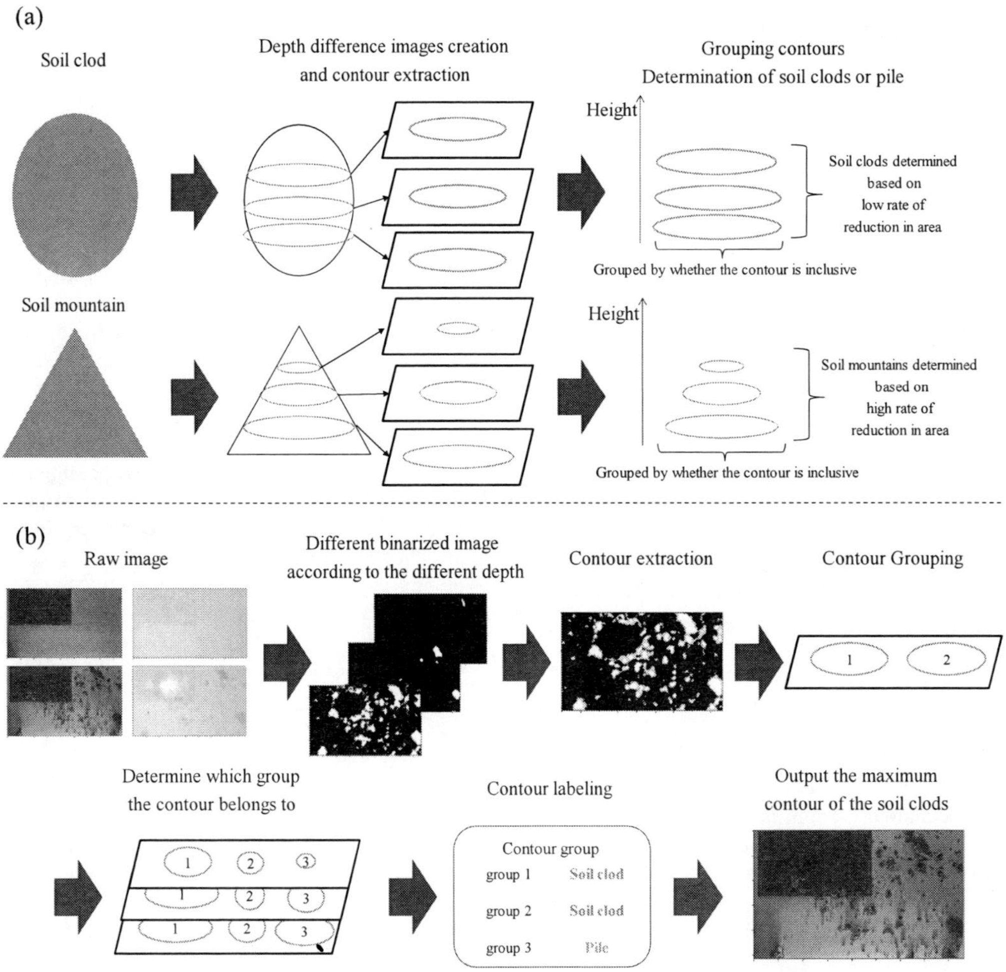

Figure 2. Soil clods recognition: a) conceptual diagram and b) flowchart of the algorithm

3. Results and Discussion

3.1 Accuracy of soil clod recognition method

To confirm the accuracy of the proposed soil clod recognition method, soil clod samples were created and recognized. Figure 3 shows the location of the sample soil clod. The size of the soil clod is defined by its length along the two axes. As shown in the figure, the x- and y-axis were set up, and their respective axial lengths were measured. The scattering position of the clod was determined by comparing the center of geometry of the clod with that of the tilling claw. The scattering distance was determined for both the x- and y-axis. First, the soil clod was measured with a ruler. Its lengths in the x-axis and y-axis directions were 53 mm and 45 mm, respectively. The distance from the center of the claw was 120 and 90 mm in the x-axis and y-axis directions, respectively.

Figure 4 and Table 4 show the measurement results for the soil clods recognized by the system. The lengths of the clods in the x- and y-axes were 55 and 44 mm, respectively. The distance from the center of the claw was 124 mm in the x-axis direction and 94 mm in the y-axis direction. Compared to the values measured with a ruler, the system accurately measured the size of the soil clods and detected their dispersal position and shape. Therefore, this system can accurately recognize the size and dispersal location of soil clods.

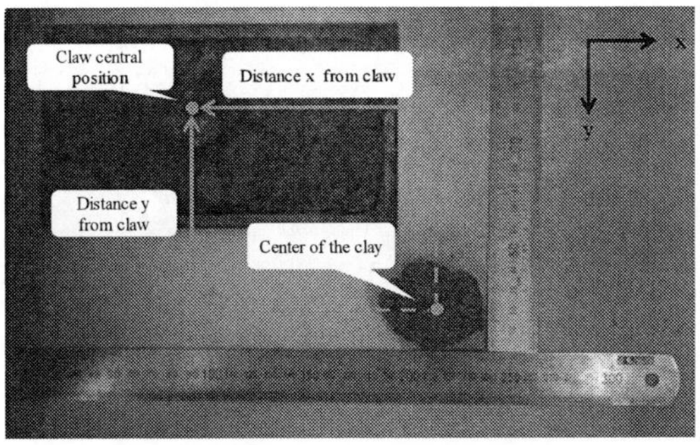

Figure 3. Sample soil clod location

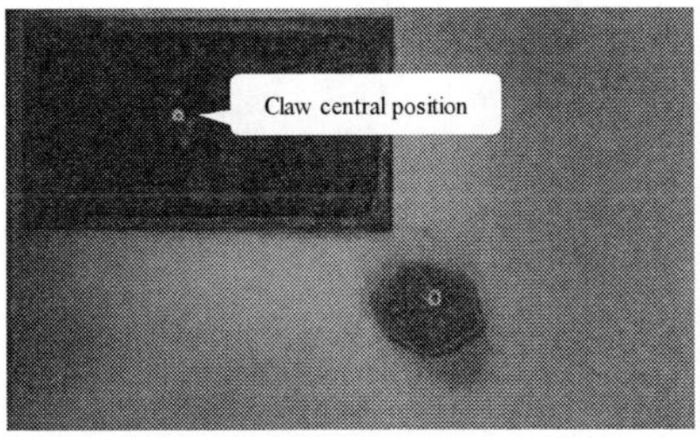

Figure 4. Recognition results of sample soil clods

Table 4. Comparison of measurement methods

	Length (mm)		Distance (mm)	
Measurement method	x	y	x	y
Measured with ruler	53	45	120	90
Proposed system	55	44	124	94

3.2 Results overview

Figure 5 shows the results of the soil clod recognition for each experimental condition. Soil clods were recognized within the recognition area on the board, except inside the container. The recognized clods are shown as red contours. As shown in the Figure 5, the soil clods were correctly recognized under almost all conditions. The soil clods were not well recognized when the rotational speed was 300 rpm, and moisture content was 8 %. The small clods scattered on top of the sediment, lining next to each other. The results indicate that sediment and small clods are more likely to occur when the moisture content is low, and the rotation speed is high.

In addition, when the rotational speed was slow, the clods were larger. When the rotation speed was high, the clods broke into small pieces. The scattering area of the clods was distributed around the container at 100 rpm. However, as the rotational speed increased, the scattering area expanded. When the moisture content is low, many fine clods and sediments are distributed. Whereas when the moisture content is high, the size of the clods is larger, and the proportion of sediment and fine clods decreases.

Figure 6 shows the distribution of the average number of pieces per soil clod size for each of the nine conditions. The figure shows that the proportion of the small clods decreased, and that of the large clods increased as the moisture content increased. Comparing the rotational speed of 200 rpm, at a moisture content of 10%, the number of clods larger than 400 mm^2 accounted for 3.0 compared to 17.2 for the total (17%). On the other hand, when the moisture content is 12%, the number of clods is 1.4 compared to 6.0 (23%), indicating that the proportion of large clods increases as the moisture content increases. As the moisture content increases, the soil clods become more adhesive, thus their ability to break the soil into smaller pieces during tilling is decreased.

In addition, it can also be seen that both the percentage and the number of small soil clods increase as the speed of rotations increases. At a moisture content of 10%, at 100 rpm, the number of soil clods larger than 400 mm^2 accounted for 1.4 out of 5.6 overall (25%), while at 200 and 300 rpm, the percentages steadily decreased to 17% and 5%, respectively. As for the number of small clods, at a water content of 10%, the number of clods smaller than 100 mm^2 increased to 2.0, 9.0, and 12.8, respectively. This could be because the increased number of rotations increased the energy given to the soil by the tilling claw.

These results are consistent with those shown in Figure 5. The results were measured and quantitatively analyzed using the recognition system proposed in this study. In the future, it will be important to improve the accuracy of soil clod recognition in sediments. Since clods are also distributed in the container, it is necessary to examine the clods around the tilled area.

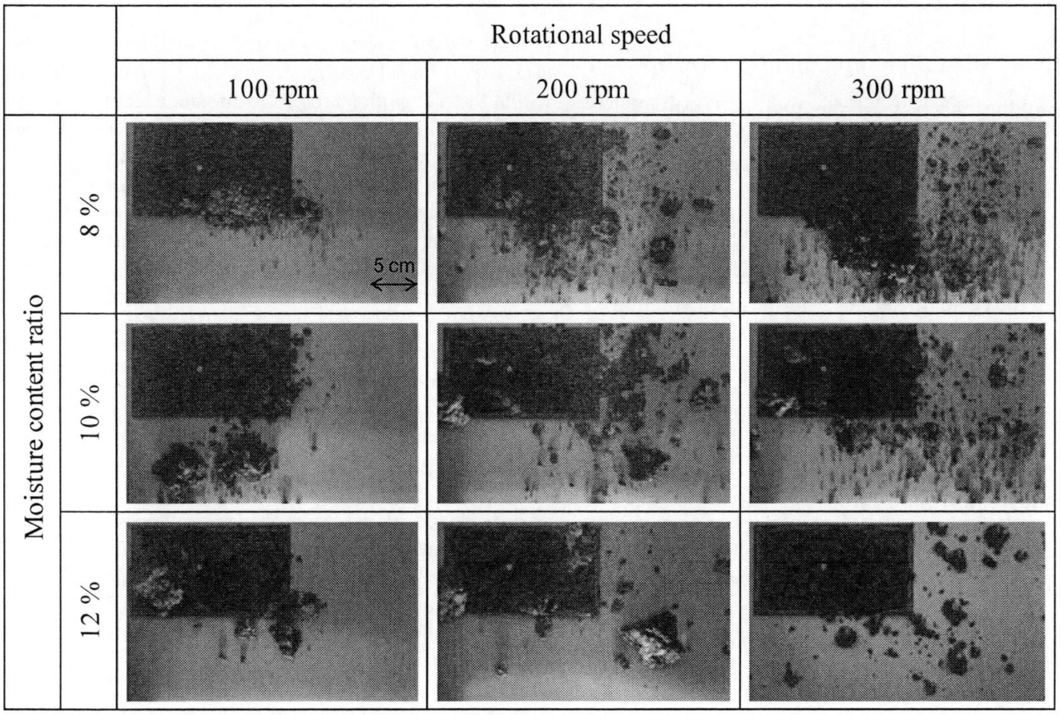

Figure 5. Soil clods recognition results for each condition

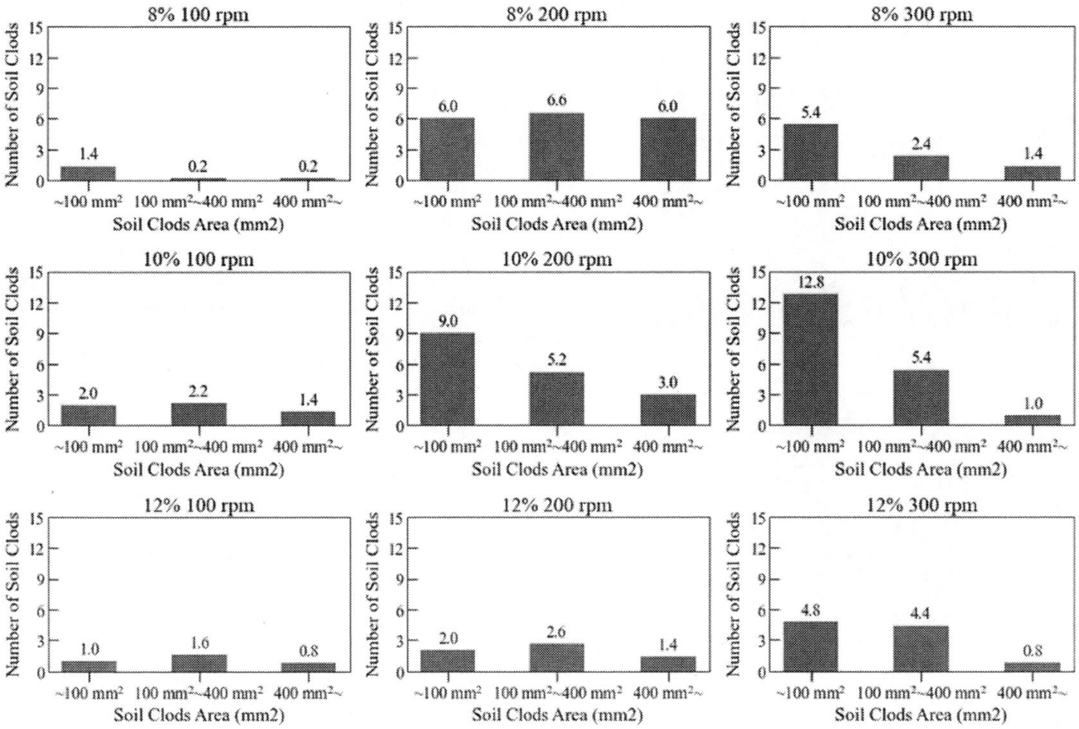

Figure 6. Size distribution of soil clods

3.3 Size and dispersal distribution of soil clods

The distribution of the number of clods per size at 10 % and 12 % moisture content, where soil clod recognition was accurate, was determined. The detection area was divided into three parts, as shown in Figure 7. The distribution of the number of clods per size in each area was determined. Each area is assumed to have a significant contribution from the following mechanisms during tilling.

Area 1: The effect of scattering backward is large.

Area 2: The effect of scattering by sliding in the direction of the axis of rotation is large.

Area 3: Both effects are large.

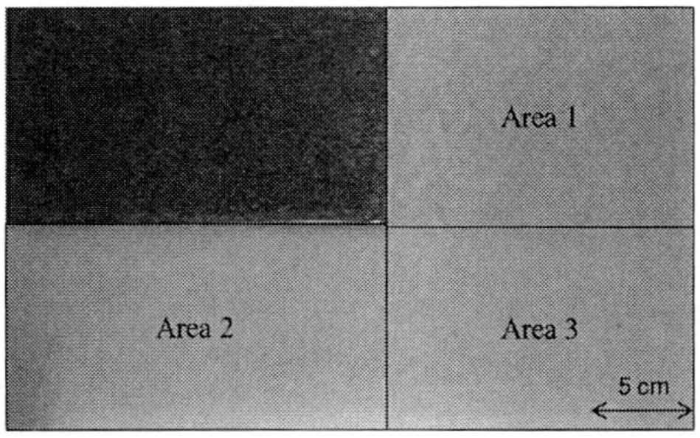

Figure 7. Detection area classification

Figure 8 shows the distribution of the size and number of clods in each area. As the rotation speed increases, the ratio of area 2 tends to decrease, and it can be confirmed that the distribution is biased in the direction of 1 and 3. When the water content ratio is 10%, area 3 accounts for 0.6 of the total 4.6 at 100 rpm (13%), and 24% and 25% at 200 rpm and 300 rpm, respectively, showing an increasing trend in the ratio to area 3. This is thought to be that at low rotational speeds, there was not enough torque to send the soil backward, causing it to slide sideways, while at high rotational speeds, the amount of soil sent to areas 1 and 3 increased because of the increased torque to send the soil in the direction of rotation. This result also agrees with the result of 3.2, confirming that the recognition system in this study can accurately quantify the amount of soil.

The difference between areas is small under all conditions, but it can also be confirmed that there is more scattering in area 2 only at 100 rpm. When the moisture content ratio was 10%, Area 2 accounted for 3.2 compared to 4.6 for the total (70%). When the moisture content was 12%, Area 2 accounted for 2.2 compared to 2.8 for the total (79%). These results clearly indicate that the amount of scatter in Area 2 is larger than that in the other areas. This indicates that at low rotational speeds, the lateral scattering effect is greater than the backward scattering effect.

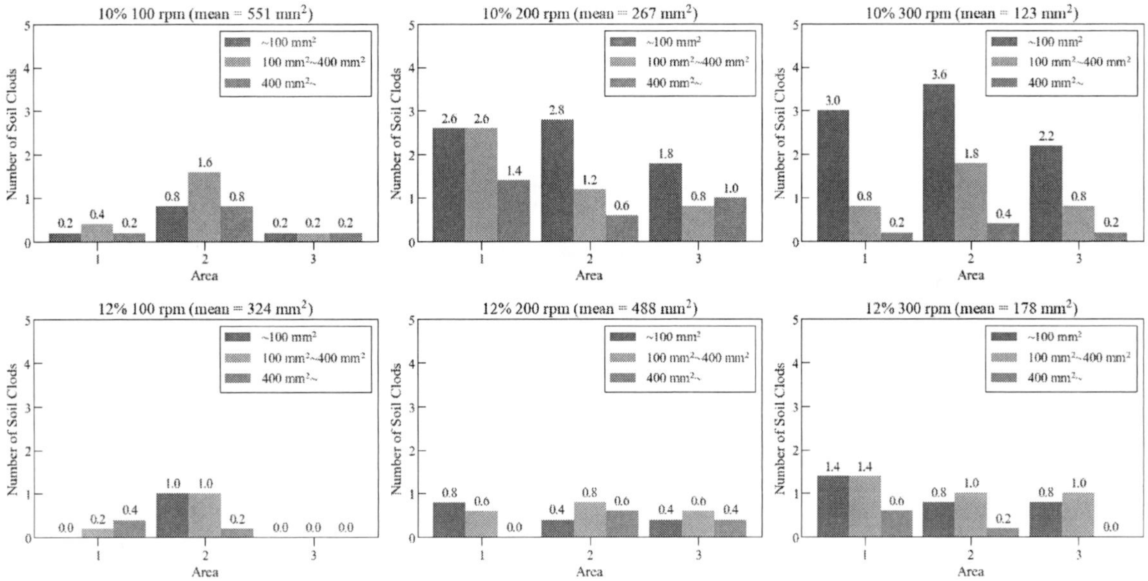

Figure 8. Size distribution of soil clods per area

3.4 Result of DEM simulation and comparison with scale-model experiment

To confirm the distribution of the location, number, and size of the scattered soil clods after tilling, we divided the area as described in section 3.3 and examined the clod distribution. Figure 9 shows the divisions of the areas in the simulation.

A model simulating the soil with water content and adhesion was created using a DEM. Soil dispersal was evaluated by changing the rotational speed of the model. Figure 10(a) shows the scattering of the soil clods in the DEM. The soil clods are colored red.

Figure 10(b) shows the distribution of the size and number of soil clods per area. The number of soil clods increased as the rotational speed increased, and the dispersal area changed from area 2 to areas 1 and 3. The total number of clods at 200 rpm was 14, while area 3 accounted for only one clod (7%). On the other hand, at 300 rpm, area 3 accounts for 10 (42%) of the total number of dispersal areas, compared to 24 for the total. This indicates that the total number of clods increased, and the direction of scattering was biased toward Area 3. This trend is consistent with the results presented in section 3.3.

However, the average size of the soil clods was smaller than that observed in the experiment. The average size of the experimental soil clods at 200 rpm was 267 mm^2 at 10% water content and 488 mm^2 at 12% water content, compared to 29 mm^2 in the DEM analysis. The number of soil clods was extremely low at 100 rpm and extremely high at 300 rpm compared to the experiment. At 100 rpm, the number of soil clods was 4.6 and 2.8 for moisture content of 10% and 12%, respectively, while it was 0 in the simulation. At 300 rpm, the number of clods was 13.0 and 7.2 for moisture content of 10% and 12%, respectively, compared to 24 in the simulation. Therefore, the parameters of Edinburgh-Elasto-Plastic-Adhesion (EEPA) need to be modified. The adhesion force between particles needs to be increased to represent the scattering of soil clods during tiling obtained in this experiment.

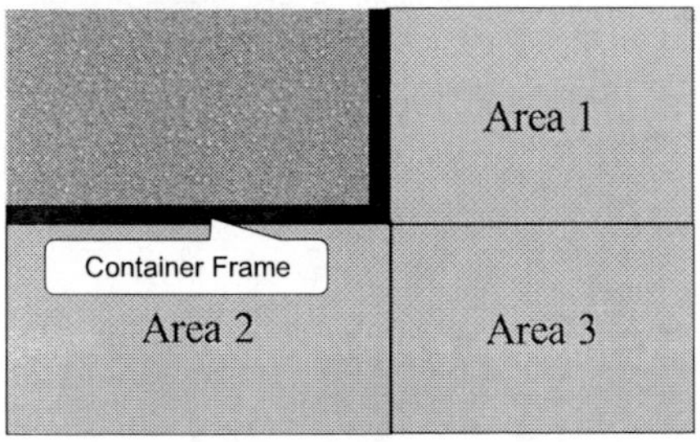

Figure 9. Detection area classification in DEM

Overall, the scattering distance of soil clods was short. The proportion of soil clods distributed in area 3 was small at a rotational speed of 200 rpm. In the experiment, this ratio is 24% at a moisture content ratio of 10% and 30% at a moisture content ratio of 12%, compared to 7% in the simulation. To resolve the difference between simulation and experiment, a model that increases particle-to-particle adhesion should be able to approximate experimental results. The energy used to destroy the soil clods could be used for dispersal.

4. Conclusion

In this study, we developed a system that can detect the presence or absence of soil clods based on depth images, which is important for the development of rotaries. A scale model of a tilling test was created. Experiments confirmed that the proposed method can accurately recognize soil clods. We also analyzed the mechanism of the proposed method when the rotation speed and moisture content changed and were able to compare the results with the DEM simulations.

As a prospect, we would like to improve recognition accuracy in situations where soil clods are difficult to recognize, such as the top of the sediment. In addition, by applying the soil cold recognition system to real-world environments, we would use the system in the evaluation of actual equipment.

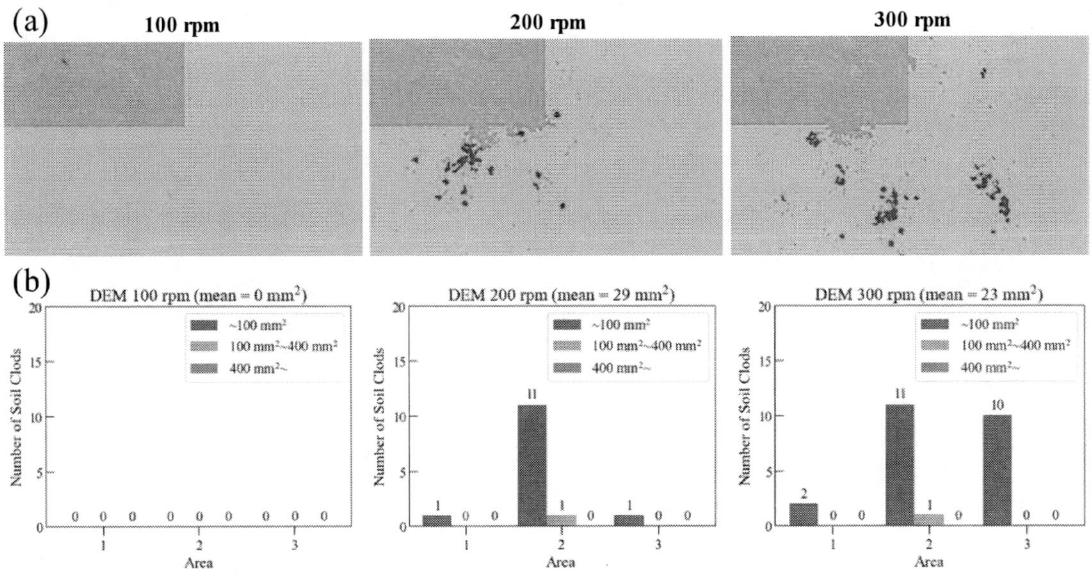

Figure 10. Soil clod dispersal: (a) appearance and (b) size distribution per area

5. References

Beucher, S. (1992) The Watershed Transformation Applied to Image Segmentation, Scanning Microscopy: Vol. 1992: No. 6, 28.

Cheng, J., Zheng, K., Xia, J., Liu, G., Jiang, L., & Li, D. (2021). Analysis of adhesion between wet clay soil and rotary tillage part in paddy field based on discrete element method. Processes, 9(5):845.

Chertkiattipol, S., & Niyamapa, T. (2010). Variations of torque and specific tilling energy for different rotary blades. International Agricultural Engineering Journal, 19(3), 1-14.

Grundy, L., Ghimire, C., & Snow, V. (2020). Characterisation of soil micro-topography using a depth camera. MethodsX, 7(September), 101144.

Guo, Q., Wang, Y., Yang, S., & Xiang, Z. (2022). A method of blasted rock image segmentation based on improved watershed algorithm. Scientific Reports, 12(1) :7143.

Hirasawa, K., Kataoka, T., Niu, H., Kubo, T. (2014). Analysis of Dynamic Phenomenon in Rotary Tilling (Part 2). Journal of the Japanese Society of Agricultural Machinery and Food Engineers, 76(1), 55-61.

Hirasawa, K., Kataoka, T., & Kubo, T. (2013). Prediction and evaluation for leveling performance in rotary tiller. In IFAC Proceedings Volumes (IFAC-Papers Online) (Vol. 4, Issue PART 1). IFAC.

Li, S., Chen, X., Chen, W., Zhu, S., Li, Y., Yang, L., Xie, S., & Yang, M. (2018). Soil-cutting simulation and parameter optimization of handheld tiller's rotary blade by Smoothed Particle Hydrodynamics modelling and Taguchi method. Journal of Cleaner Production, 179, 55-62.

Matin, M. A., Desbiolles, J. M. A., & Fielke, J. M. (2016). Strip-tillage using rotating straight blades: Effect of cutting-edge geometry on furrow parameters. Soil and Tillage Research, 155, 271-279.

Matin, M. A., Fielke, J. M., & Desbiolles, J. M. A. (2015). Torque and energy characteristics for strip-tillage cultivation when cutting furrows using three designs of rotary blade. Biosystems Engineering, 129, 329-340.

Matin, Md A., Hossain, M. I., Gathala, M. K., Timsina, J., & Krupnik, T. J. (2021). Optimal design and setting of rotary strip-tiller blades to intensify dry season cropping in Asian wet clay soil conditions. Soil and Tillage Research, 207(May 2020), 104854.

Morrisey, J. P., Thakur, S. C., & Ooi, J. Y. (2014). EDEM Contact Model: Adhesive Elasto-Plastic Model. Granul. Matter, 16(3), 383-400.

Usaborisut, P., Sukcharoenvipharat, W., & Choedkiatphon, S. (2020). Tilling tests of rotary tiller and power harrow after subsoiling. Journal of the Saudi Society of Agricultural Sciences, 19(6), 391-400.

Zhao, H., Li, H., Ma, S., He, J., Wang, Q., Lu, C., Zheng, Z., & Zhang, Chao. (2020). The effect of various edge-curve types of plain-straight blades for strip tillage seeding on torque and soil disturbance using DEM. Soil and Tillage Research, 202(17), 104674

8658 / Investigation of the Relationship between the Cone Index and the Physical and Mechanical Parameters of the Soil of Typical Surfaces of the Movement of Agricultural Tractors and Machines

Sergey Zhukov[a]*, Alexander Belyaev[b]
Vladimir Belyakov[b], and Vladimir Makarov[b],
[a] *Nizhny Novgorod State Engineering and Economic University, Knyaginino, Nizhny Novgorod region, Russia*
[b] *Nizhny Novgorod State Technical University n.a. R.E.Alekseev, Nizhny Novgorod, Russia*
**Corresponding author: ser-9.02.94@yandex.ru*

https://doi.org/10.56884/VNZQ2520

Abstract

The development of high-performance mobile transport complexes for agricultural purposes is largely based on the results of scientific research. The main direction is to study the mobility properties of robotic complexes in specific natural and climatic conditions of the area. The purpose of this work is to determine the cone index CI for typical surfaces of agricultural machinery movement, based on the physical and mechanical characteristics of agricultural soils. In 2020 year, a large-scale experimental and theoretical study of agricultural soils was conducted for 7 months using a soil penetrometer and equipment for measuring soil density and humidity. The peculiarity of this work is that the study has a spatiotemporal character of changes in the mechanical parameters of soils. Thus, the cone index CI is determined during the agricultural season for five typical plots: the first plot is fallow land located on a slope; the second plot is virgin land; the third section is a plowed field; the fourth section is a dirt road and the fifth section is a dirt road covered with broken bricks. Thus, the study showed a change in the cone index for each site during the year: the first site – 4.8-10.93 kPa; the second site – 5.13-5.32 kPa; the third site – 4.14-4.57 kPa; the fourth site - 4.43 – 11.20 kPa; the fifth site – 5.57 – 5.81 kPa. The data were also approximated and a mathematical model for calculating the cone index from the deformation modulus E, (MPa) was obtained. The obtained mathematical dependencies can be used in the future as the basis for mapping the mobility of any region of the world, as well as in the compilation of the algorithm of the mobile robotic complex.

Keywords: terrain mobility, tracked vehicle, inverse approach

1. Introduction

The effectiveness of the technological agricultural operation largely depends on the conditions of movement of the terrain. The driving conditions of agricultural machines are determined from the following components: 1) weather and climatic characteristics; 2) terrain; 3) physical and mechanical parameters of the soil base. The first and second components directly affect the mechanics of the interaction of the soil and the engine of the machine. Currently, Industry 4.0 is taking place in agricultural production and digital and robotic technologies are being integrated into the production processes. Accordingly, the introduction of autonomous robotic mobile complexes is required, for which it is necessary to develop and implement algorithms that will predict the mechanics of the interaction of the engine of the machine with the roadbed.

The mathematical model should be based on previously obtained data on the physical and mechanical state of soils. In 2020, an experimental and theoretical study was conducted and the following parameters were obtained: deformation modulus, soil specific adhesion, internal friction angle, shear modulus as soil characteristics. But these values represent an extremely large array of data, which is quite difficult to implement into the algorithm. Therefore, it was decided to identify the relationship between the known data and obtain an indicator of the resistance of the soil to the indentation of the conical indenter, which is part of the general "WES-model" and more simply and fully describes the mechanical state of the roadbed. This work is a continuation of the study [1], where the main soil base is loam and the roadbed is represented by several main routes for moving agricultural machinery.

2. Purpose and objectives of the study

This work is part of a general study aimed at increasing the productivity of the agricultural sector by introducing automatic robotic mobile complexes into the chain of technological processes of production while minimizing the risks of the human factor. Since the role of man in mobile agricultural complexes is quite high, it is necessary to achieve its reduction. By integrating intelligent systems into the movement of transport and technological machines, the role of the human operator is significantly reduced. The high performance of intelligent systems is justified by the perfection of the algorithm, which takes into account all aspects of the TTM operation within the area of operation. The purpose of our work is to determine and study the relationship between physical and mechanical parameters and to supplement the scientific description of connected soils using a cone indenter as the most practically applicable indicator for basic soils and typical routes of agricultural machinery. To achieve the goal, it is necessary to solve the following tasks:

1) Determination of physical and mechanical parameters of agricultural soils and dynamics of their changes during the agricultural season.

2) Identification of the mathematical relationship of physical and mechanical parameters through the modulus of deformation as the main value.
3) Investigation of the relationship between physical and mechanical parameters and a cone indenter for agricultural soils.

3. Experimental research

3.1 Materials and methods

As materials for the research, we use data obtained as part of an experimental and theoretical research of the physical and mechanical parameters of the soil of an agricultural area in 2020 year, as well as reference data on the mechanics of the interaction of the propulsion and the roadbed. Experimental data were obtained in an agricultural area (Figure 1).

The research was conducted during the agricultural season from April to October 2020. For the most complete representation of the specifics of the terrain, photographs of the plots were taken in April (Figure 2).

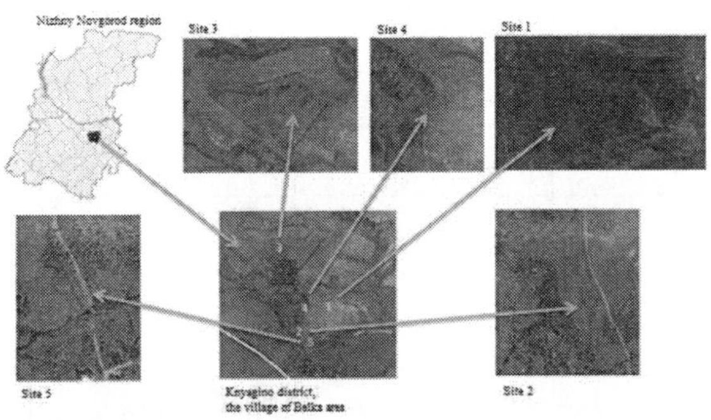

Figure 1. Map of terrain

Figure 2. Photos of the sites where the measurements were conducted.

3.2 Research methodology

In the field conditions were determined: the value of the penetrometer indicator (S, 10^{-6} m) with a PG-10 ground penetrometer; mass (M, kg) and humidity (%) of the soil by sampling (Figure 3). Further calculation was performed in the MS Excel 2020 environment to determine the following values according to the algorithm (Figure 4).

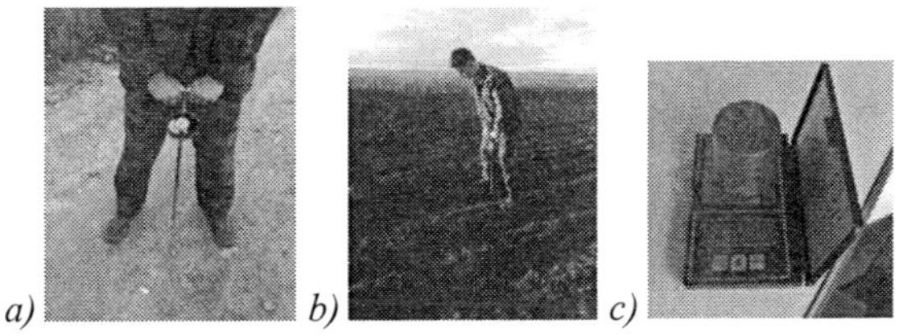

Figure 3. Determination of initial values:
a) determination of the value of the penetrometer of a dirt road; b) determination of the value of the penetrometer of arable land; c) determination of the mass of the soil sample

3.3 Processing of research results

The processing of the research results took place according to the algorithm for calculating the necessary indicators, which is considered in [Belyaev 2018, 2019, 2020, Belyaev and Makarov 2018, Makarov 2017]. The penetration force (P), the modulus of deformation of the soil (E, MPa), the angle of internal friction (φ, rad), the specific adhesion of the soil (C, MPa) are determined by the value of the indicator indicator of the penetrometer (S, microns) (Figure 4).

In the process of data processing, it turned out that some zones of the studied sites are heterogeneous in terms of humidity. Therefore, the data set was divided into ranges of soil moisture, since there was a significant difference between the values. At the end, the histograms in Figure 5 were obtained.

One of the main tasks of designing transport and technological machines is to predict mobility for any mechanical condition of the movement surface, which means it is necessary to study the dynamics of changes in physical and mechanical characteristics. A number of authors have previously proposed theoretical dependencies that allow determining the traction properties of wheeled vehicles through the use of mathematical dependencies between physical and mechanical parameters of forest soils [2],[3],[4]. Newer research suggests using the index of the cone indenter *CI* as a more optimal value for predicting mobility [5]. In our research, we use a similar technique for determining the specific adhesion of soil C (MPa) and the angle of internal friction φ (rad) through the main parameter - the deformation modulus E (MPa). The second stage of the study is most relevant for site №1 – field and site № 2 – dirt road, as the most intensive from the point of view of the movement of agricultural machinery. Sites № 3, № 4, № 5 are homogeneous in type of soil, but different in technological condition at this time and with less intensive traffic, the analysis of dependencies has an additional character and will be demonstrated in subsequent works. Thus, we obtain the following mathematical dependencies for site №1 and site №2 (Table 1) and graphs (Figure 2).

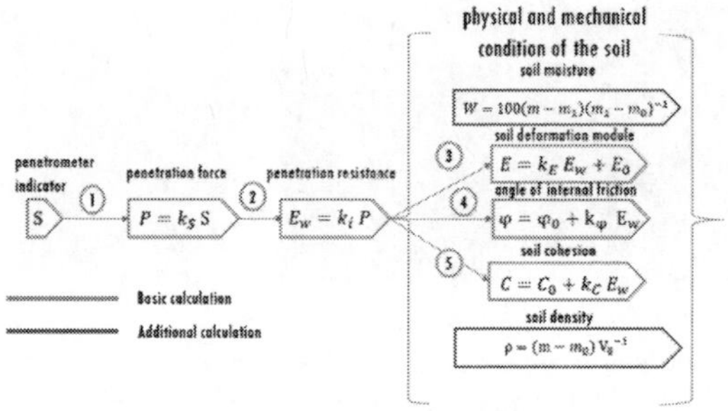

Figure 4. Algorithm for calculating physical and mechanical characteristics

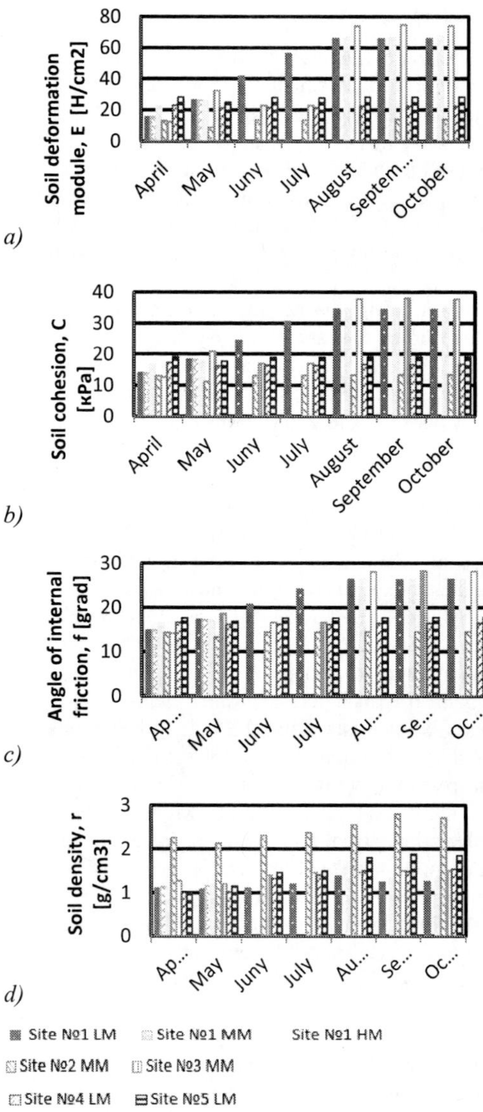

Fig. 5. Dynamics of changes in physical and mechanical characteristics during the season:
a) Soil deformation modulus, E; b) Soil adhesion, C; c) Internal friction angle; d) soil density, r [g/cm3]; e) Mark (LM – low moisture; MM – middle moisture; HM – high moisture)

Table №1 Mathematical relationship between physical and mechanical parameters

Month / Equation	**1. G [kPa] –E [MPa]** **2. C, [MPa] - E [MPa]** **3. φ [rad] - E [MPa]**
Site №1	
April	$G = 1838.2E - 10^{-12}$ $R^2 = 1$ $C = 4.7458E - 0.0357$ $R^2 = 0.8799$ $\varphi = 0.0085E - 0.0952$ $R^2 = 0.8592$
May	$G = 1838.2E - 10^{-12}$ $R^2 = 1$ $C = 0.2107E + 0.0075$ $R^2 = 0.8747$ $\varphi = 117.86 E + 11.216$ $R^2 = 1$
June	$G = 1838.2E - 7*10^{-12}$ $R^2 = 1$ $C = 0.2107E + 0.0075$ $R^2 = 0.901$ $\varphi = 117.86E + 11.216$ $R^2 = 0.995$
July	$G = 1838.2E - 7*10^{-13}$ $R^2 = 1$ $C = 0.2107E + 0.0075$ $R^2 = 0.899$ $\varphi = 117.86E + 11.216$ $R^2 = 1$
August	$G=1838.2E$ $R^2 = 1$ $C = 0.2107E + 0.0075$ $R^2 = 0.922$ $\varphi = 117.86E + 11.216$ $R^2 = 0.879$
September	$G = 1838.2E + 5*10^{-12}$ $R^2 = 1$ $C = 0.2107E + 0.0075$ $R^2 = 0.922$ $\varphi = 117.86E + 11.216$ $R^2 = 1$
October	$G = 1838.2E - 5*10^{-12}$ $R^2 = 1$ $C = 0.2107E + 0.0075$ $R^2 = 0.871$ $\varphi = 117.86E + 11.216$ $R^2 = 0.9823$
Site №2	
April	$G = 1838.2E$ $R^2 = 0.989$ $C = 0.2107E + 0.0075$ $R^2 = 0.998$ $\varphi = 117.86E + 11.216$ $R^2 = 0.9823$
May	$G = 1838.2E + 1*10^{-12}$ $R^2 = 1$ $C = 0.2107E + 0.0075$ $R^2 = 0.998$ $\varphi = 117.86E + 11.216$ $R^2 = 0.8891$
June	$G = 1838.2E - 2*10^{-12}$ $R^2 = 1$ $C = 0.2107E + 0.0075$ $R^2 = 0.998$ $\varphi = 117.86E + 11.216$ $R^2 = 1$
July	$G = 1838.2E$ $R^2 = 0.988$ $C = 0.2107E + 0.0075$ $R^2 = 0.998$ $\varphi = 117.86E + 11.216$ $R^2 = 1$
August	$G = 1838.2E - 4*10^{-12}$ $R^2 = 1$ $C = 0.2107E + 0.0075$ $R^2 = 0,998$ $\varphi = 117.86E + 11.216$ $R^2 = 0.978$
September	$G = 1838.2E + 2*10^{-12}$ $R^2 = 1$ $C = 0.2107E + 0.0075$ $R^2 = 0.998$ $\varphi = 117.86E + 11.216$ $R^2 = 1$
October	$G = 1838.2E$ $R^2 = 0.99$ $C = 0.2107E + 0.0075$ $R^2 = 0.998$ $\varphi = 117.86E + 11.216$ $R^2 = 1$

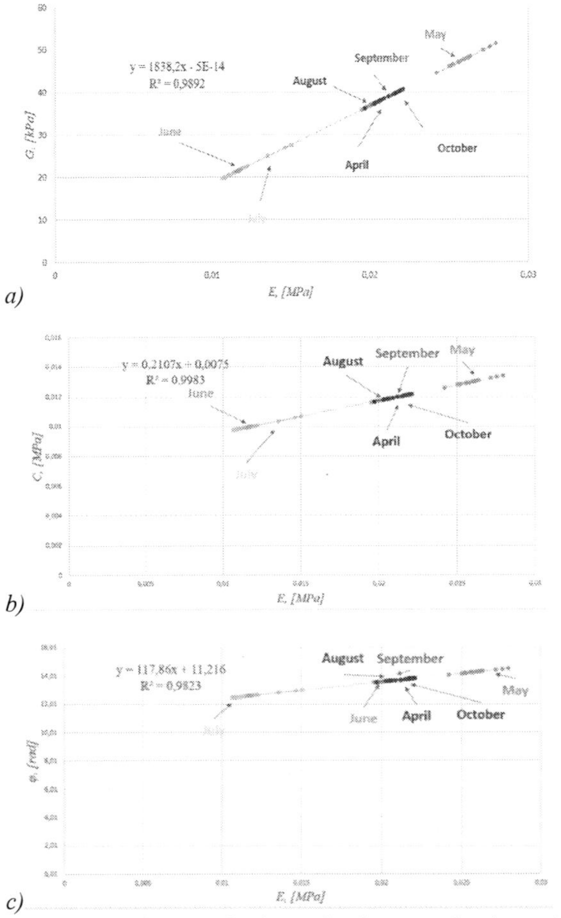

Figure 6. Mathematical relationship between the physical and mechanical parameters of the soil:
a) Dependence of the shear modulus G, [kPa] and the deformation modulus E, [MPa]
b) The dependence of the specific adhesion of the soil C, [MPa] and the modulus of deformation E, [MPa]
c) The dependence of the angle of internal friction φ, [rad] and the modulus of deformation E, [MPa]

The key feature of the obtained equations is their reliability and scale: 1) reliability is provided by the empirical method of obtaining and processing physical and mechanical parameters; 2) the scale lies in the fact that each site was surveyed for seven months in the same weather conditions. For each month, graphs of dependencies between parameters were obtained (Figures 6, 7) for site №1 and site №2.

For site №2, the physical and mechanical characteristics of the soil are as follows:

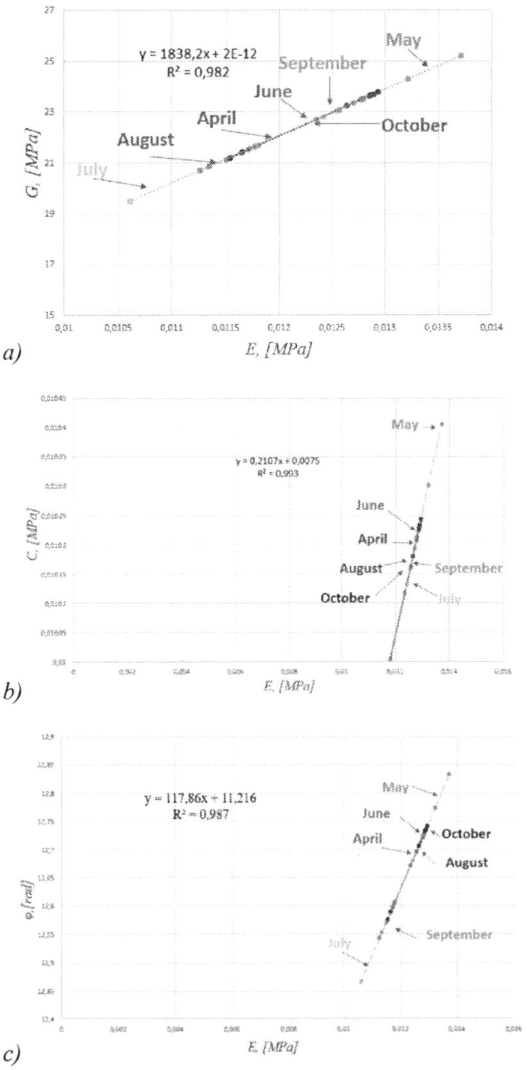

Fig. 7. Mathematical relationship between the physical and mechanical parameters of the soil:
a) Dependence of the shear modulus G, [kPa] and the deformation modulus E, [MPa]
b) The dependence of the specific adhesion of the soil C, [MPa] and the modulus of deformation E, [MPa]
c) The dependence of the angle of internal friction φ, [rad] and the modulus of deformation E, [MPa]

Graphs of physical and mechanical characteristics are quite informative and give a complete picture of the condition of the site for each month. Figures 6 and 7 show several general patterns: 1) the maximum values are reached in May due to an increase in the intensity of movement of heavy agricultural transport and technological machines; 2) the lowest value is in July, since crops enter the active growing season and the departure of agricultural machines in the field can have negative consequences; 3) on each graph there are similar values in April—the beginning of the departure of agricultural machinery in the field and in August, September and October—this is the period of the harvesting campaign and a further decrease in the activity of agricultural machinery in the field. The result of this research is to obtain a generalized mathematical dependence and the confidence value of the approximation R2 between the parameters (Table 2). The systematization of mathematical dependencies shows that the change in physical and mechanical characteristics occurs according to the same function and obeys a linear law.

Table 2. Equations of dependences of physical and mechanical characteristics

Month / Equation	G [kPa] – E [MPa] C, [MPa] - E [MPa] φ, [rad] - E [MPa]
Site №1	$G = 1838.2E - 5*10^{-14}$ $R^2 = 0.9892$ $C = 0.2107E + 0.0075$ $R^2 = 0.9983$ $\varphi = 117.86E + 11.216$ $R^2 = 0.9823$
Site №2	$G = 1838.2E + 2*10^{-12}$ $R^2 = 0.982$ $C = 0.2107E + 0.0075$ $R^2 = 0.993$ $\varphi = 117.86E + 11.216$ $R^2 = 0.987$

The third stage of the research is to determine the value of the cone indenter *CI* of the soil for agricultural use. The scientific work [5] describes in detail the essence of «WES-modeling» and the role of the cone indenter *CI*. Also in this research, the author obtained a relationship between the deformation modulus *E* and the value of the cone indenter *CI* for loamy soils [5]:

$$E = 8.914 + 5.019 \, ln \, CI \qquad (1)$$

The data of the deformation modulus *E* is already available in the data array and it is necessary to determine the value *CI* for specific surfaces of motion, for this you need to express the necessary value from the formula:

$$CI = e^{\ln CI} = e^{\frac{E-8.914}{5.019}} \qquad (2)$$

Using the MS Excel 2020, we determine the values of the cone indenter and plot the dependence graph (Figure 8).

The graph provides data on the values of the cone indenter for loamy soil, so the spread of values is indicated in Table 3.

Based on the known dependence (2) and the obtained values of the cone indenter *CI* (Figure 8), it is possible to determine the minimum and maximum values of the indicator, as well as to determine the deviation of the value from the average value. It can be seen from the graphs that the *CI* values for site № 1 and № 2 are quite close. This is due to the fact that the surface of the movement is loam and the mechanics of the soil is homogeneous. Since the dirt road under study leads to the field under study, the intensity of the impact of energy-saturated and heavy machines is approximately the same. Comparing the *CI* values with reference values, it can be seen that the obtained values correspond to loams and clays of low plasticity

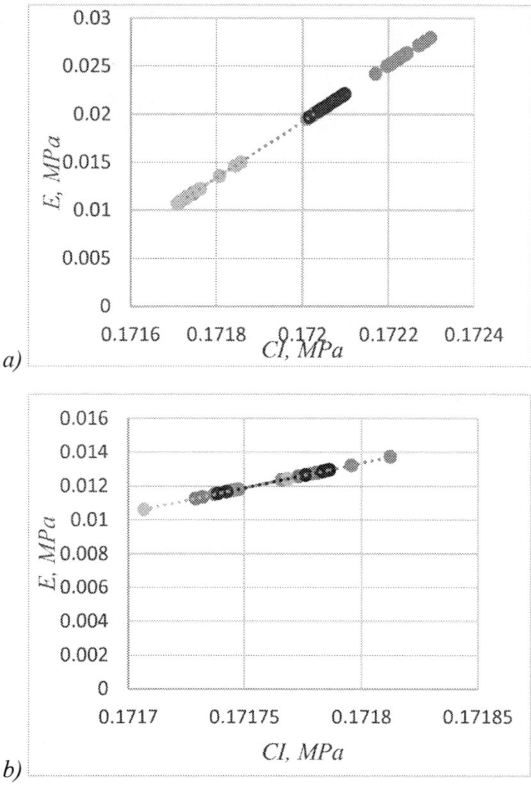

Figure 8. Dependences of the cone indenter on the modulus of loam deformation: a) site № 1; b) site #2

Table 3. Average value and deviation of CI

№	CI_{min}, MPa	CI_{max}, MPa	ΔCI, MPa	$\overline{CI}$, MPa	σ %
1.	0.1717	0.1723	0.0006	0.172	0.2
2.	0.1717	0.17181	0.00011	0.171755	0.1

4. Results

The relationship between the physical and mechanical parameters and the cone indenter during the agricultural season for loamy soil is investigated. This relationship allows us to evaluate the bearing properties of the movement surface, that is, to select the permissible mass of machinery and predict the mobility of agricultural machinery at any time of the season.

1) The physical and mechanical parameters of loam as an agricultural soil and the dynamics of their changes during the agricultural season are **determined.**

2) A linear distribution law is revealed and a mathematical relationship is determined in the form of equations for each month of the season and a general equation taking into account the space-time factor between:

a) the soil shear modulus G [kPa] and the deformation modulus E [MPa].

b) the specific adhesion of the soil C, [MPa] and the deformation modulus E [MPa].

c) the angle of internal friction, φ [rad] and the modulus of deformation E [MPa].

5. Further research

The developed methodology opens up a wide field of scientific activity in the study of mobility of transport and technological machines for any purpose. The final point of the research is the development and testing of a decision–making algorithm for autonomous robotic agricultural complexes based on artificial intelligence. To reach the final stage, you need to perform the following tasks:

1. To clarify the mathematical relationship between the deformation modulus *E*, [MPa] and the cone indenter *CI* [MPa]. To do this, it is necessary to consider site № 3, №4 and №5 according to the developed methodology, as well as additionally investigate the deformation modulus during a period of large amounts of climatic precipitation.
2. To conduct a "winter" block of research, according to which it is necessary to develop a physical and mechanical model of snow cover and a statistical model of snow cover changes as a surface of movement of transport and technological machines.
3. On the basis of points 1 and 2, develop a dynamic terrain mobility map to predict the changing conditions of the movement surface on the basis of a mathematical regularity
4. To develop a decision-making algorithm as the basis for the functioning of an adaptive chassis for autonomous robotic agricultural complexes.

6. Acknowledgements

The results of the given study have been obtained with financial support of the grants of the President of the Russian Federation № MK-336.2022.4.

7. References

Zhukov S.S. Study of trafficability conditions of typical agricultural soils in the Nizhny Novgorod region (Russian Federation), S.S. Zhukov, V.V. Belyakov, V.S. Makarov, Proceedings of the 20th International and 9th Americas Conference of the International Society for Terrain-Vehicle Systems, ISTVS 2021, 2021

Investigation of the coefficient of resistance to the movement of wheeled forest vehicles [Studies of rolling resistance coefficient of wheeled forestry machines] / I. V. Grigor'ev [et al.] // Vestnik MGUL – Lesnoj vestnik [Herald MSFU – Forestry Gazette]. – 2014. – No. S2. – P. 36-41.

Posharnikov F. V. Improvement of technical means for forest nurseries [Improvement of technical equipment for forest nurseries] / F. V. Posharnikov, V. S. Popov, V. G. Sviridov // Lesotehnicheskii jurnal [Forestry Journal]. – 2011. – № 4. – P. 110-117.

Calculation of traction mine-technical properties of a wheeled skidding tractor using data from foreign colleagues [Calculation the tractive performance of a wheeled skidder using foreign colleagues data] / S. E. Rudov, E. G. Khitrov, M. E. Rudov, V. V. Ustinov // Aktual'nye napravlenija nauchnyh issledovanij XXI veka: teorija i praktika [Current research trends of the XXI century: Theory and Practice]. – 2015. – T. 3. – No. 1 (12). – P. 223-228.

Khitrov, E. G. Investigation of the connection of the cone index and the modulus of deformation of various types of soils [Text] / E.G. Khitrov, A.M. Khakhina, V.A. Luhminsky,

D. P. Kazakov // Resources and Technologies, volume 14 (2017), number 4, 1 - 16 P.

Ageikin, Ya.S., 1972. All-terrain wheels and dual propellers. Theory and Design. Mashinostroenie. Moscow, 232 p.

Ageikin, Ya.S. and Volskaya, N. S., 2008. Theory of car. Moscow, MGIU, 318 p.

Ageikin, Ya.S., Volskaya, N. S. and Chichekin, I.V., 2010. Trafficability car. Moscow, MGIU, 275 p.

Barakhtanov, L V et al. 1996. Trafficability car. Nizhny Novgorod, NNSTU, 200 p.

Belyaev, A. M. and Makarov, V. S. 2018. Method of assessment of special wheel chassis mobility in cases of sand-gravel bases crossing. Paper presented at the IOP Conference Series: Earth and Environmental Science 194(2) doi:10.1088/1755-1315/194/2/022019

Belyaev, A. M. et al. 2018. Analysis of the main bases of the coastal zone, as a pathway for the movement of special vehicle and monitoring complexes. Paper presented at the IOP Conference Series: Materials Science and Engineering 386(1) doi:10.1088/1757-899X/386/1/012002

Belyaev, A. M. et al. 2019 Field research of drag-and-traction characteristics of mobile robotic system in coastal zone. Paper presented at the Journal of Physics: Conference Series 1177(1) doi:10.1088/1742-6596/1177/1/012050

Belyaev, A. et al. 2020. Experimental Theoretical Study of the Mobile Robotic System Movement with Caterpillar-modular Propulsion on the Beach Line Terrain. In Proceedings of the 6th International Conference on Vehicle Technology and Intelligent Transport Systems (VEHITS 2020), 567-572. DOI:10.5220/0009794705670572

Bekker, M. 1960. Theory of land locomotion. University of Michigan Press, 520 p.

Belyakov, V.V. and Kulyashov, A.P. (Eds.), 2004. All-terrain and technological vehicles. Basics of motion theory. Scientific and technical literature (Monograph). TALAM. N. Novgorod 560 p.

Makarov, V., et al. 2017. Study of trafficability conditions of typical soils of coastal zones of Sakhalin Island (Russian Federation). Paper presented at the 19th International and 14th European-African Regional Conference of the ISTVS

Volskaya, N. S., 2017. Assessment of patency of wheeled vehicles when driving on uneven dirt surface. MGIU, 215p.

Wong, J. Y. 2010. Terramechanics and Off-Road Vehicle Engineering. Elsevier. 463 p.

9352 / 3D-DEM Simulation and Post-Process Method of Wheel-Terrain Interaction for Planetary Rovers

Qingning Lan, Zhengyin Wang, Huaiguang Yang*, Liang Ding, and Haibo Gao
State Key Laboratory of Robotics and System, Harbin Institute of Technology, Harbin 150001, China
**Corresponding author: yanghuaiguang@hit.edu.cn*

https://doi.org/10.56884/WWUX2021

Abstract

Understanding the physical essence of wheel-terrain interaction is critical to establish the terramechanics model. The discrete element method (DEM) can be used to simulate a series of wheel-terrain interaction quickly and efficiently, which can be convenient to analyze the physical information of the interaction. This paper simulates the interaction of slipping wheel running on soil simulant using 3D-DEM. The information of particles and wheel is obtained from the wheel-terrain interaction. The consequent velocity and stress field of soil simulant is obtained by smoothening the velocity of discrete soil particles and contact interaction force between particles. A method to calculate the fraction field of soil simulant in the particle bed with boundary is proposed to obtain the consequent fraction field distribution. The simulation results are consistent with experimental results. This study can provide instructions for the establishment of physical terramechanics model.

Keywords: terramechanics, DEM simulation, wheel-terrain interaction

1. Introduction

In recent years, several countries have launched planetary probes, setting off a new trend of space exploration. Planetary rovers are widely used in the exploration of Mars and the Moon, which is efficient and reliable mobile research platforms in complex environments. The terramechanics model for the rover's wheel is an important part of the planetary rover design and control research, aiming to model the wheel's stress distribution along the rim during the motion of the wheel on the surface of the extraterrestrial planet. The terramechanics model can be used to optimize the design of the planetary rover's wheel (Niksirat et al., 2020), improve the movement control effect of the planetary rover (Recuero et al., 2017), and improve the accuracy of the simulation of planetary rover (Zheng et al., 2018).

Most of the current terramechanics models of planetary vehicles are semi-empirical models with less research on the motion of soil particles, which only focus on the motion state of the wheels and the soil's parameters. Comprehensive analysis of the relationship between velocity, stress and fraction field distribution of the soil during the wheel-terrain interaction can provide new solution ideas for the wheel-terrain interaction and help to improve the accuracy of the terramechanics model.

The most common method to study soil particle flow is the discrete element method (DEM). The planetary soil is simulated as thousands of discrete particles, and the microscopic information of the wheel-terrain interaction can be obtained by recording the motion information of each discrete particle. DEM simulation research is mostly used to predict the sinkage and drawbar pull of a single wheel at different slip ratios, so as to achieve rapid iteration and optimize wheel design (Jiang et al., 2018; Shin et al., 2020). Some studies focus on the flow of planetary soil and the stress distribution of the wheel to study the wheels' mobility and sinkage. For instance, Pang et al. (2021) simulated the mobility of the bionic wheel in the soil simulant using DEM, observed the velocity of single particles in part of soil simulant, and optimized the design of the bionic lugs. Zhang et al. (2020) conducted a single-wheel test simulation of an irregular legged wheel, compared the force and sinkage of the wheel between the test and the simulation, and optimized the simulation system. Ravula et al. (2021) analyzed the force of the wheel and force chain distribution of soil simulant at different times, which is helpful for the study of the dynamic process of the moving wheel. The current research generally analyzes the motion and force of a large number of discrete particles in the soil. The distribution of microscopic states such as velocity distribution and force chain of particles are relatively chaotic in the local. It is difficult to see the local and integral continuous properties of the particles, which means the analysis of the overall performance of the soil is not intuitive and incomplete. Planetary soil is a kind of granular medium exhibiting the properties of fluid or solid in different motion states. It can be regarded as a continuous medium when it flows under the intrusion of wheels (Li et al., 2013). Regarding granular medium as continuum can facilitate the study of its physical properties. Therefore, it is necessary to analyze the planetary soil in the wheel-terrain interaction as a continuum, which is needed to calculate and analyze the regular and continuous cloud diagrams of the velocity field, stress field and fraction field of the planetary soil.

In this paper, DEM simulation is used to carry out series of single-wheel test simulation with a fixed wheel sinkage and use a smoothing algorithm to study the continuous physical field of the soil simulant's velocity field and pressure field in the wheel-terrain interaction. The fraction field of soil simulant with particle object boundary is calculated using a convex hull-based fraction algorithm. It provides an analytical basis for understanding wheel-terrain interaction from the perspective of soil simulant.

2. DEM simulation of wheel-terrain interaction

This section used the DEM simulation to simulate the single-wheel test of the planetary wheel with fixed sinkage. The DEM simulation simulates the overall behavior of the granular medium by simulating the collisions between thousands of tiny particles. In this paper, the discrete element simulation software EDEM is used to conduct 3D-DEM simulation experiments of wheel-terrain interaction.

2.1 Contact model

In the DEM simulation, the soil simulant is composed of many independent tiny particles, and a single particle is a discrete element unit. The simulation calculates the motion and force of all particles in the simulated space, including the position, velocity, contact force and other information of the particles. The forces between different particles include normal force F_n, tangential force F_t and rolling friction torque $\boldsymbol{M}$. Normally, both normal and tangential forces are described as a kind of elastic damping spring contact, and there will be a force only when two particles are in contact, as shown in Figure 1 (Feng et al., 2019). The contact model between particles used in this paper is the standard Hertz-Mindlin model (Tsuji et al., 1993) with the friction coefficient between particles μ. The rolling friction between particles is also considered with the rolling friction coefficient μ_r (Sakaguchi et al., 1993). The specific contact force expressions are shown as Eq. (1)~(3).

$$F_n = k_n\delta_n^{3/2} + \gamma_n\delta_n^{1/4}v_n^{rel} \tag{1}$$

$$F_t = \begin{cases} k_t\delta_n^{1/2}\delta_t + \gamma_t\delta_t^{1/4}v_t^{rel} & \text{if } |F_t| < \mu|F_n| \\ \mu F_n & \text{else} \end{cases} \tag{2}$$

$$\boldsymbol{M} = -\mu_r F_n d\hat{\boldsymbol{\omega}} \tag{3}$$

where δ_n and δ_t are the relative displacements in the normal and tangential directions at the contact point of the two particles, respectively. v_n^{rel} are v_t^{rel} the relative velocities in the normal and tangential directions at the contact point of the two particles, respectively. d is the displacement of the particle contact point relative to the particle center of mass, and $\hat{\omega}$ is the unit vector in the direction of the angular velocity vector at the point of contact.

The specific function expressions of parameters are shown in Table 1 where k_n, k_t, γ_n, γ_t are functions of particle diameter d, Young's modulus E, Poisson's ratio v and coefficient of restitution e. G is the shear modulus of the particle. m is the mass of a single particle, i.e., $m = \rho_g \pi d^3/6$, where ρ_g is the density of particle.

The choice of the time step has a great impact on the time consumption and accuracy of simulation. If the time step is too large, the particles may move irregularly, while it may cost more time if the time step is too short. For the DEM simulation of granular medium in quasi-static conditions, the time step can be empirically selected to be 20% of the Rayleigh time step T, i.e., the update time of simulation is selected to be about 20% of the period of the elastic wave propagating between particles (Paulick et al., 2015), as shown below:

$$T = \frac{\pi R\left(\rho_g/G\right)^{1/2}}{0.1631v + 0.8766} \tag{4}$$

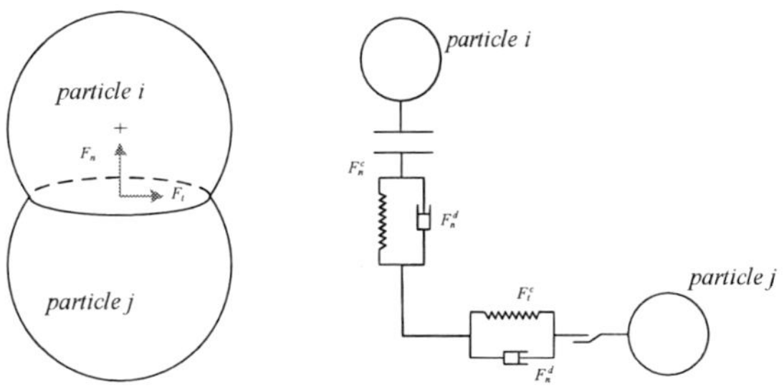

Figure 1. Schematic diagram of the force between particles

Table 1. Parameter expressions of the force model in the simulation

Parameter	Particle-particle interaction	Particle-object interaction
k_n	$\frac{E}{3(1-v^2)}\sqrt{d}$	$\frac{2\sqrt{2}E}{3(1-v^2)}\sqrt{d}$
k_t	$\frac{2G}{2-v}\sqrt{d}$	$\frac{4\sqrt{2}G}{2-v}\sqrt{d}$
γ_n	$-\sqrt{\frac{5}{2}}\beta\sqrt{mk_n}$	$-\sqrt{5}\beta\sqrt{mk_n}$
γ_t	$-\sqrt{\frac{5}{3}}\beta\sqrt{mk_t}$	$-\sqrt{\frac{10}{3}}\beta\sqrt{mk_t}$

Note: $\beta = \ln e/\sqrt{\ln^2 e + \pi^2}$

2.2 Establishment of simulation system

Since the research of granular medium physics is generally under the working condition of fixed depth. The wheel-terrain interaction is simulated under the condition of fixed wheel sinkage by EDEM. The motion information of particle and wheel is recorded during the wheel movement, which is beneficial for model analysis and post-process of granular medium.

Slip ratio is an important concept in terramechanics, which is used to describe the degree of wheel slip and skid. In order to facilitate the analysis of the two working conditions, this paper defines the slip ratio as follows (Ding, 2009):

$$s = \begin{cases} \omega r/v - 1 & (\omega r < v) \\ 1 - v/\omega r & (\omega r \geq v) \end{cases} \quad (5)$$

where ω is the angular velocity of the wheel. v is the translational velocity of the wheel, and r is the wheel radius. If s is less than zero, it is at a status of skid that often occurs in the situation of downhill and break. If s is larger than zero, it is at a status of slip that often occurs in the situation of uphill and acceleration.

It can be known from Eq. (5) that when the wheel's translational velocity is fixed and the slip ratio is small ($s \leq 0.2$), the wheel's angular velocity ω increases almost linearly with the increase of the slip ratio s, and when the slip ratio s is large ($s > 0.2$), the wheel's angular velocity ω begins to increase rapidly and nonlinearly with the increase of the slip ratio s. Because the degree of slip of the wheel is not large when it is working normally, the range of the slip ratio selected in this paper is -1 to 0.6.

The diameter and Young's modulus of the granular medium are the quantities that influence simulation results significantly. The smaller the diameter of the granular medium or the larger the Young's modulus, the closer to the real situation, but the simulation time will be greatly increased. In this paper, DEM simulation parameters were selected to make the response angle of the granular medium as a typical value. The soil simulant is composed with many same single particles to accelerate simulation. This paper chose the shear modulus G and Poisson's ratio v commonly used in the literature, whose values are 10MPa and 0.3 respectively (Guillard et al., 2014). The Young's modulus E can be determined to be 26MPa. The diameter of the particle d was chosen to be smaller than 1/20 of the minimum size of the object. The wheel slip ratio simulated was series typical value selected from -1.0 to 0.6. The wheel's translational velocity was selected as a typical value of about $v = \sqrt{2gd_g}/5$, and the fixed wheel sinkage was selected so that the wheel entrance angle θ_1 is close to 30°, which is a typical value of entrance angle. The friction coefficient between particles was selected that the test result of the response angle is about 38°, which is a typical value for simulating the Martian soil. The specific parameter configuration used in the DEM simulation is shown in Table 2.

The specific steps of the simulation are as follows:

(1) DEM simulation parameters are determined and defined. The parameter calibration program provided by the EDEM software was used for parameter calibration.
(2) Particle samples are generated, and the granular medium adopts a single sphere model. This single sphere model can reduce the simulation time and obtain good results.
(3) The container for the particle bed is created, and the width direction of the container is set as the periodic boundary, so that the influence of the lateral boundary of the container is as small as possible.
(4) The particles bed is piled up by dropping particles under gravity and then resting.
(5) The wheel is set to initial position. The simulation wheel adopts a wheel without lugs to explore the essential characteristics of the wheel-terrain interaction. The wheel's size is very close to that of China's Lunar rover "Yutu-2" (Ding et al., 2022). Create and move the wheel above the initial target position of the wheel and slowly drop it into the target sinkage of granular medium.
(6) The sinkage of wheel in the vertical direction is fixed, and the translational and rotational motion of the wheel are defined to fix the slip ratio of the wheel.
(7) The appropriate simulation step is set to simulate.

In the simulation, the wheel ran for a total of 15s, and the translational displacement was 1050 mm. The initial and final states of the system in the simulation are shown in Figure 2.

Table 2. DEM simulation parameter configuration

Parameters	Values
Wheel slip ratio s	-1.0, -0.6, -0.3, 0.0, 0.3, 0.6
Wheel radius r (mm)	135
Wheel width b (mm)	150
Wheel translational velocity v (mm/s)	70
Wheel sinkage z (mm)	18
Particle diameter d (mm)	6
Particle density ρ (kg/m^3)	2660
Particle Poisson's ratio v	0.3
Particle shear modulus G (MPa)	10
Coefficient of restitution e	0.3
Friction coefficient between particles μ_{gg}	0.7
Friction coefficient between particles and objects μ_{gi}	0.9
Rolling friction coefficient μ_r	0.15
Size of particle bed (mm)	1600×250×300

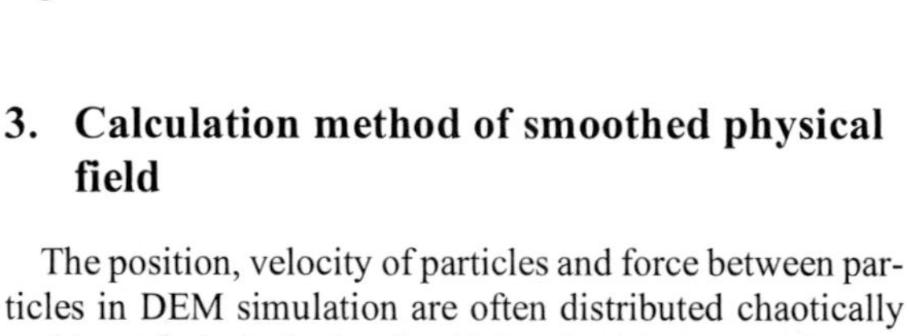

a) initial state *b) end state*

Figure 1. Simulation initial and end state

3. Calculation method of smoothed physical field

The position, velocity of particles and force between particles in DEM simulation are often distributed chaotically and irregularly in the local, which only exists a certain continuity in a large spatial range. Since the granular medium can be regarded as a continuum similar to solid and liquid in the wheel-terrain interaction, it is necessary to regard the planetary soil in the wheel-terrain interaction as a continuum for physical field analysis. It can provide new ideas to model the wheel-terrain interaction.

The wheel is symmetrical in the direction of width in the wheel-terrain interaction. This paper mainly focuses on the physical field distribution of the granular medium at the center section of the wheel for post-process, including the velocity, stress and fraction field distribution of granular medium.

In this paper, the velocity and stress field distributions of the soil simulant are calculated based on the physical field smoothing algorithm of DEM proposed by Goldhirsch and Andreotti et al. (Glasser and Goldhirsch, 2001; Goldhirsch, 2010). The algorithm is used to convert the physical quantities of discrete particles into a uniformly distributed physical field by averaging the velocity, stress and density over a spherical region around the target point of the computation. In order to obtain continuous and smoothed physical field distribution, a regular grid is divided for the calculated area and the physical field on each grid point is calculated separately.

Figure 3 is the schematic diagram of the smoothing algorithm. For each point to be calculated, a spherical area with radius R is chosen to average physical quantities. The velocity and stress of continuous medium at this point are averaged weightily by all the particles in this spherical area. The radius R of the spherical area is equivalent to the threshold value of the particles to be averaged. Only the particles within the area will be averaged, so the size of the spherical area has little effect on the results. R is chosen to be 3/2 the diameter of the particle d from reference (Guillard et al., 2014).

$$\begin{cases} \mathbf{v}(\mathbf{r}) = \dfrac{\sum_{i=1}^{N} \mathbf{v}_i \times \mathrm{W}(\mathbf{r}_i, \mathbf{r})}{\sum_{i=1}^{N} \mathrm{W}(\mathbf{r}_i, \mathbf{r})} \\ \boldsymbol{\sigma}(\mathbf{r}) = \dfrac{1}{V_{smooth}} \sum_{i,j \in V(\mathbf{r})} \mathbf{f}_{ij} \otimes \mathbf{r}_{ij} \mathrm{W}(\mathbf{r}_{ij}, \mathbf{r}) \end{cases} \quad (6)$$

$$\mathrm{W}(\mathbf{r}_i, \mathbf{r}) = \begin{cases} 0 & \|\mathbf{r}_i - \mathbf{r}\| \geq R + d/2 \\ 1 & \|\mathbf{r}_i - \mathbf{r}\| \leq R - d/2 \\ \dfrac{R + d/2 - \|\mathbf{r}_i - \mathbf{r}\|}{R - d/2} & others \end{cases} \quad (7)$$

where $\mathbf{r}$ is the displacement vector of the target point. $\mathbf{r}_i$ is the displacement vector of the ith particle. $\mathbf{v}_i$ is the velocity vector of the ith particle. $\mathbf{f}_{ij}$ and $\mathbf{r}_{ij}$ are the force and displacement between the ith particle and the jth particle, respectively, i.e., $\mathbf{r}_{ij} = \mathbf{r}_i - \mathbf{r}_j$. $\mathrm{W}(\mathbf{r}_i, \mathbf{r})$ is the contribution weight of a single particle to the target point. Since the working condition studied in this paper has a boundary and the diameter of the granular medium is not much smaller than the interval of the requested grid point, the segmented linear function in Eq. (7) is used to weight each particle by the distance from the particle to target point, and the particle completely within the spherical region is regarded as a complete contribution to the target point. V_{smooth} is the volume of the spherical smooth region. It can be found that the closer to the target point, the greater the contribution of that particle to the velocity and stress at the target point.

It is worth noting that when the scale of the granular medium particle is not negligibly small relative to the calculational region, the fraction on the boundary of the granular medium calculated by the smoothing algorithm is smaller than the real value. The reason is that the algorithm calculates the fraction based on particles in a given spherical region and the region is larger than real situation near the boundary significantly. The selected spherical region will exceed the boundary of the granular medium. It expands the boundary of the granular medium outward, making the spherical region larger relative to the actual area occupied by the particles, which causes the calculated fraction of the granular medium smaller than real.

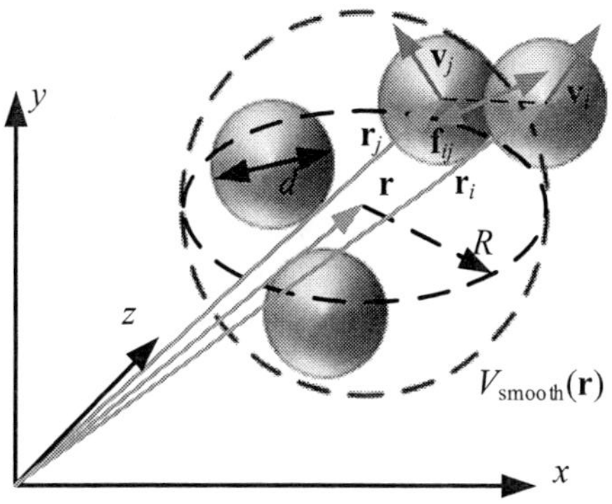

Figure 3. Schematic diagram of smoothing algorithm

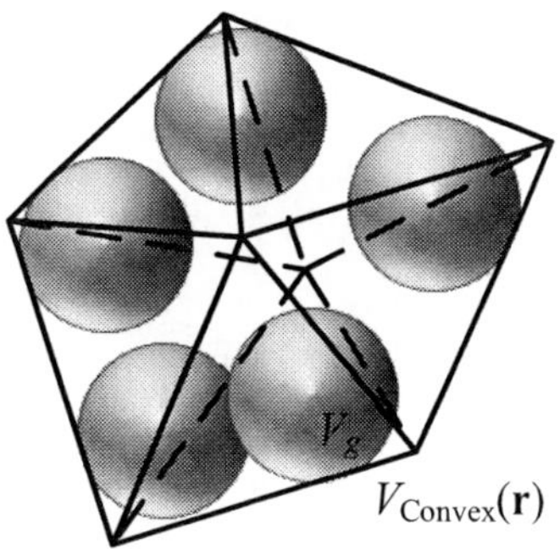

Figure 4. Schematic diagram of fraction algorithm based on convex hull

In order to solve the problems mentioned above, this paper proposed an algorithm to calculate the fraction field of granular medium based on the convex hull of 3D point cloud. This method calculates the fraction of granular medium under the condition of fixed space. As shown in Figure 4, in order to calculate the fraction of the granular medium at the target point, the N particles closest to the requested grid point are identified first, and then a number of points are added uniformly on the surface of these particles. The convex hull of these points on the surface of all particles is used to approximate the smallest polyhedron that can contain these N particles. Thus, the volume of the convex hull can be used to approximate the entire space that can contain these N particles. The fraction of this target point is the ratio of the volume of these particles to the volume of the convex hull. Therefore, the fraction of target point can be calculated by the following equation.

$$\phi(\mathbf{r}) = \frac{NV_g}{V_{\text{Convex}}(\mathbf{r})} \tag{8}$$

where V_g is the volume of a single particle, i.e., $V_g = \pi d^3/6$. $V_{\text{Convex}}(\mathbf{r})$ is the volume of the convex hull containing the nearest N particles at that point. Usually, the spherical region with a radius of $3/2d$ can completely include 13 particles with diameter of d at most. So N is selected as 13 in this paper.

4. Simulation Result

The diagram of the normal force and drawbar pull of the wheel versus the displacement in the simulation of the fixed sinkage condition is shown in Figure 5. After the wheel's displacement is greater than 400 mm, the wheel enters the steady state at most slip ratios, and this paper focuses on the steady state. When the wheel slips, the wheel generates a positive drawbar pull, which is consistent with terramechanics theory. In this paper, the period after the wheel's force curve enters the steady state is selected for post-process.

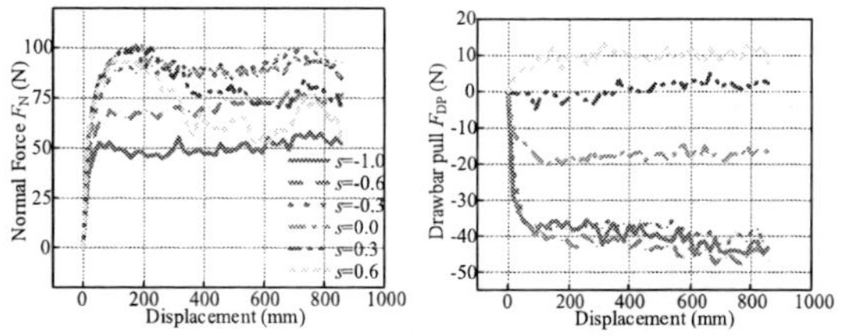

a) Normal force *b) Drawbar pull*

Figure 5. The force of wheel versus displacement

As shown in Figure 6, the velocity field distribution of the soil simulant at different slip ratios is shown. When the wheel slip ratio is -1, the wheel's angular velocity is zero, and there is only translational motion. The soil simulant piled up in front of the wheel, and the dividing line between the moving and non-moving parts of the soil simulant is almost horizontal, because the soil simulant is pushed by wheel to move forward horizontally. When the wheel starts to rotate, the accumulation of soil simulant in front of the wheel decreases with the increase of the slip ratio. When the wheel slip ratio is -0.3, the dividing line between the moving and non-moving parts of the soil simulant gradually shifts downward for the soil simulant rotating driven by the wheel. The direction of particle flow changes from horizontal to diagonal downward. The characteristics of the particle flow are consistent with the results obtained using PIV measurements in the literature, as shown in Figure Figure 7a (Senatore and Iagnemma, 2014). When the wheel shifts to a slip state, the accumulation of soil simulant in front of the wheel became almost negligible. The flow region of soil simulant under the wheel changes to two sectors, forward and backward respectively. However, the velocity of the particles is significantly smaller than in the skid state. When the wheel slip ratio is larger, as shown in Figure 6d, the soil simulant under the wheel moves almost only backward. The overall performance of soil simulant shows a tendency to move backward, while the sector area where the particles move forward almost disappears since the transportation effect of the wheel on the soil simulant gradually becomes significant. The experimental test of the slip condition shown in Figure 7b also shows that the velocity field of the granular medium under the wheel moving forward is very small. The particles almost only transport backward. Comparing the velocity field diagrams of the skid and slip conditions in Figure 6 and Figure 7, the particles move forward in the skid and backward in the slip on the whole. The diagrams have the same characteristics, indicating the reliability of the simulation.

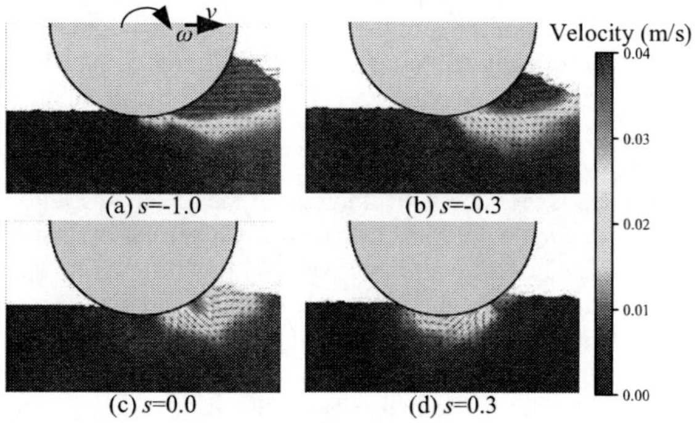

Figure 6. The velocity field of the soil simulant at different slip ratios obtained by simulation

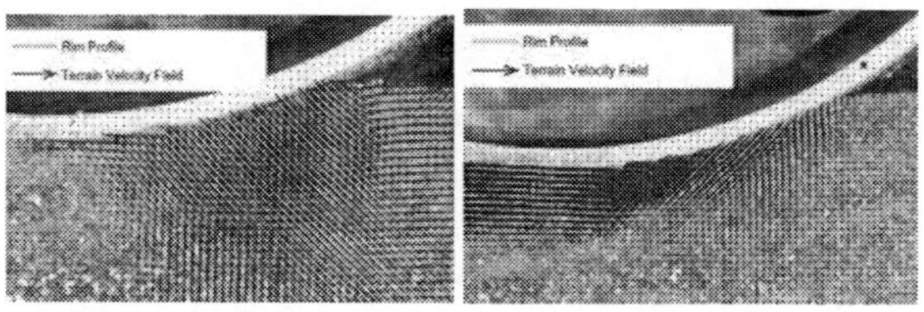

a) s = -0.3 *b) s = 0.3*

Figure 7. The velocity field of the soil at different slip ratios measured by PIV (Senatore and Iagnemma, 2014)

Figure 8 shows the fraction field distribution of soil simulant at different slip ratios based on the convex hull algorithm. The wheel pushes the soil simulant in front and disturbs it when slipping, causing the fraction of the soil simulant in front of the wheel getting smaller than the initial fraction. With the increase of wheel rotation, the area with smaller fraction gradually expands downward, as shown in Figure 8b, which is almost the same as the area with larger velocity in the slip state in Figure 6. It is due to the fact that only the particles disturbed can change their fraction. When the wheel enters the slip state, as shown in Figure 8d, the particles transfer from the front of the wheel to the rear of the wheel due to the transportation effect of the wheel on the soil simulant, the wheel squeezes the soil simulant in front and makes the fraction of the soil simulant in front of the wheel increases compared with the skid state. The soil simulant behind the wheel comes from the front of the wheel through the transportation effect, and the fraction of the soil simulant behind the wheel reduces due to the disturbance of the movement.

Figure 9 is the pressure field distribution of the soil simulant under different slip ratio of the wheel. The definition of pressure is the same as that in the fluid, which is the average value of normal stress in three directions, as shown in the following:

$$p = \left(\sigma_{xx} + \sigma_{yy} + \sigma_{zz}\right)/3 \quad (9)$$

where σ_{ii} is the diagonal terms of stress tensor of the particle continuum after the being smoothed by algorithm at that point, i.e., the normal stress in the three directions. The pressure can reflect the stress distribution and force chain distribution inside soil simulant to a certain extent. It can be seen that no matter what the wheel slip ratio is, the pressure of the soil simulant is concentrated in the middle part of the contact area between the wheel and the soil simulant, while the pressure of the particles close to the surface of the particle bed accounts for a small proportion. Due to the accumulation of soil simulant in front of the wheel in the skid, the contact area between the wheel and the soil simulant is larger, but the stress inside the soil simulant is small. The force chain of the soil simulant under the wheel extends from the bottom of the wheel to the front of the wheel. The distribution area of the force chain is concentrated in the oblique front of the wheel. When the wheel is in slip state, the direction of transportation of the particles also changes from forward to backward. The force chain of the soil simulant under the wheel also starts to extend to the rear of the wheel, and it will extend both forward and backward at this time.

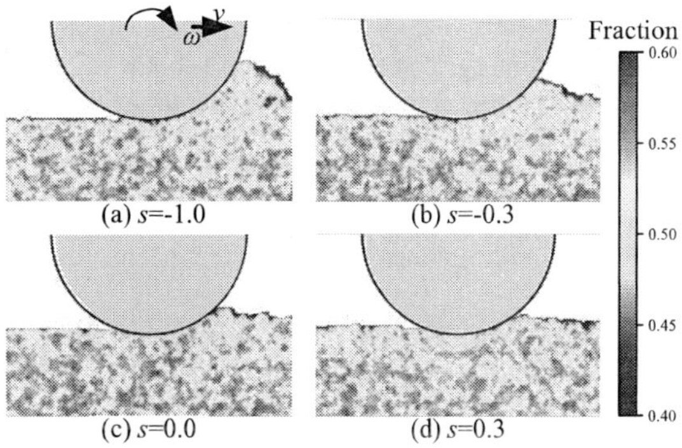

Figure 8. The fraction field of soil simulant at different slip ratios

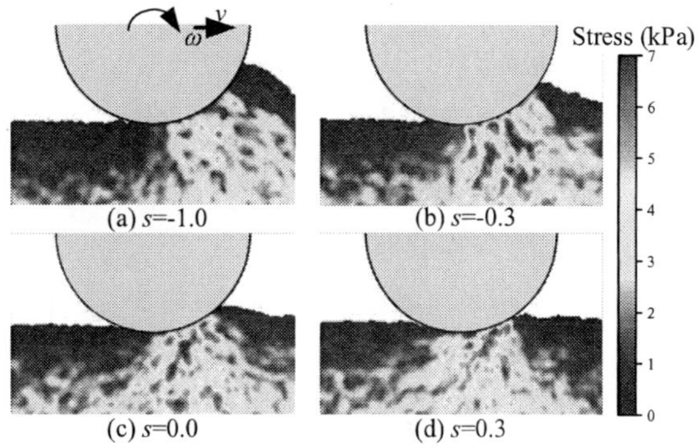

Figure 9. The pressure field of soil simulant at different slip ratios

5. Conclusions

In this paper, the 3D-DEM simulation is used to simulate the wheel-terrain interaction with fixed wheel sinkage. The physical field smoothing algorithm is used to obtain the regular and continuous velocity and stress field distribution of the granular medium regarding as continuum. A new method to calculate fraction field of the granular medium with boundary is proposed. The simulation obtained the wheel force-displacement relation, the fraction, velocity and pressure field distribution of the soil simulant. As the slip ratio of the wheel increases, less soil simulant accumulates in front of the wheel. The transportation velocity of the soil simulant shifts from oblique front to rear. The force chain extends from forward to behind the wheel, and the fraction of the soil simulant gradually increases. The smoothed and continuous physical field distribution of soil will help to visualize the process of wheel-terrain interaction, analyze the motion of soil, understand the essence of wheel-terrain interaction, and help to establish a physical model of terramechanics from the perspective of soil physical properties.

6. References

Zheng, J., Gao, H., Yuan, B., Liu, Z., Yu, H., Ding, L., Deng, Z., 2018. Design and terramechanics analysis of a Mars rover utilising active suspension. Mech Mach Theory 128, 125-149.

Niksirat, P., Daca, A., Skonieczny, K., 2020. The effects of reduced-gravity on planetary rover mobility. Int J Robot Res 39, 797-811.

Recuero, A., Serban, R., Peterson, B., Sugiyama, H., Jayakumar, P., Negrut, D., 2017. A high-fidelity approach for vehicle mobility simulation: Nonlinear finite element tires operating on granular material. J Terramechanics 72, 39-54.

Jiang, M., Dai, Y., Cui, L., Xi, B., 2018. Experimental and DEM analyses on wheel-soil interaction. J Terramechanics 76, 15-28.

Shin, Y.-J., Jeong, J.-S., Jun, C.-W., Sohn, J.-H., 2020. Interacting analysis between wheel and sand particles based on DEM and its validation with experiments. J Mech Sci Technol 34, 4537-4544.

Pang, H., Zhang, R., Ge, P., Liu, F., Wang, C., Dong, W., Li, T., Li, J., 2021. 3D DEM analysis on tractive trafficability of a lunar rover wheel with bionic wheel lugs. Rendiconti Lincei. Scienze Fisiche e Naturali 32, 377-387.

Zhang, R., Pang, H., Dong, W., Li, T., Liu, F., Zhang, H., Hu, Z., Li, J., 2020. Three-dimensional discrete element method simulation system of the interaction between irregular structure wheel and lunar soil simulant. Adv Eng Softw 148, 102873.

Ravula, P., Acar, G., Balachandran, B., 2021. Discrete element method-based studies on dynamic interactions of a lugged wheel with granular media. J Terramechanics 94, 49-62.

Li, C., Zhang, T., Goldman, D.I., 2013. A terradynamics of legged locomotion on granular media. Science 339, 1408-1412.

Feng, Y., Blumenfeld, R., Liu, C., 2019. Support of modified Archimedes' law theory in granular media. Soft matter 15, 3008-3017.

Tsuji, Y., Kawaguchi, T., Tanaka, T., 1993. Discrete particle simulation of two-dimensional fluidized bed. Powder Technol 77, 79-87.

Sakaguchi, H., Ozaki, E., Igarashi, T., 1993. Plugging of the flow of granular materials during the discharge from a silo. Int J Mod Phys B 7, 1949-1963.

Paulick, M., Morgeneyer, M., Kwade, A., 2015. Review on the influence of elastic particle properties on DEM simulation results. Powder Technol 283, 66-76.

Ding, L., 2009. Wheel-soil interaction terramechanics for lunar/planetary exploration rovers: modeling and application, Harbin Institute of technology.

Guillard, F., Forterre, Y., Pouliquen, O., 2014. Lift forces in granular media. Phys Fluids 26, 043301

Ding, L., Zhou, R., Yuan, Y., Yang, H., Li, J., Yu, T., Liu, C., Wang, J., Li, S., Gao, H., 2022. A 2-year loco-motive exploration and scientific investigation of the lunar farside by the Yutu-2 rover. Sci Robot 7, eabj6660.

Goldhirsch, I., 2010. Stress, stress asymmetry and couple stress: from discrete particles to continuous fields. Granul Matter 12, 239-252.

Glasser, B., Goldhirsch, I., 2001. Scale dependence, correlations, and fluctuations of stresses in rapid granular flows. Phys Fluids 13, 407-420.

Senatore, C., Iagnemma, K., 2014. Analysis of stress distributions under lightweight wheeled vehicles. J Terramechanics 51, 1-17.

9768 / Design and Traction Performance Test of Bionic Paddy Wheel Based on Cattle Hoof

Lan Li[a], Jing Li[a,b], Baofeng Xie[a], Fei Lin[a], and Long Xue[a,b]*

[a]*Jiangxi Agricultural University, Nanchang, Jiangxi, China*
[b]*Key Lab of Modern Agricultural Equipment, Nanchang, Jiangxi, China*
**Corresponding author: ultimata@163.com*

https://doi.org/10.56884/OTPF6196

Abstract

In order to improve the traction performance of the micro-tiller wheel on the paddy soil surface, a bionic paddy wheel was designed with a cattle hoof as the bionic prototype, and its diameter and wheel width were 0.46 m and 0.08 m, respectively. The traction performance test was carried out in a soil bin test-bed with a moisture content of 36 %. The vertical loads were 82.57 N, 131.40 N and 179.42 N, respectively. The driving speeds were 0.3 m/s, 0.5 m/s and 0.7 m/s, respectively. The drawbar pull was in the range of 10 – 120 N. The results showed that at the driving speed of 0.7 m/s, with the increase of the vertical load, the driving torque and the drawbar pull are increasing. The vertical load has a significant effect on the change of driving torque and maximum drawbar pull. Under the vertical load of 179.42 N and different driving speeds, when the slip ratio is less than 0.37, the efficiency coefficient begins to grow rapidly, and the greater the driving speed is, the greater the growth rate is. When the slip ratio is about 0.37, the efficiency coefficient reaches the maximum and then begins to decrease. Driving speed has a significant effect on the maximum efficiency coefficient of wheels. This paper can provide a reference for the traction performance of the micro-tiller wheel on the paddy soil surface and the design of the new bionic paddy wheel.

Keywords: paddy soil, traction performance, structure design, micro-tiller wheel, slip ratio

1. Introduction

Rice is one of the world's major food crops because its growth characteristics must be planted in paddy soils. Paddy soil is a typical soft ground type with obvious stratification, which is composed of a plough layer and plow layer. The main power source of paddy tillage machinery in hilly areas of southern China is micro-tillers. When the wheels of conventional micro-tillers travel on the surface of paddy soil and operate normally, serious vehicle subsidence and slippage often occur, and there are problems such as poor traction performance, large energy consumption, and even unable to travel, which seriously restrict the development of rice planting in hilly areas of southern China. Thus, it is of great significance to carry out the design and traction performance research of the paddy soil micro-tiller wheel to improve its passing ability on the paddy soil surface.

In order to improve the passing ability of vehicles on soft ground, many scholars have conducted a large number of traction performance tests on soft ground and explored the wheels by using bionic engineering technology. Tested the wheel with grousers, studied the influence of wheel radius and grousers height on wheel performance (Iizuka et al., 2011). Used single-wheel testbed to test the wheel of planetary exploration vehicle with different radius, width, lug height, lug number, and lug inclination under different slip rates to evaluate the traction performance of the wheel in deformed soil (Ding et al., 2011). Conducted Traction performance tests by controlling tire and soil parameters to understand tire-soft soil interaction (Naranjo et al., 2014). Obtained the stress distribution under the wheel with grooves at each angle during wheel rotation by measuring the three-dimensional force acting on the three positions of the grousers and the wheel surface, respectively. The test showed that the shear stress almost did not affect the wheel surface between each grouser (Higa et al., 2016). Studied the effects of four types of grouser (parallel, inclined, V-shaped, and offset V-shaped) on the traction performance of wheels in the sand. The results show that the inclined shape is the optimal design of the grouser for wheeled roamers on lunar and planetary soils (Nagaoka et al., 2019). Simulated the interaction between camel hoof and sand and developed a kind of bionic tire for deserts. The test results showed that the bionic camel hoof tire had a high sand passing ability (Li et al., 1999). Proposed a deployable wheel structure to realize the change of 1 – 2.1 times wheel diameter. Based on the wheel-soil interaction mechanical model of stress distribution, the obstacle-crossing performance of the wheel was analyzed. It was concluded that using a large diameter relative to a small diameter can reduce resistance and increase the drawbar pull and driving torque (Wei and Cheng, 2013). Studied the traction performance of wheeled tractors in rice production, discussed soil utilization, environmental impact, and planting methods, and linked them to tractor operations (Keen et al., 2013). Proposed a new type of wheel mechanism with an active driving lug, which can generate stable traction by actively adjusting the subsidence length of the lug (Yang et al., 2014). Proposed a new type of agricultural machinery rigid wheel suitable for mud paddy soil. The test results show the arc wheel has good driving performance in wheel sinking and soil reaction force (Chen et al., 2020). Developed a new type of grousers wheel, so that the grousers can be telescopic in various regions around the wheel. Especially when the wheel rotates, the grousers can always be extended in front of the wheel and retracted under the wheel, explaining how the thrust is mainly generated by the front grousers (Nassiraei and Skonieczny, 2020). Designed and tested four bionic walking wheels by imitating the structure and posture of ostrich foot walking on the sand. The experimental results show that the bionic walking wheel designed by multi-pose wheel lugs can reduce soil resistance and sand disturbance, and obtain superior sand walking performance (Zhang et al., 2020). Designed a bionic wheel surface monomer according to the structural characteristics and plantar geometry of the ostrich plantar convex ball. The experimental results show that the bionic convex ball structure greatly improves the friction performance of the wheel surface (Pang et al., 2021).

The above research mainly focuses on wheel traction performance and bionic wheel design in sand or simulated lunar soil. However, there are few studies on the traction performance of the wheel in paddy soil and the bionic wheel in paddy soil. In addition, in order to solve the problem of restricting the development of the rice industry in southern hilly areas due to the passing of tillage machinery. Thus, in this paper, a bionic paddy wheel is designed based on the surface morphology of cattle hoof. In the single wheel soil bin test-bed, the traction performance test is carried out by controlling the vertical load, driving speed and drawbar pull on the wheel to evaluate its traction performance in paddy soil. At the same time, it can provide a reference for the design of a new bionic paddy wheel.

2. Methods

2.1 Design of bionic paddy wheel

Buffalo often walks in paddy soil. In the long-term interaction with the soil, the cattle hoof has gradually formed an excellent geometric shape through continuous evolution, which has the characteristics of strong passing ability, small movement resistance and high action efficiency in paddy soil. In this paper, a bionic paddy wheel was designed based on the excellent geometric shape of the cattle hoof and the previous research and design experience of the paddy wheel of micro-tiller, according to the conventional paddy wheel size, wheel width and blade number. Its diameter and wheel widths were 0.46 m and 0.08 m, respectively. The number of bionic wheel lugs was 10, as shown in Figure 1.

The bionic paddy wheel is composed of a bionic wheel lug and wheel ring. The bionic wheel lug is fixed in the wheel ring by bolts. When the bionic paddy wheel works, the bionic wheel lugs interact with the soil, resulting in force and reaction force, so that the micro-tiller moves forward. The bionic wheel lug surface morphology is to use the three-dimensional scanner (ATOS II Triple Scan, GOM, Germany) to scan the three-dimensional point cloud data of the cattle hoof surface, and then reconstruct the cattle hoof model by Geomagic Wrap reverse software, extract the shape characteristic curve, and finally convert it into a 3D model by UG software. The process is shown in Figure 2.

Figure 1. Bionic paddy wheel

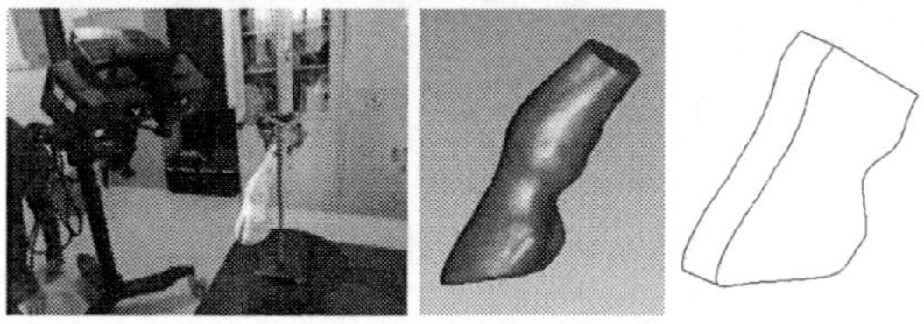

Figure 2. 3D Point cloud data acquisition (left), and model reconstruction (center), and 3D model (right)

The obtained shape characteristic curve is imported into CAD software. The selected characteristic curve data are extracted by CAD drawing software, and the point coordinates are generated. The coordinate points obtained from CAD software are input into Origin software for drawing, as shown in Figure 3. Analysis of buffalo walking in paddy soil, cattle hoofs into the soil, unearthed when the action, will be extracted in Figure 3 shape characteristic curve of cattle hoofs is divided into soil curve section AB, soil curve section BC, transition curve section CD, bulldozing curve section DE.

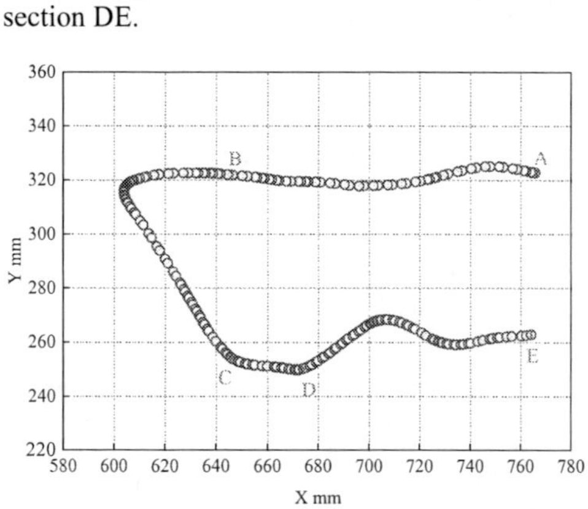

Figure 3. Shape characteristic curve of cattle hoof

By analyzing the changing trend of soil curve AB, it can be seen that the curve section is similar to a slope. Combined with the fluidity of paddy soil, the slope has the effect of accelerating the separation of paddy soil and bionic wheel lugs. The BC section of the soil-entry curve is a cone, which is conducive to the penetration of the bionic lug into paddy soil when it works and reduces the resistance of the bionic lug when it enters the soil. The transition curve section CD increases the thickness of the bionic wheel lug in the longitudinal direction, thereby increasing the shear strength of the bionic wheel lug. At the same time, it increases the contact with paddy soil in the transverse direction and increases the bulldozing resistance. When the DE of the bulldozing curve is concave inward, the contact area between the bionic wheel lug and the paddy soil increases, the bulldozing resistance increases, and the drawbar pull force of the bionic paddy wheel is increased, and the traction performance in the paddy soil is improved.

By fitting the above four curves in Origin software using the least square method, the general fitting equation is shown in Eq.1, the fitting parameters are shown in Table .

$$y = a_1x^5 + a_2x^4 + a_3x^3 + a_4x^2 + a_5x + a_6 \quad (1)$$

It can be seen from the above fitting that the regression coefficient R2 of curve fitting is greater than 0.99, and the relationship of the BC section is obtained by curve fitting after a coordinate transformation. According to the size of the paddy wheel of the conventional micro-tiller, the 3D model of the bionic wheel lug was scaled down. At the same time, in order to ensure the accuracy of the surface morphology of the bionic wheel lug, the polyurethane material is selected to process the bionic wheel lug through CNC. The external size of the bionic wheel lug is 0.08m×0.08m×0.04 m.

Table 3. Fitting parameters

Type	Curved portion			
	AB	BC	CD	DE
a_1	2E-06	3E-06	-1E-08	-2E-08
a_2	-0.0055	-0.0096	7E-05	9E-05
a_3	7.9283	12.65	-0.115	-0.1184
a_4	-5727.4	-8309.8	95.764	82.287
a_5	2E+06	3E+06	-38901	-28581
a_6	-3E+08	-4E+08	6E+06	4E+06
R^2	0.9977	0.9998	0.9965	0.9972

2.2 Experimental setup

The experiment was carried out on a self-designed single wheel soil bin test-bed (5.00m×1.20m×1.20m) to explore the traction performance of a bionic paddy wheel in paddy soil. The paddy soil in the soil bin was taken from Lianwei Town, Xinjian district in Jiangxi Province, and its physical parameters were shown in Table 2.

The single wheel soil bin test-bed is mainly composed of a soil bin, soil leveling device, test vehicle and data acquisition system, as shown in Figure 4. The moisture content of paddy soil in the soil trough was 36 %, and the test wheel was the bionic paddy wheel designed above. The leveling of paddy soil was referred to in the previous study (Zeng et al., 2022).

Figure 5 shows the structure diagram of the test vehicle, where the encoder A (E6B2, OMRON, Japan) contacts the optical axis guide rail, and the actual driving distance of the wheel is recorded in the test. Encoder B contacts the motor shaft, and the theoretical driving distance of the wheel is recorded in the test. The theoretical driving speed and actual driving speed are calculated by Eq.2.

$$v = \frac{s}{t} \quad (2)$$

where v is the driving speed (m/s), s is the driving distance (m), t is the driving time (s).

Test factors are vertical load, driving speed, which vertical load was 82.57N, 131.40N and 179.42N, driving speed were 0.3m/s, 0.5m/s and 0.7m/s, drawbar pull 10-120N, the change is realized by adding and subtracting weighter to achieve. The specific drawbar pull is obtained by pre-test, such as the vertical load is 82.57N, the drawbar pull range is 10-55N. When the vertical load is 131.40N, the drawbar pull range is 10-95N. When the vertical load is 179.42N, the drawbar pull is10-120N.

In this paper, the drawbar pull and efficiency coefficient were used as test indexes to evaluate the traction performance of the bionic paddy wheel on paddy soil. The driving torque is measured directly by a dynamic torque sensor (JN-DN, Bengbu Sensor Systems Engineering Co Ltd, China), and the efficiency coefficient is calculated by *Eq. 3*. The slip ratio is calculated by the actual speed of the wheel and the theoretical speed of the wheel by *Eq. 4*.

$$\eta_{TE} = \frac{DPv_a}{Tv_t / r_a} \quad (3)$$

$$s = \frac{v_t - v_a}{v_t} \quad (4)$$

where η_{TE} for the efficiency coefficient, DP is the drawbar pull (N), T is the drive torque (N · m), v_t is the theoretical speed (m/s), v_a is the actual speed (m/s), and r_a is the wheel drive radius (m).

Table 4. Physical parameters of the tested soil

Clay particles [%]	Soil particles [%]	Sand particles [%]	Plastic limit [%]	Liquid limit [%]	Specific gravity [g/cm³]
10.84	83.87	5.29	21.78	34.17	2.57

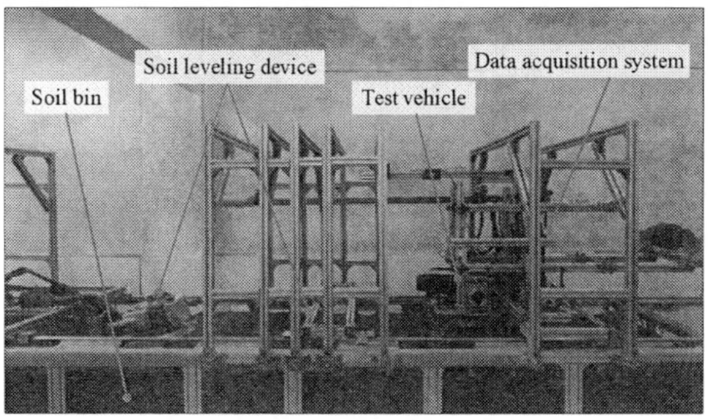

Figure 4. The single wheel soil bin test-bed

Figure 5. The structure diagram of the test vehicle

3. Results and discussion

3.1 Influence of vertical load on traction performance of bionic paddy wheel

The driving torque of the wheel is one of the evaluation indexes of vehicle traction performance and understanding its variation law is helpful to study the traction performance of the wheel. Figure 6 shows the relationship between different vertical loads and wheel driving torque at a speed of 0.7 m/s obtained by polynomial fitting. It can be seen from Figure 6 that with the increase in slip ratio, the driving torque gradually increases, and finally tends to be flat. The reason is that the maximum shear strength of paddy soil has been reached under this test condition, and the weight is further increased in the test to increase the drawbar pull, and the wheel will slip in situ. At the same slip rate, with the increase of vertical load, the reaction force obtained by wheel shearing the paddy soil breaking the ring increases, and the driving torque increases. When the driving speed is 0.7 m/s, the maximum driving torque of the wheel is 47.04 N· m when the vertical load is 179.42 N, which is increased by 55.71 % compared with the minimum driving torque. At the same time, the maximum driving torque of the wheel is increased by 135.67 % and 35.91 % compared with the vertical load of 82.57 N and 131.40 N.

Figure 7 shows the relationship between the vertical load and the maximum drawbar pull when the driving speed is 0.7m/s. It can be seen from the figure that as the vertical load increases, the maximum drawbar pull increases. When the vertical load is 179.42 N, the maximum drawbar pull of the wheel is 120 N, which increases by 166.67% and 33.33% respectively compared with the vertical load of 82.57 N and 131.40 N.

Figure 8 is the curve of the relationship between the wheel drawbar pull and slip ratio under different vertical loads when the driving speed is 0.7 m/s. It can be seen from Figure that the drawbar pull increases with the increase in slip ratio. In the range of 0.15-0.50 slip ratio, the drawbar pull increases by 3.5-11 times.

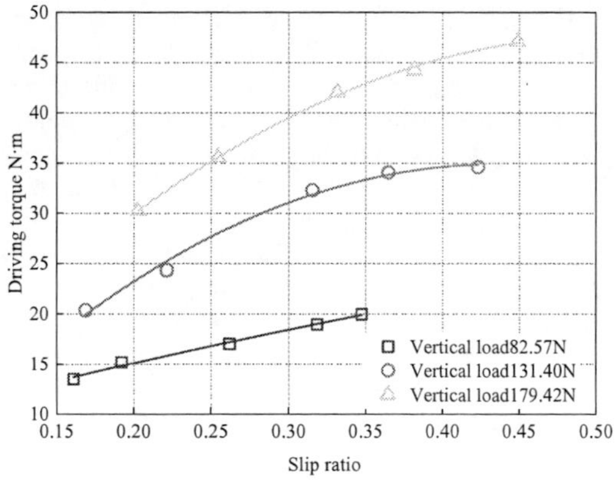

Figure 6. Influence of vertical load on the wheel driving torque

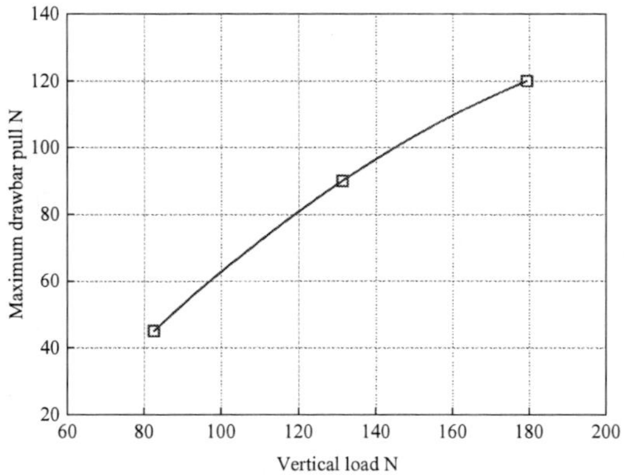

Figure 7. Relation curve between vertical load and wheel maximum drawbar pull

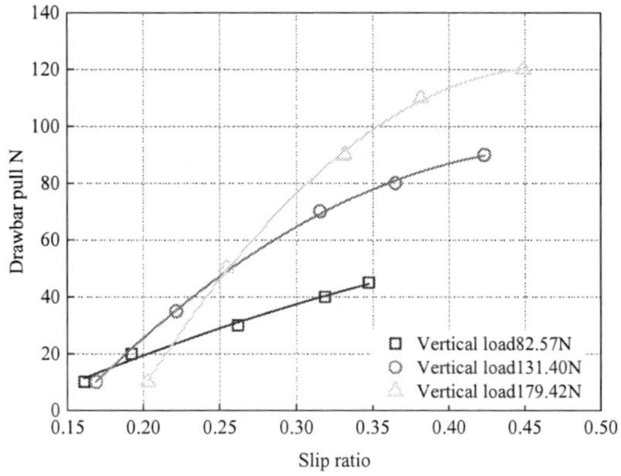

Figure 8. Influence of vertical load on wheel drawbar pull

3.2 Influence of driving speed on traction performance of bionic paddy wheel

The efficiency coefficient refers to the ratio of the output power to the input power on the driving wheel, which is calculated by the drawbar pull, driving torque, theoretical speed, actual speed and other parameters by Eq. 3. When the vertical load is 179.42 N, the influence of different driving speeds on the efficiency coefficient is shown in Figure 9.

Figure 9 shows that the maximum efficiency coefficient increases with the increase of wheel driving speed. Under the same vertical load and driving speed, the efficiency coefficient first increases and then decreases with the increase of slip ratio. When the slip ratio is less than 0.37, the efficiency coefficient begins to grow rapidly, and the greater the driving speed is, the greater the growth rate is. When the slip ratio is about 0.37, the efficiency coefficient reaches the maximum, and then begins to decrease. When the vertical load is 179.42 N, the maximum efficiency coefficient of the wheel with a driving speed of 0.7 m/s is 554 % and 98 % higher than that of the wheel with a driving speed of 0.3 m/s and 0.5 m/s, respectively.

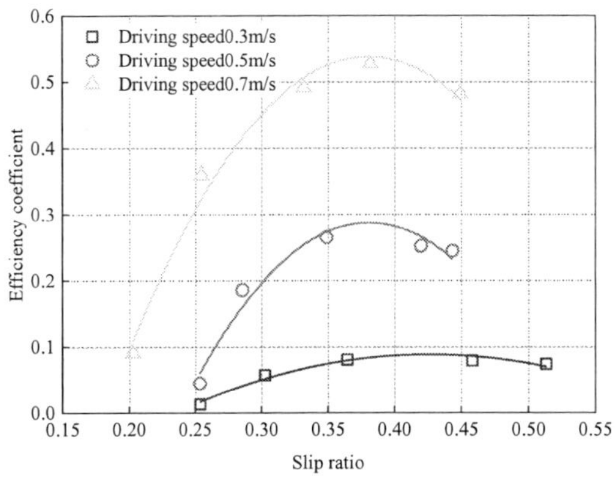

Figure 9. Influence of driving speed on efficiency coefficient

4. Conclusions

Based on the structure and morphology of cattle hoof, a bionic paddy wheel is designed by using bionic engineering technology and according to the design index of the soft ground walking mechanism. Through the traction performance test in a single wheel soil bin test-bed, the influence of vertical load on the traction performance of the bionic paddy wheel is analyzed. At the driving speed of 0.7 m/s, under different vertical loads, with the increase of slip ratio, the driving torque increases gradually, and finally tends to be gentle. Maximum drawbar pull increases as vertical load increases, when the driving speed is 0.7 m/s and the vertical load is 179.42 N, the maximum drawbar pull of the wheel is 120 N, which increases by 166.67 % and 33.33 % respectively compared with the vertical load of 82.57 N and 131.40 N. The influence of driving speed on traction performance of bionic paddy wheel is analyzed, under the vertical load of 179.42 N, with the increase of slip ratio, the efficiency coefficient first increases and then decreases, and reaches the maximum when the slip ratio is about 0.37.

5. Acknowledgements

This work was funded by the National Natural Science Foundation of China (51865018), the Natural Science Foundation of Jiangxi Province (20192BAB206025).

6. References

Chen, Z., Gu, J., Yang, X., 2020. A novel rigid wheel for agricultural machinery applicable to paddy field with muddy soil. Journal of terramechanics 87, 21-27.

Ding, L., Gao, H., Deng, Z., Nagatani, K., Yoshida, K., 2011. Experimental study and analysis on driving wheels' performance for planetary exploration rovers moving in deformable soil. Journal of Terramechanics 48, 27-45.

Higa, S., Sawada, K., Nagaoka, K., Nagatani, K., Yoshida, K., 2016. Measurement of stress distributions of a wheel with grousers traveling on loose soil, 2016 IEEE International Conference on Robotics and Automation (ICRA), pp. 2828-2833.

Iizuka, K., Yoshida, T., Kubota, T., 2011. Effect of tractive given by grousers mounted on wheels for lunar rovers on loose soil.

Keen, A., Hall, N., Soni, P., Gholkar, M., Cooper, S., Ferdous, J., 2013. A review of the tractive performance of wheeled tractors and soil management in lowland intensive rice production. Journal of Terramechanics 50, 45-62.

Li, J., Zhuang, J., Wei, D., 1999. Design and Test of the Bionic Camel Foot Rubber Tire for Desert Vehicles. Agricultural Engineering, 38-42.

Nagaoka, K., Sawada, K., Yoshida, K., 2019. Shape effects of wheel grousers on traction performance on sandy terrain. Journal of Terramechanics 90.

Naranjo, S.D., Sandu, C., Taheri, S., Taheri, S., 2014. Experimental testing of an off-road instrumented tire on soft soil. Journal of Terramechanics 56, 119-137.

Nassiraei, A.A.F., Skonieczny, K., 2020. Grousers improve drawbar pull by reducing resistance and generating thrust at the front of a wheel. Journal of Terramechanics 91, 73-84.

Pang, H., Zhang, H., Zhang, R., Dong, W., Li, J., 2021. Design of the bionic wheel surface based on the friction characteristics of ostrich planta. Rendiconti Lincei. Scienze Fisiche e Naturali.

Wei, Y., Cheng, Z., 2013. Design and Cross-barrier Capability Analysis of Road Condition Adaptive Variable Diameter Wheel. Journal of Mechanical Transmission 37, 125-129.

Yang, Y., Sun, Y., Ma, S., 2014. Effect of lug sinkage length to drawbar pull of a wheel with an actively actuated lug on sandy terrain, 2014 IEEE International Conference on Robotics and Automation (ICRA).

Zeng, J., Li, J., Li, L., Qiu, H., Wu, C., Bao, Y., Liu, M., Xue, L., 2022. Traction Performance Test of Micro-cultivator Wheel in Paddy Soil. Acta Agriculturae Universitatis Jiangxiensis, 1-9.

Zhang, R., Pang, H., Ji, Q., Li, G., Li, J., 2020. Structure design and traction trafficability analysis of multi-posture wheel-legs bionic walking wheels for sand terrain. Journal of Terramechanics 91, 31-43.

9913 / Acquisition of Flipper Motion of Tracked Robot in Step-Climbing Using Reinforcement Learning

Ryosuke Eto* and Junya Yamakawa
National Defense Academy of Japan, Yokosuka, Kanagawa, Japan
**Corresponding author: eto@nda.ac.jp*

https://doi.org/10.56884/SGWJ1011

Abstract

Remotely piloted robots have been expected for disaster operations to prevent secondary disasters to rescuers. The robots are required to have a high performance to overcome obstacles such as debris and bumps. Tracked robots with flipper arms on the front, back, left, and right sides can improve their ability to overcome bumps by changing the flipper angles, and are thus expected to be used as rescue robots. However, the six degrees of freedom of the left and right crawlers and four flipper arms require a high level of skill to maneuver the robot. Therefore, a semi-autonomous control system that automatically controls the flipper arms according to the terrain is expected. In this study, we proposed a method to determine the front and rear flipper angles in step-climbing using reinforcement learning with Double Deep Q Network. The input data were the step height, distance to the step, and current front and rear flipper angles. The outputs were amounts of front and rear flipper angle variations. The behavior of the robot and rewards were calculated using a quasi-static model that considers the slips between the step, floor, and crawler. Positive rewards were given for successful step stepping over steps and negative rewards for unsuccessful steps. Furthermore, the flipper motions at which slippage decreased were obtained by subtracting the sum of the squared values of the slippage rates from the rewards. As the results, it was confirmed that the robot slips less when the body is lifted by the rear flipper than when it runs over along the step shape.

Keywords: tracked robot, flipper, slip, step-climbing, reinforcement learning.

1. Introduction

In disaster operations, remotely piloted robots are effective in preventing secondary disasters to rescuers. The rescue robots are required to have a high performance to overcome obstacles such as debris and bumps and to travel on soft terrain. For this reason, tracked robots are commonly employed. Tracked robots with flipper arms on the front, back, left, and right sides can improve their performance of overcoming obstacles by changing the flipper angles and are thus expected to be used as rescue robots.

On the other hand, high skill is required to maneuver the left and right crawlers and the four flipper arms in a total of six degrees of freedom. Therefore, a semi-autonomous control system that automatically controls the flipper arms according to the terrain is expected. As shown in Fig. 1, the control plan for the flipper angles relative to a step when climbing a step can be divided into two methods: one in which the robot body is controlled to maintain a horizontal position and the other in which the robot is controlled to move along the step shape. The latter method is effective on soft ground because the rear flipper can increase the ground contact area. However, these control methods determine the flipper angles based on geometric relationships and do not consider the slip state of the crawler. The objective of this study is to obtain a motion plan for the flippers that reduces the slip to prevent the robot from slipping down from the step on hard ground.

Reinforcement learning is a method for determining the robot's action plan. Reinforcement learning is a method of acquiring optimal action rules for a robot through repeated trial and error. Reinforcement learning for flipper arms has been studied (Isabell, 2012, Totani, 2019), however, they only judge whether the robot was able to climb the steps and do not consider the state of slip of the robot.

In this study, reinforcement learning with Deep Q Network (DQN) is adapted to determine the angles of the front and rear flippers that minimizes slip and enables the robot to climb a step. Repeating the trial-and-error process necessary for reinforcement learning through experiments takes a lot of time. Therefore, in this study, we aim at acquiring the flippers motion by learning through a step-climbing simulation using a quasi-static model that considers slips between the floor and the crawler and between the step and the crawler.

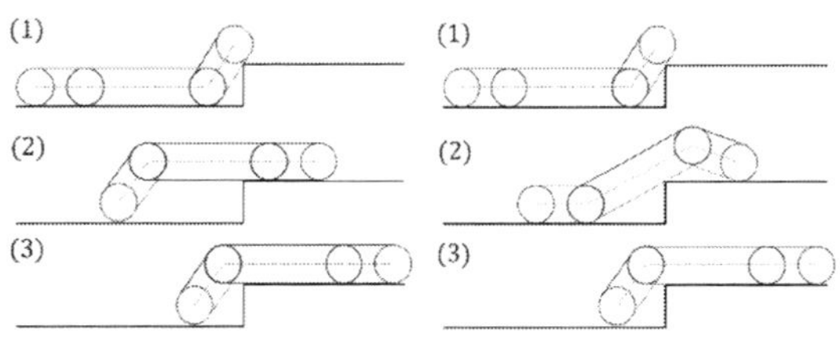

(a) keeping body horizontal (b) along a step shape
Figure 1 Difference in flipper angles when climbing a step

2. Quasi-Static Model

2.1 Calculation of slip ratio and normal load

The quasi-static model used in this study assumes that the crawler robot moves at extremely low speeds and calculates the statics balance at each position of the robot (Eto, 2021). The crawler is not deformed and the circumferential speeds of the main crawler and subcrawler are assumed to be the same.

Figure 2 shows the dynamic model of a crawler robot without grousers climbing a step. F_b, N_b, and R_b are the driving force, vertical drag force, and driving resistance acting at the contact point between the rear subcrawler and the floor, and F_a, N_a and R_a are the driving force, vertical drag force, and driving resistance acting at the contact point between the step and crawler. X, m, g are the distance between the robot and the step, the mass of the entire robot, and the acceleration of gravity, respectively. ϕ_f, ϕ_r, θ_x are the front flipper angle, the rear flipper angle, and contact angle, respectively. The angles are positive in the counterclockwise direction.

In this model, the vertical forces N_a, N_b and the slip ratios s_a, s_b are calculated by solving four simultaneous equations: (1) the equilibrium of the horizontal forces, (2) the equilibrium of the vertical forces, (3) the equilibrium of the moments, and (4) the relationship between the slip ratios at the contact point between the floor and the step. The conditions that make it impossible to climb a step in this method are when the center of gravity of the robot cannot pass directly above the step angle and when the solution of the simultaneous equations cannot be obtained. Changing the flipper angles changes the pitch angle and the location of center of gravity of the robot, allowing the robot to pass directly above the step angle and changing the critical height for climbing a step.

The slip ratio s_i is the ratio of the slip velocity to the crawler circumferential velocity v_t and is defined by the following equation.

$$s_i = \frac{v_t - v_i}{v_t} \tag{1}$$

where i is the contact point and v_i is the velocity of movement at each contact point. The relationship of the slip ratio at each contact point is obtained from the relation of the angular velocity ω_c around the instantaneous center O in Fig. 2 as follows.

$$l_a(1 - s_b) = l_b(1 - s_a) \tag{2}$$

The relationship between vertical force and driving force can be expressed by the following equation using the coefficient of friction.

$$F_i = \mu_i N_i \tag{3}$$

The friction coefficient μ_i is expressed by the following equation using the sliding ratio s_i.

$$\mu_i = \mu_{i0}(1 - \exp(-K_i|s_i|)) \cdot \mathrm{sign}(s_i) \tag{4}$$

where μ_{i0} is the maximum coefficient of friction and K_i is a positive constant, determined by the floor, the step, and the crawler material.

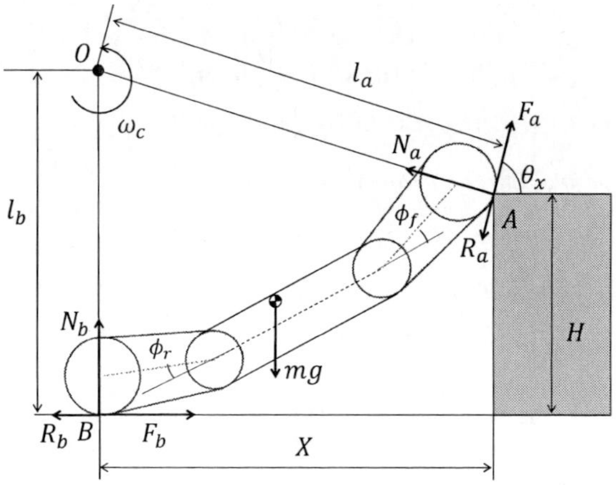

Figure 2. Dynamic model of stepping over a step

2.2 Model with grousers

Figure 3 shows the dynamic model of a crawler robot with grousers. In the case of the state in which the main crawler is in contact with a step, the moving velocity of the contact point $A(x_a, y_a)$ between the grouser and the step can be expressed as follows.

$$\dot{x}_a = \dot{x}_b + \dot{l}_X(\cos\theta_c - \tan\theta_d \sin\theta_c) \\ -\omega_c\{l_r\sin\theta_r + l_X\sin\theta_c - (r - l_X + h_{gr})\cos\theta_c\} \tag{5}$$

$$\dot{y}_a = \omega_c\{(l_r\cos\theta_r + l_X\cos\theta_c) \\ +(r - l_X\tan\theta_d)\sin\theta_c + h_{gr}\sin\theta_x\} \\ +\dot{l}_X(\tan\theta_d\cos\theta_c + \sin\theta_c) = 0 \tag{6}$$

where x, y are the horizontal and vertical positions, respectively. $\dot{x}_b$ is the horizontal velocity of the contact point $B(x_b, y_b)$, ω_c is the pitch angular velocity of the robot, l_r is the rear crawler length, l_X is the distance to the main crawler contact point from the rear wheel of main crawler, r is the main crawler wheel diameter, θ_r is the rear flipper pitch angle, θ_c is the robot pitch angle, θ_d is the angle between the crawler centerline and the belt ($\theta_d = 0$ for the main crawler), and h_{gr} is the height of the grouser. The vertical position of contact point A is equal to the step height, so $\dot{y}_a = 0$. The moving velocity $\dot{x}_b$ of contact point B can be obtained from equation (1) as follows.

$$\dot{x}_b = (1 - s_b)(r_r + h_{gr})\omega_r \tag{7}$$

where r_r is the diameter of the rear flipper idler wheel and ω_r is the rotation speed of the rear flipper idler wheel. Since the circumferential speed of the main crawler can be obtained from equation (7), $\dot{l}_X$ is given by

$$\dot{l}_X = -\frac{\dot{x}_b r_r \cos\theta_d}{(r_r + h_{gr})(1 - s_b)} \tag{8}$$

The pitch angular velocity ω_c is obtained by substituting $\dot{l}_X$ into equation (6). And, the velocity $\dot{x}_a$ is obtained by substituting ω_c, $\dot{l}_X$, and $\dot{x}_b$ into Eq. (5).

$$\dot{x}_a = \dot{x}_b - a_1 a_2 \frac{\dot{x}_b r_r}{(r_r + h_{gr})(1 - s_b)} + a_3 \frac{\dot{x}_b r_r}{(r_r + h_{gr})(1 - s_b)} \tag{9}$$

$$a_1 = \frac{\sin\theta_d\cos\theta_c + \cos\theta_d\sin\theta_c}{l_r\cos\theta_r + l_X\cos\theta_c + (r - l_X\tan\theta_d)\sin\theta_c + h_{gr}\sin\theta_x} \tag{10}$$

$$a_2 = l_r\sin\theta_r + l_X\sin\theta_c - (r - l_X + h_{gr})\cos\theta_c \tag{11}$$

$$a_3 = \cos\theta_d\cos\theta_c + \sin\theta_d\sin\theta_c \tag{12}$$

Since $\dot{x}_a$ is the sliding velocity at contact point A, dividing $\dot{x}_a$ by the crawler belt circumferential velocity horizontal component yields the following slip ratio relationship.

$$s_a = b_1(1 - s_b) + b_2 \quad (13)$$

Where, b_1, b_2 are values obtained from the robot specifications and pitch angle. When considering grousers, the vertical forces N_a, N_b and the slip ratios s_a, s_b are obtained by solving the simultaneous equations using the relationship in Eq. (13) instead of Eq. (2).

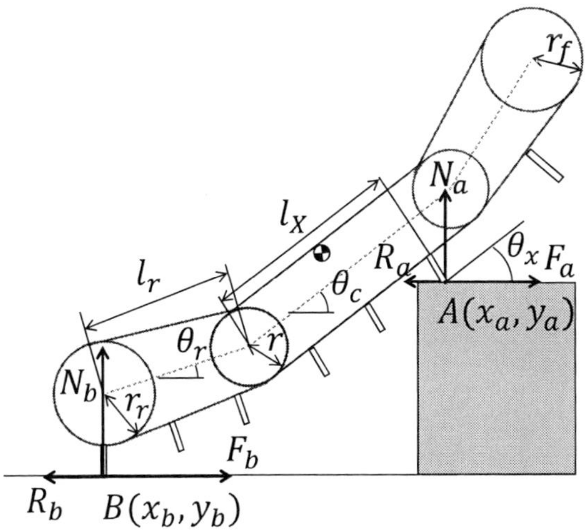

Figure 3. Dynamics model considering grousers

3. Acquisition of flipper motion

3.1 Reinforcement Learning with DQN

In this study, we used reinforcement learning with DQN as a method of acquiring flipper motion. DQN is a method that uses deep neural network (DNN) model for Q-learning that is a typical method of reinforcement learning (Volodymyr, 2015). In Q-learning, the agent learns a strategy for selecting the action that maximizes the action value Q, when the agent selects action a in state s. The Q value is updated using the action value function $Q(s_t, a_t)$ as follows.

$$Q(s_t, a_t) \leftarrow Q(s_t, a_t) + \alpha(r_t + \gamma \max_{a'} Q(s_{t+1}, a') - Q(s_{t+1}, a_{t+1})) \quad (14)$$

where s_t, a_t, and r_t are the state, action, and reward at step t, respectively. α is the learning rate and γ is the discount rate.

DQN is a method that uses a DNN model as a function approximator of the action value function $Q(s_t, a_t)$. In DQN, the target value function q_n are set as follows for each of the total N experience data (s_n, a_n, r_n, s'_n), $n \in \{1, \ldots, N\}$ in the historical data.

$$q_n = r_n + \gamma \max_{a'} Q(s'_n, a') \quad (15)$$

Where, s'_n is the state after the transition. The parameters of the DNN model are updated so that the loss function for the mean squared error is smaller. the loss function η is defined as follows.

$$\eta = \frac{1}{N}\sum_{n=1}^{N}(Q'(s_n, a_n) - q_n)^2 \quad (16)$$

Where, Q' is a network after parameter optimization.

3.2 Learning Conditions

The robot targeted for learning is Dr Robot's Jaguar V6 with Arm shown in Fig. 4. The driving resistance in the quasi-static model was set to 0, the maximum friction coefficient μ_{i0} in the friction model was 1, and K was 20.

The targets of the acquired motion are the front flipper angle and the rear flipper angle. There are four states: step height H, distance to step X, front flipper angle ϕ_f, and rear flipper angle ϕ_r. The actions were set to three changing angles for each flipper relative to the distance the robot travels: 2.5 deg/cm, 0 deg/cm, and -2.5 deg/cm. Therefore, there are 9 actions in each step. The initial front and rear flipper angles were set to 10 deg. and -10 deg., respectively. The step height was set to 20 cm.

The reward was defined as follows using the slip ratios.

$$r_n = -20(s_{an}^2 + s_{bn}^2) \quad (17)$$

In addition, we added a reward of 10 for a successful climbing and -10 for an unsuccessful. Since learning is performed to maximize the reward, it is expected that the robot will acquire the behavior that allows it to successfully climbing a step with a small slip. The learning rate α was 0.05 and the discount rate γ was 0.99.

The max operator in DQN, in Eq. (15), uses the same values to select and evaluate an action. This makes it more likely to select overestimated values, resulting in overoptimistic value estimates. In this study, the Double DQN method (Hado, 2016) was employed to prevent the overestimation. Double DQN uses the action value estimate Q' based on the latest parameters for action selection and the target network based on previous parameters for action evaluation. The target value in Eq. (15) is replaced as follows.

$$q'_n = r_n + \gamma \max_{a'} Q\left(s'_n, \operatorname{argmax}_{a'} Q'(s'_n, a')\right) \quad (18)$$

The action is selected by the ε-greedy policy (Volodymyr, 2015). The greedy policy is $a' = \operatorname{argmax}_{a'} Q'(s'_n, a')$. In this study, ε was updated by the following equation.

$$\varepsilon_n = 0.995^{n-1} \quad (19)$$

The minimum value of ε was 0.01.

The DNN model is a four-layer network model consisting of an input layer with 4 neurons, an output layer with 9 neurons, and two intermediate layers which have 30 and 60 neurons, respectively. The numbers of neurons of input and output layers are determined by the number of states and the number of actions in learning. The intermediate layer was determined by trial and error. The ReLU function was used as the activation function.

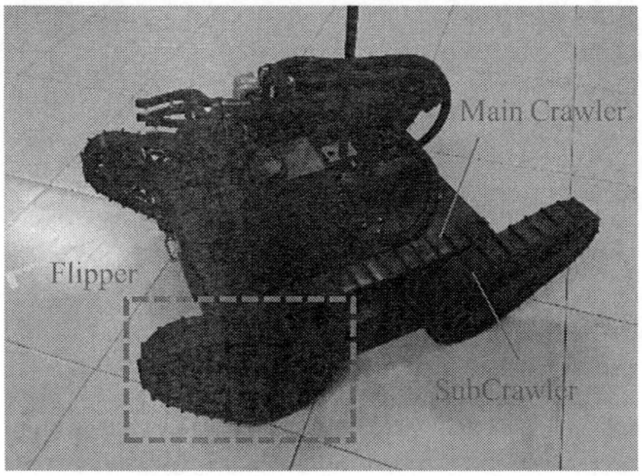

Figure 4. Tracked robot

Table 1. Specification of robot

Parameter	Value
Body mass	50 [kg]
Flipper mass	1 [kg]
Main crawler length	0.46 [m]
Subcrawler length	0.22 [m]
Driving wheel diameter	0.115 [m]
Idler wheel diameter	0.07 [m]
Grouser height	0.01 [m]

3.3 Results

Figure 5 shows the obtained reward for each episode. The gray line is the unprocessed data, and black line is the smoothed data with 15 points. In the case of the tracked robot without grousers, the reward was relatively stable after 500 episodes. In the case of the tracked robot with grousers, the increase in reward stopped when the episode exceeds 1500. In some episodes, rewards were low even when the number of episodes was a lot due to the random selection of actions by the ε-greedy policy. The reason why the reward does not increase to nearly 10 that is a reward for a successful climbing is that it is not possible to climb a step with zero slip. The reason why the reward of the robot with grousers is smaller is that the forward slip at the point of contact with the step is larger than without grousers during the first half of climbing the step.

Table 2 shows the integral of the sum of the squares of the slip ratios at the two contact points and the success or failure of the climbing. Figure 6 also shows the acquired flipper angles. When the flipper angle is not changed (episode number 0), the robot's center of gravity is not able to move above the step and the robot cannot climb it. However, with repeated learning, as shown in Figure 6(a) and (b), the robot makes the pitch angle of the main body closer to horizontal by lowering the rear flipper and moves its center of gravity forward, regardless of the presence or absence of grousers and can climb a step. The learning process also reduces the slip ratio. After the learning, the rear flipper was lowered to a maximum angle of 45 degrees immediately after the start of driving, and the robot climbed a step while maintaining that pitch angle. Therefore, to reduce the total slip ratio up to the time when the robot climbs a step, it is better to run the robot while keeping the body close to the horizontal. However, the front flipper angle dose not changed from the initial angle after learning. Since the impact of landing on a step can be reduced by lowering the front flipper when climbing up a step, it is thought that the motion of the front flipper can be acquired by adding a parameter related to this to the reward.

As described in the introduction, another motion for the robot to climb a step is to drive along the step shape. In order to compare this method with the acquired motion, slip ratios of the robot are analyzed when the robot climbs a step with the motion plan of the rear flipper as shown in Fig. 7 with red lines. In the acquired motion, the rear flipper is lowered after the start of the drive, but in the drive along a step shape, the rear flipper is raised in the first half of the run and lowered in the second half for climbing the step. The front flipper remains unchanged. In this method, the pitch angle of the robot is larger than that of the acquired motion, as shown in Fig. 8. The black dots in Fig. 8 indicate the center of gravity of the robot. Finally, the center of gravity is located above the step in both methods, and the robot succeeds in climbing the step.

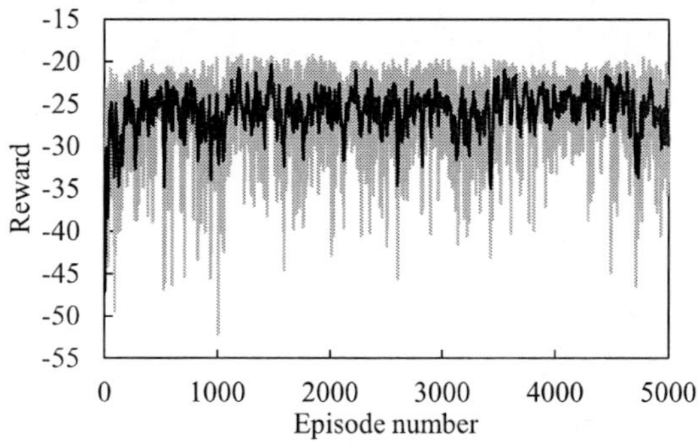

a) without grouser

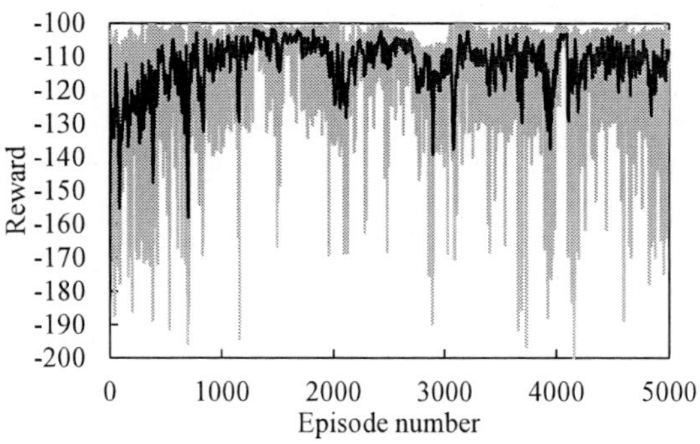

b) with grouser

Figure 5. Rewards in learning

Table 2. Results of tests

Episode number		Without grouser	With grouser
0	Slip ratio	0.0071	0.068
	Success/Failure	Failure	Failure
1000	Slip ratio	0.0062	0.066
	Success/Failure	Success	Failure
5000	Slip ratio	0.0061	0.048
	Success/Failure	Success	Success

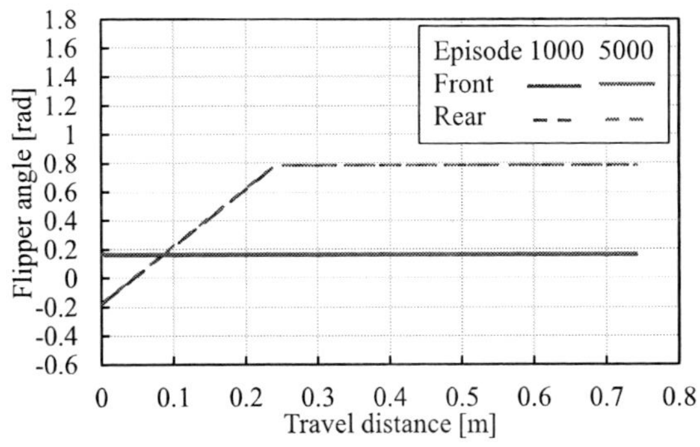

a) with grouser

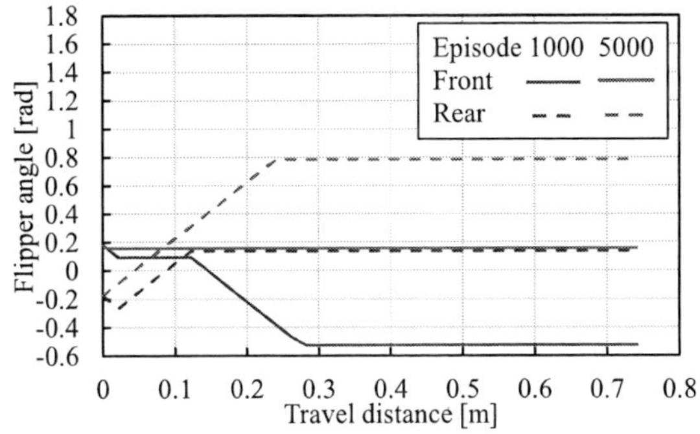

b) without grouser
Figure 6. Acquired flipper motions

Figure 9 shows the results of the slip ratios in the cases of driving with the acquired motion and driving along a step shape. The differences in slip ratios are small up to a travel distance of 0.25 m. However, when the travel distance exceeds 0.25 m and the main crawler contacts a step, both slip ratios at the contact point with the step become larger when traveling along the step shape. This is because the pitch angle of the main body is large, and a larger driving force is required to balance the vertical forces in the case without grousers. In the case with grousers, the horizontal component of the moving velocity of the main crawler's grouser becomes small, and the difference between it and the travel velocity at the contact point with the floor becomes large, which causes forced slip. Therefore, regardless of the presence or absence of the grousers, the slip can be reduced by lifting the main body to bring its posture closer to the horizontal. When the pitch angle is large, slippage is likely to occur at the contact point with the step if there is no grouser and is likely to occur on the floor if there are grousers.

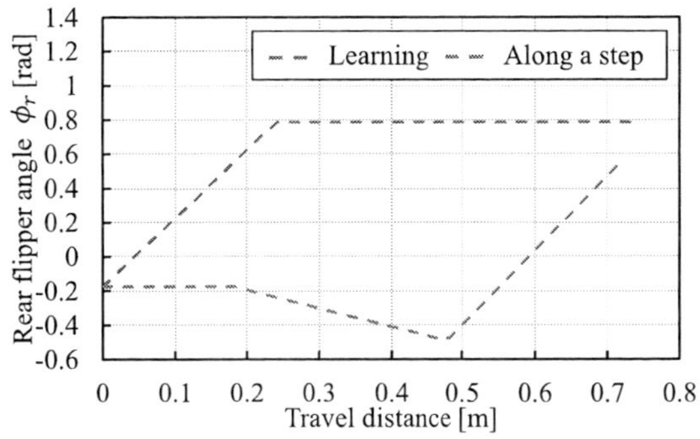

Figure 7. Rear flipper angles

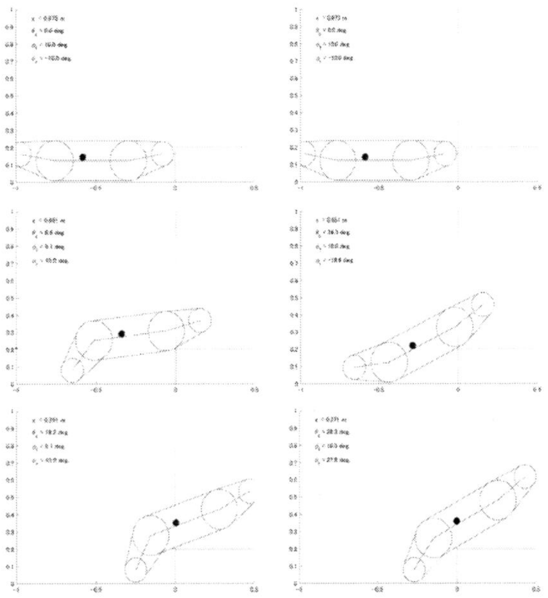

a) Learning *(b) Along a step shape*
Figure 8. States of robot in step-climbing

4. Conclusions

In this study, we proposed a method for acquiring a motion when climbing a step using reinforcement learning with Double DQN for a tracked robot with flippers. The proposed method can acquire the motion of climbing a step with small slip by analyzing a quasi-static model of climbing a step. It was confirmed that the robot slips less when the rear part of body is lifted by the rear flipper than when the robot drives along the step shape.

In the future, we will further study the initial conditions and step height. Further studies are also needed for more complex terrain and the case where the crawler rotation speed is also considered.

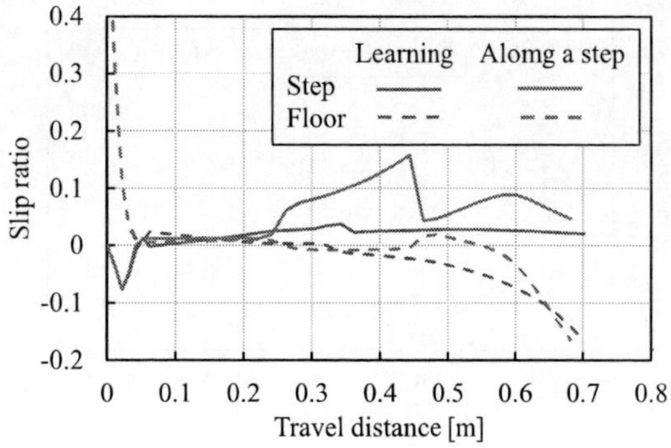

a) with grouser

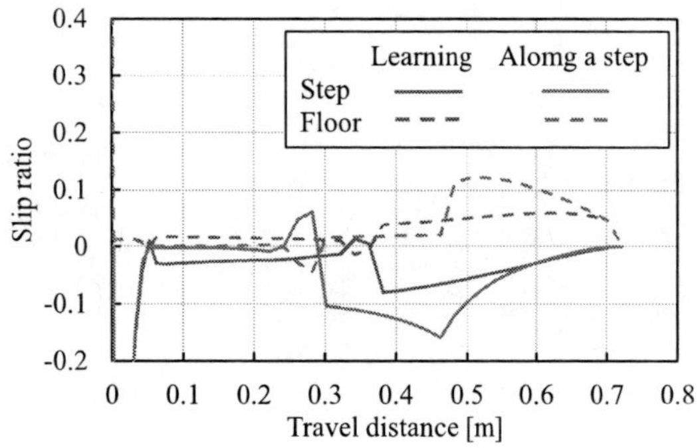

b) without grouser

Figure 9. Results of slip ratios on step and floor

5. References

Eto, R., Yamakawa, J., 2021. Evaluation of step-climbing performance of tracked robot with flippers. Proceedings of the 20th International and 9th Americas Conference of the ISTVS. 039, 1-9.

Hado, v. H., Arthur, G., David, S., 2016. Deep reinforcement learning with double Q-Learning. Proceedings of the Thirtieth AAAI Conference on Artificial Intelligence. 2094-2100.

Isabelle, V., Qiao, S., 2012. A combined reactive and reinforcement learning controller for an autonomous tracked vehicle. Robotics and Autonomous Systems. 60, 599-608.

Totani, M., Sato, N., Morita, Y., 2019. Step climbing method for crawler type rescue robot using reinforcement learning with Proximal Policy Optimization. Proceedings of the 12th International Workshop on Robot Motion and Control. 154-159.

Volodymyr, M., Koray, K., et al., 2015. Human-level control through deep reinforcement learning. Nature. 518, 529-533.

ISTVS

Abstract-only submissions

0196 / Analysis on Climbing Mode of Zhurong Rover

Zhen Chen, Meng Zou, Dong Pan, Baofeng Yuan, and Lining Chen
Video: https://vimeo.com/showcase/9710424/video/747934001

Due to the complex situation on the surface of Mars and the long delay of signal transmission between Mars and the Earth, many rovers of different countries have experienced the situation of reduced mobility or even failed exploration missions due to climbing process on Mars. In order to avoid such a situation, this paper takes 'Zhurong' Rover as an example. Four kinds of climbing modes that it can use are analyzed namely normal climbing, Z-type climbing, oblique climbing, bionic wriggle climbing. The advantages and disadvantages of four kinds of climbing modes and their applicable scope are compared. When the slope is small, normal climbing should be used. When the slope is moderate, Z-type and oblique climbing should be used. When the slope is large, bionic wriggle climbing should be used. Through the experiment, the critical value of the rover climbing slope is 23°, when the rover climbing slope is greater than 23°, the normal can not climbing normally which should be backward downhill. In addition, the current criterion other than the slope criterion is given to improve the climbing efficiency and safety of the rover under different climbing modes. The critical value of the current is 1.92 A. the research results can provide experimental data and reference for the in-orbit use of 'Zhurong' Rover.

1704 / Comparative Analysis of Hydrodynamical Efficiency of Full-Submerged Archimedes Screws of Rotary-Screw Propulsion Units of Snow and Swamp-Going Amphibious Vehicles with Single and Tandem Propulsor Design

Svetlana Karaseva, Aleksey Papunin, Vladimir Belyakov, Vladimir Makarov, and Dmitry Malahov
Video: https://vimeo.com/showcase/9710424/video/748218893

Abstract. The paper considers the problems of increasing of efficiency of amphibious vehicles with rotary-screw propulsion units (RSP). The results of computer simulation of hydrodynamical interaction of full-submerged Archimedes screws of rotary-screw propulsion units with water area in travelling and mooring modes are presented. The computer simulation was done for geometrically similar Archimedes screws with single and tandem design. During simulation the hydrodynamical characteristics of Archimedes screws of RSP with the most typical for snow and swamp-going vehicles geometrical characteristics were studied. The three-run Archimedes screws with helix angles in the range from 24 to 39 degrees were considered. The non-dimensional dependences of thrust coefficient, torque coefficient and propulsive coefficient on advance ratio for travelling mode (performance curves) and of relative quality coefficient on rotation speed of Archimedes screw for mooring mode got by the results of computer simulation are given and analyzed. The advantages of tandem design of RSP in comparison with single design with relation to realization of overwater characteristics of snow and swamp-going amphibious vehicles are highlighted and quantitative estimated.

2398 / Modeling Tire-Soil Compression Resistance on Artificial Soil Using a Scaling Law of Pressure—Soil Sinkage Relationship

Pius Jjagwe, Mehari Tekeste, and Nisreen Alkalifa
Video: https://vimeo.com/showcase/9710424/video/750588095

Semi-empirical traction models utilize soil parameters estimated from ASABE cone and flat plate soil sinkage data; however, limited studies exist that apply the scaling law of the soil measurement tool to a full-scale tire to soil system at various initial soil conditions. This study investigated the effects of three scaled plates (size and shape) and two density conditions on pressure-sinkage relationships on an artificial soil. Elliptical shape tire-soil and contact area of 484 cm2 were estimated from a vertical loading a tire (235/75R15, tire inflation pressure of 179 kPa, and 8 kN vertical load) soil bin test on an artificial soil. Pressure-sinkage data was collected on an artificial soil column at 1.21 Mg/m3 (loose) and 1.41 Mg/m3 (dense, 75% of Proctor Density) initial conditions using circular, rectangular, and square, each at three scaled areas ($n = 0.5$, $n = 0.25$, and $n = 0.125$, where $n = 1$ is the tire-soil estimated area from the single tire soil bin test). A scaling law with a strong correlation was established between the geometry scale and the energy in compressing the soil. The plate pressure data from $n = 0.125$ exhibited an asymptotically decreased pressure with depth, similar to the soil cone penetrometer data. The pressure-sinkage data from $n = 0.5$ exhibited similar trend as Reece-Bekker-Wong models but varied by the initial soil bulk density. Using ground-contact pressure of the tested tire and pressure-sinkage prediction model ($n = 0.5$), the loose soil resulted in a 38% increase in simulated sinkage compared to the dense soil. The study demonstrated applying a scaling law to simulate the tire-soil system on soft and dense soils. Semi-empirical prediction of motion resistance ratio using data from single-tire soil bin on soft and dense soils will be presented.

2681 / Research on High Traction Bionic Wheel Based on Traction Characteristics of Jerboa Foot

Hao Pang, Liangliang Zhao, Huailiang Wang, and Rui Zhang

The lunar surface is covered with loose lunar soil, and the bearing capacity and shear resistance of the lunar soil are relatively low. Therefore, when the wheels roll on the lunar soil, the lunar soil particles flow easily, which reduces the passing performance of the mobile system. Jerboa has superior traction in sandy terrain. In this paper, the 3D reconstruction of the jerboas foot is firstly carried out, and three models of jerboas feet with different postures are established according to the posture changes of the jerboas foot. The discrete element numerical simulation of the sand jumping process was carried out to explore the interaction mechanism of the jerboas feet and sand when jumping in the sand.The simulation shows that jerboas toenails play a role in improving traction during the sand jumping process, and the trapezoidal toe gap structure with narrow upper and lower lower feet plays a role in sand-fixing and current-limiting in the process of foot-sand interaction. The key structure to improve the traction performance of jerboas foot was extracted through simulation analysis, and a high-traction bionic lunar rover wheel was designed by using the jerboa's foot as a bionic prototype. The wheel structure was optimized by discrete element numerical simulation. Taking Yutu wheel as the control wheel, the traction performance of the bionic wheel was verified by the wheel soil test on the bench.

2928 / Bearing Capacity Analysis of Bionic Walking Wheel for Manned Lunar Rover.

Rui Zhang, Liangliang Zhao, Yupei Du, Bin Zhao, Hao Pang, Lige Wen, Weijun Wang, Hua Zhang, Zhenyu Hu, Zhenyu Hu, and Meng Zou

The special curved shape of African ostrich sole can limit current and fix sand and improve the bearing capacity of wheel. In the case of limited size and mass of lunar rover wheels, the use of flexible wheels can reduce the occurrence of slippage and subsidence on the soft surface of the lunar rover, and the bearing capacity of the existing flexible wheels cannot meet the requirements of heavy load. Therefore, an innovative rigid-flexible coupling bionic walking wheel was proposed, which was composed of the bionic main monomer and the integrated rim hub formed by the array of rigid-flexible coupling sub-monomers of the flexible imitation ostrich sole morphology and rigid limit structure. In order to fully analyze the mechanical properties of the rigid-flexible coupling bionic walking wheel under different loads, the Apollo elastic wheel of the same size was proposed as the control group. Based on the lunar soil thrust received by the wheel when driving on the lunar surface, the lateral force received when turning and the huge impact force received when driving, The static load and impact load of vertical load, tangential force and lateral force are simulated by finite element method. The thickness of the sheet (1.5mm) was determined by vertical load simulation, taking into account elasticity and stiffness. Through the simulation of lateral force, tangential force and impact load, it is found that the control wheel outside the bionic wheel has plastic deformation. The results show that the rigid-flexible coupling bionic walking wheel can bear larger load and impact. This research has certain reference value for the follow-up development of heavy flexible metal wheel.

2960 / Terramechanics Model Augmentation Using Machine Learning

Eric Karpman, Jozsef Kovecses, and Marek Teichmann

Thus far, terramechanics simulations for real time applications have focused largely on the well-established semi-empirical models as these are the only ones that are fast enough to be run in real time. For wheel-soil interaction these are the models hallmarked by the names of Bekker and Wong, and for other earthworking maneuvers the Fundamental Earthmoving Equation (FEE) proposed by Reece is commonly used. One major drawback to these models is that they are formulated with a steady state assumption which makes their accuracy in dynamic simulations questionable. More flexible simulation alternatives include the Finite Element Method (FEM) and the Discrete Element Method (DEM) - both of which are extremely computationally demanding which can be prohibitive in many practical applications. In this work, we propose a method for utilizing advances in machine learning to augment the forces predicted by existing models in an effort to improve their accuracy in dynamic simulations. This is done by running virtual experiments using DEM simulations and comparing the soil reaction forces at each time step to those that would be predicted by the semi-empirical models. The difference between the actual recorded forces and those predicted by the models are used to train a neural network. The objective is for the neural network to learn how much to add or subtract from the forces predicted by the steady state model under any given dynamic conditions. Once trained, the neural network could be used in combination with the semi-empirical models to run more accurate terramechanics simulations without the need to run computationally expensive DEM or FEM simulations. As a proof of concept for this approach, we will illustrate how it can be applied to the augmentation of the Fundamental Earthmoving Equation in the simulation of a blade cutting soil, and also for the augmentation of the Bekker-Wong wheel-soil model.

3004 / Obstacle Avoidance of Mobile Robots Using Modified Artificial Potential Field Algorithm Based on Vortex and PID Adjustment

Zhuoran Sheng and Song Wang

In recent years, autonomous mobile robot has become a key research direction in academia due to their good application prospects. The ability to autonomously avoid obstacles and reach a preset target point is the basic requirement of an intelligent mobile robot. As a common obstacle avoidance algorithm for mobile robots, artificial potential field algorithm (APF) has defects such as local minima problem and GNRON problem. Aiming at the shortcomings of the traditional APF algorithm, this paper proposes an improved artificial potential field algorithm based on point vortex and PID adjustment. The implementation of the improved algorithm is based on the two following main points. Firstly, an irrotational point vortex flow field is added to the original repulsion field of the obstacle to form a composite potential field. Properties of point vortex are used to provide additional virtual deflection force to the robot to avoid getting stuck. Secondly, aiming at GNRON problem, the strength of vortex and the range of action of the vortex repulsive composite potential field are adjusted in real time by means of PID adjustment. In order to verify the feasibility of the improved algorithm proposed in this paper, it is compared with the traditional APF algorithm in 7 cases of environment with different obstacle layouts. The results show that the improved algorithm can be effectively applied to a variety of obstacle environments, and successfully solve the local minima problem and GNRON problem of the traditional APF algorithm. The mobile robot can flexibly and efficiently avoid obstacles and reach the end point.

3290 / Why we need alternative ground robots to traverse sandy and rocky extraterrestrial terrain, and how we can progress towards them

Chen Li and Kevin Lewis

Video: https://vimeo.com/showcase/9710424/video/748992412

Robotic spacecrafts and vehicles have helped expand human's reach in many planetary exploration missions. Most mobile ground robotic vehicles for planetary exploration use wheeled or modified wheeled platforms, and they have been extraordinarily successful at completing intended mission goals. However, due to the limitations of wheeled locomotion, these robotic vehicles have been largely limited to relatively benign, solid terrain with sparse obstacles, and they have avoided extreme terrain with loose soil/sand and cluttered large rocks. Unfortunately, such challenging terrain is often among the most scientifically interesting for planetary geology. Many animals traverse such challenging terrain seemingly at ease, but robots have not matched their locomotor performance and robustness. This lack is largely due to a lack of understanding of the fundamental principles of how effective locomotion (or lack thereof) is generated from controlled interaction with complex terrain, on the same level of flight aerodynamics and underwater vehicle hydrodynamics. Fundamental understanding of legged and limbless locomotor-ground interaction gained over the past few decades has already enabled stable and efficient bio-inspired robot locomotion on relatively flat ground with small, scattered obstacles. Recent progress in the new field of terradynamics of locomotor-terrain interaction has also elucidated the principles of legged and limbless locomotion on loose soil/sand via a diversity of surface and subsurface locomotor modes, as well as over cluttered large obstacles via destabilizing locomotor transitions across different modes overcoming obstacles comparable to or larger than robots themselves. Such bio-inspired multi-legged and limbless robots using terradynamic principles will provide versatile, robust alternative platforms for traversing scientifically important extreme extraterrestrial terrain and expand the reach in planetary exploration.

4136 / DEM Analysis of the Dynamics of Granular Media Considering with Interparticle Forces under Low Gravity Condition

Takuru Nishino, Kenta Takase, Shingo Ozaki, Takao Maeda, Mitsuhisa Baba, and Masatsugu Otsuki

Video: https://vimeo.com/showcase/9710424/video/750291615

For future satellite and planetary explorations, it is important to understand the dynamic interaction between the soft ground and the mechanical system under low gravity. One of the key points in understanding the interactions in low gravity environments is the competition between gravity and the forces acting between particles. The typical forces between particles or between particle and object are considered to be electrostatic and van der Waals forces. These forces are very small compared to gravity under 1G environment on Earth and have little effect on the dynamics. However, it has been suggested that in low gravity environments on small bodies, these forces can compete with gravity and affect the dynamics of granular media. In this study, the analysis using discrete element method (DEM) of rod penetration into granular media and pad impact was performed. First, various parameters were determined for the interparticle forces. The parameters were determined by considering the relationship between the force and the particle size. Next, DEM analyses were performed under four different interparticle force conditions (none, constant, electro-static, and van der Waals) and several levels of gravity environments, to compare reaction forces, sinkage, and particle dispersal. Comparison of the reaction forces, sinkage, and particle dispersal obtained from the analysis confirms the influence of interparticle forces under low gravity condition.

4170 / Development and Traveling Performance Analysis of Driving Test Rover in Sandy Terrain

Masataku Sutoh, Yuji Katsumata, and Sachiko Wakabayashi

Video: https://vimeo.com/showcase/9710424/video/747931452

The interests on the lunar exploration and utilization are recently increasing in worldwide. The Japan Aerospace Exploration Agency, JAXA, also plans to conduct a lunar polar exploration mission, in which lunar volatiles such as water ice are investigated using a rover, and various activities toward human lunar exploration in the future. Towards these missions, it is essential to analyze the traveling performances of vehicles operating in an environment similar to the surface of the Moon which is covered with loose regolith. In this study, the development and performance analysis of a driving test rover are described. We first developed a driving test rover for motion analysis of vehicles traveling on sandy terrain. This four-wheeled rover has a driving system compatible with ground vehicles on Earth. In addition to the lunar exploration, this rover is also used to analyze the motion behaviors of the vehicles on Earth. Subsequently, we conducted traveling tests using the rover in an indoor sand field having an area of approximately 400 m2. The motion of the rover was three-dimensionally measured on flat and inclined terrains using a motion capture system equipped in the field. Furthermore, numerical simulations were conducted using the dynamic model of the rover based on the wheel-terrain interaction. From the experimental and simulation results, a similar tendency was observed regarding the motion behaviors of the rover.

4323 / Development and Demonstration of Robotic Investigation Tool for Terrain Mechanical Property

Yuzuki Morita and Genya Ishigami

Video: https://vimeo.com/showcase/9710424/video/746407755

The Artemis Program led by NASA and the Global Exploration Roadmap summarized by the International Space Exploration Coordination Group has been focusing on the construction of lunar infrastructure in the coming decades. Terrain parameter estimation is essential for the design phase of infrastructure construction in rough terrain. A robotic platform should perform the estimation because of the harsh environment on the moon. While the classical terramechanics model holds several terrain parameters such as cohesion, internal friction, or shear modulus, the resistive force theory needs one lumped parameter called scaling factor, which is basically identified with a flat-plate sinkage test. We developed a handheld investigation tool for the terrain scaling factor that can also be mounted on a mobile robot. The tool consists of a flat plate with a precise linear actuator for intruding the plate into the soil. A force sensor attached between the plate and the actuator measures the resistive force generated during the plate intrusion. The scaling factor is then determined by the relationship between the force and intrusion depth. In our work, we first conducted a laboratory control test of the tool and verified that the tool consistently identifies the scaling factor under different density and grain size distribution conditions. Subsequently, we mounted the tool on a four-wheeled robotic mobile platform and remotely operated the robot deployed on a sandy field, where the robot experimentally demonstrated a terrain parameter investigation. The demonstration confirmed that all of the scaling factors determined by the robotic tool were within the range of the data obtained in the laboratory control experiment.

5636 / Numerical Analysis of Multi-Pass Effect Based on the Extended Terramechanics Theory

Shingo Nakano and Shingo Ozaki

Video: https://vimeo.com/showcase/9710424/video/750254517

In recent years, various off-road vehicles have been used for disaster response and space exploration. In the future, it is expected that off-road vehicles will become even more important. Along with this, there is also a need to improve the traveling performance of off-road vehicles. Currently, multibody dynamics analysis is promising as a preliminary examination method that is indispensable for research, however, to improve the analysis accuracy, the interaction between the ground and wheels, such as deformation of the ground surface shape and multipath effect, have to be appropriately considered. Therefore, in this study, we focus on the terramechanics theory and propose an extended model that combines cellular automata and conventional theory. In the proposed model, the stress distribution between the road surface and the wheel is evaluated by the conventional Bekker-Wong-Reece model, and change in the bulk density and soil property is evaluated based on the cellular automaton. The purpose of this study is to establish a methodology for MBD analysis that can take into account the deformation of the terrain surface and the change in soil state variables due to wheel traveling. Then, we performed the numerical analysis of traveling on flat ground with the single-wheel model and small rover model. Here, we used the general-purpose simulation software Simscape, and implemented the proposed model into it. As a result, we successfully showed the effectiveness of the proposed methodology by the systematic analysis.

5943 / Research on Bionic Anti-Skid Tires Suitable for Icy and Snow-Covered Roads

Rui Zhang, Yu Han, and Hao Pang

In the icy and snowy environment, the tire surface of the vehicle, as a key component of the vehicle's interaction with the icy and snow-covered ground, plays a vital role in improving the vehicle's passability. Reindeer run on ice and snow all year round, and their excellent foot attachment characteristics are obvious. A bionic tire tread structure was designed according to the structural characteristics of the reindeer's sole and the geometry of the reindeer's sole, and the tire structure was optimized by the finite element method (CAE). The simulation results show that the hard ice-breaking part of the tire will cut into the ice layer and generate greater friction with the ice surface, and the soft material part of the tire adheres to the snow, reducing the effect of the snow layer on the tire's traction. Finally, the CAE models of the bionic wheel and the conventional snow wheel are established. Wheel simulations show that the driving traction of a wheel with a bionic tread is 85.4% greater than that of a conventional snow tire. The above studies show that the bionic structure greatly improves the adhesion performance of the wheel surface. In addition, the bionic hair can play a better role in combination with the bionic wheel contour based on the curve of the reindeer's foot. This paper can provide new inspiration and methods for the design of tires on icy and snowy roads.

7349 / Research on Energy-Saving Underactuated Bionic Biped Robot

Youhao Diao, Zhiqiang Zhuang, and Xuebo Wang

There is still a certain gap between the current biped walking machines and humans in terms of walking efficiency and walking naturalness. important scientific value. Taking the human lower limb musculoskeletal system as the learning model, the bionic design of the biped robot based on the passive walking principle is carried out, and the physical prototypes of the fully passive walking machine and the underactuated walking machine are developed. Completed the development of a knee-free completely passive bionic walker with bionic knee joint, bionic foot, bionic subtalar joint and metatarsophalangeal joint, and carried out Adams-based simulation analysis for the virtual prototype to verify the feasibility of its walking. The experimental results show that the walking efficiency of the new kneeless fully passive bionic walking machine is increased by 50%, while achieving continuous and stable walking. In this paper, through the in-depth analysis of the human body's delicate and complex muscle-skeletal system and high-efficiency and energy-saving movement patterns, the inspiration is obtained, and then the mechanical body of the biped walking machine based on the passive walking principle is finely and subtly designed. Design brings bionic design inspiration.

7739 / Modeling of the Shear Displacement for Tracked Vehicles in Transient Maneuvers

Yang Jiao and Jozsef Kovecses

Many design specifications of tracked vehicles are given in terms of the steady-state characteristics. They are indeed quite important to tracked vehicles because of the nature of their operations. However, in some situations such as starting or steering, the transient state performance of these tracked vehicles could also be of great significance. The main trend solution to model the traction force of off-road vehicles is to use shear stress and shear displacement curve by assuming the track soil interaction behaves like the one in the direct shear box test. Therefore, the approximation of shear displacement is quite important in order to accurately predict the shear stress which is directly related to the traction performance of a tracked vehicle. In this work, we propose a general representation of the shear displacement along a discretized track model. This is done by analyzing the shear displacement accumulation at each special point along the track in every time step. Then a new shear displacement model is proposed based on the general representation. The performance of this model is compared with Wong's shear displacement model and Solis and Longoria's (SL) shear displacement model. Wong's model has the simplest form but has difficulty in predicting transient state shear displacement. SL model achieves a better performance in transient state by estimating the shear displacement at each discretized track element with an ordinary differential equation. In order to validate the new model and compare the performance of the three models, a simplified track soil interaction DEM model is developed by using a commercial DEM simulation software 'EDEM'. The shear force obtained from the DEM simulation is used as a reference for the results obtained by using semi-empirical shear force model with different shear displacement models. As proved by the comparison, the new method shows a clear advantage in dynamic shear displacement prediction over Wong's method. The difference between the SL model and the new model is not as large but also obvious in some particular test cases.

8104 / Determination of the Necessary Indicators of Vehicles and the Movement Surface for Calculating the Mobility Criterion

Alina Markovnina, Umar Vakhidov, Vladimir Belyakov, and Vladimir Makarov

This article discusses wheeled vehicles moving on mixed pavements consisting of various types of soil, and indicates the driving conditions typical for such pavements. The main tasks of traffic control are analyzed and it is revealed that the main indicator of the possibility of a vehicle moving in difficult road conditions is the reserve of traction force. The paper presents the general aspects of the methodology for calculating the mobility criterion, according to which it is possible to determine the possibility or impossibility of overcoming a terrain by a given machine. A system of indicators is described that characterizes the possibility of driving a car in given road conditions. The schemes of interaction of the wheel mover with the track bed during rectilinear, curvilinear motion and during maneuvering when bypassing discrete and water obstacles are given. The paper shows various patterns of surface deformation when passing by several axles of a vehicle. The influence of profile obstacles on the cross-country ability and the value of the loss of mobility in terms of cross-country ability are considered. The relationship between the magnitude of the traction force and the magnitude of the obstacle that the vehicle can overcome is determined.

8362 / A High-Fidelity Dynamics Simulation Method for Legged Robot Based on Foot–Terrain Interaction Model

Jianghua Ge, Jianghu Wu, and Xiaofei Zhu

The usage of legged robot for extraterrestrial exploration is a research hotspot at present, and the emulation method to simulate the landing and roving process can effectively improve the design efficiency and verify the control, but due to the complex terrain environment on the surface of the celestial body, it is a challenge to realize the unification of simulation accuracy and efficiency, therefore, the foot–terrain interaction process is an important factor to limit the simulation effect. In this paper, an ADAMS-based simulation system for legged robot dynamics is proposed to achieve high precision simulation of legged robot by introducing a foot–terrain interaction model. The main work includes three aspects: the derivation of dynamic impact model, static pressure model and tangential walking model for legged robot rover in complex terrain based on foot–terrain interaction mechanics model, the development of foot–terrain contact interaction solution module for legged robot rover based on the aforementioned three models and the secondary development of foot–terrain interaction model solution module based on ADAMS user subroutine. The simulations of landing and patrols were carried out to verify that the designed legged robot can effectively meet the landing and patrols functions. Compared with the CONTACT simulation method of ADAMS, the ADAMS high-fidelity dynamics simulation system can verify the probe dynamics requirements with higher accuracy and provide reference for foot pad design and motion control optimization of the legged robot.

8685 / Design and Testing of Parallel Embedded Six-Axis Force Sensors for Paddy Walking Wheels

Zaiman Wang, Jianfei He, Erli Zhang, Miao Su, Yue Huang, Wenwu Yang, Minghua Zhang, Weiqin Jia, Yuanli Huang, Yifan Ma, Dongyang Yu, Peizhao Zhong, Zhihao Zeng, and Ziyou Guo

The operational load spectrum of paddy power machinery can effectively reflect the machine-paddy soil interactions. In response to the problems of difficult operational performance data measurement and lack of observation means for paddy power machinery under complex working conditions in paddy fields, a paddy wheel parallel embedded six-axis force sensor test system is developed and mounted on the field soil through test device and paddy power chassis for performance testing. The test system consists of six-axis force sensors, angle sensors, signal amplifiers, slip ring components, a data acquisition system, data analysis software, and cables. During operation, the five six-axis force sensors transmit electrical signals from the real-time measurement of six-axis forces to the signal amplifier and slip-ring components, and the slip-ring transmits the signals transmitted by the signal amplifier to the cable, which in turn transmits them to the data acquisition system, where the measurement data are processed and displayed in real-time by the data processing software on the host computer. To verify the accuracy of the measurement data of the parallel embedded six-axis force sensor of the paddy wheel in the complex working environment of a paddy field, the bench performance experiment of the field soil test device was carried out to reveal the relationship between the driving force and rolling resistance of the paddy wheel under different rotational speed and different depth of soil entry conditions, and to investigate the influencing factors affecting the operating performance of the paddy field power chassis. The test results show that the measurement data of the six-axis force sensor test system of the paddy walking wheel is accurate, the relative error is less than 0.2%FS of the measurement accuracy, and the principle and method of the test system are reliable; the six-axis force sensor test system can be used for the load spectrum data acquisition of paddy power machinery field operation.

9092 / Examining the Simulation-to-Reality Gap of a Wheel Loader Interacting with Deformable Terrain

Koji Aoshima, Daniel Lindmark, and Martin Servin

Video: https://vimeo.com/showcase/9710424/video/750588946

Simulators are essential for developing autonomous control of off-road vehicles and heavy equipment. They allow automatic testing under safe and controllable conditions, and the generation of large amounts of synthetic and annotated training data necessary for deep learning to be applied. Limiting factors are the computational speed and how accurately the simulator reflects the real system. When the deviation is too large, a controller transfers poorly from the simulated to the real environment. On the other hand, a finely resolved simulator easily becomes too computationally intense and slow for running the necessary number of simulations or keeping realtime pace with hardware in the loop. We investigate how well a physics-based simulator can be made to match its physical counterpart, a full-scale wheel loader instrumented with motion and force sensors performing a bucket filling operation. The simulated vehicle is represented as a rigid multibody system with nonsmooth contact and driveline dynamics. The terrain model combines descriptions of the frictional-cohesive soil as a continuous solid and particles, discretized in voxels and discrete elements. Strong and stable force coupling with the equipment is mediated via rigid aggregate bodies capturing the bulk mechanics of the soil. The results include analysis of the agreement between a calibrated simulation model and the field tests, and of how the simulation performance and accuracy depend on spatial and temporal resolution. The system's degrees of freedom range from hundreds to millions and the simulation speed up to ten times faster than realtime. Furthermore, it is investigated how sensitive a deep learning controller is to variations in the simulator environment parameters.

9219 / Penetration Dynamics of Asteroid Exploration Landing Based on DEM

Yongbin Wang, Shiqing Wu, Shutong Chen, Xinghua Liu, Huan Liu, He Jia, Xuyan Hou, Shaomin Liang, Shunying Ji, and Zijian Zhang

Asteroid exploration is one of the hot areas of deep space exploration in recent years. In this paper, a landing attachment system for asteroid landing detection is proposed, which can complete the landing and attachment of the lander on the asteroid surface. Based on the discrete element method, this paper analyzes the landing impact and penetration dynamics in the process of land penetration. In this paper, the triaxial compression performance of simulated lunar soil is tested, and the physical and mechanical parameters of simulated lunar soil are studied, so as to realize the parameter matching between meso parameters of bulk medium and macro mechanical properties of soil. The penetration process of the detached body is studied. The penetration characteristics under different incident angles, velocities and shapes of the spear body are studied. The characteristics of the depth, velocity and acceleration of the anchor body with time are obtained. The landing impact dynamic characteristics of the anchoring device are obtained. This study provides a new analysis method for the study of asteroid detection anchoring and attachment, and has reference significance for the design of asteroid and other planetary landing and attachment detection.

9260 / Experimental Analysis and Numerical Modeling of Bulldozing Force with Varied Soil Moisture Content

Naohiro Sato and Genya Ishigami

Video: https://vimeo.com/showcase/9710424/video/748991406

Bulldozing/earthmoving works on loose soil are typical operations of construction machines. The bulldozing force acting on a machine blade from soil often impedes the machine mobility. The interaction forces depend on the mechanical property and/or bulk characteristics of soil such as moisture content (i.e., dry or wet sand). The Distinct Element Method (DEM) has been widely employed in precise and high-fidelity simulation for varied types of soil particles interacting with machines. However, the DEM requires accurate V&V (verification and validation), and in particular, such a process for a soil having different moisture content remains an open issue. Therefore, the main scope of this paper is to reveal key parameters in the DEM simulation that appropriately corresponds to specific moisture content. First, we conduct bulldozing experiments under different levels of soil moisture and investigate how the bulldozing force varies with moisture content. The experimental results show that the bulldozing force significantly increases as the moisture content saturation exceeds around 60 %. Subsequently, we employ a DEM simulation to calculate the bulldozing force under different moisture content. Here, we focus on the adhesive forces acting between soil particles because past research suggested that the moisture in the soil is largely subject to the adhesive forces. The adhesive force model of the DEM consists of two parameters, i.e., a scaling magnitude w.r.t. particle weight and the maximum adhesive distance of two soil particles. The values for those two parameters are then adjusted such that the maximum bulldozing force calculated in the DEM coincides with that of the experiments in different moisture content. Finally, we reveal the quantitative relationship of the key parameters corresponding to the various moisture content in the soil. The results imply that the DEM simulation is applicable to moist soil environments by carefully adjusting the adhesive force parameters.

9273 / Force China Analysis on Wheel-Soil Interaction Using Photoelastic Method

Yuto Yoshida, Sota Yuasa, and Kenji Nagaoka

Video: https://vimeo.com/showcase/9710424/video/748991406

This paper presents a force chain analysis of soil-wheel interaction using a photoelastic method. So far, soil-wheel interaction has been experimentally studied by several approaches; such as an analysis of the empirical relationship between wheel's mobility performance and wheel design parameters, measurement of soil reaction acting the wheel surface, the soil flow analysis based on a particle image velocimetry method. However, the stress distribution of a soil particle layer is difficult to be directly observed. In this study, we propose a photoelastic method to visualize the stress distributions by which photoelastic particles are used as terrain, simulating soil particles. The photoelastic disks and their layer which simulate soil are shown in Fig. 1. The photo elastic particles can visualize the applied stresses as interference fringes of light (see Fig. 1). As a result, dynamic change of stress distribution and stress propagation in the terrain can be observed. In particular, the fringe and the stress propagate between the particles and form force chains in the terrain. We developed a single-wheel testbed as an experimental apparatus for wheel-soil interaction using the photoelastic method, as shown in Fig. 2. This apparatus can quantitatively provide the relationship between the particle stresses and the interference fringes based on the brightness gradient. Moreover, we investigated the effects of mixing with different filler particles to simulate various wheel slip-sinkage states. In this paper, we first show the force chain analyses of particles beneath a single traveling wheel. Fig. 3 shows an example of the force chain visualization. Then, orientational orders of the force chains are quantitatively evaluated for small and large wheel slippage. The experimental results confirmed the photoelastic method shows the dynamic change of the stress distributions and propagation in terrain.

International Society for Terrain-Vehicle Systems
www.istvs.org

ISBN 979-8-3313-1248-0

CURRAN ASSOCIATES INC.
proceedings.com

Maintenance and Operations 2018: Parts 1 and 2

Transportation Research Record: Journal of the Transportation Research Board Volume 2672, Issue 12

SAGE Publishing